Olefin Metathesis

and

Ring-Opening Polymerization
of Cyclo-Olefins

V. Drăguţan

Centre of Organic
Chemistry, Bucharest
ROMANIA

A. T. Balaban

Department of Chemistry,
The Polytechnic, Bucharest
ROMANIA

M. Dimonie

Department of Chemistry,
The Polytechnic, Bucharest
ROMANIA

Olefin Metathesis

and

Ring-Opening

Polymerization

of Cyclo-Olefins

A Wiley-Interscience Publication

EDITURA ACADEMIEI
Bucureşti

JOHN WILEY & SONS LIMITED
Chichester ● New York ● Brisbane
Toronto ● Singapore

Ist Edition published in 1981 by Editura Academiei
IInd revised edition published in 1985 by Editura Academiei
and John Wiley & Sons, Ltd.

Copyright © 1985 by John Wiley & Sons, Ltd.

Library of Congress Cataloging in Publication Data:

Drăguțan, Valerian.

Olefin metathesis and ring-opening polymerization of cyclo-olefins.

Revised translation of: Metateza olefinelor și polimerizarea prin deschidere de inel a
cicloolefinelor.

'A Wiley-Interscience publication.'
Includes bibliographies and index.
1. Olefins. 2. Polymers and polymerization.
I. Balaban, Alexandru T. II. Dimonie, Mihai.
III. Title.

QD305.H7D713   1985   547'.412   83-10189

ISBN.0 471 90267 5 (Wiley)

British Library Cataloguing in Publication Data:

Drăguțan, V.

Olefin metathesis and ring-opening polymerization of cyclo-olefins. — 2nd rev. ed.
1. Coordination compounds 2. Olefins
I. Title. II. Balaban, Alexandru T.
III. Dimonie, M.
546'.3   QD474

ISBN 0 471 90267 5

PRINTED IN ROMANIA

Contents

8

Preface

This book covers olefin metathesis and ring-opening polymerization of cyclo-olefins, an important topic of recent interest. It discusses both the scientific and the applied aspects of this field.

This new, unexpected, and interesting reaction discovered less than twenty years ago (in heterogeneous catalysis in 1964, and with homogeneous catalysts in 1967) has become soon afterwards a large and independent research area.

The metathesis of olefins (alkenes) is the reaction converting two identical or different olefin molecules into two other olefins by formal permutation of the two double bonds in a four-centred array, under the action of certain catalytic systems. Cyclo-olefins undergo the same reaction affording macrocyclic molecules or polymers, according to the reaction conditions. The metathesis was also applied to acetylenes, and was generalized to a large variety of organic unsaturated compounds. Its main technical applications are in the synthesis of olefins and functionalized olefins, of monomers, in the formation of plastics and elastomers, and in petrochemistry.

Research in this area continues at a rapid pace, and new practical applications emerge at the same time as mechanistic details are elucidated. It is significant to follow the steep rise in the number of patents and papers published during the last fifteen years : whereas the patent literature increased markedly during the first part of this period, the number of papers increased more steeply during the last years. The major interest manifested for this area is apparent from the organization of a series of successful international symposia on olefin metathesis (Mainz, 1976, Amsterdam, 1977, Lyons, 1979, Belfast, 1981, Graz, 1983). Olefin metathesis is present in traditional current symposia on catalysis, organometallics, polymers, etc.

The present revised, expanded, and updated English version contains more than 2000 literature citations (while the Romanian edition which was published in 1981 had only half this number of references, and did not contain the extensive table listing all reactions described so far). The literature has been reviewed exhaustively through 1982, with occasional more recent references.

The first chapter of the book describes briefly the history of this reaction, defines the metathesis reaction and classifies the various known reaction types pertaining to this process. The classification introduced by the authors is based on the nature of the unsaturated compounds undergoing the reaction. The second chapter deals with the reaction conditions, namely the catalytic systems (either the heterogeneous catalysts used mainly for olefin disproportionation and alkenolysis, or the homogeneous catalytic systems used mainly for ring-opening polymerization or for alkene metathesis) and the reaction parameters (temperature, pressure, solvents, etc.). Chapter 3 presents a systematic description of known metathesis reactions, namely for acyclic and cyclic alkenes (with or without substituents), dienes, acetylenes. Separate sections of this chapter are devoted to the synthesis of carbocycles, catenanes, knots or rotaxanes through metatheses and cometatheses which involve two different molecules, and which have applicability either as alkenolysis or as ring scission reactions. Chapter 4 describes the ring-opening polymerization of cyclo-olefins, ordered according to the ring size of the monomers and the nature (mono-, bi- or polycyclic) of the olefins; a special emphasis is placed on the ring-opening polymerization of cyclopentene leading to polypentenamers, a reaction with important practical applications. Chapters 5 and 6 discuss the thermodynamic aspects, and the kinetics and the mechanism of the reaction, respectively. A brief historical survey of controversial mechanisms is presented, together with an extensive discussion of the currently accepted mechanism involving metallacyclobutanes formed *via* metallacarbenes. The seventh chapter presents the reaction stereochemistry both for the metathesis of acyclic olefins and for the ring-opening polymerization of cyclo-olefins; stereochemical features of the reaction are correlated with mechanistic aspects. Chapter 8 is devoted to practical applications of the reaction : olefin metatheses, petrochemical uses, syntheses of uncommon alkenes, catenanes, rotaxanes, knots, macrocycles, polyalkenamers, monomers, etc; many of these procedures are accompanied by technological schemes. The last chapter provides a comprehensive tabular survey of unsaturated substrates (alkenes, cyclo-alkenes and alkynes) and the metathetical catalytic systems for which metathesis reactions and ring-opening polymerizations have been reported.

The book is addressed to advanced students and teaching staff in organic chemistry, chemical technology, petrochemistry, and macromolecular chemistry, as well as to research chemists and chemical engineers in organic or macromolecular chemistry and petrochemistry.

The authors are deeply indebted to Professors C. Boelhouwer, B. A. Dolgoplosk, H. Höcker, K. Hummel, T. J. Katz and J. Levisalles as well as to Drs. L. Bencze, W. J. Feast, L. Hocks, F. W. Küpper, J. C. Mol, K. W. Scott, K. Seyferth, F. Stelzer and R. Streck for kindly providing reprints and preprints of their publications in this field. One of the authors (V. D.) expresses his thanks to Professor J. M. Basset and Dr. R. Mutin, Professors K. J. Ivin and J. J. Rooney, and Professor K. Hummel and Dr. F. Stelzer for their kindness during the symposia in Lyons, Belfast and Graz.

Bucharest, 1983

Valerian DRĂGUȚAN
Alexandru T. BALABAN
Mihai DIMONIE

Chapter 1

INTRODUCTION

The metathesis of olefins is the reaction in which the molecules of these hydrocarbons are formally fragmented at their double bonds and new olefin molecules result by recombination of fragments originating from different molecules (1.1).

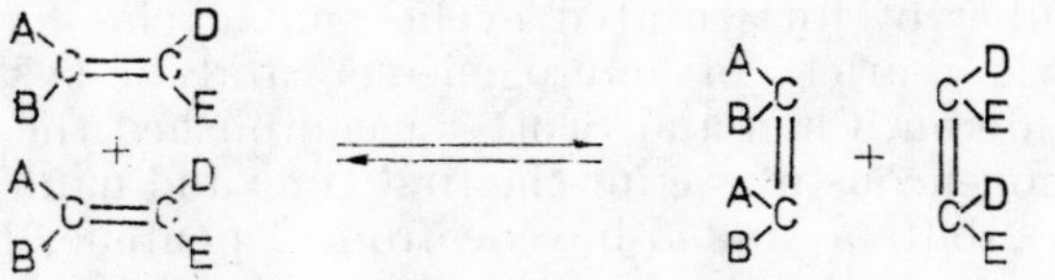

$$
\begin{array}{c}
\underset{B}{\overset{A}{>}}C = C\underset{E}{\overset{D}{<}} \\
+ \\
\underset{B}{\overset{A}{>}}C = C\underset{E}{\overset{D}{<}}
\end{array}
\;\; \rightleftharpoons \;\;
\begin{array}{c}
\underset{B}{\overset{A}{>}}C \\
\parallel \\
\underset{B}{\overset{A}{>}}C
\end{array}
+
\begin{array}{c}
C\underset{E}{\overset{D}{<}} \\
\parallel \\
C\underset{E}{\overset{D}{<}}
\end{array}
\tag{1.1}
$$

The ring-opening polymerization of cyclo-olefins, closely related to metathesis, is the reaction by which such molecules are cleaved at the double bond producing unsaturated linear polymers of the polyalkenylene type, called polyalkenamers (1.2).

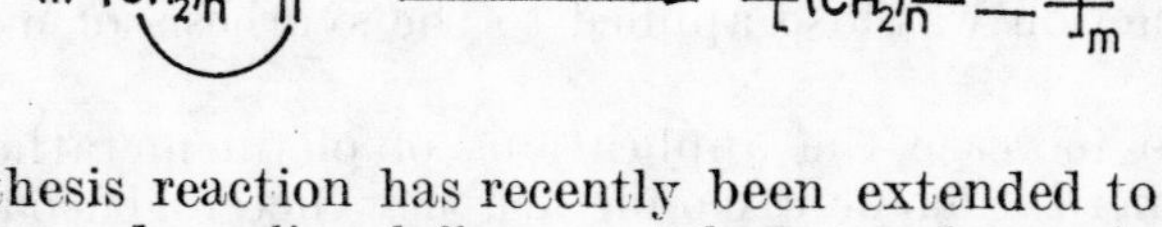

$$
m \; (CH_2)_n \big| \big| \longrightarrow \text{---}\big[(CH_2)_n\text{---}=\text{---}\big]_m
\tag{1.2}
$$

The metathesis reaction has recently been extended to a large number of acyclic and cyclic olefins, acetylenes and unsaturated organic compounds. Similarly, the ring-opening polymerization has been applied to simple or substituted mono-, di-, and polycyclic olefins.

Soon after their discovery, these two types of reaction proved to be of commercial use. The metathesis is now in use on an industrial scale and is accepted as a method which has considerable possibilities in the production of olefins for the petrochemical industry, such as ethene, propene and butene, and of olefin mixtures used to obtain high octane gasoline, high-grade mineral oils, monomers, or other raw materials for chemical industry. The ring-opening polymerization of cyclo-olefins is used to synthesize polyalkenamers with useful properties, such as elastomers and plastics. One of these, polypentenylene (known also as polypentenamer) is a valuable competitor of natural rubber and has considerable economic importance.

1.1. Historical Survey

The non-catalytic disproportionation of olefins has been known since 1931, when Schneider and Frölich[1] obtained ethene and 2-butene by pyrolyzing propene at 852°C. The catalytic disproportionation of olefins was discovered later. In 1964, Banks and Bailey[2] obtained ethene and 2-butene, with high selectivity, by the disproportionation of propene in the presence of heterogeneous catalytic systems consisting of alumina-supported molybdenum hexacarbonyl, tungsten hexacarbonyl or molybdenum oxide. These authors also carried out this reaction with butene, pentene, hexene and octene. Later, the new type of reaction was extended to a large number of unsaturated cyclic and acyclic hydrocarbons and to this end a wide variety of heterogeneous catalytic systems was used.

In 1967, Calderon, Chen and Scott[3] accomplished the disproportionation in the homogeneous phase for the first time and named it metathesis. They obtained 2-butene and 3-hexene from 2-pentene, by employing a catalytic system formed from tungsten hexachloride, ethanol and ethylaluminium dichloride. In the same year, Calderon et al.[4] applied the metathesis to cyclo-olefins, synthesizing, with high selectivity, polyalkenamers by ring-opening polymerization of cyclo-olefins.

Soon after the discovery of olefin metathesis, Pennella, Banks and Bailey[5a] carried out the metathesis of acetylenes; Heckelsberg, Banks and Bailey[5b] reported the metathesis of dienes; and other groups of chemists described the metathesis of a large number of simple olefins[6] and substituted olefins bearing functional groups,[7] cyclic olefins,[8] or acetylenes.[9] The reaction was also applied to the synthesis of many polyalkenamers.[10]

In order to increase the applications of olefin metathesis and ring-opening polymerization of cyclo-olefins, new heterogeneous and homogeneous catalytic systems have been developed. The heterogeneous catalysts based on molybdenum and tungsten oxides, carbonyls, sulphides or salts, which were initially described by Banks, Bailey and Heckelsberg,[11] were diversified.[12,13] Similarly, the homogeneous catalytic systems described by Calderon,[3] Zuech,[14] and Wang and Menapace[15] were extended to a wide range of transition metal complexes.[16,17] By employing modern means of investigation, it was possible to obtain data on the structure and the formation reactions of the active species for these catalysts. Mention should be made of the recent contributions of Bencze,[18-20] Boelhouwer,[21,22] Casey,[23,24] Dolgoplosk[25] and other investigators[26-29] in obtaining and characterizing several new catalytic systems. At present, the known catalysts suitable for metathesis and ring-opening polymerization contain a combination of elements from almost all groups in the periodic system.

Since 1967, the reaction kinetics and mechanism have been studied intensively. In the disproportionation of 1- and 2-butene with heterogeneous catalytic systems, Bradshaw, Howman and Turner [50] obtained results which suggested a reaction mechanism through a four-centred cyclobutanic intermediate; Calderon et al. also proposed this mechanism for metathesis in the homogeneous phase [31] and extended it to ring-opening polymerization [32]; Mango and Schachtschneider [33] elaborated a theoretical basis for the cyclobutanic reaction mechanism and developed a new theory for symmetry-forbidden cyclo-additions catalyzed by transition-metal complexes. Other similar theoretical interpretations were subsequently reported. [34-36] The mechanism through cyclobutanic intermediate also seemed to be supported by the studies performed by Crain, [37] Boelhouwer [38,39] and others [40,41] on the reaction in homogeneous phase.

MECH I

The first kinetic studies on propene metathesis in the heterogeneous phase were done by Begley and Wilson [42] and on 2-pentene metathesis in the homogeneous phase by Hughes. [43] Although the results obtained by Begley and Wilson indicate a first order kinetics for the reaction of propene, contradicting the mechanism through a cyclobutanic intermediate, a series of further investigations carried out in heterogeneous catalysis [44-50] seemed to indicate a second order kinetics. The real kinetics of the process is still uncertain in view of several more recent results obtained from the reaction of propene [51] or 2-pentene [51,53] which indicated a first-order kinetics.

In 1970, Hérisson and Chauvin [54] suggested a chain mechanism for metathesis and ring-opening polymerization in the homogeneous phase, through carbenic intermediates. This suggestion has stimulated a new series of investigations on the mechanism and stereochemistry of the reaction [55-67] and nowadays this idea is favoured by most investigators.

MECH II

Meanwhile the metathetic mechanism of cyclo-olefin polymerization, formulated initially by Calderon and Scott [32] for the reaction of cyclo-octene, has been generalized by the studies carried out by Dall'Asta et al. [68] on a large number of simple and substituted cyclo-olefins. However, the carbenic chain mechanism suggested by Hérisson and Chauvin has been supported by the more recent results obtained by Amass, [69] Calderon, [70] Dolgoplosk, [71] Katz, [72] Ivin [73] and Rooney. [74]

It has been less than three years since the discovery of the metathesis in the heterogeneous phase, but the process has been applied on an industrial scale to synthesize olefins such as ethene, propene or butene. [75-77] Similarly, the ring-opening polymerization of cyclo-olefins has become a simple and economical method of manufacturing polyalkenamers, such as polypentenamer [10]; this polymerization is at present ready for industrial scale application in several countries.

Interest in the metathesis of olefins and ring-opening polymerization of cyclo-olefins for their use in organic synthesis, petrochemistry, elastomer and plastics industries has stimulated the publication of a series of reviews on these reactions since 1969. The reader may find further information on metathesis in the reviews published by Bailey,[12] Banks,[13] Bencze and Markó,[78] Calderon,[79] Drăguţan,[16] Grübbs,[80] Hocks,[81] Haines and Leigh,[17] Hughes,[82] Katz,[83a] Khidekel,[83b] Mol and Moulijn,[84] Rooney,[85] Streck,[86] Taube and Seyferth[87a] and Weber.[87b] For ring-opening polymerization, reviews by Calderon,[88] Dall'Asta,[89] Dimonie,[90] Dolgoplosk,[91] Drăguţan,[92] Günther[93] and Höcker[94] should be consulted.

1.2. Definitions and Reaction Types

The olefin metathesis belongs to the general class of disproportionation reactions which include those reactions in which : i. two identical molecules of the same olefin can form two different olefin molecules, one larger and one smaller than the initial olefin

$$2 \ C_nH_{2n} \longrightarrow C_{n-m}H_{2(n-m)} + C_{n+m}H_{2(n+m)} \tag{1.3}$$

where $n \geqslant 3$ and $n - m \geqslant 2$,
ii. two identical molecules of the same olefin can form two different molecules of new hydrocarbons, one with a higher and the other with a lower degree of un saturation than the initial olefin,

$$2 \ C_nH_{2n} \longrightarrow C_nH_{2n+2} + C_nH_{2n-2} \tag{1.4}$$

iii. two identical molecules of the same olefin form a diene and a hydrogen molecule by hydrogen transfer.

$$2 \ C_nH_{2n} \longrightarrow C_{2n}H_{4n-2} + H_2 \tag{1.5}$$

Of these three disproportionation reactions, only the first, equation (1.3), is related to the metathesis reaction.

The metathesis of olefins is a disproportionation reaction which proceeds formally by cleavage of two molecules at the double bonds and by recombination of the alkylidene groups ((1.6), which constitutes a special case of equation (1.1)).

$$
\begin{array}{c}
R' \quad R' \\
\| + \| \longrightarrow \\
R'' \quad R''
\end{array}
\qquad
\begin{array}{c}
R' \longequal R' \\
+ \\
R'' \longequal R''
\end{array}
\tag{1.6}
$$

The metathesis reaction has been given several different names. It was discovered in the heterogeneous phase by Banks and Bailey [2] who named it disproportionation; later, Bradshaw, Howman and Turner [30] called it dismutation, Crain [37] mutual cleavage, and Calderon, Chen and Scott [3] used the term metathesis for the reaction in a homogeneous phase. Nowadays, for the reaction in a heterogeneous phase, the term disproportionation is more commonly used, while metathesis is preferred for the reaction in a homogeneous phase.

Since, from the viewpoint of the reaction products, there are no essential differences between the two types of processes, the term disproportionation may be considered as synonymous with metathesis. In the following, we shall use only the term metathesis for process (i).

For the class of acyclic olefins, two main types of metathesis can be distinguished:

a) simple metathesis or homometathesis, because actually disproportionation should also encompass processes (ii) and (iii), which will not be discussed here, and b) cross metathesis or cometathesis.

a) simple metathesis is the reaction which takes place between two molecules of the same olefin. In this reaction, it is possible to form two molecules of new olefins as in the general case expressed by equations (1.1) and (1.6) or to regenerate the initial olefin (as illustrated in equation (1.7)).

$$
\begin{array}{c}
\overset{R^I}{\underset{R^{II}}{\vert\vert}} + \overset{R^{II}}{\underset{R^I}{\vert\vert}} \longrightarrow \begin{array}{c} R^I \!-\!\!=\!\!-\! R^{II} \\ + \\ R^{II} \!-\!\!=\!\!-\! R^I \end{array}
\end{array} \qquad (1.7)
$$

Equation (1.7) represents a degenerate metathesis or intermolecular automerization [95,96] which always accompanies the simple metathesis and can be proved to occur by isotopic labelling. In the case when $R' = R'' = R$, the reaction is totally degenerate, and the initial olefin is regenerated as the unique reaction product.

$$
\begin{array}{c}
\overset{R}{\underset{R}{\vert\vert}} + \overset{R}{\underset{R}{\vert\vert}} \longrightarrow \begin{array}{c} R \!-\!\!=\!\!-\! R \\ + \\ R \!-\!\!=\!\!-\! R \end{array}
\end{array} \qquad (1.8)
$$

This metathesis reaction, also called intermolecular topomerization, [97] occurs in symmetrically substituted internal olefins (i.e. those in which the double bond occupies a central position with identical substituents on both sides).

α-Olefins, i.e. alkenes with a terminal double bond, yield through metathesis ethene and an internal olefin (1.9 A).

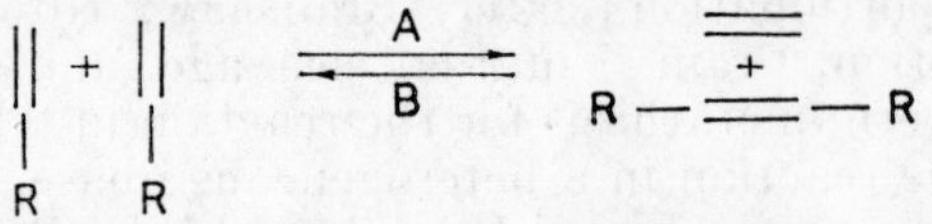

$$\begin{array}{c} || \\ | \\ R \end{array} + \begin{array}{c} || \\ | \\ R \end{array} \quad \underset{B}{\overset{A}{\rightleftharpoons}} \quad R - \overset{+}{=\!=} - R \qquad (1.9)$$

The reverse reaction (1.9 B) is known as ethenolysis.

β-Olefins yield, through metathesis, 2-butene and an internal olefin.

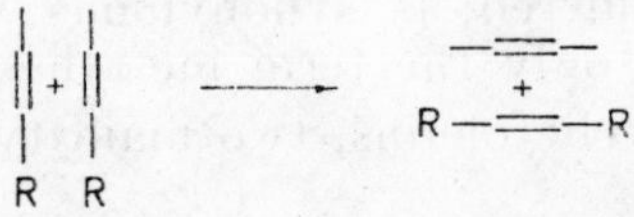

$$\begin{array}{c} || \\ | \\ R \end{array} + \begin{array}{c} || \\ | \\ R \end{array} \quad \longrightarrow \quad R - \overset{+}{=\!=} - R \qquad (1.10)$$

The metathesis of internal olefins yields different products, according to the structure of the particular olefin. Symmetrically substituted olefins undergo a totally degenerate reaction, as it has been shown above (1.8), whereas non-symmetrically substituted olefins undergo both a degenerate reaction (1.7) and a non-degenerate (productive) reaction forming new symmetrically substituted olefins (1.6).

α,ω-Dienes yield, through intermolecular metathesis, simple alkenes and linear trienes or polyenes (1.11)

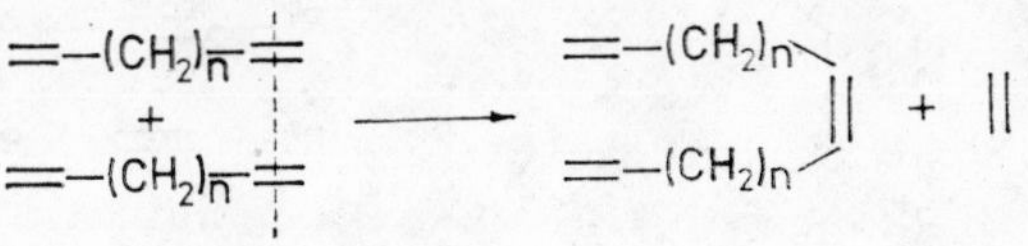

$$\begin{array}{c} =\!-(CH_2)_{\overline{n}}\!=\!\!= \\ + \\ =\!-(CH_2)_{\overline{n}}\!=\!\!= \end{array} \quad \longrightarrow \quad \begin{array}{c} =\!-(CH_2)_n \\ =\!-(CH_2)_n \end{array} || \quad + \quad || \qquad (1.11)$$

or as a result of subsequent inter- and intra-molecular reactions, cyclic polyenes.

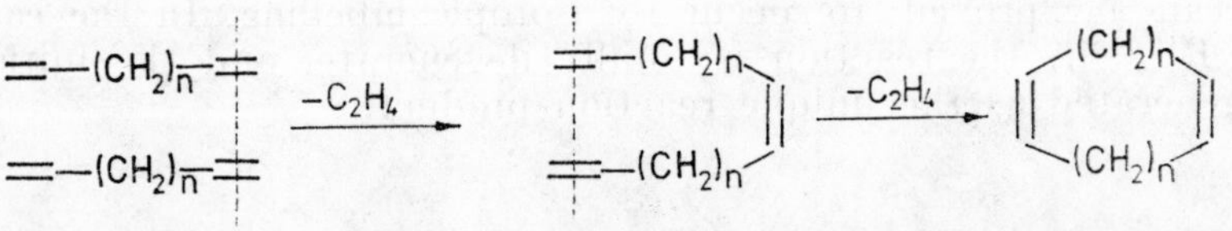

$$\begin{array}{c} =\!-(CH_2)_{\overline{n}}\!=\!\!= \\ =\!-(CH_2)_{\overline{n}}\!=\!\!= \end{array} \quad \xrightarrow{-C_2H_4} \quad \begin{array}{c} =\!-(CH_2)_n \\ =\!-(CH_2)_n \end{array} || \quad \xrightarrow{-C_2H_4} \quad \begin{array}{c} (CH_2)_n \\ (CH_2)_n \end{array} \qquad (1.12)$$

The two double bonds of larger-chain α,ω-acyclic dienes can also react intramolecularly, yielding cyclo-olefins and ethene.

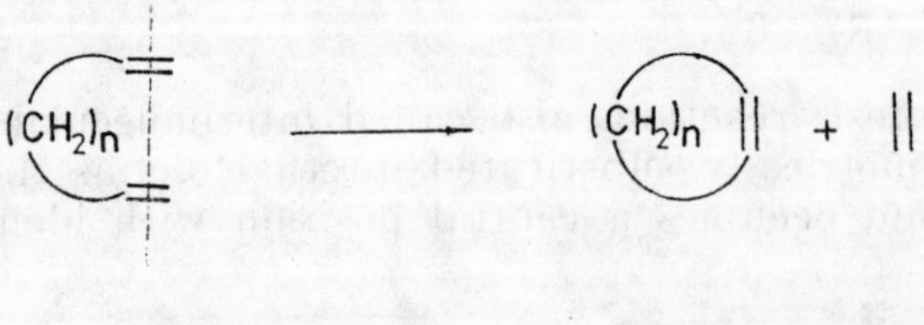

$$(CH_2)_n \quad \longrightarrow \quad (CH_2)_n \, || \quad + \quad || \qquad (1.13)$$

b) The cross metathesis, for which we shall use the term cometathesis, is the reaction between two or more different olefin molecules, yielding other new olefins.

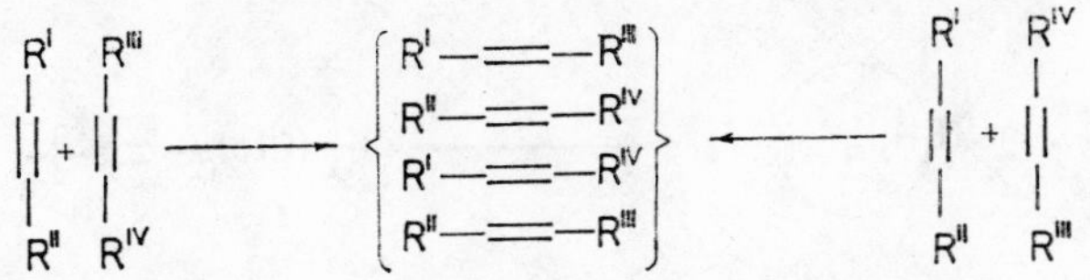

$$
\begin{matrix} R^{I} & R^{III} \\ || & + & || \\ R^{II} & R^{IV} \end{matrix}
\longrightarrow
\left\{
\begin{matrix}
R^{I} - = - R^{III} \\
R^{II} - = - R^{IV} \\
R^{I} - = - R^{IV} \\
R^{II} - = - R^{III}
\end{matrix}
\right\}
\longleftarrow
\begin{matrix} R^{I} & R^{IV} \\ || & + & || \\ R^{II} & R^{III} \end{matrix}
\tag{1.14}
$$

The cometathesis between a cyclo-olefin and ethene is called ethenolysis (1.15) which is (1.9 B) i.e. the reverse of equation (1.9 A).

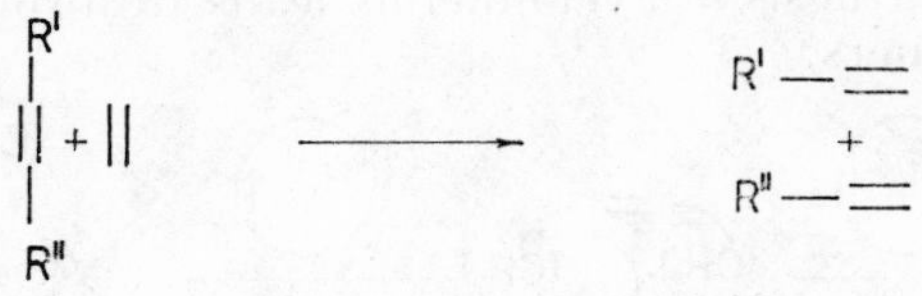

$$
\begin{matrix} R^{I} \\ || \\ \\ R^{II} \end{matrix} + ||
\longrightarrow
\begin{matrix} R^{I} - = \\ + \\ R^{II} - = \end{matrix}
\tag{1.15}
$$

The cometathesis reaction with propene is called propenolysis.

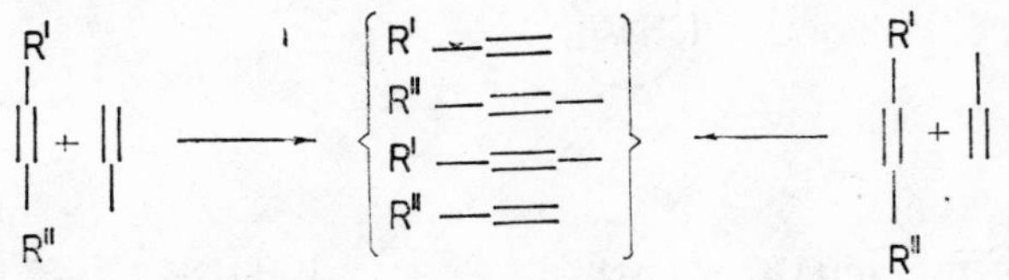

$$
\begin{matrix} R^{I} \\ || \\ R^{II} \end{matrix} + ||
\longrightarrow
\left\{
\begin{matrix}
R^{I} - = \\
R^{II} - = - \\
R^{I} - = - \\
R^{II} - =
\end{matrix}
\right\}
\longleftarrow
\begin{matrix} R^{I} \\ || \\ R^{II} \end{matrix} + ||
\tag{1.16}
$$

In general, when an unspecified alkene is employed in cometathesis, as in equation (1.14), the reaction is called alkenolysis. Alkenolysis leads to a cleavage of the acyclic polyene chain. Thus, unsaturated polymers can be converted, by alkenolysis, into fragments with a lower molecular weight.

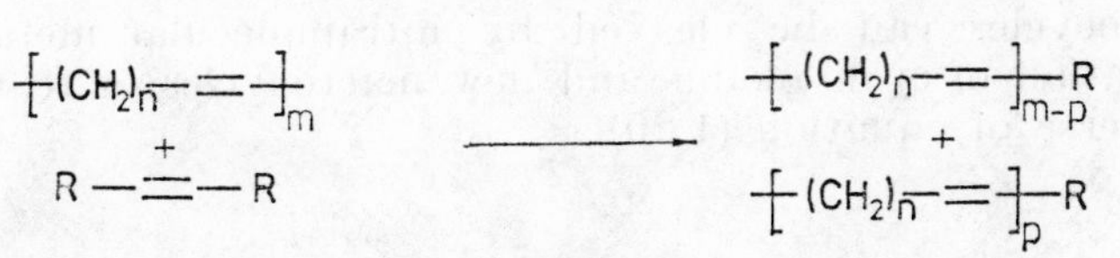

$$
\begin{matrix} \left[(CH_2)_n - = \right]_m \\ + \\ R - = - R \end{matrix}
\longrightarrow
\begin{matrix} \left[(CH_2)_n - = \right]_{m-p} R \\ + \\ \left[(CH_2)_n - = \right]_p R \end{matrix}
\tag{1.17}
$$

The metathesis of cyclo-olefins takes place in a completely different way. Through the metathesis of the double bonds of two cyclo-olefin molecules, cyclo-olefins ought to yield cyclic dimers with two double bonds.

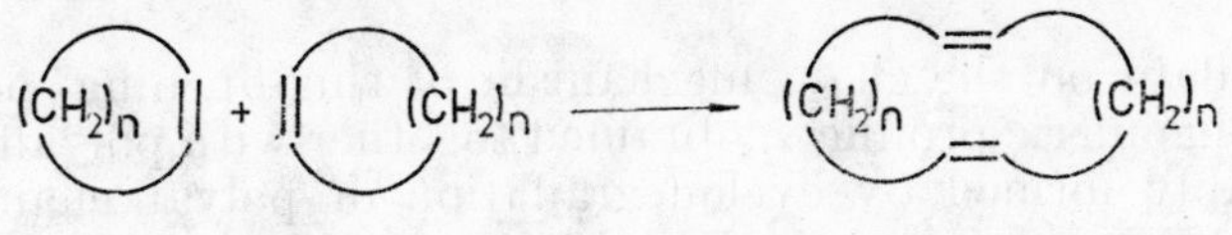

$$
(CH_2)_n \, || \; + \; || \; (CH_2)_n
\longrightarrow
(CH_2)_n \overset{=}{\underset{=}{\bigcirc}} (CH_2)_n
\tag{1.18}
$$

Actually, however, such a metathesis of cyclo-olefins occurs only in a few cases. In most reactions, in the presence of metathesis catalysts, a ring-opening polymerization of cyclo-olefins takes place, with formation of polyalkenamers.

$$m \ (CH_2)_n \ || \longrightarrow -[(CH_2)_n - = -]_m \qquad (1.19)$$

The presence of cyclo-olefins oligomers and of unsaturated macrocycles along with polyalkenamers was interpreted as supporting the idea that the ring-opening polymerization of cyclo-olefins proceeds by successive metathesis steps initially forming dimers (1.18), then trimers, etc. Therefore, the metathesis of cyclo-olefins leads to formation of cyclic and macrocyclic oligomers.

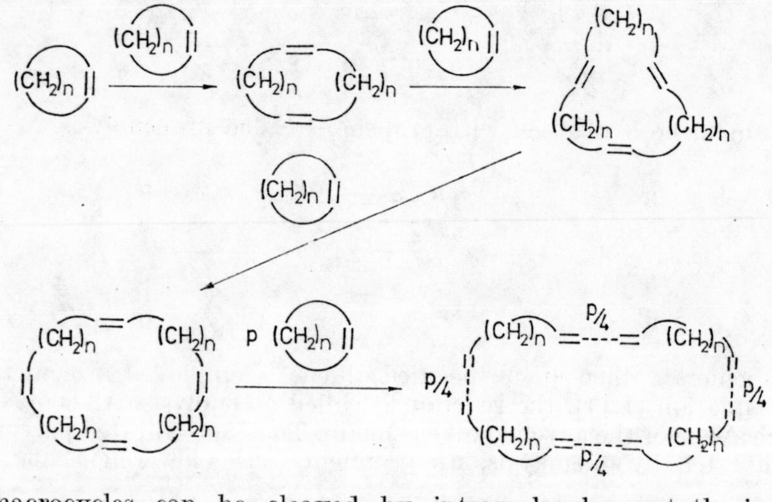

$$(1.20)$$

The macrocycles can be cleaved by intramolecular metathesis (1.21) with formation of cyclo-olefins and new macrocycles, a reaction related to the reverse of equation (1.20)

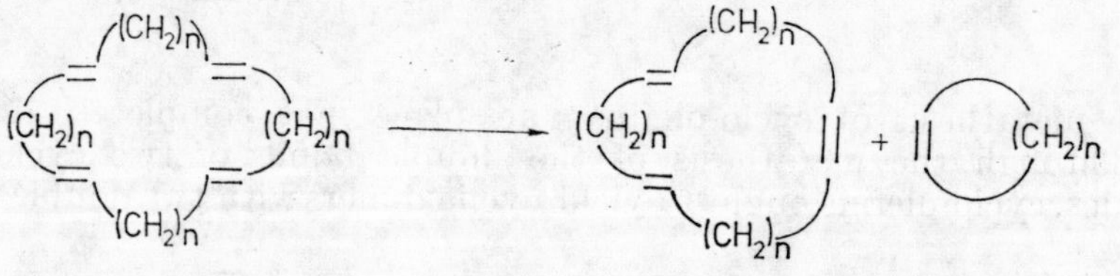

$$(1.21)$$

Recent data on the chain mechanism of ring-opening polymerization show that oligomeric products, obtained together with polyalkenamers, are more probably formed by cyclodegradation of polyalkenamers.

If, during the oligomerization of cyclo-olefins through metathesis, the macrocyclic ring becomes twisted, it is possible to obtain catenanes or knots, depending on the number of half-twists.

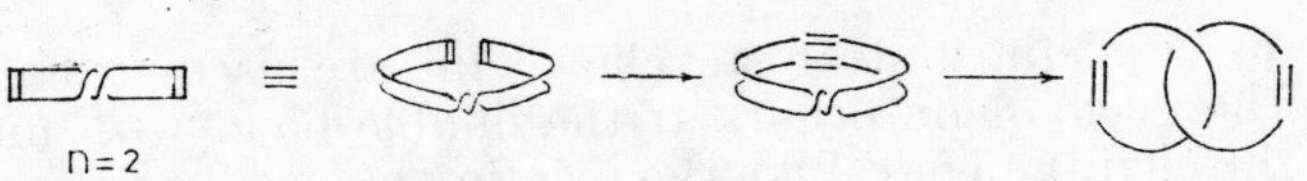

$$\text{(1.22)}$$

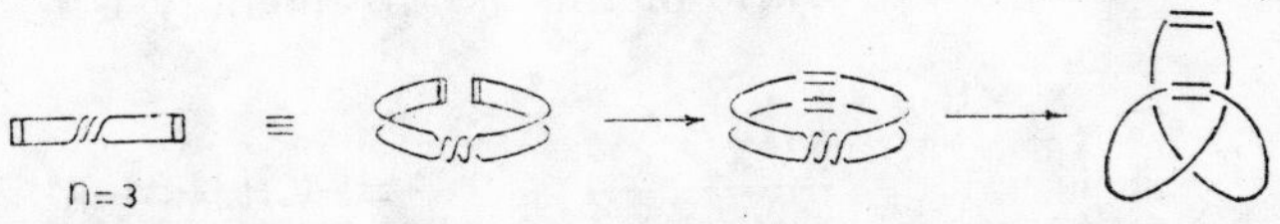

$$\text{(1.23)}$$

The catenanes are formed when the macrocycles undergo n = 2 half-twists and the knots when the macrocycles undergo n=3 half-twists.

When the number of half-twists is zero (n = 0), the macrocycles cleave into two cyclo-olefins (1.24)

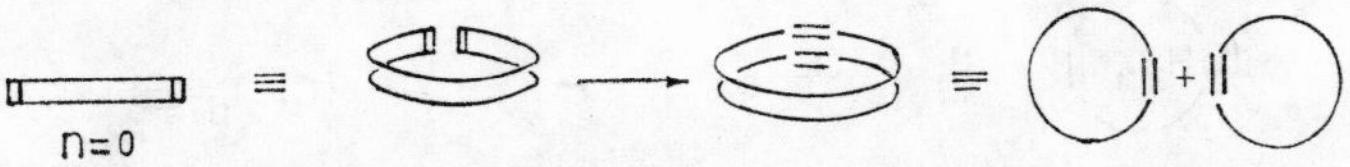

$$\text{(1.24)}$$

a reaction which represents the reverse of equation (1.18). For one half-twist (n = 1) a degenerate reaction occurs when the macrocycle is reformed.

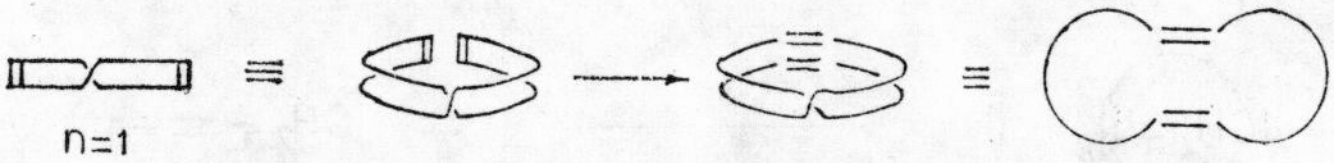

$$\text{(1.25)}$$

In this case the double bonds merely change their positions in the cycle.

For technical applications of ring-opening polymerizations, the poly-saturated macrocycle (1.20) is alkenolyzed with a controlled amount of alkene (e.g. 2-butene) resulting in an elastomer with a desired molecular weight, as will be shown below (1.32).

The cometathesis reaction between two different cyclo-olefins leads to codimers, co-oligomers and/or copolyalkenamers, according to a general scheme, similar to simple olefin metathesis.

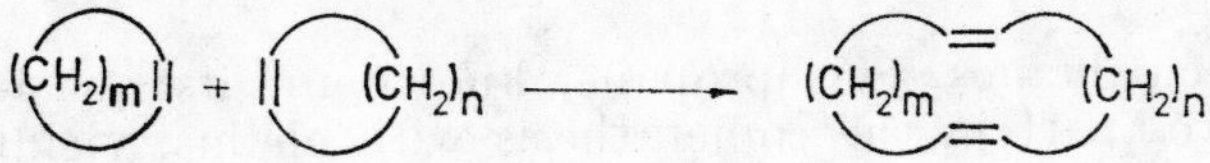

$$(CH_2)_m + (CH_2)_n \longrightarrow (CH_2)_m \quad (CH_2)_n \qquad \text{(1.26)}$$

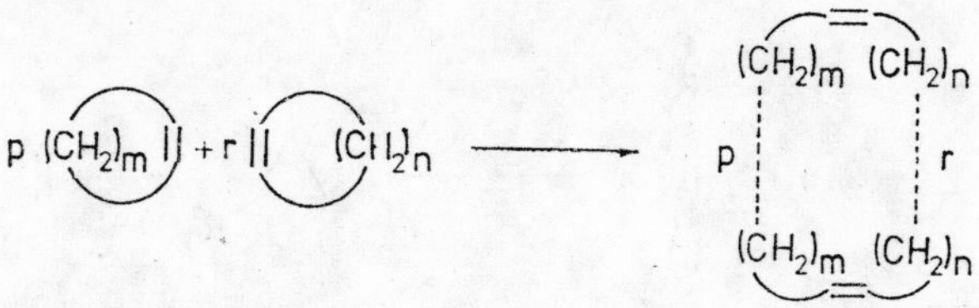

$$p\ (CH_2)_m + r\ (CH_2)_n \longrightarrow p \begin{array}{c} (CH_2)_m\ (CH_2)_n \\[4pt] \\ (CH_2)_m\ (CH_2)_n \end{array} r \qquad \text{(1.27)}$$

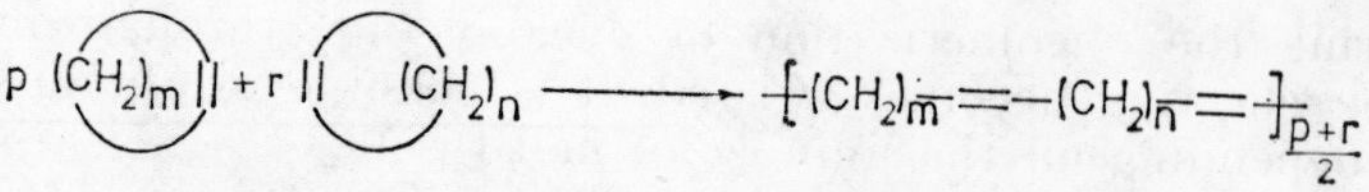

$$p \; (CH_2)_m \|| + r \; \| (CH_2)_n \longrightarrow \left[(CH_2)_{\overline{m}} = \!\! = \!\! -(CH_2)_{\overline{n}} = \!\! = \right]_{\frac{p+r}{2}} \quad (1.28)$$

When a cyclo-olefin and an acyclic olefin are used in cometathesis, cleavage of the cyclo-olefin occurs at the double bond with formation of linear (straight-chain) dienes. Like the reaction of acyclic olefins (see 1.14), the cometathesis reaction of cyclo-olefins is an alkenolysis.

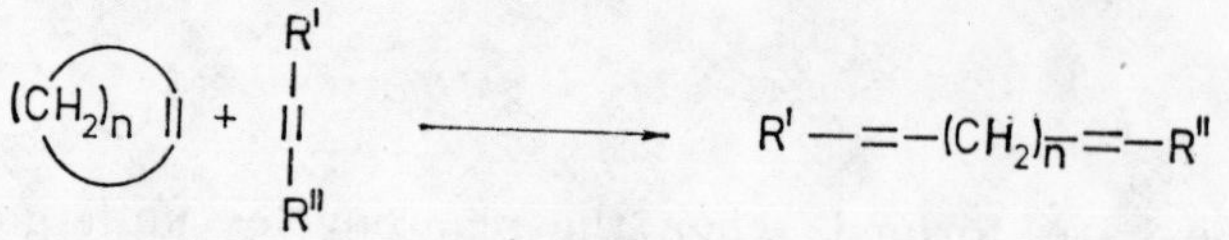

$$(CH_2)_n \|| + \underset{R''}{\overset{R'}{\|}} \longrightarrow R' - \!\! = \!\! -(CH_2)_{\overline{n}} = \!\! = \!\! -R'' \quad (1.29)$$

The ethenolysis of cyclo-olefins yields α, ω-dienes with terminal double bonds (1.30);

$$(CH_2)_n \|| + \| \longrightarrow = \!\! -(CH_2)_{\overline{n}} = \!\! = \quad (1.30)$$

this reaction represents the reverse of reaction (1.13).

By alkenolysis with an α-olefin, cyclo-olefins give dienes with only one terminal double bond.

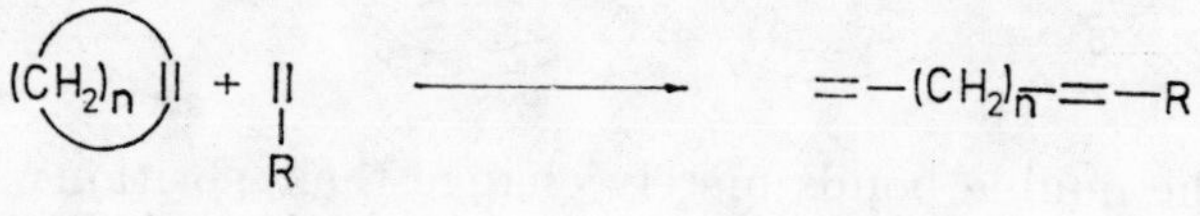

$$(CH_2)_n \|| + \underset{R}{\|} \longrightarrow = \!\! -(CH_2)_{\overline{n}} = \!\! = \!\! -R \quad (1.31)$$

The cleavage of macrocycles with acyclic olefins yielding polyalkenamers is related to the alkenolysis of cyclo-olefins.

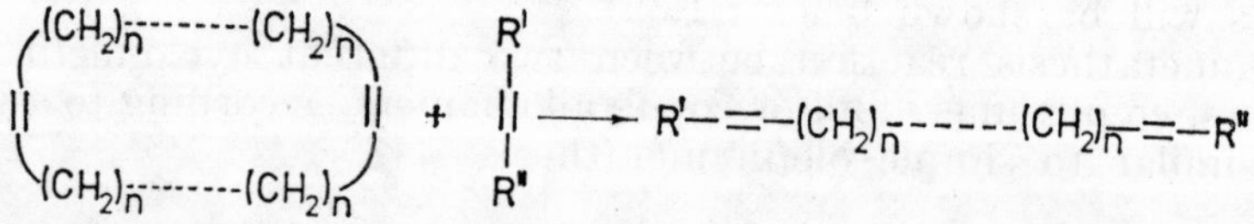

$$\left(\begin{array}{c} (CH_2)_{\overline{n}} \text{------} (CH_2)_{\overline{n}} \\ \| \qquad\qquad \| \\ (CH_2)_{\overline{n}} \text{------} (CH_2)_n \end{array} \right) + \underset{R''}{\overset{R'}{\|}} \longrightarrow R' - \!\! = \!\! -(CH_2)_{\overline{n}} \text{-----} (CH_2)_{\overline{n}} = \!\! = \!\! -R'' \quad (1.32)$$

Among cycloalkanes, cyclopropane, known for its quasi-unsaturated character, takes part in the cometathesis with olefins, yielding disubstituted cyclopropane. [98]

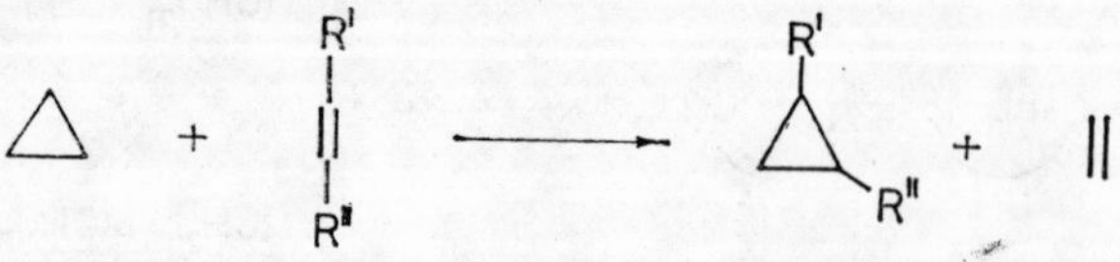

$$\triangle + \underset{R''}{\overset{R'}{\|}} \longrightarrow \underset{R''}{\overset{R'}{\triangle}} + \| \quad (1.33)$$

Acetylenes (alkynes) undergo metathesis reactions similar to those of olefins. Through such reactions acetylenes are cleaved at the triple bond yielding new acetylenes by recombination of the alkyne fragments.

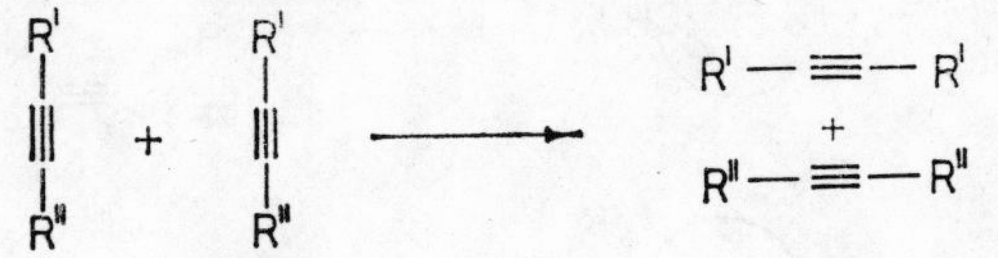

$$R^{I} - \equiv - R^{I} \quad + \quad R^{II} - \equiv - R^{II} \tag{1.34}$$

The reaction is accompanied by a degenerate metathesis which can be evidenced through isotopic labelling.

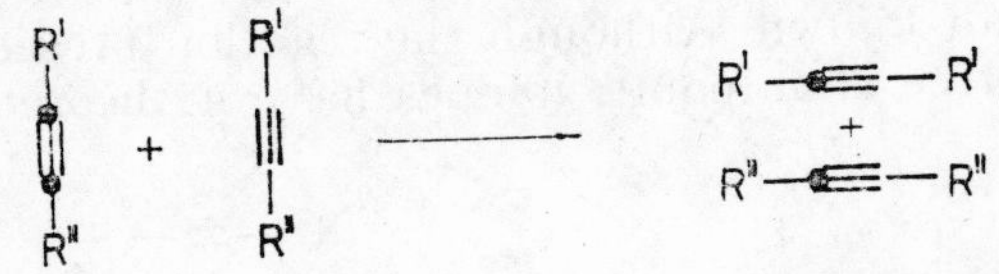

$$\begin{aligned} R^{I} - \equiv - R^{I} \\ + \\ R^{II} - \equiv - R^{II} \end{aligned} \tag{1.35}$$

When the acetylene is symmetrical, a completely degenerate intermolecular topomerization occurs.

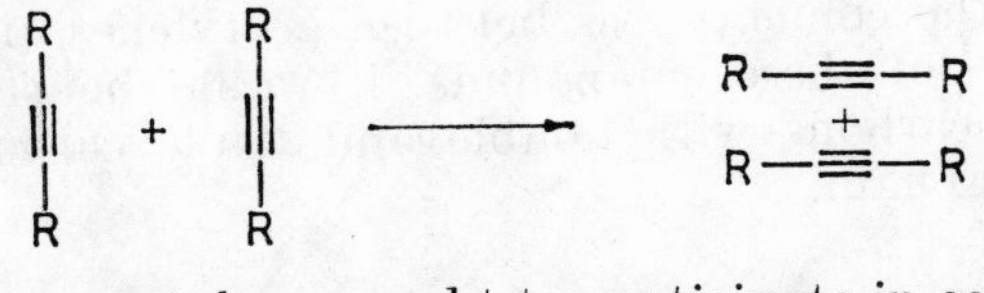

$$\begin{aligned} R - \equiv - R \\ + \\ R - \equiv - R \end{aligned} \tag{1.36}$$

Like the olefins, acetylenes ought to participate in cometatheses,

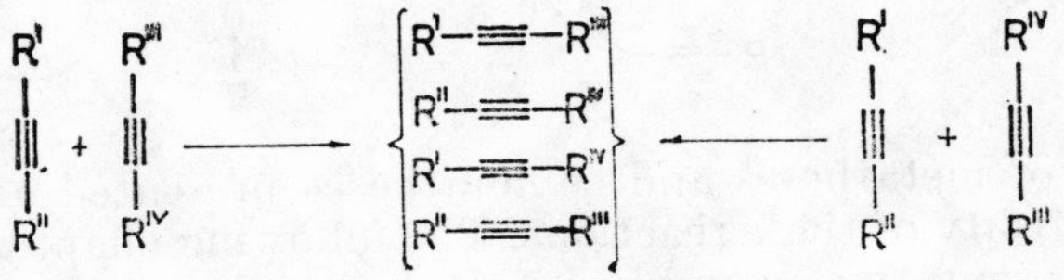

$$\tag{1.37}$$

but, so far, no examples of such reactions have been described.

It is known that if, in the following formulae $n > 6$, the cyclic acetylenes exist; their metathesis could yield cyclic diynes or polyalkynamers.

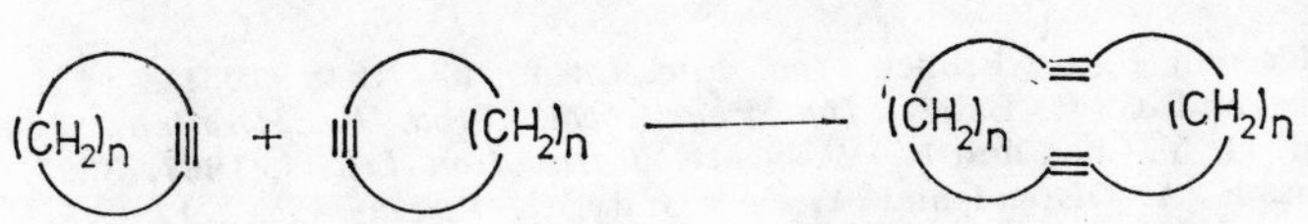

$$\tag{1.38}$$

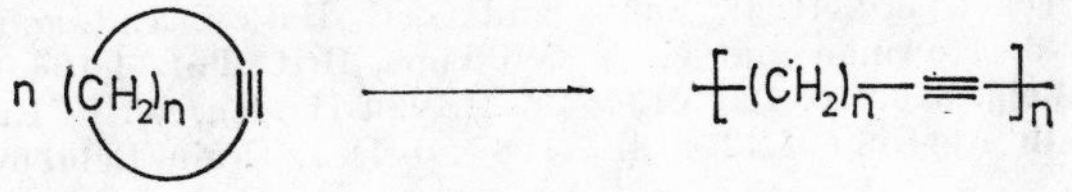

$$\tag{1.39}$$

Similarly, the cometathesis of cyclic acetylenes could form cyclic diynes or copolyalkynamers.

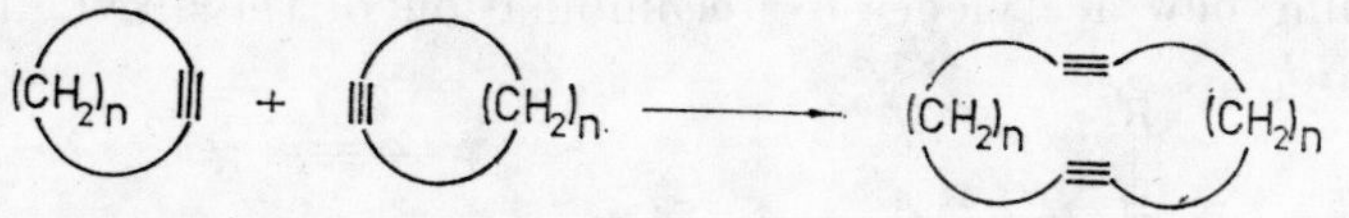

$$(1.40)$$

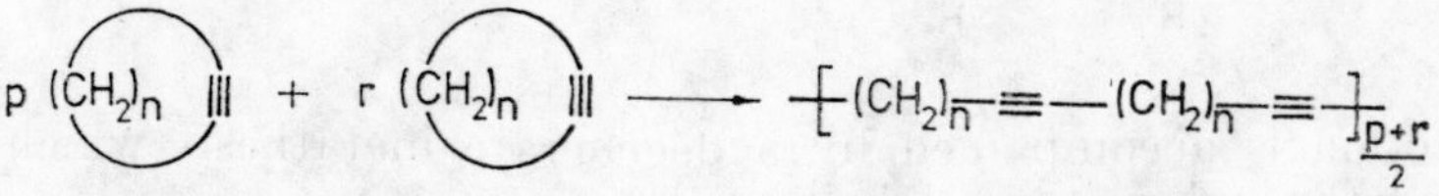

$$(1.41)$$

The cometathesis reaction which might occur between an acetylene and an olefin is not known. Although the reaction presents little interest from the preparative standpoint, it does have a theoretical significance.

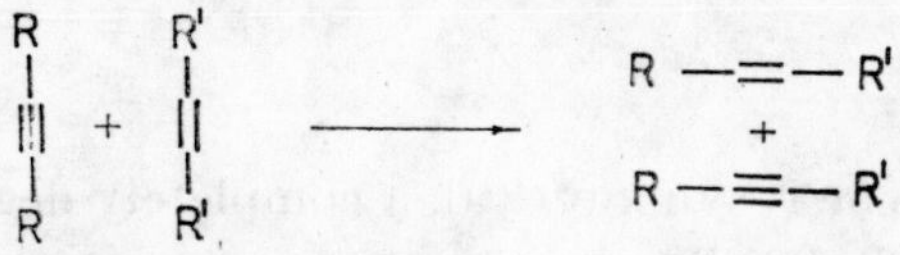

$$(1.42)$$

By extending the cometathesis between acetylenes and olefins to the cyclic counterparts of these compounds, it would be possible to obtain unsaturated hydrocarbons with double and triple carbon-carbon bonds in the same linear chain.

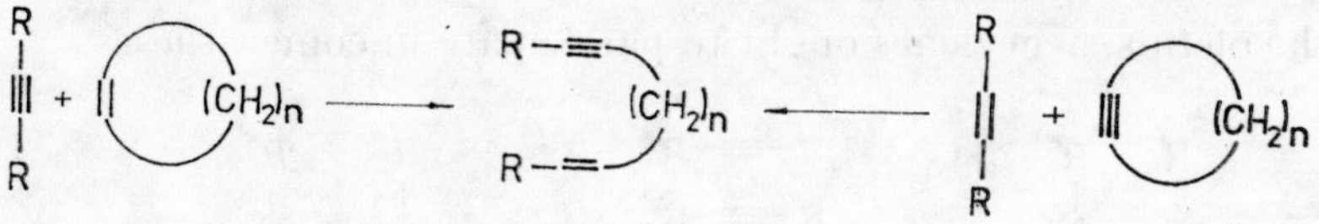

$$(1.43)$$

The types of metatheses and cometatheses presented above illustrate the great versatility of these reactions. The following chapters will describe the most representative examples of such reactions.

References

1. V. Schneider and P. K. Frölich, *Ind. Eng. Chem.*, **23**, 1045 (1931).
2. R. L. Banks and G. C. Bailey, *Ind. Eng. Chem., Prod. Res. Develop.*, **3**, 170 (1964).
3. N. Calderon, H. Y. Chen and K. W. Scott, *Tetrahedron Letters*, **1967**, 3327.
4. N. Calderon, E. A. Ofstead and W. A. Judy, *J. Polymer Sci., A. 1*, **5**, 2209 (1967).
5. a. F. Pennella, R. L. Banks and G. C. Bailey, *J. Chem. Soc., Chem. Commun.*, **1968**, 1548; b. L. F. Heckelsberg, R. L. Banks and G. C. Bailey, *J. Catalysis*, **13**, 99 (1969).
6. a. L. Turner, E. J. Howman and K. V. Williams, **Brit. Pat.** 1,106,015, 23.07.1965; b. R. M. Marsheck, **Belg.** 678,016, 19.09.1966; c. R. van Helden, W. F. Engel, A. J. Alkema and P. Krijger, **Neth. Appl.** 6,608,427, 18.12.1967; d. D. L. Crain, **Belg.** 691,711, 23.06.1967;

e. R. P. Arganbright, **U.S.** 3,702,827, 14.11.1972; f. C. P. C. Bradshaw, **Ger. Offen.** 1,900,490, 31.07.1969; g. British Petroleum Co., **Fr.** 1,554,287, 17.01.1969; h. E. J. Howman and B. P. McGrath, **Brit.** 1,123,500, 14.08.1968; i. R. E. Reusser, **U.S.** 3,511,890, 12.05.1970; j. H. J. Alkema and R. van Helden, **U.S.** 3,536,777, 27.10.1970; k. H. W. Ruhle, **Ger. Offen.** 2,062,448, 16.09.1971; l. H. R. Menapace, **U.S.** 3,636,126, 18.01.1972; m. K. Ichikawa and K. Fukuzumi, *J. Org. Chem.*, **41**, 2633 (1976).

7. a. P. B. van Dam, M. C. Mittelmeijer and C. Boelhouwer, *J. Chem. Soc., Chem. Commun.*, **1972**, 1221; b. P. B. van Dam, M. C. Mittelmeijer and C. Boelhouwer, *J. Amer. Oil Chem. Soc.*, **51**, 389 (1974); c. P. B. van Dam, M. C. Mittelmeijer and C. Boelhouwer, *Fette, Seifen, Anstrichm.*, **67**, 264 (1974); d. S. R. Wilson and D. E. Schalk, *J. Org. Chem.*, **41**, 3928 (1976); e. S. I. Beilin *et al.*, *Izvest. Akad. Nauk S.S.S.R., Ser. Khim.*, **1974**, 947; f. R. Nakamura, S. Matsumoto and E. Echigoya, *Chem. Letters*, **1976**, 1019; g. R. Nakamura and E. Echigoya, *Chem. Letters*, **1977**, 1227; h. A. N. Bashkirov, R. A. Fridman, L. G. Liberov and S. M. Nosakova, **U.S.S.R.** 564,301, 5.07.1977.

8. a. N. Calderon and W. A. Judy, **Brit.** 1,124,456, 21.08.1968; b. D. L. Crain, **Fr.** 1516853, 15.03.1968.

9. a. J. A. Moulijn, H. J. Reitsma and C. Boelhouwer, *J. Catalysis*, **25**, 434 (1972); b. A.V. Mushegyan, V. Kh. Ksipteridis and G. A. Chukhadzhyan, *Arm. Khim. Zh.*, **28**, 674 (1975); *Chem. Abstr.*, **84**, 16693 (1976); c. A. Mortreux, N. Dy and M. Blanchard, *J. Mol. Catalysis*, **1**, 101 (1976).

10. a. G. Dall'Asta and G. Carella, **Brit.** 1,062,367, 11.02.1965; b. K. Nützel, K. Dinges and F. Haas, **Ger. Offen.** 1,720,797, 7.03.1968; c. G. Günther, W. Oberkirch and G. Pampus, **F.** 2003536, 7.11.1969; d. C. P. C. Bradshaw, **Brit.** 1,252,799, 10.11.1971; e. R. Strck and H. Weber, **Ger. Offen.** 2,028,935, 16.12.1971; f. A. J. Amass, *Brit. Polymer J.*, **4**, 327 (1972).

11. L. F. Heckelsberg, R. L. Banks and G. C. Bailey, *Ind. Eng. Chem., Prod. Res. Develop.*, **8**, 259 (1969).

12. G. C. Bailey, *Catalysis Revs.*, **3**, 37 (1969).

13. a. R. L. Banks, *Fortschr. Chem. Forsch.*, **25**, 39 (1972); b. R. L. Banks, *J. Mol. Catalysis*, **8**, 269 (1980).

14. E. A. Zuech, *J. Chem. Soc., Chem. Commun.*, **1968**, 1182.

15. J. L. Wang and H. R. Menapace, *J. Org. Chem.*, **33**, 3794 (1968).

16. V. Drăguţan, *Studii cercetări chim., Acad. R.S.R.*, **20**, 1171 (1972).

17. R. J. Haines and G. J. Leigh, *Chem. Soc. Revs.*, **4**, 155 (1975).

18. L. Bencze and L. Markó, *J. Organomet. Chem.*, **69**, C19 (1974).

19. E. K. Tarcsi, F. Ratkovics and L. Bencze, *J. Mol. Catalysis*, **1**, 435 (1975/76).

20. L. Bencze, J. Pallos and L. Markó, *J. Mol. Catalysis*, **2**, 139 (1977).

21. P. B. van Dam and C. Boelhouwer, *React. Kin. Catalysis Letters*, **1**, 165 (1974).

22. A. A. Olsthoorn and C. Boelhouwer, *J. Catalysis*, **44**, 197, 207 (1976).

23. C. P. Casey and T. J. Burkhardt, *J. Amer. Chem. Soc.*, **95**, 5833 (1973).

24. C. P. Casey, L. D. Albin and T. J. Burkhardt, *J. Amer. Chem. Soc.*, **99**, 2533 (1977).

25. B. A. Dolgoplosk, K. L. Makovetskii, E. I. Tinyakova and I. A. Oreshkin, *Doklady Akad. Nauk S.S.S.R.*, **228**, 1109 (1976).

26. M. T. Mocella, R. Rovner and E. L. Muetterties, *J. Amer. Chem. Soc.*, **98**, 4689 (1976).

27. a. G. Pampus, J. Witte and M. Hoffmann, *Rev. Gen. Caout. Plast.*, **43**, 1343 (1970); b. G. Pampus, G. Lehnert and D. Maertens, *Polymer Prepr. ACS, Div. Polymer Chem.*, **13**, 880 (1972).

28. a. R. Taube and K. Seyferth, *Z. Chem.*, **14**, 410 (1974); b. R. Taube and K. Seyferth, *Z. anorg. allgem. Chem.*, **437**, 213 (1977); c. R. Taube, K. Seyferth, L. Bencze and L. Markó, *J. Organomet. Chem.*, **111**, 215 (1976).

29. J. L. Bilhou and J. M. Basset, *J. Organomet. Chem.*, **132**, 395 (1977).

30. C. P. C. Bradshaw, E. J. Howman and L. Turner, *J. Catalysis*, **7**, 269 (1967).

31. N. Calderon, E. A. Ofstead, J. P. Ward, W. A. Judy and K. W. Scott, *J. Amer. Chem. Soc.*, **90**, 4133 (1968).

32. K. W. Scott, N. Calderon, E. A. Ofstead, W. A. Judy and J. P. Ward, *Adv. Chem. Ser.*, **91**, 399 (1969).

33. a. F. D. Mango and J. H. Schachtschneider, *J. Amer. Chem. Soc.*, **89**, 2484 (1967); b. F. D. Mango and J. H. Schachtschneider, in "Transition Metals in Homogeneous Cataly-

sis", G. N. Schrauzer (Editor), M. Dekker, New York, 1971, pp 223 ; c. F. D. Mango, *Top. Curr. Chem.*, **45**, 39 (1974) ; d. F. D. Mango, *Coord. Chem. Rev.*, **15**, 109 (1975).

34. K. Fukui and S. Inagaki, *J. Amer. Chem. Soc.*, **97**, 4445 (1975).
35. a. G. L. Caldow and R. A. McGregor, *J. Chem. Soc.*, A, **1971**, 1654 ; b. R. G. Pearson, *Top. Curr. Chem.*, **41**, 75 (1973).
36. a. D. R. Armstrong, *Rec. Trav. Chim.*, **96**, M17 (1977) ; b. A. A. Andreev, R. M. Edreva-Kardjieva and N. M. Neshev, *Rec. Trav. Chim.*, **96**, M23 (1977).
37. D. L. Crain, *J. Catalysis*, **13**, 110 (1969).
38. J. C. Mol, J. A. Moulijn and C. Boelhouwer, *J. Chem. Soc., Chem. Commun.*, **1968**, 633.
39. J. C. Mol, L. A. Moulijn and C. Boelhouwer, *J. Catalysis*, **11**, 87 (1968).
40. C. T. Adams and S. G. Brandenberger, *J. Catalysis*, **13**, 360 (1969).
41. A. Clark and C. Cook, *J. Catalysis*, **15**, 420 (1969).
42. J. W. Begley and R. T. Wilson, *J. Catalysis*, **9**, 375 (1967).
43. W. B. Hughes, *J. Amer. Chem. Soc.*, **92**, 532 (1970).
44. M. J. Lewis and G. B. Wills, *J. Catalysis*, **15**, 140 (1969).
45. A. J. Moffat and A. Clark, *J. Catalysis*, **17**, 264 (1970).
46. C. J. Lin, A. W. Aldag and A. Clark, *J. Catalysis*, **45**, 287 (1976).
47. E. V. Pletneva, Yu. N. Usov and E. V. Skortsova, *Neftekhim.*, **15**, 347 (1975).
48. E. S. Davie, D. A. Whan and C. Kemball, *J. Catalysis*, **24**, 272 (1972).
49. R. C. Luckner, G. E. McConchie and G. B. Wills, *J. Catalysis*, **28**, 63 (1973).
50. U. R. Hattikudur and G. Thodos, *Adv. Chem. Ser.*, **133**, 80 (1974).
51. F. H. M. van Rijn and J. C. Mol, *Rec. Trav. Chim.*, **96**, M96 (1977).
52. V. I. Mar'in and V. I. Labunskaya, *Issled. Obl. Sint. Katal. Org. Soed.*, **1975**, 46.
53. a. A. Ismayel-Milanovic, J. M. Basset, H. Praliaud, M. Dufaux and L. de Mourgues, *J. Catalysis*, **31**, 408 (1973) ; b. J. L. Bilhou, J. M. Basset, R. Mutin and W. F. Graydon, *J. Amer. Chem. Soc.*, **99**, 4083 (1977).
54. J. L. Hérisson and Y. Chauvin, *Makromol. Chem.*, **141**, 161 (1970).
55. a. R. H. Grubbs, D. D. Carr, C. R. Hoppin and L. P. Burk, *J. Amer. Chem. Soc.*, **98**, 3478 (1976) ; b. R. H. Grubbs and C. R. Hoppin, *J. Chem. Soc., Chem. Commun.*, **1977**, 634 ; c. R. H. Grubbs, *Inorg. Chem.*, **18**, 2623 (1979) ; d. R. H. Grubbs and C. R. Hoppin, *J. Amer. Chem. Soc.*, **101**, 1499 (1979) ; e. R. H. Grubbs and S. J. Swetnic, *J. Mol. Catalysis*, **8**, 25 (1980) ; f. T. R. Howard, J. B. Lee and R. H. Grubbs, *J. Amer. Chem. Soc.*, **102**, 6876 (1980).
56. a. J. McGinnis, T. J. Katz and S. Hurwitz, *J. Amer. Chem. Soc.*, **98**, 605 (1976) ; b. T. J. Katz and J. McGinnis, *J. Amer. Chem. Soc.*, **99**, 1903 (1977) ; c. T. J. Katz and W. H. Hersh, *Tetrahedron Letters*, **1977**, 585 ; d. T. J. Katz, S. J. Lee, M. Nair and E. B. Savage, *J. Amer. Chem. Soc.*, **102**, 7940 (1980) ; e. T. J. Katz, E. B. Savage, S. J. Lee and M. Nair, *J. Amer. Chem. Soc.*, **102**, 7942 (1980).
57. a. E. L. Muetterties, *Inorg. Chem.*, **14**, 951 (1975) ; b. M. T. Mocella, M. A. Busch and E. L. Muetterties, *J. Amer. Chem. Soc.*, **98**, 1293 (1976) ; c. E. L. Muetterties and E. Band, *J. Amer. Chem. Soc.*, **102**, 6572 (1980).
58. a. C. P. Casey and H. E. Tuinstra, *J. Amer. Chem. Soc.*, **100**, 2270 (1978) ; b. C. P. Casey and A. J. Shusterman, *J. Mol. Catalysis*, **8**, 1 (1980).
59. a. R. R. Schrock, S. Rocklage, J. H. Wengrovius, G. Rupprecht and J. Fellmann, *J. Mol. Catalysis*, **8**, 73 (1980) ; b. J. H. Wengrovius, R. R. Schrock, M. R. Churchill, J. R. Missert and W. J. Youngs, *J. Amer. Chem. Soc.*, **102**, 4515 (1980).
60. a. F. N. Tebbe, G. W. Parshall and D. W. Ovenall, *J. Amer. Chem. Soc.*, **101**, 5074 (1979) ; b. U. Klabunde, F. N. Tebbe, G. W. Parshall and R. L. Harlow, *J. Mol. Catalysis*, **8**, 37 (1980).
61. a. P. G. Gassman and T. H. Johnson, *J. Amer. Chem. Soc.*, **98**, 6055 (1976) ; b. P. G. Gassman and T. H. Johnson *J. Amer. Chem. Soc.*, **98**, 6057 (1976).
62. a. W. S. Greenlee and M. F. Farona, *Inorg. Chem.*, **15**, 2129 (1976) ; b. V. W. Motz and M. F. Farona, *Inorg. Chem.*, **16**, 2545 (1977) ; c. M. F. Farona and R. L. Tucker, *J. Mol. Catalysis*, **8**, 85 (1980).
63. a. R. J. Puddephatt, M. A. Quyser and C. F. H. Tipper, *J. Chem. Soc., Chem. Commun.*, **1976**, 626 ; b. R. J. Al-Essa, R. J. Puddephatt, M. A. Quyser and C. F. H. Tipper, *J. Amer. Chem. Soc.*, **101**, 364 (1979) ; c. R. J. Al-Essa, R. J. Puddephatt, P. J. Thomson and C.F.H. Tipper, *J. Amer. Chem. Soc.*, **102**, 7546 (1980).

64. a. M. Ephritikhine and M. L. H. Green, *J. Chem. Soc., Chem. Commun.*, **1976**, 926;
b. G. J. A. Adam, S. G. Davies, K. A. Ford, M. Ephritikhine, P. F. Todd and M. L. H. Green,
J. Mol. Catalysis, **8**, 15 (1980); c. M. L. H. Green, *Ann. N. Y. Acad. Sci.*, **333**, 229
(1980).

65. a. J. M. Basset, J. L. Bilhou, R. Mutin and. A. Theolier, *J. Amer. Chem. Soc.*, **97**,
7376 (1975); b. J. L. Bilhou, R. Mutin, M. Leconte and J. M. Basset, *Rec. Trav. Chim.*, **96**,
M5 (1977); c. M. Leconte and J. M. Basset, *Nouveau J. Chim.*, **3**, 429 (1979); d. M. Leconte, Y. Ben Taarit, J. L. Bilhou and J. M. Basset, *J. Mol. Catalysis*, **8**, 263 (1980);
e. M. Leconte and J. M. Basset, *Ann. N. Y. Acad. Sci.*, **333**, 165 (1980).

66. a. P. Krausz, F. Garnier and J. E. Dubois, *J. Organomet. Chem.*, **146**, 125 (1978);
b. F. Garnier, P. Krausz and J. E. Dubois, *J. Organomet. Chem.*, **170**, 195 (1979); c. F. Garnier and P. Krausz, *J. Mol. Catalysis*, **8**, 91 (1980); d. F. Garnier, P. Krausz and H. Rudler, *J. Organomet. Chem.*, **186**, 77 (1980).

67. a. H. Rudler, *J. Mol. Catalysis*, **8**, 53 (1980); b. J. Levisalles, H. Rudler and D. Villemin, *J. Organomet. Chem.*, **192**, 195 (1980); c. J. Levisalles, H. Rudler and D. Villemin, *J. Organomet. Chem.*, 192, 375 (1980); d. E. Verkuijlen, *J. Mol. Catalysis*, **8**, 107
(1980).

68. G. Dall'Asta, G. Motroni and L. Motta, *J. Polymer Sci., Part Al*, **10**, 1601 (1972).

69. a. A. J. Amass and C. N. Tuck, *European Polymer J.*, **14**, 817 (1978); b. A. J. Amass
and J. A. Zurimendi, *J. Mol. Catalysis*, **8**, 243 (1980); c. A. J. Amass and T. A. Mc-
Gourtey, *European Polymer J.*, **16**, 235 (1980).

70. a. N. Calderon, J. P. Lawrence and E. A. Ofstead, *Adv. Organomet. Chem.*, **17**, 449 (1979);
b. E. A. Ofstead, J. P. Lawrence, M. L. Senyek and N. Calderon, *J. Mol. Catalysis*, **8**,
227 (1980).

71. a. T. G. Golenko, B. A. Dolgoplosk, K. L. Makovetskii and I. Ya. Ostrovskaya, *Doklady
Akad. Nauk S.S.S.R.*, **220**, 863 (1975); b. B. A. Dolgoplosk, K. L. Makovetskii, Yu. V.
Korshak, I. A. Oreshkin, E. I. Tinyakova and V. A. Yakolev, *Rec. Trav. Chim.*, **96**,
M35(1977); c. B. A. Dolgoplosk, K. L. Makovetskii, E. I. Tinyakova, T. G. Golenko and
I. A. Oreshkin, *Izvest. Akad. Nauk S.S.S.R.*, *Ser. Khim.*, **1976**, 1084; d. I. A. Kopieva,
I. A. Oreshkin, E. I. Tinyakova and B. A. Dolgoplosk, *Doklady Akad. Nauk S.S.S.R.*,
249, 374 (1979); e. I. A. Oreshkin et al., *Vysokomol. Soedineniya*, **19B**, 55 (1977); f. B.A.
Dolgoplosk, I. A. Oreshkin and S. A. Smirnov, *European Polymer J.*, **15**, 237 (1979);
g. B. A. Dolgoplosk, I. A. Kopieva, I. A. Oreshkin and E. I. Tiniyakova, *European Polymer
J.*, **16**, 547 (1980).

72. a. T. J. Katz, S. J. Lee and N. Acton, *Tetrahedron Letters*, **1976**, 4247; b. T. J. Katz
and N. Acton, *Tetrahedron Letters*, **1976**, 4251; c. T. J. Katz, S. J. Lee and M. A.
Shippey, *J. Mol. Catalysis*, **8**, 219 (1980); d. T. J. Katz, S. J. Lee, M. Nair and E.B.
Savage, *J. Amer. Chem. Soc.*, **102**, 7940 (1980).

73. a. K. J. Ivin, D. T. Laverty and J. J. Rooney, *Makromol. Chem.*, **178**, 1545 (1977);
b. K. J. Ivin, D. T. Laverty and J. J. Rooney, *Makromol. Chem.*, **179**, 253 (1978);
c. K. J. Ivin, J. H. O'Donnell, J. J. Rooney and C. D. Stewart, *Makromol. Chem.*, **180**,
1975 (1979); d. K. J. Ivin, D. T. Laverty, J. H. O'Donnell, J. J. Rooney and C. D.
Stewart, *Makromol. Chem.*, **180**, 1989 (1979); e. K. J. Ivin, G. Lapienis, J. J. Rooney
and C. D. Stewart, *J. Mol. Catalysis*, **8**, 203 (1980).

74. D. T. Laverty, M. A. McKervey, J. J. Rooney and A. Stewart, *J. Chem. Soc., Chem.
Commun.*, **1976**, 193.

75. Phillips Petroleum Company, *Hydrocarbon Processing*, **46**, 232 (1967).

76. P. H. Johnson, *Hydrocarbon Processing*, **46**, 149 (1967).

77. R. S. Logan and R. L. Banks, *Hydrocarbon Processing*, **47**, 135 (1968).

78. a. L. Bencze and L. Markó, *Magy. Ken., Lapja*, **27**, 213 (1972); b. L. Markó, *J. Hung.
Chem. Soc.*, **27**, 253 (1972).

79. a. N. Calderon, *Accounts Chem. Res.*, **5**, 127 (1972); b. N. Calderon, E. A. Ofstead and
W. A. Judy, *Angew. Chem.*, **88**, 433 (1976).

80. a. R. H. Grubbs, *Progr. Inorg. Chem.*, **24**, 1 (1978); b. R. H. Grubbs in "Comprehensive
Organometallic Chemistry", G. Wilkinson (Ed.), Pergamon Press, Oxford, 1981, Vol. 5.

81. L. Hocks, *Bull. Soc. chim. France*, **1975**, 1893.

82. a. W. B. Hughes, *Organomet. Chem. Synth.*, **1**, 341 (1972); b. W. B. Hughes, *Adv. Chem. Ser.*, **132**, 192 (1974); c. W. B. Hughes, *Ann. N. Y. Acad. Sci.*, **295**, 271 (1977).

83. a. T. J. Katz, *Adv. Organomet. Chem.*, **16**, 283 (1977); b. M. L. Khidekel, A. D. Shebaldova and I. V. Kalechits, *Usp. Khim.*, **40**, 1416 (1971).

84. J. C. Mol and J. A. Moulijn, *Adv. Catalysis*, **24**, 131 (1975).

85. J. J. Rooney and A. Stewart, *Catalysis*, **1**, 277 (1977).

86. R. Streck, *Chem.-Ztg.*, **99**, 397 (1975).

87. a. R. Taube and K. Seyferth, *Wiss. Z. TH Leuna-Merseburg*, **17**, 350 (1975); b. H. Weber, *Chem. Unserer Zeit*, **11**, 22 (1977).

88. N. Calderon, *J. Macromol. Sci.-Rev. Macromol. Chem.*, **C7**, 105 (1972).

89. G. Dall'Asta, *Rubber Chem. Technol.*, **47**, 511 (1974).

90. a. M. Dimonie, C. Oprescu and G. Hupcă, *Mat. plastice*, **10**, 239 (1973); b. M. Dimonie, C. Oprescu and G. Hupcă, *Mat. plastice*, **10**, 417 (1973).

91. B. A. Dolgoplosk, K. L. Makovetskii, Yu. V. Korshak, I. A. Oreshkin, E. I. Tinyakova and V. A. Iakovlev, *Vysokomol. Soedineniya*, **19A**, 2464 (1977).

92. a. V. Drăguțan, *Studii cercetări chim.*, *Acad. R. S. R.*, **22**, 293 (1974); b. V. Drăguțan, *Studii cercetări chim.*, *Acad. R. S. R.*, **22**, 455 (1974).

93. P. Günther, F. Haas, G. Marwede, K. Nützel, W. Oberkirch, G. Pampus, N. Schön and J. Witte, *Angew. Makromol. Chem.*, **16/17**, 27 (1971).

94. J. Höcker, W. Reimann, L. Reif and K. Riebel, *J. Mol. Catalysis*, **8**, 191 (1980).

95. A. T. Balaban and D. Fărcașiu, *J. Amer. Chem. Soc.*, **89**, 1958 (1967).

96. A. T. Balaban, *Rev. Roumaine Chim.*, **22**, 243 (1977).

97. G. Binsch, E. L. Eliel and H. Kessler, *Angew. Chem.*, **83**, 618 (1971); *Internat. Ed.*, **10**, 570 (1971).

98. P. G. Gassman and T. H. Johnson, *J. Amer. Chem. Soc.*, **98**, 6058 (1976).

Chapter 2

REACTION CONDITIONS

The metathesis reaction proceeds quite readily in the presence of a specific catalytic system. The process can be carried out under heterogeneous phase conditions, at optimum temperature, pressure, and reactant flow rate or in homogeneous phase where these parameters are less critical.

In order to explore the scope and limitations of metathesis and ring-opening polymerization reactions, numerous studies have been carried out. Owing to the importance of the working conditions for the yield and selectivity of the reaction, the present chapter will discuss the catalytic systems and the influence of physical parameters of the process, such as temperature, pressure, reactant flow, and reaction medium.

2.1. Catalytic Systems

Numerous heterogeneous and homogeneous catalytic systems are known to initiate the metathesis reactions of olefins and the ring-opening polymerization of cyclo-olefins. These catalytic systems consist of one, two, or several components, the main one being a compound of a transition metal, especially from groups VI or VII of the periodic system (Mo, W, Re).

2.1.1. Heterogeneous Catalysts

Heterogeneous catalysts are employed on a large scale for the metathesis of olefins and, to a smaller extent, for ring-opening polymerization of cyclo-olefins. These catalysts consist of a promoter, which can be an oxide, a metallic carbonyl, a sulphide or a transition metal salt supported on a carrier, especially alumina, silica, metal oxide, or salt.

2.1.1.1. Types of Heterogeneous Catalysts for Metathesis

Several types of metathesis heterogeneous catalysts have been described according to the chemical compounds employed as promoter and supporting material. The classification of heterogeneous systems adopted in the present review takes into account the chemical structure of the promoter, since the compounds used as support are fewer than those constituting the promoter. According to the chemical structure of the carrier, there exist heterogeneous catalysts based on oxides, carbonyls, sulphides and metal salts.

A. Catalysts based on metallic oxides. The most important heterogeneous catalytic systems, particularly for industrial applications, consist of metallic oxides, especially of molybdenum, tungsten or rhenium, deposited on inert materials.

Molybdenum oxide catalysts (MoO_3). Even in the first studies carried out on propene disproportionation, Banks and Bailey [1] employed catalysts consisting of molybdenum oxide supported on alumina. These catalysts, containing up to 13 % molybdenum oxide, led to a conversion of propene into ethene and 2-butene of maximum $11-12$ %. Later, molybdenum oxide was deposited on other carriers, particularly on oxides such as silicon dioxide (silica), titanium dioxide (titania), zirconium dioxide (zirconia) or mixtures of these oxides, and variable degrees of conversion in the disproportionation of propene were thus obtained.[2] Inorganic salts, such as phosphates, silicates, tungstates, etc. were also used, but most of these carriers gave only very low conversions.[2]

The results obtained by Banks and Bailey [1] and Heckelsberg [2] using supported molybdenum oxide catalysts in the disproportionation of propene are given in Table 2.1. It may be seen that silica is a better carrier than alumina.

TABLE 2.1. Disproportionation of propene over supported MoO_3 catalysts [a]

Catalyst	MoO_3/Al_2O_3	MoO_3/SiO_2	$MoO_3/AlPO_4$	MoO_3-CoO/Al_2O_3
Temperature, °C	50	538	538	205
Product composition (mole%):				
Ethene	—	38.7	40.0	27.4
1-Butene	—	19.1	22.0	8.0
trans-2-Butene	—	24.4	22.0	32.7
cis-2-Butene	—	17.8	16.0	18.4
C_5^+	—	traces	traces	13.4
Conversion, %	$11-12$	28	5	42.6
Selectivity to ethene and butenes, %	100	>95	>95	~94

[a]Data from ref. [1, 2].

The catalytic systems containing other metallic oxides besides molybdenum oxide (such as those of cobalt, chromium, titanium, etc.) have a higher activity than those containing only molybdenum oxide. The former systems are supported on the same carriers as those used for simple molybdenum oxide: i.e. alumina, silica or metallic salts. Banks and Bailey [1] employed, with good results, catalytic systems formed from molybdenum oxide and cobalt oxide to disproportionate a large number of α-olefins, including propene, 1-butene, 1-pentene, 1-hexene and 1-octene. The conversions obtained in experiments with propene are high (over 42%), close to the equilibrium conversion (Table 2.1).

Cobalt oxide can be used in these catalytic systems in a proportion of up to 20%, but, preferably, its concentration should range between 1 and 5%.[3] The molar ratio cobalt : molybdenum should be generally about 1:3. A decrease in the cobalt oxide content lowers the catalytic activity, while an increase gives rise to side-reactions with decreasing metathesis selectivity. In this system, cobalt oxide also reduces the amount of coke deposition on the catalyst. To a certain extent, it also favours the isomerization of α-olefins to internal olefins.

Chromium oxide proved to be a good activator of molybdenum oxide in propene disproportionation. Atlas et al.[4] obtained a propene conversion of 36%, by employing alumina-supported catalyst containing 5% Cr_2O_3 and 11% MoO_3. In this reaction, chromium oxide led to higher selectivity than that of cobalt, repressing to a greater extent the isomerizations and other side-reactions. Titanium dioxide was also used along with molybdenum oxide in catalytic systems to disproportionate olefins. The results thus obtained by Olivé[5] showed how the molar ratio Ti : Mo influenced the catalyst activity. For instance, in the disproportionation of propene, by varying the molar ratio Ti : Mo between 0 and 19, the authors obtained an increase in conversion from 0.7 to 7.8% and in the disproportionation of 1-butene, from 0.3 to 22.4% (Table 2.2).

TABLE 2.2 Disproportionation of propene and butene over the catalytic system MoO_3-TiO_2/Al_2O_3 [a]

Olefin	$Mo \times 10^3$ g-atom/g cat	Ti : Mo molar ratio	Olefin moles minx g-at.Mo	Temp. °C	Conversion %
Propene	0.1	0.0	0.21	68	0.8
	0.1	9.5	0.21	68	4.5
	0.1	19.0	0.21	68	7.8
	0.8	3.3	0.23	68	37.8
1-Butene	0.1	0.0	0.24	68	0.3
	0.1	1.9	0.24	68	3.4
	0.1	9.5	0.24	68	11.5
	0.1	19.0	0.24	68	22.4
	0.8	3.3	0.03	55	59.5

[a]Data from ref. [5].

Although the activity of many catalytic systems based on molybdenum oxide was increased by adding certain activators of the types mentioned above, the selectivity of these systems was unsatisfactory, particularly for practical applications. Their selectivity was improved significantly by the addition of inorganic compounds which contain metals from the first or second group of the periodic system.[6,7] Thus, Heckelsberg [8] prepared a catalytic system of molybdenum oxide supported on silica in the presence of small quantities of sodium hydroxide. By employing this catalyst, he obtained higher conversions and selectivities in the disproportionation of propene, 1-butene, and 2-butene. High selectivity in the disproportionation of 3-heptene, 1-octene and 2-octene was obtained by Crain[9] when treating the alumina-deposited catalytic system based on molybdenum oxide with potassium hydroxide. Bradshaw et al.[10] used solutions of sodium hydrogen carbonate to anihilate the acidic centres of the cobalt molybdate catalyst, eliminating thereby the isomerization of 1-butene to 2-butene which otherwise accompanies the disproportionation of 1-butene. Likewise, Ray and Crain [11] and Mango [12] obtained high selectivity with catalytic systems of cobalt molybdate for cyclo-olefin cleavage, by treating the molybdate with aqueous solutions of potassium hydroxide. Van Helden and co-workers [13] prepared cobalt molybdate catalysts by introducing potassium, sodium, rubidium or caesium ions into the system, during the preparation of the catalysts.

Tungsten oxide catalysts (WO_3). Among the most active heterogeneous catalysts known for the metathesis of olefins are those containing tungsten oxide as a promoter.[14] Tungsten oxide is used in a low concentration, usually 1 to 10%; it is deposited on silica or alumina, and leads readily to high conversions of olefins (Table 2.3).

TABLE 2.3. Disproportionation of propene over the catalytic system WO_3/SiO_2 [a]

WO_3, %	1	2	3	4.5	6	8
Conversion, %						
After 1 hr	13.5	33.6	29.7	41.5	37.8	41.6
After 2 hrs	15.1	31.0	29.0	41.8	37.8	42.4

[a]Data from ref. [14].

As can be seen in Table 2.3., the catalysts with 4 to 8% tungsten oxide are very active in the disproportionation of propene. Nevertheless, the degree of conversion depends appreciably on the working conditions. Thus, a catalyst containing only 3% tungsten oxide on silica, at 427°C, led to 44.8% conversion in propene and 2-butene, a value very close to the equilibrium value for this reaction (45%).

Catalysts of this type have their optimum activity at higher temperatures, namely within the range of 300 to 500°C, than those based on molybdenum oxide. These temperatures have the advantage that the catalyst maintains constant activity during a longer period of operation through a continuous regeneration, the temperature range being rather high. However an accidental overheating may cause a rapid deactivation through coke deposition on the catalyst surface.

The activity and selectivity of tungsten oxide catalysts were continuously improved by additions of alkali metals or of other compounds. Heckelsberg [15] employed a catalyst of tungsten oxide supported on silica with 0.15% sodium content for the metathesis of propene, 2-butene, 1,3-butadiene and 1,5-hexadiene. Regier [16] treated a catalyst of tungsten oxide with ammonium fluoride, obtaining higher conversions in the reaction of 1-butene. Banks [17] obtained very active catalytic systems by adding 1—2% metallic oxides, such as Nd_2O_3, Pr_2O_3 or Sm_2O_3 into a tungsten-based catalytic system; with such a catalyst containing 1% Sm_2O_3 the conversion increased from 45 to 80%. Similarly, Takahashi improved the activity of this type of catalysts by post-treating the system, after preparation, with ammonia [18] or amines. [19]

Owing to its high activity, tungsten oxide was tested for deposition on a large number of carriers. [2] As well as alumina and silica, other metallic oxides or mixtures of these oxides, metallic phosphate or other salts, organic or inorganic polymers, etc., were also used. In many cases, the conversion depends very much on the carrier and the results obtained with these catalytic systems are interesting from a preparative standpoint.

Catalysts of rhenium oxide (Re_2O_7). A class of heterogeneous catalysts, also of interest industrially, are the catalysts containing rhenium oxide as a promoter. [20-30] Rhenium oxide is deposited on an alumina or silica carrier in a ratio between 1 and 30%, preferably between 14 and 20%, and leads to high conversions when working at lower temperatures than those for molybdenum or tungsten oxide catalysts. Rhenium oxide catalysts are usually employed at temperatures of 20 to 100°C, depending on the promoter concentration and on the nature of the carrier. The advantages of the lower temperatures consist in (i) reducing the number of side-reactions, (ii) improving the selectivity and (iii) lengthening the period of catalyst operation as a result of the reduction of coke and polymer deposition. The conversions of propene obtained with these catalysts under various working conditions can exceed 30%.

Besides alumina and silica, other carriers were also used for rhenium oxide catalysts. Thus, catalytic systems are known with rhenium oxide deposited on titanium, zirconium, thorium or tin dioxides in concentrations ranging from 1 to 20%. [31] These catalysts led to variable conversions of propene, depending on the nature of the carrier and the working conditions (Table 2.4). For instance, tin dioxide favours reaction at lower temperatures, whereas thorium dioxide gives better results at higher temperature.

Numerous improvements have been made to rhenium oxide catalysts. Palmer and McGrath [32] prepared alumina-deposited catalysts containing 10% rhenium oxide and anions such as sulphate, phosphate or fluoride. Such catalysts increased the disproportionation activity and at the same time, by the inclusion of cations of alkali metals such as sodium or potassium, the polymerization side reaction has decreased. These

authors increased the degree of conversion and the yield of the metathesis
of propene or 2-butene with a catalyst containing 2% sodium sulphate.
Van Helden and co-workers [33] showed that the activity of alumina-deposited rhenium oxide catalyst is conserved for longer periods by adding a
mixture of rare metal oxides (La, Nd, Pr, Sm, Gd, Ce, Y) in 25%
concentration.

TABLE 2.4. Conversion of propene over rhenium heptoxide
at various temperatures [a]

Catalyst	Conversion, %		
	50°C	150°C	250°C
$Re_2O_7 \cdot SnO_2$	15.0	2.0	0.0
$Re_2O_7 \cdot ThO_2$	6.8	12.0	2.1
$Re_2O_7 \cdot ZrO_2$	1.9	2.1	0.2

[a] Data from ref. [31].

Catalysts of other metallic oxides. Although the most active heterogeneous catalytic systems for the metathesis of olefins are those based on
molybdenum, tungsten or rhenium oxides, active catalytic systems containing a large range of other metallic oxides as promotors have also been
reported. A first category of such oxides is that of metals belonging to
the 5th group of the periodic system, namely vanadium, niobium and
tantalum. Vanadium pentoxide, V_2O_5, deposited on silica even at 1%
vanadium content, led to a good conversion of olefins : at 538°C and with
a volumetric flow of 250 L/hr, Banks [34] obtained a 13.2% conversion of
propene into ethene and 2-butene. Silica-deposited catalysts of niobium
and tantalum led to 18—20% conversions of propene.

Another category is that of rhodium, iridium and osmium oxides.[35]
This class of catalysts exhibited a marked dependence on the working
temperature. The highest conversions were obtained at 350°C. Rhodium
oxide displayed higher activity in the disproportionation of propene than
iridium or osmium oxides.

The third category of oxides is that of tellurium, lanthanum and tin.
Such catalytic systems deposited on silica or alumina afforded variable
conversions, depending on the promotor concentration ; even a low concentration (3%) of silica-deposited tellurium oxide led to 20% conversion of propene.

There exist oxides of other metals which possess high activity in the
metathesis of olefins. Thus, magnesium oxide treated with a flow of a
reducing gas (either hydrogen or carbon oxide) at high temperature, was
employed by Banks [36] to disproportionate propene.

B. **Catalysts based on metallic carbonyls.** The catalytic systems with
which Banks and Bailey discovered the disproportionation of olefins were
heterogeneous catalysts consisting of molybdenum hexachloride supported

on alumina.[1] These catalysts had a high activity for the disproportionation of a large number of α-olefins, in several cases leading to conversions approaching 50% (Table 2.5).

TABLE 2.5. Disproportionation of olefins over the $Mo(CO)_6/Al_2O_3$ catalytic system [a]

Olefin	Propene	1-Butene	1-Pentene	1-Hexene
% Conversion	25.0	10.0	60.0	54.0
Product composition (mole%)				
Ethene	42.0	8.0	2.0	5.0
Propene	—	34.0	21.0	13.0
Butene	55.0	—	27.0	12.0
Pentene	2.0	18.0	—	15.0
Hexene	1.0	32.0	27.0	—
C_7^+	—	8.0	23.0	55.0

[a] Data from ref. [1].

Under the given working conditions, the molybdenum hexacarbonyl catalyst led to extensive isomerization of the reaction products, particularly in the disproportionation of 1-butene, 1-pentene and 1-hexene. Thus, in the reaction of 1-butene, the ratio 1-butene : 2-butene was 84 : 16 and in the reactions of 1-pentene and 1-hexene the analogous ratios were 33 : 67 and 52 : 48, respectively. Under similar conditions, ethene led to a reaction product consisting of 8% cyclopropane and 12% methylcyclopropane, with 2.7% conversion.

The catalytic system of molybdenum hexacarbonyl supported on alumina behaves similarly to that consisting of tungsten hexacarbonyl. Banks and Bailey [1] obtained conversions of 7% in the reaction of 1-butene and of 44% in the reaction of 1-pentene.

Later on, Heckelsberg, Banks and Bailey [2] obtained more active catalytic systems than those supported on alumina, by employing silica as support for molybdenum or tungsten hexacarbonyl. Such systems allowed the reaction to be carried out at higher temperatures, while maintaining high activity for long periods.

Alumina-supported pentacarbonyl catalysts possess moderate activity in the metathesis of olefins. With such a catalyst, containing 10% rhenium pentacarbonyl, Williams and Turner [37] obtained a 12% conversion of propene, working at 120°C. Such catalysts are usually employed at lower temperatures than those of molybdenum and tungsten carbonyls, in order to achieve a reasonable life for the catalyst.

C. **Catalysts based on metallic sulphides.** Supported molybdenum and tungsten sulphides proved to be efficient catalysts for the metathesis of olefins [2] when silica and alumina were used as supports. The promotor is usually deposited in low concentration, up to 10%. Heckelsberg [38] obtained propene conversions of about 20% with catalytic systems consisting of 5% silica-supported tungsten or molybdenum sulphides (WS_2, MoS_2) (Table 2.6). When the same sulphides were supported on alumina the conversions of propene were lower.

TABLE 2.6. Disproportionation of propene over metal sulphides [a]

Catalyst	5% $WS_2 \cdot SiO_2$	5% $MoS_2 \cdot SiO_2$
Reaction conditions		
Temperature, °C	538.0	538.0
Flow, L/1.hrs	10.0	10.0
Product composition (mole%)		
Ethene	6.7	3.4
Propene	81.7	80.9
1-Butene	3.2	1.7
trans-2-Butene	4.7	12.2
cis-2-Butene	3.7	1.8
Conversion, %	18.3	19.1

[a] Data from ref. [38].

The working temperatures of these catalysts depend on the nature of the carrier. Those supported on alumina lead to optimum conversions at temperatures of 65—300°C, while those supported on silica operate best at temperatures of 200—500°C. High temperatures are not advisable for these systems because sulphides easily promote dimerization and polymerization side-reactions.

D. **Catalysts based on metallic salts.** A wide range of active heterogeneous catalytic systems for metathesis consists of salts of transition metals alone or in combination with organometallic compounds, supported on a carrier. Halides, oxyhalides, sulphates or phosphates were used most often. Alumina, silica, mixtures of oxides or phosphates were used as carriers and recently, in many cases, organic polymers (such as polyethylene, polyisoprene, etc.) were selected as supports.

Alkema and co-workers [39] prepared a series of catalysts consisting of W, Mo, Ti, Zr, Nb, Ta or Re salts supported on an oxide or phosphate of one of the metals in groups IIa—IVa or IVb of the periodic system, which exhibited high activities and selectivities in the metathesis of acyclic olefins. Such a system comprising tungsten hexachloride on a carrier consisting of 75% SiO_2 and 25% Al_2O_3 afforded a 35% conversion of 2-pentene. The selectivity of the system was improved by adding 0.01 to 1% lithium hydroxide. The activity of the system was also increased by treating it with triethylaluminium or with diethylaluminium chloride.

Bradshaw and Blake [40] obtained binary catalysts of tungsten hexachloride and organometallic compounds deposited on alumina or silica which led to higher degree of conversion of 2-pentene when compared to analogous homogeneous systems. Ethylaluminium dichloride, tri-n-butyl and tri-t-butylaluminium or tetrabutyltin were used as organometallic compounds.

Catalytic systems of tungsten hexachloride or molybdenum pentachloride supported on organic polymers were employed by Khidekel and co-workers; [41-43] polyethylene, polyisoprene or polycyclopentadiene were used as supports. Besides tungsten or molybdenum halides, small quantities of Lewis acids ($AlCl_3$, $SbCl_5$, $TiCl_4$, $ZnCl_2$) were added, as well as organometallic compounds (n-C_4H_9Li) or Grignard compounds (n-C_4H_9MgCl, $C_7H_{11}MgBr$).

Smith and co-workers [44] obtained heterogeneous catalytic systems based on organometallic complexes of molybdenum or tungsten supported on silica or alumina. Metal alkyl derivatives or metal acetates were used, and the catalysts obtained proved to have a high activity in the reaction of propene at room temperature. Such a system, consisting of hexamethyltungsten and silica [45] with 7.8% tungsten content, exhibited an activity comparable to that of the molybdenum hexacarbonyl system (with 1.5% molybdenum) on the same carrier.

Another series of heterogeneous catalysts contains π complexes of transition metals supported on silica or alumina. Thus, Morris and co-workers [46] employed compounds with the formula R_mMX_n (where R is an allyl or crotyl radical, M is molybdenum, tungsten or rhenium and X is a halogen) for the metathesis of 1-butene, 1-hexene, 3-heptene, 3,3-dimethyl-1-butene, cyclopentene and 2-norbornene. The reaction of 1-hexene, for instance, yielded a 73% conversion of the starting material.

2.1.1.2. Heterogeneous Catalysts for Ring-Opening Polymerizations

Although most of the catalytic systems used for the ring-opening polymerization of cyclo-olefins are homogeneous [47] ring-opening polymerizations catalyzed by heterogeneous systems are also reported. The heterogeneous systems employed for these reactions belong to the class of metallic oxides supported on alumina or silica, on other metallic oxides or salts as carriers.

Eleuterio [48-49] effected the ring-opening polymerization of several cyclo-olefins employing catalysts based on chromium, molybdenum or uranium oxides supported on alumina, titania (TiO_2) or zirconia (ZrO_2) which had been first reduced with hydrogen at temperatures between 350 and 800°C, and then treated with metallic hydrides (hydrides of alkali metals, of earth alkaline metals, of boron or aluminium) at temperatures between 20 and 300°C. The polymerizations took place in the liquid phase, using a suspension of 0.01 to 10% catalyst in a hydrocarbon which is a solvent for polymer. By working at 100°C, with a catalytic

system consisting of alumina-supported molybdenum oxide which was
then reduced with lithium-aluminium hydride, Eleuterio obtained, after 6 hours of reaction, a 4% conversion of cyclopentene to polypentenamer. Later, Alkema and Van Helden [50] attempted the polymerization
of cyclo-octene with a catalyst consisting of 3.2% cobalt oxide and 9.3%
molybdenum oxide supported on alumina; this is known to be a very
active catalyst for the metathesis of cyclic olefins. The catalytic system
had a low selectivity in the reaction of cyclo-octene, yielding dimers and
oligomers of cyclo-octene with a degree of oligomerization between 3 and 6.

A series of heterogeneous catalysts, consisting of cobalt and molybdenum oxides, supported on alumina, silica, thoria (ThO_2) or on a phosphate of Al, Zr, Ca, Mg or Ti was used by Crain [51] to polymerize cyclopentene, cyclohexene, cyclo-octene and 1,5-cyclo-octadiene. Such a catalyst consisting of 3.4% cobalt oxide and 11% molybdenum oxide supported on alumina led to a 54% conversion of cyclo-octene with 10.5%
selectivity with respect to polymer.

Catalysts based on rhenium oxide proved to be more active in ring-
opening polymerization of cyclo-olefins. Turner and Bradshaw [52] obtained
high yields of polyalkenamers from cyclo-octene and cyclododecene by
employing a catalytic system of alumina-supported rhenium heptoxide,
while Toyoda and co-workers [53] obtained polypentenamer from cyclopentene in the presence of a similar catalytic system.

2.1.1.3. Physical Characteristics

The activity of heterogeneous catalysts, and hence the degree of olefin
conversion, depends to a great extent on the surface characteristics of
the catalysts. We shall review briefly the most important ones: the specific surface, the pore size and the form of the grains.

An increase in the specific surface of a heterogenous catalyst results in
an increase of its activity. Usually, one prepares metathesis catalysts
with a specific surface of over 100 m²/g. When alumina is used as a carrier, it is possible to obtain good activities within the range of 100—
—300 m²/g. Silica-supported catalysts usually have a higher specific
surface, between 300 and 500 m²/g. Nevertheless, there exist cases when
silica catalysts led to low conversions because they had a low specific
surface of 2 m²/g.[14]

The pore size is estimated by reference to the average diameter and
the volume of pores. In order to obtain high activities, it is necessary to
have a diameter of 50—200 Å, but it is possible to obtain satisfactory
results also with smaller sizes. Generally, the volume of the pores is
below 1 cm²/g, but dimensions over this limit are sometimes accepted.

The form of the particles greatly depends on the method used for
preparing the carrier. In general, one uses pellets obtained through extrusion, or grains formed through precipitation. In the case of alumina,

particles of 20—40 mesh are chosen, whereas the finer silica particles can
go down to 60 mesh.

Table 2.7 shows the main physical characteristics of several catalysts
based on metallic oxides.

TABLE 2.7. Physical characteristics of catalytic systems based on metal
oxides

Catalyst	MoO_3/Al_2O_3	$MoO_3 \cdot CoO/Al_2O_3$	Re_2O_7/Al_2O_3	WO_3/SiO_2
Specific surface, m^2/g	178	284	255	345
Pore volume, cm^3/g	—	0. 58	—	0. 98
Average pore diameter, Å	107	82	58	114
Volume density, g/cm^3	—	0. 61	—	—
Particle size (in.)	—	1/16	—	—

The physical characteristics of heterogeneous catalysts depend to a
great extent on the method of preparing the carrier and of applying on it
the promoter, on the activation and regeneration conditions, especially
the temperature and the processing to which the catalyst is subjected
afterwards.

2.1.1.4. Composition of Heterogeneous Catalysts

Depending on the working conditions and the required conversion degree, heterogeneous catalysts for metathesis can have variable concentrations of the promoter deposited on the carrier. Banks and Bailey [1]
investigated how the content of alumina-supported molybdenum oxide
influences the conversion of propene. The catalyst proved to be active
at a content between 1 and 13.2%, but the conversion of propene reached
its maximum value for promoter concentrations between 7 and 12%. In
the catalytic system formed from CoO. MoO_3 and Al_2O_3, the cobalt oxide
is used in a concentration of 2 to 4%. [1] Optimum conversions of propene
were obtained at 3.4% cobalt oxide. The influence of the tungsten oxide
concentration in the catalytic system $WO_3 \cdot SiO_2$ on the degree of conversion of propene was studied by Heckelsberg, Banks and Bailey. [14] For
concentrations of tungsten oxide between 1 and 75% these authors obtained a range of propene conversions (Table 2.8).
Catalysts with high tungsten oxide concentrations lead to good conversions in relatively short reaction times, but these catalysts become deac-

TABLE 2.8. Effect of WO_3 concentration in SiO_2-supported catalysts on the conversion of propene [a]

% WO_3	Conversion, %			
Reaction time, hrs	1	2	3	4
1	13.5	15.1	—	—
2	33.6	31.0	30.4	32.4
3	29.7	29.0	29.2	30.0
4.5	41.5	41.8	36.3	37.8
6	37.8	37.8	—	—
8	41.6	42.4	42.8	39.7
Reaction time, min :		15		
10		47.0		
25		42.0		
50		18.0		
75		8.0		

[a] Data from ref. [14].

tivated more rapidly. Low promoter concentrations maintain constant activity of the system, even if the conversion is lower. Under these conditions, higher selectivities in metathesis products are also obtained.

2.1.1.5. Structure of Heterogeneous Catalysts

A key factor in the metathesis reaction catalyzed by heterogeneous systems is the structure of the active centres existing on the catalyst surface. Nowadays, a significant number of studies try to elucidate the structure of heterogeneous metathesis systems consisting of molybdenum, tungsten or rhenium compounds on alumina or silica. These investigations refer mainly to the determination of the surface composition and the structure of the active centers, the number and nature of active centers, and to the coordination or valence state of the transition metal. This is important especially as the surface structures are continuously changing during activation, evacuation, reaction or other treatments effected.

The structure of conventional catalysts of molybdenum trioxide supported on alumina has been widely studied by various chemical and spectroscopic methods [54-73] which evidenced the existence of several structural species on the catalyst surface. It is likely that on the alumina surface, molybdenum trioxide crystallites exist along with combinations of molybdena and alumina. According to Massoth, [55-58] the molybdena is well dispersed over the alumina support in the form of a molybdena-alumina interaction, having a variable range of bond strengths. Using structural studies on oxidized and reduced molybdena-alumina catalysts by various chemical and physical techniques, Giordano *et al.*[65-67] evidenced the presence of monolayer deposition in catalysts containing up to $15-20\%$ MoO_3 activated at $500°C$. They consider that the chemisorbed molybdenum species, present in the uncalcined, dried catalyst, interact with the carrier to form stable alumina-molybdenic acid complexes at the surface, having octahedral oxomolybdenum configuration in the MoO_3-poorest catalyst and a polyanionic mixed tetrahedral-octahedral structure in the medium composition range; at high MoO_3 concentration, part of the molybdenum is incorporated in octahedral interstices of bulk Al_2O_3 while $Al_2(MoO_4)_3$ is formed and unreacted MoO_3 is left behind as clusters. These authors suggest that the octahedral $[MoO_6]$ species are possible precursor of the $Al_2(MoO_4)_3$ phase.

Stork *et al.*[70] also evidenced, by means of luminescence spectroscopy, the existence of aluminium molybdate, $Al_2(MoO_4)_3$ on the catalyst surface. From the sharpness of the emission bands it is inferred that this compound is formed in more than a monolayer and is reasonably well crystallized. In the formation of aluminium molybdate, Mo^{VI} changes from octahedral to tetrahedral coordination. Similarly, Medema and co-workers [71] found, by infra-red spectroscopy, an aluminium molybdate monolayer on the catalyst surface. In addition, Olsthoorn and Moulijn [73] followed the intensity change of infra-red bands as a function of molybdenum trioxide concentration and demonstrated the formation of a continuous layer of a two-dimensional molybdenum compound covering the support surface.

In catalytic systems of cobalt molybdate on alumina, $CoO \cdot MoO_3 \cdot Al_2O_3$, widely used for heterogeneous metathesis, Okamoto *et al.*[74] found, by X-ray photoelectron spectroscopy, a high content of Co_3O_4 and pseudo-$CoAlO_4$ on the catalyst surface. On the basis of the photoelectron transition intensity ratios, they came to the conclusion that the surface structure of the catalyst was highly sensitive to the preparation method. Thus, bilayer structures are proposed for the catalysts prepared by sequential impregnations, while a separate phase structure is suggested to be plausible for the catalysts prepared by simultaneous impregnation.

Molybdenum trioxide catalysts supported on silica do not differ formally when compared to those on alumina.[75-77] Castellan *et al.*[75] concluded, from electronic spectra and from the results they obtained by titration techniques, that in $MoO_3 \cdot SiO_2$ systems dimolybdate, polymolybdate and silico-molybdic acid are present. From Raman spectroscopic studies and temperature programmed reduction determinations they found that the surface of molybdenum trioxide systems consists of a mixture of crystalline trioxide and molybdenum compounds. The ratio of crystalline trioxide and surface compounds is a function of molybdenum oxide content. The amount of crystalline trioxide increases with the molybdenum concentration and is independent of the impregnation technique. It is interesting to note that Thomas and co-workers [77] found that the activity of $MoO_3 \cdot SiO_2$ systems correlates with the amount of surface compounds and not with that of crystalline trioxide. Moreover, these authors observed that dilution with silica resulted in a considerable improvement in the performance of the catalysts having high molybdenum concentrations.

The surface structure of tungstena-alumina catalysts is still debated. Stork and Pott [78] established by phosphorescence spectroscopy and X-ray diffraction that the catalyst surface contained a significant amount of aluminium tungstate, $Al_2(WO_4)_3$, formed by interaction between promoter and support. They correlated this structure with the activity of the catalytic system in metathesis (Fig. 2.1).

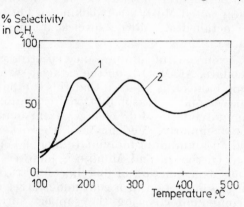

Fig. 2.1. — Dependence of activity of tungsten catalysts on temperature in the metathesis of propene. Curve $1 = Al_2(WO_4)_3$ which had been heated at $550°C$; curve $2 = WO_3/\gamma—Al_2O_3$. (Data from ref. [78].)

Similar behaviour was observed by de Vries and Pott [79] in the study of the processes occurring during catalyst activation or treatment in propene metathesis. These authors showed that aluminium tungstate, $Al_2(WO_4)_3$, can be formed in one or more layers on the carrier as a function of loading and preparation conditions. They concluded that the atomic environment on the surface was best represented by structures I and II, corresponding to hydrated and dehydroxylated $Al_2(WO_4)_3$.

$$\text{(2.1)}$$

I   II

Furthermore, they postulated that during the break-in period the active sites are formed according to equation (2.2)

$$\xrightarrow{\;C_3H_6\;}$$

$$\text{(2.2)}$$

III

In contrast to these results, Thomas *et al.*[80] recently found by laser Raman spectroscopy that tungsten-alumina catalysts do not contain aluminium tungstate on the surface. This finding is confirmed by Kerkhof in temperature-programmed reduction measurements. [81]

Several structural investigations were reported for tungsten-silica catalysts.[82-84] From Raman spectra, Thomas *et al.*[82] inferred that, at low tungsten coverages, tungsten is mainly present as surface compounds, while at high coverages it is mainly present as crystalline WO_3. Further, Kerkhof *et al.*[83] showed, by X-PES measurements, that the formation of tungsten silicate in these systems is improbable. From reduction rates of the catalysts they concluded that along with crystalline WO_3, a hardly reducible surface compound is present; these authors correlate the catalytic activity with this surface compound. A structure like IV is suggested resulting from a reaction between the silanol groups on the surface and tungsten trioxide.

$$\equiv Si-O-H \atop \equiv Si-O-H \quad + \quad O=W{\stackrel{\textstyle O}{\diagdown O}} \quad \longrightarrow \quad {\equiv Si-O \atop \equiv Si-O}{\diagup}W{\stackrel{\textstyle O}{\diagdown O}} \quad + \quad H_2O \tag{2.3}$$

IV

Like the molybdena-alumina catalysts, rhenium heptoxide catalysts on the same carrier have been described. By infra-red spectroscopy, Olsthoorn and Boelhouwer [85,86] evidenced the formation of a monolayer of Re_2O_7 on the catalyst surface. The continuity of this monolayer depends on the rhenium heptoxide concentration. At small content of rhenium, alumina aggregates are exposed. These results are confirmed by the activity studies of Kapteijn [87] and Kerkhof.[88] In addition, Andreev *et al.*[89,90] studying the metathesis of alkenes over $Re_2O_7 \cdot Al_2O_3$, by means of reflectance and EPR spectra observed the formation of a surface aluminium mesoperrhenate structure indicating the existence of anion vacancies at active centers via hydration of the surface at high temperature.

The structure of molybdenum hexacarbonyl supported on alumina has been the subject of many investigations.[91-103] For this system, Davie, Whan and Kemball [91,92] pointed out a remarkable correlation existing between catalyst treatment, catalytic activity and infra-red spectra. Thus, the unactivated catalyst has a single sharp carbonyl frequency corresponding to unchanged hexacarbonyl on the support. The activated catalyst loses this sharp band and presents two wider and broader new bands, while the dead catalyst obtained after exposure to air shows no absorption in the carbonyl region. Davie and co-workers concluded that, during activation, molybdenum hexacarbonyl loses one or more carbonyl groups, passing to a lower symmetry. Similarly, Howe *et al.*[95-97] by infra-red studies on molybdenum hexacarbonyl heated *in vacuo*, evidenced the decomposition of adsorbed $Mo(CO)_6$ during the thermal treatment.

Brenner and Burwell[98-102] succeeded in analysing quantitatively the molybdenum carbonyl species under various conditions. These authors observed a reversible decarbonylation of $Mo(CO)_x$, where $3 \leqslant x \leqslant 6$. The "$Mo(CO)_3$-ads" proved to be stable enough at these temperatures, in particular on alumina which was not extensively dehydroxylated. These compounds are readily transformed, however, in the presence of oxygen; even at room temperature, they are transformed into "$Mo(CO)_2O_2$-ads". In contrast, at temperatures above 300°C complete irreversible decarbonylation of $Mo(CO)_6$ was observed. The decarbonylation is accompanied by the reaction of molybdenum with the hydroxyl groups of the support resulting in reoxidation of molybdenum. Similar results were reported by Olsthoorn and Moulijn [73] in their infra-red spectroscopic study of the $Mo(CO)_6 \cdot Al_2O_3$ system. These authors confirmed the decarbonylation of molybdenum hexacarbonyl during heating and evacuation. Furthermore, they correlated the catalyst activity with pretreatment conditions and the degree of hydroxylation of the carrier.

Tungsten hexacarbonyl on alumina is found to be formally related to the molybdenum analogue. Brenner and Hucul [103] showed that under thermal conditions W^{VI} was transformed into a lower valency state on losing some of its carbonyl groups.

2.1.1.6. Deactivation and Poisoning of Heterogeneous Catalysts

Heterogeneous catalysts can lose their activity under the action of external physical or chemical factors. A primary factor sensibly modifying their activity is the temperature. Heterogeneous catalysts function optimally in metathesis reaction within a definite range of temperature. Below the lower limit of temperature the system is inactive, while above the upper limit it deactivates rapidly. The deactivation at temperatures above the limit may be chiefly a consequence of the modification of the crystalline configuration at the surface of the catalytic system. This modification involves the destruction of the active centres for metathesis, and may give rise to new centres more or less active for side-reactions of the starting material. These side-reactions may lead to the formation of either polymers or coke which will cover all the active centres of the catalyst. Likewise, the modification of the crystalline configuration may determine changes of geometry of the catalyst surface accompanied by variation in the specific surface or changes in the size and form of the catalyst pores. These variations lead to modification of the diffusion and adsorption coefficient (chemisorption included) accompanied by modification of the reaction kinetics and hence of the catalyst activity.

Another factor which modifies the activity of heterogeneous catalysts is the working pressure. An increase of pressure in metathesis reactions promotes side-reactions such as olefin oligomerization and polymerization. These side-reactions may yield new compounds with higher molecular weight, which remain adsorbed on the active surface of the catalyst for a longer period of time, thus reducing the activity of the catalytic system. Through a mechanism similar to that of the increase in temperature, the polymers formed at high pressures yield significant amounts of coke which become deposited on the active centres of the catalyst.

Pronounced deactivation of the heterogeneous catalytic system for metathesis can be caused by the medium in which the reaction occurs. The reaction medium, through its components, can deactivate the catalyst slightly or even poison it. Both phenomena occur through the chemisorption of various substances present in the medium in lower or higher concentration, such substances having a higher affinity for the active centres than the olefins. The chemisorption of traces of oxygen, water, or carbon dioxide may thus block some of the active centres, decreasing the activity of the catalyst. Such deactivations are manifested by many heterogeneous catalytic systems. Thus, the ternary catalytic system of cobalt oxide-molybdenum oxide-alumina is deactivated by traces of oxygen, water, carbon dioxide, hydrogen sulphide, acetylene, propadiene or other dienes or acetylenes existing in the starting material.[3] Some of these substances can be removed during the usual activation of the catalyst, under suitable conditions, whereas others cannot be removed even using special physical or chemical treatments, the catalyst, in the latter case, being poisoned.

The poisoning of metathesis catalysts is generally caused by impurities in the starting material which, through strong chemisorption or even through chemical reactions, block the active centres of the catalyst. In the case of the catalytic system of silica-supported tungsten oxide, substances such as water, acetone, methanol, amines, etc., readily poison the catalyst.[14] Regeneration of the poisoned catalysts is a difficult problem and only rarely succeeds (when employing special chemical treatments).

2.1.1.7. Improvements in Heterogeneous Catalysts

The activity and selectivity of heterogeneous catalysts have been steadily improved by special treatments applied to the carrier or the catalytic system, during or after the application of the promoter. Thus, Banks and Bailey[1] obtained higher conversions of olefins with alumina-supported molybdenum oxide when, prior to impregnating the carrier with the promoter, the former had been treated with organic or inorganic acids. Howman and McGrath,[104] before impregnating γ-alumina with rhenium oxide, washed the alumina with 0.1 N hydrochloric acid and then brought the pH to the value of 5.4. Williams and Turner[37] applied this treatment before impregnating γ-alumina with rhenium carbonyl.

A larger number of improvements were effected on catalytic systems during the impregnation of the carrier with the promoter or even afterwards. Bol'shakov et al.[105] prepared silica-supported tungsten oxide catalytic systems with higher stability than the known catalysts of this type, performing the impregnation of the carrier with the promoter in successive steps accompanied by partial calcination after each impregnation. In another procedure, the same authors used for impregnation a solution of ammonium tungstate containing 13.5—18% hydrogen peroxide and 2.7—6.3% oxalic acid. [106] They also used preformed solutions of tungsten salt and silicagel.[107-108] The activity of some catalytic systems containing silica-supported tungsten oxide was improved by treating the catalytic system with hydrochloric acid or with chlorinated compounds before metathesis. Thus, Pennella [109] observed a considerable increase in the conversion of propene in a series of experiments with catalysts of this type, by treating them with hydrochloric acid, vinyl chloride, chloropropane, or chlorobenzene.

The selectivity of many heterogeneous catalysts for metathesis was increased by adding to the system small amounts of alkali or alkaline earth metal cations. These were introduced, in most cases, as inorganic compounds such as oxides, hydroxides or salts either during the carrier impregnation phase or after this phase. This resulted in neutralization of the acid centres from the carrier, diminishing the side-reactions of olefin isomerization and oligomerization promoted by these centres. Bradshaw and co-workers [10] prepared a cobalt molybdate catalyst by treating it with sodium hydrogen carbonate solutions of variable concentration. With such a system impregnated with sodium ions, these authors studied, in the metathesis of 1-butene, the variation of the degree of conversion, the selectivity in metathesis products, and the isomerization of 1-butene to 2-butene. These authors noticed that on increasing sodium ion into the catalyst, although the catalyst activity decreases, the selectivity in metathesis products increases much more due to the inhibition of the isomerization of 1-butene to 2-butene (see Fig. 2.2).

Crain [9] treated an alumina-supported molybdenum oxide catalyst with potassium hydroxide and obtained high selectivity in the metatheses of 3-heptene, 1-octene, or 2-octene and in the ethenolysis of 2-octene or of 2,3-dimethyl-2-butene. Heckelsberg [8,110] prepared catalytic systems consisting of silica-supported molybdenum or tungsten oxides, modified with 0.05—1% alkali or alkaline earth metal ions such as Li, Na, Rb, Ce, Ca, Sr, or Ba in the form of oxides, hydroxides, carbonates, dicarbonates, sulphates, halides, nitrates or even acetates. The addition of these modifiers was performed in most cases before the calcination of the catalyst. Similarly, Alkema and Van Helden [111] improved the selectivity in metathesis products of α-olefins by incorporating potassium, rubidium or caesium in the catalytic systems based on molybdenum, tungsten, rhodium, vanadium, iridium, niobium, or tantalum oxide. In another procedure, the same authors [112] increased the yield of olefin metathesis and extended the life of the cobalt molybdate catalyst by adding 1.2% potassium hydroxide. Ray and Crain [11] as well as Mango [12] improved the selectivity of cyclo-olefin cleavage via alkenolysis with cobalt molybdate catalysts. Arganbright [113] treated catalytic systems consisting of alumina-supported molybdenum or rhenium oxides with metallic fluorides such as NaF, KF, CdF_2 or ZnF_2; the highest selectivity in the reaction of internal olefins was obtained with the KF-treated catalytic system.

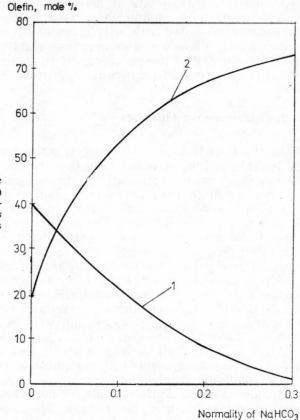

Olefin, mole %

Fig. 2.2. — Variation of the conversion of 1-butene (1) and of the selectivity in metathesis products (2) *versus* the amount of sodium ions in $CoO.MoO_3/Al_2O_3$.[10]

Normality of $NaHCO_3$

The activity of tungsten oxide catalysts was improved by treating the catalytic system with a large range of compounds. Schmidt *et al.*[114] employed alkali metal oxides in a concentration of 0.01—0.4% or halides of these metals in a concentration of 0.1—1.3%, increasing thereby the degree of conversion of propene at high temperatures and pressures. Regier[16] increased the conversion of 1-butene from 50.8 to 58.2% by treating a catalytic system of tungsten oxide with ammonium fluoride. Takahashi[18] improved the activity of this system by adding ammonia or primary, secondary and tertiary amines,[19] or chlorinated compounds such as carbon tetrachloride or chloroform.[115]

Catalysts based on molybdenum, tungsten, or rhenium carbonyl were also modified by the addition of alkali or alkaline earth metal compounds to the system. The procedure patented by British Petroleum Company [116] exemplifies how the treatment of alumina-supported molybdenum hexacarbonyl catalyst with sodium hydrogen carbonate leads to higher conversions of 1-butene, 2-butene, and isobutene.

At present, such improved heterogeneous catalytic systems are exclusively used. Since the number of these systems is large and cannot be exhaustively presented here, only a few more representative patents will be mentioned. These are heterogeneous catalytic systems treated with alkali metals,[117-123] fluoride anions,[124-126] phosphorus compounds,[127,128] derivatives of silver,[129] magnesium[130] or zinc.[131]

2.1.2. Homogeneous Catalysts

When the first homogeneous catalyst was obtained by Calderon, Chen and Scott[132] in 1967, it was noticed that the reaction occurred in a homogeneous phase under particularly mild conditions of temperature and pressure. Furthermore, the reaction in the homogeneous phase proceeds with very high selectivity, because some of the side-reactions of the heterogeneous phase are eliminated. The homogeneous catalytic system of metathesis also made it possible to carry out ring-opening polymerizations of cyclo-olefins [132,133] through a new route with high yields and selectivities, to give polyalkenamers, which are polymers with valuable properties. As a consequence of the versatile possibilities when effecting these reactions in homogeneous phase and of the advantages of carrying them out under mild conditions, investigations concerning both metathesis and ring-opening polymerization were extended considerably; various new fields were opened, such as : how to obtain new homogeneous catalytic systems; the kinetics and mechanism of the reaction; the synthesis of organic compounds with complicated structures; the preparation of macromolecular compounds.

Homogeneous catalytic systems for metathesis and ring-opening polymerization are of three main types. They contain a transition metal compound, particularly of molybdenum, tungsten or rhenium, used either as such or in combination with an organometallic compound. In some cases, they may also contain a third component, usually a compound with oxygen, halogen, nitrogen, etc. In the present chapter, the homogeneous catalytic systems will be classified according to the number of components into the following main classes : one-component (unary), two-component (binary) and three-component (ternary) catalysts. Each class is subdivided according to the chemical nature of the components.

2.1.2.1. One-Component (Unary) Catalysts for Metathesis

One-component (unary) catalysts for metathesis consist of a transition metal compound without any cocatalyst; this compound is generally a coordination complex.

Lewandos and Pettit [134] used a one-component catalyst consisting of toluenetricarbonyltungsten, $CH_3C_6H_4W(CO)_3$, with which they carried out the metathesis of 4-nonene. By working with a molar ratio olefin : tungsten of 4 :1, at 98°C and using heptane as a solvent, they obtained a 28% conversion.

The metathesis took place with this catalyst only after carbon monoxide had been removed from the system; in an enclosed system, the reaction did not occur. Molar ratios olefin : tungsten higher than 4 : 1 or an extra amount of toluene interrupted the reaction.

Cardin, Doyle and Lappert [135] obtained one-component catalytic systems based on rhodium complexes, active for the metathesis of electron-rich olefins. They carried out the cometathesis of tetrasubstituted olefins with the catalytic system $Rh(PPh_3)_2LCl$ (where the ligand L is CO or PPh_3), obtaining high conversions.

The cometathesis of norbornene with a series of acyclic olefins such as 2-butene, isobutene, 2-methyl-2-butene, 1-pentene and 2-pentene was carried out by Rossi et al.[136] by employing a complex consisting of iridium, di-μ-chlorochlorobis(cyclo-octene)iridium. By working with molar rations alkene : norbornene of about 2 : 1 they obtained yields of 11—12% in the reactions of 2-butene and 2-pentene.

The catalytic activity of phenyltrichlorotungsten, $C_6H_5WCl_3$, in the absence of any cocatalyst, was noticed by Opitz, Thiele, Bencze and Markó [137] in the metathesis of 2-pentene. Under the working conditions used by these authors, 2-butene and 3-hexene were obtained as metathesis products in 10% yield. Since the catalyst was only partially soluble in pentane which was used as solvent, it is possible that the activity of this catalyst might be much higher in a totally homogeneous system.

A new category of one-component catalysts which have been developed lately is that of the carbenic complexes of transition metals. Diphenyl-pentacarbonyltungsten-carbene, $(C_6H_5)_2C = W(CO)_5$, was the first carbenic complex used as an initiator for the metathesis of unsymmetrically substituted olefins. Katz and co-workers first used this catalyst for carrying out the cometathesis of 1-octene with 1-hexene [138] and afterwards for the ring-opening polymerization of cyclo-olefins.[139] They obtained very high product selectivities in these reactions. Later, Katz and co-workers used also other carbenic complexes of this type. They carried out the metathesis of 2-pentene [140] with phenyl-methoxy-pentacarbonyl-tungsten-carbene, $C_6H_5(CH_3O) C = W(CO)_5$, and the metathesis of 1-methylcyclo-octene with diphenyl-pentachlorotungsten-carbene $(C_6H_5)_2C = WCl_5$. Casey and Burkhardt [141] demonstrated the activity of the catalytic system diphenyl-pentacarbonyltungsten-carbene mentioned above, in the cyclopropanation of unsymmetrically substituted olefins. With another carbenic complex, ditolylpentacarbonyltungsten-carbene, $(CH_3C_6H_4)_2C = W(CO)_5$, Casey et al.[142] observed the transfer of the ditolyl-carbenic ligand to an alkylidene group in the reaction of the following olefins : 1-pentene, isobutene, 2-butene, 2-methyl-2-hexene and styrene. The cyclopropanation of styrene was carried out by Casey et al.[143] employing a new complex, phenyl-methyl-pentacarbonyltungsten-carbene, $C_6H_5(CH_3)C = W(CO)_5$. This complex proved to be particularly active for this type of reaction.

2.1.2.2. One-Component Catalysts for Ring-Opening Polymerization

The catalytic activity of one-component systems in the ring-opening polymerization of cyclo-olefins was noticed before the discovery of metathesis. Thus, Dall'Asta and Carella[144] obtained polyalkenamers in the polymerization of cyclo-olefins with WCl_6, WBr_6 or $WOCl_4$, without any cocatalyst. The yields were low but the stereospecificity of the polymer was high.

Later, Oshika and Tabuchi,[145] using $MoCl_5$, $ReCl_5$ or WCl_6 in CCl_4 or CS_2 as reaction media, polymerized norbornene, exo-trimethylenenorbornene and endo-di(cyclopentadiene) to the corresponding polyalkenamers, obtaining good yields. The stereospecificity of the reaction product was high. The polymerization of norbornene with these catalytic systems was studied recently by Laverty et al.[146] who observed the formation of dimers and oligomers along with the polyalkenamer.

Another category of catalysts of this class is represented by ruthenium, osmium and iridium halides. Natta, Dall'Asta et al.[147-149] polymerized cyclobutene and 3-methylcyclobutene with $RuCl_3$ in a highly polar medium, such as water or ethanol. A polyalkenamer with a higher content of trans configuration was obtained when ethanol was used as reaction medium. Michelotti and Keaveney[150] noticed that the hydrates of ruthenium, osmium or iridium trichloride in ethanol are particularly active in the polymerization of norbornene. The catalyst activity in the reaction decreased in the order $Ir^{3+} > Os^{3+} > Ru^{3+}$. The osmium-based system yielded a polyalkenamer with a high content of cis configuration, whereas the iridium-based system yielded a polyalkenamer with a high content of trans configuration.

Several of iridium or ruthenium complexes were shown to lead to selective ring-opening polymerization of cyclo-olefins. These complexes are formed from a cyclo-olefin with one or two double bonds and one iridium or ruthenium halide.[151,152] Porri et al.[151] employed di-μ-chlorochlorobis(cyclo-octene)iridium for the polymerization in good yield of norbornene.

Among aryltungsten compounds, phenyltrichlorotungsten $C_6H_5WCl_3$ catalyzes the ring-opening polymerization of cyclo-olefins in the absence of cocatalysts. By employing this compound, Grahlert et al.[153] obtained appreciable polyalkenamer yields in the polymerization of cyclopentene.

The carbenic complexes which have recently proved active in the metathesis of olefins also initiate the ring-opening polymerization of small-ring and medium-ring cyclo-olefins. For instance, Katz and co-workers[139] used diphenyl-pentacarbonyltungsten-carbene, $(C_6H_5)_2C = W(CO)_5$, as a catalyst to polymerize 1-methylcyclobutene, cycloheptene and norbornene, affording the corresponding polyalkenamers. The catalytic system displayed during these reactions a pronounced control of stereochemistry.

Various catalysts based on π complexes of transition metals were used by Körmer et al.[154] to polymerize a large number of cyclo-olefins such as cyclobutene, cyclopentene, cyclo-octene, cyclo-octadiene, cyclododecatriene and norbornene. These catalysts proved to be particularly active, leading to very high yields of vinylic polymers or polyalkenamers. It is to be noted that the systems based on tungsten or molybdenum led to ring-opening polymerization, while those based on nickel, cobalt, palladium or rhodium led only to vinylic polymerization (Table 2.9).

TABLE 2.9. Polymerization of cyclo-olefins with
π-al-lylic systems [a]

Catalytic system	Opening way, %	
	Opening at the double bond	Ring opening
$(\pi\text{-allyl})_4 Zr$	70	30
$(\pi\text{-allyl})_3 Cr$	70	30
$(\pi\text{-allyl})_4 Mo$	0	100
$(\pi\text{-allyl})_4 W$	0	100
$(\pi\text{-allyl})_2 Ni$	100	0
$(\pi\text{-allyl})_3 Co$	100	0
$(\pi\text{-allyl } PdX)_2$	100	0
$(\pi\text{-allyl})_2 RhX_2$	100	0

[a] Data from ref. [154].

2.1.2.3. Two-Component (Binary) Catalysts for Metathesis

The two-component (binary) catalysts for metathesis contain a salt of a transition metal from groups VI or VII of the periodic system, usually a halide or a complex, and a cocatalyst. In most cases, the cocatalyst is an organic compound of a metal in groups Ia — IVa of the periodic system, but it is also possible to use as cocatalyst a Lewis acid or another similar compound. Depending on the nature of the cocatalyst, this class of catalytic systems is divided into Ziegler-Natta type and Friedel-Crafts type two-component catalysts. In the first case, the cocatalyst is an organometallic compound or a precursor thereof, while in the second case the cocatalyst is a Lewis acid.

A. **Two-component catalysts of the Ziegler-Natta type.** Calderon and co-workers [155] used, in their first studies concerning the metathesis of 2-butene with 2-butene-d_8, the two-component catalytic system $WCl_6 \cdot C_2H_5AlCl_2$. In comparison with the tri-component catalyst containing, in addition ethanol, the two-component system gave a lower conversion of the starting material. Later, this two-component catalyst was used also by other authors in the metathesis of terminal or internal olefins. Immediately after the $WCl_6 \cdot C_2H_5AlCl_2$ system was reported as a catalyst for metathesis, Wang and Menapace [156] used a new two-component catalytic system in which the aluminium compound was replaced by butyllithium, viz. $WCl_6 \cdot n\text{-}C_4H_9Li$. These authors obtained conversions up to 50% for 2-pentene with this catalyst, using a molar ratio Li : W = 2. Simultaneously, Zuech [157] employed, instead of tungsten hexachloride, nitrosyl complexes of molybdenum or tungsten in the form $L_2Cl_2(NO)_2M$ (where the metal M is Mo or W and the ligand is Ph_3P, C_5H_5N, Ph_3PO etc); the cocatalyst was an organometallic compound. e.g. ethylaluminium dichloride $C_2H_5AlCl_2$, diethylaluminium chloride $(C_2H_5)_2AlCl$, aluminium methylsesquichloride $(CH_3)_3AlCl_3$ and aluminium ethylsesquichloride $(C_2H_5)_3Al_2Cl_3$. The systems based on molybdenum were more active than those based on tungsten.

The three types of two-component catalysts, $WCl_6 \cdot C_2H_5AlCl_2$, $WCl_6 \cdot n\text{-}C_4H_9Li$ and $L_2Cl_2(NO)_2M \cdot C_2H_5AlCl_2$ represent the basis for developing new binary catalytic systems which have been used for the metathesis of a large variety of olefins.

Catalysts of the type $WCl_6 \cdot R_mMX_{3-m}$. The first catalyst of this class is the system $WCl_6 \cdot C_2H_5AlCl_2$ which, as has been mentioned above, was used by Calderon and co-workers [155] in the metathesis of 2-butene with 2-butene-d_8. The catalytic activity of this binary system was low, however. Kothari and Tazuma[158] demonstrated the alkylating activity of the system when working in an aromatic medium. Thus, the reaction of 2-pentene carried out with $WCl_6 \cdot C_2H_5AlCl_2$ in benzene yielded a mixture of phenylpentane, phenylbutane and phenylhexane. The last two compounds are formed at the 10% level as a consequence of the alkylation of benzene with the metathesis products of 2-pentene.

Höcks, Hubert and Teyssié [159-161] determined quantitatively the effect of the reaction parametres on the competition between metathesis and alkylation when the reaction of 2-pentene was carried out in benzene, toluene, xylene and mesitylene in the presence of the system $WCl_6 \cdot C_2H_5AlCl_2$. The alkylation-promoting activity of this catalytic system was also noticed by Uchida et al.[162] in the reaction of terminal olefins such as 1-butene, 1-pentene and 1-octene or of trisubstituted olefins such as 2-methyl-2-butene. In an aliphatic medium, the metathesis prevails, but it is accompanied by side-reactions of isomerization and disproportionation type. When the reaction of 1-octene or 1-undecene in the presence of the system $WCl_6 \cdot C_2H_5AlCl_2$ was carried out in hexane, Hümmel and Ast [163] obtained a large number of α-olefins as reaction products, in the former case olefins with a $C_7—C_{15}$ skeleton, in the latter with $C_8—C_{21}$ skeleton.

A more active catalyst was obtained on replacing the ethylaluminium dichloride by methylaluminium sesquichloride, $(CH_3)_3Al_2Cl_3$. Nakamura et al.[164] effected the metathesis of esters, nitriles, ketones and unsaturated amides with the system $WCl_6 \cdot (CH_3)_3Al_2Cl_3$, obtaining good yields. The systems WCl_6 and $(C_2H_5)_2AlCl$ or $(C_2H_5)_3Al_2Cl_3$ proved less active than those based on $(CH_3)_2Al_2Cl_3$ and $C_2H_5AlCl_2$.[165]

Catalytic systems where the alkylaluminium halide was replaced by similar compounds of other metals also proved active in the metathesis reaction. Boelhouwer and co-workers [166] obtained high conversions of 2-pentene with $WCl_6 \cdot (C_6H_5)_3SnCl$ and $WCl_6 \cdot (C_6H_5)_2SnCl_2$ catalysts. Unlike the arylstannic halide, the alkylstannic halides did not always lead to good results : whereas Boelhouwer [167] succeeded in carrying out the metathesis of unsaturated fatty esters with the system $WCl_6 \cdot (CH_3)_3SnCl$, Takagi et al. [168] failed to obtain any reaction of 2-heptene with $WCl_6 \cdot (C_4H_9)_3SnCl$ or $WCl_6 \cdot (C_4H_9)_2SnCl_2$.

The systems consisting of WCl_6 and Grignard compounds of the type C_4H_9MgI, related to the above catalysts, led to high conversions of olefins. In the reaction of 2-heptene, Tagaki and co-workers [169] obtained 50% conversion working with a molar ratio Mg : W = 2. In comparison with $WCl_6 \cdot C_2H_5AlCl_2$, this system promoted alkylation to only a very low extent (below 3%), when working in benzene as a reaction medium.[170]

Numerous catalytic systems of this class are known in which the tungsten halide is replaced by halides or other derivatives of various transition metals. Thus, Uchida et al. [171] prepared catalysts in which WCl_6 was replaced by $MoCl_5$ or $ReCl_5$. These authors used the two-component catalytic systems $MoCl_5 \cdot (C_2H_5)_2AlCl$ and $ReCl_5 \cdot (C_2H_5)_2AlCl$ to carry out the metathesis of 1-pentene, 2-pentene and 2-hexene [171] with high conversion in the presence of oxygen. The system $Mo(OC_2H_5)_2Cl_3 \cdot (CH_3)_3Al_2Cl_3$[164] led to a good conversion of substituted olefins with functional groups. Similarly, a tungsten alkoxide such as $WCl_6[OCH(CH_2Cl)_2]_2$ or $(C_4H_9)_2AlCl$ proved to be a very selective system for the cometathesis of 2-butene and 1-pentene. Greenlee and Farona [172] obtained high conversions of terminal and internal olefins by employing a system formed from $Re(CO)_5Cl$ and $C_2H_5AlCl_2$. Other rhenium derivatives, such as $ReOBr_3$, $ReOCl_3$ or $ReCl_4$ were used by Kittleman and Zuech [173] in binary and ternary systems for the metathesis of 2-pentene, giving good conversions. These authors used also binary catalysts consisting of lanthanide halides and methylaluminium sesquichloride.[174,175] Though fairly active, lanthanide halides led to low conversions of olefins.

Two-component catalysts of the type $WCl_6 \cdot R_mM$. The first catalysts of this type were those in which the alkylaluminium halide in Calderon's two-component systems was replaced by other alkylaluminium compounds. Uchida et al.[176,177] carried out the metathesis of 2-pentene in good yields by using the system $WCl_6 \cdot (C_2H_5)_3Al$. Surprisingly, this system had low activity in the metathesis of α-olefins,[171] but traces of oxygen brought about an important increase in its activity.[171] Other similar catalysts are systems consisting of $MoCl_5 \cdot (C_2H_5)_3Al$ and $ReCl_5 \cdot (C_2H_5)_3Al$.[171] Nakamura et al.[165] used the compound $(C_8H_{17})_3Al$ together with WCl_6 in the metathesis of unsaturated esters, such as methyl oleate, oleyl acetate and ethyl vinylacetate, and in the cometathesis of these esters with ethene and 1-hexene. In these reactions, instead of tungsten halides, the above authors also used molybdenum and rhenium catalysts, $Mo(C_2H_5O)_2Cl_2 \cdot (C_8H_{17})_3Al$ and $ReCl_5 \cdot (C_8H_{17})_3Al$, respectively.

New systems were obtained on replacing the alkylaluminium compound by organic derivatives of other metals. Tin compounds were most commonly used, but good results were also obtained with compounds of the metals from groups IIa—IVa in the periodic table.

Bradshaw [178] carried out the metathesis of 2-pentene with 50% conversion by means of the $WCl_6 \cdot Bu_4Sn$ system. Later, systems of this type were used by several other investigators. Ichikawa et al.[179] obtained high yields in the metathesis of α-olefins with the catalytic systems $WCl_6 \cdot (C_4H_9)Sn$ and $\cdot WCl_6 \cdot (C_2H_5)_4Sn$. Compounds such as alkyl acetates, acetonitrile, phenylacetylene, di(cyclopentadiene) or various esters raised the selectivity in these reactions. In the metathesis of 1-hexene with the catalytic system $WCl_6 \cdot R_4Sn$, Bespalova et al.[180] noticed that the activity of the catalyst decreases in the order $R = C_2H_5 > C_4H_9 > C_6H_5 > CH_3 > (CH_3)_3 Si$. Boelhouwer and co-workers [166] prepared a series of systems containing phenyl derivatives of tin, lead, antimony, bismuth or mercury as cocatalysts. As can be seen from Table 2.10, most of these systems were very active. In the metathesis of 2-pentene, the highest conversions were recorded for $(C_6H_5)_4Sn$, $(C_6H_5)_4Pb$, $(C_6H_5)_2SnCl$ and $(C_6H_5)_6Sn_2$.

TABLE 2.10 Metathesis of 2-pentene with two-component catalysts $WCl_6 \cdot MPh_mX_n$

Catalytic system	Reaction ime, hrs	Conversion, %
$WCl_6 \cdot Ph_4Sn$	5	48
$WCl_6 \cdot Ph_3SnCl$	5	49
$WCl_6 \cdot Ph_2SnCl_2$	24	14
$WCl_6 \cdot Ph_6Sn_2$	24	5
$WCl_6 \cdot Ph_4Pb$	24	48
$WCl_6 \cdot Ph_3PbCl$	24	38
$WCl_6 \cdot Ph_6Pb_2$	24	10
$WCl_6 \cdot Ph_3Sb$	24	5
$WCl_6 \cdot Ph_3Bi$	24	23
$WCl_6 \cdot Ph_2Hg$	24	3

Ichikawa et al.[181] prepared other active catalytic systems with aryl derivatives of bismuth, lead or tin. With such systems these authors obtained high conversions in the metathesis of 2-heptene. By employing metal alkyl derivatives instead of aryl derivatives, they found a good correlation between the catalyst activity and the molar ratio alkyl derivative : tungsten halide.[182] In two-component (binary) catalysts of this type, WCl_6 can be replaced by $MoCl_5$ or $ReCl_5$. Boelhouwer and co-workers [166] obtained 11% conversion of 2-pentene with the $MoCl_5 \cdot (C_6H_5)_4Sn$ system. However, a similar catalyst, $ReCl_5 \cdot (C_4H_9)_4Sn$, used by the same authors, led to a much higher conversion of 2-pentene.

Catalysts of the type $WCl_6 \cdot RLi$ *or* $WCl_6 \cdot RNa$. Catalysts $WCl_6 \cdot$n-C_4H_9Li were examined by Wang and Menapace[156] shortly after the discovery of the metathesis reaction. The activity of related systems in the metathesis of 2-pentene decreases in the order n-butyl > sec-butyl > tert-butyl, the last one actually being inactive. By employing the $WCl_6 \cdot$-n-BuLi catalyst in the metathesis of 2-heptene, Ichikawa *et al.* showed that there is a close dependence on the molar ratio n-C_4H_9Li : WCl_6 of the activity of the system. The proportion of alkylation reactions of the aromatic solvents decreased with increase in the above ratio, being very low as compared to that of other systems.

Boelhouwer and co-workers[183] carried out reactions of 2-pentene, 3-hexene and 4-octene with $WCl_6 \cdot$n-C_4H_9Li and $WCl_6 \cdot$n-C_3H_7Li. In the reaction of 2-pentene equilibrium conversion was readily obtained, whereas 3-hexene and 4-octene yielded only by-products resulting from disproportionation of the alkyl radicals of the catalytic system.

Ichikawa *et al.*[184] employed cyclopentadienylsodium and phenylethylsodium as cocatalysts in two-component (binary) systems of this type. The metathesis of 2-heptene occurred with maximum conversions for molar ratios Na:W between 1.5 and 3.0. The authors noticed the high selectivity of this catalytic system for metathesis, the very low extent of Friedel-Crafts reactions (below 1%) and the high stability towards poisoning agents as compared with other similar catalysts.

Catalysts of the type $WCl_6 \cdot LiAlH_4$. Chatt and Leigh[185] obtained two-component catalytic systems with an activity comparable to that of the systems containing Grignard or alkyllithium compounds from WCl_6 and $LiAlH_4$. The highest activity of these systems was recorded when their components were mixed prior to the addition of the olefin. In the presence of aromatic solvents, this type of catalyst did not lead to side-reactions (Friedel-Crafts alkylations). However, the formation of isomerization by-products was noticed; indeed, it is known that $LiAlH_4$ isomerizes internal olefins to α-olefins.[186] The replacement of tungsten hexachloride by molybdenum or rhenium pentachlorides produced active catalytic systems of this type. The use of lithium or calcium halides as cocatalyst failed to give any results.

The two-component catalytic systems formed from WCl_6 and $LiAlH_4$ were extensively employed by Rossi *et al.*[187,188] in the metathesis of a large number of α or β-olefins, such as 1-pentene, 2-pentene and 2-octene or of cyclo-olefins such as cyclo-octene or cyclononene. In most cases, the reactions afforded high yields of products.

Catalysts of the type $ML_2(NO)_2Cl_2 \cdot R_mMX_{3-m}$. The first systems of this class were catalysts consisting of $ML_2(NO)_2Cl_2$ (where M is Mo or W, L is Ph_3P, C_5H_5N, Ph_3PO etc.) and $C_2H_5AlCl_2$, $(CH_3)_3Al_2Cl_3$, $(C_2H_5)_3Al_2Cl_3$ or $(C_2H_5)_2AlCl$ obtained by Zuech. These catalysts afforded very high yields in the reaction of 2-pentene. After the discovery of these catalysts, Zuech et al.[189] found a wide range of molybdenum compounds, particularly complexes, which exhibit catalytic activity in combination with aluminium derivatives after being treated with nitrogen oxide. A series of such compounds is presented in Table 2.11.

TABLE 2.11. Metathesis of 1-pentene and 1-octene with catalysts of molybdenum compounds treated with NO [a]

Molybdenum compound	Olefin	Reaction time, min	% Conversion
$MoCl_3(PhCo_2)_2$	1-pentene	27	60
$MoO_2(acac)_2$	1-pentene	60	18
$MoOCl_3$	1-pentene	30	55
$MoCl_5$	1-pentene	30	43
$MoCl_5 + Py$	1-pentene	60	38
$MoCl_5 + Ph_3PO$	1-pentene	60	45
$MoCl_5 + n-Oct_3PO$	1-pentene	30	17
$MoCl_5 + n-Bu_3P$	1-pentene	60	58
$MoCl_4(PPh_3)_2$	1-octene	30	48
$MoCl_4Py_2$	1-octene	90	47
$MoCl_4(n-PrCN)_2$	1-octene	60	40
$C_5H_5Mo(CO)_3I$	1-octene	240	17

[a] Data from ref. [189].

Mononitrosyl complexes of molybdenum were used by Taube and Seyferth [190] in two-component catalytic systems for the metathesis of 2-pentene. They obtained $54-55\%$ conversions of 2-pentene with systems consisting of $Mo(NO)Cl_3$ or $Mo(NO)Cl_2L_2$ [$L=OP(C_6H_5)_3$] and $C_2H_5AlCl_2$. In these complexes, the authors used a wide range of ligands[191] such as propionitrile, acetone, pyridine, dimethylformamide, dimethylsulphoxide etc.

New catalytic systems based on tungsten in which the nitrosyl group was replaced by carbonyl were prepared by Bencze and Markó.[192,193] These authors obtained a 50% conversion of 2-pentene with $W(CO)_3Cl_2(PPh_3)_2 \cdot C_2H_5AlCl_2$. The addition of carbon monoxide to this system or to other complexes such as $W(C_5H_5N)_2Cl_4 \cdot C_2H_5AlCl_2$ and $W[C_2H_4(PPh_3)_2]_2Cl_3 \cdot C_2H_5AlCl_2$ led to a considerable increase in the activity and selectivity of metathesis. Farona and Greenlee [172] obtained very high conversions by using two-component catalytic systems consisting of $Re(CO)_4X$ (X=Cl, Br) and $C_2H_5AlCl_2$. Unlike the tungsten and molybdenum complexes mentioned above, these systems proved particularly stable, being active for several days.

A series of particularly active catalysts consisting of carbonyl complexes of transition metals were obtained by Doyle [194] starting from complexes with the formula $R_4N^+[M(CO)_5X]^-$ (where M is W, Cr and X is Br, Cl) and from CH_3AlCl_2. With the catalyst, $Bu_4N[Mo(CO)_5Cl]$ ·- CH_3AlCl_2 [194], this author obtained a 56% conversion of 1-pentene. In another series of catalytic systems Kroll and Doyle [195] replaced the monometallic carbonyl complex with a similar bimetallic complex with the formula $A^+[(CO)_5M—M'(CO)_5]^-$ (where M—M' is Mo—Mo, W—W, Mo—Re, Mo—Mn and A=Et_4N, Bu_4N, Na etc). The system $(Bu_4N)[(CO)_5Mo—Mo(CO)_5]$ · CH_3AlCl_2 had the highest activity; it gave a 72.6% [195] conversion of 1-pentene and produced cyclohexene from 1,7-octadiene with a conversion of 44%.[196]

Catalysts from carbenic complexes. Kroll and Doyle prepared a series of catalysts formed from ionic carbenic complexes with the formula $R_4N^+[M(CO)_5COR']^-$ (where M=Mo, W; R' = CH_3,C_6H_5) and CH_3AlCl_2 or $(CH_3)_3Al_2Cl_3$. By using these systems, they obtained high conversions in the metathesis of 1-pentene, 2-pentene, 4-methyl-1-pentene, 1-octene and 1,7-octadiene.[197 198] The order of reactivity in the reaction of 1-pentene was : Mo >W and C_6H_5 >CH_3. Lower activity was shown by the similar system obtained from the neutral carbenic complexes $W(CO)_5 —$ $—C(OCH_3)C_2H_5$ and $W(CO)_5[CN(CH_3)_2]CH_3$. However, on adding ammonium salts, e.g. Bu_4NCl, the activity of these systems became even higher than that of the ionic carbenic complexes mentioned above.

B. Friedel-Crafts two-component catalysts.

The catalytic systems of this class contain a Lewis acid as a cocatalyst along with the transition metal compound.

Catalysts of the type $WX_4 · AlX_3$. Dall'Asta and Carella [144] were the first to show that the catalytic activity of the systems containing WCl_6, $WOCl_4$, WCl_2 or WBr_5 is increased by adding a Lewis acid, e.g. $AlCl_3$ or $AlBr_3$. They also noticed that the activity of WCl_6-based catalytic systems is higher if the catalysts are prepared by reducing WCl_6 with aluminium. Later, Singleton [199,200] obtained particularly active catalytic systems comprising WCl_4 or $MoCl_4$ and AlX_3 (X = Cl, Br). The WCl_4 halide was prepared from WCl_6 by reduction with hydrogen and $MoCl_4$ by reduction with $SnCl_2$. With a system consiting of $WCl_4 · AlCl_3$, it was possible to obtain 50% conversion of 2-pentene.

Catalysts of the type $WO_3 · AlX_3$. Hérisson and Chauvin [201] prepared homogeneous and heterogeneous catalysts formed from tungsten compounds and AlX_3 (where X is Cl or Br). These catalysts contain a tungsten compound such as oxide, carbonyl, halide, oxyhalide, acid, isopolyacid, heteropolyacid or other complex salt. The reaction of α-olefins, such as 2-pentene, 2-hexene and 2-octene gave good yields of metathesis products, whereas the reaction of cyclo-olefins always yielded polymers.

Catalysts of the type $ML_2(CO)Cl_2 · AlX_3$. Bencze and Markó [202] showed that adducts such as $W(CO)_3(PPh_3)_2Cl_2 · AlCl_3$ have properties resembling those of similar complexes with $C_2H_5AlCl_2$ and are active in olefin metathesis. Consequently, Bencze and co-workers [203,204] studied a series of properties such as infra-red absorption and conductance of certain systems of the type $W(CO)_3L_2Cl_2 \cdot AlX_3$ (where X is Cl, Br) in methylene chloride, correlating these results with the catalytic activity of the same complexes in metathesis.

Catalysts of the type $WBn_{4-n}X_n \cdot AlX_3$. Aryltungsten derivatives with the formula WBn_4, WBn_3X, WBn_2X_2 and $WBnX_3$ (Bn = benzyl, X = Cl, Br, I) in combination with AlX_3 (where X is Cl, Br) were used by Bencze, Thiele and Marquardt [205] in the metathesis of 2-pentene. The highest catalytic activity was manifested by WBn_3X systems. In this series, the catalytic activity increases in the order Cl<Br<I.

2.1.2.4. Binary Catalysts for Polymerization

Binary catalysts for polymerization like those for metathesis, consist of a transitional metal salt and a cocatalyst.

Depending on the nature of the catalyst, they can be divided into Ziegler-Natta and Friedel-Crafts types. The former ones generally contain an organometallic compound as a cocatalyst, while the latter contain a Lewis acid.

A. Ziegler-Natta binary catalysts. The first ring-opening polymerization of cyclo-olefins was carried out by Merckling and Anderson [206] who employed for norbornene polymerization a catalyst consisting of titanium tetrachloride, $TiCl_4$ and tetraheptyllithium-aluminium, $(C_7H_{15})_4AlLi$. In this reaction they obtained a mixture of vinylic polymers and polyalkenamers; but the activity of the catalytic system was low. However, Ziegler-Natta catalysts continued to be used.

Truett et al. [207] studied the polymerization of norbornene with binary catalytic systems consisting of $TiCl_4$ and organo-aluminium compounds. These authors investigated the influence of the molar ration Al : Ti on catalyzing the two kinds of norbornene polymerization, vinylic and ring-opening. They noted that, for molar ratios Al : Ti > 1, the reaction proceeds through vinylic polymerization and that for molar ratios Al : Ti < 1, the reaction proceeds through ring-opening.

Later, Natta, Dall'Asta et al. [208,209] obtained a large number of binary catalytic systems with which they carried out the polymerization of cyclobutene. On using vanadium and titanium catalysts, these authors noticed that cyclobutene, in the presence of $VCl_4 \cdot (C_6H_{13})_3Al$ and $V(acac)_3 \cdot (C_2H_5)_2AlCl$, yields mainly vinylic polymers, whereas the systems $TiCl_4 \cdot (C_2H_5)_3Al$ and $TiCl_3 \cdot (C_2H_5)_3Al$ lead mostly to polybutenamer.[208] Starting from this observation, Natta and co-workers [209] studied the behaviour of binary systems consisting of transition metal salts from groups V—VII in the periodic table and of organoaluminium compounds, $(C_2H_5)_3Al$ or $(C_2H_5)_2AlCl$ in the reaction of cyclobutene. These authors found that titanium-, vanadium-, chromium- or tungsten-based systems are the most active ones, those based on molybdenum are less active, and those based on ytterbium, uranium, manganese, iron or cobalt are inactive. Among the first group, those with titanium, tungsten, or molybdenum yield polybutenamer along with vinylic polymers and those of vanadium or chromium yield only vinylic polymers (Table 2.12). In all cases, the structure of the polybutenamer was a mixture of *cis* and *trans* configurations with a certain preference for one or the other configuration.

TABLE 2.12. Polymerization of cyclobutene with two-component Ziegler-Natta systems [a]

Catalytic system	Conversion, %	Polymer structure, %		
		% Polycyclo-butylenamer	% Poly-butenamer	
			cis	*trans*
$3TiCl_3 \cdot AlCl_3 \cdot Et_3Al$	100	40	20	40
$TiCl_4 \cdot Et_3Al$	100	5	30	65
$V(acac)_3 \cdot Et_2AlCl$	100	100	—	—
$VCl_4 \cdot Et_3Al$	100	99	0	1
$Cr(acac)_3 \cdot Et_2AlCl$	100	100	—	—
$MoCl_3 \cdot Et_3Al$	3	10	60	30
$MoCl_5 \cdot Et_3Al$	5	30	30	30
$MoO_2(acac)_2 \cdot Et_2AlCl$	55	10	45	45
$WCl_6 \cdot Et_3Al$	100	40	30	30

[a] Data from ref. [209].

Natta and co-workers [210,211] then carried out the polymerization of cyclopentene with a large number of similar transition metal binary catalysts. In this case, the catalytic systems based on vanadium and chromium led to vinylic polymers, those based on molybdenum and tungsten to polypentenamer and those based on titanium to both types of polymers. Among these catalysts, those consisting of $MoCl_5$ or WCl_6 and $(C_2H_5)_3Al$ or $(C_2H_5)_2AlCl$ proved to be the most active ones. Molybdenum pentachloride led to *cis*-polypentenamer with high stereospecificity and WCl_6 led to *trans*-polypentenamer.

Natta and co-workers [212–214] carried out the polymerization of cycloolefins with seven or more carbon atoms in the ring with catalytic systems based on WCl_6 and $MoCl_5$. These authors used as cocatalysts organic derivatives or hydrides of Al, Zn, Mg, Be, or Ca. Tungsten catalysts proved to be more active than those of molybdenum, the catalytic activity decreasing with the increase of the number of carbon atoms in the cyclo-olefin. In most cases, the resulting polyalkenamers had the *trans* configuration at the double bond.

A catalytic system of this type, particularly active for the ring-opening polymerization of cyclo-olefins, [133] is $WCl_6 \cdot C_2H_5AlCl_2$, a binary system used by Calderon and co-workers for the metathesis of acyclic olefins.[155] With this catalytic system, Calderon and co-workers polymerized cyclo-octene, 3-methylcyclo-octene, 3-phenylcyclo-octene,1,5-cyclo-octene, cyclo-decene and 1,5,9-cyclododecatriene; they gave the corresponding polyalke-namers with the *trans* configuration at the double bond. In most cases, the conversions were over 75%, working under mild conditions at normal temperature and pressure.

A large range of binary catalysts based on tungsten halides was used by Günther *et al.* [215-217] to polymerize cyclopentene. The first binary systems prepared by these authors led to polypentenamer, with relatively low yields, up to 35% (Table 2.13).

TABLE 2.13. Polymerization of cyclopentene with various two-component catalytic systems [a]

Catalytic system	Yield, %	% *trans*-Poly-pentenamer
$WCl_6 \cdot i\text{-}Bu_3Al$	35	90.4
$WF_6 \cdot Et_3Al_2Cl_3$	34	17.2
$WCl_6 \cdot Na_3WPh_5$	5	24.4
$WCl_6 \cdot (HSiOCH_3)_4$	8	65.1
$WCl_6 \cdot (HSiOCH_3)_8$	11	55
$WCl_6 \cdot Li_3WEt_6$	4	88.2
$WCl_6 \cdot (allyl)_3Cr$	6	27.4
$WCl_6 \cdot (allyl)_4W_2$	18	60.3
$WCl_6 \cdot HSnEt_3$	35	90.2
$WCl_6 \cdot Ca$	21	90.0
$WCl_6 \cdot Li$	38	80.0
$WCl_5OX \cdot Ca$	66	83.0
$WCl_6 \cdot Al \cdot AlCl_3$	45	80.0
$WBr_5 \cdot Ca$	30	89.0
$Li_3WPh_6 \cdot BCl_3$	52.0	92.0
$Li_3MoPh_6 \cdot BF_3$	43.0	62.0
$Pr_4W \cdot SnCl_4$	20.0	92.0
$WCl_6 \cdot AlCl_3$	45.0	80.0
$Ph_3Cr \cdot WCl_6$	62.0	45.0
$Li_2VPh_6 \cdot MoCl_5$	32.0	32.0

[a] Data from ref. [215—217].

Afterwards, the authors improved the activity and stability of the catalytic systems by using activators (see below, ternary systems, 2.1.2.6).

As can be seen from Table 2.13, polypentenamer with the *trans* configuration at the double bond was obtained in most cases. However, the use of tungsten hexafluoride [217] in the catalytic system $WF_6 \cdot (C_2H_5)_2AlCl$ produced a polypentenamer with 82% *cis* configuration, thus demonstrating that a totally different stereospecificity is generated by tungsten hexafluoride as compared to the hexachloride.

The catalytic systems containing trialkylaluminium exhibit a different type of activity, leading to different stereospecificities, depending on the nature of the alkyl radical and of the cyclo-olefinic substrate. Thus, the system $WCl_6 \cdot (C_2H_5)_3Al$ [211] led, with $40-60\%$ yields, to *trans*-polypentenamer with a high content of *trans* stereoisomer (97%), whereas the catalytic systems $WCl_6 \cdot (i\text{-}C_3H_7)_3Al$ [218] and $WCl_6 \cdot (i\text{-}C_4H_9)_3Al$ [215] had lower activity and led to polypentenamer with a lower content of *trans*-stereoisomer. With the catalytic system $WCl_6 \cdot (i\text{-}C_4H_9)_3Al$, Calderon and Morris [219] obtained, in the polymerization of cyclo-octene, a polyoctenamer with a variable content of *trans* and *cis* configuration depending on the duration of the polymerization.

Binary catalysts of this type were extended also to include other tungsten compounds or other transition metals. With a catalytic system consisting of $W(PhO)_4Cl_2$ and CH_3AlCl_2, [220] it was possible to obtain a polypentenamer in 69% yield containing 84% *trans* configuration at the double bond. In these systems, the tungsten halide can be replaced by $MoCl_5$, [221] $ReCl_5$ [222] or $TaCl_5$. [223] Günther *et al.* [222] obtained a polypentenamer containing 96% *cis* structure and having the properties of natural rubber by employing a catalyst consisting of $ReCl_5$ and $(i\text{-}C_4H_9)_3Al$.

B. Friedel-Crafts binary catalysts. On replacing the organometallic compounds of the Ziegler-Natta binary catalysts with Lewis acids, new Friedel-Crafts catalytic systems were obtained, which proved to have a high activity in the ring-opening polymerization of cyclo-olefins. Dall'Asta and Carella [144] used tungsten-based catalytic systems consisting of WCl_6, $WOCl_4$, WCl_2 or WBr_5 and $AlCl_3$ or $AlBr_3$ to polymerize cyclopentene or other cyclo-olefins and obtained polyalkenamers rich in the *trans* configurations. Marshall and Ridgewell [224] carried out the polymerization of a large number of cyclo-olefins such as cyclopentene, cycloheptene, cyclo-octene, cyclododecene, 1,5-cyclo-octadiene and 1,5,9-cyclododecatriene with catalysts consisting of WCl_6 or $MoCl_5$ and $AlBr_3$. The use of aluminium bromide instead of aluminium chloride in these systems afforded a higher efficiency for the process.

A large variety of catalysts formed from AlX_3 ($X = Cl$ or Br) and inorganic compounds of tungsten (such as tungsten trioxide, tungstic acid, isopolyacids, heteropolyacids or the salts of these acids) was used by Hérisson and Chauvin [201] to polymerize cyclopentene. By employing chlorobenzene as a reaction medium, these authors obtained high yields of polypentenamer, although the products did not possess a simple structure.

Good results in cyclopentene polymerizations were obtained by Nützel *et al.*[225] with binary systems formed from Lewis acids and organic compounds of transition metals from groups $IV-VI$ of the Periodic System. They used especially tungsten, molybdenum, titanium, chromium or vanadium aryl compounds and boron, tin, tungsten or molybdenum halides (Table 2.13). It can be seen that Friedel-Crafts systems also present high stereospecificity in the reaction of cyclopentene as those of the

Ziegler-Natta type. Catalysts such as $Li_3WPh_6 \cdot BCl_3$ and $Pr_4W \cdot SnCl_4$ led to a polypentenamer with 92 % content of *trans* stereoisomer.

Judy [226-229] used several catalytic systems which were very active for the polymerization and copolymerization of cyclo-olefins such as cyclopentene, cyclo-octene, cyclo-octadiene and cyclododecatriene; he obtained a polyoctenamer in 74.5 % yield by using a catalyst of WCl_6, $AlCl_3$ and powdered aluminium. Instead of tungsten hexachloride, Judy has also used WF_6, $MoCl_5$, MoF_6, MoF_4, WOF_4, $WOCl_4$ or $MoOCl_4$.[226] He also obtained polyoctenamer in 82 % yield with another system consisting of $KWCl_6$ and $AlCl_3$ or $AlBr_3$; he [226] also successfully used carbonyl complexes. One such system, namely $W(CO)_4$ (*o*-phenanthroline) $\cdot AlBr_3$, with cyclo-octene afforded a 86.3 % yield of polyoctenamer.

C. **Binary catalysts with π-complexes.** Günther, Nützel *et al.*[215-230] carried out the polymerization of cyclopentene with catalysts containing tungsten hexachloride and allyltungsten or allylchromium. They obtained *trans*-polypentenamer but the yields were low.

Catalytic systems consisting of allyltungsten and aluminosilicates of trichloroacetic acid were employed by Oreshkin *et al.*[231] to polymerize *trans, trans, cis*-1,5,9-cyclododecatriene.

This reaction yielded 1,4-polybutadiene with a certain degree of stereospecificity, the *trans* : *cis* ratio being close to that of the initial monomer.

Tungsten and molybdenum π-complexes in combination with aluminium or gallium halides led to very active binary catalysts. Yufa and co-workers [232] carried out the ring-opening polymerization of 1,5-cyclo-octadiene with (π-allyl)$_4$W and GaBr$_3$ to 1,4-polybutadiene in high yields. The resulting polymer had 40 % *trans* and 60 % *cis* stereoisomer content. However, when employing the system (π-crotyl)$_4$W and AlBr$_3$ to polymerize *trans, trans, cis*-1,5,9-cyclododecatriene, [233,234] these authors obtained polybutadiene with 60 % *trans* and 40 % *cis* stereoisomer content, respectively.

D. **Carbenic binary catalysts.** Carbenic systems, which have recently come into use with good results for the metathesis of olefins have also found applications in a series of binary systems for ring-opening polymerization of cyclo-olefins. Catalysts containing complexes such as $RR'C=M(CO)_5$ [$R = C_3H_5O$, CH_3O; $R' = CH_3$, C_6H_5, $M = W$, Mo] or $Cl_2(Ph_3P)Pd=C(OCH_3)NHC_6H_5$ and $AlCl_3$ or C_2H_5AlCl were prepared by Chauvin *et al.*[235] They obtained polypentenamer in 70 % yield by using such a system. Commereuc *et al.*[236] have also used titanium tetrachloride, $TiCl_4$, along with tungsten carbenic compounds, $(CO)_5W=C(OC_2H_5)R$.

2.1.2.5. Ternary Catalysts for Metathesis

In many cases a third component is added to binary catalysts for metathesis to act either as an activator, promotor, stabilizer of the catalytic system, or as an inhibitor of side-reactions. This substance is usually an organic compound which readily becomes complexed with the basic binary system. The ternary systems thus obtained are very active in metathesis, they are highly selective and possess high stability and show good reproducibility. They can be classified according to chemical composition, as for binary systems.

The first ternary catalytic system consisting of WCl_6, C_2H_5OH and $C_2H_5AlCl_2$ was employed by Calderon and co-workers [132] when they discovered metathesis in homogeneous phase. This system, used initially with 2-pentene, was then extended by Calderon and co-workers to the metathesis of acyclic [155,237] and cyclic olefins [238,239] and by other research groups to various other metathesis reactions. [240-243] Calderon and Chen [237] carried out the metathesis of β-olefins such as 2-pentene, 2-hexene, 2-heptene, or 2-octene and obtained 2-butene and internal olefins. In these reactions, equilibrium conversions were reached. The preferred molar ratio for the catalytic components $WCl_6 : C_2H_5OH : C_2H_5AlCl_2$ varies between $1:1:2$ and $1:1:4$. With such a system, Calderon [238,239] prepared macrocyclic compounds by the metathesis of cyclo-octene, while Wasserman [240] and Wolovsky [241] prepared large-ring carbocycles, catenanes and knots through the metathesis of cyclo-octene and cyclododecene. Küpper and Streck [242,243] obtained cyclopentene of high purity through the metathesis of 1,6-cyclodecadiene with the ternary catalyst $WCl_6 \cdot C_2H_5OH \cdot C_2H_5AlCl_2$ in the molar ratio $1:3:5$.

On replacing ethanol in Calderon's catalytic system by another compound containing oxygen, it was possible to obtain very active catalytic systems for metathesis. One example is that of systems containing acetic acid. Medema et al. [244] obtained high conversions of 2-pentene, close to the equilibrium conversion, depending on the acetic acid content of the catalyst. In this case, the molar ratio between the catalytic system components was rather different ($WCl_6 : CH_3COOH : C_2H_5AlCl_2 = 1:1:8$) in comparison with the system containing ethanol. With such a catalyst, other internal olefins such as 2-hexene, 2-octene and 2-decene also gave the corresponding metathesis products. [245] In another series of ternary catalysts, benzoic acid was used as a complexing agent. Thus, Hughes and Zuech [246,247] disproportionated various β-olefins with ternary catalytic systems consisting of vanadium or niobium halides, benzoic acid and methylaluminium sesquichloride ($Me_3Al_2Cl_3$). However, the activity of these catalysts was low : the reaction of 2-pentene in the presence of the system $NbCl_5 \cdot C_6H_5COOH \cdot (CH_3)_3Al_2Cl_3$ reached a conversion of only 2 %.

Ternary metathesis catalysts which contain amines or ammonium salts have also been described. Wang, Brown and Menapace [248a] carried out the metathesis of α-olefins such as propene and 1-pentene, using ternary catalysts based on WCl_6 or $WCl_6[O(CH_3OOC)C_6H_4O]WCl_5$, $C_2H_5AlCl_2$ or $(C_2H_5)_3Al_2Cl_3$ and aniline. With systems containing metallic complexes $L_nMo(CO)_{6-n}$[L=CO, amine (pyridine), n = 1−3], aluminium compounds, $(CH_3)_3Al_2Cl_3$ or $(CH_3)_2Al_2Cl_4$ and ammonium salts

$R_4NCl(R=C_3H_7, C_4H_9)$, Ruhle [248b] obtained high yields in the metathesis of simple α- and β-olefins such as 1-pentene, 2-hexene, of substituted α- and β-olefins such as 2-phenyl-1-butene, 3-methyl-1-butene and of dienes, such as 1,7-octadiene. From the metathesis of 1,7-octadiene Ruhle obtained quantitative yields of cyclohexene (along with ethene) in the presence of the $(C_5H_5N)_2Mo(CO)_4 \cdot (C_3H_7)_4NCl \cdot (CH_3)_3Al_2Cl_3$ system. Similarly, the catalytic systems obtained by Doyle [194] from monometallic carbonyl complexes $R_4N^+[M(CO)_5X]^-$ and CH_3AlCl_2 or those obtained by Kroll and Doyle [195] from bimetallic carbonylic complexes $R_4N^+[(CO)_5M-M'(CO)_5]^-$ and CH_3AlCl_2 may be considered as ternary systems of the above type when used in α-olefin metathesis taking into account that the preparation of such carbonyl complexes involves the respective carbonylic compounds and ammonium salts. Kubicek and Zuech [249] used pyridine as a complexing agent for zirconium, titanium and hafnium catalytic system. One such system, $TiCl_4 \cdot 2C_5H_5N \cdot C_2H_5AlCl_2$, proved active for the metathesis of 2-pentene.

Catalysts containing triphenylphosphine along with transition metal salts and organoaluminium compounds proved to be active in β-olefin metathesis.[250,251] Kittleman and Zuech [173] used rhenium, manganese or technetium halides, triphenylphosphine and ethylaluminium dichloride for the metathesis of 2-pentene. Among rhenium compounds, the complexes $ReCl_4 \cdot P(C_6H_5)_3$, $ReOCl_3 \cdot P(C_6H_5)_3$ and $ReOBr_3 \cdot P(C_6H_5)_3$ formed active catalysts. The latter complexes led to a conversion of 2-butene amounting to about 30%. Kubicek and Zuech obtained such catalysts, with low activity, from copper, silver and gold salts : they obtained conversions of 0.6% in the metathesis of 2-pentene [251] with a catalyst consisting of $Cu_2Cl_2[P(C_6H_5)_3]$ and $C_2H_5AlCl_2$, and a conversion of 0.2% in the metathesis of 2-heptene [250] with another catalyst, $CuCl[P(C_6H_5)_3]_3$ and $[(CH_3)_3Al_2Cl_3]$.

Zuech [252,253] diversified the binary systems based on nitrosyl complexes [157] by employing a third component in order to extend the range of utilization of these systems. He obtained such ternary systems which can be employed for the simultaneous metathesis and dimerization of olefins. By using such a catalyst formed from $(Ph_3P)_2Mo(NO)_2Cl_2$, $(PH_3P)_2NiCl_2$ and $C_2H_5AlCl_2$, the metathesis and dimerization of propene yielded, with very high conversions, the fractions C_4, C_5, C_6 and C_7 used in petrochemistry.

If aluminium chloride or bromide is added to Wang and Menapace's binary system consisting of WCl_6 and n-C_4H_9 [156] a considerable increase in the activity and selectivity of the metathesis system is observed. Such ternary systems proved their efficiency in the reaction of 2-pentene [254] for molar ratios $AlX_3 : WCl_6$ lower than 1 (Table 2.14). The increase of aluminium halide over this value, although leading to an increase in conversion, considerably reduces the selectivity in 2-butene and 3-hexene. Aluminium chloride proved to be more active than aluminium bromide, but the latter maintained high selectivity in metathesis products.

TABLE 2.14. Metathesis of 2-pentene with $WCl_6 \cdot nBuLi \cdot AlX_3$ (X = Cl, Br) [a]

$AlCl_3$, mmole	0	0.05	0.1	0.2	0.4
Conversion, %	15	42	48	54	76
Selectivity, %	92	86	93	86	54
$AlBr_3$, mmole	0	0.05	01	0.2	0.4
Conversion, %	15	51	40	51	58
Selectivity, %	92	100	100	88	54

[a] Data from ref. [254].

2.1.2.6. Ternary Catalysts for Ring-Opening Polymerization

As in the case of ternary systems for metathesis, ternary systems for the ring-opening polymerization of cyclo-olefins have been described which have, in addition to the transition metal complex and the organometallic compound, a third component, usually containing oxygen, halogen, sulphur or nitrogen. Although the detailed mechanism through which the third component acts has not fully been elucidated, experiments show that this component increases the activity or selectivity, improves the reproducibility, and modifies the solubility of the catalytic system, and it can even initiate the polymerization or control the molecular weight of the polymer. This type of catalytic system was discovered by Dall'Asta and Carella [255] in 1965 and since then it has been very much diversified, numerous investigators having created new systems with high activity and special uses. There are many studies and patents concerning the application of these systems in polyalkenamer syntheses, particularly for obtaining *trans*-polypentenamer, as well as for the synthesis of copolyalkenamers,[256] interpolymers, block polymers and graft polymers.[257]

The first catalytic system of this type obtained by Dall'Asta and Carella [255] consisted of tungsten or molybdenum salts, organometallic compounds and an oxygenated compound such as peroxide, hydroperoxide, alcohol, phenol, molecular oxygen or water. The salts they employed were WCl_6, $WOCl_4$ or $MoCl_5$, the organometallic compounds were $(C_2H_5)_2AlCl$, $(C_2H_5)_3Al$, $(C_6H_{13})_3Al$, $(i\text{-}C_4H_9)_2AlCl$ or $(C_2H_5)_2Be$, and the oxygenated compounds were benzoyl peroxide (Bz_2O_2), t-butyl peroxide, cumyl hydroperoxide, hydrogen peroxide, ethanol, phenol, molecular oxygen, or water. The molar ratio between the components of the system was W : Al : O $=1$: $(0.5-100)$: $(0.5-1)$. With such a ternary system, $WCl_6 \cdot (C_2H_5)_2AlCl \cdot Bz_2O_2$, Dall'Asta and Carella [255] obtained in 40% yield a polypentenamer containing 80% *trans* configuration at the double bond. Later, Dall'Asta, Motroni and Carella [258] employed ternary catalytic systems to polymerize and copolymerize a large number of cyclo-olefins, such as cycloheptene, cyclo-octene or cyclopentadiene.

After the discovery of the metathesis of olefins in homogeneous phase, Calderon and co-workers [133,259] carried out the polymerization of cyclo-olefins with the ternary system $WCl_6 \cdot C_2H_5OH \cdot C_2H_5AlCl_2$. They obtained polyalkenamers with *trans* configuration at the double bond in high yields in the reaction of cyclo-octene, 3-methyl-cyclooctene, 3-phenylcyclo-octene, 1,5-cyclo-octadiene and 1,5,9-cyclododecatriene. Instead of ethanol, Calderon and Judy [260] employed also methanol, allyl alcohol, cumyl alcohol, glycol, phenol, thiophenol or cumyl hydroperoxide.

Ofstead [261] obtained particularly active ternary catalytic systems for the polymerization of cyclo-olefins to polyalkenamers by using tungsten or molybdenum carbonyl complexes. High yields in polyoctenamer were obtained in the polymerization of cyclo-octene (in solution or bulk) by using systems such as $(CO)_4(1.5\text{-}COD)W$ or $(CO)_4(NBD)Mo$, $C_2H_5AlCl_2$ and oxygen, bromine, iodine or cyanogen bromide *. More recently, Ofstead prepared new active systems for cyclopentene polymerization which contain, along with WCl_6 and $C_2H_5AlCl_2$, a hydroxynitrile (cyanohydrin) such as $HOCH_2CH_2CN$ [262-265] or a chlorosilane of the type $ClCH_2CH_2OSiMe$. [266]

A wide range of ternary systems for cyclo-olefin polymerization was prepared by Günther and co-workers, particularly for the polymerization of cyclopentene to *trans*-polypentenamer. [215] In these systems, they used especially oxygenated compounds such as alcohols, epoxides, peroxides, hydroperoxides, acetates, phenols or ethers, nitrogen compounds such as amides, amines or nitroderivatives, halogenated derivatives of these compounds or even inorganic peroxides such as sodium or barium peroxides. Pampus et al. [267-273] employed chloroethane or epichlorohydrin in systems consisting of WCl_6 and $(C_2H_5)_2AlCl$ or $(i\text{-}C_4H_9)_3Al$. With such ternary systems, they obtained *trans*-polypentenamer, copolymers of cyclopentene with butadiene or graft polymers with good elastomeric properties.

2-Cyclopentenehydroperoxide increased the activity of the systems formed from WCl_6 and $(i\text{-}C_4H_9)_3Al$ [274] or from tungsten salts and Lewis acids [225] and the resulting ternary catalysts were used by Nützel and co-workers to polymerize cyclopentene. The systems thus obtained proved to be particularly stable and increased the conversion to *trans*-polypentenamer. The same authors also used nitro-aromatic compounds [275] or inorganic peroxides such as Na_2O_2 or BaO_2 [276] with the former systems mentioned above.

Witte et al. [277,278] showed that, in cyclopentene polymerization, α-halogenated alcohols or halophenols impart higher stability to catalytic systems based on WCl_6 and $(i\text{-}C_4H_9)_3Al$ or $(C_2H_5)_2AlCl$ relative to other similar compounds. Examples of such halogenated alcohols are 2-chloro-ethanol, 2-bromo-ethanol, 1,3-dichloro-2-isopropanol, 2-chlorocyclohexanol, 2-iodo-cyclohexanol, o-chlorophenol. With a catalytic system containing $WCl_6 \cdot ClCH_2OH$ and $(i\text{-}C_4H_9)_3Al$, these authors obtained *trans*-polypentenamer in very high yields and with 94 % *trans* configuration at the double bond. Acetals such as $CH_2(OCH_3)_2$, $CH_2(OCH_2CH_2Cl)_2$, $CH_3CH(OC_2H_5)_2$,

---

* 1,5—COD = 1,5—cyclo-octadiene ; NBD=norbornadiene.

$Cl_3CCH(OCH_3)_2$ or $C_6H_5CH(OC_2H_5)_2$ employed by Schön [279] also proved to be good stabilizers for the systems WCl_6 and $(C_2H_5)_2AlCl$ in the polymerization of cyclopentene. The same authors also used various epoxides such as etheneoxide or 1-buteneoxide for increasing the activity of the systems containing WCl_6 and $(i-C_4H_9)_3Al$. By employing a catalyst formed from equimolar amounts of WCl_6, C_2H_4O and $(i-C_4H_9)_3Al$, they obtained in 82% yield a polypentenamer having 90.5% *trans* configuration at the double bond.[280]

Oberkirch *et al.*[281,282] used an unsaturated halogenated hydrocarbon such as vinyl chloride in catalytic systems formed from WCl_6 or $TaCl_5$ and organoaluminium compounds. These compounds increase the activity of the system but, at the same time, they make possible the adjustment of the molecular weight of the polyalkenamer. By employing the system $WCl_6 \cdot (i-C_4H_9)_3Al \cdot CH_3{=}CHCl$, they obtained in 78% yield *trans*-polypentenamer having an intrinsic viscosity of 3.27 dl/g (toluene 25°C). Similarly, Streck and Weber,[283–285] by using tungsten- or molybdenum-based systems containing a halide or an unsaturated ether succeeded in adjusting the molecular weight of the polypentenamer to the desired value.

Küpper *et al.*[286,287] obtained polymers with *cis* configuration at the double bond by using organic acids or their salts in the catalytic systems formed from WCl_6 and $C_2H_5AlCl_2$. Cyclo-octene gave polyoctenamer with 63% *cis* configuration at the double bond in the presence of the catalyst $WCl_6 \cdot CH_3COOH \cdot C_2H_5AlCl_2$.

Finally, ternary catalytic systems of the type described above were also used by other research groups to polymerize monocyclic and polycyclic cyclo-olefins: mention should be made of Kurosawa,[288–290] Ueshima,[291–293] Uraneck,[294–297] and other authors.[298,299]

2.2. Reaction Parameters

The reactions of metathesis and ring-opening polymerization in heterogeneous or homogeneous catalysis are processes of industrial importance so it is of particular interest to know the influence exerted by the reaction parameters on the course of the reaction. In order to achieve maximum selectivities and yields, optimization of the process is necessary, exploiting the correlation between the reaction parameters and the resulting output. The parameters which will be studied are the temperature, the pressure and the nature of the reactive medium.

2.2.1 Reaction Temperature

A first important factor which considerably influences the course of the process in heterogeneous or homogeneous catalysis is the working temperature.

In heterogeneous catalysis, a very wide range of working temperatures is used, depending on the catalytic system. Banks and Bailey [1] studied the disproportionation of propene with a catalyst consisting of molybdenum oxide — cobalt oxide — alumina within a temperature range of 90 and 260°C. Although the conversion increases with rise in temperature, optimum yields of products are obtained only when working at lower temperatures. The effect of temperature on the conversion of propene, on the product yield, and on the composition of the reaction products is given in Table 2.15. At high temperatures, the conversion and yield decrease considerably owing to the appearance of side-reactions.

TABLE 2.15. Disproportionation of propene over CoO-MoO_3/Al_2O_3 at various temperatures [a]

Temperature, °C	93	163	205	300
Product composition, %				
Ethene	6	10.0	11.6	1.9
Propene	82.1	62.8	57.4	85.4
Propane	traces	0.1	0.2	1.5
1-Butene	0.1	1.9	3.4	2.5
trans-2-Butene	7.3	14.6	13.9	2.7
cis-2-Butene	3.4	7.5	7.8	1.5
Isobutane	—	traces	traces	0.1
1-Pentene	—	traces	0.2	0.2
2-Pentene	—	0.8	1.7	0.1
2-Methylbutene	—	0.4	0.5	0.2
n-Pentane	—	0.1	traces	—
C_6^+	1.1	1.8	3.3	2.9
Conversion, %	17	37.9	42.6	14.6
Yield, %	93	91.4	86.2	58.9

[a] Data form ref. [1].

It can be seen that with increasing temperature both the percentage of hydrocarbons with more than 4 carbon atoms and the amount of propane or of other saturated hydrocarbons increase as a result of hydrogenation reactions. Furthermore, the amount of residues and of coke deposited on the catalyst increases continuously with increasing temperatures.

Ogata and Kamiya [300,301] effected a comparative study on the disproportionation of propene with the catalytic system molybdenum oxide — alumina in gas and liquid phase. The selectivity in ethene and butene decreased with the increase of temperature in both cases while the quantity of 1-butene in the butene fraction increased (Table 2.16). It is important to follow the variation of the ratio trans : cis-2-butene with temperature. For the reaction in the gaseous phase, this ratio has a higher value than the equilibrium one and it decreases towards the equilibrium value at higher temperatures. It can be noticed that the reaction exhibits a more pronounced kinetic control at lower temperatures.

TABLE 2.16. Disproportionation of propene over the MoO_3/Al_2O_3 catalyst in gas or liquid phases [a, b]

Reaction phase	Reaction temperature °C	Conversion %	Selectivity to ethene and butenes %	*trans-: cis*-butene mol/mol
Gas	0	15.0	99.93	2.37
	20	15.0	99.89	2.26
	35	15.0	99.80	2.20
	50	15.0	99.67	2.12
	85	15.0	99.60	1.98
Liquid	0	9.2	99.85	1.80
	20	13	99	1.83
	35	21	98	1.93
	60	22	97	1.96
	85	22	95	2.02

[a] Reaction in gas phase was carried out at constant conversion (15%)
[b] Data from ref. [301].

The results obtained by Milanović *et al.*[302] in the kinetics of propene disproportionation with a catalytic system of $CoO \cdot MoO_3 \cdot Al_2O_3$ correspond to the effect of temperature on the activity and selectivity of the reaction previously observed by the preceding research group.

Bradshaw *et al.*[10] have drawn interesting conclusions by studying the disproportionation of 1-butene with a catalyst of molybdenum oxide — cobalt oxide — alumina in the temperature range between 122 and 245°C. The conversion increased with the temperature, reaching a maximum of 41.4% at 180°C (Table 2.17). Selectivity towards the primary disproportionation product, $C_2 + C_6$, decreased on increasing the temperature. On

TABLE 2.17. Effect of temperature on the disproportionation of 1-butene [a]

Temperature, °C	122	140	160	180	200	222	245
Product coxposition,							
C_2H_4	25.2	21.4	15.0	8.9	6.0	4.3	4.1
C_3H_6	25.2	29.7	37.2	46.1	48.6	49.1	48.3
C_5H_{10}	27.0	28.9	31.4	32.8	35.5	36.5	38.3
C_6H_{12}	20.3	17.7	12.5	9.2	7.4	7.0	6.4
C_7H_{14}	1.4	1.2	1.8	1.9	1.5	1.5	1.3
C_8H_{16}	0.8	0.6	1.4	0.4	0.9	1.2	1.5
C_9H_{18}	0.1	traces	0.7	0.2	0.1	0.4	0.1
Conversion, %	18.3	23.1	31.9	41.4	35.2	34.4	23.4
Selectivity to C_2+C_6%	45.5	39.5	27.5	16.1	13.4	11.3	10.5
Selectivity to C_3+C_5,%	52.2	58.6	68.6	78.9	84.1	85.6	86.6

[a] Data from ref. [10].

the other hand, the extent of isomerization (the content of 2-butene in the butene fraction) increased and, consequently the amount of disproportionation by-products, namely $C_3 + C_5$, increased on raising the temperature.

In the homogeneous phase, the reaction proceeds very easily at room temperature, but variations of temperature considerably modify the conversion and the yield of the process. Hughes,[303] using the complex $Py_2Mo(NO)_2Cl_2.EtAlCl_2$ as a catalytic system, studied the influence of temperature on 2-pentene metathesis. This author showed that there is only a restricted temperature range within which the homogeneous catalytic system maintains a convenient level of activity.

Wang and Menapace [304] found that, in the metathesis of 2-pentene with a catalytic system consisting of WCl_6 and n-BuLi, it is possible to obtain high conversions if temperatures higher than room temperature are used. They obtained maximum conversions and optimum yields within the range of 61 to 130°C (Table 2.18). Considerable deactivation of the

TABLE 2.18. Effect of temperature on the metathesis of 2-pentene [a, b]

Temp., °C	25			61			94			120		
Reaction time (seconds)	1800	3600	7200	60	120	240	60	120	240	60	120	240
Conversion, %	23	40	43	40	43	43	28	47	48	20	23	37
Selectivity, %	100	100	98	100	100	88	100	100	100	100	97	98
trans-: cis-2-pentene	1.5	2.6	3.0	2.4	2.1	1.8	1.4	2.2	2.2	1.1	1.2	1.5
trans-:cis-2-butene	0.9	1.4	1.6	1.2	1.4	1.2	1.0	1.4	1.5	1.0	1.0	1.2

[a] Catalyst: $WCl_6 \cdot$ n-BuLi;   [b] Data from ref. [304]

catalytic system was noticed with increase of temperature above the higher threshold. Surprisingly, the ratio trans : cis-2-butene (having the value of 1.6) did not vary with the reaction temperature within the range 61—130°C. This fact indicates pronounced kinetic control of the reaction within this temperature range.

2.2.2. Reaction Pressure

The pressure at which the metathesis or ring-opening polymerization occurs is another important factor which determines the development of the process. Taking into account the fact that the diffusion of molecules is important in metathesis, especially in the gaseous phase, the reaction pressure will modify the reaction kinetics under these conditions and will therefore also influence the outcome of the reaction. Because of this, the maintenance of a constant pressure when carrying out the reaction in the gaseous phase becomes an essential factor for the optimum development of the process.

The effect of pressure on conversion, within a pressure range between 1 and 30 atm was studied by Heckelsberg and Bailey [14] for the disproportionation of propene in the presence of a silica-supported tungsten oxide catalyst. By working at 260°C and having the flow rate adjusted in such a way as to obtain a constant contact time, they obtained a significant variation of the conversion with change in pressure. At atmospheric pressure the conversion of propene is very low and after reaching an optimum at 10 atm the value of the conversion decreases again.

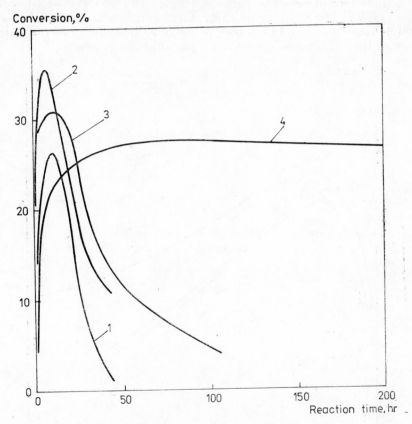

Fig. 2.3. — Effect of pressure on the disproportionation of propene in gass phase (catalyst MoO_3/Al_2O_3) : 1 — 130 kg/cm^2 ; 2 — 30 kg/cm^2 ; 3 — 15 kg/cm^2; 4 — 1.03 kg/cm^2. [255]

Sodesawa, Ogata and Kamiya [305] studied the influence of pressure on the disproportionation of propene within the range 1 to 130 atm, performing the reaction in both gaseous and liquid phases. They used an alumina-

supported molybdenum oxide catalyst. These authors observed the important effect exerted by the variation of pressure on the conversion of propene under the given working conditions. In both phases, the conversion
increased with increase of pressure, reaching a maximum value at about
15—30 atm. For the reaction in the gas phase, at high pressure, the conversion decreased suddenly with extension of the reaction time, convenient
conversions being obtained for short reaction times (Fig. 2.3). Thus, the
increase of pressure leads to rapid deactivation of the catalyst.
However, at low pressures, e.g. 1.03 atm, it was possible to maintain high
catalytic activities for an indefinite period of time. For the reaction in liquid
phase, the effect of pressure varies, owing to the presence of the solvent
in the system. The activity of the catalyst may remain high for a long
period of time (Fig. 2.4.) because the solvent (heptane) removes from the
surface of the catalyst compounds which cause inhibition.

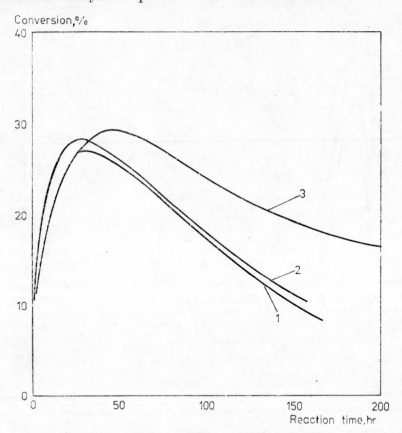

Fig. 2.4. — Effect of pressure on the disproportionation of propene in liquid phase (catalyst MoO_3/Al_2O_3): 1 — 50 kg/cm²; 2 — 30 kg/cm²;
3 — 15 kg/cm².[255]

Similar results are known also for the disproportionation of 2-pentene. Milanović *et al.*,[302] carrying out the reaction of 2-pentene in the gas phase with the catalytic system $CoO \cdot MoO_3 \cdot Al_2O_3$, noticed rapid deactivation of the catalyst with increase of pressure. Figure 2.5 represents the change

$v \times 10^6$ mol. s^{-1}. g^{-1}

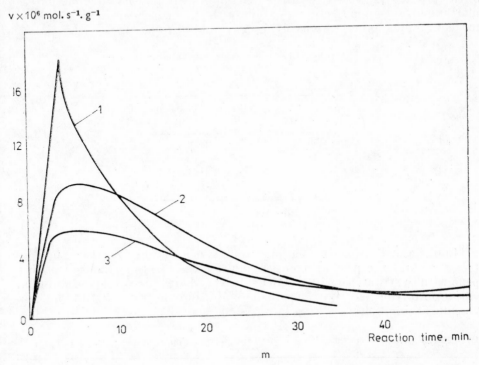

m

Fig. 2.5. — Effect of pressure on the disproportionation of 2-pentene : 1—300 Torr ; 2—156 Torr ; 3 — 55 Torr.[252]

of the reaction kinetics for various pressures versus time, working at 70°C. The effect of pressure on the disproportionation of 2-pentene was interpreted by assuming the occurrence of side-reactions (promoted by increased pressure) which deactivate the catalyst. Among these side-reactions, the polymerization of the olefin on the acid centres of the carrier and the reduction of the active centres of the catalyst by 2-pentene are the most plausible.

2.2.3. Reactant Flow

The flow of reagents is another factor which influences the metathesis when operating a continuous process with heterogeneous catalytic systems. Heckelsberg, Banks and Bailey [14] obtained a decrease in the conversion of propene on increasing the feeding flow of propene (Table 2.19). This increase of the feeding flow leads to a decrease in the contact time, which consequently reduces the degree of conversion of the starting material. The modification of the contact time can cause changes in the composition of the product mixture and in the reaction selectivity. This fact has particular importance for the optimization of the metathesis process, especially for continuous processes involving gas-solid interfaces.

TABLE 2.19. Conversion of propene versus the space rate of starting material [a, b]

Space rate, WHSV g/L·hr	Propene conversion, %
20	45.5
40	44.4
60	44.2
80	41.8

[a] Catalyst : WO_3/SiO_2 ;
[b] Data from ref. [14].

Banks and Bailey [1] studied in detail the correlations between the contact time and the product composition, for the disproportination of propene with the catalytic system $CoO \cdot MoO_3 \cdot Al_2O_3$. They observed different variation of the selectivity in ethene and 2-butene from that of the conversion (Table 2.20).

TABLE 2.20. Effect of contact time on the disproportionation of propene [a,b]

Contact time, seconds	1	2	3	8	14	30
Propene conversion, %	6.9	17.3	26.2	37.2	42.9	41.0
Efficiency in ethene and butene, %	98.0	95.9	93.6	91.2	94.1	86.6
Product composition, %						
Ethene	35.5	35.0	30.2	32.4	29.8	23.7
1-Butene	0.5	0.9	1.9	1.7	3.6	6.6
trans-2-Butene	38.0	36.0	40.0	36.9	39.9	37.0
cis-2-Butene	24.0	24.0	21.5	20.2	20.8	19.3
C_5^+	2.0	4.1	6.4	8.8	5.9	13.4

[a] Catalyst : $CoO-MoO_3/Al_2O_3$;   [b] Data from ref. [1]

Banks [3] also obtained variable conversions by changing the contact time (Table 2.21) in the disproportionation of 1-butene with a catalytic system of the same type as mentioned above. Under the given working conditions, side-reactions of isomerization and disproportionation compete with the main reaction, leading to appreciable amounts of propene, 2-butene, pentenes and other higher homologues.

TABLE 2.21. Effect of contact time on the disproportionation of 1-butene [a, b]

Contact time, minutes	30	60	120
1-Butene conversion,%	21.7	48.0	63.3
Product composition, %			
Ethene	0.1	3.5	4.2
Propene	2.7	8.4	14.8
1-Butene	78.3	52.0	36.7
trans-2-Butene	3.7	5.3	4.9
cis-2-Butene	5.6	7.0	4.3
1-Pentene	0.1	0.2	0.4
trans-2-Pentene	1.9	6.9	12.4
cis-2-Pentene	0.8	2.7	5.0
C_6^+	6.8	14.0	17.3

[a] Catalyst: $CoO-MoO_3/AlO_3$;   [b] Data from ref. [3].

The ratio olefin : catalyst also modifies the degree of conversion and the product yield in the disproportonation of 1-butene over $CoO \cdot MoO_3 \cdot Al_2O_3$ catalyst. By maintaining constant flow of 1-butene and by gradually increasing the amount of catalyst, Banks [3] obtained different degrees of conversion (Table 2.22).

TABLE 2.22. Effect of the ratio 1-butene : catalyst on the product composition [a,b]

1-Butene : catalyst ratio	68	33	12	7.6
1-Butene conversion, %	5.5	9.8	41.7	48.0
Product composition, %				
Ethene	0.06	0.6	1.5	3.5
Propene	0.04	0.5	4.9	8.4
1-Butene	94.5	90.2	58.3	52.0
trans-2-Butene	1.9	2.3	9.6	5.3
cis-2-Butene	2.5	3.5	10.0	7.0
1-Pentene	traces	—	0.1	0.2
trans-2-Pentene	traces	0.2	4.2	6.9
cis-2-Pentene	traces	0.1	1.7	2.7
C_6^+	1.0	2.5	9.7	14.0

[a] Catalyst: $CoO-MoO_3/Al_2O_3$;   [b] Data from ref. [3].

For the reactions in the homogeneous phase, which usually occur in a discontinuous process, the ratio olefin : catalyst allows one to obtain optimum, well defined yields for a fairly wide variation range in most cases. Thus, in the metathesis of 2-pentene with the catalytic system $WCl_6 \cdot Bu_4Sn$, Ichikawa [306] noticed a considerable change of the yield relative to the ratio 2-heptene : WCl_6. The reaction proceeded readily for a ratio within the range from 50 to 200 mole/mole. In the reaction of 1-octene, the same authors obtained maximum values of the yield for ratios 1-octene : WCl_6 between 20 and 400 mole/mole.

2.2.4. Reaction Medium

In heterogeneous catalysis, the reaction was carried out mostly in biphasic systems, in the presence of solid catalysts, with the olefin in the gas or liquid phase ; a triphasic system was also used, the olefin being in gaseous phase, and liquid diluent being used as reaction medium. In homogeneous catalysis, the reaction medium is an inert solvent which solvates the catalytic system and dissolves both the reagents and the reaction products. It is important that the reaction medium, regardless of the state of aggregation, should not contain any trace of impurities such as water, oxygen, polar organic compounds, etc., which act as inhibitors for the catalytic system.

2.2.4.1. Diluents for Heterogeneous Catalysis

The reaction in the gas phase proceeds with higher efficiency if one adds to the olefin flow, in a certain proportion, a gaseous diluent which is inert towards the catalytic system. As a gaseous diluent, one may use an inert gas, (nitrogen, argon), saturated hydrocarbon with small molecules such as methane, ethane, propane or butane, or another gas (hydrogen, carbon dioxide, etc.). In some cases, it is possible to use gaseous diluents which are also activators for the catalytic system, such as carbon monoxide or even oxygen. Sometimes, the olefin itself can play the part of the diluent. Thus, in order to obtain high efficiency in the process, it is necessary to work with high olefin flows, i.e. conditions which afford a low conversion of the starting material but which extend the working period of the catalyst. Gaseous diluents such as saturated hydrocarbons are, in most cases, present in the olefin because the olefin is obtained from a petroleum fraction. Such compositions already containing diluent can be used directly in metathesis. In other cases, the diluent is introduced in a desired proportion into the olefin flow.

In the disproportionation of propene with the catalytic system $CoO \cdot MoO_3 \cdot Al_2O_3$, Banks and Bailey[1] obtained different conversions and selectivities according to the ratio propene : propane in the starting material. When propane constituted 40% of the input flow, higher conversions of olefin were obtained. Alkema and van Helden[112] used hydrogen as a gaseous diluent, in a proportion of 10—60%, in the metathesis of propene, 1-butene and 2-butene in the presence of the catalytic system $CoO \cdot MoO_3Al_2O_3$, obtaining higher yields than without hydrogen. An increase in the molar ratio hydrogen : propene resulted in a raise in the conversion degree of propene. Hydrogen also led to extension of the catalyst life and diminished the inhibiting effect of dienes e.g. butadiene, which may be present in the starting material.

Gaseous diluents play several roles in metathesis. First, they moderate the thermal regime, absorbing the heat generated by possible overheating of the catalytic system and thus help to maintain the activity of the latter. Secondly, they make possible the removal from the catalyst surface of heavy product deposits, which would destroy the active centres of the catalytic system. In this way, they contribute to activation and regeneration of the catalytic system, extending the operating lifetime of the latter. Third, the gaseous diluents diminish the effect of the inhibitors which cannot be totally removed from the starting material, decreasing their concentration below the threshold and thus avoiding poisoning of the catalyst.

The metathesis of olefins with heterogeneous catalysts was carried out also in a three-phase system, employing liquid diluent as a reaction medium. These liquid diluents are generally inert relative to the catalytic system, e.g. saturated hydrocarbons such as heptane, octane, etc. Their role is similar to that of the gaseous diluents, namely to moderate the thermal effects of the reaction and to extend the lifetime of the catalyst. The activation and regeneration effects of this type of diluents on the catalyst are greater in the present case. The liquid diluent can protect better the active centres of the catalyst through solvation of the heavy by-products of the reaction. These products would otherwise cover the catalyst surface and rapidly deactivate it. Likewise, the liquid diluents can remove the coke or polymer deposits more easily from the surface of the catalyst. Thus, it is important for these diluents to have, besides inertness towards the catalytic system, the quality of solvating and efficiently removing the higher molecular weight products or polymers formed during the reaction.

Kamiya [307] and Ogata [308] studied the effect of heptane or benzene as reaction medium in the conversion of propene with the catalytic system $MoO_3 \cdot Al_2O_3$. In the case of heptane, they noticed a substantial increase of the conversion with an increase in the ratio n-heptane : propene. Benzene gave good results only for a short reaction time and for lower benzene : propene ratios. Extension of the reaction time as well as increase in the ratio benzene : propene led to large amounts of alkylation products of benzene with propene. Furthermore, benzene proved to be a poorer solvent for the polymers formed in the reaction.

2.2.4.2. Solvents for Homogeneous Catalysis

The reactions of metathesis and ring-opening polymerization in the homogeneous phase, generally take place in an adequate solvent which allows solvation of the starting material, of the reaction products and of the catalytic complex. Several conditions have to be fulfilled by the solvents used in these reactions. First, they should be inert relative to the catalytic complex, neither deactivating nor consuming it ; also, the solvent should not undergo any transformation in the presence of the catalyst. Secondly, it is necessary that the solvent should be well purified from traces of water, air, or other substances which could inhibit the catalyst. Furthermore, the solvent should have a high solvating power, especially in reactions of metathesis or ring-opening polymerization of cyclo-olefins which yield oligomers or polymers with low solubility. Hence, the solvent should easily remove the reaction products from the region of the active centres, avoiding the formation of deposits on the catalyst surface. The solvent thus plays an important role in maintaining high activity of the catalyst for long period of time.

The main types of solvent employed in metathesis and ring-opening polymerization are non-polar unsaturated hydrocarbon compounds such as n-hexane, n-heptane, n-octane, etc. non-polar or scarcely polar aromatic hydrocarbon compounds such as benzene, toluene, xylene, mesitylene, etc. and scarcely polar or weakly polar chlorinated hydrocarbon compounds such as monochlorobenzene, m- and p-dichlorobenzene, 1,2,4-trichlorobenzene, methylene chloride, chloroform, carbon tetrachloride, trichloroethene, tetrachloroethene etc. For ring-opening polymerization of cyclo-olefins, carbon disulphide or highly polar media such as water and ethanol were also used with certain catalysts.

Most of the metathesis reactions taking place in the homogeneous phase use saturated hydrocarbons (hexane, heptane etc.) as reaction media. Aromatic compounds are employed only in certain cases : since the most widely used catalytic systems for metathesis are based on tungsten and molybdenum halides, which have a marked electrophilic character, they tend to promote alkylations of aromatic solvents with the olefin. Kothari and Tazuma [158] showed that the metathesis of 2-pentene in benzene or toluene mostly yields alkylation products of the solvent with the starting material or with its metathesis products. That is why aromatic solvents should be used with caution in such reactions.

A systematic study concerning the influence of the medium on the metathesis of 2-pentene was carried out by Hocks, Hubert and Teyssié.[159-161] They showed how the amount and nature of the aromatic solvent influence competition between the metathesis of the olefin and the alkylation of the solvent. An increase in the molar ratio benzene : 2-pentene diminishes considerably the yield of metathesis products, in favour of alkylation products. Among aromatic hydrocarbons, it is benzene which favours the metathesis, and the metathesis/alkylation ratio decreases in the order benzene > toluene > p-xylene > m-xylene > mesitylene. The yield of alkylation products of the solvent increases in the same order.

Ichikawa and co-workers [309] studied the influence of numerous chlorinated solvents on the degree of conversion of 2-pentene, in the presence of the catalytic systems WCl_6 and Et_3Al or Ph_4Sn. They showed that some polychlorinated solvents lead to higher conversions of 2-pentene than are obtained in benzene or chlorobenzene. Such solvents are trichloroethene, tetrachloroethene, *m*- and *p*-dichlorobenzene and 1,2,4-trichlorobenzene. Di- and trichlorinated derivatives of benzene totally eliminate the competing Friedel-Crafts alkylation reactions.

Solvents used for ring-opening polymerizations (besides the fact that they should be inert towards the catalytic system) have to ensure good dissolution of the resulting polymer, allowing the free access of the monomer to the catalytic complex, and at the same time they must have as low a polarity as possible so as to ensure the normal propagation of the reaction chain. Furthermore, it is necessary that, as in the case of metathesis, the solvent should be of high purity, because traces of water or oxygen can rapidly deactivate the catalytic system. Aliphatic saturated hydrocarbons such as hexane, heptane, octane etc. are the most efficient solvents used for ring-opening polymerizations; among aromatic compounds, benzene and chlorobenzene are most often used. Höcker and Musch [310] showed that, in the reaction of cyclo-octene, the use of chlorobenzene as a solvent instead of n-heptane led to an increase in the amount of oligomers in the reaction product.

Höcker and Jones [311] studied the effect of a large number of solvents on the polymerization of cyclo-octene. Among the aromatic solvents, chlorobenzene favoured the formation of Friedel-Crafts products to a greater extent than did benzene. Among the chlorinated aliphatic compounds, carbon tetrachloride does not achieve complete solvation of the catalytic system, so that the system remains heterogeneous. However, cyclo-octene is quantitatively polymerized to *trans*-polyoctenamer in this reaction medium. The reaction in methylene chloride proceeded differently according to the reaction conditions. When the catalyst was prepared in benzene and the metathesis reaction was continued in methylene chloride, polyoctenamer was obtained almost quantitatively (94% yield). When methylene chloride was employed as a medium both for preparing the catalyst and for the metathesis reaction, side-reactions occurred, which diminished the polyoctenamer yield below 60%.

The polarity of the solvent has a significant influence on the ring-opening polymerizations of cyclo-olefins with catalytic systems consisting of ruthenium, osmium or iridium halides.[312] Thus, in the case of cyclobutene polymerization with $RuCl_3$, the structure of the resulting polybutenamer was modified when the reaction was carried out in water or in ethanol. On employing water as a reaction medium, a 50 : 50 mixture of *cis*- and *trans*-polybutenamer is obtained. When working in ethanol a polymer with totally *trans* configuration results.

2.2.4.3. Inhibitors

Heterogeneous and homogeneous catalytic systems for metathesis and ring-opening polymerization are affected by a wide range of substances which may be present in the reaction medium and   which act as inhibitors. Most of these substances rapidly deactivate the catalytic systems, stopping the metathesis of the ring-opening polymerization. The inhibitors of these reactions are organic or inorganic substances which generally contain oxygen, e.g. alcohols, ethers, acids, esters, nitriles, thiols, amines, phosphines, etc. Unsaturated hydrocarbons such as dienes, mono- and diacetylenes, and some aromatic compounds also act as inhibitors.

Banks [3] showed that, in the disproportionation of propene with alumina-supported molybdenu moxide, the conversion is substantially decreased if the starting material contains the following impurities in concentrations between 300 and 200 ppm : oxygen, water, carbon dioxide, hydrogen sulphide, diethyl sulphide, ethyl methyl sulphide, acetylene, methylacetylene, propadiene and butadiene. By removing these impurities and by

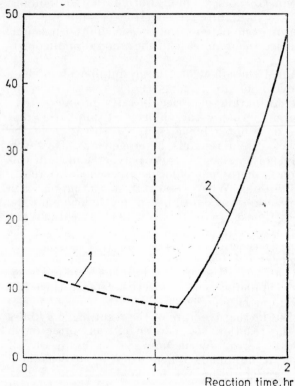

Fig. 2.6. — Variation of the conversion of propene *versus* reaction time : 1 — moist propene ; 2 — dry propene.[14]

reactivating the catalyst with air and an inert gas, it is possible to revive its normal activity. Silica-supported tungsten oxide catalysts have better resistance towards inhibitors because they are used at higher temperatures. However, if water, air, acetone, carbon monoxide, hydrogen or methanol are present in the olefin, they deactivate this catalytic system even at high temperatures.[14] As a result of the removal of these impurities from the olefin flow, the catalyst easily regains its initial activity. An example of how water deactivates the disproportionation of propene with $WO_3 \cdot SiO_2$ catalyst is furnished by Heckelsberg, Banks and Bailey (Fig. 2.6). In this example, propene saturated with water was passed over a catalyst at 427°C and 30 atm. The conversion of propene decreased below 10%. After one hour's operation, the feed was changed to dry propene. The conversion increased gradually up to 44—45% after another hour of reaction. Similar results were obtained in the disproportionation of propene containing air.

Homogeneous catalytic systems, particularly those consisting of tungsten or molybdenum halides and organometallic compounds react vigorously with water or alcohols yielding hydrated compounds or alkoxides, respectively. These systems can easily be destroyed by traces of water, alcohols or other similar substances present in the solvent or in the olefin. Amines, phosphines, and organic compounds with strong nucleophilic character can form stable combinations with most of the homogeneous catalytic systems. Similarly, the homogeneous catalysts for metathesis easily give complexes with dienes, acetylenes and aromatic hydrocarbons, thus inhibiting the metathesis of olefins.

References

1. R. L. Banks and G. C. Bailey, *Ind. Eng. Chem., Prod. Res. Develop.*, **3**, 170 (1964).
2. L. F. Heckelsberg, R. L. Banks and G. C. Bailey, *Ind. Eng. Chem., Prod. Res. Develop.*, **8**, 259 (1969).
3. R. L. Banks, **U.S.** 3,261,879, 19.07.1966; *Chem. Abstr.*, **65**, 12105e (1966).
4. V. V. Atlas, I. I. Pis'man and A. M. Bakhshi-Zadé, *Khim. Prom.*, **45**, 734 (1969).
5. G. Henrici-Olivé and S. Olivé, *Angew. Chem.*, **85**, 148 (1973).
6. G. C. Bailey, *Catalysis Revs.*, **3**, 37 (1969).
7. R. L. Banks, *Top. Curr. Chem.*, **25**, 39 (1972).
8. L. F. Heckelsberg, **Fr.** 1,562,397, 4.04.1969; *Chem Abstr.*, **71**, 93358 (1969).
9. D. L. Crain, *J. Catalysis*, **13**, 110 (1969).
10. C. P. C. Bradshaw, E. J. Howman and L. Turner, *J. Catalysis*, **7**, 269 (1967).
11. G. C. Ray and D. L. Crain, **Belg.** 694,420, 22.08.1967.
12. F. D. Mango, **U.S.** 3,424,811. 29.01.1969; *Chem. Abstr.*, **70**, 106042 (1969).
13. R. van Helden *et al.*, **Neth. Appl.** 6,608,427, 18.12.1967; *Chem. Abstr.* **69**, 433891 (1968).
14. L. F. Heckelsberg, R. L. Banks and G. C. Bailey, *Ind. Eng. Chem., Prod. Res. Develop.*, **7**, 29 (1968).
15. L. F. Heckelsberg, **S. African** 68,01,408, 31.07.1968; *Chem. Abstr.*, **70**, 69924 (1969).
16. R. B. Regier, **U.S.** 3,761,427, 25.09.1973; *Chem. Abstr.*, **80**, 41318 (1974).
17. R. L. Banks, **U.S.** 3,883,606, 13.05.1975; *Chem. Abstr.*, **83**, 58066 (1975).
18. T. Takahashi, **Japan. Kokai** 7668,508, 14.06.1976; *Chem. Abstr.*, **85**, 176801 (1976).
19. T. Takahashi, **Japan. Kokai** 7632,502, 19.03.1976; *Chem. Abstr.*, **85**, 93775 (1976).
20. British Petroleum Company, **Brit.** 1,054,864, 8.09.1964; *Chem. Abstr.*, **64**, 19408 (1966).
21. British Petroleum Company, **Brit.** 1,064,829, 19.11.1964; *Chem. Abstr.*, **65**, 13539 (1966).
22. British Petroleum Company, **Brit.** 1,105,563, 23.04.1965; *Chem. Abstr.*, **66**, 48080 (1967).

23. British Petroleum Company, **Brit.** 1,105,564, 23.04.1965 ; *Chem. Abstr.*, **66**, 48081 (1967).
24. L. Turner, E. J. Howman and K. V. Williams, **Brit.** 1,106,015, 23.07.1965 ; *Chem Abstr.*, **67**, 53607 (1967).
25. L. Turner, E. J. Howman and K. V. Williams, **Brit.** 1,106,016, 23.07.1965 ; *Chem. Abstr.*, **67**, 53607 (1967).
26. L. Turner and C. P. C. Bradshaw, **Brit.** 1,103,976, 21.02.1968 ; *Chem. Abstr.*, **68**, 86834 (1968).
27. C. P. C. Bradshaw, **Ger. Offen**. 1,900,490, 31.07.1969 ; *Chem. Abstr.*, **71**, 90763 (1969).
28. L. Turner, E. J. Howman and C. P. C. Bradshaw, **Brit. Ammend.** 1,054,864, 11.12.1970 ; *Chem. Abstr.*, **75**, 140216 (1972).
29. L. F. Heckelsberg, **Brit.** 1,282,592, 19.07.1972 ; *Chem. Abstr.*, **77**, 100701 (1972).
30. British Petroleum Company, **Fr.** 1,485,859, 23.06.1967 ; *Chem. Abstr.*, **68**, 61420 (1968).
31. British Petroleum Company, **Brit.** 1,093,784, 25.08.1965 ; *Chem. Abstr.*, **67**, 53624 (1967).
32. N. M. Palmer and B. P. McGrath, **Ger. Offen**. 1,945,282, 14.04.1970 ; *Chem. Abstr.*, **72**, 136911 (1970).
33. R. van Helden, C. P. Kohll and P. A. Verbrugge, **Ger. Offen**. 2,109,011, 16.09.1971 ; *Chem. Abstr.*, **75**, 140218 (1971).
34. R. L. Banks, **U.S.** 3,442,969, 6.05.1969 ; *Chem. Abstr.*, **71**, 38315 (1969).
35. L. Turner and K. V. Williams, **Brit.** 1,096,200, 20.12.1967 ; *Chem. Abstr.*, **68**, 77691 (1968).
36. R. L. Banks, **U.S.** 3,546,313, 8.12.1970 ; *Chem. Abstr.*, **74**, 78085 (1971).
37. K. V. Williams and L. Turner, **Brit.** 1,116,243, 6.06.1968 ; *Chem. Abstr.*, **69**, 29085 (1968).
38. L. F. Heckelsberg, **U.S.** 3,340,322, 5.09.1967 ; *Chem. Abstr*, **67**, 116529 (1967).
39. A. J. Alkema, D. Medema and F. Wattimena, **Ger. Offen**. 1,929,136, 11.12.1969 ; *Chem. Abstr.*, **72**, 81244 (1970).
40. C. P. C. Bradshaw and J. I. Blake, **Brit.** 1,266,340, 8.03.1972 ; *Chem. Abstr.*, **76**, 126352 (1972).
41. M. L. Khidekel *et al.*, **U.S.S.R.** 591,397, 15.11.1975 ; *Chem. Abstr.*, **84**, 65758 (1976).
42. A. D. Shebaldova *et. al.*, *Neftekhim.*, **16**, 533 (1976).
43. A. D. Shebaldova *et. al.*, *Izvest. Akad. Nauk S.S.S.R.*, *Ser. Khim.*, **1976**, 2509.
44. J. Smith *et al.*, *J. Chem. Soc.*, *Dalton Trans.*, **1974**, 1742.
45. W. Mowat, J. Smith and C. A. Whan, *J. Chem. Soc.*, *Chem. Commun.*, **1974**, 34.
46. A. Morris, H. Thomas and C. J. Attridge, **Ger. Offen.** 2,213,948, 21.12.1972 ; *Chem. Abstr.*, **79**, 65769 (1973).
47. V. Drăguțan, *Studii Cercetări Chim.*, **22**, 293 (1974).
48. H. S. Eleuterio, **U.S.** 3,074,918, 20.06.1957.
49. H. S. Eleuterio, **Ger. Offen**. 1,072,811, 7.01.1960 ; *Chem. Abstr.*, **55**, 16005 (1961).
50. A. J. Alkema and R. van Helden, **Brit.** 1,118,517, 3.07.1967 ; *Chem. Abstr.*, **69**, 96066 (1968).
51. D. L. Crain, **U.S.** 3,575,947, 20.04.1971 ; *Chem. Abstr.*, **75**, 6607 (1971).
52. L. Turner and C. P. C. Bradshaw, **Brit.** 1,105,565, 6.03.1968 ; *Chem. Abstr.*, **68**, 87758 (1968).
53. A. Toyoda, T. Matsui and S. Shiozaki, **Japan. Kokai**, 7349,900 ; 13.07.1973 ; *Chem. Abstr.*, **80**, 15494 (1974).
54. J. M. J. P. Lipsch and G. C. A. Schuit, *J. Catalysis*, **15**, 174 (1969).
55. F. E. Massoth, *J. Catalysis*, **30**, 204 (1973).
56. J. Sonnemans and P. Mars, *J. Catalysis*, **31**, 209 (1973).
57. G. C. A. Schuit and B. C. Gates, A. I. Ch. E. J, **19**, 147 (1973).
58. F. E. Massoth, *J. Catalysis*, **36**, 164 (1975).
59. W. K. Hall and M. LoJacono, *Proc. Sixth Int. Congr. Catalysis*, **1976**, 246.
60. E. A. Lombardo, M. LoJacono and W. K. Hall, *J. Catalysis*, **51**, 243 (1978).
61. H. Knozinger and H. Jeriarowski, *J. Phys. Chem.*, **82**, 2002 (1978).
62. L. Wang and W. K. Hall, *J. Catalysis*, **66**, 251 (1980).
63. W. H. J. de Beer, M. J. M. van der Aalst, C. G. Machiels and G. C. A. Schuit, *J. Catalysis*, **43**, 78 (1976).
64. A. Janibello and F. Trifiro, *Z. anorg. allgem. Chem.*, **413**, 293 (1975).
65. N. Giordano, J. C. J. Bart, A. Vaghi, A. Castellan and G. Martinotti, *J. Catalysis*, **36**, 81 (1975).
66. N. Giordano, A. Castellan, J. C. J. Bart, A. Vaghi and F. Campadelli, *J. Catalysis*, **37**, 204 (1975).

67. N. Giordano, M. Padovan, A. Vaghi, J. C. J. Bart and A. Castellan, *J. Catalysis*, **38,** 1 (1975).
68. F. R. Brown, L. E. Makovsky and R. H. Rhee, *J. Catalysis*, **50,** 162 (1977).
69. F. R. Brown, L. E. Makovsky and R H. Rhee, *J. Catalysis*, **50,** 382 (1977).
70. W. H. J. Stork, J. G. P. Coolegem and G. T. Pott, *J. Catalysis*, **32,** 497 (1974).
71. J. Medema, C. van Stam, V.H.J. de Beer, A. J. A. Konings and D. C. Koningsberg *J. Catalysis*, **53,** 386 (1978).
72. H.-C. Liu and S. W. Weller, *J. Catalysis*, **66,** 65 (1980).
73. A. A. Olsthoorn and J. A. Moulijn, *J. Mol. Catalysis*, **8,** 147 (1980).
74. Y. Okamoto, T. Imanaka and S. Teranishi, *J. Catalysis*, **65,** 448 (1980).
75. A. Castellan, J. C. J. Bart, A. Vaghi and N. Giordano, *J. Catalysis*, **42,** 162 (1976).
76. A. Vaghi, A. Castellan, J. C. J. Bart and N. Giordano, *J. Catalysis*, **42,** 381 (1976).
77. R. Thomas, J. A. Moulijn, W. H. J. de Beer and J. Medema, *J. Mol. Catalysis*, **8,** 161 (1980).
78. W. H. J. Stork and G. T. Pott, *Rec. Trav. chim.*, **96,** M105 (1977).
79. J. L. K. F. de Vries and G. T. Pott, *Rec. Trav. chim.*, **96,** M115 (1977).
80. R. Thomas, F. P. J. M. Kerkhof, J. A. Moulijn, J. Medema and W. H. J. de Beer, *J. Catalysis*, **61,** 559 (1980).
81. F.P.J.M. Kerkhof, M. C. Mittelmeijer-Hazeleger and J. A. Moulijn, cited after 80.
82. R. Thomas, J. A. Moulijn and F. P. J. M. Kerkhof, *Rec. Trav. chim.*, **96,** M134 (1977).
83. F. P. J. M. Kerkhof, R. Thomas and J. A. Moulijn, *Rec. Trav. chim.*, **96,** M121 (1977).
84. P. Biloen and G. T. Pott, *J. Catalysis*, **30,** 169 (1973).
85. A. A. Olsthoorn and C. Boelhouwer, *J. Catalysis*, **44,** 197 (1976).
86. A. A. Olsthoorn and C. Boelhouwer, *J. Catalysis*, **44,** 207 (1976).
87. F. Kapteijn, L. H. G. Bredt and J. C. Mol, *Rec. Trav. chim.*, **96,** M139 (1977).
88. F. P. J. M. Kerkhof, J. A. Moulijn and R. Thomas, *J. Catalysis*, **56,** 279 (1979).
89. R. M. Edreva-Kardjieva and A. A. Andreev, *React. Kin. Catal. Letters*, **5,** 465 (1976).
90. A. A. Andreev, R. M. Edreva-Kardjieva and N. M. Neshev, *Rec. Trav. chim.*, **96,** M23 (1977).
91. E. S. Davie, D. A. Whan and C. Kemball, *J. Chem. Soc., Chem. Commun.*, **1969,** 1430.
92. E. S. Davie, D. A. Whan and C. Kemball, *J. Catalysis*, **24,** 272 (1972).
93. D. A. Whan, M. Barber and P. Swift, *J. Chem. Soc., Chem. Commun.*, **1972,** 198.
94. E. S. Davie, D. A. Whan and C. Kemball, *Proc. Fifth. Int. Congr. Catalysis*, **1972,** 86.
95. R. F. Howe, D. E. Davidson and D. A. Whan, *J. Chem. Soc., Faraday Trans. I*, **68,** 2266 (1972).
96. R. F. Howe and I. R. Leith, *J. Chem. Soc., Faraday Trans. I*, **69,** 1967 (1973).
97. J. Smith, R. F. Howe and D.A. Whan, *J. Catalysis*, **34,** 191 (1974).
98. A. Brenner and R. L. Burwell, Jr., *J. Catalysis*, **52,** 353 (1978).
99. A. Brenner and R. L. Burwell, Jr., *J. Catalysis*, **52,** 364 (1978).
100. R. L. Burwell, Jr. and A. Brenner, *J. Mol. Catalysis*, **1,** 77 (1976).
101. A. Brenner and R. L. Burwell, Jr., *J. Amer. Chem. Soc.*, **97,** 2565 (1975).
102. A. Brenner, *J. Mol. Catalysis*, **5,** 157 (1979).
103. A. Brenner and D. A. Hucul, *J. Catalysis*, **61,** 216 (1980).
104. E. J. Howman and B. P. McGrath, **Brit.** 1,123,500, 14.08.1968 ; *Chem. Abstr.*, **69,** 95908 (1968).
105. D. A. Bol'shakov *et al.*, **U.S.S.R.** 536,837, 30.11.1976 ; *Chem. Abstr.*, **86,** 79426 (1977).
106. D. A. Bol'shakov *et al.*, **U.S.S.R.** 525,467, 25.08.1976 ; *Chem. Abstr.*, **85,** 182916 (1976).
107. D. A. Bol'shakov *et al.*, **U.S.S.R.** 480,442, 15.08.1975 ; *Chem. Abstr.*, **83,** 153213 (1975).
108. D. A. Bol'shakov *et al.*, **U.S.S.R.** 514,626, 25.05.1976 ; *Chem. Abstr.*, **85,** 83669 (1976).
109. F. Pennella, **S. African,** 68,01,533, 6.08.1968 ; *Chem. Abstr.*, **70,** 67555 (1969).
110. L. F. Heckelsberg, **S. African,** 68,01,707, 15.08.1968 ; *Chem. Abstr.*, **70,** 67557 (1969).
111. A. J. Alkema and R. van Helden, **U.S.** 3,536,777, 27.10.1970.
112. A. J. Alkema and R. van Helden, **Brit.** 1,117,968, 26.06.1968 ; *Chem. Abstr.*, **69,** 95906 (1968).
113. R. P. Arganbright, **U.S.** 3,697,613, 10.10. 1972 ; *Chem. Abstr.*, **78,** 15471 (1973).
114. F. K. Schmidt et al., **U.S.S.R.** 429,049, 25.05.1974 ; *Chem. Abstr.*, **81,** 77442 (1974).
115. T. Takahashi, **Japan. Kokai** 7626,805, 5.03.1976 ; *Chem. Abstr.*, **85,** 93773 (1976).
116. British Petroleum Company, **Fr.** 1,554,287, 17.01.1969 ; *Chem. Abstr.*, **71,** 60655 (1969).

117. L. Turner, E. J. Howman and C. P. C. Bradshaw, **Brit.** 1,056,980, 1.02.1967; *Chem. Abstr.*, **66,** 75651 (1967).
118. D. L. Crain, **Belg.** 691,711, 23.06.1967.
119. R. E. Reusser, **U.S.** 3,511,890, 12.05.1970.
120. R. E. Reusser, **S. African** 68,01,879, 20.08.1968; *Chem. Abstr.*, **70,** 77280 (1969).
121. R. E. Reusser and S. D. Turk, **U.S.** 3,786.112, 15.01.1974; *Chem. Abstr.*, **80,** 95210 (1974).
122. Shell Internat. Res. Maatschapij., **Neth. Appl.** 7,111,882, 25.05.1972; *Chem. Abstr.*, **77,** 125930 (1972).
123. Y. Morita, H. Yoshiura and T. Isaka, **Japan. Kokai** 7366,086, 11.09.1973; *Chem. Abstr.*, **81,** 13074 (1974).
124. R. P. Arganbright, **U.S.** 3,792.108, 12.02.1974; *Chem. Abstr.*, **80,** 95212 (1974).
125. R. P. Arganbright, **U.S.** 3,702,827, 14.11.1972; *Chem. Abstr.*, **78,** 110518 (1973).
126. R. B. Regier, **U.S.** 3,923,920, 2.12.1975; *Chem. Abstr.*, **84,** 58594, (1976).
127. S. Malinovski *et. al.*, **Pol.** 71,930, 15.01.1975; *Chem. Abstr.*, **84,** 43274 (1976).
128. M. J. Lawrenson, **Ger. Offen.** 2,340,081, 28.02.1974; *Chem. Abstr.*, **81,** 54837 (1974).
129. A. F. Ellis and E. S. Sabourin, **Ger. Offen.** 2,306,813, 30.08.1973; *Chem. Abstr.*, **79,** 125796 (1973).
130. L. F. Heckelsberg, **S. African** 68,01,620, 15.08.1968; *Chem. Abstr.*, **79,** 67560 (1969).
131. R. B. Regier, **U.S.** 3,792,106, 12.02.1974; *Chem. Abstr.*, **80,** 95211 (1974).
132. N. Calderon, H. Y. Chen and K. W. Scott, *Tetrahedron Letters*, **1967,** 3327.
133. N. Calderon, E. A. Ofstead and W. A. Judy, *J. Polymer Sci.*, $A-1$, **5,** 2209 (1967).
134. G. S. Lewandos and R. Pettit, *J. Amer. Chem. Soc.*, **93,** 7087 (1971).
135. D. J. Cardin, M. J. Doyle and M. F. Lappert, *J. Chem. Soc., Chem. Commun.*, **1972,** 927.
136. R. Rossi, P. Diversi, A. Lucherini and L. Porri, *Tetrahedron Letters*, **1974,** 879.
137. R. Opitz, K.-H. Thiele, L. Bencze and L. Markó, *J. Organomet. Chem.*, **96,** C53 (1975).
138. J. McGinnis, T. J. Katz and S. Hurwitz, *J. Amer. Chem. Soc.*, **98,** 605 (1976).
139. a) T. J. Katz, J. McGinnis and C. Altus, *J. Amer. Chem. Soc.*, **98,** 606 (1976); b) T.J. Katz, S. J. Lee and N. Acton, *Tetrahedron Letters*, **1976,** 4247.
140. T. J. Katz and N. Arton, *Tetrahedron Letters*, **1976,** 4251.
141. C. P. Casey and T. J. Burkhardt, *J. Amer. Chem. Soc.*, **96,** 7808 (1974).
142. C. P. Casey, H. E. Tuinstra and M. C. Saeman, *J. Amer. Chem. Soc.*, **98,** 608 (1976).
143. C. P. Casey, L. D. Albin and T. J. Burkhardt, *J. Amer. Chem. Soc.*, **99,** 2533 (1977).
144. G. Dall'Asta and G. Carella, **Ital.** 784,307, 2.08.1966.
145. T. Oshika and H. Tabuchi, *Bull. Chem. Soc. Jap.*, **41,** 211 (1968).
146. D. T. Laverty *et al.*, *J. Chem. Soc. Chem. Commun.*, **1976,** 193.
147. G. Natta, G. Dall'Asta, and L. Porri, *Makromol. Chem.*, **81,** 253 (1965).
148. L. Porri, G. Natta and M. Gallazzi, *Chimica e Industria*, **46,** 428 (1964).
149. G. Natta, G. Dall'Asta and G. Motroni, *J. Polymer Sci.*, **B,** **2,** 349 (1964).
150. F. W. Michelotti and W. P. Keaveney, *J. Polymer Sci.*, **A,** **3,** 895 (1965).
151. L. Porri, R. Rossi, P. Diversi and A. Lucherini, *Makromol. Chem.*, **175,** 3097 (1974).
152. L. Porri, P. Diversi, A. Lucherini and L. Rossi, *Makromol. Chem.*, **176,** 3121 (1975).
153. W. Grahlert, K. Milowski, U. Langbein and E. Taeger, *Plaste und Kautschuk*, **22,** 229 (1975).
154. V. A. Körmer, I. A. Poleyatova and T. L. Yufa, *J. Polymer Sci.*, $A-1$, **10,** 251 (1972).
155. N. Calderon, E. A. Ofstead, J. P. Ward, W. A. Judy and K. W. Scott, *J. Amer. Chem. Soc.*, **90,** 4133 (1968).
156. J. L. Wang and H. R. Menapace, *J. Org. Chem.*, **33,** 3794 (1968).
157. E. A. Zuech, *J. Chem. Soc., Chem. Commun.*, **1968,** 1182.
158. V. M. Kothari and J. J. Tazuma, *J. Org. Chem.*, **36,** 2951 (1971).
159. L. Hocks, A. J. Hubert and Ph. Teyssié, *Tetrahedron Letters*, **1972,** 3687.
160. L. Hocks, A. J. Hubert and Ph. Teyssié, *Tetrahedron Letters*, **1973,** 2719.
161. L. Hocks, A. J. Hubert and Ph. Teyssié, *Tetrahedron Letters*, **1974,** 877.
162. A. Uchida *et al.*, *Ind. Eng. Chem.*, *Prod. Res. Develop.*, **10,** 372 (1971).
163. K. Hummel and W. Ast, *Naturwiss.*, **57,** 245 (1970).
164. R. Nakamura *et. al.*, *Chem. Letters*, **1976,** 1019.
165. R. Nakamura *et. al.*, **Japan. Kokai** 76125,1317, 1.11.1976; *Chem. Abstr.*, **86,** 170899 (1977).
166. P. B. van Dam and C. Boelhouwer, *React. Kin. Catalysis Letters*, **1,** 165 (1974).

167. P. B. van Dam, M. C. Mittelmeijer and C. Boelhouwer, *Fette, Seifen, Anstrichm.*, **67**, 264 (1974).
168. T. Takagi *et al.*, *Yukagaku*, **24**, 377 (1975).
169. T. Takagi *et al.*, *J. Chem. Soc., Chem. Commun.*, **1972**, 838.
170. T. Takagi *et al.*, *Yukagaku*, **24**, 518 (1975).
171. Y. Uchida *et al.*, *Bull. Chem. Soc. Japan*, **45**, 1158 (1972).
172. W. S. Greenlee and M. Farona, *Inorg. Chem.*, **15**, 2129 (1976).
173. E. T. Kittleman and E. A. Zuech, **Fr.** 1,561,025, 21.03.1969 ; *Chem. Abstr.*, **72**, 31193 (1970).
174. E. T. Kittleman and E. A. Zuech, **U.S.** 3,554,924, 12.01.1971 ; *Chem. Abstr.*, **74**, 99440 (1971).
175. E. T. Kittleman and E. A. Zuech, **U.S.** 3,708,551, 2.01.1973 ; *Chem. Abstr.*, **78**, 71382 . (1973).
176. A. Uchida *et al.*, *Ind. Eng. Chem., Prod. Res. Develop.*, **10**, 369 (1971).
177. A. Uchida *et al.*, *Ind. Eng. Chem., Prod. Res. Develop.*, **11**, 389 (1972).
178. C. P. C. Bradshaw, **Ger. Offen.** 1,909,951, 16.10.1069 ; *Chem. Abstr.*, **72**, 12047 (1970).
179. K. Ichikawa and K. Fuzukumi, *J. Org. Chem.*, **41**, 2633 (1976).
180. N. B. Bespalova *et al.*, *Doklady Akad. Nauk S.S.S.R.*, **225**, 1071 (1975).
181. K. Ichikawa *et al.*, *Trans. Met. Chem.*, **1**, 183 (1976).
182. K. Ichikawa *et al.*, *J. Catalysis*, **44**, 416 (1976).
183. P. B. van Dam, M. C. Mittelmeijer and C. Boelhouwer, *React. Kin. Catalysis Letters*, **1**, 481 (1974).
184. K. Ichikawa *et al.*, *Bull. Chem. Soc. Japan*, **49**, 750 (1976).
185. J. Chatt, R. J. Haines and G. J. Leigh, *J. Chem. Soc., Chem. Commun.*, **1972**, 1202.
186. S. A. Matlin and R. G. Sammes, *J. Chem. Soc., Chem. Commun.*, **1973**, 174.
187. R. Rossi and R. Giorgi, *Chimica e Industria*, **58**, 517 (1976).
188. R. Rossi and R. Giorgi, *Tetrahedron Letters*, **1976**, 1721.
189. E. A. Zuech *et al.*, *J. Amer. Chem. Soc.*, **29**, 528 (1970).
190. R. Taube and K. Seyferth, *Z. Chem.*, **14**, 284 (1974).
191. R. Taube and K. Seyferth, *Z. Chem.*, **14**, 410 (1974).
192. L. Bencze and L. Markó, *J. Organomet. Chem.*, **28**, 261 (1971).
193. L. Bencze, A. Rédey and L. Markó, *Hung. J. Ind. Chem.*, **1**, 453 (1973).
194. G. Doyle, **Ger. Offen.** 2,047,270, 1.04.1971 ; *Chem. Abstr.*, **75**, 5202 (1971).
195. W. R. Kroll, G. Doyle and W. H. Ruhle, **U.S.** 3.691.095, 12.09.1972 ; *Chem. Abstr.*, **78**, 110511 (1973).
196. W. R. Kroll and G. Doyle, *J. Catalysis*, **24**, 356 (1972).
197. W. R. Kroll and G. Doyle, *J. Chem. Soc., Chem. Commun.*, **1971**, 839.
198. W. R. Kroll and G. Doyle, **U.S.** 3,689,433, 5.09.1972 ; *Chem. Abstr.*, **77**, 151431 (1972).
199. D. M. Singleton, **Fr.** 2,013,672, 25.08.1968.
200. D. M. Singleton, **U.S.** 3,530,196, 29.01.1970 ; *Chem. Abstr.*, **72**, 89734 (1970).
201. J. L. Hérisson and Y. Chauvin, **Ger. Offen.** 2,008,246, 8.10.1970 ; *Chem. Abstr.*, **74**, 3306 (1971).
202. L. Bencze and L. Markó, *J. Organomet. Chem.*, **69**, C19 (1974).
203. E.-K. Tarcsi, F. Ratkovics and L. Bencze, *J. Mol. Catalysis*, **1**, 435 (1975/76).
204. L. Bencze, J. Pallos and L. Markó, *J. Mol. Catalysis*, **2**, 139 (1977).
205. L. Bencze, K.-H. Thiele and V. Marquardt, *Rec. Trav. chim.*, **96**, M8, (1977).
206. A. W. Anderson and N. C. Merckling, **U.S.** 2,721,189, 18.10.1955 ; *Chem. Abstr.*, **68**, 3656h (1968).
207. W. L. Truett *et al.*, *J. Amer. Chem. Soc.*, **82**, 2377 (1960).
208. G. Dall'Asta, G. Mazzanti, G. Natta and L. Porri, *Makromol. Chem.*, **56**, 224 (1962).
209. G. Natta, G. Dall'Asta, G. Mazzanti and G. Motroni, *Makromol. Chem.*, **69**, 163 (1963).
210. G. Natta, G. Dall'Asta and G. Mazzanti, **Fr.** 1,394,380, 2.04.1965 ; *Chem. Abstr.*, **68**, 3656h (1968).
211. G. Natta, G. Dall'Asta and G. Mazzanti, *Angew. Chem.*, **76**, 765 (1964).
212. G. Natta, G. Dall'Asta, I. W. Bassi and G. Carella, *Makromol. Chem.*, **91**, 87 (1966).
213. G. Natta, G. Dall'Asta and G. Mazzanti, **Belg.** 667,392, 16.11.1965 ; *Chem. Abstr.*, **65**, 3996c (1966).

214. G. Natta, G. Dall'Asta and G. Mazzanti, **Fr.** 1,425,601, 21.01.1966; *Chem. Abstr.*, **65,** 10692f (1966).
215. P. Günther *et al.*, *Angew. Makromol. Chem.*, **14,** 87 (1970).
216. P. Günther, W. Oberkirch and G. Pampus, **Fr.** 2,065,300, 1.10.1969.
217. W. Oberkirch, P. Günther and G. Pampus, **Ger. Offen.** 1,957,025, 19.05.1971.
218. F. Haas, K. Nützel, G. Pampus and D. Theisen, *Rubber Chem. Technol.*, **43,** 1126 (1970).
219. N. Calderon and M. C. Morris, *J. Polymer Sci.*, *A—2*, **5,** 1283 (1967).
220. Shell Internat. Res. Maatschapij., **Neth. Appl.** 7,014,132, 28.03.1972; *Chem Abstr.*, **77.** 49806 (1972).
221. G. Dall'Asta, **Ger. Offen.** 2,161,878, 6.07.1972; *Chem. Abstr.*, **77,** 127242 (1972).
222. P. Günther, W. Oberkirch and G. Pampus, **Ger. Offen.** 1,812,338, 18.06.1970.
223. W. Oberkirch, P. Günther and G. Pampus, **U.S.** 3,787,381, 22.01.1974.
224. P. R. Marshall and B. J. Ridgewell, *European Polymer J.*, **5,** 29 (1969).
225. K. Nützel, K. Dinges and F. Haas, **Ger. Offen.** 1,720,797, 7.03.1968.
226. W. A. Judy, **Fr.** 2,016,360, 26.08.1968; *Chem. Abstr.*, **72,** 112067 (1970).
227. W. A. Judy, **Fr.** 2,016,364, 26.08.1968; *Chem. Abstr.*, **72,** 112039 (1970).
228. W. A. Judy, **Fr.** 2,029,747, 31.01.1969; *Chem. Abstr.*, **73,** 88654 (1970).
229. W. A. Judy, **U.S.** 3,746,696, 17.07.1973; *Chem. Abstr.*, **79,** 126938 (1973).
230. K. Nützel, F. Haas and G. Marwede, **Ger. Offen.** 1,919,047, 15.04.1969.
231. I. A. Oreshkin *et al.*, *Izvest. Akad. Nauk S.S.S.R.*, *Ser. Khim.*, **1971,** 1123.
232. T. L. Yufa *et al.*, **Belg.** 770,823, 2.02.1972; *Chem. Abstr.*, **77,** 76404 (1972).
233. T. L. Yufa *et al.*, **Brit.** 1,338,155, 21.11.1973; *Chem. Abstr.*, **80,** 122125 (1974).
234. T. L. Yufa *et al.*, **Ger. Offen.** 2,133,905, 20.01.1972; *Chem. Abstr.*, **76,** 142093 (1972).
235. Y. Chauvin *et al.*, **Ger. Offen.** 2,151,662, 4.05.1972; *Chem. Abstr.*, **77,** 25343 (1972).
236. D. Commereuc, Y. Chauvin and D. Cruypelnick, *Makromol. Chem.*, **177,** 2637 (1976).
237. N. Calderon and N. Y. Chen, **Brit.** 1,125,529, 28.08.1968; *Chem. Abstr.*, **69,** 195851 (1968).
238. N. Calderon, **U.S.** 3,439,056, 15.04.1969; *Chem. Abstr.*, **71,** 38438 (1969).
239. N. Calderon, **U.S.** 3,439,057, 15.04.1969; *Chem. Abstr.*, **71,** 80807 (1969).
240. E. Wasserman, D. A. Ben-Efraim and R. Wolovsky, *J. Amer. Chem. Soc.*, **90,** 3286 (1968).
241. R. Wolovsky, *J. Amer. Chem. Soc.*, **92,** 2132 (1970).
242. F. W. Küpper and R. Streck, *Makromol. Chem.*, **175,** 2055 (1974).
243. F. W. Küper and R. Streck, **Ger. Offen.** 2,326,196, 12.12.1974; *Chem. Abstr.*, **82,** 155528 (1975).
244. D. Medema *et al.*, **Brit.** 1,193,943, 3.06.1970; *Chem. Abstr.*, **73,** 76637 (1970).
245. Shell Internat Res. Maatschapij., **Fr.** 2,132,099, 22.12.1972.
246. W. B. Hughes and E. A. Zuech, **U.S.** 3,562,178, 9.02.1971; *Chem. Abstr.*, **75,** 10868 (1971).
247. W. B. Hughes and E. A. Zuech, **U.S.** 3,691,253, 12.09.1972; *Chem. Abstr.*, **78,** 3657 (1973).
248. a) J. L. Wang, M. Brown and H. R. Menapace, **Ger. Offen.** 2,158,990, 6.07.1972; b) H. W. Ruhle, **Ger. Offen.** 2,062,448, 16.09.1971; *Chem. Abstr.*, **75,** 151341 (1971).
249. D. H. Kubicek and E. A. Zuech, **U.S.** 3,670,043, 13.01.1972; *Chem. Abstr.*, **77,** 100716 (1972).
250. D. H. Kubicek and E. A. Zuech, **U.S.** 3,558,520, 26.01.1971; *Chem. Abstr.*, **74,** 99441 (1971).
251. D. H. Kubicek and E. A. Zuech, **U.S.** 3,703,561, 21.11.1972; *Chem. Abstr.*, **78,** 71378 (1973).
252. E. A. Zuech, **Fr.** 1,575,778, 25.07.1969; *Chem. Abstr.*, **72,** 99988 (1970).
253. E. A. Zuech, **U.S.** 3,865,892, 11.02.1975; *Chem. Abstr.*, **82,** 139267 (1975).
254. J. L. Wang, H. R. Menapace and M. Brown, *J. Catalysis*, **26,** 455 (1972).
255. G. Dall'Asta and G. Carella, **Brit.** 1,062,367, 11.02.1965; *Chem. Abstr.*, **66,** 19005q (1967).
256. G. Dall'Asta and G. Motroni, **Ger. Offen.** 2,237,353, 15.02.1973; *Chem. Abstr.*, **79,** 61765 (1973).

257. K. W. Scott and N. Calderon, **Ger. Offen.** 2,058,198, 9.07.1971; *Chem. Abstr.,* **75,** 118801 (1971).
258. G. Dall'asta, G. Motroni and G. Carella, **Fr.** 1,545,643, 15.11.1968; *Chem. Abstr.,* 13733 (1969).
259. K. W. Scott, N. Calderon, *et al., Adv. Chem. Ser.,* **91,** 399 (1969).
260. N. Calderon and W. A. Judy, **Brit.** 1,124,456, 21.08.1968; *Chem. Abstr.,* **69,** 77970 (1968)
261. E. A. Ofstead, **U.S.** 3,597,403, 21.10.1969; *Chem. Abstr.,* **75,** 141671 (1971).
262. E. A. Ofstead, **U.S.** 3,997,471., 14.12.1976; *Chem. Abstr.,* **86,** 56559 (1977).
263. E. A. Ofstead, **U.S.** 4,010,113, 1.03.1977; *Chem. Abstr.,* **86,** 140694 (1966).
264. E. A. Ofstead, **Ger. Offen.** 2,509,578, 2.10.1975; *Chem. Abstr.,* **84,** 74875 (1976)
265. E. A. Ofstead, **Ger. Offen.** 2,509,577, 4.12.1975; *Chem. Abstr.,* **84,** 90793 (1976).
266. E. A. Ofstead, **U.S.** 4,020,254, 16.04.1977; *Chem. Abstr.,* **87,** 24528 (1977).
267. G. Pampus and J. Witte, **Ger. Offen.** 1,770,143, 6.04.1968.
268. G. Pampus, J. Witte and M. Hoffmann, **Ger. Offen.** 1,954,092, 28.12.1969.
269. G. Pampus, *et al.,* **Ger. Offen.** 1,961,865, 10.12.1969.
270. G. Pampus *et al.,* **Ger. Offen.** 2,009,740, 30.09.1971.
271. G. Pampus, J. Witte and M. Hoffmann, **Ger.Offen.** 2,016,471, 28.10.1971.
272. G. Pampus, J. Witte and M. Hoffmann, **Ger.Offen.** 2,016,840, 21.10.1971.
273. G.Pampus, G. Lehnert and J. Witte, **Ger.Offen.** 2,101,684, 20.06.1972.
274. K. Nützel *et al.,* **Ger.Offen.** 1,720,791, 24.02.1968.
275. K. Nützel, F. Haas and G. Marwede, **Ger.Offen.** 1,919,048, 15.04.1969.
276. K. Nützel and F. Haas, **Fr.** 2,008,180, 9.05.1968.
277. J. Witte, N. Schön and G. Pampus, **Fr.** 2,014,139, 17.04.1970.
278. J. Witte, G. Pampus and D. Maertens, **Ger.Offen.** 2,203,303, 2.08.1973.
279. N. Schön, G. Pampus, J. Witte and D. Theisen, **Ger.Offen.** 2,006,776, 19.08.1971.
280. N. Schön, G. Pampus and J. Witte, **Ger.Offen.** 2,012,675, 20.03.1970.
281. W. Oberkirch, P. Günther and G. Pampus, **Belg.** 742,391, 29.11.1968.
282. W. Oberkirch, P. Günther and G. Pampus, **Ger.Offen.** 1,811,653, 17.09.1970.
283. R. Streck and H. Weber, **Ger. Offen.** 2,028,716, 16.12.1971; *Chem. Abstr.,* **76,** 113903 (1972).
284. R. Streck and H. Weber, **Ger.Offen.** 2,028,905, 20.01.1972; *Chem. Abstr.,* **76,** 127817 (1972).
285. R. Streck and H. Weber, **Ger.Offen.** 2,028,935, 16.12.1972; *Chem. Abstr.,* **76,** 113891 (1972).
286. F. W. Küpper, R. Streck and K. Hummel, **Ger.Offen.** 2,051,798, 24.04.1972; *Chem. Abstr.,* **77,** 52529 (1972).
287. F. W. Küpper, R. Streck, H. T. Heims and H. Weber, **Ger.Offen.** 2,051,799, 27.04.1972; *Chem. Abstr.,* **77,** 62528 (1972).
288. S. Kurosawa, T. Ueshima and S. Kobayashi, **Japan. Kokai** 75110,000, 29.08.1975; *Chem. Abstr.,* **84,** 31737 (1976).
289. S. Kurosawa *et al.,* **Japan. Kokai** 7565,500, 27.05.1975; *Chem. Abstr.,* **83,**179939 (1975).
290. S. Kurosawa *et al.,* **Ger. Offen.** 2,529,963, 15.01.1976; *Chem. Abstr.,* **84,** 122656 (1976).
291. T. Ueshima, S. Kobayashi and M. Matsuoka, **Ger.Offen.** 2,316,087, 4.10.1973; *Chem. Abstr.,* **80,** 83932 (1974).
292. T. Ueshima, Y. Tanaka and S. Kobayashi, **Japan. Kokai** 75128,799, 11.10.1975; *Chem. Abstr.,* **84,** 90770 (1976).
293. T. Ueshima and S. Kobayashi, **Japan. Kokai** 7763,298, 25.05.1977; *Chem. Abstr.,* **87,** 102850 (1977).
294. C. A. Uraneck and W. J. Trepka, **Fr.** 1,542,040, 11.10.1968; *Chem. Abstr.* **71,** 22762 (1969).
295. C. A. Uraneck and W. J. Trepka, **Belg.** 705,673, 26.04.1968.
296. C. A. Uraneck and J. E. Burgleigh, **U.S.** 3,681,300, 1.08.1972; *Chem. Abstr.,* **77,** 127236 (1972).

297. J. E. Burgleigh and C. A. Uraneck, U.S. 3,640,986, 8.02.1972; *Chem. Abstr.*, **76**, 141537 (1972).
298. H. Ikeda *et al.*, **Ger. Offen.** 2,447,687, 17.04.1975; *Chem. Abstr.*, **83**, 44046 (1975).
299. F. Arai *et al.*, **Japan. Kokai** 76101,060, 7.09.1976; *Chem. Abstr.*, **85**, 178525 (1976).
300. E. Ogata and Y. Kamiya, *Chem. Letters*, **1973**, 603.
301. E. Ogata and Y. Kamiya, *Ind. Eng. Chem., Prod. Res. Develop.*, **13**, 226 (1974).
302. A. Ismayel-Milanović *et al.*, *J. Catalysis*, **31**, 408 (1973).
303. W. B. Hughes, *J. Amer. Chem. Soc.*, **92**, 532 (1970).
304. J. L. Wang and H. R. Menapace, *J. Catalysis*, **28**, 300 (1973).
305. T. Sodesawa, E. Ogata and H. Kamiya, *Bull. Chem. Soc. Japan*, **50**, 998 (1977).
306. K. Ichikawa, T. Tagaki and K. Fukuzumi, *Trans. Met. Chem.*, **1**, 54 (1976).
307. Y. Kamiya and E. Ogata, **Japan. Kokai** 7569,002, 9.06.1975.
308. E. Ogata, T. Sodesawa and Y. Kamiya, *Bull. Chem. Soc. Japan*, **49**, 1317 (1976).
309. K. Ichikawa, T. Takagi and K. Fukuzumi, *Yukagaku*, **25**, 136 (1976).
310. H. Höcker and R. Musch, *Makromol. Chem.*, **176**, 3117 (1975).
311. H. Höcker and F. R. Jones, *Makromol. Chem.*, **161**, 251 (1972).
312. G. Natta, G. Dall'Asta and L. Porri, *Makromol. Chem.*, **81**, 253 (1965).

Chapter 3

METATHESIS REACTIONS

The metathesis reaction discovered by Banks and Bailey,[1] in heterogeneous catalysis for α-olefins and by Calderon and Scott[2] in homogeneous catalysis for β-olefins, was later on extended to a large number of linear olefins, both simple or substituted,[3-12] to cyclo-olefins,[13-15] to linear and cyclic dienes,[16,17] to acetylenes[18] and even to unsaturated macromolecuar compounds.[19-20] The reaction is used to prepare organic compounds belonging to various classes, such as acyclic olefins,[13-12,21] cyclo-olefins,[17] unsaturated acids and esters,[22,23] unsaturated nitriles,[24] unsaturated compounds containing the ammonium group,[25] macrocycles,[26] knots and catenanes,[27] natural compounds[28,29] and macromolecular compounds having the properties of elastomers or plastic materials.[30] The field into which metathesis has penetrated during less than ten years since its discovery is very vast, and the reaction provides large possibilities of being applied in petrochemistry, synthesis of basic organic chemicals, of natural and macromolecular compounds.

The present chapter describes the main metathesis reactions of linear olefins, dienes, cyclo-olefins and of acetylenes. Also included are the cometathesis as well as the reactions of olefins bearing functional groups. The ring opening polymerization of cyclo-olefins, mechanistically related to metathesis, will be treated in the next chapter. Several aspects of the metathesis reaction, such as its kinetics and thermodynamics, its mechanism and stereochemistry, will be treated in separate chapters. For further information on the various aspects concerning the metathesis reaction, the reader may consult general reviews published by Bailey,[31] Banks,[32] Calderon,[33] Drăguțan,[34] Grubbs[35a] Hocks[35b] Haines and Leigh,[36] Hughes[37] Mol and Moulijn,[38] Taube and Seyferth[39] and other authors.[40]

3.1. Metathesis of Acyclic Olefins

As it has been shown in the first chapter, through metathesis, acyclic olefins disproportionate at the double bond with the formation of two new olefins, one with a shorter chain and the other with a longer chain than in the initial olefin (3.1).

$$
\begin{array}{c}
R'\text{---}=\text{---}R'' \\
+ \\
R'\text{---}=\text{---}R''
\end{array}
\quad\rightleftarrows\quad
\begin{array}{c}
\underset{R'}{\overset{R'}{\|}} \;+\; \underset{R''}{\overset{R''}{\|}}
\end{array}
\qquad (3.1)
$$

According to the nature of the radicals R' and R'', the following three main types of metathesis reactions can be distinguished in the class of acyclic olefins :

a) The metathesis of α-olefins, when $R' = H$ and $R'' = R$ (hydrocarbon radical), yielding ethene and an internal olefin (3.2).

$$
\begin{array}{c}
H_2C\text{==}CHR \\
+ \\
H_2C\text{==}CHR
\end{array}
\quad\rightleftarrows\quad
\begin{array}{c}
\underset{CH_2}{\overset{CH_2}{\|}} \;+\; \underset{CHR}{\overset{CHR}{\|}}
\end{array}
\qquad (3.2)
$$

b) The metathesis of unsymmetrical internal olefins, yielding two symmetrical internal olefins. The metathesis of β-olefins that leads to 2-butene and to a symmetrical internal olefin is an example (3.3).

$$
\begin{array}{c}
\text{---}=\text{---}R \\
+ \\
\text{---}=\text{---}R
\end{array}
\quad\rightleftarrows\quad
\begin{array}{c}
\| \;+\; \underset{R}{\overset{R}{\|}}
\end{array}
\qquad (3.3)
$$

γ-Olefins yield 3-hexene and a symmetrical internal olefin whereas δ-olefins afford 4-octene and a symmetrical internal olefin.

c) The metathesis of symmetrical internal olefins that regenerates the initial olefin, this reaction being an example of a totally degenerate rea tion [41] or of topomerization.[42]

$$
\begin{array}{c}
R\text{---}=\text{---}R \\
+ \\
R\text{---}=\text{---}R
\end{array}
\quad\rightleftarrows\quad
\begin{array}{c}
\underset{R}{\overset{R}{\|}} \;+\; \underset{R}{\overset{R}{\|}}
\end{array}
\qquad (3.4)
$$

When the metathesis reaction is carried out under homogeneous catalysis, the process takes place according to equations (3.1)−(3.4), depending on the starting material; it is selective and leads to the reaction products mentioned above, with the exception of the special cases when less selective catalysts or solvents promoting side reaction are used. If the reaction is carried out under heterogeneous catalysis, the process is more complex as these systems promote, in an important amount, side reactions of izomerization, oligomerization and polymerization of olefins, with a decrease of the selectivity in the metathesis products. The isomerization products can readily take part in reactions of cometathesis with the initial olefin, leading to other new reaction products. The metathesis of acyclic olefins in heterogeneous phase is thus a rather complex process of disproportionation that leads to a greater variety of reaction products.

3.1.1. Metathesis of Unsubstituted Olefins

Among the simple, unsubstituted olefins, propene was mostly studied under heterogeneous catalysis and 2-pentene in the homogeneous reaction. The interest in the metathesis reaction for processing a wide range of olefins and for establishing the reaction mechanism imposed the extension of the reaction to practically all unsubstituted simple olefins which are commonly used in laboratory or in industry.

This chapter discusses the metathesis of propene and¹ 2-pentene and briefly reviews known data about other representatives of the series.

3.1.1.1. Metathesis of Ethene

Ethene, the simplest olefin, yields again ethene through metathesis (3.5).

$$
\begin{array}{ccc}
CH_2 & CH_2 & CH_2{=}CH_2 \\
\| \; + \; \| & & + \\
CH_2 & CH_2 & CH_2{=}CH_2
\end{array}
\qquad (3.5)
$$

This reaction is a simple example of a totally degenerate reaction or topomerizarion.[42]

The disproportionation reaction of ethene was studied by Bank and Bailey,[1] immediately after they had discovered the reaction with propene. Under the conditions employed by them, ethene led, with low yields, to cyclopropane and methylcyclopropane. The formation mechanism of these products was interpreted much later on the basis of the carbene intermediates that may appear during the reaction. Taking into consideration the simple, symmetrical structure of the ethene molecule, its metathesis reaction could be evidenced only by isotopic labelling. Olsthoorn[43] studied the cometathesis reaction of a mixture of ethene and ethene-d_4. He obtained, together with the two initial olefins, a new labelled olefin, ethene-d_2.

These results showed that the reaction had taken place according to equation (3.6).

$$
\begin{array}{c}
CH_2 \\ \| \\ CH_2
\end{array}
+
\begin{array}{c}
CD_2 \\ \| \\ CD_2
\end{array}
\quad \rightleftharpoons \quad
\begin{array}{c}
CH_2{=\!=}CD_2 \\ + \\ CH_2{=\!=}CD_2
\end{array}
\tag{3.6}
$$

3.1.1.2. Metathesis of Propene

Through metathesis, propene can form normal products of disproportionation, ethene and 2-butene (3.7), or it may also lead back to propene

$$
\tag{3.7}
$$

via an automerization reaction (partially degenerate rearrangement) (3.8).

$$
\tag{3.8}
$$

Propene is the olefin for which Banks and Bailey [1] obtained, for the first time, the disproportionation reaction to ethene and 2-butene, in heterogeneous catalysis. The reaction, originally carried out with catalysts of molybdenum hexacarbonyl, tungsten hexacarbonyl and molybdenum oxide deposited on alumina was carefully studied with a wide range of catalytic systems and under different reaction conditions. For instance, by using a catalyst of molybdenum oxide and cobalt oxide deposited on alumina and working at temperatures between 93°C and 300°C, Banks and Bailey obtained conversions ranging from 20 % to 43 % and selectivities in ethene and butene up to 91 — 94 %. The degree of conversion of propene, as well as the selectivity in ethene and butene, are considerably influenced by the nature of the promoter, its concentration on the support and the nature of this support.

The disproportionation reaction of propene used various heterogeneous catalytic systems constituted mainly of carbonyl compounds, oxides or metallic sulphides deposited on alumina or silica, mixtures of oxides, or phosphates. [1,44-60] Among these systems, only a restricted number present a good activity which permits to obtain reasonable conversions of propene into metathesis products. Many of the catalysts with a low activity eliminate a number of undesirable side reactions leading to the reaction products with high selectivities. A number of active catalysts for the disproportionation of propene are given in Table 3.1.

TABLE 3.1. Heterogeneous catalysts for propene disproportionation

Catalyst	% Conversion	% Selectivity	References
$Mo(CO)_6/Al_2O_3$	25.0	97	[1, 44]
MoO_3/Al_2O_3	11.0	100.0	[1, 3, 45, 46]
$CoO-MoO_3/Al_2O_3$	42.9	94.1	[1, 47]
$MoO_3/AlPO_4$	5.0	95	[47]
MoO_3/SiO_2	28.0	95	[47, 48]
MoS_2/Al_2O_3	1.3	100	[47, 49]
MoS_2/SiO_2	9.1	100	[47, 49]
$MoO_3-Cr_2O_3/Al_2O_3$	36.0	97	[50, 51]
WO_3/Al_2O_3	7.4	100	[3]
$WO_3/AlPO_4$	34.0	82	[47]
WO_3/SiO_2	44.8	97.8	[47, 48]
WS_2/Al_2O_3	1.0	100	[47, 49, 52]
WS_2/SiO_2	18.3	100	[47, 49, 52]
$Re_2(CO)_{10}/Al_2O_3$	20.4	100	[53]
Re_2O_7/Al_2O_3	19.2	100	[54, 46]
Re_2O_7/SiO_2	4.0	100	[47, 55]
Re_2O_7/ZrO_2	2.1	100	[55]
Re_2O_7/ThO_2	12.0	100	[55]
Re_2O_7/SnO_2	15.0	100	[55]
Re_2O_7/TiO_2	13.0	100	[55]
Re_2O_7/Fe_2O_3	0.4	100	[56]
Re_2O_7/MoO_3	3.9	100	[56]
Re_2O_7/WO_3	0.75	100	[56]
V_2O_5/SiO_2	11.2	44	[57]
Nb_2O_5/SiO_2	3.7	90	[57]
Ta_2O_5/SiO_2	8.3	56	[57]
Al_2O_3	0.9	100	[58, 59]
MgO	1.5	100	[60]

For each catalytic system, there exists an optimum range of temperatures within which maximum conversions of propene can be obtained. Molybdenum oxide containing cobalt oxide supported on alumina leads to good conversions within the temperature range 150—200°C. An increase in the temperature over 200°C leads to a rapid deactivation of the catalyst (see Fig. 3.1).

The temperature of the reaction also modifies the composition of the disproportionation products. At lower temperature, the reaction products are unitary, resulting from a simple metathesis between two molecules of propene. On increasing the temperature, owing to the competition of the isomerization and dimerization reactions, a reaction of cometathesis between molecules of different olefins determines the formation of higher olefins and of other heavier products. The increase in temperature from 100°C to 300°C completely modifies the range of products obtained via propene disproportionation, which is rendered evident when using a catalyst of $CoO \cdot MoO_3 \cdot Al_2O_3$ (Table 3.2).

Banks and Bailey [1] have studied the influence of the contact time on the conversion of propene and the distribution of the reaction products, using a catalytic system formed of $CoO \cdot MoO_3 \cdot Al_2O_3$. They have noticed that, for a long contact time, higher conversions of propene are

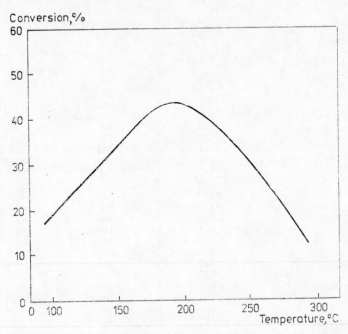

Fig. 3.1. — Variation of the conversion of propene *versus* temperature.[1]

obtained but the increase of the contact time over a certain value leads to the decrease of the process selectivity. The composition of the reaction product is much affected by the increase of the contact time. For a short

TABLE 3.2. Effect of temperature on the disproportionation of propene [a, b]

Reaction tempe-rature °C	Product composition, wt %					Conver-sion %	Selecti-vity %
	Ethene	Butenes	Pentenes	C_6^+	Paraffins		
93	33.5	60.4	—	6.1	—	17.9	93.9
163	27.0	64.5	3.2	4.8	0.5	37.9	91.4
205	27.2	59.0	5.6	7.7	0.5	42.6	86.2
300	13.0	45.9	10.2	19.9	11.0	14.6	58.9

[a] Catalyst : $CoO-MoO_3/Al_2O_3$;   [b] Data from ref. [1].

contact time (e.g. below 2 seconds), ethene and 2-butene are obtained as main products. The increase of the contact time leads to the increase of the quantity of 1-butene and of higher olefins among the reaction products and to the change of the ratio of *trans* to *cis* isomers of 2-butene (Table 3.3).

TABLE 3.3. Effect of contact time on the disproportionation of propene [a,b]

Contact time, seconds	3	8	14	30
Conversion, %	26.2	37.2	42.9	41.0
Selectivity, %	93.6	91.2	94.1	86.6
Product composition, wt %				
Ethene	30.2	32.4	29.8	23.7
1-Butene	1.9	1.7	3.6	6.6
trans-2-Butene	49.0	36.9	39.9	37.0
cis-2-Butene	21.5	20.2	20.8	19.3
C_5^+	6.4	8.8	5.9	13.4

[a] Catalyst: $CoO-MoO_3/Al_2O_3$; [b] Data from ref. [1].

It was noticed that, for a short contact time, the ratio of *trans* to *cis* isomers of 2-butene is smaller than the equilibrium value, but this ratio increases with the increase of the contact time and tends towards the thermodynamic equilibrium value.

The disproportionation of isotopically labelled propene represented one of the methods used by several research groups in order to elucidate the mechanism of the metathesis reaction. Mol, Moulijn and Boelhouwer [61,62] carried out the reaction of the isotopically labelled propene with [14]-C in all of the three positions of the molecule in the presence of heterogeneous catalysts constituted of Re_2O_7 and Al_2O_3.

A first series of expriments, in which 1-^{14}C-propene was used, yielded non-radioactive butene and ethene possessing the double specific radioactivity relative to that of the starting material. Consequently, the reaction took place according to equation 3.9.

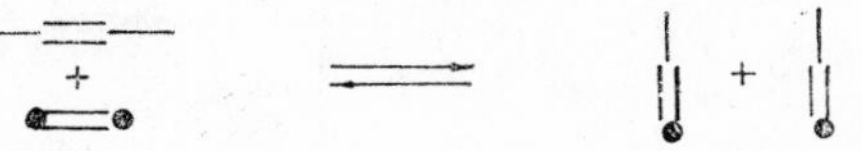

$$\underset{\bullet=\!=\!\bullet}{\overset{-\!=\!-}{+}} \quad\rightleftharpoons\quad \| \;+\; \| \tag{3.9}$$

Starting from 2-^{14}C-propene, non-radioactive ethene and butene with a specific radioactivity double that of the initial propene were obtained (3.10).

$$\| \;+\; \| \quad\rightleftharpoons\quad \underset{-\!\bullet\!=\!\bullet\!-}{\overset{=}{+}} \tag{3.10}$$

Products in which the methyl group preserved their identity during the reaction were formed during the experiments with 3-^{14}C-propene (3.11):

$$\text{(3.11)}$$

Isagulyants and Rar [59] studied the mixed metathesis of labelled and nonlabelled propene with alumina catalysts. Non-radioactive ethene and radioactive butene have been obtained in the reaction between non-radioactive propene and 2-^{14}C-propene, (3.12), while in the reaction be-

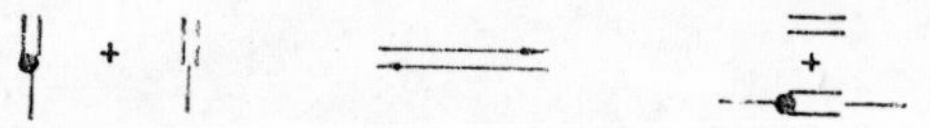

$$\text{(3.12)}$$

tween 1,3-^{14}C$_2$-propene and non-radioactive propene, both reaction products contain half of the initial radioactivity of the starting material (3.13).

$$\text{(3.13)}$$

Clark and Cook [63] studied the disproportionation of 1-^{14}C-and 2-^{14}C-propene by means of the catalytic system CoO· MoO$_3$· Al$_2$O$_3$, under conditions differing from those used by the authors mentioned above. Working at a temperature of 60°C, they obtained normal metathesis products, similar to those obtained by Mol et al.[61, 62] At temperatures higher than 60°C, the isomerization of 1-^{14}C-propene to 3-^{14}C-propene was an important side-reaction, so that it led to a mixture of simple metathesis products of the initial olefins and of the mixed metathesis of the initial olefin with its isomer (3.14).

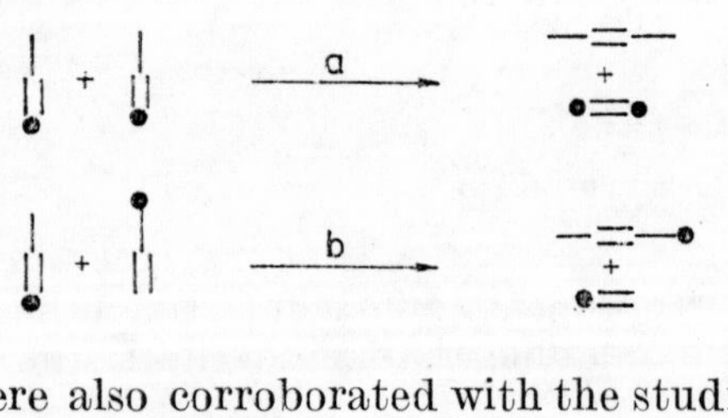

$$\text{(3.14)}$$

These results were also corroborated with the studies of Woody, Lewis and Wills [64] on the metathesis of 1-^{14}C-propene.

Mol, Visser and Boelhouwer [65] disproportonated 2-deuteropropene using as a catalyst alumina-supported rhenium oxide. Working at a temperature of 85°C, undeuterated ethene and 2,3-dideutero-2-butene were obtained as reaction products, according to equation (3.15).

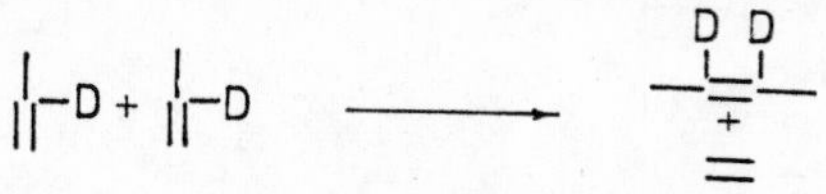

$$\text{(3.15)}$$

The results obtained in the disproportionation of isotopically label-led propene constitute an important evidence for the mechanism of the metathesis reaction. These aspects will be discussed in greater detail in the next chapter which will deal with the mechanism of the metathesis reaction.

3.1.1.3. Metathesis of Butenes

According to the position of the double bond (1- or 2-butene) or to the structure of the chain (n-butene or isobutene) butene may lead to diffe-rent metathesis products. Thus, 1-butene forms ethene and 3-hexene (3.16),

$$\text{(3.16)}$$

while 2-butene leads back to 2-butene by a totally degenerate reaction (3.17).

$$\text{(3.17)}$$

Isobutene leads, via metathesis, to ethene and 2,3-dimethyl-2-butene (3.18).

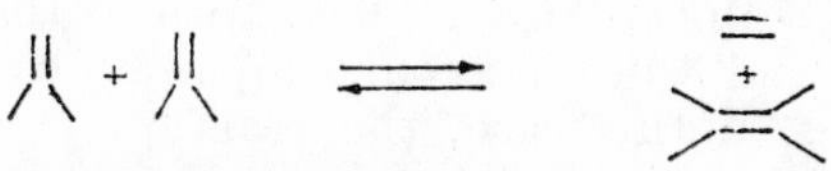

$$\text{(3.18)}$$

but, at the same time forms isobutene by a totally degenerate reaction (3.19).

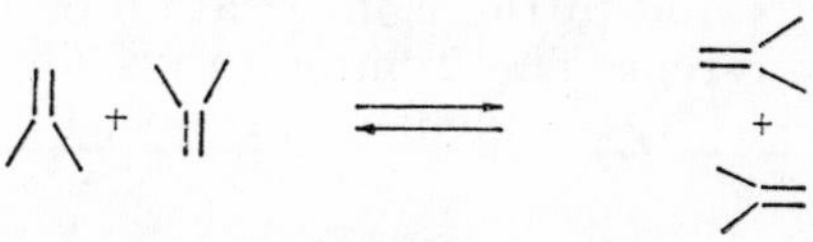

$$\text{(3.19)}$$

The disproportionation of 1-butene and 2-butene was for the first time studied by Banks and Bailey[1] who used the heterogeneous cata-lytic system $CoO \cdot MoO_3 \cdot Al_2O_3$. It was noted that 1-butene disproportionates more readily than 2-butene, the former reaching the maximum conversion (55%) at 150–175°C, while the latter reaches it only at 200°C (30%) (see Fig. 3.2).

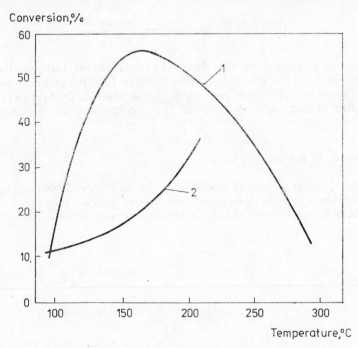

Fig. 3.2. — Variation of the conversion of 1-butene (1) and 2-butene (2) *versus* temperature.[1]

Later on, Bradshaw, Howman and Turner [6] studied in greater detail the disproportionation of 1- and 2-butene with heterogeneous catalytic systems based on $CoO \cdot MoO_3 \cdot Al_2O_3$. They noticed that the metathesis of 1-butene to 2-butene and 3-hexene is accompanied by an isomerization of 1-butene to 2-butene. In this case, the reaction product is much more complex : at different working temperatures, a mixture of $C_2—C_9$ olefins was obtained in the ratio indicated in Table 3.4.

It should be noted that a larger amount of 2-butene was found in the reaction product; this is due to the isomerisation of 1-butene in the presence of the catalytic system. The 2-butene thus formed reacts with the initial 1-butene and leads, via mixed metathesis, to propene and 2-pentene (3.20).

$$\begin{array}{c}|| \\ | \\ |\end{array} \quad + \quad \begin{array}{c}| \\ ||\end{array} \quad \rightleftharpoons \quad \begin{array}{c}=\!\!= \\ = \\ --\!\!=\!\!-\end{array} \hspace{3cm} (3.20)$$

TABLE 3.4. Disproportionation of 1-butene at various temperatures [a,b]

Reaction temperature, °C	122	140	160	180	200	222	245
Products (mole%)							
C_2H_4	24.2	21.9	15.0	8.9	6.0	4.3	4.1
C_3H_6	25.2	29.7	37.2	16.1	48.6	49.1	48.3
C_5H_{10}	27.0	28.9	31.4	32.8	35.5	36.5	38.3
C_6H_{12}	20.3	17.7	12.5	9.2	7.4	7.0	6.4
C_7H_{14}	1.4	1.2	1.8	1.9	1.5	1.5	1.3
C_8H_{16}	0.8	0.6	1.4	0.9	0.9	1.2	1.5
C_9H_{18}	0.1		0.7	0.2	0.1	0.4	0.1
Conversion, %	18.3	23.1	31.9	41.4	35.2	34.4	23.4
Selectivity to $C_2 + C_6$, %	52.2	58.6	68.6	78.9	84.1	85.6	86.6
Selectivity to $C_3 + C_5$, %	45.5	39.5	27.5	18.1	13.4	11.3	10.5

[a] Catalyst: $CoO-MoO_3/Al_2O_3$; GHSV; 2000; Reaction time: 30 minutes.
[b] Data from ref. [6].

Similarly, the disproportionation of 2-butene, in the presence of the catalytic system mentioned above does not proceed normally; 2-butene should have yielded back 2-butene via metathesis (in a totally degenerate reaction). Actually, starting from 2-butene, Bradshaw, Howman and Turner obtained the same complex mixture of reaction products as in the case of 1-butene, but these products appeared in different ratios. The catalytic system promoted the isomerization of 2-butene and the latter, via simple metathesis (see equation 3.16) and via cometathesis with 2-butene (see equation 3.20), led to the formation of ethene, propene, pentene and hexene in the reaction mixture.

By studying the effect of temperature on the degree of conversion of 2-butene and on the selectivity in reaction products, the above authors noticed that, within the temperature range of 122°C to 245°C, the conversion reaches a maximum at about 180°C and, unexpectedly, the selectivity in ethene and hexene decreases much above 160°C, while the selectivity in propene and pentene increases steadily.

The flow ratio of 1-butene has also a significant influence on the degree of conversion and on the selectivity in reaction products. It was noticed that, for high 1-butene flow ratios, although the conversion decreases considerably, the selectivity of the reaction in ethene and hexene increases; this is due to the decrease of the importance of the isomerization reaction of 1-butene to 2-butene and, consequently, to the decrease of the cometathesis between 1-butene and 2-butene.

On inhibiting or promoting the isomerization reaction of 1-butene to 2-butene, by adding suitable compounds to the catalytic system, Bradshaw, Howman and Turner showed that the disproportionation process of butene could be directed either towards the formation of ethene and

3-hexene as main products, or even of a mixture of olefins containing ethene, propene, pentene and hexene. For this purpose, they have studied how the treatment of the catalytic system $CoO \cdot MoO_3 \cdot Al_2O_3$ with sodium ions affects the disproportionation products of 1-butene. The results

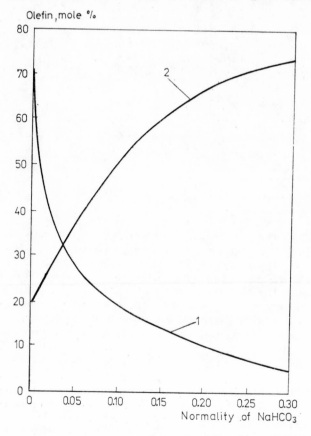

Fig. 3.3. — Disproportionation and isomerization of 1-butene with $CoO \cdot MoO_3/$ $/Al_2O_3$ poisoned by sodium ions : 1 — isomerization of 1-butene to 2-butene ; 2— selectivity in disproportionation products. [6]

obtained in this reaction, as a function of the degree of catalyst poisoning, are given in Figure 3.3. It can be noticed that with the decrease of the conversion degree, the content of 2-butene in the butene fraction also decreases, thus proving that the effect of the poisoning agent is similar for the metathesis and isomerization reactions.

3.1.1.4. Metathesis of Pentenes

1-Pentene forms by metathesis ethene and 4-octene while 2-pentene forms 2-butene and 3-hexene, according to equations (3.21) and (3.22). The metathesis reaction of 1-pentene with catalysts formed of $CoO \cdot MoO_3 \cdot Al_2O_3$

was studied by Banks and Bailey [1] for the temperature range 100—300°C. For 1-pentene, a maximum conversion of 78% was obtained at the tem-

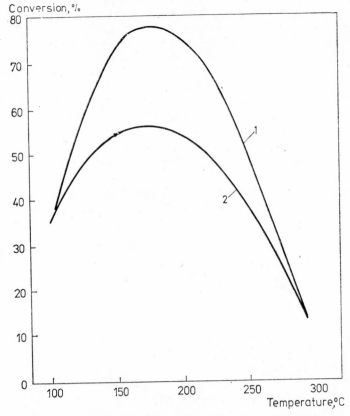

$$(3.21)$$

$$(3.22)$$

perature of 175°C, and for 2-pentene a maximum conversion of 58% at about 175—180°C (see Fig. 3.4.). In most of the experiments, the ratio

Fig. 3.4. — Variation of the conversion of 1-pentene (1) and 2-pentene (2) *versus* temperature. [1]

of *trans*-2-butene to *cis*-2-butene in the reaction products was lower than the equilibrium ratio, while the ratio of *trans*-2-pentene to *cis*-2-pentene was higher than the corresponding equilibrium ratio.

In homogeneous phase, the metathesis reactions of 1-pentene and 2-pentene were studied by several research groups. Calderon, Chen and Scott [2] carried out, for the first time, the metathesis of 2-pentene to 2-butene and 3-hexene, using the homogeneous catalytic system $WCl_6 \cdot EtAlCl_2$. Shortly afterwards, Zuech [66] obtained a new series of homogeneous catalysts constituted of nitrosyl complexes of molybdenum and tungsten with alkylaluminium halides, which proved particularly active for the metathesis of 1-pentene and 2-pentene, while Wang and Menapace [67] prepared a system consisting of WCl_6 and n-butyllithium which was used in the metathesis of 2-pentene.

Uchida et al.[68] employing a series of binary Friedel—Crafts catalysts based on WCl_6 and $EtAlCl_2$ or Et_3Al, proved that, in benzene medium, 1-pentene leads to alkylbenzenes via alkylation, and to 1-pentene oligomers via oligomerization. The same authors showed that, unlike 1-pentene, 2-pentene leads to the normal metathesis products, 2-butene and 3-hexene, in the presence of the catalysts mentioned above.

High conversions in the metathesis of 2-pentene were obtained by Boelhouwer and co-workers [69-71] with a large number of catalysts based on WCl_6, $MoCl_5$ or $ReCl_5$ and organometallic compounds. These authors used, besides n-alkyllithium, phenyl derivatives of tin, lead, antimony, bismuth and mercury. The results obtained with some of these systems are given in Table 3.5.

TABLE 3.5. Metathesis of 2-pentene with binary systems based on organom etallic compounds [a]

Catalytic system	Conversion, %
$WCl_6 \cdot n\text{-}C_4H_9Li$	52
$WCl_6 \cdot n\text{-}C_3H_7Li$	52
$WCl_6 \cdot (C_6H_5)_4Sn$	48
$WCl_6 \cdot (C_6H_5)_3SnCl$	49
$WCl_6 \cdot (C_6H_5)_2SnCl_2$	14
$WCl_6 \cdot (C_6H_5)_4Pb$	48
$WCl_6 \cdot (C_6H_5)_3PbCl$	38
$WCl_6 \cdot (C_6H_5)_3Sb$	5
$WCl_6 \cdot (C_6H_5)_3Bi$	23
$WCl_6 \cdot (C_6H_5)_2Hg$	3
$MoCl_5 \cdot (C_6H_5)_4Sn$	11
$MoCl_5 \cdot (C_6H_5)_4Pb$	3

[a] Data from ref. [70, 71].

Bencze et al.[72-76] studied the activity of catalytic systems based on carbonyl compounds of tungsten and molybdenum or of σ-organotungsten derivatives in the metathesis of 2-pentene.

Taube and Seyferth [77-78] carried out the metathesis of 2-pentene with a catalytic system based on nitrosyl complexes of molybdenum.

They obtained high conversions for a very short reaction time, namely of several minutes.

Finally, Basset and co-workers [79–84] studied the stereochemistry of the metathesis reaction of cis-2-pentene with a series of catalytic systems based on tungsten. They noticed that the isomerization reaction of cis-pentene to trans-2-pentene takes place by a metathesis reaction (3.23).

$$ \qquad\qquad (3.23)$$

3.1.1.5. Metathesis of Hexenes

The three positional isomers of hexene behave differently in the metathesis reaction. Thus, metathesis of 1-hexene yields ethene and 5-decene according to equation

$$ \qquad\qquad (3.24)$$

but, at the same time it reforms 1-hexene through automerization, according to equation

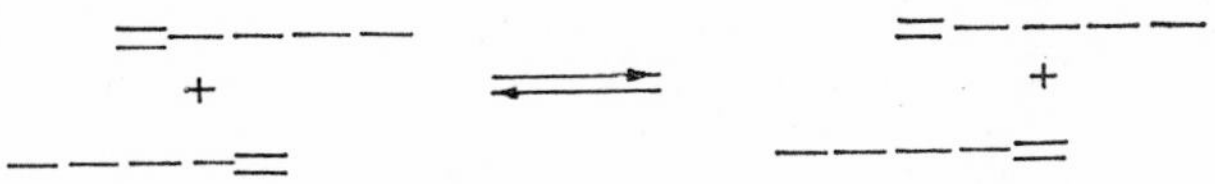

$$ \qquad\qquad (3.25)$$

2-Hexene, besides undergoing automerization, yields the metathesis products, 2-butene and 4-octene

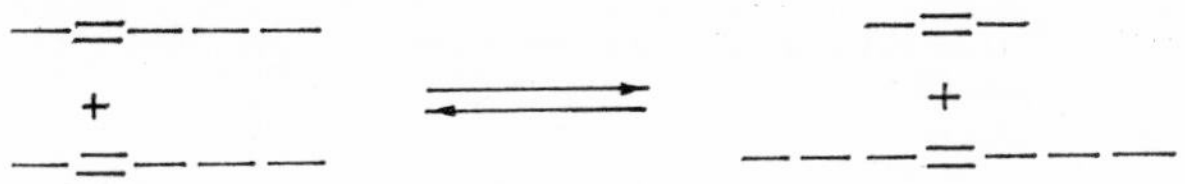

$$ \qquad\qquad (3.26)$$

while 3-hexene undergoes a totally degenerate reaction.

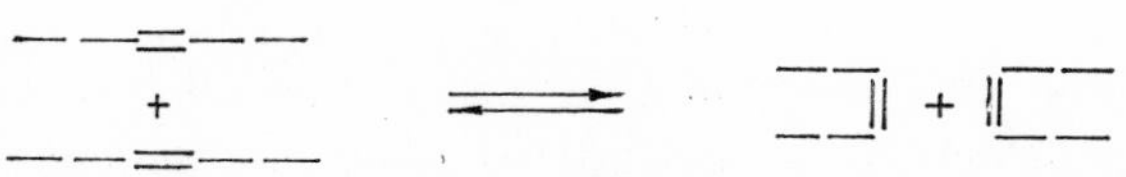

$$ \qquad\qquad (3.27)$$

In the metathesis reaction of 1-hexene, carried out by Banks and Bailey [1] with the heterogeneous catalytic system $CoO \cdot MoO_3 \cdot Al_2O_3$, propene, butene, pentene and heptene were obtained instead of the expected ethene and decene (see equation 3.24). This proves that the rate of the isomerization reactions is greater than that of metathesis. The process is a complex disproportionation invoving cometathesis between the hexenes formed through isomerization. Although the conversion of 1-hexene was over 55 % within the range 150°C—200°C, the formation of the preceding products and not of those which would be expected from the direct metathesis indicated that the metathesis of α-olefins takes place with greater difficulty than that of internal olefins.

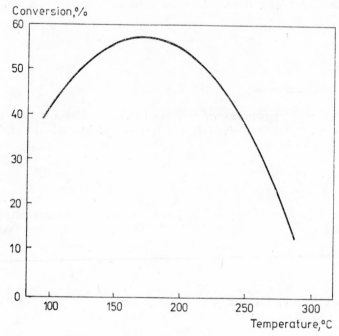

Fig. 3.5.—Variation of the conversion of 1-hexene *versus* temperature.[1]

The temperature dependence of the conversion of 1-hexene displays an optimum point, as indicated by the variation of the conversion with temperature (Fig. 3.5). A marked resemblance was obtained between 1-hexene and 1-butene conversion values within the temperature range 100°C—300°C; it is noticeable that the values of the conversion for the two olefins are very close.

The metathesis reaction of 1-hexene in the homogeneous phase yielded ethene and 5-decene as the main products. In the presence of the catalytic systems formed from $WCl_6 \cdot Bu_4Sn$ and n-propyl acetate or acrylonitrile, Ichikawa and Fukuzumi[85] obtained 5-decene with yields of about 40%. The content in the reaction mixture of the *cis* isomer was 20%, close to the equilibrium value for the *cis* and *trans* isomers of 5-decene.

3.1.1.6. Metathesis of Heptenes

The three isomeric heptenes, 1-heptene, 2-heptene and 3-heptene form, by metathesis, the reaction products indicated in equations

$$(3.28)$$

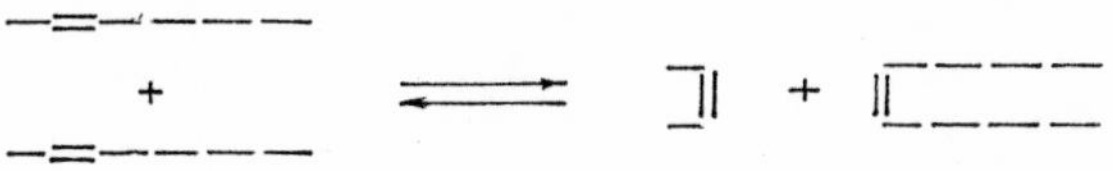

$$(3.29)$$

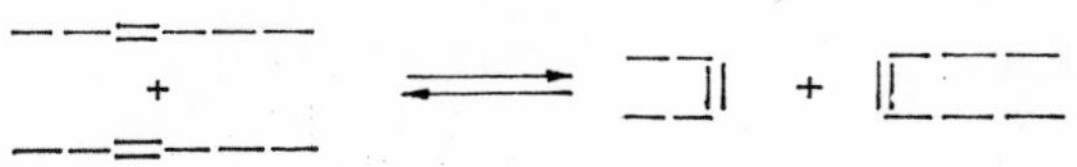

$$(3.30)$$

Under heterogeneous catalysis, the process is usually a complex disproportionation, in which simple metathesis, automerization, isomerization and cometathesis reactions between isomeric heptenes take place. Only the most selective heterogeneous catalytic systems can direct the reaction towards a simple metathesis and can diminish or prevent the other isomerizations or cometatheses. An example is the metathesis of 3-heptene with a selective catalyst ($MoO_3 \cdot Al_2O_3$ treated with potassium hydroxide) effected by Crain.[86] When carried out under reflux, in a fixed catalyst layer, the reaction of 3-heptene led to a conversion of 65% and a selectivity over 90% for 3-hexene and 4-octene (Table 3.6).

TABLE 3.6. Disproportionation of 3-heptene over the catalytic system MoO_3/Al_2O_3 + KOH[a]

Product Composition (mole %)	C_3H_6 1.2	C_5H_{10} 2.5	C_6H_{12} 45.6	C_8H_{16} 45.6	C_9H_{18} 3.2	$C_{10}H_{20}$ 1.9

[a] Data from ref. [86].

However, when carried out under homogeneous catalysis, the reaction of heptene directly yielded simple metathesis products, as in the above reactions. Thus, 1-heptene afforded ethene and 6-dodecene with yields between 20 and 30%, when employing ternary catalytic systems, $WCl_6 \cdot Bu_4Sn \cdot CH_3CO_2C_3H_7$ or $WCl_6 \cdot Bu_4Sn \cdot CH_3CN$, and working in trichloroethylene as solvent. [85] The *cis* isomer content of the product was 32—33%. The presence of additives in these catalytic systems prevented the migration of the double bond in heptene or dodecene, but must have allowed the geometric isomerization to occur since the ratio between the *cis* and *trans* isomers of the final olefin was close to the equilibrium value.

3.1.1.7. Metathesis of Octenes

Octene presents a greater number of positional isomers than the lower homologues previously discussed. 1-Octene, 2-octene and 3-octene can yield, by metathesis, the reaction products indicated in the following equations

$$(3.31)$$

$$(3.32)$$

$$(3.33)$$

but 4-octene is reformed via a totally degenerate reaction

$$(3.34)$$

In the studies carried out by Banks and Bailey [1] on the reactions of 1-octene under heterogeneous catalysis using $CoO \cdot MoO_3 \cdot Al_2O_3$, a very high conversion was reached, as much as 75 % under optimum conditions; nevertheless, the products thus obtained were totally different from those expected for the simple metathesis of 1-octene (Fig. 3.6).

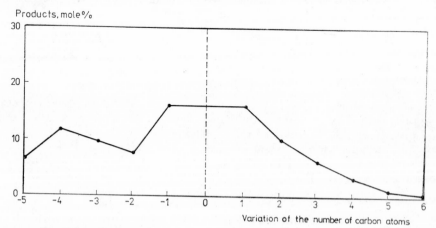

Fig. 3.6. — Product distribution in the disproportionation of 1-octene.[1]

The isomerization rate was high in this case too, the process being complicated by the various cometathesis reactions occurring between the positional isomers of octene. The simple metathesis of 1-octene which yields ethene and 5-decene, according to reaction (3.31), takes place with difficulty.

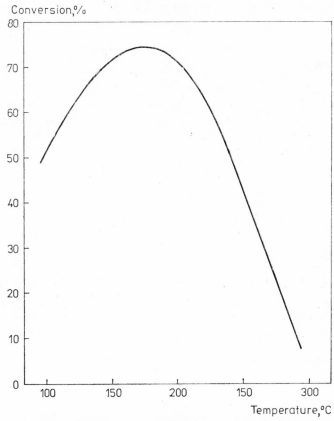

Fig. 3.7. — Variation of the conversion of 1-octene *versus* temperature.[1]

The conversion of 1-octene, under the above catalytic system, undergoes a marked variation within the temperature range of 90°C and 315°C. The conversion reaches its maximum of about 75 % at 175—200°C and decreases suddenly towards 300°C. (Fig. 3.7).

The more pronounced decrease of the conversion in the disproportionation reaction of 1-octene is explicable on the basis that, in this case, polymerization products with a higher molecular weight, which deactivate the active centres of the catalyst, are more readily formed.

High selectivity was obtained in the metathesis of 1-octene to 7-tetra-decene by Crain[86] in the reaction carried out over a catalytic system form-ed from $MoO_3 \cdot Al_2O_3$ treated with potassium hydroxide. The treatment effected on this catalytic system led to the elimination of the isomeriza-tion side reactions of 1-octene to 2-octene, 3-octene or 4-octene. The metathesis of 1-octene in the homogeneous phase proceeds differently from the reaction in the heterogeneous phase. When effected with the homogeneous system $Et_3Al \cdot WCl_6$ in chlorobenzene, it yields ethene and 7-tetradecene. When carried out with the catalytic system $Et_3Al \cdot WCl_6 \cdot O_2$, a mixture of olefins ranging from ethene to tetradecene is obtained. If benzene is used as a reaction medium instead of chlorobenzene then, along with the metathesis and oligomerization products, Friedel-Crafts reactions of olefins also yield alkylbenzenes.

The metathesis of 1-octene takes place with good selectivity to give ethene and 7-tetradecene when using ternary catalytic systems consti-tuted of $WCl_6 \cdot Bu_4Sn$ and additives, such as alkyl acetates, acetonitrile, phenylacetylene, dicyclopentadiene, ethyl ether, propyl ether, tetrahy-drofuran, 1-propanol.[85] When working in trichloroethylene, the side-re-actions of alkylation are eliminated, and the isomerization and oligome-rization reactions of olefins are reduced.

The metathesis of 2-octene was carried out by Crain [86] using the hete-rogeneous catalytic system based on MoO_3 and Al_2O_3, mentioned above. The reaction took place with high selectivity (92%), leading to 2-butene and 6-dodecene.

3.1.1.8. Metathesis of Decene and Higher Homologues

The reaction of decene in heterogeneous phase yields a large number of products, because of the great number of isomers of decene which are formed in the presence of the catalytic system and which take part in simple or cometathesis reactions. The process is less selective and leads to a rapid deactivation of the metathesis catalytic system as a conse-quence of polymerization and cracking.

In the homogeneous phase, the reaction of decene takes place with high selectivity to normal metathesis products, the isomerization and co-metathesis reactions being eliminated. Thus, the metathesis of 1-decene in the presence of catalytic systems formed from WCl_6, Bu_4Sn and aceto-nitrile or propyl acetate yields 9-octadecene with a yield of 44—45%.[87]

The metathesis of the higher homologues of decene, namely of unde-cene, dodecene, tridecene, tetradecene, etc. have been less studied. Obvi-ously, on increasing the number of carbon atoms in the olefin chain, the number of isomers increases and the reaction becomes more complex and less selective. The activity of heterogeneous catalysts will be rapidly dimi-nished and the process will proceed with greater difficulty. However, the reactions in the homogeneous phase of these hydrocarbons are interesting because in this medium the side reactions, especially those of isomeriza-tion, are eliminated.

3.1.2. Metathesis of Substituted Olefins

31.2.1. Olefins Substituted with Alkyl or Alkenyl Groups

Isobutene. A tetrasubstituted alkene, 2,3-dimethyl-2-butene can be easily obtained through metathesis of 2-methylpropene (isobutene).

$$\text{(3.35)}$$

The metathesis of isobutene has been carried out more frequently in the heterogeneous phase, with binary or ternary catalytic systems, based on molybdenum or tungsten oxides. Arganbright [88] obtained high selectivities in metathesis products by employing alumina-supported molybdenum oxide catalytic systems and by inhibiting the dimerization and oligomerization side reactions with fluorine ions. Heckelsberg [89] improved the selectivity of the metathesis reaction of isobutene to 2,3-dimethyl-2-butene in the presence of silica-supported tungsten oxide catalytic systems, adding to the latter small quantities of barium, caesium or potassium oxides. The metathesis of isobutene with other heterogeneous catalytic systems is less selective. Catalysts formed from WO_3, SiO_2 and MgO lead to the isomerization of 2,3-dimethyl-2-butene resulting from the primary reaction[90] and those formed from CoO, Mo_2O_3 and Al_2O_3 yield oligomerization products.[91] The reaction of isobutene is even more complex over catalysts formed from aluminosilicates.[92,93] In their presence, hydrogen and alkyl group transfers are predominant, leading to branched oligomerization products.

2-Methyl-1-butene. 2-Methyl-1-butene, like isobutene, should form ethene and 3,4-dimethyl-3-hexene, via metathesis.

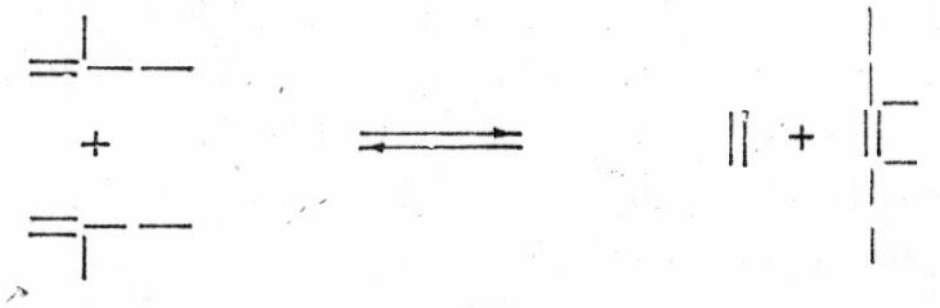

$$\text{(3.36)}$$

In the presence of catalytic systems such as $CoO \cdot MoO_3 \cdot Al_2O_3$,[3,94] although the conversion of 2-methyl-1-butene was high, a mixture of olefins $C_6 - C_8$ and heavier compounds was obtained, indicating low selectivity in primary reaction products.

3-Methyl-1-butene. Through the metathesis of 3-methyl-1-butene (isopentene), ethene and 2,5-dimethyl-3-hexene are formed.

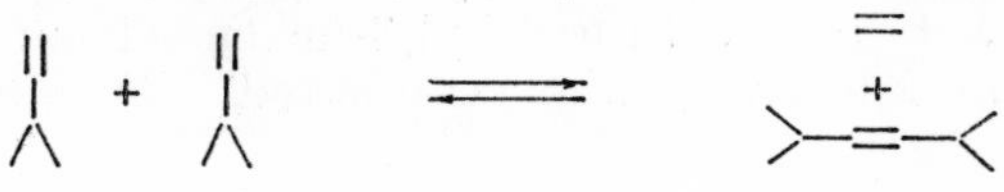

$$\text{(3.37)}$$

Although the substituent is in the β-position relative to the double bond, and therefore would be expected to exert only minor influence, the metathesis of 3-methyl-1-butene takes place in a totally different way from that of 1-butene. The reaction of 3-methyl-1-butene in the presence of catalytic systems formed from $CoO \cdot MoO_3 \cdot Al_2O_3$[3] is non-selective, yielding isomerization products of the starting material, 2-methyl-1-butene and 2-methyl-2-butene, and linear and branched C_6-C_8 olefins as well.

2-*Methyl-2-butene.* 2-Methyl-2-butene should afford 2-butene and 2,3-dimethyl-2-butene by metathesis, in the presence of suitable catalytic systems.

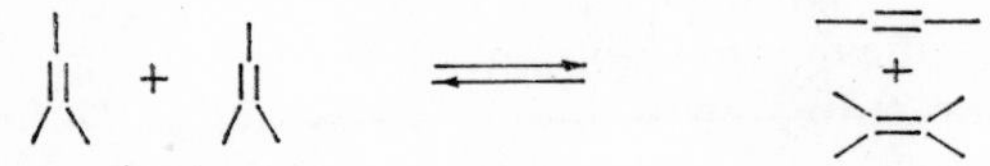

$$\text{(3.38)}$$

The reaction of 2-methyl-2-butene carried out by Banks[3] with the catalytic systems $CoO \cdot MoO_3 \cdot Al_2O_3$ yielded, with a very low conversion, C_6-C_8 products. The formation of 2,3-dimethyl-2-butene by this method, is thus more difficult than by the similar reaction of isobutene.

2-*Methyl-2-pentene.* The normal metathesis products of 2-methyl-2-pentene are 2,3-dimethyl-2-butene and 3-hexene.

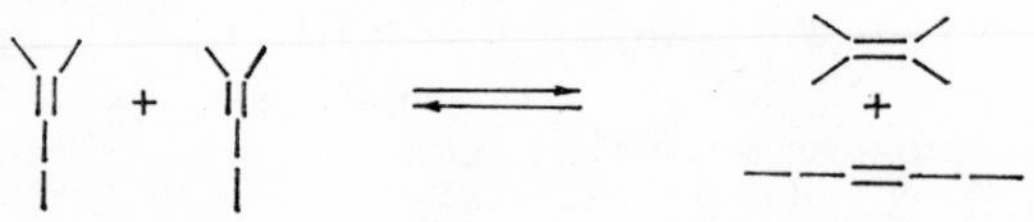

$$\text{(3.39)}$$

In the reaction of 2-methyl-2-pentene, in the presence of homogeneous catalytic systems formed from WCl_6, ethanol and $EtAlCl_2$, Calderon and Chen[95] obtained a mixture of 2-methyl-2-butene, 3-methyl-2-pentene, 3-ethyl-2-pentene, 3-methyl-3-hexene and 3-ehtyl-3-hexene, thus proving that the metathesis reaction is accompanied, in these conditions, by isomerizytion and transalkylation side reactions.

3-*Methyl-2-pentene.* 3-Methyl-2-pentene ought to yield 2-butene and 3,4-dimethyl-3-hexane, via metathesis.

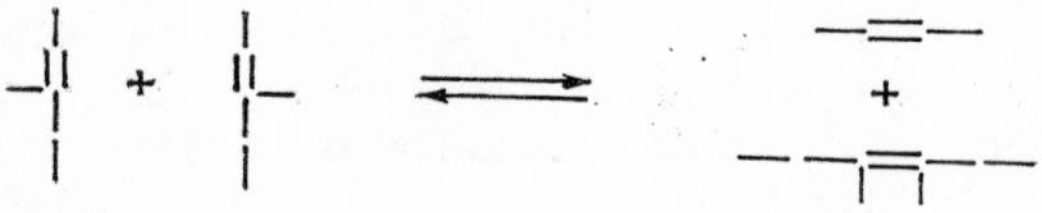

$$\text{(3.40)}$$

By employing particularly active catalytic systems containing molybdenum or tungsten, of the type $R_4N^+[M(CO)_5Cl]^-$, where R is an alkyl group, Doyle[96] obtained 3,4-dimethyl-3-hexene from 3-methyl-2-pentene with very low selectivity when compared with the similar reaction of 1-pentene.

3,3-*Dimethyl-1-butene*. The metathesis of 3,3-dimethyl-1-butene (neo-xene) is a simple method for obtaining 1,2-di-t-butylethene.

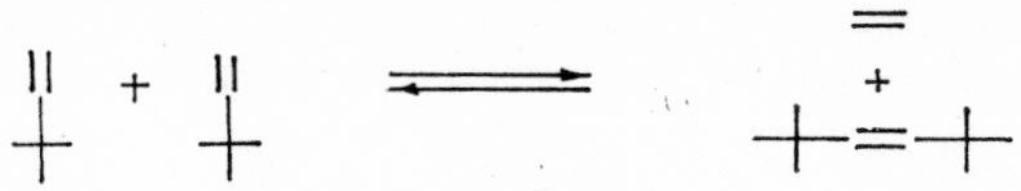

$$(3.41)$$

The reaction took place with good selectivity in 1,2-di-t-butylethene, in the presence of heterogeneous catalytic systems formed from WO_3, SiO_2 and MgO [97] or with heterogeneous and homogeneous systems based on π complexes of tungsten or molybdenum.[98,99]

Vinylcyclohexane. Finkel'stein and co-workers [100] obtained, via metathesis, 1,2-dicyclohexylethene with heterogeneous catalytic systems derived from Re_2O_7 and Al_2O_3

$$(3.42)$$

Vinylcyclohexene. The metathesis of 4-vinylcyclohexene takes place at the exocyclic double bond, with formation of 1,2-dicyclohexenylethene.

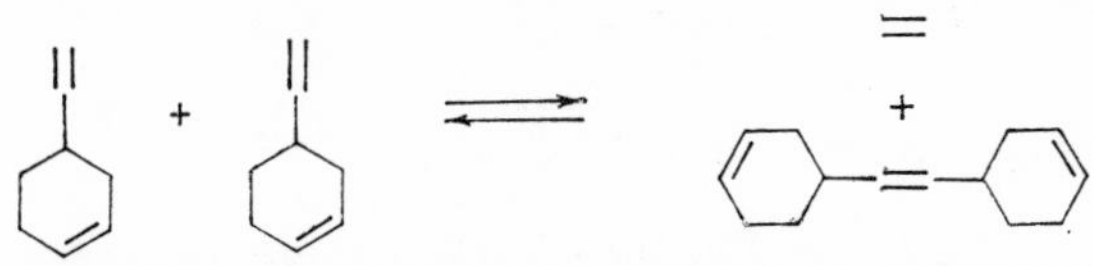

$$(3.43)$$

The reaction was carried out by Crain,[101,102] who used heterogeneous catalytic systems of alumina-supported molybdenum or tungsten oxides. In all cases, the double bond of the cyclohexene ring was not affected.

3.1.2.2. Olefins Substituted with Aryl Groups

Styrene. Through metathesis, styrene yields ethene and stilbene.

$$(3.44)$$

In the presence of heterogeneous catalytic systems based on tungsten, molybdenum or rhenium oxides,[103] the equilibrium is shifted towards the formation of styrene from stilbene and ethene, but, under homogeneous catalysis, with systems based on $Mo(CO)_6$ or $Mo(CO)_2(Tol)$ [104] the reaction takes place with formation of stilbene.*

---

* Abbreviations: Tol = *para*-tolyl; Oct = n-octyl

ω-Arylolefins. The metathesis reaction applied to ω-arylolefins is a simple method for the synthesis of ω,ω'-diarylolefins

$$
\begin{array}{c}
C_6H_5-(CH_2)_n-CH=CH_2 \\
+ \\
C_6H_5-(CH_2)_n-CH=CH_2
\end{array}
\quad \longrightarrow \quad
\begin{array}{c}
C_6H_5-(CH_2)_n-CH \\
\parallel \\
C_6H_5-(CH_2)_n-CH
\end{array}
\; + \;
\begin{array}{c}
CH_2 \\
\parallel \\
CH_2
\end{array}
\qquad (3.45)
$$

where n = 1 − 3.

By employing catalytic systems from $MoCl_2(NO)_2$ $(PPh_3)_2$ and $EtAlCl_2$, Descotes, Chevalier and Sinou [21,105] obtained ω,ω'-diarylolefins with yields between 50 and 90%. The reaction was also carried out with substituted ω-arylolefins, where the aryl is *para*-chlorophenyl.

2,2'-Divinylbiphenyl. 2,2'-Divinylbiphenyl yields, through metathesis, ethene and phenanthrene.

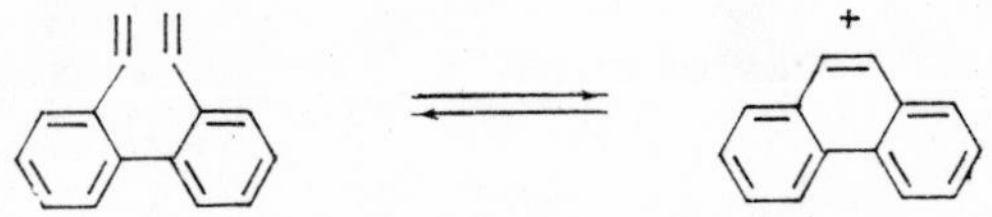

$$(3.46)$$

The reaction was carried out by Katz and Rotchild [106] in homogeneous phase with binary catalytic systems consisting of $MoCl_2(NO)_2(Oct_3P)_2$ or $WCl_2(NO)_2(Oct_3P)_2$ and $Me_2Al_2Cl_3$.

3.1.2.3. Olefins Substituted with Functional Groups

The metathesis reaction has been recently extended to olefins substituted with halogen, amino, nitrile, acid, ester or other functional groups.

Fridman and co-workers [107] carried out the reaction of allyl chloride, allyl bromide, crotyl chloride, crotyl bromide and 5-bromo-1-hexene over alumina-supported heterogeneous catalytic systems prepared from Re_2O_7 treated with Bu_4Sn. The reaction was also applied to other halo-olefins [108] and shown to constitute a general synthetic route to dihalo-olefins,

$$
\begin{array}{c}
XCRR'(CH_2)_mCH=CH_2 \\
+ \\
XCR''R'''(CH_2)_nCH=CH_2
\end{array}
\quad \longrightarrow \quad
\begin{array}{c}
CH_2 \\
\parallel \\
CH_2
\end{array}
\; + \;
\begin{array}{c}
CH(CH_2)_mCRR'X \\
\parallel \\
CH(CH_2)_nCR''R'''X
\end{array}
\qquad (3.47)
$$

where X = Cl, Br; R, R^I, R^{II}, R^{III} = H, Me; m, n = 0 − 3. The metathesis of unsaturated amines $CH_2CH(CH_2)_n$ NRR' [109] leads to unsaturated diamines in the presence of catalytic systems based on $W(CO)_3(Mes)$ and $EtAlCl_2$.

$$
\begin{array}{c}
CH_2=CH(CH_2)_n NRR' \\
+ \\
CH_2=CH(CH_2)_n NRR'
\end{array}
\quad \longrightarrow \quad
\begin{array}{c}
CH_2 \\
\parallel \\
CH_2
\end{array}
\; + \;
\begin{array}{c}
CH(CH_2)_nNRR' \\
\parallel \\
CH(CH_2)_nNRR'
\end{array}
\qquad (3.48)
$$

The higher the molar ratio Al :W in the catalytic system, the better the yield. The reactions carried out with hydrochlorides gave higher conversions than those with free amines.

Acrylonitrile reacts with propene over a silica-supported tungsten oxide catalytic system, yielding ethene and crotonic nitrile.

$$CH_2{=}CH{-}CN + CH_2{=}CH{-}CH_3 \rightleftharpoons \begin{array}{c} CH_2 \\ \| \\ CH_2 \end{array} + \begin{array}{c} CH{-}CN \\ \| \\ CH{-}CH_3 \end{array} \qquad (3.49)$$

The reaction takes place with 24% selectivity in nitrile.

Among nitriles of higher olefins, the reaction of oleic nitrile is known[110] which, over binary catalytic systems formed from alkyl aluminium compounds and tungsten, molybdenum or rhenium compounds yields the corresponding dinitrile.

$$2\ CH_3(CH_2)_7CH{=}CH(CH_2)_7CN \longrightarrow \begin{array}{c} CH_3(CH_2)_7\,CH \\ \| \\ CH_3(CH_2)_7\,CH \end{array} + \begin{array}{c} CH(CH_2)_7CN \\ \| \\ CH(CH_2)_7CN \end{array} \qquad (3.50)$$

Through the metathesis of unsaturated esters, it is possible to obtain esters of dicarboxylic acids.

$$CH_2{=}CH(CH_2)_nCOOR + CH_2{=}CH(CH_2)_nCOOR \longrightarrow \begin{array}{c} CH_2 \\ \| \\ CH_2 \end{array} + \begin{array}{c} CH(CH_2)_nCOOR \\ \| \\ CH(CH_2)_nCOOR \end{array} \qquad (3.51)$$

The reaction of methyl esters of lower unsaturated acids such as acrylic, methacrylic, 1- and 2-butenoic acids, 1- and 2-pentenoic acids was carried out by Verkuijlen *et al.*[111] with homogeneous catalytic systems formed from WCl_6 and Me_4Sn. The same authors [112] also used in the reaction of methyl 1-pentenoate a heterogeneous catalytic system consisting of $Re_2O_7 \cdot Al_2O_3$ and Me_4Sn.

The corresponding esters of dicarboxilic acids can be obtained from the esters of unsaturated fatty acids with a double bond. Thus, methyl oleate, in the presence of homogeneous catalytic systems formed from WCl_6 and Me_4Sn [23,113,114] disproportionates to the corresponding olefin and dicarboxylic ester.

$$CH_3(CH_2)_7CH{=}CH(CH_2)_7COOMe + CH_3(CH_2)_7CH{=}CH(CH_2)_7COOMe \longrightarrow \begin{array}{c} CH(CH_2)_7\,CH_3 \\ \| \\ CH(CH_2)_7CH_3 \end{array} + \begin{array}{c} CH(CH_2)_7COOMe \\ \| \\ CH(CH_2)_7COOMe \end{array} \qquad (3.52)$$

Other catalytic systems formed from alkylaluminium derivatives and tungsten, molybdenum or rhenium [110,115] were also used for the metathesis of methyl oleate. Butyl oleate reacted similarly in the presence of the above systems. Methyl erucate yielded the corresponding dicarboxylic ester and olefin with binary systems formed from WCl_6 and Me_4Sn, Ph_4Sn or Ph_4Pb.[114] The unsaturated dicarboxylic acid thus obtained may be used in the synthesis of polyesters or polyamides.

The metathesis of unsaturated fatty acids with several C=C double bonds affords a more complex mixture of products. During the reaction, the double bonds may participate inter- and intramolecularly with formation of polymers, mono- or dicarboxylic esters and even cyclo-olefins. Such compounds were obtained in the reactions of methyl linoleate or inolenate over binary catalytic systems of WCl_6 and Me_4Sn.[116 a]

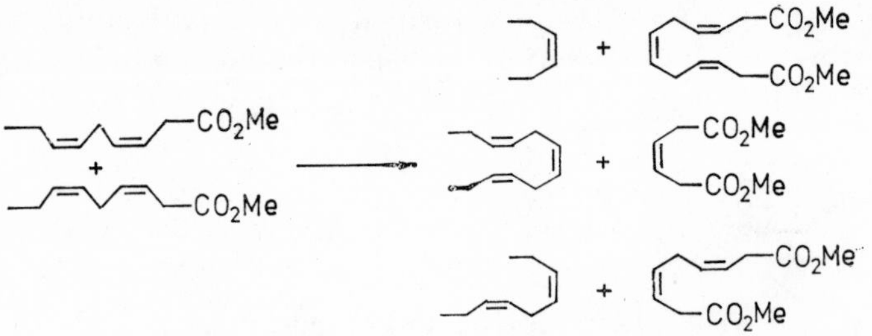

$$(3.53)$$

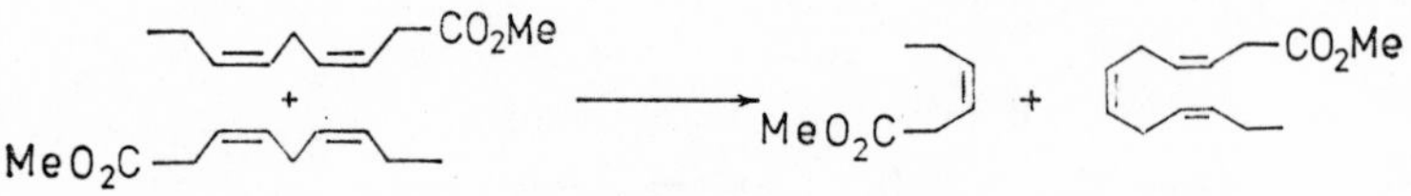

$$(3.54)$$

The polyene compounds obtained from methyl linoleate or linolenate yield, via intramolecular metathesis, cyclic polyenes and new esters of mono- or di-carboxylic acids.

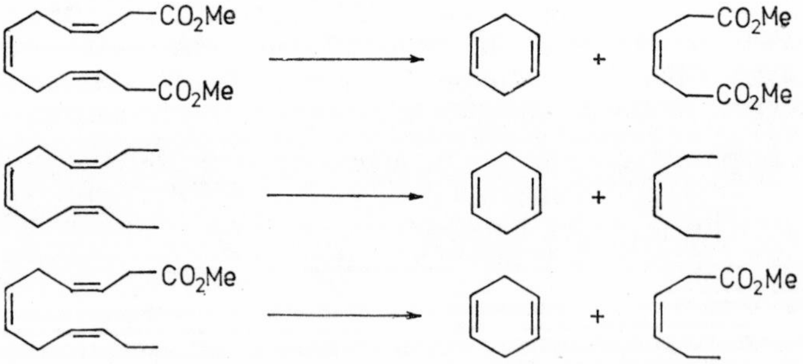

$$(3.55)$$

The conversion and the composition of the products obtained during the metathesis of methyl linoleate or linolenate over WCl_6 and Me_4Sn are presented in Table 3.7.

TABLE 3.7. Conversion and composition of metathesis products of methyl linoleate and linolenate with $WCl_6 \cdot Me_4Sn$ catalysts [a]

Starting material	% Conversion	Monocarbo-xylic esters (mole %)	Dicarboxylic esters (mole %)	Acyclic poly-enes (mole %)	Cyclic polyenes (mole %)
Methyl linoleate	84	46	23	23	8
Methyl linolenate	95	30	15	15	40

[a] Data from ref. [116b].

3.2. Metathesis of Acyclic Dienes

The metathesis was successfully applied to acyclic dienes. This is an important pathway for processing dienes in organic synthesis. Depending on the position of the double bond in dienes they can participate in inter-molecular or intramolecular reactions, yielding different reaction products in the two cases.

Short-chain dienes, with conjugated or non-conjugated double bonds, usually form intermolecular metathesis reactions. Thus, in the reaction of 1,3-butadiene or isoprene over tungsten oxide heterogeneous catalysts, Heckelsberg, Banks and Bailey [16] obtained metathesis products resulting from the participation of only one of the double bonds of the molecule.

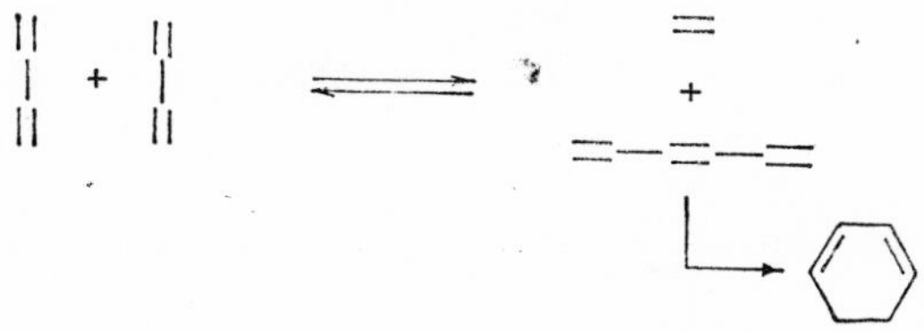

$$(3.56)$$

Cyclohexadiene was readily formed by the cyclization of hexatriene. The analysis of the reaction products showed that the reaction mixture was more complex, some side-reactions of cyclo-addition and isomerization of dienes taking place under these conditions (Table 3.8).

TABLE 3.8. Disproportionation of 1,3-butadiene over the catalytic system WO_3/SIO_2 [a]

Reaction products, mole %	
Ethene	29
Propene	4
Butene	12
Cyclohexadiene	28
Vinylcyclohexene	27
Conversion, %	3—5

[a] Data from ref. [16].

1,3-Pentadiene reacted more readily than butadiene, but in the reaction mixture a large variety of olefins and dienes was present because of side-reactions.

The similar reaction of 1,4-pentadiene has not been studied. Under heterogeneous catalysis, the product will be a mixture of olefins similar to that obtained in the reaction of 1,3-pentadiene, because of the isomerization side-reactions. Under homogeneous catalysis, by suppressing any side-reactions, the metathesis of 1,4-pentadiene can yield ethene and 1,4,7-octatriene or cyclopropene through, respectively, intermolecular and intramolecular participation of the double bonds.

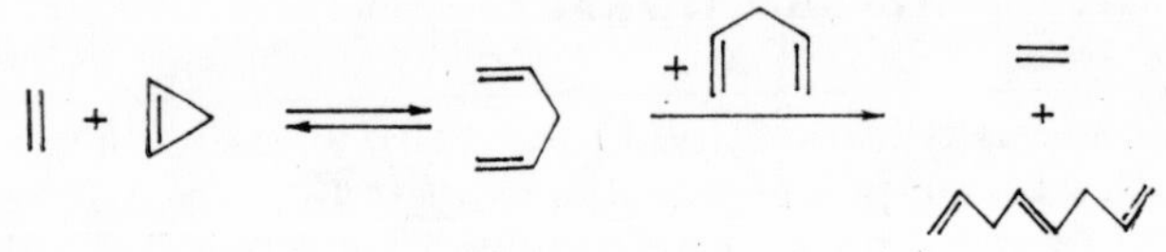

$$\text{(3.57)}$$

Zuech et al.[116b] carried out the reaction of 1,5-hexadiene with a homogeneous catalytic system based on a molybdenum nitrosyl complex and aluminium methylsesquichloride. 1,5,9-Decatriene was obtained as the main reaction product, with traces of 1,5,9,13-tetradecatetraene and 1,5,9,13,17-octadecapentaene. No traces of cyclobutene could be detected. These results prove that the reaction took an intermolecular course, and that the intramolecular cyclization by metathesis was excluded.

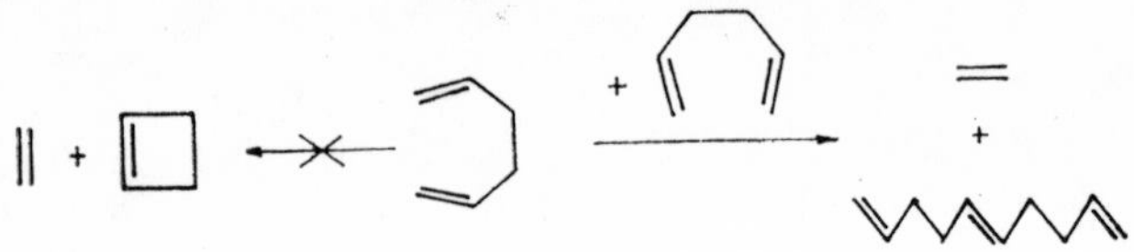

$$\text{(3.58)}$$

Unlike 1,5-hexadiene, 1,7-octadiene led, under similar reaction conditions, to the formation of cyclohexene, as a main product, with traces of 1,7,13-tetradecatriene.

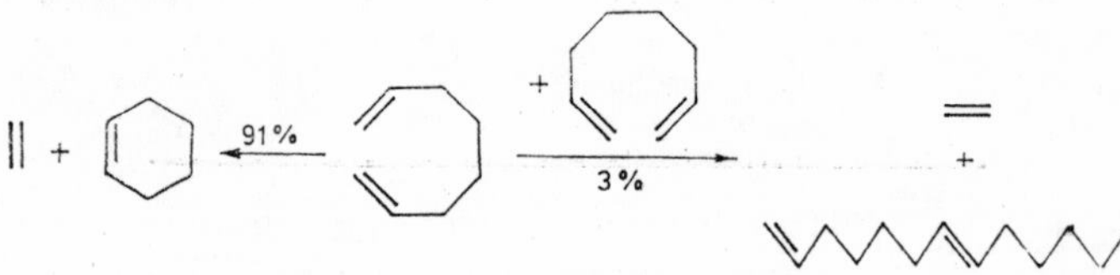

$$\text{(3.59)}$$

The reaction took mostly an intramolecular course, favoured also by the high stability of cyclohexene, unlike that of the smaller rings which are not formed in the similar reactions of lower homologues.

It is very likely that in the reaction of 1,6-heptadiene, the double bonds will participate in comparable amounts in the intramolecular and intermolecular metathesis with formation of cyclopentene and 1,6,11-decatriene but, so far, this reaction has not been studied.

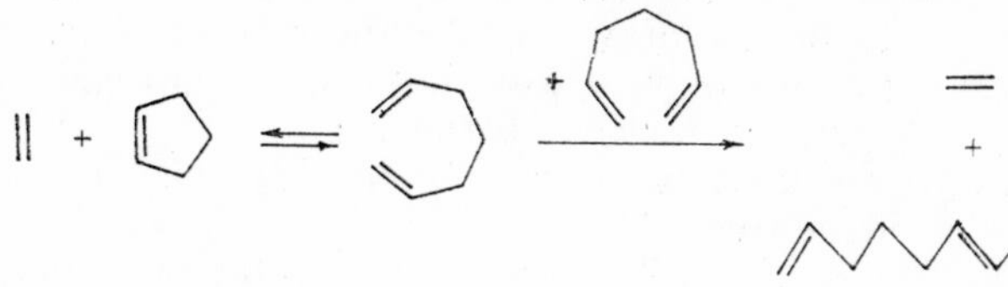

$$\| + \bigcirc \rightleftharpoons \bigcirc \xrightarrow{\quad +\ \text{(diene)}\quad} \quad = \quad (3.60)$$

The above mentioned examples of diene metathesis prove the usefulness of this reaction in the synthesis of polyenic or cyclic olefins starting from acyclic dienes. In the general case of α,ω-dienes, the metathesis can be a preparative method for cyclo-olefins or long linear chain polyenes.

3.3. Metathesis of Cyclo-Olefins

Monocyclic olefins with a single double bond yield, by metathesis, cyclic diolefins having twice as many carbon atoms in the molecule.

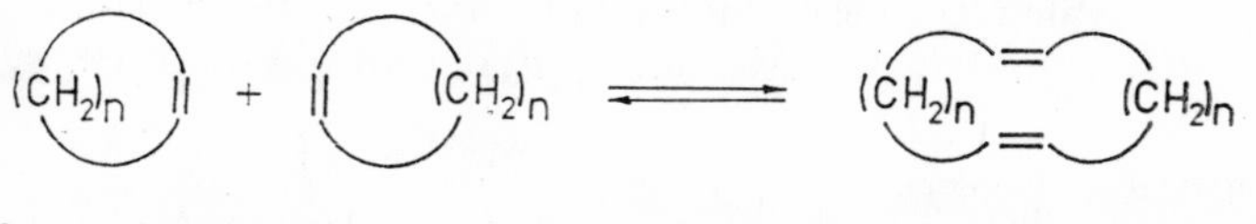

$$(CH_2)_n\ \| \ + \ \| \ (CH_2)_n \ \rightleftharpoons \ (CH_2)_n \overset{=}{\underset{=}{\bigcirc}} (CH_2)_n \qquad (3.61)$$

Therefore, the reaction is a dimerization accompanied by cleavage of the double bonds of the starting olefin. The diolefin thus formed may yield, through a further reaction with another molecule, a trimer,

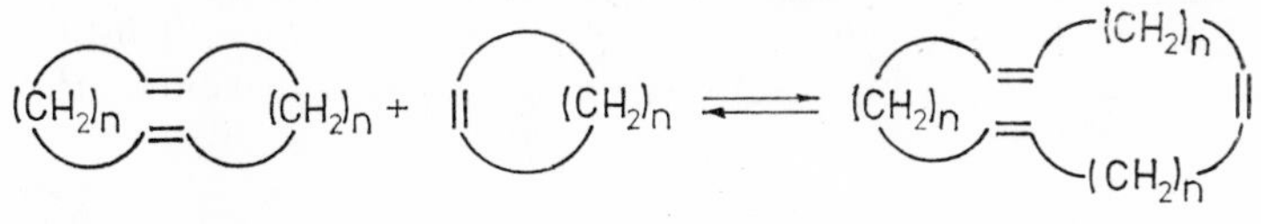

$$(CH_2)_n \overset{=}{\underset{=}{\bigcirc}} (CH_2)_n + \| \ (CH_2)_n \ \rightleftharpoons \ (CH_2)_n = \overset{(CH_2)_n}{\underset{(CH_2)_n}{\bigcirc}} \| \qquad (3.62)$$

or, through a new dimerization, a tetramer.

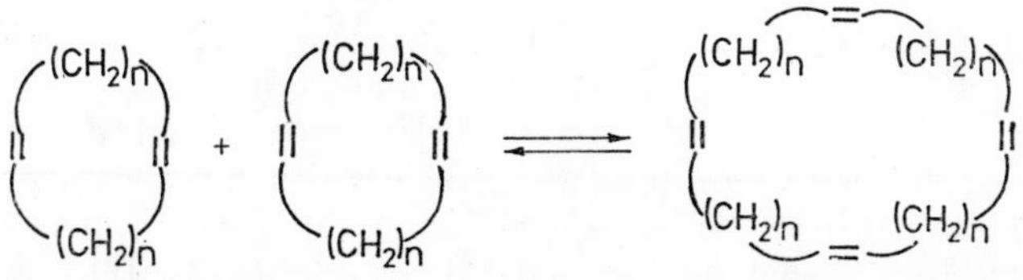

$$\begin{pmatrix} (CH_2)_n \\ \| \quad \| \\ (CH_2)_n \end{pmatrix} + \begin{pmatrix} (CH_2)_n \\ \| \quad \| \\ (CH_2)_n \end{pmatrix} \rightleftharpoons \begin{pmatrix} (CH_2)_n \overset{=}{} (CH_2)_n \\ \| \qquad \| \\ (CH_2)_n \underset{=}{} (CH_2)_n \end{pmatrix} \qquad (3.63)$$

The reaction chain may continue with the formation of higher molecular weight cyclic oligomers.

On starting from a monocyclic diolefin, the reaction may proceed either towards dimerization or towards cleavage (through reverse metathesis or retrometathesis) with the formation of mono-olefins. Cyclo-olefins with several double bonds may show similar behaviour.

So far the following types of reactions have been noticed for cyclo-olefins with catalytic systems for metathesis:

i) dimerizations, trimerizations and oligomerizations, according to the reactions presented above;

ii) cleavage of cyclic diene and polyenes to cyclic mono-olefins, the reactions constituting the reverse of those mentioned above;

iii) polymerization through ring-opening yielding polyalkenamers.

Dimerizations of cyclo-olefins are not selective; they are accompanied by trimerizations, tetramerizations and oligomerizations, so that the result of the reaction is a mixture of dimers, trimers and oligomers. In most instances, metathesis catalytic systems yield polyalkenamers accompanied by a small amount of cyclic oligomers.

3.3.1. Metathesis of Monocyclic Olefins

The majority of monocyclic olefins, with most of the metathesis catalytic systems, yield polyalkenamers by ring-opening polymerization. However, an appreciable number of monocyclic mono-, di- or tri- olefins yield, through reverse metathesis reactions, cyclic mono-olefins.

3.3.1.1. Oligomerization Reactions

Of all monocyclic olefins, the behaviour of small and medium rings was mostly studied in metathesis.

Cyclobutene. In the presence of metathesis catalytic systems, cyclobutene readily yields polybutenamer. Even its dimer, 1,5-cyclo-octadiene, gives oligomerization and polymerization reactions, the reaction being practically irreversible.

$$\square \longrightarrow \bigcirc \longrightarrow -\cdots- \longrightarrow \left[-\!\!-\!\!=\!\!- \right]_n \qquad 3.64)$$

Natta and co-workers [117,118] carried out the polymerization of cyclobutene with a large range of catalysts formed from transition metal salts and organometallic compounds. These authors obtained polymers and no formation of dimers or oligomers was noticed during these reactions. The results of these studies are presented in a separate chapter (Chapter 4).

1-Methylcyclobutene together with a series of catalytic systems, e.g. catalysts of tungsten salts and organoaluminium compounds,[119] forms saturated cyclization products resulting by oligomerization, while with other catalytic systems, e.g. the diphenylpentacarbonyltungsten carbe nic complex,[120] it forms polymers with polyisoprenic structure. These reactions will be discussed in more detail later on (p. 158).

Cyclopentene. Cyclopentene ought to form cyclodecadiene through dimerization as a first step in metathesis. However, when starting from

$$
\text{(3.65)}
$$

cyclopentene, it is not usually possible to obtain the dimer or other oligo- mers, but rather polypentenamer is formed directly. The tendency of cy- clopentene to form predominantly a polymer is due, among other factors, to the steric strain within the unsaturated five-membered ring. Thus, the reverse metathesis of 1,6-cyclodecadiene affords a high yield of cyclopen- tene, the method being a simple way to prepare high purity cyclopent- ene.[121]

1-Methylcyclopentene [122] usually forms polymeric products with me- tathesis catalytic systems.

The reactions of unsubstituted and substituted cyclopentenes to form polyalkenamers in the presence of metathesis catalysts will be discussed at length in the next chapter.

Cyclohexene. Few studies on six-membered rings have been published. It is well known that such cycles have a low tendency towards ring-open- ing, because of their high stability. In fact, a mixture of un defined oli- gomers and polymers was obtained in low yield in reactions carried out with cyclohexene over certain alumina-supported oxides of transition metals or over catalysts based on WCl_6. In the oligomerization and poly- merization reactions, the equilibrium shifted towards cyclohexene; in- deed, reverse metathesis with the formation of cyclohexene from 1,7-cy- clododecadiene is favoured.

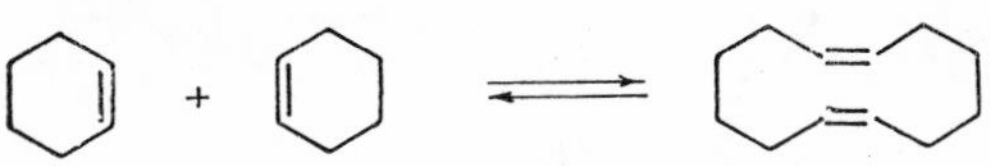

$$
\text{(3.66)}
$$

Cycloheptene. In the reaction of cycloheptene with metathesis catalytic systems, only polyalkenamers are obtained.[123] The reactivity of cyclo- heptene is lower than that of other cyclo-olefins, therefore it is likely that compounds of dimeric and oligomeric nature are also present among the reaction products.

Cyclo-octene. Cyclo-octene readily polymerizes by ring-opening under the influence of metathesis catalytic systems with formation of *trans-* or *cis*-polyoctenamers.[124] However, in the polymerization reactions of cyclo-octene, appreciable amounts of cyclo-octene dimers and trimers were formed together with the polymers.[13] The dimers and trimers were the result of metathesis reactions between two molecules of monomer or

of dimer and monomer, respectively. Successive metathesis reactions
yield macrocycles and polymers, depending on the reaction conditions.

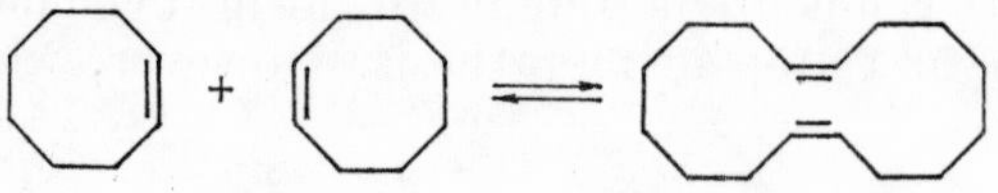

$$\text{(3.67)}$$

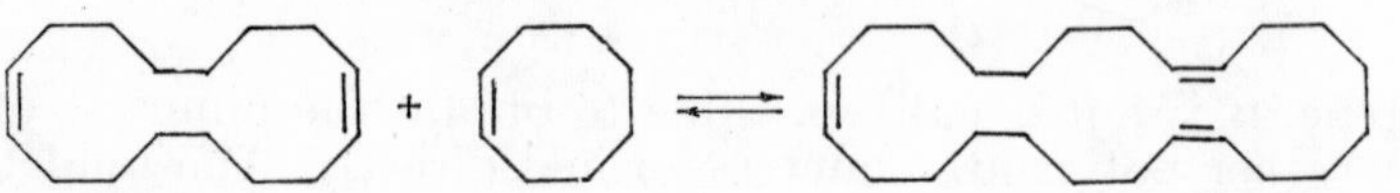

$$\text{(3.68)}$$

1,5-*Cyclo-octadiene*. 1,5-Cyclo-octadiene can undergo intramolecular
metathesis, leading to cyclobutene,

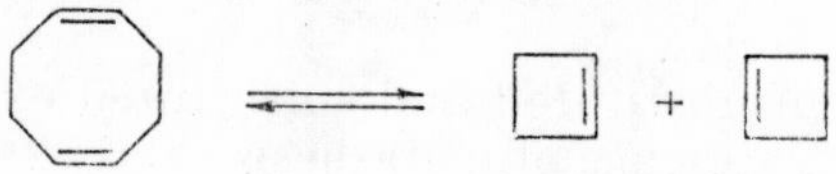

$$\text{(3.69)}$$

or intermolecular metathesis with formation of dimers and trimers, as
in the case of cyclo-octene, but with a higher degree of unsaturation.

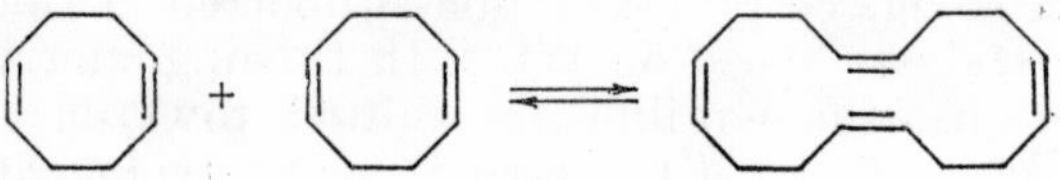

$$\text{(3.70)}$$

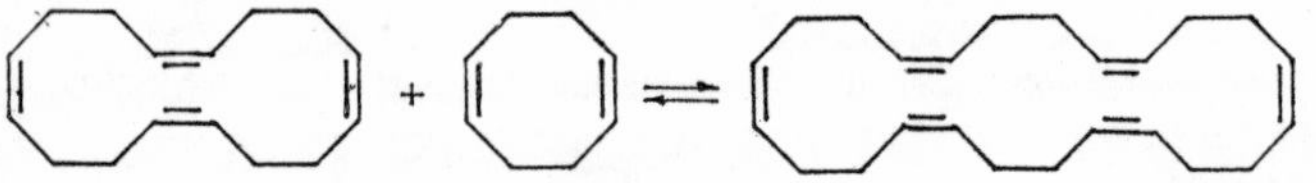

$$\text{(3.71)}$$

The equilibrium in the intramolecular metathesis (3.69) is totally
shifted towards the formation of 1,5-cyclo-octadiene because of the strain
within the four-membered ring of cyclobutene so that, when starting from
1,5-cyclo-octadiene, practically no reaction to cyclobutene takes place.
On the other hand, 1,5-cyclo-octadiene easily forms dimers and trimers
together with oligomers and polyalkenamers with butenamer monomeric
units. 1,5-Cyclo-octadiene dimers and trimers of a higher unsaturation
degree undergo further, along with intermolecular metathesis reactions, intramolecular reactions leading to compounds which contain the

cyclo-octadiene unit for a fractional number of times. Through such a reaction, the dimer of 1,5-cyclo-octadiene affords 1,5,9-cyclododecatriene

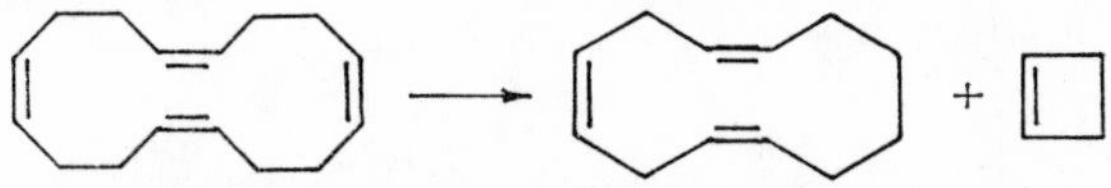

$$(3.72)$$

and the trimer yields the cyclo-olefin $C_{20}H_{30}$ in a similar way.

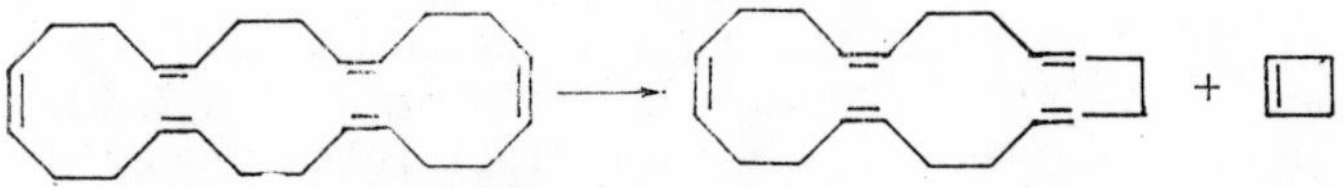

$$(3.73)$$

Scott and others evidenced the formation of 1,5,9-cyclododecatriene [13] and cyclo-olefins $C_{20}H_{30}$, together with other oligomers and polyalkenamer in the reaction of 1,5-cyclo-octadiene over the catalytic system $WCl_6 \cdot EtOH \cdot EtAlCl_2$.

Cyclononene. Cyclononene, similarly to cyclo-octene, forms polyalkenamers in the presence of metathesis catalytic systems.[29] This monomer has been less studied.

1,5-Cyclononadiene. 1,5-Cyclononadiene can yield cyclopentene and cyclobutene through an intramolecular metathesis reaction.

$$(3.74)$$

Actually, the formation of cyclopentene and cyclobutene does not occur; the equilibrium is likely to be totally shifted towards 1,5-cyclononadiene, in parallel with the reaction of 1,5-cyclo-octadiene. Rossi and Giorgi [125] demonstrated the exclusive formation of polyalkenamer from 1,5-cyclononadiene in the presence of the binary catalytic system WCl_6 and $LiAlH_4$.

Cyclodecene. Cyclodecene forms polydecenamer [126] through ring-opening polymerization over catalytic systems based on tungsten and alkylaluminium compounds. It is very likely that cyclodecene oligomers are formed together with polydecenamer, but this has not been proved. Cyclodecene can form 1,10-cyclo-eicosadiene through simple dimerization with catalytic systems for metathesis.

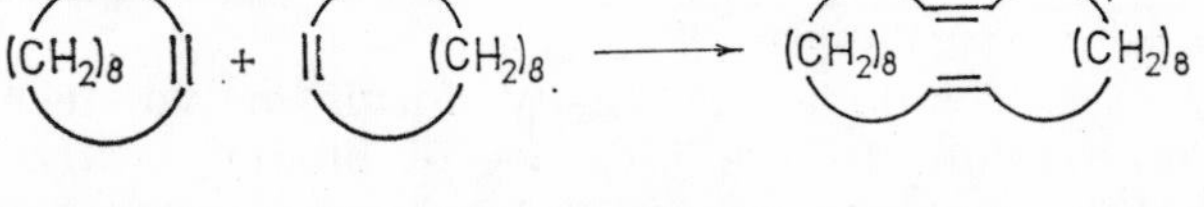

$$(3.75)$$

1,10-Cyclo-eincosadiene, through successive metathesis reactions with cyclodecene or with itself, further leads to trimers, tetramers and oligomers of cyclodecene. These oligomers are of great interest because the synthesis of carbocycles by alternative reactions is difficult.

1,5-*Cyclodecadiene*. 1,5-Cyclodecadiene can yield cyclohexene and cyclobutene through metathesis.

$$\text{(3.76)}$$

Although cyclohexene is particularly stable, the equilibrium of this reaction is totally shifted towards 1,5-cyclodecadiene, because cyclobutene possesses appreciable ring strain. In the presence of binary catalytic systems of WCl_6 and $EtAlCl_2$, Hocks and others [127] obtained polyalkenamer with cyclohexenic-cyclobutenic copolymer structure formed directly from 1,5-cyclodecadiene through ring-opening polymerization. For a long reaction time, they demonstrated the formation of cyclohexene from polyalkenamer, the structure of the latter becoming richer in butenamer units with reaction time.

1,6-*Cyclodecadiene*. As has been mentioned above, 1,6-cyclodecadiene reacts intramolecularly in the presence of binary catalytic system of WCl_6 and $EtAlCl_2$,[17,121] yielding cyclopentene.

$$\text{(3.77)}$$

The reaction takes place with conversions over 20 %, giving pure cyclopentene. Among the possible conformers of 1,6-cyclodecadiene, only *cis*, *cis*-cyclodecadiene reacts through the above intramolecular metathesis, the *cis,trans* and *trans,trans* conformers yielding polypentenamer.

Cyclo-undecene. Cyclo-undecene can form polyundecenamer over metathesis catalytic systems but the reactivity of this cyclo-olefin is probably low.

1,6-*Cyclo-undecadiene*. Cycloundecadiene can form cyclohexene and cyclopentene through an intramolecular reaction.

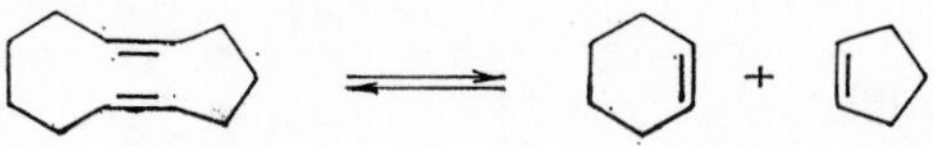

$$\text{(3.78)}$$

The stability of the reaction products could act as a driving force, favouring their formation in this reaction. No studies have been reported on the metathesis of 1,6-cyclo-undecadiene.

Cyclododecene. Cyclododecene easily forms polydodecene via ring-opening polymerization. In the presence of binary catalytic systems of WCl_6 and $EtAlCl_2$, the reaction led to carbocycles, catenanes and knots arising from cyclododecene by oligomerization.[27,128-131] The formation of these compounds will be discussed below in more detail.

1,7-Cyclododecadiene. 1,7-Cyclododecadiene may yield cyclohexene, through intramolecular reaction.

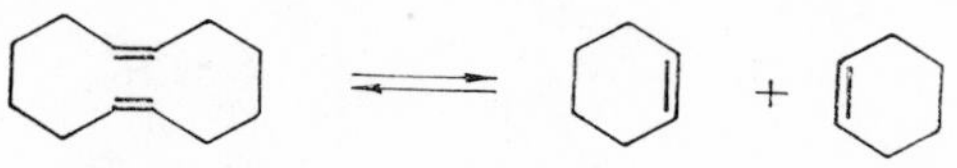

$$(3.79)$$

The very high stability of cyclohexene shifts the equilibrium of this reaction entirely towards the formation of cyclohexene. Though the reaction has not been studied in detail it is very likely that pure cyclohexene may be obtained, starting from 1,7-cyclododecadiene in the presence of WCl_6 and $EtAlCl_2$, i.e. the same catalytic system which was employed by Küpper and Streck [17,121] for the similar preparation of pure cyclopentene.

3.3.2. Syntheses of Carbocycles, Catenanes, Knots and Rotaxanes

The presence of a fraction with lower molecular weight along with the polyalkenamers is a general feature of ring-opening polymerization of cyclo-olefins. This fraction, named by Scott *et al.* [13] "extractable" can easily be separated from the polymer by extraction, fractionated into components by gas chromatography, and analyzed structurally by suitable physical or chemical methods. The structure determination of these products was a first step in elucidating the mechanism of cyclo-olefin ring-opening polymerization. It was found that these products are oligomers of the initial cyclo-olefins and that they belong to several classes of substances. These compounds can be obtained, in some cases, as main products in the metathesis of cyclo-olefins. Their formation gives information on the mechanism of the two processes, metathesis and ring-opening polymerization.

3.3.2.1. Synthesis of Carbocycles

A first category of oligomeric compounds which is obtained in the metathesis or ring-opening polymerization of cyclo-olefins consists of carbocycles.

The "extractable" fraction obtained by Scott and others [13] through the polymerization of cyclo-octene and 1,5-cyclo-octadiene in the presence of catalytic systems containing WCl_6, EtOH and $EtAlCl_2$ was formed from unsaturated carbocycles with molecular weight up to 1200. By NMR, gas-chromatography analysis and mass spectroscopy, the above authors established that these unsaturated carbocycles are oligomers, i.e., dimers, trimers and tetramers of the initial cyclo-olefins. Their formation was accounted for by dimerization, trimerization and tetramerization reactions during the successive steps of metathesis with participation of the double bonds from the cyclo-olefins. In the case of cyclo-octene, in the first step, an unsaturated carbocycle, 1,9-cyclohexadecadiene is formed. It leads, through subsequent metathesis reactions with the initial cycloolefin or with itself to cyclo-octene trimer or tetramer, respectively.

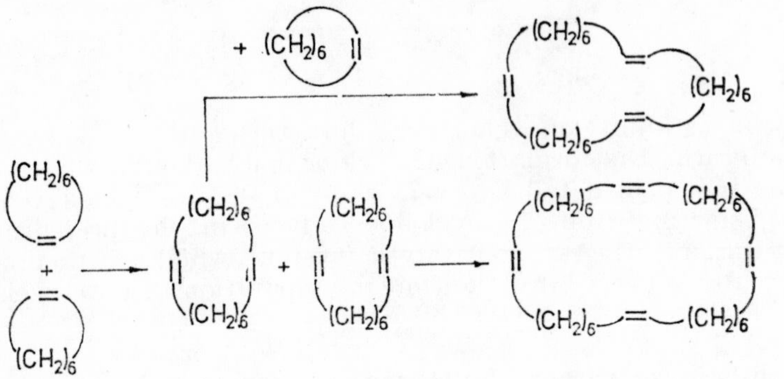

$$(3.80)$$

This stepwise metathesis gives rise to macrocycles with higher molecular mass.

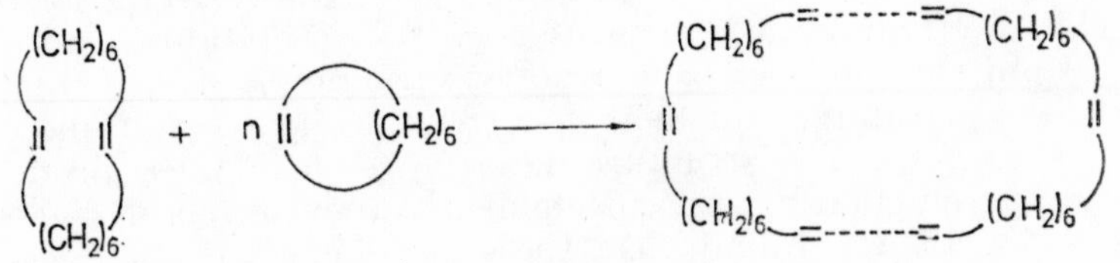

$$(3.81)$$

Carbocycles with up to 120 carbon atoms were thus synthesized by Wasserman, Ben-Efraim and Wolovsky [26] by the metathesis of cyclo-octene in the presence of catalytic systems containing WCl_6, EtOH and $EtAlCl_2$. The products formed through metathesis corresponded to cyclic oligomers of cyclo-octene which, after gas chromatographic separation and hydrogenation, led to saturated carbocycles. Unsaturated carbocycles of the same type were also obtained by Höcker and Musch [132] through metathesis of cyclo-octene with catalytic systems based on WCl_3 and $EtAlCl_2$. The distribution of the oligomers in the reaction mixture depends on the concentration of the catalyst and is related to the molar ratio between the catalyst components.

Wolovsky [27] obtained carbocycles with up to 168 carbon atoms, through the metathesis of cyclododecene. The reaction was carried out at room temperature in n-octane using the same catalytic system based on tungsten as in the case of cyclo-octene. The structure of the reaction products, established by mass spectrometry, corresponded mainly to C_{24}, C_{36}, C_{48}, C_{72} compounds and, in smaller amount, to higher oligomers. Together with the non-branched carbocycles, Wolovsky also obtained interlocked ring structures belonging to the class of catenanes and knots.

The unsaturated compounds arising from successive intermolecular reactions of cyclo-olefins can further undergo intramolecular metathesis reactions leading to new carbocycles with a different number of carbon atoms.

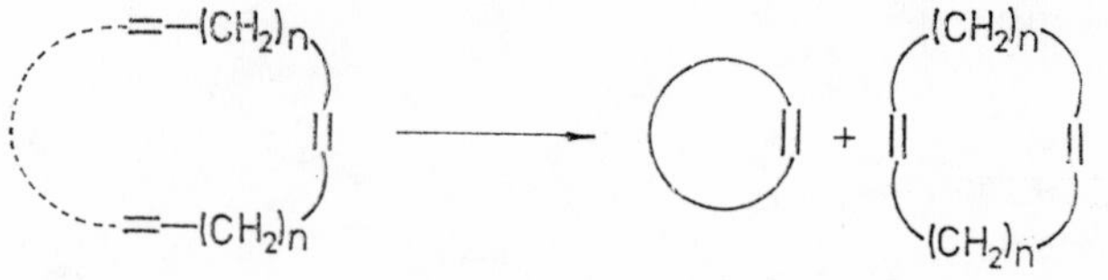

$$(3.82)$$

In this case, smaller unsaturated carbocycles result leading to a wide molecular mass dispersion of the products. Metathesis between such unsaturated carbocycles and acyclic olefins can lead even to linear open chain polyalkenamers.

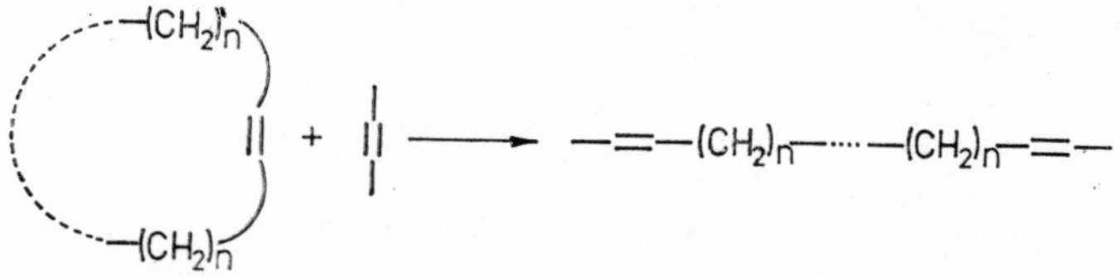

$$(3.83)$$

The size of the carbocycles which are formed in the metathesis of cyclo-olefins can be deduced from the molecular mass of the polyalkenamers obtained through ring-opening polymerization of cyclo-olefins with acyclic alkenes.

3.3.2.2. Catenanes, Knots and Rotaxanes

Wolovsky [27] showed that the metathesis of cyclododecene with catalytic systems consisting of WCl_6, EtOH and $EtAlCl_2$ afforded, along with unsaturated oligomeric carbocycles, other compounds with interlocked ring structures belonging to the class of catenanes. By studying the metathesis of cyclododecene with the same type of catalysts, Ben-Efraim, Batich and Wasserman [128] proved that the structure of the resulting macrocyclic compounds was more complex, various interlocked cyclic structures being formed together with non-branched carbocycles, catenanes and knots. Later, Wolovsky [130,131] obtained new topological isomers belonging to the class of catenanes, nectinodanes and plectanes, by carrying out the metathesis of cyclododecene under various conditions with catalytic systems based on WCl_6 and $EtAlCl_2$.

The formation of compounds of these types from cyclo-olefins was rationalized only after elucidating the mechanism of the metathesis reaction. In the case of catenanes and knots, the formation mechanism can be explained as follows : the macrocycles arising from cyclo-olefins undergo a series of twists between successive metathesis reactions, following the "Möbius strip" principle. The structure of catenanes or knots is determined by the number of these twists. According to the "Möbius strip" principle, for a number of half-twists n = 2 from a carbocycle, a catenane of the order 2 with $\alpha = 1$ results by intramolecular metathesis (α = the folding topological number).

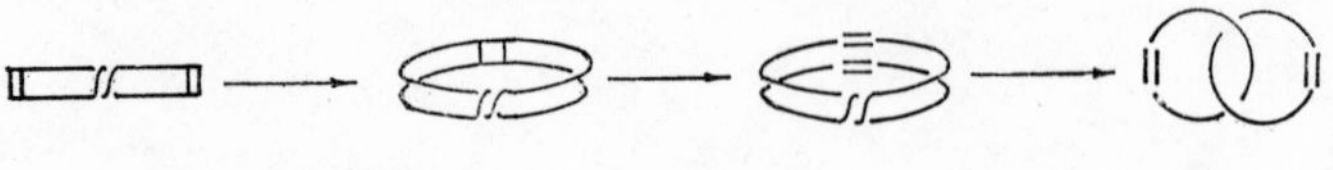

$$(3.84)$$

For a number of half-twists n = 3, it is possible to form a knot with N = 3 (N = the interlocking number).

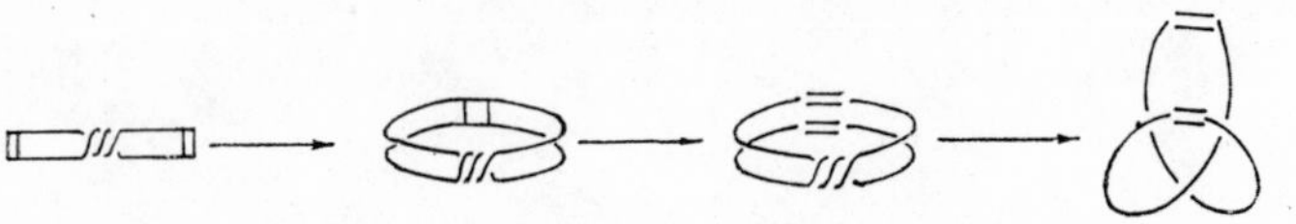

$$(3.85)$$

Generally, for a variable number of half-twists of the carbocycle during metathesis, it is possible to obtain catenanes and knots with different structures according to the number n of the half-twists.

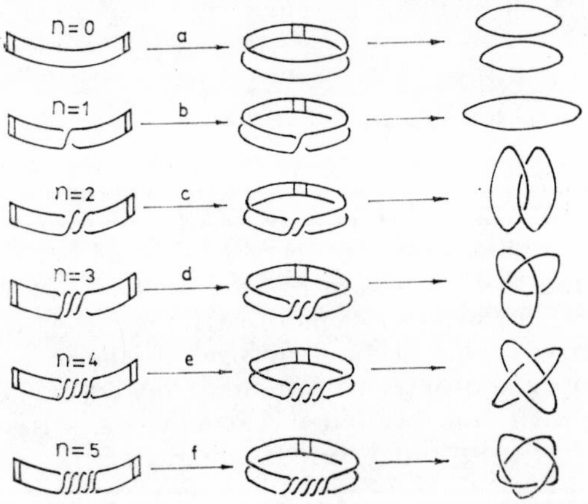

$$(3.86)$$

The entropy freedom of the system, which depends on the dilution, can shift towards the formation of some of these products. The method can thus become a new way of synthesizing natural products which are difficult to prepare by known, conventional methods.[133-135]

Starting from catenanes, it is possible to obtain rotaxanes, via metathesis, by opening one of the rings by reaction with an acyclic olefin.

$$\text{|||(CH}_2\text{)}_n\text{(CH}_2\text{)}_n\text{|||} \quad + \quad \begin{matrix} R \\ || \\ R \end{matrix} \quad \longrightarrow \quad \text{|||(CH}_2\text{)}_n\text{(CH}_2\text{)}_n \qquad (3.87)$$

Thus, in a catenane of order 2, with the folding topological number 1,
it is possible to open a ring through metathesis with a linear olefin, yielding
a rotaxane whose ring axis has as terminal groups the two moieties of the
acyclic olefin.[136] It is highly probable that these compounds may be formed
in the metathesis reactions of cyclo-olefins along with carbocycles, catenanes and knots, but their presence in these systems has not been proved as yet.

3.4. Metathesis of Acetylenes

As with olefins, acetylenes can be disproportionated by metathesis in
heterogeneous or homogeneous catalysis with formation of new acetylenes.

$$\begin{matrix} R^I \\ ||| \\ R^{II} \end{matrix} \quad + \quad \begin{matrix} R^I \\ ||| \\ R^{II} \end{matrix} \quad \rightleftharpoons \quad \begin{matrix} R^I - \!\equiv\! - R^I \\ + \\ R^{II} - \!\equiv\! - R^{II} \end{matrix} \qquad (3.88)$$

Along with metathesis, automerization may also occur.

$$\begin{matrix} R^I \\ ||| \\ R^{II} \end{matrix} \quad + \quad \begin{matrix} R^{II} \\ ||| \\ R^I \end{matrix} \quad \rightleftharpoons \quad \begin{matrix} R^I - \!\equiv\! - R^{II} \\ + \\ R^{II} - \!\equiv\! - R^I \end{matrix} \qquad (3.89)$$

The last mentioned reaction can be shown to occur by labelling with
radioactive isotopes.

The automerization study carried out by Bauer and Jeffers in 1965
on radioactively labelled acetylene (3.90) has drawn attention to the meta-

thesis of acetylene. The same reaction was observed later by Shida *et al.*[137] in the mercury photosensitized reaction of acetylene.

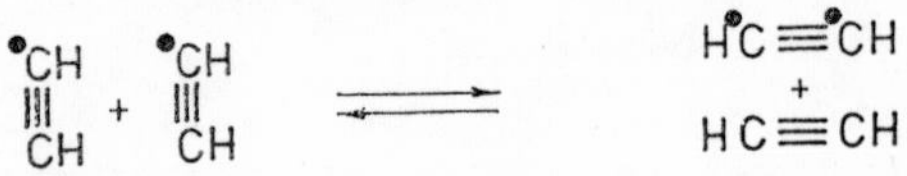

$$
\begin{array}{c}
{}^{\bullet}\mathrm{CH} \\ \|| \\ \mathrm{CH}
\end{array}
+
\begin{array}{c}
{}^{\bullet}\mathrm{CH} \\ \|| \\ \mathrm{CH}
\end{array}
\ \rightleftharpoons\
\begin{array}{c}
\mathrm{H}\overset{\bullet}{\mathrm{C}}\equiv\mathrm{CH} \\ + \\ \mathrm{HC}\equiv\mathrm{CH}
\end{array}
\tag{3.90}
$$

Pennella, Banks and Bailey[18] carried out the metathesis of 2-pentyne to 2-butyne and 3-hexyne in 1968, using heterogeneous catalysts which were active in the metathesis of olefins.

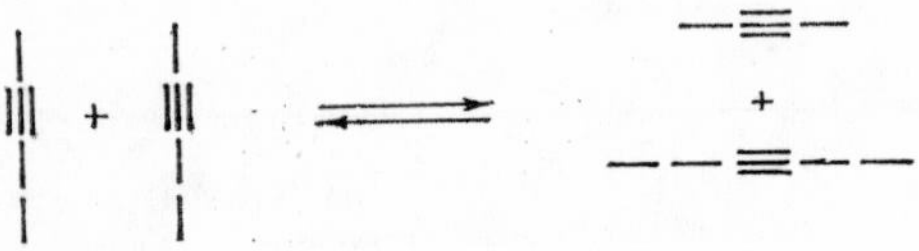

$$
\begin{array}{c}
| \\ \|| \\ |
\end{array}
+
\begin{array}{c}
| \\ \|| \\ |
\end{array}
\ \rightleftharpoons\
\begin{array}{c}
-\!\!\equiv\!- \\ + \\ -\!-\!\equiv\!-\!-
\end{array}
\tag{3.91}
$$

Operating at atmospheric pressure and within a temperature range of 200°C to 450°C, they obtained 2-butyne and 3-hexyne in a 1:1 molar ratio. As an example, using 6.8 % WO_3 on silica as catalyst at 350°C, and a feed flow rate of 12 L mixture/L. reactor. hour, the conversion of 2-pentyne was 44 % and the selectivity in the reaction products was 53 %.

Moulijn, Reitsma and Boelhouwer[138] have studied in more detail the metathesis of a large number of alkynes, employing a heterogeneous catalytic system formed from 5.5 % WO_3 on silica. They noticed dual behaviour of the catalytic system at temperatures between 100°C and 500°C; in most cases, metathesis reactions occurred together with cyclotrimerizations of alkynes. These authors also noticed that α-alkynes disproportionated less readily than β-alkynes, α-alkynes showing a marked tendency to give cyclotrimerization reactions. The results obtained in these studies are given in Table 3.9.

TABLE 3.9. Reaction of alkynes over the catalytic system WO_3/SiO_2 [a]

Alkyne	Temperature, °C	% Metathesis	% Cyclotrimerization
Ethyne	250—500	—	50
Propyne	90—145	0	90
	465	1	50
	495	7	50
1-Butyne	110—260	0	99
	460—500	3	50
1-Pentyne	250—270	0	84
	270—360	1.5	74
	360—490	9	54
2-Pentyne	294	1.5	—
	366	26	—
	446—450	11	—

[a] Data from ref. [138].

The metathesis of acetylene could not be achieved under these operating conditions. Only the cyclotrimerization of acetylene to benzene in 50% yield was observed.

In the reaction of propyne at temperatures below 145°C, mesitylene formed by trimerization was obtained as the main product. On increasing the temperature, the amount of trimerization reaction decreased to 40%, while the parallel metathesis to acetylene and 2-butyne increased.

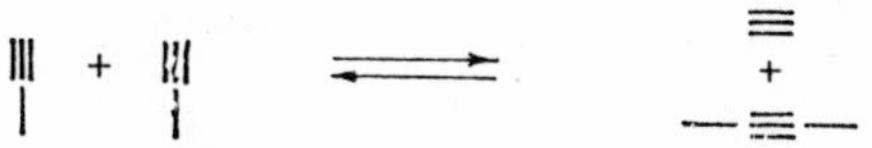

$$\text{(3.92)}$$

However, the selectivity for metathesis products remains relatively low, reaching 7% at 495°C. The formation of xylenes and other trimethylbenzenes together with metathesis products was noticed above 465°C.

1-Butyne exhibits a similar behaviour in this reaction. At temperatures between 110°C and 260°C, only cyclotrimerization products were obtained. At higher temperatures, the degree of cyclotrimerization decreased to 50%, metathesis occurring also along with other reactions in a proportion of 3%.

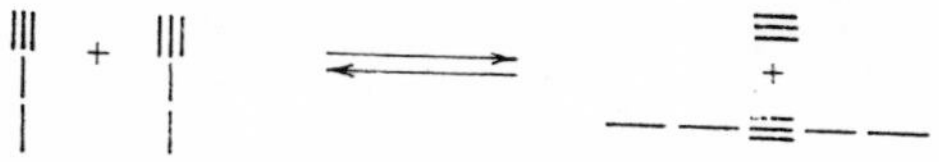

$$\text{(3.93)}$$

The hexyne thus formed was identified by mass spectroscopy.

The metathesis of 1-pentyne took place in a proportion of about 9% under optimum temperature conditions.

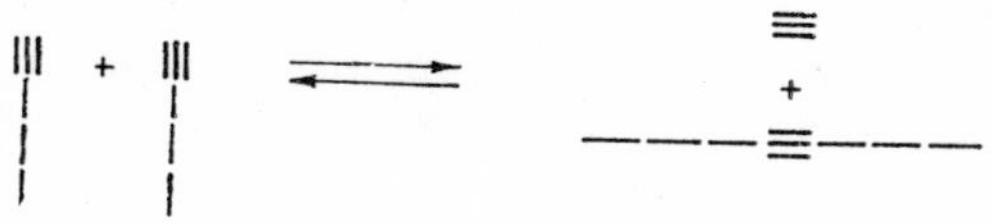

$$\text{(3.94)}$$

Unlike 1-pentyne, 2-pentyne, under similar reaction conditions, yielded metathesis products : 2-butyne and 3-hexyne with a selectivity of 26%.

The metathesis of 1-hexyne was carried out by Mushegyan et al.[139] using a heterogeneous catalyst of silica-supported MoO_3, 10%.

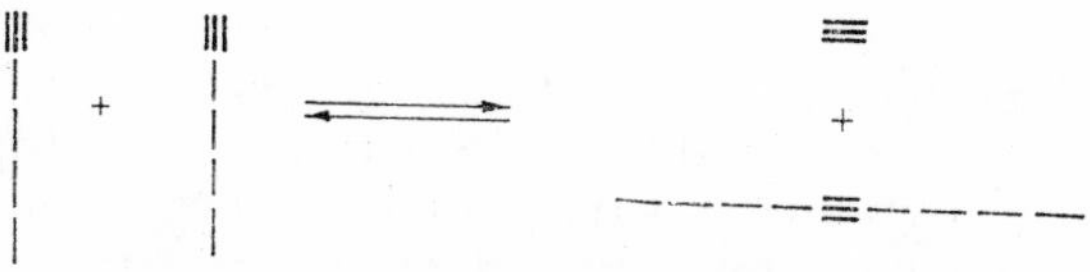

$$\text{(3.95}$$

Acetylene and 5-decyne were obtained with a selectivity of 30—40%, operating in the temperature range between 300°C and 350°C. The authors reported that the yield in metathesis products could be raised to 70% when cyclohexene was used as diluent within the temperatures 400—500°C.

The addition of cobalt oxides also leads to an increase of the catalyst activity in the reaction.

Among aromatic acetylene, the metathesis of phenylacetylene is to be mentioned first. Mushegyan obtained diphenylacetylene and acetylene in a 30—40% yield, by employing a heterogeneous catalyst consisting of silica-supported MoO_3 and containing 1% cobalt oxide.

$$
\begin{array}{c}
C_6H_5-C\equiv CH \\
+ \\
C_6H_5-C\equiv CH
\end{array}
\rightleftharpoons
\begin{array}{c}
C_6H_5-C \\
\parallel \\
C_6H_5-C
\end{array}
+
\begin{array}{c}
CH \\
\parallel \\
CH
\end{array}
\tag{3.96}
$$

Under optimum operating conditions the yield in the reaction products attained 60—70%.

More recently, Mortreux et al.[140] carried out the metathesis of 1-phenyl-1-hexyne labelled with ^{14}C in position 1, in the presence of a catalytic system formed from $Mo(CO)_6$ and phenol. The authors obtained diphenylacetylene and 5-decyne with a conversion of 44%, the radioactivity being entirely found in the diphenylacetylene.

$$
\begin{array}{cc}
\overset{Ph}{\underset{|}{\overset{\bullet}{C}}} & \overset{Ph}{\underset{|}{\overset{\bullet}{C}}} \\
\parallel & \parallel \\
\underset{|}{C} & \underset{|}{C} \\
Bu & Bu
\end{array}
\rightleftharpoons
\begin{array}{c}
Ph-\overset{\bullet}{C}\equiv\overset{\bullet}{C}-Ph \\
+ \\
Bu-C\equiv C-Bu
\end{array}
\tag{3.97}
$$

The simultaneous participation of a double bond and a triple bond in the metathesis reaction was noticed by Mushegyan et al.[139b] in the metathesis of a vinylacetylene. Thus, starting from isopropenylacetylene and operating in the range 350—400°C, with a catalytic system based on MoO_3 and SiO_2, they obtained the olefinic and acetylenic metathesis products in 50—60% overall yield.

$$
\begin{array}{c}
\equiv-\underset{|}{\overset{|}{=}}-\equiv \\
+ \\
=
\end{array}
\rightleftharpoons
\begin{array}{cc}
-\underset{\parallel}{\overset{\parallel}{|}} & +\overset{\parallel}{\underset{\parallel}{|}}-
\end{array}
\rightleftharpoons
\begin{array}{c}
\underset{=}{\overset{|}{=}}-\underset{=}{\overset{|}{\equiv}}-\overset{|}{=} \\
+ \\
\equiv
\end{array}
\tag{3.98}
$$

Among acetylene possessing functional groups, propargyl alcohol was studied under metathesis conditions. Starting from propargyl alcohol, Mushegyan et al.[139b] obtained 1,4-butyndiol and acetylene with a heterogeneous catalyst of silica-supported MoO_3 at a temperature of 630°C.

$$
\begin{array}{cc}
\underset{|}{\overset{\parallel\parallel}{|}} & + & \underset{|}{\overset{\parallel\parallel}{|}} \\
CH_2OH & & CH_2OH
\end{array}
\rightleftharpoons
\begin{array}{c}
\equiv \\
+ \\
HOCH_2-\equiv-CH_2OH
\end{array}
\tag{3.99}
$$

3.5. Cometathesis Reaction

Since the first metathesis reaction was reported in a homogeneous system for 2-pentene, by Calderon, Chen and Scott,[2] it has been known that the reaction is reversible and that the same equilibrium composition is obtained whether starting from 2-pentene or from an equimolecular mixture of 2-butene and 3-hexene

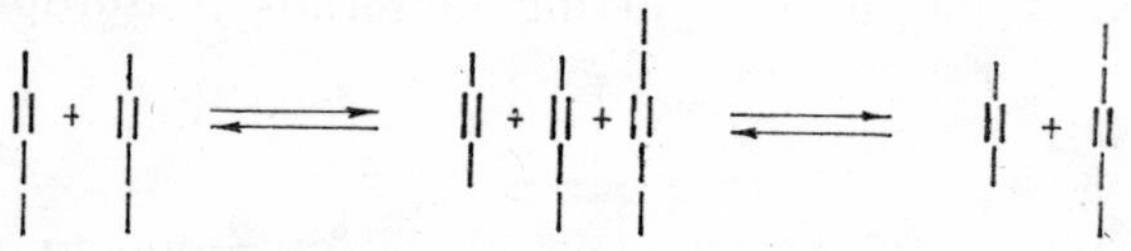

$$
\| + \| \rightleftharpoons \| + \| + \| \rightleftharpoons \| + \| \tag{3.100}
$$

The reaction between 2-butene and 3-hexene is therefore the first known case of cometathesis or mixed metathesis. Considering the unlimited possibilities of varying the two reacting olefins, the number of possible cometatheses will be much larger. The structural diversity of olefins that can be employed in cometathesis facilitates the study of the reaction mechanism, stereochemistry and kinetics.

Cometathesis reactions involve splitting of one double bond in an alkene or a cyclo-alkene with the aid of another alkene and they are, therefore, also called alkenolyses. With smaller alkenes, ethenolysis propenolysis, butenolysis, etc., reactions are known. Cyclo-alkene cleavage with linear alkenes is an important diversification route with great possibilities for industrial application. The fragmentation reaction of unsaturated polymer chains by cometathesis with alkenes is also an application for this type of reactions.

3.5.1. Ethenolysis

Ethenolysis is the cleavage reaction of a linear or cyclic olefin with ethene. The simplest possible ethenolysis is the cleavage of an isotopically labelled ethene molecule by an unlabelled one. This reaction, as has been mentioned in a previous section (p. 89), was carried out by Olsthoorn[43] using a heterogeneous catalytic system consisting of Re_2O_7 and Al_2O_3.

$$
\begin{array}{c} CD_2 \\ \| \\ CD_2 \end{array} + \begin{array}{c} CH_2 \\ \| \\ CH_2 \end{array} \rightleftharpoons \begin{array}{c} CH_2{=}CD_2 \\ + \\ CH_2{=}CD_2 \end{array} \tag{3.101}
$$

Through ethenolysis, α-olefins yield the same olefins and ethene. This is an example of a totally degenerate reaction and can be made evident only by isotope labelling.

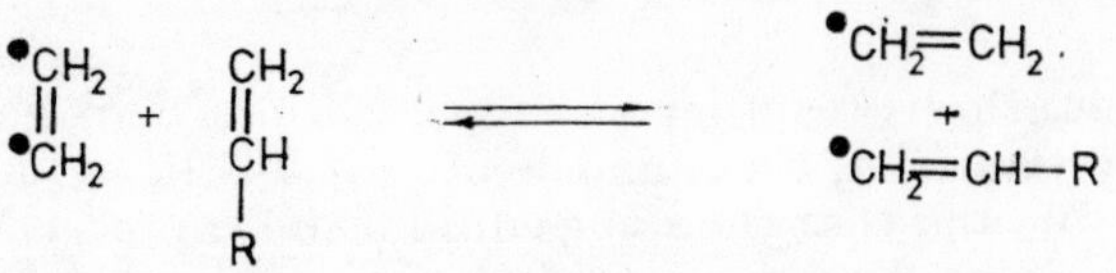

$$
\begin{array}{c}
\overset{\bullet}{C}H_2 \\
\parallel \\
\overset{\bullet}{C}H_2
\end{array}
+
\begin{array}{c}
CH_2 \\
\parallel \\
CH \\
\mid \\
R
\end{array}
\rightleftharpoons
\begin{array}{c}
\overset{\bullet}{C}H_2{=}CH_2 \\
+ \\
\overset{\bullet}{C}H_2{=}CH{-}R
\end{array}
\qquad (3.102)
$$

A simple example of such a totally degenerate reaction is the ethenolysis of propene which yields propene and ethene; the ethenolysis reaction can be observed if one of the starting materials is isotopically labelled.

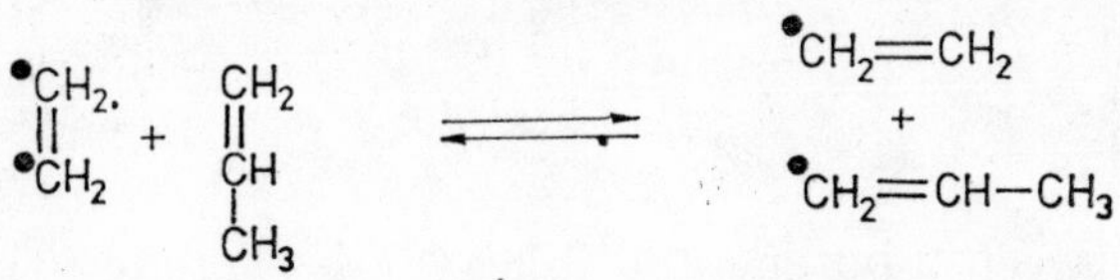

$$
\begin{array}{c}
\overset{\bullet}{C}H_2 \\
\parallel \\
\overset{\bullet}{C}H_2
\end{array}
+
\begin{array}{c}
CH_2 \\
\parallel \\
CH \\
\mid \\
CH_3
\end{array}
\rightleftharpoons
\begin{array}{c}
\overset{\bullet}{C}H_2{=}CH_2 \\
+ \\
\overset{\bullet}{C}H_2{=}CH{-}CH_3
\end{array}
\qquad (3.103)
$$

From the practical point of view, more important is the scission of internal olefins with ethene, yielding two lower α-olefins.

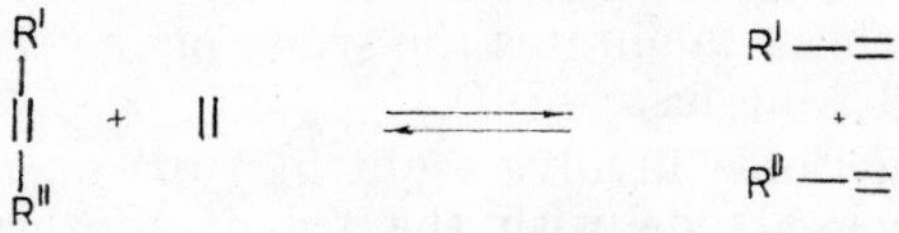

$$
\begin{array}{c}
R^{I} \\
\mid \\
\parallel \\
\mid \\
R^{II}
\end{array}
+
\parallel
\rightleftharpoons
\begin{array}{c}
R^{I}{-}{=} \\
+ \\
R^{II}{-}{=}
\end{array}
\qquad (3.104)
$$

The first reaction of this type, the cleavage of 2-butene with ethene, was the first ethenolysis studied by Bradshaw, Howman and Turner.[6]

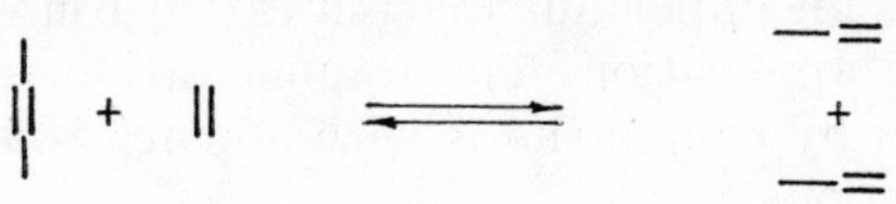

$$
\parallel + \parallel \rightleftharpoons
\begin{array}{c}
{-}{=} \\
+ \\
{-}{=}
\end{array}
\qquad (3.105)
$$

The reaction was carried out with the heterogeneous catalytic system $CoO \cdot MoO_3 \cdot Al_2O_3$.

The reaction of ethene with 2-butene is the reverse metathesis reaction of propene.

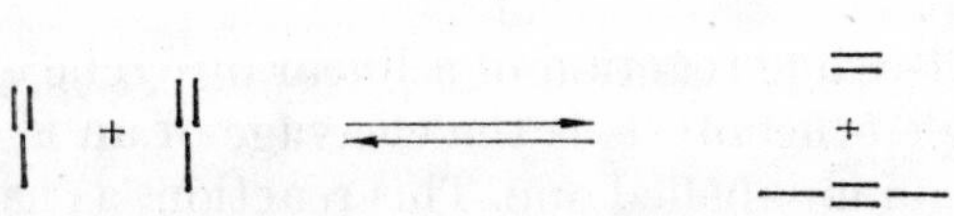

$$
\parallel + \parallel \rightleftharpoons
\begin{array}{c}
{=} \\
+ \\
{-}{=}{-}
\end{array}
\qquad (3.106)
$$

Bradshaw, Howman and Turner actually showed that the ethenolysis of 2-butene with a $CoO \cdot MoO_3 \cdot Al_2O_3$ catalyst at 180°C yields propene as a main reaction product together with the unreacted ethene and 2-butene. The only by-product was 2-pentene formed by the cometathesis of 2-butene with 1-butene, the latter resulting from isomerization.

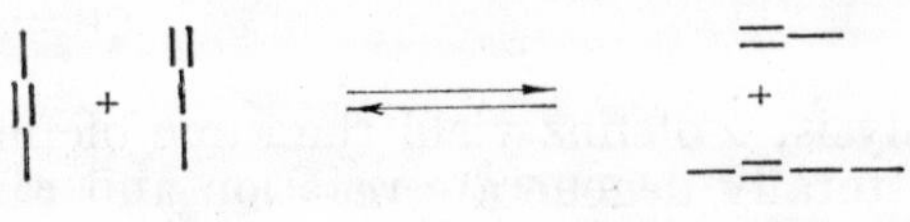

$$
\parallel + \parallel \rightleftharpoons
\begin{array}{c}
{=}{-} \\
+ \\
{-}{=}{-}{-}
\end{array}
\qquad (3.107)
$$

The small amount of 2-pentene indicated that the isomerization reaction occurred to a low extent, the main reaction being that of ethenolysis and not the metathesis of 1-butene or of 2-butene.

The ethenolysis of 2,3-dimethyl-2-butene which can yield isobutene as the only reaction product seems quite interesting.

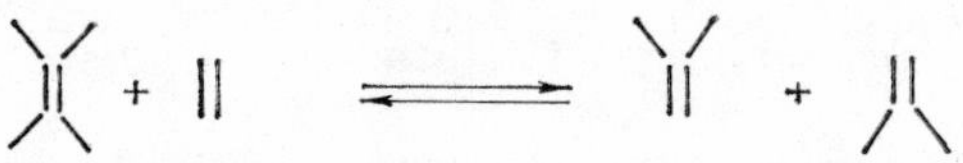

$$\text{)(} + \text{||} \ \rightleftharpoons \ \text{Y} + \text{|}\text{\textbackslash} \qquad (3.107a)$$

Crain[86] studied the ethenolysis of this olefin in heterogeneous catalysis, using $MoO_3 \cdot Al_2O_3$ treated with potassium hydroxide as a catalytic system. When working at 165°C and atmospheric pressure, a reaction mixture containing 10.8% isobutene, 72.3% 2,3-dimethyl-2-butene and 16.8% 2,3-dimethyl-1-butene was obtained.

Unlike 2-butene, higher β-olefins yield, via ethenolysis, propene and different α-olefins depending on the initial β-olefin.

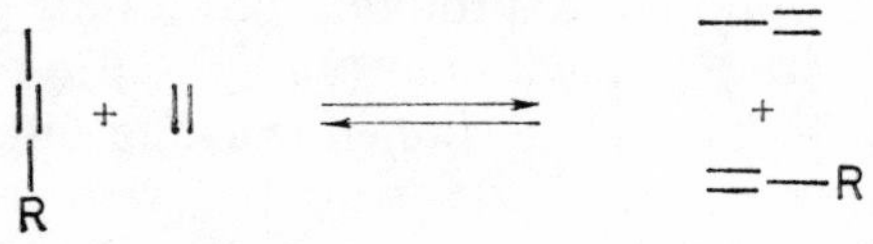

$$\underset{R}{\overset{|}{\text{||}}} + \text{||} \ \rightleftharpoons \ \overset{-\!\!=}{\underset{=\!-R}{+}} \qquad (3.108)$$

The reaction can be considered as a general method for preparing α-olefins from β-olefins. For instance, propene and 1-butene can be obtained by the ethenolysis of 2-pentene.

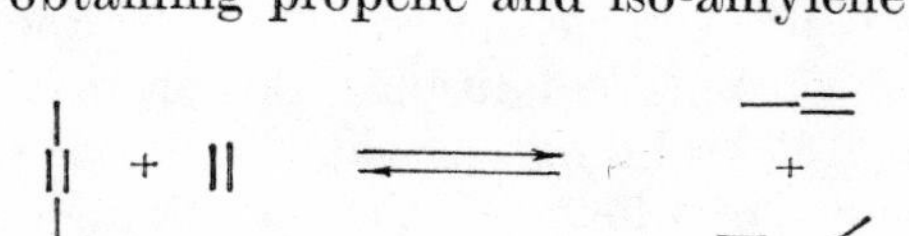

$$\text{||} + \text{||} \ \rightleftharpoons \ \overset{-\!\!=}{\underset{=\!-\!-}{+}} \qquad (3.109)$$

Bradshaw, Howman and Turner [6] carried out the ethenolysis of 4-methyl-2-pentene with the heterogeneous catalytic systems $CoO \cdot MoO_3 \cdot Al_2O_3$ and $Re_2O_7 \cdot Al_2O_3$, obtaining propene and iso-amylene as main products.

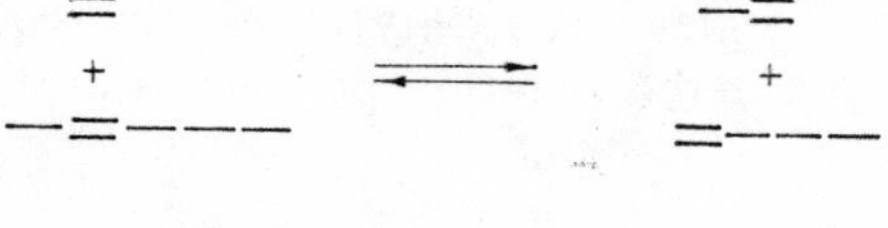

$$\underset{\text{\textbackslash}}{\overset{|}{\text{||}}} + \text{||} \ \rightleftharpoons \ \overset{-\!\!=}{\underset{=\!\!<}{+}} \qquad (3.110)$$

Similarly, by ethenolysis of 2-hexene, it is possible to obtain propene and 1-pentene,

$$\overset{=}{\underset{-\!=\!-\!-\!-}{+}} \ \rightleftharpoons \ \overset{-\!=}{\underset{=\!-\!-\!-}{+}} \qquad (3.111)$$

while from 2-heptene, propene and 1-hexene

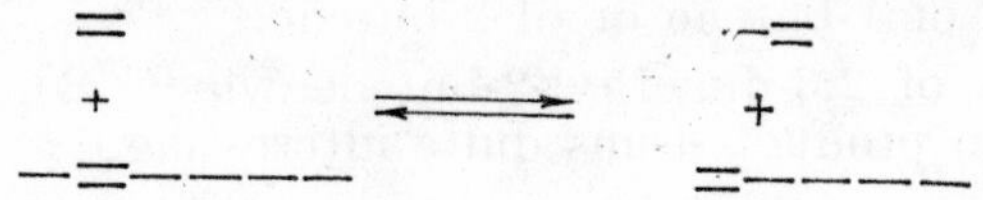

$$
\begin{array}{ccc}
= & & -= \\
+ & \rightleftharpoons & + \\
-=---- & & =----
\end{array}
\tag{3.112}
$$

and from 2-octene, propene and 1-heptene.

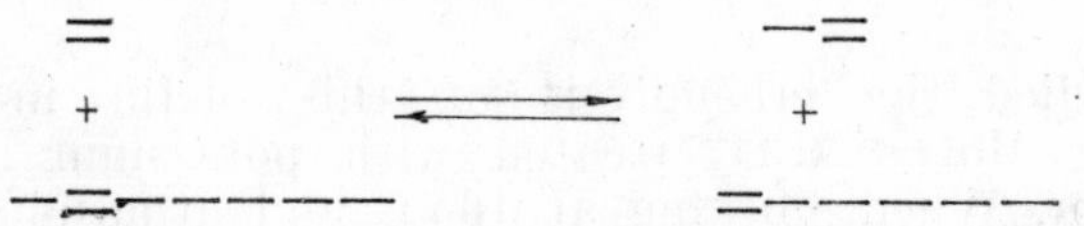

$$
\begin{array}{ccc}
= & & -= \\
+ & \rightleftharpoons & + \\
-=---- & & =----
\end{array}
\tag{3.113}
$$

The reaction of 2-heptene with ethene was carried out by Zuech [116] in homogeneous catalysis; the catalytic system he employed was the complex $(Ph_3P)_2Cl_2(NO)_2Mo$ and aluminium methylsesquichloride $Me_2Al_2Cl_3$. After an hour at room temperature, using chlorobenzene as a reaction medium, he obtained a product containing propene (6%), 2-butene (3%), 1-hexene (19%), 2-heptene (66%) and 5-decene (7%). It can be seen that, under the above mentioned working conditions, the primary cometathesis products propene and 1-hexene were converted, by metathesis, into 2-butene and 5-butene, respectively. Obviously, under proper conditions, propene and 1-hexene might have been obtained with high selectivity.

Crain [86] studied the ethenolysis of 2-octene with a heterogeneous catalytic system containing MoO_3 and Al_2O_3 treated with potassium hydroxide. At 125°C and 50 atm pressure, for a 29% conversion of octene, Crain obtained a product with the following molar composition (%): C_3H_6 (1.7), C_4H_8 (20.1), C_5H_{10} (2.8), C_7H_{14} (71.7), C_9H_{18} (0.4), $C_{11}H_{22}$ (0.2) and $C_{12}H_{24}$ (3.0). In this case, the primary products of cometathesis, propene and 1-heptene, were also converted, via metathesis, into 2-butene and 6-dodecene, respectively, immediately after their formation. The obviously low amount of propene observed in the product was due to the fact that, under the working conditions, the propene was carried away in the ethylene flow. It can be noted that a series of side reactions of isomerization and cometathesis between the resulting isomers also occurred, yielding small amounts of pentene, nonene, decene and undecene. The heptenes in the reaction product consisted of 98.4% 1-heptene, the remainder being 2- and 3-heptenes.

Among higher olefins, the ethenolysis of dodecene was studied in more detail. By using a heterogeneous catalytic system consisting of $Mo(CO)_6$ and Al_2O_3, Crain [141] obtained a reaction product containing C_8-C_{11} olefins with an 86% content of olefins.

3.5.2. Propenolysis, Butenolysis and Other Cleavage Reactions of Internal Olefins

Propenolysis is the cleavage of an acyclic or cyclic olefin with propene, via metathesis. Examples of propenolysis reactions are similar to those described previously for ethenolysis, propene being empolyed as a cleavage agent instead of ethene.

Depending on the nature of the olefin used as a substrate, the following types of propenolysis can be distinguished : a) the reaction of an α-olefin with propene yielding a β-olefin and ethene

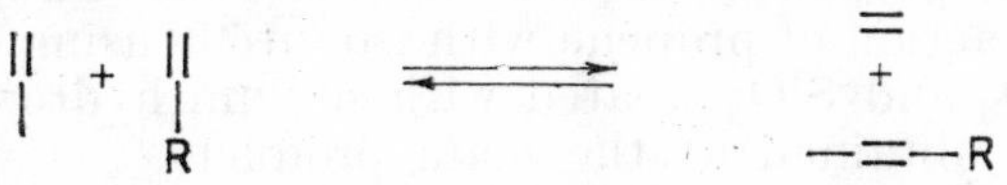

$$\text{(3.114)}$$

b) the reaction of a β-olefin with propene when two possibilities arise :
(1) when the productive reaction (3.115) yields an α-olefin and 2-butene

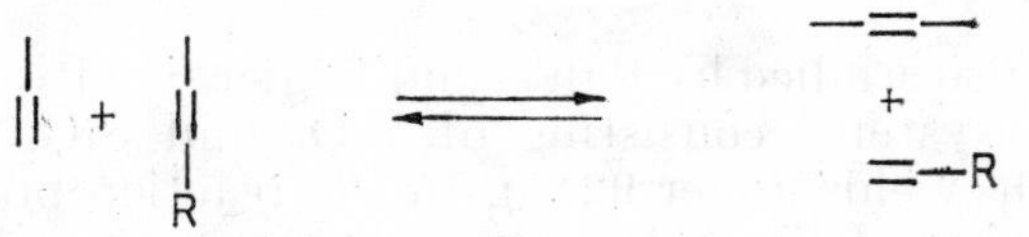

$$\text{(3.115)}$$

or (2) by a degenerate reaction (3.116) which yields the starting materials

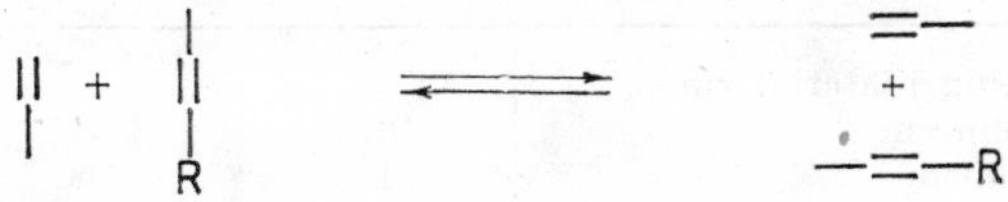

$$\text{(3.116)}$$

If both (1) and (2) occur simultaneously, the conversion is lowered ("partly degenerate reaction").
c) the reaction of propene with an internal olefin yielding an α-olefin and a β-olefin.

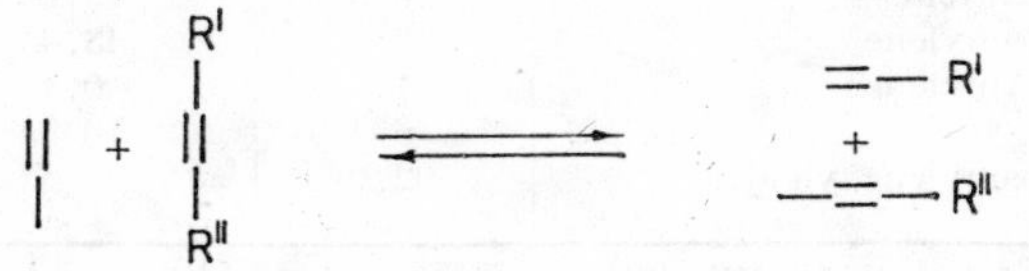

$$\text{(3.117)}$$

The simplest propenolysis is the reaction mentioned above between propene and ethene, which takes place by a totally degenerate reaction and can be evidenced by isotopic labelling. The conversion between isotopically labelled propene and non-labelled propene can be considered as a propenolysis reaction which can be monitored by isotopic labelling.

The reaction studied by Isaguliantz and Rar,[59] who used either 2-^{14}C-propene alone or 1,3-[14]C$_2$-propene together with non-labelled propene, is an example of this category of reactions.

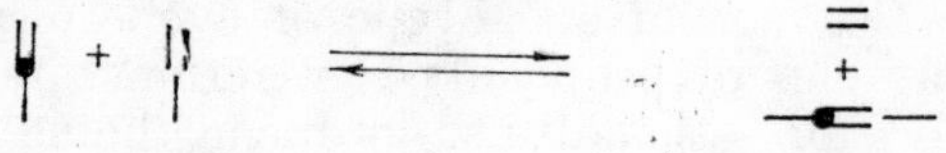

$$\text{(3.118)}$$

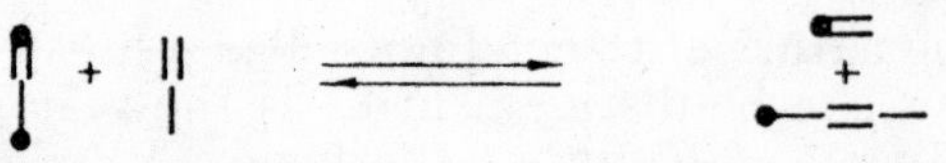

$$\text{(3.119)}$$

More complex propenolyses are also known. Thus Heckelsberg[142] carried out the reaction of propene with isobutene using catalytic systems consisting of WO_3 and SiO_2 treated with sodium hydroxide. Ethene and iso-amylene were obtained as the main products.

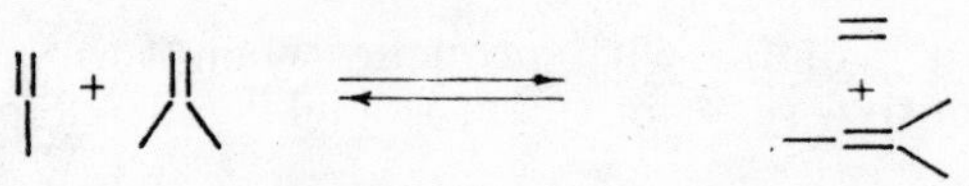

$$\text{(3.120)}$$

This reaction was also studied by Banks and Regier[143] with the same heterogeneous catalytic systems consisting of WO_3 and SiO_2. They obtained isoamylene in high yields (over 92%). In the reaction product, n-butene, n-amylene and higher olefins arising from isomerization and disproportionation reactions were also found (Table 3.10).

TABLE 3.10. Cometathesis of isobutene with propene [a,b]

Starting material (mole%)		
Isobutene	48.7	61.3
Propene	51.3	38.7
Conversion, %		
Isobutene	24.3	22.9
Propene	46.8	51.1
Product composition (mole %)		
Ethene	28.0	26.0
n-Butene	26.5	15.0
Isoamylene	38.1	48.4
n-Amylenes	1.0	0.1
C_5^+	6.4	10.5
Isoamylene yield, %	92.0	92.8

[a] Catalyst: WO_3/SiO_2; reaction temperature: 450°C; reaction pressure: 20 atm. [b] Data from ref. [143].

As in ethenolysis and propenolysis, butenolysis is the cleavage of a linear or cyclic alkene with butene, via metathesis. 1-Butene, 2-butene or isobutene can be used as the cleavage agent and any acyclic or cyclic olefin can serve as substrate.

The cleavage reactions with 1-butene are similar to those described above for propenolysis. Thus, depending on the olefin used as a substrate, α-, β- or γ- and other internal olefins can be obtained in this way.

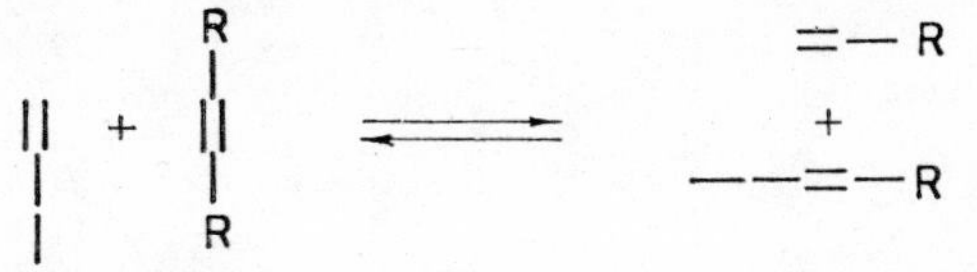

$$(3.121)$$

The reactions of 1-butene with ethene, propene and 1-butene were described above. The reaction of 1-butene with 2-butene yields, via cometathesis, propene and 2-pentene.

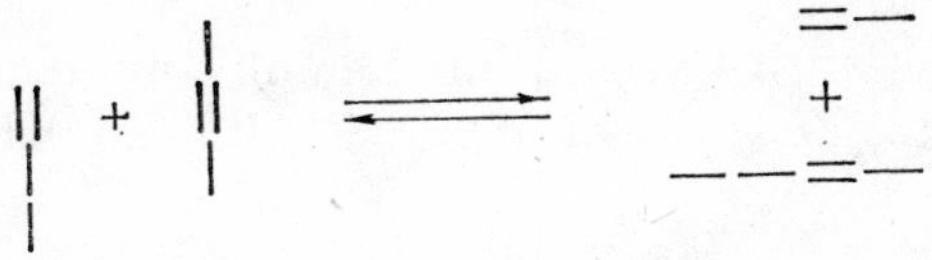

$$(3.122)$$

This reaction was carried out by Bradshaw, Howman and Turner [6] employing heterogeneous catalytic systems consisting of $CoO \cdot MoO_3 \cdot Al_2O_3$.

The cometathesis reactions with 2-butene give rise to α-olefins, β-olefins, γ-olefins, etc. depending on the olefin undergoing the cleavage. Two different new β-olefins result from a non-symmetrically substituted internal olefin.

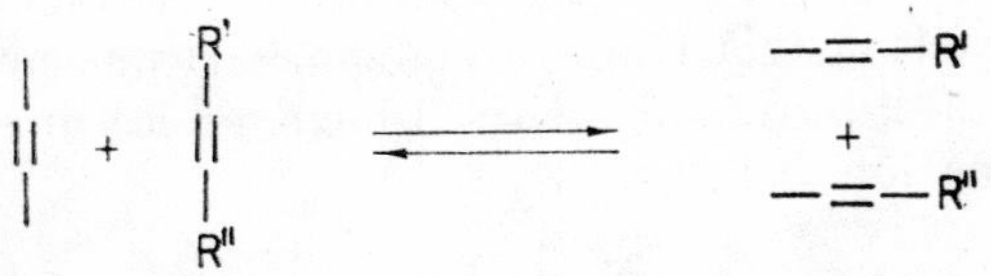

$$(3.123)$$

The cometathesis of 2-butene with isobutene carried out by Heckelsberg with a catalytic system consisting of WO_3 and SiO_2 treated with sodium hydroxide yielded mainly propene and isoamylene.

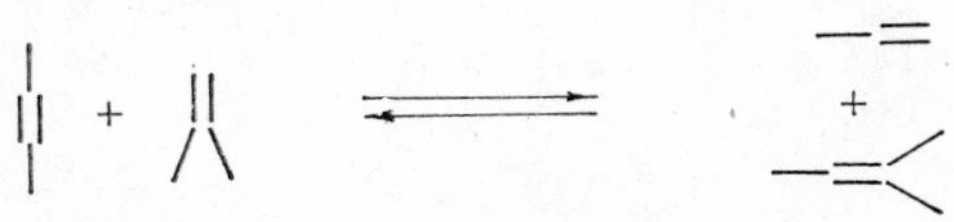

$$(3.124)$$

Later on, Banks and Regier [143] with a view to obtaining higher yields of isoamylene, studied the cometathesis between 2-butene and isobutene, as well as that of several mixtures of isobutene, 1-butene and 2-butene, by using heterogeneous catalytic systems consisting of WO_3 and SiO_2. Isoamylene was obtained in 68—80% yield (Table 3.11).

The cometathesis of 2-butene with a series of symmetrically substituted internal olefins was studied in the homogeneous phase as well. Calderon and co-workers [8] obtained only the transalkylidenation compound as the

TABLE 3.11. Cometathesis of *iso*-butene with butenes [a,b]

Starting material (mole %)					
Isobutene	59.2	49.5	49.0	61.5	63.1
2-Butene	40.8	45.8	38.2	35.1	38.0
1-Butene	—	4.7	12.8	3.4	8.9
Product composition (mole %)					
Ethene	3.4	5.2	4.5	5.0	6.3
Isoamylene	50.0	39.6	34.6	43.8	39.6
n-Amylenes	1.4	5.1	11.0	5.5	7.1
C_5^+	14.9	19.1	20.2	16.2	20.6
Isoamylene yield, %	80.8	68.8	70.4	69.6	66.4

[a] Catalyst: WO_3/SiO_2: temperature: 450°C; [b] Data from ref. [143].

product of the reaction between 2-butene and 2-butene-d_8 in the presence of the ternary catalytic system $WCl_6 \cdot EtAlCl_2 \cdot EtOH$.

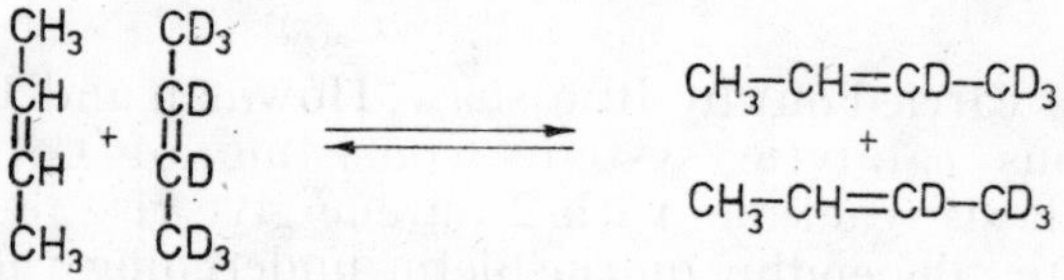

$$
\begin{array}{c}
CH_3 \\ | \\ CH \\ || \\ CH \\ | \\ CH_3
\end{array}
+
\begin{array}{c}
CD_3 \\ | \\ CD \\ || \\ CD \\ | \\ CD_3
\end{array}
\rightleftharpoons
\begin{array}{c}
CH_3{-}CH{=}CD{-}CD_3 \\ + \\ CH_3{-}CH{=}CD{-}CD_3
\end{array}
\qquad (3.125)
$$

Among higher olefins, 3-hexene can be easily cleaved with 2-butene, affording 2-pentene, as has been shown above. The fact that 2-pentene is formed in this reaction, via transalkylidenation, was proved by Calderon in the reaction of 2-butene-d_8 with 3-hexene, leading to 2-pentene-d_4, according to equation 3.126.

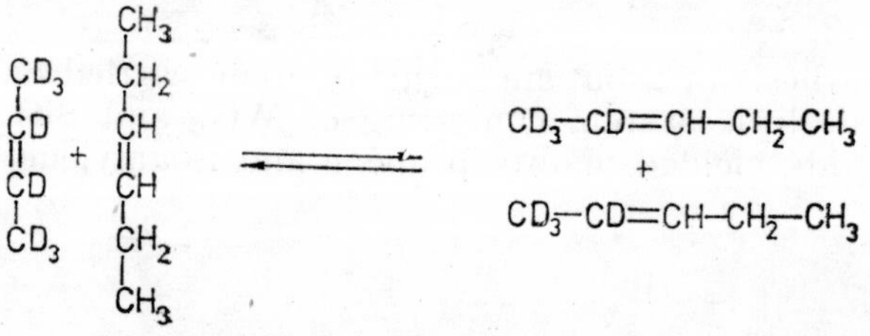

$$
\begin{array}{c}
CD_3 \\ | \\ CD \\ || \\ CD \\ | \\ CD_3
\end{array}
+
\begin{array}{c}
CH_3 \\ | \\ CH_2 \\ | \\ CH \\ || \\ CH \\ | \\ CH_2 \\ | \\ CH_3
\end{array}
\rightleftharpoons
\begin{array}{c}
CD_3{-}CD{=}CH{-}CH_2{-}CH_3 \\ + \\ CD_3{-}CD{=}CH{-}CH_2{-}CH_3
\end{array}
\qquad (3.126)
$$

Stilbene reacts with 2-butene in the presence of the above ternary catalytic system yielding β-methylstyrene by a similar reaction.

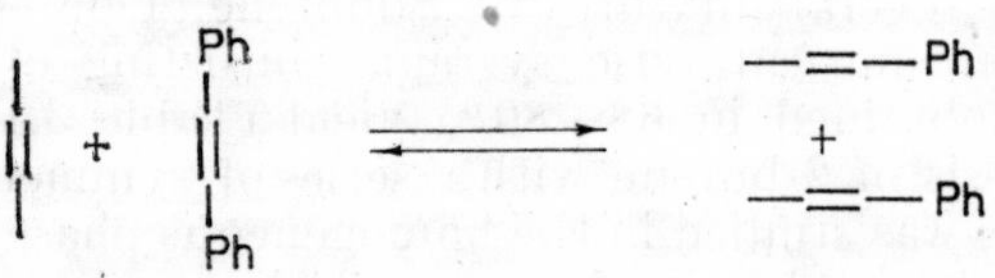

$$
\begin{array}{c}
| \\ || \\ |
\end{array}
+
\begin{array}{c}
Ph \\ | \\ || \\ | \\ Ph
\end{array}
\rightleftharpoons
\begin{array}{c}
{-}{=}{-}Ph \\ + \\ {-}{=}{-}Ph
\end{array}
\qquad (3.127)
$$

The reaction of unbranched linear olefins with isobutene is a general method for preparing olefins with three substituents at the double bond, two of which are *gem*-dimethyl groups.

$$\text{(3.128)}$$

In this way Banks and Regier [143] prepared *iso*-amylene, effecting the co-metathesis of isobutene with propene or 2-butene as described above (see equation 3.124).

As in the case of the cleavage reaction with ethene, propene and butene described above, other higher olefins can be employed as cleavage agents. Thus, internal olefins in which the position of the double bond depends on the olefin used as cleavage agent can be obtained. Such reactions are the cleavage of acyclic olefins with pentene, hexene, heptene or other higher homologues. Among these the cometathesis of 6-dodecene with 2-pentene is an example; the reaction was carried out by Calderon and co-workers [8]; they obtained 2-butene, 3-hexene, 2-octene and 3-nonene.

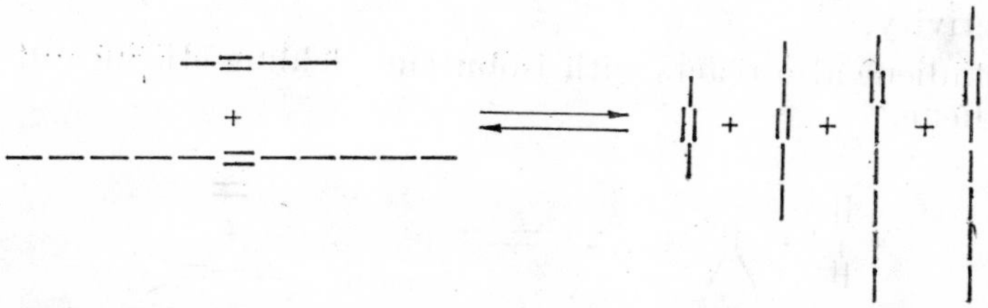

$$\text{(3.129)}$$

The alkenolysis of branched olefins or of olefins with functional groups is of interest for organic syntheses. The reactions of styrene with 2-pentene and of 4-methyl-2-pentene with 2-hexene in the presence of the catalytic system $WCl_6 \cdot EtOH \cdot EtAlCl_2$ are such examples. Other useful reactions from this point of view are the cleavage of esters of fatty acids with acyclic olefins presented in chapter 8.

3.5.3. Cometathesis of Dienes with Mono-Olefins

By ethenolysis, butadiene undergoes a totally degenerate reaction, in agreement with the general course of the corresponding reaction of α-olefins. The reaction can be made apparent only by isotope labelling.

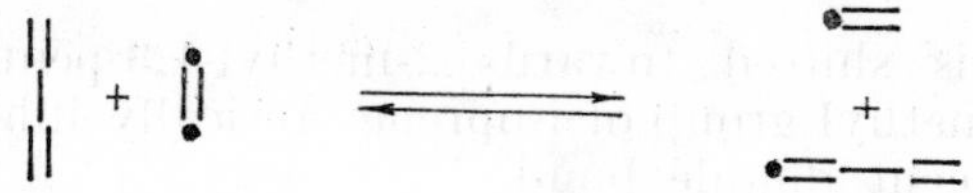

$$\text{(3.130)}$$

The propenolysis of butadiene was carried out by Heckelsberg, Banks and Bailey [16] in heterogeneous catalysis; by using WO_3 supported on SiO_2, at 538°C, they obtained ethene and 1,3-pentadiene with 46% selectivity.

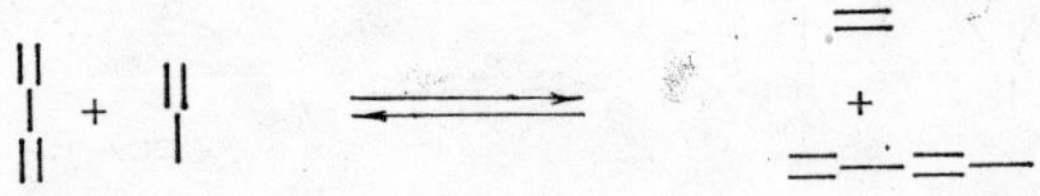

$$\text{(3.131)}$$

Besides propenolysis, sidereactions of isomerization and dimerization of the starting material and of the products also occurred.

Similar results were obtained by Heckelsberg, Banks and Bailey [16] in the butenolysis of 1,3-butadiene.

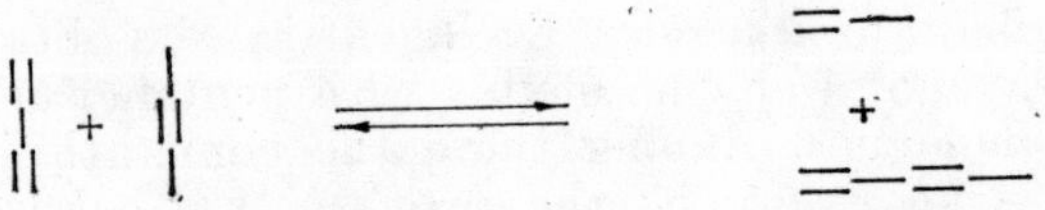

$$\text{(3.132)}$$

Working under conditions similar to those utilized for propenolysis, the reaction with *trans*-2-butene yielded propene and 1,3-pentadiene with 77% selectivity.

1,3-Butadiene also reacts with isobutene yielding ethene and 4-methyl-1,3-pentadiene.[16]

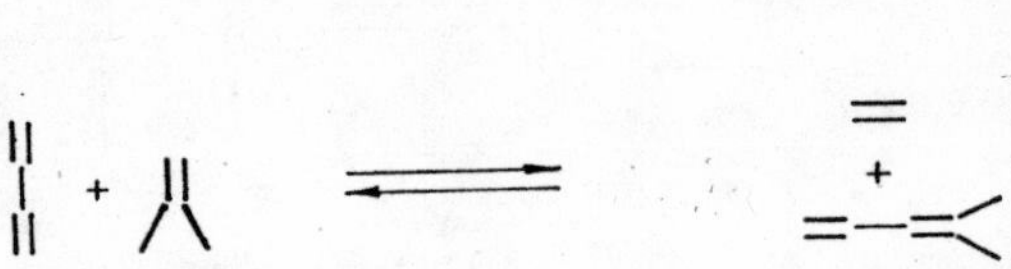

$$\text{(3.133)}$$

Although the conversion is only 5% for one cycle over the catalyst (because the reaction is partly degenerate), selectivity with respect to ethene and 4-methyl-1,3-pentadiene is high, hence the reaction is useful for the synthesis of the substituted diene.

In the reaction of isoprene with *trans*-2-butene, the participation in metathesis of any of the two double bonds of the conjugated olefin was demonstrated because 2-methyl-1,3-pentadiene and 3-methyl-1,3-pentadiene were obtained as the main reaction products.[16]

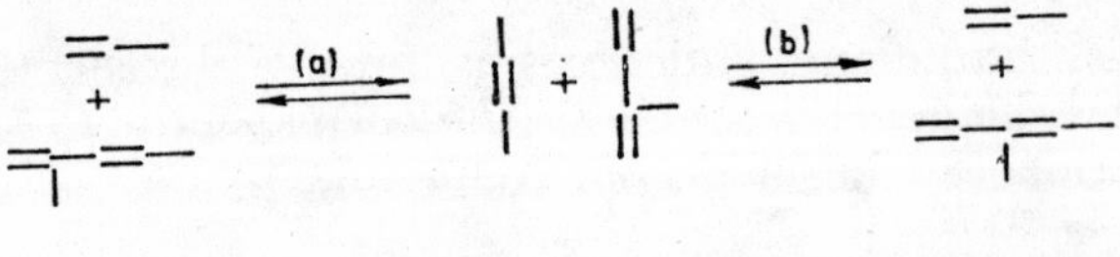

$$\text{(3.134)}$$

The equilibrium is shifted towards 2-methyl-1,3-pentadiene (pathway (a)) because the methyl group of isoprene sterically inhibits the involvement of the adjacent double bond.

3.5.4. Cometathesis of Cyclo-olefins

There are two types of cometathesis reactions of cyclo-olefins. A first class consists in the cleavage of cyclo-olefins at the double bond with an alkene; this reaction is known as alkenolysis. Depending on the number of the double bonds in the initial cyclo-olefin, linear dienes or polyenes may be obtained by this reaction. Monocyclo-olefins yield acyclic dienes,

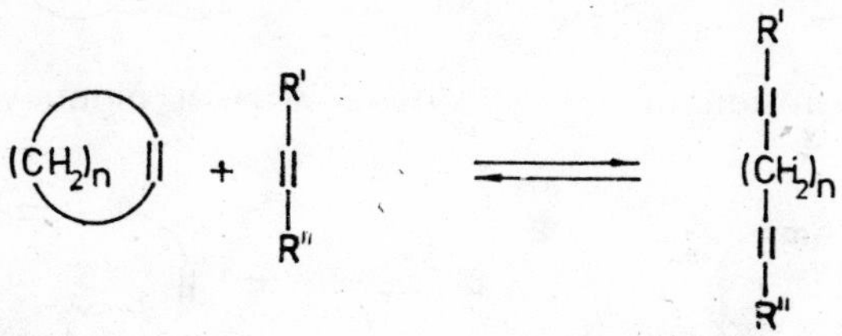

$$(3.135)$$

and cyclic dienes form linear dienes or trienes, if the reaction is bimolecular or trimolecular.

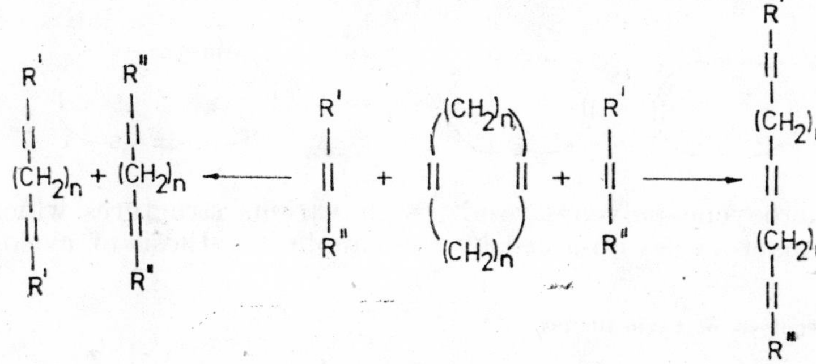

$$(3.136)$$

For cyclic polyenes, a cleavage of the cycle can occur at the double bond

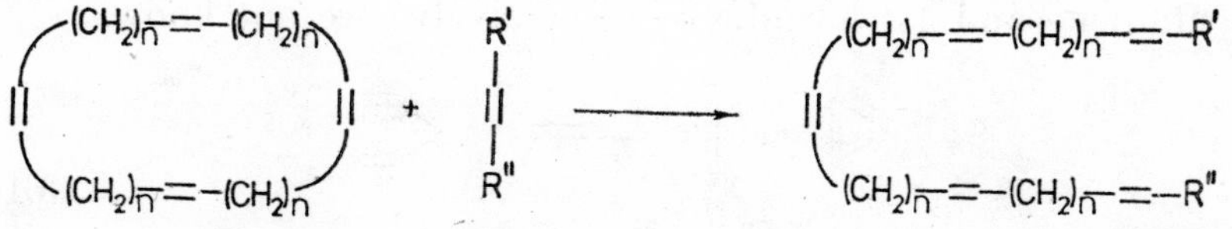

$$(3.137)$$

or even a fragmentation of the cyclic polyene to lower linear dienes

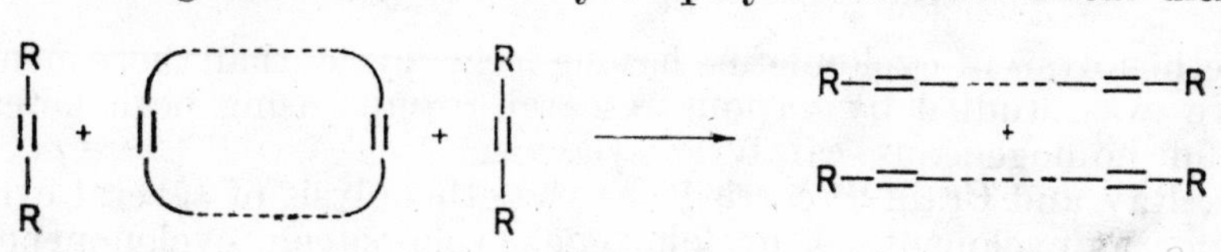

$$(3.138)$$

A second class of cometathesis of cyclo-olefins is the reaction between two different cyclo-olefins. Depending on the number of double bonds in the starting cyclo-olefins, it is possible to obtain cyclic dienes, trienes or polyenes. Thus, the cometathesis reaction of two cyclo-olefins having one double bond yields a cyclic diene.

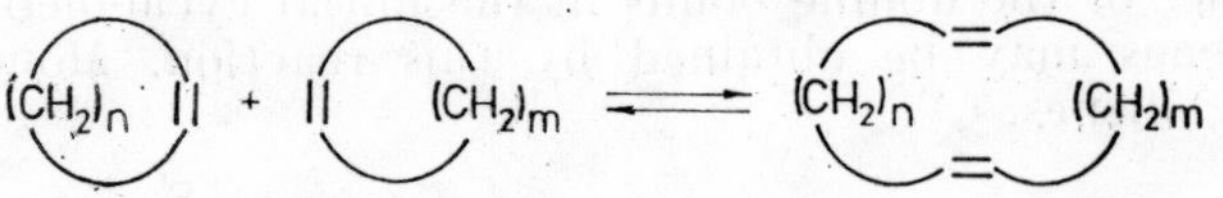

$$(CH_2)_n \; || \; + \; || \; (CH_2)_m \; \rightleftharpoons \; (CH_2)_n \qquad (CH_2)_m \tag{3.139}$$

Similarly, the reaction of a cyclic diene with a monocyclo-olefin yields a cyclic triene

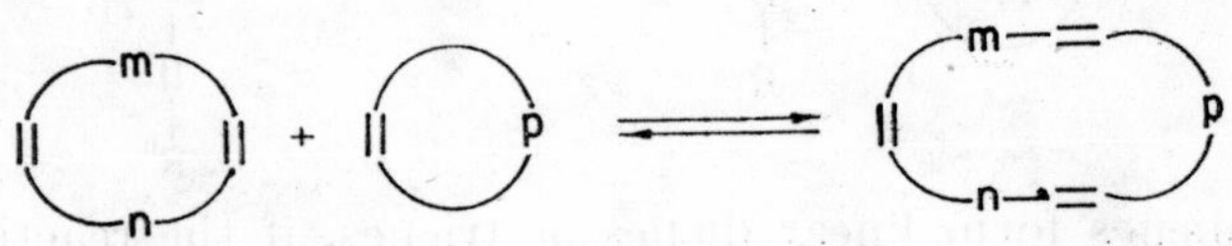

$$\tag{3.140}$$

and the cometathesis between two cyclic dienes yields a cyclic tetraene.

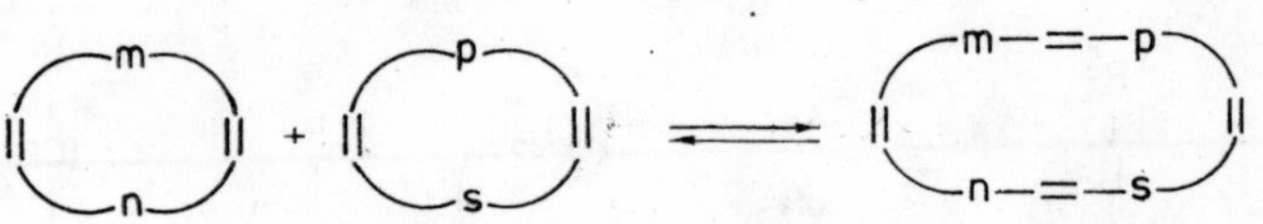

$$\tag{3.141}$$

Thus, macrocyclic polyenes result, with various structures which differ from the macrocycles obtained by the simple metathesis of cyclo-olefins.

3.5.4.1. Ethenolysis of Cyclo-Olefins

Among the alkenolyses of cyclo-olefins, the simplest reactions are those of ethenolysis. The ethenolysis of cyclopropene and cyclobutene leading to 1,3-pentadiene and 1,4-hexadiene, respectively, are not known.

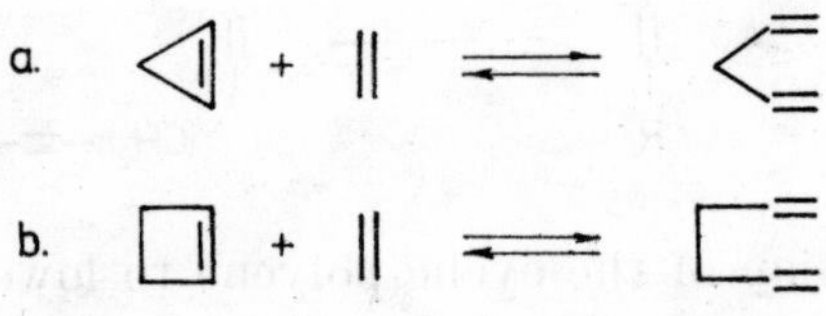

$$a. \qquad \triangleleft \; + \; || \; \rightleftharpoons \; \lessgtr$$

$$\tag{3.142}$$

$$b. \qquad \square \; + \; || \; \rightleftharpoons \; \sqsubset$$

The ethenolysis of cyclo-olefins having larger cycles than those mentioned above was studied by various research groups using both heterogeneous and homogeneous catalytic systems.

Thus, Ray and Crain [144] carried out the ethenolysis of several cycloalkenes, such as cyclopentene, cyclohexene, cyclo-octene, cyclononene and cylodecene with heterogenous catalysts consisting of alumina-supported

MoO_3 or $Mo(CO)_6$. From the reaction of cyclopentene with ethene, in the presence of catalytic systems based on $Mo(CO)_6$ at 125°C and a pressure of 69 atm, 1,6-heptadiene was obtained, in a 9—10% yield.

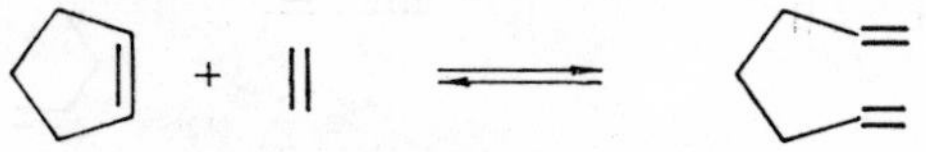

$$\text{(structure)} \quad + \quad || \quad \rightleftharpoons \quad \text{(structure)} \qquad (3.143)$$

Ray and Crain, using the same catalytic system, prepared 1,7-octadiene in similar yield by ethenolysis of cyclohexene.

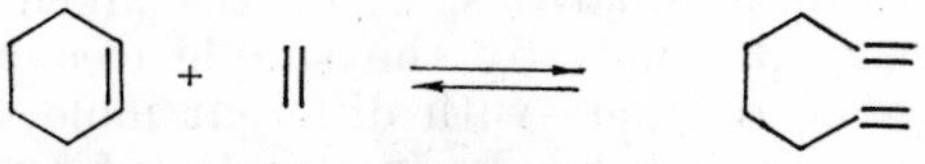

$$\text{(structure)} \quad + \quad || \quad \rightleftharpoons \quad \text{(structure)} \qquad (3.144)$$

and similarly, 1,9-decadiene, 1,10-undecadiene or 1,11-dodecadiene by ethenolysis of cyclo-octene, cyclononene or cyclodecene, respectively.

In homogeneous catalysis, the ethenolysis of cycloalkenes gave alkadienes in higher yields. Indeed, Zuech and co-workers [116] obtained by the ethenolysis of cyclo-octene, 1,9-decadiene in about 20% yield, when using the catalytic system formed from $(Ph_3P)_2Cl_2(NO)_2Mo$ and $(CH_3)_3Al_2Cl_3$; with this catalytic system, the same authors prepared 1,13-tetradecadiene by the similar ethenolysis of cyclododecene.

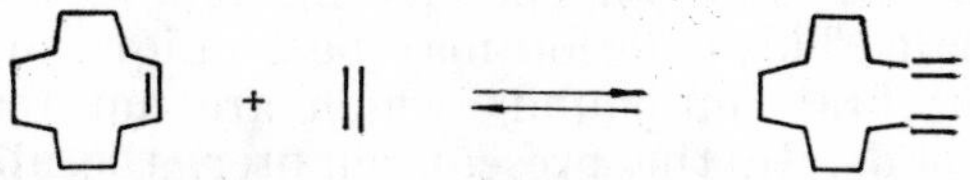

$$\text{(structure)} \quad + \quad || \quad \rightleftharpoons \quad \text{(structure)} \qquad (3.145)$$

As mentioned above, cyclodienes yield linear trienes by ethenolysis. The ethenolysis of 1,5-cyclo-octadiene, carried out both by Ray and Crain [144] with the heterogeneous catalytic system consisting of alumina-supported $Mo(CO)_6$ and by Mango [15] with cobalt molybdate, gave 1,5,9-decatriene.

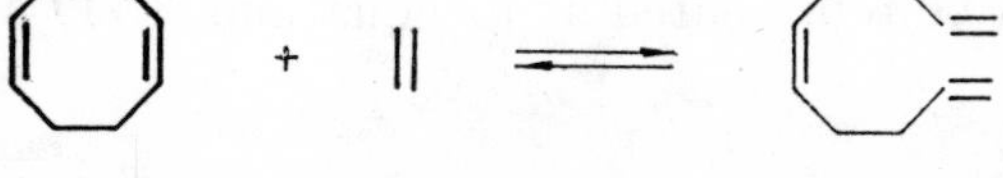

$$\text{(structure)} \quad + \quad || \quad \rightleftharpoons \quad \text{(structure)} \qquad (3.146)$$

Among bicyclic olefins, norbornene and norbornadiene were cleaved with ethene in the presence of heterogeneous catalysts of cobalt molybdate; 1,3-divinylcylopentane was obtained from norbornene

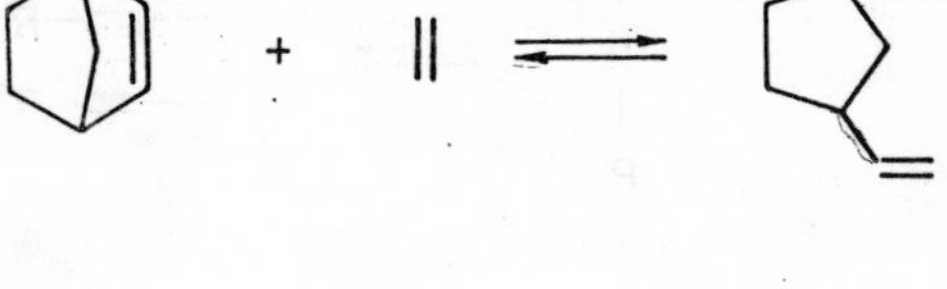

$$\text{(structure)} \quad + \quad || \quad \rightleftharpoons \quad \text{(structure)} \qquad (3.147)$$

and 3,5-divinylcyclopentene from norbornadiene.

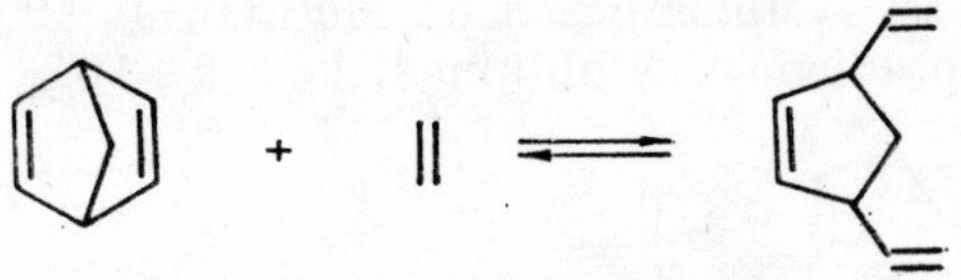

$$\text{(3.148)}$$

3.5.4.2. Cometathesis of Cyclo-Olefins with Ethene Homologues

Unlike ethenolysis, the alkenolysis of cyclo-olefins with ethene homologues exhibits a series of peculiarities. First, the alkenolysis does not afford dienes as primary products by the simple cleavage of the olefinic ring but rather polyenic telomers with different molecular weights resulted from one mole of alkene and several moles of cyclo-olefin. In fact, the reaction is a combination of ring-opening polymerization and of alkenolysis of the resulted macrocycle. Secondly, these cometathesis reactions usually proceed more slowly than ethenolysis and are also slower than the simple metathesis of either cyclo-olefins or alkenes separately. Also in the alkenolysis of cyclo-olefins, there exist differences in reactivity between various cyclo-olefins or alkenes; for example, cyclopentene and cyclooctene proved to be more reactive, while cyclohexene and cycloheptene are less reactive; among alkenes, β-olefins are more reactive than α-olefins. Thirdly, alkenolysis represents a vast class of reactions by which it is possible to obtain dienes or polyenes having double bonds in the desired position in the chain. This reaction may be a unique method for the synthesis of many natural compounds which are not readily available by conventional methods. In the present chapter, the alkenolysis of cyclo-olefins will be dealt with following a classification based on the cyclo-olefin to which the reaction is applied.

No alkenolysis of cyclopropene or cyclobutene is known. It is likely that these alkenolyses proceed very easily, as a consequence of the tendency of small rings to open owing to their angle strain, but such processes may be accompanied by polymerization reactions which diminish the selectivity of the alkenolysis. Schematically, the reactions of cyclopropene and cyclobutene can be represented as in equations 3.149 and 3.150, respectively:

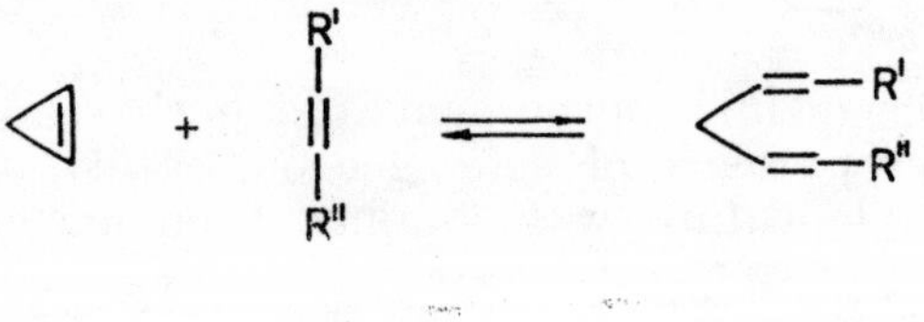

$$\text{(3.149)}$$

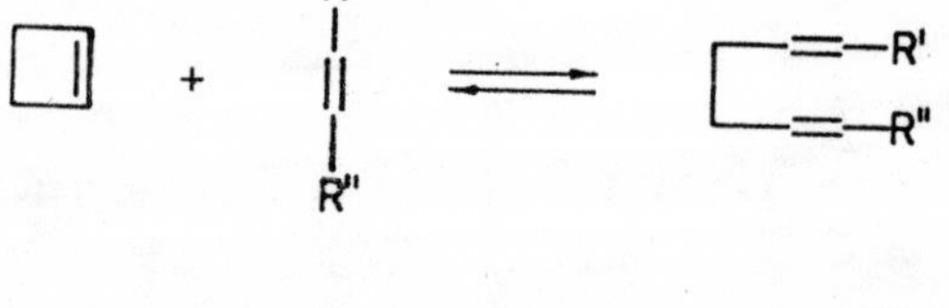

$$\text{(3.150)}$$

On the other hand, numerous alkenolyses of cyclo-olefins with medium and large rings are known. Thus, Hérisson and Chauvin [145a] studied the telomerization reactions of cyclopentene, cyclohexene, cyclo-octene, 1,5-cyclo-octadiene and 1,5,9-cyclododecatriene with various olefins in the presence of homogeneous catalytic systems based on $WOCl_4$ and Bu_4Sn or Et_2AlCl.

In the reaction of cyclopentene with 2-butene, catalyzed by the system formed from $WOCl_4$ and Bu_2Sn, four types of polyenes were obtained as main products, C_9H_{16}, $C_{14}H_{24}$, $C_{19}H_{32}$ and $C_{24}H_{40}$, in the molar proportion of 109, 76, 24, 10. It is significant to note that the direct metathesis product C_9H_{16} predominates in the reaction mixture.

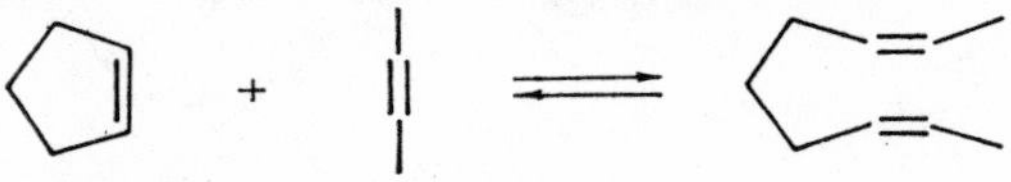

$$(3.151)$$

Unlike the reaction with 2-butene, the reaction between cyclopentene and 2-pentene, catalyzed by $WOCl_4$ and Bu_4Sn or Et_2AlCl yielded four groups of compounds, each group being a "triad", corresponding to the four types of polyenes mentioned above (Table 3.12).

TABLE 3.12. Molar composition of reaction products between cyclopentene and 2-pentene [a]

Starting material	Molar composition, mmol		
$2-C_5H_{10}$	58		47
C_5H_8	58		58
Catalyst	$WOCl_4 \cdot Sn(C_4H_9)_4$	$WOCl_4 \cdot Al(C_2H_5)_2Cl$	
Reaction time	7 min	14 min	24 hrs
Reaction products	Molar composition, mmol		
C_5H_8	32	35	108
C_4H_8	9.3	18.7	} 159
C_5H_{10}	58	31.3	
C_6H_{12}	8.8	17.7	
C_9H_{16}	4.5 } 18.3	8.1 } 34.4	16.8 } 70.4
$C_{10}H_{18}$	9.5	18.1	36.8
$C_{11}H_{20}$	4.3	8.2	16.8
$C_{14}H_{24}$	2.0 } 8.3	3.6 } 13.9	8.9 } 36.0
$C_{15}H_{26}$	4.3	6.9	18.5
$C_{16}H_{28}$	2.0	3.4	8.6
$C_{19}H_{32}$	1.0 } 4.8	1.5 } 6.0	5.0 } 18.6
$C_{20}H_{34}$	2.0	3.0	8.7
$C_{21}H_{36}$	1.0	1.5	4.9
$C_{24}H_{40}$	—	0.8 } 3.1	—
$C_{25}H_{42}$	—	1.6	—
$C_{26}H_{44}$	—	0.7	—

[a] Data from ref. [145 a].

The first triad is composed of the reaction products of cyclopentene with 2-pentene and with the metathesis products of the latter, 2-butene and 3-hexene.

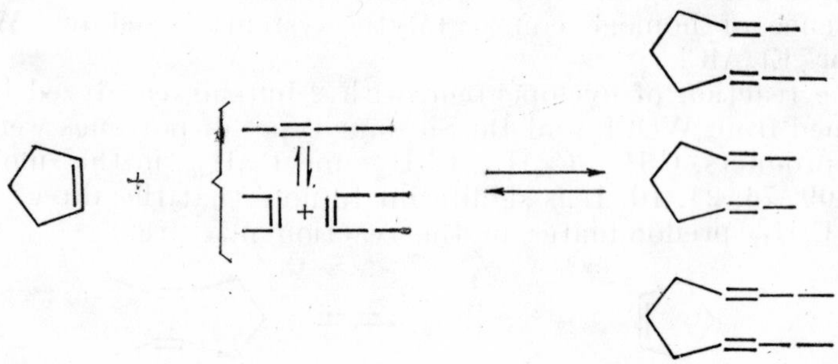

$$(3.152)$$

The subsequent triads are similarly formed out of the reaction products of the cyclopentene dimer (trimer, tetramer, etc.) with 2-butene, 2-pentene and 3-hexene.

$$C_{10}H_{16} \quad + \quad \begin{cases} C_4H_8 \longrightarrow \\ C_5H_{10} \longrightarrow \\ C_6H_{12} \longrightarrow \end{cases} \quad \begin{matrix} C_{14}H_{24} \\ C_{15}H_{26} \\ C_{16}H_{28} \end{matrix} \qquad (3.153)$$

$$C_{15}H_{24} \quad + \quad \begin{cases} C_4H_8 \longrightarrow \\ C_5H_{10} \longrightarrow \\ C_6H_{12} \longrightarrow \end{cases} \quad \begin{matrix} C_{19}H_{32} \\ C_{20}H_{34} \\ C_{21}H_{36} \end{matrix} \qquad (3.154)$$

$$C_{20}H_{32} \quad + \quad \begin{cases} C_4H_8 \longrightarrow \\ C_5H_{10} \longrightarrow \\ C_6H_{12} \longrightarrow \end{cases} \quad \begin{matrix} C_{24}H_{40} \\ C_{25}H_{42} \\ C_{26}H_{44} \end{matrix} \qquad (3.155)$$

The results obtained by Hérisson and Chauvin in the reaction between cyclopentene and 2-hexene or 3-heptene agreed with the behaviour of cyclopentene in the telomerization reaction with 2-pentene discussed above. Moreover, these data were also corroborated with the studies carried out by Kelley and Calderon [145b] dealing with the cometathesis of cyclopentene with 1-pentene.

Cyclohexene, a cyclo-olefin known to be inert under metathesis conditions, when reacted with 2-pentene in the presence of catalytic systems based on $WOCl_4$ led to a dimer ($C_{12}H_{20}$) and a trimer ($C_{18}H_{30}$) along with a product of the cometathesis with 2-pentene ($C_{11}H_{20}$). The structures of these compounds are not exactly known, but it is very likely that they are the products of simple metathesis of cyclohexene and of cometathesis of this cyclo-olefin with 2-pentene.

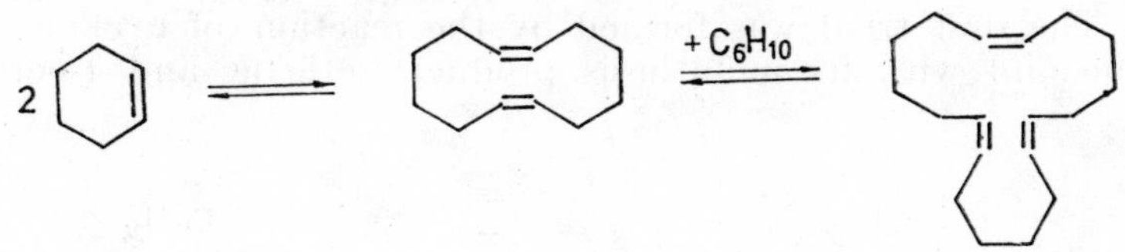

$$(3.156)$$

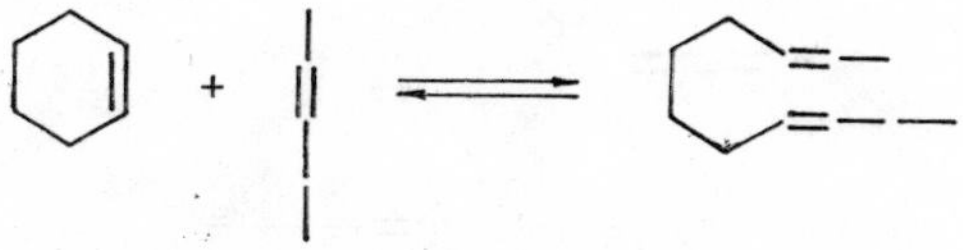

$$(3.157)$$

Hérisson and Chauvin also studied in detail the reaction of cyclo-octene with α-olefins. In the reaction with propene, in the presence of the catalytic system consisting of $WOCl_4$ and Bu_4Sn, cyclo-octene yielded a mixture of two triads, $C_{10}-C_{12}$ and $C_{18}-C_{20}$. The first triad arises from the reaction of a molecule of cyclo-octene with propene and with the metathesis products of the latter, ethene and 2-butene

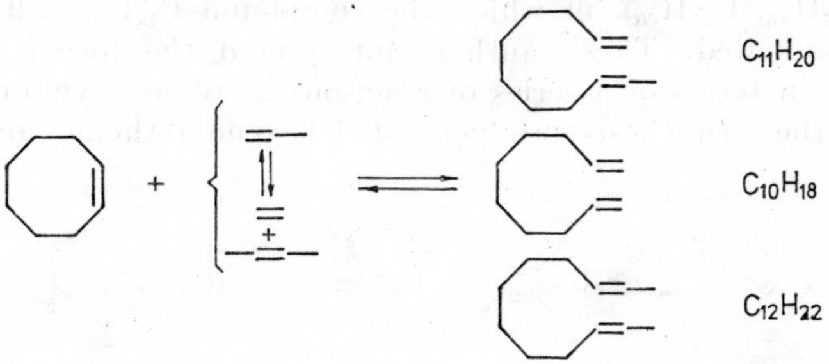

$$(3.158)$$

and the second triad, by the reaction of cyclo-octene dimer with propene, ethene and 2-butene.

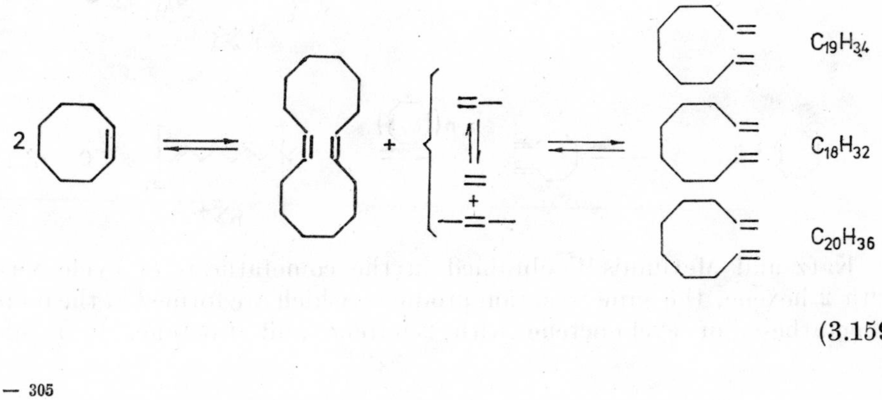

$$(3.159)$$

Under similar conditions, the reaction of cyclo-octene with 1-pentene yielded a mixture of triads $C_{10}H_{18}$, $C_{13}H_{24}$, $C_{16}H_{30}$ and $C_{18}H_{32}$, $C_{21}H_{38}C_{24}H_{44}$. The first triad was formed by the reaction of cyclo-octene with 1-pentene and with its metathesis products, ethene and 4-octene.

$$ (3.160) $$

Using a variety of ternary catalysts based on WCl_6, Lal and Smith [146] obtained in the cometathesis of cyclo-octene with 1-hexene linear unconjugated polyenes belonging to three different homologous series, namely : A ($C_{14}H_{26}$, $C_{22}H_{40}$, $C_{30}H_{54}$), B ($C_{18}H_{34}$, $C_{26}H_{48}$, $C_{34}H_{62}$) and C ($C_{10}H_{18}$, $C_{18}H_{32}$, $C_{26}H_{46}$, $C_{34}H_{60}$), in which the compound $C_{14}H_{26}$ (1,9-tetradecadiene) predominated. These authors interpreted the formation of these compounds in terms of a series of reactions involving cyclo-octene, 1-hexene and the metathesis products of 1-hexene (ethene and 5-decene)

$$ A \quad (3.161) $$

$$ B \quad (3.162) $$

$$ C \quad (3.163) $$

Katz and McGinnis [147] obtained in the cometathesis of cyclo-octene with 2-hexene, the same reaction products which are formed in the double cometathesis of cyclo-octene with 2-butene and 4-octene. It is clear that cyclo-octene reacted with 2-hexene and with the metathesis products of the latter.

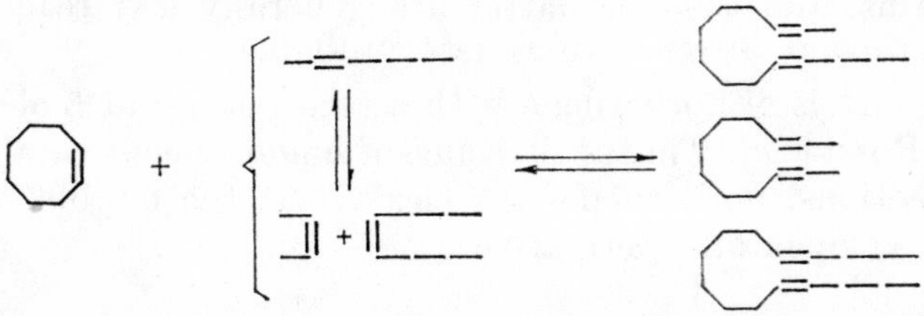

$$(3.164)$$

The reaction of 1,5-cyclo-octadiene with 2-pentene in the presence of the catalytic system $WOCl_4$ and $EtAlCl_2$ yielded a large number of triads as a result of the cometathesis between the cyclic diolefin and 2-pentene or with the metathesis products of the latter. [145a] The groups of triads were much more complex in this reaction and it was not possible to establish a simple quantitative relationship between their components.

Pinazzi and Reyx [148] studied the reaction of 1,5-cyclo-octadiene with 4-octene in the presence of the homogeneous catalytic system $WCl_6 \cdot EtAlCl_2 \cdot EtOH$ and obtained 4,8-dodecadiene and 4,8,12-hexadecatriene as main products, the structure of both olefins being a multiple structure of the polybutadiene unit. The formation of these products can be accounted for by cometathesis of 1,5-cyclo-octadiene with one or two molecules of 4-octene

$$(3.165)$$

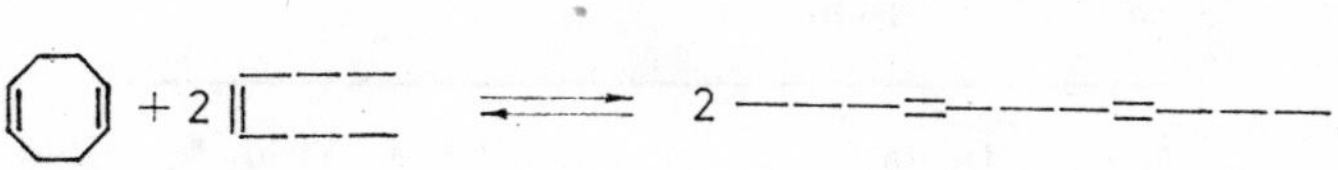

$$(3.166)$$

Unlike the preceding reaction, the similar reaction of 1,5-cyclo-octadiene with 4-methyl-4-octene yielded as main products 4,9-dimethyl-4,8-dodecadiene, 4-methyl-4-dodecadiene and 4,8-dodecadiene. These compounds could arise either from the reaction of 1,5-cyclo-octadiene with 4-methyl-4-octene according to the two possible orientations of the acyclic alkene,

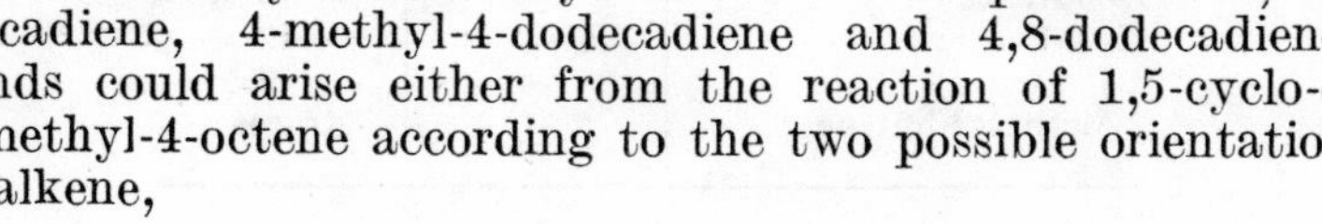

$$(3.167)$$

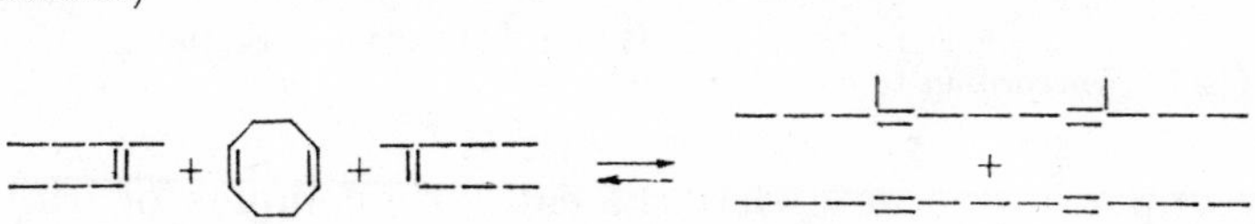

$$(3.168)$$

or from the reaction of cyclo-octene with 4-methyl-4-octene and with its metathesis products, 4-octene and 4,5-dimethyl-4-octene. Considering the fact that the latter possibility involves simultaneously tri- and tetra-substituted olefins, and that the latter are generally less reactive in metathesis, this second alternative is less probable.

The alkenolysis of norbornene with a series of α-and β-olefins was carried out by Rossi et al.[149] in the presence of homogeneous catalytic systems. In these reactions, norbornene was cleaved at the double bond yielding 1,3-divinyl-cyclopentane derivatives.

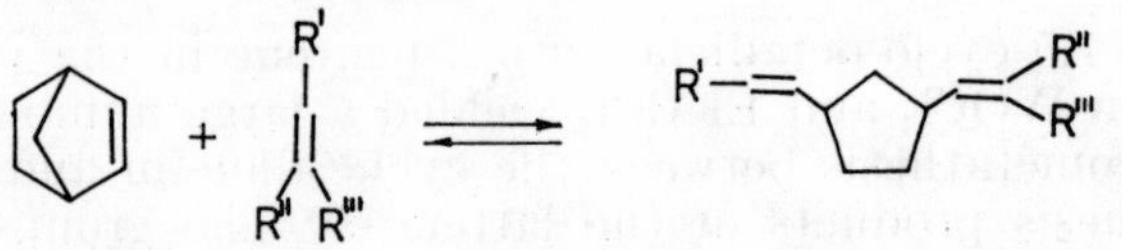

$$\text{(3.169)}$$

a. R'=R''=Me, R'''=H   d. R'=H, R''=R'''=Me

b. R'=Me, R''=H, R'''=Et   e. R'=R''=H, R'''=n-Pr

c. R'=R'''=Et, R''=H   f. R'=R''=R'''=Me

The results obtained in the reactions of norbornene with 2-butene, isobutene, 1-pentene, 2-pentene and 2-methyl-2-butene in the presence of μ-chlorochlorobis(cyclo-octene)iridium catalyst are presented in Table 3.13.

TABLE 3.13. Alkenolysis of norbornene in the presence of μ-chlorochlorobis (cyclo-octene) iridium [a]

Olefin	Product [b]	Yield, %
cis-2-Butene	a	11
Isobutene	d	5
cis-2-Pentene	a+b+c	12
1-Pentene	e	2
2-Methyl-2-butene	f	traces

[a] Data from ref. [149];   [b] Product structure according to equation 3.169.

It is very important to note that the catalytic complex of iridium which is not active in the usual olefin metathesis, presents in this reaction of norbornene cometathesis, an even higher activity than the known metathesis catalyst $WCl_6 \cdot Me_4Sn$. Furthermore, the iridium system in the presence of norbornene became active also for the metathesis of 2-pentene because the reaction mixture consisted in the cometathesis products of norbornene with 2-butene, 2-pentene and 3-hexene.

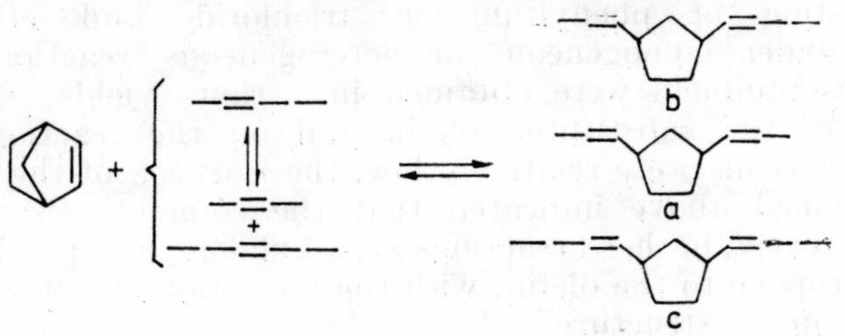

$$(3.170)$$

3.5.5. Cometathesis of Cyclopropane with Olefins

Cyclopropane shows unique behaviour in metathesis. Unlike the other cyclo-alkanes, cyclopropane behaves like an unsaturated hydrocarbon, yielding cometathesis products with a series of substituted olefins.

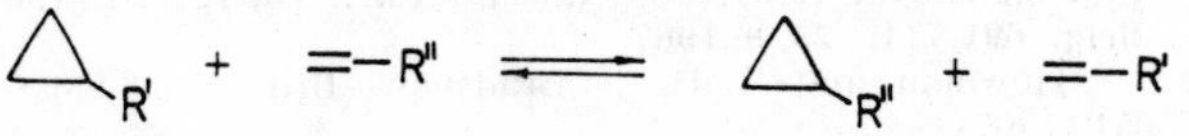

$$(3.171)$$

TABLE 3.14. Cometathesis of ethylcyclopropane with substituted olefins using chlorobenzene or hexane as solvents [a]

Olefin	Product	Yield, %	
		$PhWCl_3\cdot\text{-EtAlCl}_2$ chlorobenzene	$PhWCl_3\cdot\text{EtAlCl}_2$ hexane
$CH_2 = CHCO_2C_2H_5$	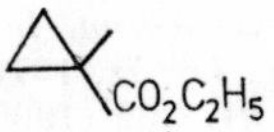 (cyclopropane with $CO_2C_2H_5$)	14	12
H—$CO_2C_2H_5$ / CH_3 H	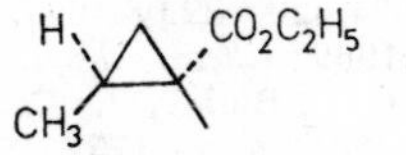 (cyclopropane, CH_3, H, $CO_2C_2H_5$)	12	10
$CH_2 = CHCN$	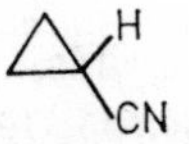 (cyclopropane with H, CN)	9	7
H—$CO_2C_2H_5$ / C_6H_5 H	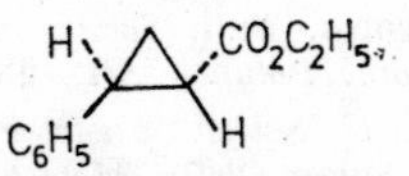 (cyclopropane H, C_6H_5, $CO_2C_2H_5$, H)	2	21
CH_3—CO_2CH_3 / CH_3 H	(cyclopropane CH_3, CH_3, CO_2CH_3, H)	11	8
H—CH_3 / H CO_2CH_3	(cyclopropane H, H, CH_3, CO_2CH_3)	13	10

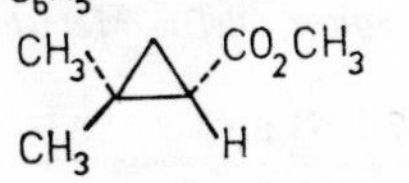

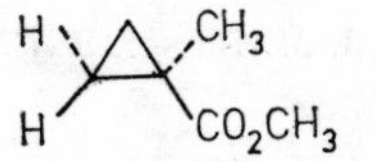

[a] Data from ref. [150].

Gassman and Johnson [150] carried out the cometathesis of ethylcyclopropane with several substituted olefins in the presence of catalytic systems consisting of phenyltungsten trichloride and ethylaluminium dichloride, under homogeneous or heterogeneous reaction conditions. Cometathesis products were obtained in various yields, depending on the nature of the substituted olefin and on the reaction conditions. Although the yields were relatively low, the isolation of the reaction products mentioned above indicated that the cometathesis reactions did occur. Apparently, in these reactions an alkylidene group was transferred from cyclopropane to the olefin, with the formation of a new olefin and a new cyclopropane structure.

References

1. R. L. Banks and G. C. Bailey, *Ind. Eng. Chem., Prod. Res. Develop.*, **3**, 170 (1964).
2. N. Calderon, H. Y. Chen and K. W. Scott, *Tetrahedron Letters*, **1967**, 3327.
3. R. L. Banks, U.S. 3,261,879, 19.07.1966; *Chem. Abstr.*, **65**, 121052 (1966).
4. D. L. Crain, **Belg.** 691,711, 23.06.1967.
5. L. Turner, E. J. Howman and C. P. C. Bradshaw, **Brit.** 1,056,980, 1.02.1967; *Chem. Abstr.*, **66**, 75651 (1967).
6. C. P. C. Bradshaw, E. J. Howman and L. Turner, *J. Catalysis*, **7**, 269 (1967).
7. C. P. C. Bradshaw, **Ger.Offen.** 1,900,490, 31.07.1969; *Chem. Abstr.*, **71**, 90763 (1969).
8. N. Calderon *et. al.*, *J. Amer. Chem. Soc.*, **90**, 4133 (1968).
9. L. F. Heckelsberg, **Fr.** 1,562,396, 4.04.1969; *Chem. Abstr.*, **71**, 93357 (1969).
10. J. L. Hérisson and Y. Chauvin, **Ger.Offen.** 2,008,246, 8.10.1970; *Chem. Abstr..* **74**, 3306 (1971).
11. a) P. H. Phung and G. Lefebvre, **Fr.** 1,594,582, 17.07.1970; *Chem. Abstr.*, **74**, 87316g (1971); b)T. J. Katz and J. McGinnis, *J. Amer. Chem. Soc.*, **99**, 1903 (1977).
12. a) A. Uchida *et al.*, *Ind. Eng. Chem., Prod. Res. Develop.*, **10**, 372 (1971); b) R. Nakamura, S. Matsumoto and E. Echigoya, *Chem. Letters*, **1976**, 1019; c) R. Nakamura, S. Matsumoto and K. Komatsu, *Chem. Letters*, **1976**, 253.
13. K. W. Scott *et al.*, *Adv. Chem. Ser.*, **91**, 399 (1969).
14. G. C. Ray and D. L. Crain, **Belg.** 694,420, 22.08.1967.
15. F. D. Mango, U.S. 3.424.811, 28.01.1969; *Chem. Abstr.*, **70**, 106042 (1969).
16. L. F. Heckelsberg, R. L. Banks and G. C. Bailey, *J. Catalysis*, **13**, 99 (1969).
17. F. W. Küpper and R. Streck, *Makromol. Chem.*, **175**, 2055 (1974).
18. F. Pennella, R. L. Banks and G. C. Bailey, *J. Chem. Soc., Chem. Commun.* **1968**, 1548.
19. K. Hummel and W. Ast, *Makromol. Chem.*, **166**, 39 (1973).
20. K. Hummel *et. al.*, *Makromol. Chem.*, **166**, 45 (1973).
21. P. Chevalier, D. Sinou and G. Descotes, *Bull. Soc. chim. France*, **1975**, 2254.
22. K. Hummel and Y. Imamoglu, *Colloid. Polymer Sci.*, **253**, 225 (1975).
23. E. Verkuijlen and C. Boelhouwer, *Fette, Seifen, Anstrichm.*, **78**, 444 (1976); D. G. Daly and M. A. McKervey, *4th Internat. Symp. Olefin Metathesis*, Belfast, August 30 – September 4, 1981.
24. G. Foster, **Ger.Offen.** 2,063,150, 8.07.1971; *Chem. Abstr.*, **75**, 63172 (1971); J. C. Mol, *J. Mol. Catalysis*, **15**, 35 (1982).
25. J. P. Laval, A. Lattes, R. Mutin and J. M. Basset, *J. Chem. Soc., Chem. Commun.*, **1977**, 502.
26. E. Wasserman, D. A. Ben-Efraim and R. Wolovsky, *J. Amer. Chem. Soc.*, **90**, 3286 (1968).
27. R. Wolovsky, *J. Amer. Chem. Soc.*, **92**, 2132 (1970).

28. a) S. R. Wilson and D. E. Shalk, *J. Org. Chem.*, **41**, 3928 (1976) ; b) R. Streck, *Chem.-Ztg.*, **99**, 395 (1973).
29. R. Rossi and R. Giorgi, *Chimica et Industria*, **58**, 517 (1976) ; D. Villemin, *Tetrahedron Letters*, **1980**, 1715 ; D. Villemin and J. Levisalles, *3rd Internat. Symposium Metathesis*, Lyons, 1979, p. 70.
30. G. Dall'Asta, *Rubber Chem. Technol.*, **47**, 511 (1974).
31. G. C. Bailey, *Catalysis Revs.*, **3**, 37 (1969).
32. R. L. Banks, *Top. Curr. Chem.*, **25**, 39 (1972).
33. a. N. Calderon, *Accounts Chem. Res.*, **5**, 127 (1972) ; b. N. Calderon, E. A. Ofstead and W. A. Judy, *Angew. Chem.*, **88**, 433 (1976) ; c. N. Calderon, J. P. Lawrence and E. A. Ofstead, *Adv. Organomet. Chem.*, **17**, 449 (1979).
34. V. Drăguțan, *Studii cercetări chim.*, *Acad. R.S.R.*, **20**, 1171 (1972).
35. a. R. H. Grubbs, *Progr. Inorg. Chem.*, **24,** 1 (1978) ; b. L. Hocks, *Bull. Soc. chim. France*, **1975**, 1893.
36. R. J. Haines and G. J. Leigh, *Chem. Soc. Revs.* **4**, 155 (1975).
37. a. W. B. Hughes, *Organomet. Chem. Synth.*, **1**, 341 (1972) ; b. W. B. Hughes, *Adv. Chem. Ser.*, **132**, 192 (1974) ; c. W. B. Hughes, *Ann. N. Y. Acad. Sci.*, **295**, 271 (1977).
38. a. T. J. Katz, *Adv. Organomet. Chem.*, **16**, 283 (1977) ; b. J. C. Mol and J. A. Moulijn, *Adv. Catalysis*, **24**, 131 (1975) ; c. J. J. Rooney and A. Stewart, *Catalysis*, **1**, 277 (1977).
39. R. Taube and K. Seyferth, *Wiss. Z. TH Leuna-Merseburg*, **17**, 350 (1975).
40. a. L. Bencze and L. Markó, *Magy. Kem. Lapja*, **27**, 213 (1972) ; b. M. L. Khidekel, A. D. Shebaldova and I. V. Kalechits, *Usp. Khim.*, **40**, 1416 (1971) ; *Russ. Chem. Rev. (Eng. Transl.)*, **40**, 669 (1971) ; c. L. Markó, *J. Hung. Chem. Soc.*, **27**, 253 (1972) ; d. R. Streck, *Chem.-Ztg.*, **99**, 397 (1975) ; e. H. Weber, *Chem. Unserer Zeit*, **11**, 22 (1977).
41. A. T. Balaban, *Rev. Roumaine Chim.*, **22**, 243 (1977).
42. G. Binsch, E. L. Eliel and H. Kessler, *Angew. Chem.*, **83**, 618 (1971) ; *Internat. Ed.*, **10**, 570 (1974).
43. A. A. Olsthoorn, cited after reference 32.
44. R. L. Banks, **Belg.** 633,418, 10.12.1968 ; *Chem. Abstr.*, **61**, 1690d (1964).
45. L. Turner, E. J. Howman and K. V. Williams, **Brit.** 1,106,015, 23.07. 1965 ; *Chem. Abstr.*, **67**, 53607t (1967).
46. L. Turner, E. J. Howman and K. V. Williams, **Brit.** 1,106,016, 23.07. 1965 ; *Chem. Abstr.*, **67**, 53607t (1967).
47. L. F. Heckelsberg, R. L. Banks and G. C. Bailey, *Ind. Eng. Chem.*, *Prod. Res. Develop.*, **8**, 259 (1969).
48. L. F. Heckelsberg, R. L. Banks and G. C. Bailey, *Ind. Eng. Chem.*, *Prod. Res. Develop.*, **7**, 29 (1968).
49. L. F. Heckelsberg, U.S.Pat. 3.340.322, 5.09.1967 ; *Chem. Abstr.*, **67**, 116529 (1967).
50. V. V. Atlas, I. I. Pis'man and A. M. Bakhshi-Zadé, *Khim. Prom.*, **45**, 734 (1969).
51. V. V. Atlas, I. I. Pis'man and A. N. Bakhshi-Zadé, **U.S.S.R.** 253,797, 7.10.1969 ; Chem Abstr., 72, 66355 (1970).
52. L. F. Heckelsberg, Belg. Pat. 713.186, 1968.
53. K. V. Williams and L. Turner, **Brit.** 1,116,243, 6.06.1968 ; *Chem. Abstr.*, **69**, 29085 (1968).
54. British Petroleum Company, **Brit.** 1,105,564, 23.04.1965 ; *Chem. Abstr.*, **66**, 48081 (1967).
55. British Petroleum Company, **Brit.** 1,093,784, 25.08.1965 ; *Chem. Abstr.*, **67**, 53624 (1967).
56. L. Turner and K. V. Williams, **Brit.** 1,096,250, 20.12.1967 ; *Chem. Abstr.*, **68**, 77693. (1968).
57. R. L. Banks, **U.S.** 3,443,969, 6.05.1969 ; *Chem. Abstr.*, **71**, 38315 (1969).
58. L. F. Heckelsberg, **U.S.** 3,395,196, 30.07.1968 ; *Chem. Abstr.*, **69**, 60530 (1968).
59. Isagulyants and L. F. Rar, *Izvest. Akad. Nauk S.S.S.R.*, *Ser. Khim.*, **1969**, 1362.
60. R. L. Banks, **U.S.** 3,546,313, 8.12.1970 ; *Chem. Abstr.*, **74**, 78085 (1971).
61. J. C. Mol, J. A. Moulijn and C. Boelhouwer, *J. Chem. Soc., Chem. Commun.*, **1968**, 633.
62. J. C. Mol, J. A. Moulijn and C. Boelhouwer, *J. Catalysis*, **11**, 87 (1968).
63. A. Clark and C. Cook, *J. Catalysis*, **15**, 420 (1969).
64. F. L. Woody, M. J. Lewis and J. B. Wills, *J. Catalysis*, **14**, 389 (1969).
65. J. C. Mol, F. R. Visser and C. Boelhouwer, *J. Catalysis*, **17**, 114 (1970).
66. E. A. Zuech, *J. Chem. Soc., Chem. Commun.*, **1968**, 1182.

67. J. L. Wang and H. R. Menapace, *J. Org. Chem.*, **33**, 3794 (1968).
68. A. Uchida *et al.*, *Ind. Eng. Chem.*, *Prod. Res. Develop.*, **10**, 369 (1971).
69. J. A. Moulijn and C. Boelhouwer, *J. Chem. Soc.*, *Chem. Commun.*, **1971**, 1170.
70. P. B. van Dam and C. Boelhouwer, *React. Kin. Catalysis Letters*, **1**, 481 (1974).
71. P. B. van Dam and C. Boelhouwer, *React. Kin. Catalysis Letters*, **1**, 165 (1974).
72. L. Bencze and L. Markó, *J. Organomet. Chem.*, **28**, 271 (1971).
73. L. Bencze, A. Rédey and L. Markó, *Hung. J. Ind. Chem.*, **1**, 453 (1973).
74. R. Opitz, L. Bencze, L. Markó and F.-H. Thiele, *J. Organomet. Chem.*, **71**, C3 (1974).
75. L. Bencze, L. Markó, R. Opitz and K.-H. Thiele, *Hung. J. Ind. Chem.*, **4**, 15 (1976).
76. L. Bencze and L. Markó, Hung. Teljes, 8.274, 28.06.1974; *Chem. Abstr.*, **82**, 35379 (1975).
77. R. Taube and K. Seyferth, *Z. Chem.*, **14**, 284 (1974).
78. R. Taube and K. Seyferth, *Z. Chem.*, **14**, 410 (1974).
79. J. M. Basset *et. al.*, *J. Catalysis*, **34**, 152 (1974).
80. J. M. Basset *et. al.*, *J. Catalysis*, **34**, 196 (1974).
81. J. M. Basset *et al.*, *J. Amer. Chem. Soc.*, **97**, 7376 (1975).
82. J. L. Bilhou, J. M. Basset and R. Mutin, *J. Organomet. Chem.*, **87**, C4 (1975).
83. J. L. Bilhou, J. M. Basset, R. Mutin and W. F. Graydon, *J. Chem. Soc.*, *Chem. Commun.*, **1976**, 970.
84. J. L. Bilhou, J. M. Basset, R. Mutin and W. F. Graydon, *J. Amer. Chem. Soc.*, **99**, 4083 (1977).
85. K. Ichikawa and K. Fukuzumi, *J. Org. Chem.*, **41**, 2633 (1976).
86. D. L. Crain, *J. Catalysis*, **13**, 110 (1969).
87. A. Uchida, K. Kobayashi and S. Matsuda, *Ind. Eng. Chem.*, *Prod. Res. Develop.*, **11**, 389 (1972).
88. R. P. Arganbright, U.S. 3,792,108, 12.02.1974; *Chem. Abstr.*, **80**, 95212 (1974).
89. L. F. Heckelsberg, S. African 68,01,797, 15.08.1968; *Chem. Abstr.* **70**, 67557 (1969).
90. R. E. Reusser, U.S. 3,729,524, 24.04.1973; *Chem. Abstr.*, **79**, 18045 (1973).
91. R. J. Sampson and C. Jackson, Ger.Offen. 1,808,459, 12.06.1969; *Chem. Abstr.*, **71**, 38314 (1969).
92. A. L. Lapidus *et al.*, *Izvest. Akad. Nauk S.S.S.R.*, *Ser. Khim.*, **1973**, 1261.
93. A. L. Lapidus *et al.*, *Izvest. Akad. Nauk S.S.S.R.*, *Ser. Khim.*, **1973**, 1266.
94. R.L. Banks, Belg. 620,440, 21.01. 1967; *Chem. Abstr.*, **59**, 8590d (1963).
95. N. Calderon and H. Y. Chen, Brit. 1,125,529, 28.08.1968; *Chem. Abstr.*, **69**, 195851 (1968).
96. G. Doyle, *J. Catalysis*, **30**, 118 (1973).
97. R. E. Reusser and S. D. Turk, U.S. 3,760,026, 18.09.1973; *Chem. Abstr.*, **79**, 125789 (1973).
98. A. Morris, H. Thomas and C.J. Attridge, Ger.Offen. 2,213,948, 21.12.1972; *Chem. Abstr.*, **79**, 65769 (1973).
99. ICI, Fr. 2.120.509, 8.12.1972; *Chem. Abstr.*, **79**, 115104 (1973).
100. E. Sh. Finkel'stein *et al.*, *Neftekhim.*, **15**, 667 (1975).
101. D. L. Crain, Fr. 1,515,037, 1.03.1968; *Chem. Abstr.*, **70**, 87137 (1969).
102. D. L. Crain, Belg. Pat. 688,784, 23.06.1967.
103. P. D. Montgomery, R. N. Moore and R. W. Knox, U.S. 3,965,206, 22.06.1976; *Chem. Abstr.*, **85**, 123540 (1976).
104. G. S. Lewandos and R. Pettit, *J. Amer. Chem. Soc.*, **93**, 7087 (1971).
105. G. Descotes, P. Chevalier and D. Sinou, *Synthesis*, **1974**, 564.
106. T. J. Katz and R. Rotchild, *J. Amer. Chem. Soc.*, **98**, 2519 (1976).
107. R. A. Fridman *et. al.*, *Doklady Acad. Nauk S.S.S.R.*, **234**, 1354 (1977).
108. A. N. Bashkirov *et al.*, U.S.S.R. 564,301, 5.07.1977; *Chem. Abstr.* **87**, 117556 (1977).
109. R. Nouguier *et al.*, *Rec. Trav. chim.*, **96**, M91 (1977).
110 Japan Synthetic Rubber Company, Fr. 2,252,314, 20.06.1975; *Chem. Abstr.*, **86**, 43190 (1976).
111. E. Verkuijlen, R. L. Dirks and C. Boelhouwer, *Rec. Trav. chim.*, **96**, 86 (1977).
112. E. Verkuijlen, F. Kapteijn, J. C. Mol and J. C. Boelhouwer, *J. Chem. Soc.*, *Chem. Commun.*, **1977**, 198.
113. P. B. van Dam, M. C. Mittelmeijer and C. Boelhouwer, *J. Chem. Soc.*, *Chem. Commun.*, **1972**, 1221.

114. P. B. van Dam, M. C. Mittelmeijer and C. Boelhouwer, *Fette, Seifen, Anstrichm.*, **76**, 264 (1974).
115. R. Nakamura *et. al.*, **Japan. Kokai** 76 125,317, 1.11.1976; *Chem. Abstr.* **86**, 170599 (1977).
116. a) E. Verkuijlen and C. Boelhouwer, *Riv. Ital. Sostanze Grasse*, **53**, 234 (1976); b) E. A. Zuech *et. al.*, *J. Amer. Chem. Soc.*, **92**, 528 (1970).
117. G. Dall'Asta, G. Mazzanti, G. Natta and L. Porri, *Makromol. Chem.*, **53**, 224 (1964).
118. G. Natta, G. Dall'Asta, G. Mazzanti and G. Motroni, *Makromol. Chem.*, **69**, 163 (1963).
119. G. Dall'Asta and R. Manetti, *Atti Acad. naz. Lincei, Rend., Classe Sci. fis. mat. nat.*, **41**, 351 (1966).
120. T. J. Katz, J. McGinnis and C. Altus, *J. Amer. Chem. Soc.*, **98**, 606 (1976).
121. F. W. Küpper and R. Streck, **Ger. Offen.** 2,326,296, 12.12.1974; *Chem. Abstr.*, **82**, 155528 (1975).
122. P. Günther *et al.*, *Angew. Makromol. Chem.*, **16/17**, 27 (1971).
123. G. Natta, G. Dall'Asta, I. W. Bassi and G. Carella, *Makromol. Chem.*, **91**, 87 (1966).
124. N. Calderon and W. A. Judy, **Brit.** 1,124,456, 21.08.1968; *Chem. Abstr.*, **69**, 77970 (1968).
125. R. Rossi and R. Giorgi, *Tetrahedron Letters*, **1976**, 1721.
126. G. Dall'Asta and R. Manetti, *European Polymer J.*, **4**, 145 (1968).
127. L. Höcks, D. Berck, A. J. Hubert and Ph. Teyssié, *J. Polymer Sci., Polymer Letters*, **13**, 391 (1975).
128. D. A. Ben-Efraim, C. Batich and E. Wasserman, *J. Amer. Chem. Soc.*, **92**, 2133 (1970).
129. R. Wolovsky, *Synthesis*, **1972**, 134.
130. R. Wolovsky, **Ger.Offen.** 2,108,965, 9.09.1971; *Chem. Abstr.*, **76**, 3466 (1972).
131. R. Wolovsky, **Israeli** 33,970, 31.05.1973; *Chem. Abstr.*, **79**, 52876 (1973).
132. H. Höcker and R. Musch, *Makromol. Chem.*, **157**, 201 (1972).
133. E. Wasserman, *J. Amer. Chem. Soc.*, **82**, 4433 (1960).
134. H. L. Frisch and E. Wasserman, *J. Amer. Chem. Soc.*, **83**, 3789 (1961).
135. G. Schill, "Catenanes, Knots and Rotaxanes", Academic Press, New York, 1971, p. 31.
136. G. Schill and A. Lüttringhaus, *Angew. Chem.*, **76**, 567 (1964).
137. S. Shida, M. Tsukada and T. Oka, *J. Chem. Phys.*, **45**, 3483 (1967).
138. J. A. Moulijn, H. J. Reitsma and C. Boelhouwer, *J. Catalysis*, **25**, 434 (1972).
139. a) A. V. Mushegyan *et al.*, *Arm. Khim. Zh.*, **28**, 672 (1975); *Chem. Abstr.*, **84**, 16675 (1976); b) A. V. Mushegyan *et al.*, *Arm. Khim. Zh.*, **28**, 674 (1975); *Chem. Abstr.*, **84**, 16693 (1976).
140. A. Mortreux, F. Petit and M. Blanchard, *Tetrahedron Letters*, **1978**, 4967.
141. D. L. Crain, **Fr.** 1,516,853, 15.03.1968; *Chem. Abstr.*, **71**, 12498 (1969).
142. L. F. Heckelsberg, **Fr.** 1,562,397, 4.04.1969; *Chem. Abstr.*, **71**, 93358 (1969).
143. R. L. Banks and R. B. Regier, *Ind. Eng. Chem., Prod. Res. Develop.*, **10**, 46 (1971).
144. G. C. Ray and D. L. Crain, **Fr.** 1,511,381, 26.01.1968; *Chem. Abstr.*, **70**, 114580 (1969).
145. a) J. L. Hérisson and Y. Chauvin, *Makromol. Chem.*, **141**, 161 (1970); b) J. W. Kelley and N. Calderon, *J. Macromol. Sci., Chem.*, **A9**, 911 (1975).
146. J. Lal and R. R. Smith, *J. Org. Chem.*, **40**, 775 (1975).
147. T. J. Katz and J. McGinnis, *J. Amer. Chem. Soc.*, **97**, 1592 (1975).
148. C. Pinazzi and D. Reyx, *Compt. rend.*. **276**, 1077 (1973).
149. R. Rossi, P. Diversi, A. Lucherini and L. Porri, *Tetrahedron Letters*, **1974**, 879.
150. P. G. Gassman and T. H. Johnson, *J. Amer. Chem. Soc.*, **98**, 6058 (1976).

RING-OPENING POLYMERIZATION OF CYCLO-OLEFINS

The ring-opening polymerization of cyclo-olefins has lately proved to be one of the most attractive fields of polymer chemistry. Indeed, by this method it is possible to obtain polymers which have a large utilization in the elastomer industry, i.e. 1,4-polybutadiene and 1,4-polyisoprene, other polymers with valuable elastomer properties, such as polychloroprene and polypentenamer, graft and block copolymers, as well as plastics with special properties.

Although known for a long time, this process has witnessed a rapid development only lately, after the discovery of olefin metathesis and the elucidation of the reaction mechanism of this type of polymerization via metathesis. The value of the polymers thus obtained and the simplicity of this procedure have prompted both academic and technological research for industrial application.

The first ring-opening polymerization of cyclo-olefins was reported by Anderson and Merckling[1] in 1955. They polymerized norbornene in the presence of a catalyst formed from tetraheptyllithium-aluminium and titanium tetrachloride and obtained solid, amorphous polymers with plastic properties. Later on, this type of reaction was carried out by Eleuterio[2,3] with several cyclo-olefins, among which he included cyclopentene, using a heterogeneous catalyst formed from alumina-supported molybdenum trioxide, thus obtaining products with elastomer properties. The weak activity of the catalytic system caused this discovery not to be immediately used. After a few years, Truett and co-workers[4] resumed the study of the norbornene polymerization with catalytic systems based on organometallic compounds and titanium tetrachloride, demonstrating the occurrence of two types of polymerization for this monomer, the ring-opening and the vinyl polymerization reactions which lead to polymers with distinct properties.

Since 1962, Natta and co-workers[5-8] have carried out a series of vast studies which evidenced these two types of cyclo-olefin polymerizations for a large number of monomers; they found new Ziegler-Natta catalyst systems for these reactions and worked out suitable procedures to obtain polyalkenamers with elastomer properties. During a first series of investigations, these authors showed that cyclobutene may undergo vinylic polymerization at the double bond in the presence of catalytic systems based on vanadium yielding solid, amorphous, saturated polymer or it may form polybutenamer by ring-opening polymerization in the presence of titanium systems, an unsaturated elastomer which is identical with 1,4-polybutadiene.[5] They have prepared a large number of catalytic systems, based on transition metals, for polymerization of cyclobutene and with these catalytic systems they demonstrated the polymerization duality of this monomer.[6] Although some of these catalytic systems, i.e. those with tungsten or molybdenum, yielded polybutenamer, the preparation of highly stereospecific 1,4-butadiene is still a problem to be solved in the future. In 1964, Natta and co-workers succeeded in polymerizing cyclopentene to polypentenamer with two new catalytic systems, one based on molybdenum and the other on tungsten. They obtained, for the first time, polypentenamer with a high content of *cis* stereoisomer in the presence of molybdenum catalysts and a high content of *trans* configuration in the presence of tungsten catalysts. Both polypentenamers had elastomer properties. Later on, Natta and co-workers extended the ring-opening polymerization with these catalytic systems to higher cyclo-olefins.[8] They obtained polyalkenamers starting from cycloheptene, cyclo-octene and cyclododecene with a high content of *trans* stereoisomer.

Soon after the discovery of metathesis [9] with the homogeneous catalytic system based on tungsten which was related to the system utilized by Natta and co-workers for the ring-opening polymerization of cyclo-olefins, Calderon and co-workers [10] reported the ring-opening polymerization of cyclo-olefins with this new catalytic system of metathesis, under milder conditions and with higher yields of polyalkenamers. They applied the reaction to a large number of cyclo-olefins, obtaining polyalkenamers with a 40—70% content of *trans* configuration at the double bond. These investigators opened new routes for synthesizing polymers such as 1,4-polybutadiene from 1,5-cyclo-octadiene or 1,5,9-cyclododecatriene. Furthermore, through the polymerization of substituted cyclo-olefins, they supplied methods for preparing alternating diene copolymers and terpolymers.

Nowadays, investigations concerning the ring-opening polymerization of cyclo-olefins are directed towards improving the conversion of cyclopentene into polypentenamer,[11] the synthesis of other polyalkenamers with elastomer properties [12-14] and the synthesis of plastics and polymers with improved properties.[14,15] The achievements in this field have been presented in a series of reviews, among which we shall mention those published by Scott,[12] Calderon,[13] Dall'Asta, [14a] Dolgoplosk [14b] and Drăguţan.[15] The present chapter will deal with ring-opening polymerizations of monocyclic, bicyclic and polycyclic olefins according to the number of carbon atoms of the ring.

4.1. Polymerization of Monocyclic Olefins

Ring-opening polymerization has been carried out for most of the mono-cyclic olefins, from cyclobutene to cyclododecene and even for higher cyclo-olefins. The smaller cyclo-olefins, cyclobutene and cyclopentene, proved reactive in the presence of polymerization catalytic systems, es-pecially cyclopentene. Cyclohexene is the only cyclo-olefin which could not be polymerized easily to polyhexenamers with the known catalytic systems. The higher homologues polymerize without difficulty. In the homologous series, the reactivity decreases with the increase of the number of carbon atoms in the ring. Cyclo-olefins with substituents at the double bond polymerize with some difficulty owing to steric hindrance, while those with substituents in other positions give polyalkenamers almost as easily as the unconjugated ones. 1,5-Cyclo-octadiene and 1,5,9-cyclo-dodecatriene, can readily be polymerized, yielding polyalkenamers with elastomeric properties.

4.1.1. Four-Membered Ring Monomers

Cyclobutene. The ring-opening polymerization of cyclobutene is of special interest because through this reaction it is possible to obtain polybutena-mer, which is identical with 1,4-polybutadiene, a polymer known for its elastomer properties and which was obtained till now by the polymeri-zation of butadiene.

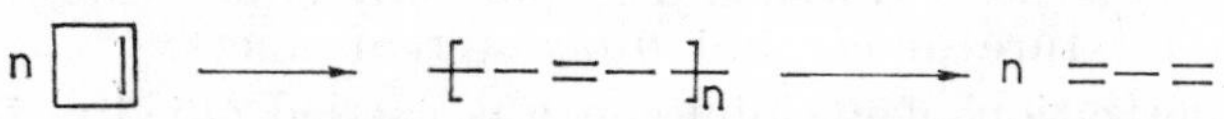

$$ n \ \square \longrightarrow \ +\!-\!=\!-\!+_n \longrightarrow n \ =\!-\!= \tag{4.1} $$

The polymerization of cyclobutene was first carried out by Natta and co-workers[5, 6] using a large range of Ziegler-Natta catalytic systems. Depending on the catalytic system, cyclobutene polymerized either at the double bond yielding the vinylic polymer, polycyclobutylenamer, or through ring-opening leading to polybutenamer, or finally through both pathways, thus affording a mixture of the two polymers.

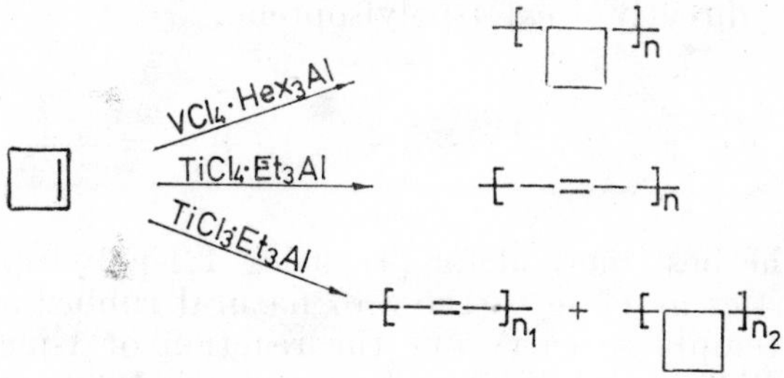

$$ \square \quad \xrightarrow[\substack{VCl_4 \cdot Hex_3Al \\ TiCl_4 \cdot Et_3Al \\ TiCl_3Et_3Al}]{} \quad \begin{array}{c} +\!\square\!+_n \\[4pt] +\!-\!=\!-\!+_n \\[4pt] +\!-\!=\!+_{n_1} + \ +\!\square\!+_{n_2} \end{array} \tag{4.2} $$

The catalytic systems utilized were formed from transition metal salts, especially vanadium, titanium, chromium, molybdenum or tungsten chlorides or acetonylacetonates, and organoaluminium compounds. Vanadium and chromium salts led to the vinylic polymer, polycyclobutylenamer, titanium tetrachloride mainly to polybutenamer, whereas the other catalysts formed both polymers.

TABLE 4.1. Polymerization of cyclobutene with catalytic systems based on transition metals[a]

Catalyst components		Molar ratio II/I	Texp. °C	Conversion %	Polymer structure, %		
					Polycyclo-butylena-mer	Polybutenamer	
I	II					cis	trans
3 TiCl$_3$·AlCl$_3$	Et$_3$Al	3	+45	100	40	20	40
TiCl$_4$	Et$_3$Al	2	−10	100	5	30	65
V(acac)$_3$	Et$_2$AlCl	5	−50	100	100	—	—
VCl$_4$	Et$_3$Al	2.5	−20	100	99	0	1
Cr(acac)$_3$	Et$_2$AlCl	7	−20	100	100	—	—
MoCl$_5$	Et$_3$Al	3	−20	5	30	30	30
MoO$_2$(acac)$_2$	Et$_2$AlCl	5	−10	55	10	45	45
WCl$_6$	Et$_3$Al	3	−20	100	40	30	30

[a] Data from ref. [6].

Although some catalysts showed reduced activity during this reaction, cyclobutene polymerized readily (in some instances with quantitative yield) in many of these systems. The polybutenamer thus obtained was in all cases a mixture of *cis* and *trans* stereoisomers.

The polymerization of cyclobutene was carried out by Natta and coworkers [16-18] also employing other salts of transition metals, even in polar solvents, such as water or ethanol. Nickel and rhodium salts led to vinylic polymers, whereas ruthenium salts yielded polybutenamer. In the latter case, a strong effect of the solvent on the steric configuration at the double bond was observed. By using ethanol as a solvent, the content of *trans* configuration in the polybutenamer was increased.

1-Methylcyclobutene. The ring-opening polymerization of 1-methylcyclobutene leads directly to 1,4-polyisoprene.

$$\tag{4.3}$$

This provides the best method for preparing 1,4-polyisoprene with structure and properties as close to those of natural rubber as possible.

The first attempts to carry out the reaction of 1-methylcyclobutene were due to Dall'Asta and Manetti,[19] as early as 1966. Catalytic systems

based on vanadium, titanium, chromium or molybdenum, which polymerize cyclobutene yielding either vinylic polymers or polyalkenamers,
proved inactive in this case. However, the polymerization of 1-methyl-
cyclobutene succeeded with catalysts based on tungsten, e.g. the following systems: $WCl_6 \cdot (C_2H_5)_3Al$ and $WCl_6 \cdot (C_2H_5)_2AlCl$. Although the
products thus obtained had a polyisoprene skeleton, they had saturated
structures. It is supposed that an intermolecular cyclization of the polyisoprene fragments, B, occurred in the presence of the catalytic systems,
with formation of condensed cyclohexane structures, C.

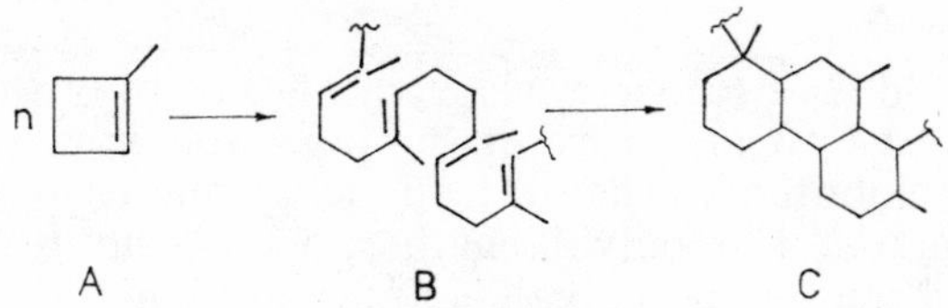

$$(4.4)$$

A   B   C

The interest presented by this method of obtaining 1,4-polyisoprene
prompted many chemists to continue the study of the ring-opening polymerization of 1-methylcyclobutene.

TABLE 4.2. Polymerization of 1-methylcyclobutene with various
catalytic systems [a]

Catalytic system	Conversion %	Polyisoprenic content %
$MoCl_2(NO)_2[(C_6H_5)_3P]_2 \cdot (CH_3)_3Al_2Cl_3$	67	20
$WCl_6 \cdot (C_6H_5)_3SnC_2H_5$	74	58
$WCl_6 \cdot n\text{-}C_4H_9Li$	83	75
$(C_6H_5)_2C=W(CO)_5$	91	94

[a] Data from ref. [20].

Katz and co-workers [20] employed a series of active catalysts in the metathesis of acyclic olefins and with some of them he obtained good results
also in the polymerization of 1-methylcyclobutene. The $MoCl_2(NO)_2(Ph_3P)_2 \cdot$
$AlMe_3 \cdot AlCl_3$ system led to a conversion of 1-methylcyclobutene of maximum
67%, with 20% polyisoprene content in the product. The systems based
on WCl_6 and $(C_6H_5)_3SnC_2H_5$ or $n\text{-}C_4H_9Li$ proved more active but the
polymer had a high content of saturated structures. The best conditions
to polymerize 1-methylcyclobutene used the carbenic catalytic system
$(C_6H_5)_2C=W(CO)_5$. In this last case, the product was polyisoprene with a
84—87% content of cis and 13—16% of trans stereoisomer. The polymer
had, along with 2-methyl-2-butene units, also 2-butene and 2,3-dimetyl-
2-butene units. These latter units probably resulted from migrations of

methyl groups in the chain under the influence of the catalytic system, after the ring-opening polymerization occurred.

3-Methylcyclobutene. The ring-opening polymerization of 3-methylcyclobutene was first effected with $RuCl_3$ in polar media,[18] and later on, with vanadium-based catalytic systems.[21] In both cases, the product was a polybutenamer with one methyl group in the monomer unit : no methyl group migration in the chain was noted.

$$n \; \square \longrightarrow \;\; \text{-}\!\!\!\text{=}\!\!-\!\!\text{-}\!\!\text{-}\!\!\text{-} \qquad\qquad (4.5)$$

If $RuCl_3$ is used as a catalyst, the structure of the resulting polymer is strongly influenced by the nature of the polar medium. As in the case of unsubstituted cyclobutene, ethanol leads to an increase of the *trans* configuration content in the polyalkenamer, while water increases the *cis* configuration content. The polyalkenamer obtained has an amorphous structure which is attributed to the lack of stereoregularity of the tertiary carbon in the chain.

Dall'Asta[21] studied the influence of various organometallic compounds added to the vanadium-based catalytic systems on the conversion and structure of polymers obtained from 3-methylcyclobutene. The most active systems proved to be those containing triethyl-aluminium ; these led to 37 % conversion of the starting material. However, the polymer thus otained had only 10 % polyalkenamer structure. On the other hand, the system with n-butyllithium, although possessing the lowest activity, led only to ring-opening polymerization. The configuration of the polymer was, in this case, 55 % *trans* and 45 % *cis*.

TABLE 4.3. Polymerization of 3-methylcyclobutene with catalysts based on vanadium chloride and organometallic compounds[a]

Catalyst components		Conversion %	Polymer structure, %		
I	II		Cyclic units	*trans-*	*cis-*
Et_3Al	VCl_4	37	90	10	0
Et_3Ga	VCl_4	25	90	6	4
Et_2Be	VCl_4	17	85	12	3
Et_2Mg	VCl_4	14	94	6	0
BuLi	VCl_4	3	0	55	45

[a] Data from ref. [21].

4.1.2. Five-membered Ring Monomers

Cyclopentene. In the presence of a large number of catalytic systems, cyclopentene can be polymerized to polypentenamer in high yields. Depending on the catalytic system used in the reaction, it is possible to obtain a polypentenamer having mainly *cis* or *trans* configuration at the double bond.

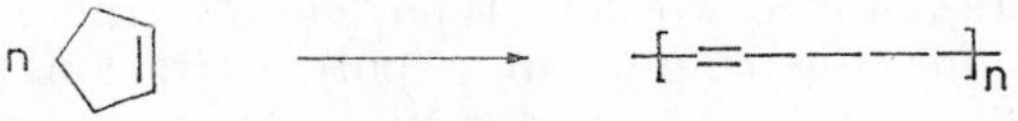

$$n \,\bigcirc \longrightarrow \quad \left[\!\!\left.=\!-\!-\!-\!-\right]\!\!\right._n \qquad (4.6)$$

Remarkably, the *trans*-polypentenamer has valuable elastomer properties similar to those of natural rubber.[14, 22]

The ring-opening polymerization of cyclopentene was carried out for the first time by Eleuterio in 1957.[2] Eleuterio employed solid catalysts based on chromium, molybdenum, tungsten or uranium, and he worked with a suspension of these catalysts in hydrocarbons.[3] The reaction occurred with low conversions (about 4%) and led to polyalkenamer with reduced stereospecificity. The polymer obtained was an amorphous elastomer but developed crystallinity on stretching.

Natta, Dall'Asta and Mazzanti [7, 22] resumed the study of cyclopentene polymerization immediately after having carried out the copolymerization of cyclopentene with ethene and propene.[23-26] They observed that a series of Ziegler-Natta type catalysts based on titanium, zirconium and vanadium with high activity in the polymerization of cyclobutene had a rather low activity towards cyclopentene, leading to conversions only of 1—3%. Unlike the reaction of cyclobutene, the reaction of cyclopentene is highly stereospecific for the above three catalytic systems, only *trans*-

TABLE 4.4. Polymerization of cyclopentene with Ziegler — Natta catalytic systems[a]

Catalytic system	Yield, %	Polymer structure	Polymer properties
$TiCl_4 \cdot Al(C_2H_5)_3$	1	*trans*	crystalline
$TiBr_4 \cdot Al(C_2H_5)_3$	2	*trans*	,,
$ZnCl_2 \cdot Al(C_2H_5)_3$	1	*trans*	,,
$VCl_4 \cdot Al(C_2H_5)_3$	1	$C_5 + C_2$	,,
$VOCl_3 \cdot Al(C_2H_5)_2Cl$	3	$C_5 + C_2$	,,
$V(acac)_3 \cdot Al(C_2H_5)_2Cl$	1.5	$C_5 + C_2$	,,
$MoCl_5 \cdot Al(C_2H_5)_3$	21	*cis*	amorphous
$WCl_6 \cdot Al(C_2H_5)_3$	39	*trans*	crystalline

[a] Data from ref. [7].

polypentenamer being obtained. Catalysts based on chromium, cobalt, uranium, manganese, iron and barium were completely inactive towards cyclopentene, but systems based on molybdenum and tungsten had high activity in the polymerization of cyclopentene. The tungsten-based catalysts proved to be the most active, leading to polypentenamer with a high content of *trans* configuration. The molybdenum-based system was less active than that of tungsten but led to a polypentenamer with *cis* configuration. The *trans*-polypentenamer obtained exhibited a high degree of crystallinity, its melting point being 23°C and having good elastomer properties. The *cis*-polypentenamer was an amorphous substance. Thus, Natta and co-workers reported, for the first time, active catalytic systems for the ring-opening polymerization of cyclopentene, obtaining products with a high degree of stereospecificity; by thorough physico-chemical studies they characterized the chemical and physical nature of the polypentenamers, demonstrating the remarkable elastomer properties, particularly for the *trans* configuration product.

The valuable properties of the polypentenamer stimulated intense research for obtaining new catalytic systems and methods for the polymerization of cyclopentene, for improving the existing procedures, for using them on industrial scale, and for perfecting the processing of the reaction products in the polymer industry. Several research groups are studying some of these aspects of cyclopentene polymerization to polypentenamer. The most outstanding results obtained by several of these groups are presented below.

Nützel [27] and Günther [28,29] carried out the polymerization of cyclopentene with a great variety of binary catalytic systems containing metal halides and organometallic compounds, and alkali or alkaline earth metal compounds. They obtained varying conversions, depending on the catalysts and on the working conditions: in most cases a polypentenamer with a high content in *trans* stereoisomer resulted. It is worth mentioning that some of these binary systems formed from tungsten halides with compounds of such metals as lithium or calcium, led to a yield of polypentenamer ranging between 21 and 66% with over 80% content of *trans* stereoisomer.

As has already been mentioned before, cyclopentene polymerizes in the presence of tungsten hexachloride and trialkylaluminium catalytic systems with different yields and stereospecificities, depending on the nature of the alkyl radical in the aluminium compound. Thus, triethylaluminium catalysts lead to higher polypentenamer yields and higher *trans* isomer content than do catalysts prepared with triisopropyl- or triisobutyl-aluminium. [27,30,31]

Though molybdenum-[32] or rhenium-based catalytic systems [33] exhibit lower activity than those of tungsten, they gained importance because of the reaction products that result. Cyclopentene polymerizes in the presence of these systems mainly to *cis*-polypentenamer, a polymer with very good performance at low temperatures. It is interesting to note that a polypentenamer with a high content of *cis* configuration is obtained on replacing chlorine by fluorine in the tungsten hexachloride from the $WCl_6 \cdot (C_2H_5)_2AlCl$ or $WCl_6 \cdot (C_2H_5)_3Al_2Cl_3$ system which yields *trans*-polypentenamer. Oberkirch [34] obtained a 34% yield of a polymer with 82.8% *cis* stereoisomer by using such a system.

Dall'Asta and Carella [35] polymerized cyclopentene with ternary catalytic systems based on tungsten or molybdenum, organometallic compounds and oxygen-containing compounds, demonstrating thereby the possibility of obtaining polypentenamer with high efficiency, mainly by employing such systems with high stability. Thus, these authors polymerized cyclopentene to *trans*-polypentenamer with catalytic systems based on tungsten and organoaluminium compounds using benzoyl peroxide, t-butyl peroxide, cumyl hydroperoxide, hydrogen peroxide, ethanol, phenol, oxygen and even water. The numerous variants of these ternary systems are continuously improving performances in the production of polypentenamer. Various research groups are working out new procedures for obtaining polypentenamer; they are attempting to diversify and improve the ternary catalytic systems for polymerizing cyclopentene by employing compounds containing oxygen, halogens, nitrogen, sulphur or phosphorus. Pampus and co-workers [36-41] use epichlorohydrin and chloroethanol in several procedures, to increase the stability of the catalytic system producing *trans*-polypentenamer. High conversions of cyclopentene were obtained by Nützel *et al.*[42-45] with cyclopentene hydroperoxide, aromatic nitroderivatives or inorganic peroxides. α-Halogenated alcohols such as 2-chloroethanol, 2-bromoethanol, 1,3-dichloro-2-isopropanol, *o*-chlorophenol, 2-chlorocyclohexanol or 2-iodocyclohexanol were employed by Witte *et al.*[45] in order to obtain some of the most stable systems, without losing their activity during cyclopentene polymerization. Acetals proved to be similar stabilizers,[46] whereas various epoxides e.g. etheneoxide and 1-buteneoxide increase to a greater extent the activity of tungsten systems utilized in cyclopentene polymerization.[47]

Together with the increase in activity, some unsaturated halogenated compounds facilitate the adjustment of the molecular weight of polypentenamer and impart useful properties for its processing. For the same purpose, Oberkirch *et al.*[48, 49] utilize vinyl chloride, and Streck and Weber [50-52] employ halides and unsaturated esters or ethers such as vinyl fluoride, vinyl acetate, allyl phenyl ether, etc.

The polymerization of cyclopentene with catalytic systems based on niobium and tantalum [53,54] made possible the preparation of polypentenamer with special properties, especially low gel content, in high yields. A system formed from $NbCl_5$ and $(i-C_4H_9)_3Al$ led to a completely gel-free polymer. Similar systems based on $MoCl_5$ yielded a polymer with 70% gel content.[53]

1-*Methylcyclopentene*. By ring-opening polymerization of 1-methylcyclopentene, it is possible to obtain a polyalkenamer whose structural unit represents the next higher homologue of that from 1,4-polyisoprene.

$$(4.7)$$

The structure of poly-1-methylpentenamer indicates elastomer properties close to those of natural rubber. Actually, its structural relationship with the polypentenamer, whose valuable elastomer properties are already known, is of the same kind as that between 1,4-polyisoprene and

1,4-polybutadiene. Unfortunately, the poor reactivity of 1-methylcyclo-pentene towards the known systems for metathesis and ring-opening poly-merization prevents the formation and characterization of the correspond-ing polyalkenamer. As in the case of 1-methylcyclobutene, the methyl group adjacent to the double bond makes the polymerization difficult. Unlike 1-methylcyclobutene, the strain in the 1-methylcyclopentene ring is much diminished, which decreases even more the reactivity of its double bond. A first attempt to polymerize 1-methylcyclopentene was made with a WCl_6 catalytic system,[27] but the results were unsatisfactory. More recently, the reaction of 1-methylcyclopentene was carried out with binary systems consisting of complexes of molybdenum nitrosyl, $MoCl_2(NO)_2 \cdot [(C_6H_5)_3P]_2 \cdot (CH_3)_3Al_2Cl_3$ when an undefined polymer was formed with low conversion.[20]

3-Methylcyclopentene. Unlike the foregoing isomer substituted at the double bond, 3-methylcyclopentene was readily polymerized to the cor-responding polyalkenamer with catalysts based on WCl_6.[27] Poly-3-methyl-pentenamer was obtained with the methyl group remaining in the mo-nomer unit, without any migration.

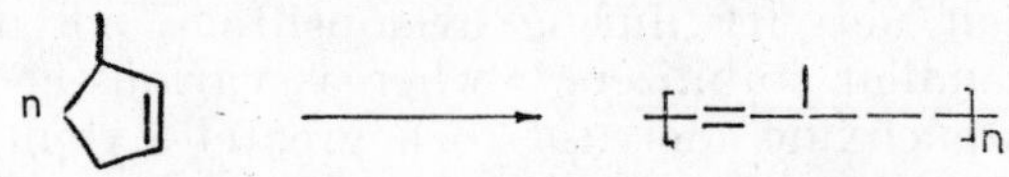

$$ n \quad \langle \text{ring} \rangle \quad \longrightarrow \quad \llbracket = - \! - - \rrbracket_n \qquad (4.8) $$

The polymer contained over 90% *trans* configuration at the double bond. The presence of the methyl group in position 3 strongly influences the properties of the polymer. An amorphous substance was obtained which had very different characteristics from those of the unsubstituted poly-pentenamer.

4.1.3. Six-Membered Ring Monomers

Cyclohexene. Cyclohexene should form, through ring-opening polymeri-zation, the next higher homologue of polypentenamer, with properties related to those of the latter.

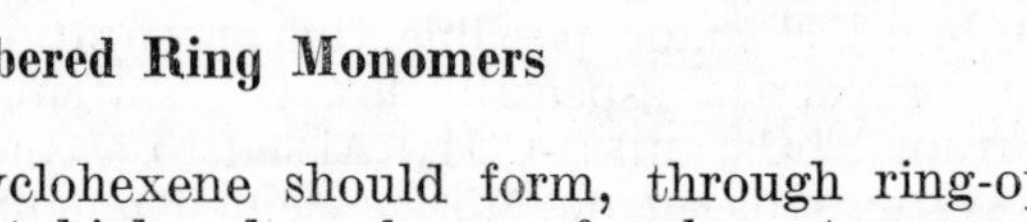

$$ n \quad \langle \text{hexagon} \rangle \quad \longrightarrow \quad \llbracket = - - - - \rrbracket_n \qquad (4.9) $$

Eleuterio[3] attempted to carry out the reaction with heterogeneous cata-lytic systems based on chromium, molybdenum, tungsten or uranium oxides. Cyclohexene scarcely reacted and only traces of oligomers were obtained. The polymerization with the homogeneous catalytic system of tungsten hexachloride carried out by Amass *et al.*[55] formed undefined polymers with a yield of only 4%.

The low reactivity of cyclohexene is explained by the great stability of the six-membered ring which shifts the equilibrium of metathesis (cyclohexene $\rightleftarrows$ dimer) towards cyclohexene and thus favours retrometathesis.

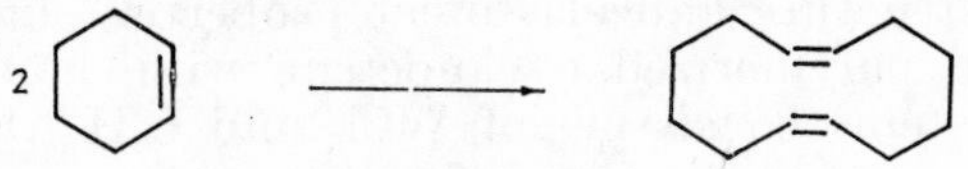

$$ 2 \quad \text{[structure]} \quad \longrightarrow \quad \text{[structure]} \qquad (4.10) $$

1,3-Cyclohexadiene. The polymerization of 1,3-cyclohexadiene was carried out by using complexes of iridium with cyclo-octene or 1,5-cyclo-octadiene.[56] Low yields of polymers with undetermined structure were obtained using a catalytic system formed from $IrCl_3 \cdot H_2O$ and cyclo-octene, in ethanol and water.

4.1.4. Seven-Membered Ring Monomers

Cycloheptene. Cycloheptene forms polyheptenamer by ring-opening polymerization, according to the reaction

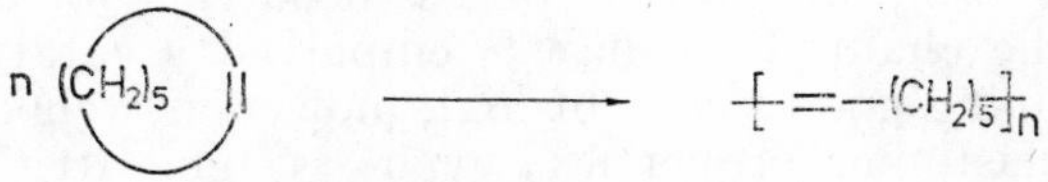

$$ n \ (CH_2)_5 \quad \text{[structure]} \quad \longrightarrow \quad \text{[structure]} \qquad (4.11) $$

The polyheptenamer is easily obtainable in the presence of tungsten- or molybdenum-based catalytic systems. Natta and co-workers [8,57] carried out the reaction of cycloheptene by employing catalytic systems constituted of WCl_6 or $MoCl_5$ and $(C_2H_5)_3Al$ or $(C_2H_5)_2AlCl$ to obtain a polyheptenamer with over 85% *trans* configuration at the double bond. By X-ray analysis it was proved that the polymer has a crystalline structure.

4.1.5. Eight-Membered Ring Monomers

Cyclo-octene. Cyclo-octene readily polymerizes in the presence of metathesis catalytic systems forming polyoctenamer.

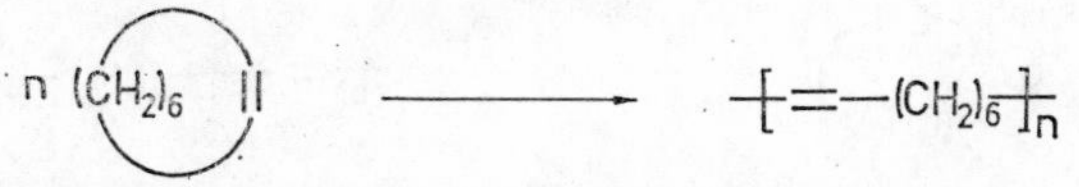

$$ n \ (CH_2)_6 \quad \text{[structure]} \quad \longrightarrow \quad \text{[structure]} \qquad (4.12) $$

The ring-opening polymerization of cyclo-octene was first achieved by Eleuterio,[3] who employed heterogeneous catalytic systems based on chromium, molybdenum, tungsten or uranium oxides. The conversions were low and polymers with a high amount of *cis* configuration at the double bond were obtained.

Natta and co-workers [8,58] obtained higher conversions in cyclo-octene polymerization, by using homogeneous catalytic systems formed from WCl_6 or $MoCl_5$ and $(C_2H_5)_3Al$ or $(C_2H_5)_2AlCl$. The configuration of the polyoctenamer at the double bond was, in this case, *trans*; the polymer had a crystalline structure and elastomer properties. Later on, Calderon and co-workers [10,59] polymerized cyclo-octene with homogeneous metathesis catalytic systems consisting of WCl_6 and $C_2H_5AlCl_2$. This catalyst was particularly active, leading with 73% conversion to polyoctenamer. The polymer exhibited a certain crystallinity and contained *cis* and *trans* configuration at the double bond. In this reaction, Calderon and co-workers noticed for the first time the formation of cyclo-octene oligomers which were separated and characterized by physico-chemical methods. These authors proved the macrocyclic nature of these oligomers formed in the reaction of cyclo-octene, a fact which had a great importance for the elucidation of the reaction mechanism.

The ethylaluminium compounds in the above catalytic systems can be successfully replaced by other organometallic compounds. Calderon and Morris [60] obtained a polyoctenamer in the presence of the catalytic system WCl_6 and $(i\text{-}C_4H_9)_3Al$ with a *trans* configuration content varying with the reaction time. The melting point of this polyoctenamer ranged between 20 and 60°C, an increase of 1°C occurring for each per cent of *trans* bond in the chain. Bradshaw [61] employed a catalyst of WCl_6 and $(C_4H_9)_4Sn$ obtaining good yields of high molecular weight polyoctenamer with excellent elastomer properties; even systems free from organometallic compounds were utilised: Marshall and Ridgewell's catalysts [62] formed from WCl_6 and $AlBr_3$ exhibited high activity during the reaction. This catalytic system is of interest for industrial applications.

Küpper *et al.* [63] obtained a polyoctenamer with 63% *cis* configuration at the double bond, by employing ternary catalytic systems constituted of WCl_6, $C_2H_5AlCl_2$ and organic acids, e.g. acetic acid. On using mineral salts of organic acids, such as lithium palmitate or alkoxides, such as aluminium s-butoxide, the content of *trans* configuration at the double bond of the polymer increased. [64]

3-Methyl- and *3-phenylcyclo-octene.* The ring-opening polymerization of these monomers led to the corresponding polyalkenamers

$$ \tag{4.13} $$

where R = methyl or phenyl. The reaction was carried out by Calderon and co-workers [10] by employing the same catalytic system as in the case of the polymerization of cyclo-octene, namely WCl_6 and $C_2H_5AlCl_2$. The catalyst activity was lower in this case than with cyclo-octene. Each monomer unit of the resulting polyalkenamer had its substituent in the α-position relative to the double bond, without any migration of alkyl or aryl groups during the polymerization. The polymers exhibited elastomer properties.

5-Methyl- and 5-phenylcyclo-octene. Calderon and co-workers,[10] by utilizing the catalytic system formed from WCl_6 and $C_2H_5AlCl_2$ which was active in cyclo-octene polymerization, carried out the polymerization of cyclo-octene substituted with alkyl or aryl groups in position 5, and obtained polyalkenamers with interesting interpolymer structures. Thus, the polyalkenamer obtained from 5-methylcyclo-octene had a ternary polymer structure, butadiene-ethene-propene and the product obtained from 5-phenylcyclo-octene had a butadiene-ethene-styrene structure,

$$n \ R-HC \underset{(CH_2)_3}{\overset{(CH_2)_2}{\big<}} \Big] \longrightarrow \Big[=-(CH_2)_3-\underset{R}{CH}-(CH_2)_2 \Big]_n \qquad (4.14)$$

where R = methyl or phenyl. The catalytic system presented a high activity in all these reactions. Substituents in position 5 relative to the double bond did not noticeably modify the monomer polymerizability. The polymers had a well-defined structure and exhibited elastomer properties.

The structure of the polyalkenamers obtained from these substituted monomers opened the way to a new and interesting method for preparing ternary polymers or interpolymers via ring-opening polymerization. The use of as widely different substituents as possible, e.g. alkyl, alkenyl, aryl, halogen, cyano or ester groups may lead to terpolymers or interpolymers which cannot be obtained by other routes and which may have valuable elastomeric or plastic properties.

1,5-Cyclo-octadiene. 1,5-Cyclo-octadiene, easily obtained from butadiene by cyclodimerization or via metathesis of cyclobutene, leads by ring-opening polymerization to the same polyalkenamer which is formed from cyclobutene, namely to polybutenamer or 1,4-polybutadiene.

$$n \ \square \longrightarrow \frac{n}{2} \ \hexagon \longrightarrow \Big[=--\Big]_n \qquad (4.15)$$

This fact is easily understandable if it is accepted that, in the first step, the polymerization of cyclobutene yields 1,5-cyclo-octadiene, via dimerization. The polymerization of 1,5-cyclo-octadiene was first carried out by Calderon and co-workers [10] by employing a catalytic system based on WCl_6 and $C_2H_5AlCl_2$. The reaction reached a conversion of over 60% and a polyalkenamer with *cis*- and *trans*-1,4-polybutadiene structure was obtained. The product presented good elastomeric properties. In order to obtain a polymer with higher steric purity, Calderon also tested other catalytic systems. Although less active than the above catalytic system, the catalyst formed from WCl_6 and $(i-C_4H_9)_3Al$ led to 1,4-polybutadiene with over 75 % content of *cis* configuration at the double bond.[65]

1-Methyl-1,5-cyclo-octadiene. The homologue of 1,5-cyclo-octadiene substituted with a methyl group at one of the double bonds, 1-methyl-1,5-cyclo-octadiene ought to yield an alternating copolymer of isoprene and butadiene by ring-opening polymerization.

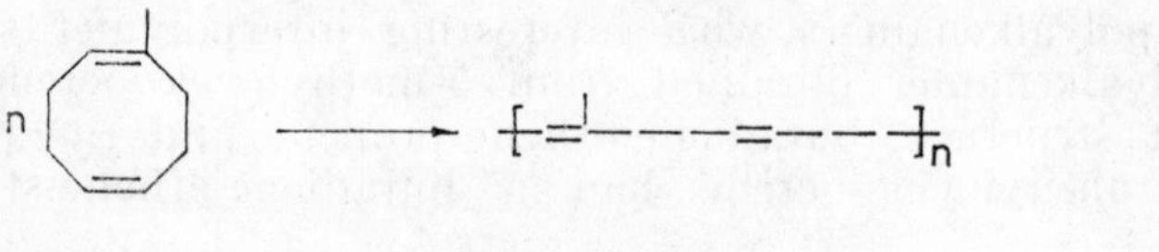

$$\text{(structure)} \qquad (4.16)$$

This polymerization was carried out by Calderon [66] with a 79 % yield, by employing a catalytic system consisting of $WCl_6 \cdot C_2H_5OH \cdot C_2H_5AlCl_2$. He obtained a polymer containing butadiene and isoprene monomer units in a uniform alternance, with very good elastomeric properties; the vitrifying temperature of the product was $-85°C$.

1-Chloro-1,5-cyclo-octadiene. Similarly to 1-methyl-1,5-cyclo-octadiene, the derivative substituted with a chlorine atom at one of the double bonds of 1,5-cyclo-octadiene, 1-chloro-1,5-cyclo-octadiene ought to form, by ring-opening polymerization, an alternating copolymer of butadiene and chloroprene.

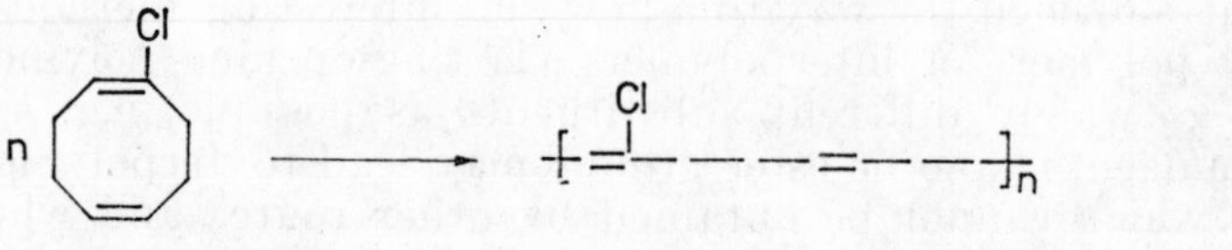

$$\text{(structure)} \qquad (4.17)$$

This polymer, which combines the properties of 1,4-polybutadiene with those of 1,4-polychloroprene, should have interesting elastomeric properties.

The polymerization of this monomer has not yet been described in the literature, although the possibility of this reaction has been mentioned by Scott *et al.*[67]

1,5-Dimethyl-1,5-cyclo-octadiene. Similarly to the formation of 1,4-poly-butadiene, via polymerization of 1,5-cyclo-octadiene, 1,5-dimethyl-1,5-cyclo-octadiene should yield 1,4-polyisoprene.

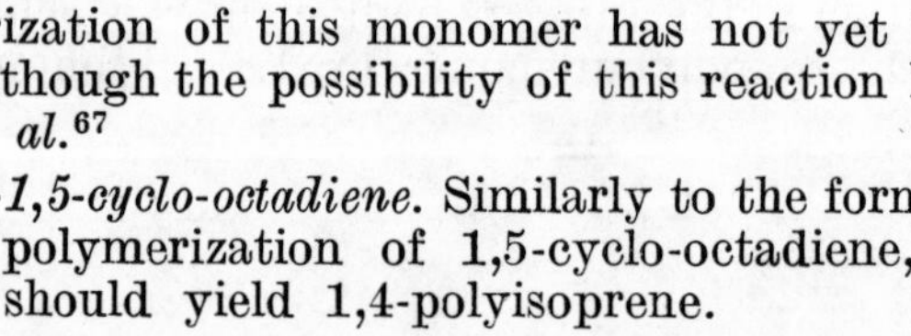

$$\text{(structure)} \qquad (4.18)$$

The polyisoprene thus obtained would be identical to that formed from 1-methylcyclobutene, due to the fact that 1,5-dimethyl-1,5-cyclo-octadiene may be formed from 1-methyl-cyclobutene by dimerization, in a first step, in the presence of the metathesis catalytic system. This reaction has not been carried out so far. The two alkyl groups at the double bond are likely to deactivate markedly the monomer in comparison with the unsubstituted 1,5-cyclo-octadiene. However, this route could constitute the most efficient method to prepare 1,4-polyisoprene with an advanced degree of steric purity and with properties very close to those of natural rubber. The synthesis of 1,4-polyisoprene by this procedure remains to be solved.

4.1.6. Nine-Membered Ring Monomers

Cyclononene. Cyclononene forms by ring-opening polymerization a polynonenamer. which is the higher homologue of polyoctenamer.

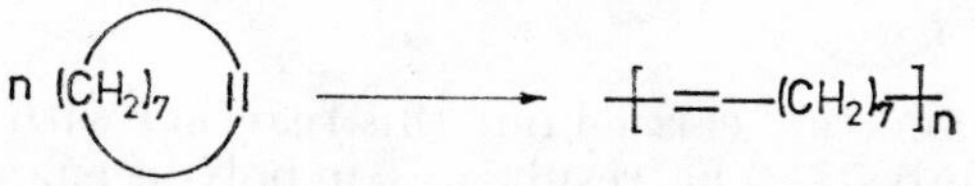

$$n \ (CH_2)_7 \quad || \quad \longrightarrow \quad \left[= -(CH_2)_7 \right]_n \qquad (4.19)$$

The polymerization of cyclononene was carried out by Natta and co-workers [8] with catalytic systems based on tungsten or molybdenum and organoaluminium compounds. A polymer with prevailing *trans* configuration at the double bond and elastomeric properties was obtained.

1,5-Cyclononadiene. The ring-opening polymerization of 1,5-cyclononadiene ought to yield an alternating copolymer with butenamer and pentenamer units.

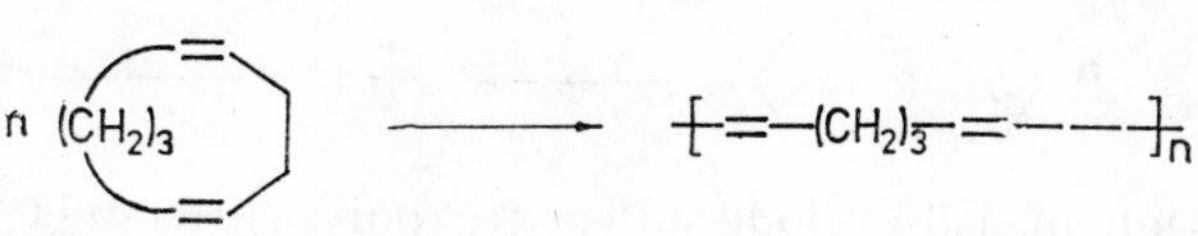

$$n \ (CH_2)_3 \quad \longrightarrow \quad \left[= -(CH_2)_3 = --- \right]_n \qquad (4.20)$$

This polymer could have good elastomeric properties close to those of 1,4-polybutadiene and polypentenamer. So far, such a copolymer with uniform alternation has not been obtained. Rossi and Giorgi [68] polymerized 1,5-cyclononadiene by employing a binary metathesis catalytic system formed from WCl_6 and $LiAlH_4$. They obtained a polymer with a higher content of butenylene units than of pentenylene ones in the chain. The formation of this polymer was interpreted through intramolecular metathesis reactions in the chain with elimination of cyclopentene, but it could also be explained by suitable mutual orientation of monomer units.

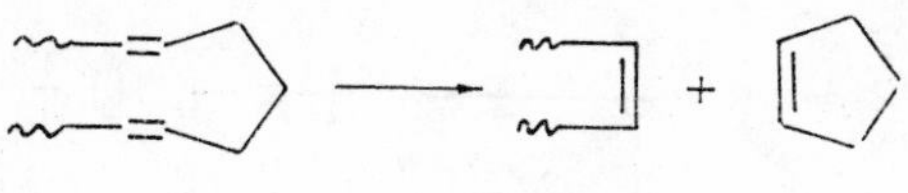

$$\qquad \longrightarrow \qquad + \qquad \qquad (4.21)$$

4.1.7. Ten-Membered Ring and Higher Monomers

The activity of the catalysts for ring-opening polymerization decreases with the increase of the olefinic ring. Among large ring cyclo-olefins, cyclodecene and cyclododecene were polymerized to macrocycles or polyalkenamers, but the yields were lower than those with the small- or medium-sized rings. Remarkably, the polymerization of nonconjugated polyenes, such as 1,5,9-cyclododecatriene or 1,9,17-cyclotetraeicosatriene was readily accomplished and the resulting polyalkenamers have good elastomeric properties.

Cyclodecene. The polymerization of cyclodecene in the presence of catalysts based on tungsten and organoaluminium compounds occurs without difficulty to give polydecenamer.

$$n \ (CH_2)_8 \ || \longrightarrow \ \left[= -(CH_2)_8 \right]_n \tag{4.22}$$

Dall'Asta and Manetti [69] carried out this reaction with WCl_6 or $WOCl_4$ and $(C_2H_5)_2AlCl$ catalysts. The result was a polydecenamer with a prevailing *trans* configuration at the double bond, in some instances over 85 %. The catalytic systems proved active for a molar ratio Al/W between 2.5 and 5.0. The activity was additionally improved by employing cumyl peroxide as an activator.

1,6-Cyclodecadiene. The ring-opening polymerization of 1,6-cyclodecadiene with metathesis catalysts offers a possible new way of obtaining polypentenamer.

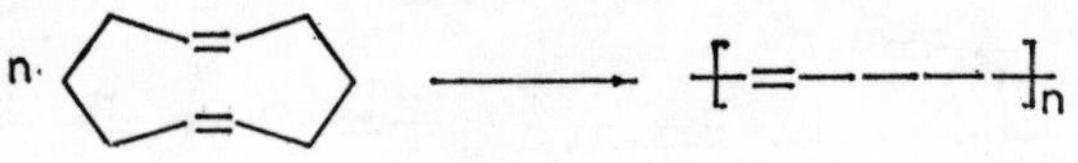

$$n \cdot \ \longrightarrow \ \left[= - - - - \right]_n \tag{4.23}$$

This reaction of 1,6-cyclodecadiene to polypentenamer has not been carried out yet, but it has been observed that 1,6-cyclodecadiene, in the presence of metathesis catalytic systems, leads to cyclopentene,[70] the equilibrium being shifted towards the formation of the cyclic mono-olefin.

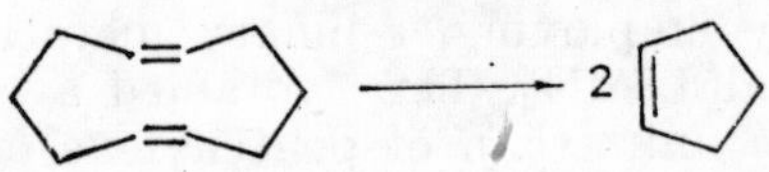

$$\longrightarrow \ 2 \tag{4.24}$$

Cyclododecene. Natta and co-workers [71] polymerized cyclododecene using tungsten- or molybdenum-based catalytic systems to obtain polydodecenamer.

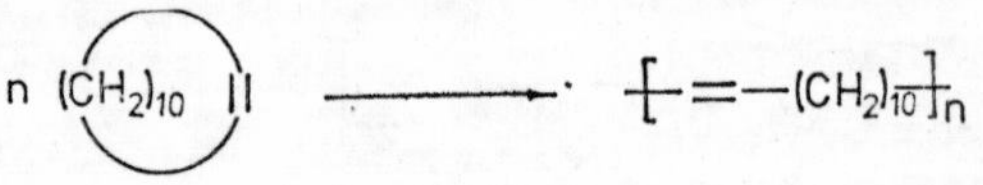

$$n \ (CH_2)_{10} \ || \longrightarrow \ \left[= -(CH_2)_{10} \right]_n \tag{4.25}$$

Catalytic systems consisting of WCl_6 or $MoCl_5$ and $(C_2H_5)_3Al$ or $(C_2H_5)_2AlCl$ are active, those of tungsten leading to the higher yields. The latter system gave a polydodecenamer with prevailing *trans* configuration at the double bond. The *trans*-polydodecenamer obtained was a crystalline substance with melting point 80°C. The characteristics of this polymer were situated between those of vinyl polymers (which have a saturated chain) and those of polymers with unsaturated chain. A *trans*-polydodecenamer with similar properties was also obtained by Calderon and co-workers [10] using the active catalytic systems for cyclo-octene polymerization, namely WCl_6 and $C_2H_5AlCl_2$.

In the presence of a catalytic system composed of WCl_6 and $(i-C_4H_9)_2AlCl$, Vardanyan *et al.*[72] polymerized cyclododecene to polydodecenamer with a variable content of *cis* and *trans* configurations. They also presented evidence for an equilibrium between the *cis* and *trans* forms in this monomer-polymer system in the presence of this catalyst.

1,5,9-Cyclododecatriene. The synthesis of 1,4-polybutadiene through more efficient routes, and the formation of a product with improved properties, revived interest in the polymerization of 1,5,9-cyclododecatriene (a trimer of butadiene or of cyclobutene) to polybutenamer, immediately after the disclosure of the catalytic systems for this type of reaction. In the presence of WCl_6 and $C_2H_5AlCl_2$, Calderon and co-workers [10] succeeded, for the first time, in obtaining polybutenamer from 1,5,9-cyclododecatriene, in high yields.

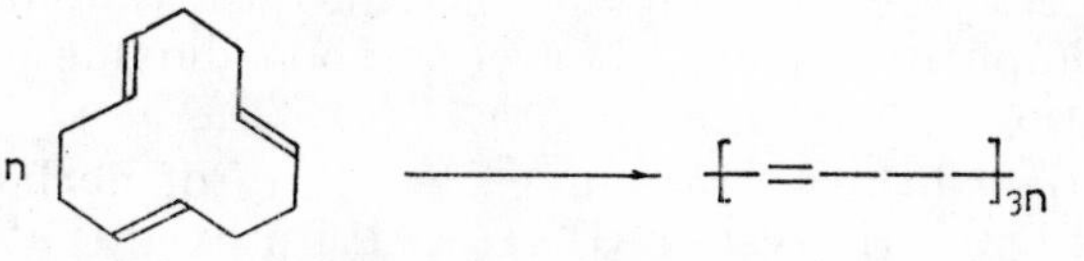

$$(4.26)$$

The reaction was carried out starting from a mixture of 40 : 60 *cis,trans,trans-* and *trans,trans,trans*-1,5,9-cyclododecatriene and resulted in 1,4-polybutadiene with a mixture of *cis* and *trans* configurations at the double bonds. It was noticed that a close correlation existed between the polymer configuration and the double bond configuration in the starting monomer. The properties of the polymer were very close to those of the polybutenamer obtained from 1,5-cyclo-octadiene.

1,9,17-Cyclotetra-eicosatriene. The ring-opening polymerization of 1,9,17-cyclotetra-eicosatriene is a new route for obtaining polyoctenamer.

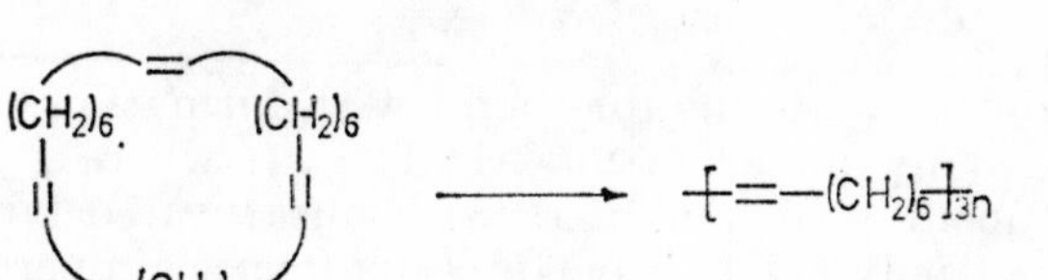

$$(4.27)$$

Ofstead and Calderon [73] carried out this reaction with a catalytic system based on WCl_6. After a short reaction time, they obtained a good conversion of 1,9,17-cyclotetra-eicosatriene, even better than that of cyclo-octene in the presence of the same catalyst.

4.2. Polymerization of Bicyclic Olefins

As already mentioned, the ring-opening polymerization was first carried out with an olefin belonging to this class, namely, bicyclo[2.2.1]heptene or norbornene. Later on, the reaction was applied to bicyclo[2.2.1]heptadiene (norbornadiene) and then extended to other bicyclic olefins containing cyclobutenic, cyclopentenic or cyclohexenic rings in their molecule.

Bicyclo[2.2.1]heptene (Norbornene). The polymerization of norbornene carried out by Anderson and Merckling [1] as early as 1955 in the presence of a Ziegler-Natta catalytic system formed from tetraheptyllithium-aluminium and titanium tetrachloride, led to a mixture of saturated and unsaturated polymers whose structure was not fully elucidated at that period. The above mentioned authors also employed (as cocatalysts, along with titanium tetrachloride) Grignard compounds, metal alkyl or aryl compounds, metal hydrides, and even alkali metals or earth alkaline metals. The activity and stability of the catalytic systems they employed were low and impurities such as water, carbon dioxide or oxygen led to rapid deactivation.

Truett and co-workers,[4] resuming the study of norbornene polymerization with the catalytic system $(C_7H_{15})_4AlLi$ proved that the polymerization in the presence of this catalyst takes place either at the double bond, as a vinylic polymerization, yielding saturated polymers, or via ring-opening yielding unsaturated polymers.

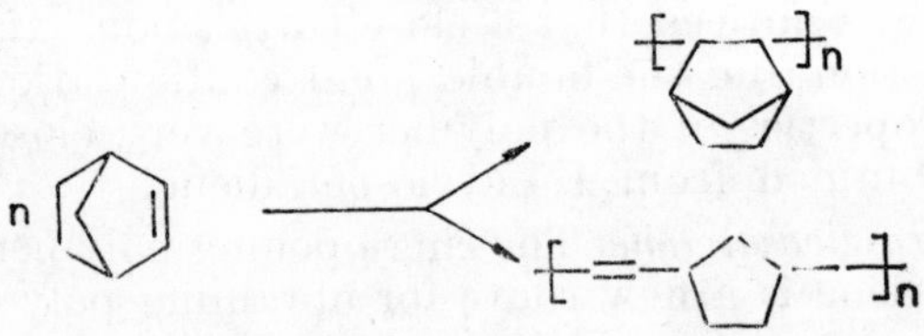

$$(4.28)$$

The polymerization at the double bond with formation of saturated polymers is favoured by molar ratios $Al : Ti < 1$, whereas ratios $Al : Ti > 1$ lead, via ring-opening, to polymers with unsaturated bonds in the chain. The unsaturated polymer had better elastomeric properties and a higher crystallinity.

Eleuterio,[3] in his studies on cyclo-olefin polymerization with hetero-geneous catalysts formed from chromium, molybdenum, tungsten or uranium oxides reduced with hydrogen and supported on alumina, titania or zirconia, demonstrated that norbornene can easily be polymerized by ring-opening with such systems. By employing γ-alumina-supported molyb-denum oxide reduced with hydrogen in combination with lithium-aluminium hydride, Eleuterio, obtained by ring-opening an unsaturated polymer with cis and trans configurations at the double bond.

Unsaturated polymers with cis configuration at the double bond were obtained by Sartori et al.[74] by using catalytic systems consisting of $MoCl_5$ and alkylaluminium compounds. However, under the same working conditions used by these authors, but with catalytic systems based on $TiCl_4$ and alkylaluminium compounds, only vinyl polymers were obtained.

The polymerization of norbornene was also reported with catalysts consisting of transition metal salts in polar solvents. Michelotti and Keaveney[75] carried out the polymerization of norbornene with $RuCl_3$, $OsCl_3$ and $IrCl_3$ hydrates in ethanol and obtained only ring-opened poly-mers. They found that the activity of these catalysts decreased in the order $Ir^{3+} > Os^{3+} > Ru^{3+}$. Although the catalytic systems were homo-geneous at the beginning of the reaction, they gradually turned hetero-geneous, suddenly modifying their activity and changing the reaction kinetics. Depending on the catalytic system, polymers with totally different physical properties were obtained working under otherwise identical con-ditions. The conversions of norbornene and the softening range of the polymers obtained with these catalysts are presented in Table 4.5.

TABLE 4.5. Polymerization of norbornene in polar media[a]

Catalyst	Reaction time	Conversion, %	Softening range, °C
Os^{3+}	4 hrs	55	58 — 90
Ru^{3+}	6.5 hrs	60	72 — 90
Ir^{3+}	7 min	73	90 — 115

[a] Data from ref. [75].

The polymer obtained with osmium catalysts had the lowest softening range and the highest content of cis configuration, whereas the polymer obtained with iridium catalysts had the highest softening range and content of trans configuration.

Similar catalytic systems of iridium and osmium salts were employed by Rinehart and Smith[76] for norbornene polymerization. The reaction products had a high degree of saturation but they did not possess a uni-form structure. In the presence of ruthenium complexes with phosphine, Hiraki and co-workers[77] obtained from norbornene both types of polymers, polyalkenamers and vinyl polymers.

The polymerization of norbornene to linear polyalkenamers with high molecular weight and uniform structure was carried out by Oshika *et al.*[78] with catalytic systems consisting of only one component, using as a reaction medium a variety of chlorinated solvents. Good conversions were obtained with molybdenum pentachloride in carbon tetrachloride, monochlorobenzene or *o*-dichlorobenzene, at normal temperature. In oxygenated compounds, such as dioxan or tetrahydrofuran or in saturated hydrocarbons such as n-heptane, no polymers were obtained with the same catalytic system. A low conversion was noticed, however, in toluene. The polymer obtained in these reactions had a *trans* configuration at the double bond. The authors noticed a gelification phenomenon, which has not yet been explained.

While extending their studies on norbornene polymerization to other catalytic systems, Oshika and Tabuchi [79] found that tungsten, molybdenum or rhenium halides in solvents such as carbon tetrachloride or carbon disulphide were particularly active. In most cases a high degree of stereospecificity was obtained. Molybdenum pentachloride yielded a polymer with a *trans* structure at the double bond, rhenium pentachloride a polymer with a *cis* structure, and tungsten hexachloride a polymer with both *cis* and *trans* configurations. On raising the temperature, the conversion and the yield of the reaction both increased. The addition, during the reaction, of tertiary amines, such as triethylamine or tributylamine, also raised the yield. However, traces of water caused a marked decrease in catalyst activity, but they did not modify the polymer structure. By elemental analysis, chlorine atoms were found in the polymer chain obtained with the molybdenum pentachloride catalyst. This fact favoured the initial interpretation of the reaction as a coordinative mechanism, with fission of the σ carbon-carbon bond in the α position relative to the double bond.

Substituted bicyclo[2.2.1]heptene. Dekking [80] noticed for the first time that norbornene substituted in position 5 with a butenyl group, yielded, in the presence of catalytic systems consisting of $TiCl_4$ and $LiAl(C_{10}H_{23})_4$, a mixture of vinylic polymers and polyalkenamers, the latter being formed by ring-opening polymerization. By spectroscopic analysis, it was noticed that the product contained 59% vinylic polymer and 41% polyalkenamer. The product had a softening temperature of 150°C and it contained no gel.

(4.29)

In their studies on the polymerization of norbornene, Michelotti and Keaveney [75] showed that the substitution of the saturated ring in position 5 influences the polymerizability of the system because when the substituent is strongly polar it forms a complex with the transition metal of the catalytic system. Less polar substituents lead to unsaturated polymers, wherein the substituent is maintained in its initial position in the cyclopentanic ring.

$$ \tag{4.30} $$

For instance, for $X = CN$ it was not possible to obtain norbornene polymers with catalytic systems such as $RuCl_3$, $OsCl_3$ or $IrCl_3$ in ethanol, but with the chloromethylene group as substituent an unsaturated polymer with low molecular weight was obtained in the presence of $IrCl_3$ as a catalyst.

Norbornene, substituted with a methyl group in position 5, in the presence of Ziegler-Natta catalytic systems formed from $TiCl_4$ and $(C_2H_5)_3Al$, led to unsaturated polymers by ring-opening. Takada, Otsu and Imoto [81] employed such a catalyst, with a molar ratio $Al : Ti = 2.5$ and obtained a polyalkenamer with the *trans* configuration at the double bond.

The polymerization of norbornene substituted with other polar groups was carried out by several authors employing binary and ternary catalytic systems which had exhibited good activity in cyclopentene polymerization. Hepworth [82] obtained, with high yields, polyalkenamers of norbornenyl acetal and 2,5-dimethylnorbornenyl carboxylate, with the ternary system WCl_6, C_2H_5OH and $(C_2H_5)_3Al_2Cl_3$. Ueshima *et al.*[83,84] and Nakamura *et al.*[85-88] obtained polymers and copolymers of norbornene derivatives, especially with cyano groups in position 5, with desirable qualities such as good mechanical strength, low gas permeability, etc. More recently, Kobayashi *et al.*[89-95] carried out the polymerization of norbornene containing a large number of substituents such as cyano, ester, ether, amide, imide, halogen or anhydride groups. They used a large range of tungsten, molybdenum, rhenium, tantalum or niobium catalysts with organic derivatives of metals from the groups Ia—IVa or IIb—IVb [89] or with inorganic compounds of these metals, such as phosphates or phosphites.[93] Kurosawa *et al.*[96-98] employed tungsten or molybdenum catalytic systems, organometallic compounds, and epoxides, peroxides, acetals, esters, organic halides or water to polymerize derivatives of norbornene such as 5-(2-pyridyl)-bicyclo[2.2.1]heptene or the *exo*-dicarboxylic anhydride. Kanega *et al.*[99] obtained good yields of thermoplastic resins from 5-cyanonorbornene, by employing WCl_6, $(C_2H_5)_2AlCl$ and triphenylphosphine, while Imaizumi *et al.*[100] obtained fireproof resins from the tribromophenylester of 2-norbornene-5-carboxylic acid with WCl_6 and $(C_2H_5)_3Al$. Fireproof resins with good properties were also obtained by Arai *et al.*[101] from norbornene derivatives having cyano, ester, ether, halogen or imide groups and using ternary systems consisting of WCl_6, $(C_2H_5)_2AlCl$ and oxygenated compounds such as 1,1-diethoxyethane. Catalysts with very high activity in the polymerization of norbornene with polar substituents were obtained by Ikeda and co-workers [102] on employing other transition metal salts such as vanadium, chromium or manganese, in addition to tungsten, molybdenum, rhenium or tantalum salts. They obtained $80-90\%$ yields of norbornene ester polymers by using chromium acetylacetonate and triethylaluminium.

Bicyclo[2.2.1]hepta-2,5-diene (Norbornadiene). The polymerization of norbornadiene with vanadium- or titanium-based catalysts, enabled Valvassori and Faina [103] to obtain unsaturated polymers via ring-opening.

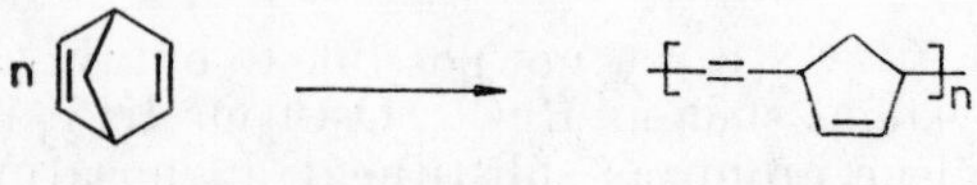

$$(4.31)$$

The polymer structure, analysed by IR spectroscopy proved to contain mostly the *trans* configuration at the double bond.

Calderon [104] employed catalysts consisting of WCl_6 with or without organometallic compounds as cocatalysts, which proved particularly active for norbornadiene polymerization. The structure of the polymers was not determined because the significant amount of gel formed on the polymer particles hindered the analysis of the reaction products. In another series of reactions, Schulz [105] employed several Ziegler-Natta catalytic systems. In most cases, both types of polymers were obtained, namely unsaturated ring-opened polymers and saturated ones by vinylic polymerization.

Bicyclo[3.2.0]hepta-2,6-diene. The polymerization of bicyclo[3.2.0]hepta-2,6-diene was carried out by Dall'Asta and Motroni [106] in the presence of Ziegler-Natta catalysts based on transition metals or in the presence of transition metal salts in polar solvents. By employing vanadium-based systems, the above authors noticed the formation of polyalkenamers together with vinylic polymers by opening the cyclobutene ring.

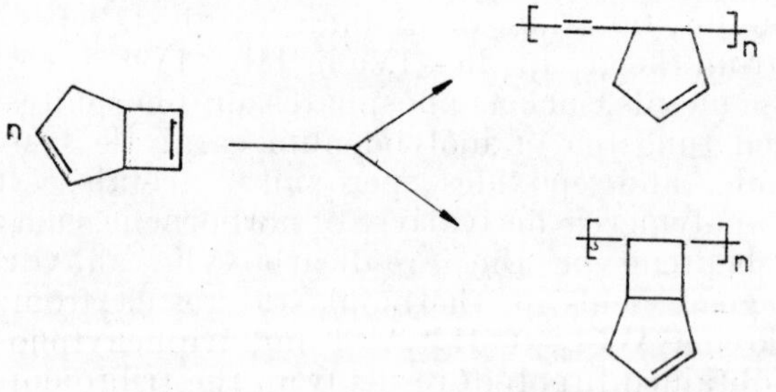

$$(4.32)$$

However, titanium-based catalysts led to polymerization by opening the cyclobutene ring. In both instances, the cyclopentene ring was not affected :

this ring is preserved in the polymer chain. The same authors attempted the polymerization of bicyclo[3.2.0]hepta-2,6-diene with $RuCl_3$ in ethanol. The resulting polymer contained transannular bonds, whose structure was not completely determined.

Bicyclo[4.2.0]oct-7-ene. The polymerization of bicyclo[4.2.0]oct-7-ene was carried out by Dall'Asta and Motroni [106] with catalytic systems based on transition metals, in hydrocarbons or polar solvents as reaction medium. According to the nature of the catalytic system, vinylic polymers, polyalkenamers or mixtures of these polymers were obtained.

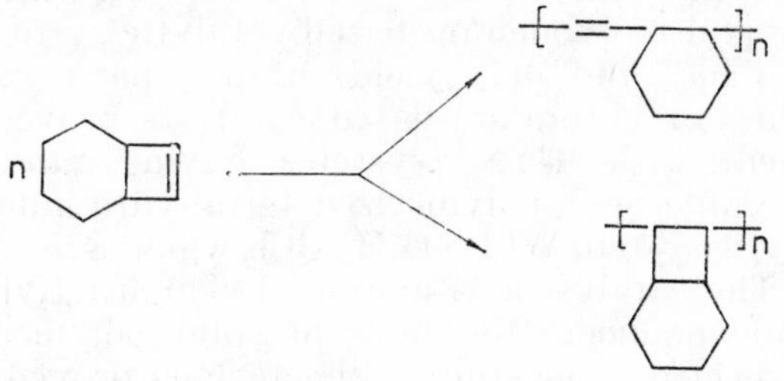

$$(4.33)$$

The reaction was carried out in the presence of Ziegler-Natta systems formed from halides or acetylacetonates of titanium, vanadium, chromium, tungsten or of unicomponent systems based on nickel, rhodium, ruthenium, iridium and palladium salts in water, ethanol, or dimethylsulphoxide as solvent. The authors noticed that the catalytic systems containing vanadium, chromium, nickel, rhodium and palladium dominantly form vinylic polymers, those with titanium, ruthenium and iridium open the cyclobutenic ring forming unsaturated polymers, whereas tungsten leads to both types of polymers. In all these cases, the molecular weight of the polymers was low.

Bicyclo[2.2.2]oct-2-ene. Ofstead [107] polymerized bicyclo[2.2.2]oct-2-ene by ring-opening with a catalytic system composed of WCl_6, C_2H_5OH, $C_2H_5AlCl_2$, obtaining the corresponding polyalkenamer.

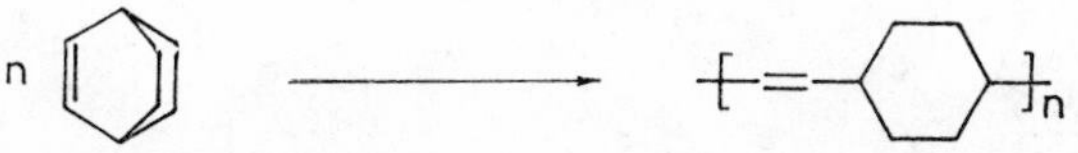

$$(4.34)$$

The above author employed a catalyst with a molar composition $W : O : Al = 1 : 1 : 4$ and obtained a polymer with *trans* configuration at the double bond.

This reaction represents one of the few examples mentioned in the literature where the cyclohexene ring undergoes ring-opening polymerization. Ofstead made several thermodynamic considerations concerning the energetics of the six-membered ring-opening polymerization in this system. The polymerizability of this monomer is mainly attributed to the conformational effects occurring in the six-membered ring of the bridge system of bicyclo[2.2.2]oct-2-ene : all the three cyclohexene or cyclohexane rings have boat conformations.

Bicyclo[4.3.0]nona-3,7-diene (Tetrahydro-indene).   Dall'Asta and Motroni [108] polymerized tetrahydro-indene in the presence of Ziegler-Natta catalytic systems to polyalkenamers.

$$n \quad \text{[structure]} \quad \longrightarrow \quad \text{[structure]}_n \qquad (4.35)$$

The resulting polymer mostly had the *trans* configuration at the double bond. Tungsten- and molybdenum-based catalytic systems were the most active, leading to 40% and 50% yields of polymers, respectively. Titanium-, vanadium- or chromium-based catalysts proved less active, the polymers formed with these systems having random structure.

Ofstead and Calderon [73] polymerized tetrahydro-indene by employing the binary catalytic system $WCl_6 \cdot C_2H_5AlCl_2$ which is active in cyclo-octene polymerization. The catalyst also exhibited a high activity in the reaction of tetrahydro-indene, moderate yields of solid polymer being obtained. On the basis of spectroscopic studies, the authors proved that the reaction occurred by exclusive ring-opening of cyclopentene ring, the six-membered ring being maintained in the polymer chain.

4.3. Polymerization of Polycyclic Olefins

Polycyclic olefins have been reported to undergo polymerization reactions in a restricted number of cases, namely tricyclic olefins such as di(cyclopentadiene) and trimethylene-norbornene. Attempts to polymerize other polycyclic systems failed till now.

Di(cyclopentadiene). Di(cyclopentadiene) polymerization may yield four types of polymers, namely two by cyclopentene ring-opening (A and B) and two by splitting the double bonds of these cycles affording vinylic polymers (C and D).

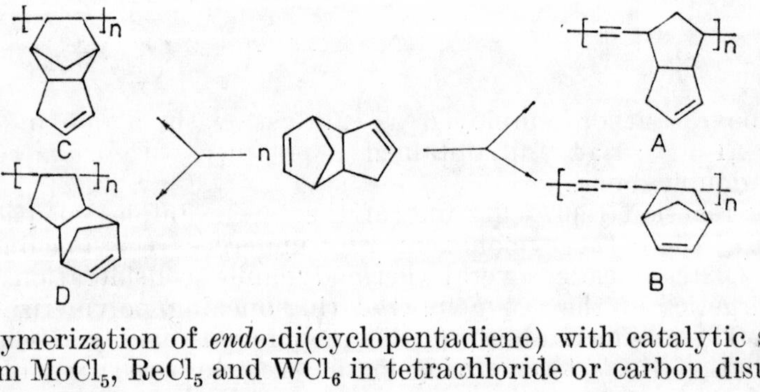

$$\text{[reaction scheme showing polymers C, D, A, B]} \qquad (4.36)$$

The polymerization of *endo*-di(cyclopentadiene) with catalytic systems formed from $MoCl_5$, $ReCl_5$ and WCl_6 in tetrachloride or carbon disulphide, carried out by Oshika and Tabuchi [79] yielded polyalkenamers of type A, with high stereospecificity : a *trans*-polymer was obtained with $MoCl_5$ catalysts, a *cis*-polymer with $ReCl_5$ and both configurations with WCl_6. In this reaction, the catalytic systems proved to be less active than in the similar reaction of norbornene.

The polymerization of di(cyclopentadiene) was carried out also by Dall'Asta and co-workers [109] with Ziegler-Natta catalytic systems composed of chromium, molybdenum or tungsten halides and organoaluminium compounds. In most cases, they obtained mixtures of type A and C polymers. The catalysts formed only from the above mentioned transition metal salts had a weak activity in this reaction.

Tricyclo[5.2.1.0]dec-8-ene (Dihydrodi(cyclopentadiene), Trimethylenenorbornene). Dihydrodi(cyclopentadiene) was polymerized, for the first time, by Eleuterio [3] who employed an alumina-supported molybdenum oxide heterogeneous catalytic system. Polymers with polyalkenamer structure were obtained, but the yield was low.

In their studies on polycyclic olefin polymerization with catalytic systems formed form $MoCl_5$, $ReCl_5$ or WCl_6, Oshika and Tabuchi [79] also carried out the polymerization of dihydrodi(cyclopentadiene). The polymer structure showed that the reaction occurred by ring-opening of the norbornenic system.

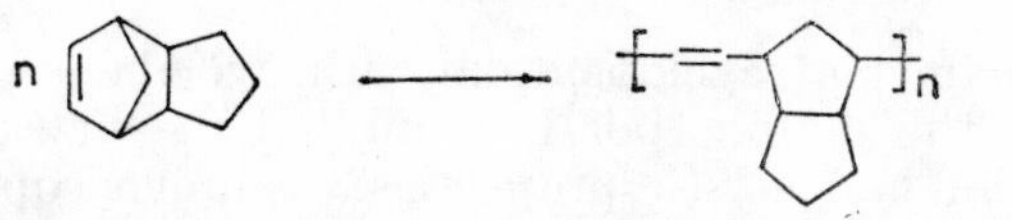

$$(4.37)$$

In the presence of these catalysts, polyalkenamers with high stereospecificity were obtained in the same way as in the reaction of norbornene. The configuration at the double bond depended on the catalytic system : $MoCl_5$ led to *trans* polyalkenamer, $ReCl_5$ to *cis* polyalkenamer and WCl_6 to a polyalkenamer having both configurations. The catalysts proved to be particularly active in the reaction of dihydrodi(cyclopentadiene), even more active than towards norbornene. In the series of polycyclic olefins, the authors established the following order of reactivity : *exo*-trimethylene-norbornene ≫ norbornene > *endo*-di(cyclopentadiene).

4.4. Copolyalkenamers from Cyclic Olefins

The copolymerization of cyclic olefins in the presence of ring-opening polymerization catalytic systems yields various copolyalkenamers, depending on the cyclic olefins employed.

$$(4.38)$$

A non-random distribution of the structural units derived from the two monomers can result only when they have closely similar reactivities. Since differences always exist between reactivities of cyclic olefins due to the nature of the ring, to its size and to the substituents it contains, it is practically impossible to obtain copolyalkenamers with a uniformly alternating distribution of units. The structure of the copolyalkenamers is also influenced by the nature of the polymerization catalyst. In many instances, these catalytic systems exhibit differences in their activity towards various cyclic olefins, regardless of the reactivity of the latter.

Cyclo-olefin copolymerization with WCl_6-based catalytic systems were studied by several research groups. Dall'Asta and co-workers [110] carried out copolymerization of mixtures of monomers such as cyclopentene, cycloheptene, cyclo-octene and cyclodecene, employing catalytic systems formed from WCl_6 or $WOCl_4$, $(C_2H_5)_2AlCl$ and benzoyl or cumyl peroxide. The reaction of cyclopentene with cycloheptene in the presence of $WCl_6(C_2H_5)_2AlCl$ and benzoyl peroxide yielded a copolyalkenamer containing 80% structural units of pentenamer in the chain. [111]

$$(CH_2)_3 \; \| \; + \; \| \; (CH_2)_5 \longrightarrow \left[= - (CH_2)_3 - = - (CH_2)_5 \right] \qquad (4.39)$$

This copolyalkenamer of cyclopentene with cycloheptene had 75% *trans* configuration at the double bond, exhibited a remarkable crystallinity on stretching, and had elastic properties. A similar copolyalkenamer was obtained by Dall'Asta and co-workers [111] from cyclopentene and cyclo-octene.

$$(CH_2)_3 \; \| \; + \; \| \; (CH_2)_6 \longrightarrow \left[= - (CH_2)_3 - = - (CH_2)_6 \right] \qquad (4.40)$$

Scott and Calderon [112] obtained copolyalkenamers of cyclo-octene and cyclododecene using catalytic systems formed from WCl_6, C_2H_5OH and $C_2H_5AlCl_2$. In fact, such polymers resulted from the initial homopolymerizations of cyclo-octene and cyclododecene separately, followed by metathesis of the resulting polyalkenamers in the presence of the same catalytic systems.[113] The above authors extended this method to the preparation of a large number of graft, block or tridimensional polymers.

To obtain copolymers with elastomeric properties, Pampus and co-workers from Farbenfabriken Bayer copolymerized monocyclic olefins with nonconjugated cyclic polyenes in the presence of various catalytic systems based on tungsten or molybdenum.[114,115] Thus, cyclopentene copolymerized with norbornene yielding a copolyalkenamer with rubber-like properties. A polyalkenamer elastomer exhibiting reduced cold flow was prepared by Donald et al.[116] by copolymerizing cyclopentene with cyclopentadiene employing a catalytic system prepared from trichloromolyb-denum dilaurate and $(C_2H_5)_2AlCl$, working in toluene at 5°C. The polymer had 89 % *trans* configuration at the double bond. Uraneck and Burgleigh [117] obtained elastomers with interesting properties by copolymerizing cyclo-pentene with di(cyclopentadiene) or 5-vinyl-2-norbornene in the presence of certain acyclic olefins such as hexene. The product obtained had good processing properties.

Copolymers of substituted polycyclo-olefins have been prepared in large number lately, because of their valuable plastic properties. Tanaka and co-workers [118] obtained copolyalkenamers from norbornene derivatives bearing various substituents e.g. cyano or phenyl groups. They obtained an unsaturated polymer with cyano and phenyl groups in the chain starting from 5-cyano-2-norbornene and 5-phenyl-2-norbornene in the presence of WCl_6 and $(C_2H_5)_2AlCl$

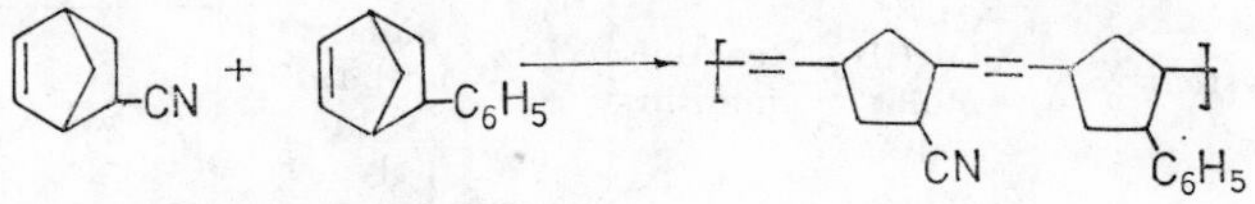

$$\text{(structure)} \quad (4.41)$$

The same research group also copolymerized norbornene derivatives with linear olefins, for instance 5-phenyl-2-norbornene with 1-hexene, obtaining copolyalkenamers with special properties.[119] Komatsu *et al.*[120] and Kothani *et al.*[121] also prepared copolyalkenamers from norbornene derivatives bearing various substituents such as cyano, ester, amide, imide, halogen or anhydride groups.

The copolymerization of cyclic olefins was successfully applied by Dall'Asta and co-workers [122] to test the four-centred mechanism for the ring-opening polymerization. By employing cyclopentene, isotopically labelled at one carbon atom of the double bond, copolymerization with cyclo-octene proved that the ring-opening occurred by splitting the double bond.

$$\text{(structure)} \quad (4.42)$$

The reaction was carried out with a catalytic system composed of $WOCl_4$, $(C_2H_5)_2AlCl$ and benzoyl peroxide. The radio-gas-chromatographic analyses of the acetylated diols obtained through polymer degradation showed that the whole radioactivity is present in pentanediol, while octanediol is non-radioactive.

For the same purpose, Dall'Asta and co-workers [123] carried out the copolymerization of cyclobutene, isotopically labelled at the double bond, with 3-methylcyclobutene, employing catalytic systems based on tungsten, molybdenum, titanium or ruthenium. They obtained a copolyalkenamer of cyclobutene and 3-methylcyclobutene which had the methyl groups in their initial position.

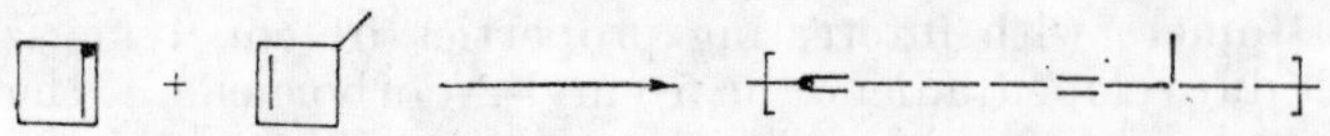

$$(4.43)$$

The yields and structure of the resulting copolyalkenamers differed, according to the catalytic system employed. Furthermore, the ratio between the amounts of the butenamer structural unit and the 3-methyl-butenamer unit was different, depending on the catalytic system.

TABLE 4.6. Copolymerization of cyclobutene (CB) with 3-methylcyclobutene (3MeCB) with transition metal catalytic systems[a]

Catalytic system	CB : 3MeCB Molar ratio	Reaction time, hrs	Yield, %	CB : 3MeCB unit ratio in polymer	Polymer configuration		
					trans	cis	saturated
$WCl_6 : (C_2H_5)_3Al(1:3)$	30 : 70	24	38	50 : 50	25	60	15
$MoO_2A_2^b : (C_2H_5)_2AlCl(1:5)$	35 : 65	15	47	40 : 60	25	70	5
$TiCl_4 : (C_2H_5)_3Al(1:2)$	26 : 74	2	100	25 : 60	55	30	15
$RuCl_3 \times H_2O : C_2H_5OH$	35 : 65	100	20	45 : 55	80	traces	20

[a] Data from ref. [123];   [b] A = acetylacetonate group.

The $TiCl_4 \cdot (C_2H_5)_3Al$ system proved to be the most active since after a reaction time of two hours it afforded a 100% yield of polymer. The ratio between the structural units butenamer : 3-methylbutenamer was 1 : 1 only for the tungsten-based catalyst, whereas for the other catalysts the ratio was in favour of 3-methylcyclobutene. Tungsten and molybdenum catalytic systems predominantly led to copolyalkenamers with the *cis* configuration at the double bond and those of titanium and ruthenium to the *trans* configuration.

References

1. A. W. Anderson and N. C. Merckling, **U.S.** 2.721.189, 18.10.1955 ; *Chem. Abstr.*, **50**, 30081 (1956).
2. H. S. Eleuterio, **U.S.** 3.074.918, 20.06.1957.
3. H. S. Eleuterio, **Ger.Offen.** 1,072,811, 7.01.1960.
4. W. L. Truett *et. al.*, *J. Amer. Chem. Soc.*, **82**, 2337 (1960).
5. G. Dall'Asta, G. Mazzanti, G. Natta and L. Porri, *Makromol. Chem.*, **56**, 224 (1962).
6. G. Natta, G. Dall'Asta, G. Mazzanti and G. Motroni, *Makromol. Chem.*, **69**, 163 (1963).
7. G. Natta, G. Dall'Asta and G. Mazzanti, *Angew. Chem.*, **76**, 765 (1964).
8. G. Natta, G. Dall'Asta, I. W. Bassi and G. Carella, *Makromol. Chem.*, **91**, 87 (1966).

9. N. Calderon, H. Y. Chen and K. W. Scott, *Tetrahedron Letters*, **1967**, 3327.

10. N. Calderon, E. A. Ofstead and W. A. Judy, *J. Polymer Sci.*, *A—1*, **5**, 2209 (1967).

11. P. Günther *et. al.*, *Angew. Chem.*, **16/17**, 27 (1971).

12. K. W. Scott, *Polymer Preprints*, **13**, 874 (1972).

13. N. Calderon, *J. Macromol. Sci.*, *Revs. Macromol. Chem.*, **C7**, 105 (1972).

14. a) G. Dall'Asta, *Rubber Chem. Technol.*, **47**, 511 (1974); b) B. A. Dolgoplosk, *Uspekhi Khim.*, **46**, 2027 (1977); c) T. J. Katz, T. J. Lee and M. A. Shippey, *J. Mol. Catalysis*, **8**, 219 (1980).

15. V. Drăguţan, *Studii Cercetări Chim.*, **22**, 293 (1974).

16. L. Porri, G. Natta and N. C. Gallazzi, *Chimica e Industria*, **46**, 428 (1964).

17. G. Natta, G. Dall'Asta and G. Motroni, *J. Polymer Sci.*, *B*, **2**, 349 (1964).

18. G. Natta, G. Dall'Asta and L. Porri, *Makromol. Chem.*, **81**, 253 (1964).

19. G. Dall'Asta and R. Manetti, *Atti Acad. naz. Lincei, Rend., Classe Sci. fis. mat. nat.*, **41**, 351 (1966).

20. T. J. Katz, J. McGinnis and C. Altus, *J. Amer. Chem. Soc.*, **98**, 606 (1966).

21. G. Dall'Asta, *J. Polymer Sci.*, *A—1*, **6**, 2397 (1968).

22. a) G. Dall'Asta and P. Scaglione, *Rubber Chem. Technol.*, **42**, 1235 (1969); b) G. Natta, G. Dall'Asta and G. Mazzanti, **Fr. Pat.** 1,394,380, 2.04.1965; *Chem. Abstr.*, **68**, 3656h (1968).

23. G. Natta *et al.*, *Makromol. Chem.*, **54**, 95 (1962).

24. G. Dall'Asta and G. Mazzanti, *Makromol. Chem.*, **61**, 178 (1963).

25. G. Natta *et. al.*, **Belg.** 619,877, 31.10.1962; *Chem. Abstr.*, **58**, 9250 (1963).

26. G. Natta *et al.*, **Fr.** 1,392,142, 12.03.1965; *Chem. Abstr.*, **63**, 13540d (1965).

27. K. Nützel, K. Dinges and F. Haas, **Ger.Offen.** 1,720,797, 7.03.1968; *Chem. Abstr.*, **72**, 79656 (1970).

28. P. Günther, W. Oberkirch and G. Pampus, **Fr.** 2,065,300, 9.10.1969; *Chem. Abstr.*, **75**, 21485 (1971).

29. P. Günther, W. Oberkirch and G. Pampus, **Fr.** 2,003,536, 7.11.1969; *Chem. Abstr.*, **72**, 79654j (1970).

30. F. Haas, K. Nützel, G. Pampus and D. Theisen, *Rubber Chem. Technol.*, **43**, 1126 (1970).

31. K. Nützel, F. Haas, K. Dinges and W. Graulich, **Ger.Offen.** 1,720,971, 24.02.1968; *Chem. Abstr.*, **72**, 913261 (1970).

32. G. Dall'Asta, **Ger.Offen.** 2,161,878, 6.07.1972; *Chem. Abstr.*, **77**, 127242 (1972).

33. P. Günther, W. Oberkirch and G. Pampus, **Ger.Offen.** 1,812,328, 18.06.1972; *Chem. Abstr.*, **73**, 56973 (1970).

34. W. Oberkirch, P. Günther and G. Pampus, **Ger. Offen.** 1,957,025, 19.05.1971; *Chem. Abstr.*, **75**, 64648 (1971).

35. G. Dall'Asta and G. Carella, **Brit.** 1,062,367, 11.02.1965; *Chem. Abstr.*, **66**, 19005q (1967).

36. G. Pampus and J. Witte, **Ger.Offen.** 1,770,143, 6.04.1968; *Chem. Abstr.*, **72**, 133851 (1970).

37. G. Pampus, J. Witte and M. Hoffmann, **Ger.Offen.** 1,954,092, 28.03.1969; *Chem. Abstr.*, **75**, 37611 (1971).

38. G. Pampus, N. Schön, W. Oberkirch and P. Günther, **Ger.Offen.** 2,009,740, 30.09.1971; *Chem. Abstr.*, **76**, 47144 (1972).

39. G. Pampus, J. Witte and M. Hoffmann, **Ger.Offen.** 2,016,840, 21.10.1971; *Chem. Abstr.*, **76**, 72998 (1972).

40. G. Pampus, J. Witte and M. Hoffmann, **Ger. Offen.** 2,016,471, 28.10.1971; *Chem. Abstr.*, **76**, 60711 (1972).

41. G. Pampus, G. Lehnert and J. Witte, **Ger.Offen.** 2,101,684, 20.06.1972; *Chem. Abstr.*, **77**, 153549 (1972).

42. K. Nützel, F. Haas and G. Marwede, **Ger.Offen.** 1,919,047, 15.02.1969; *Chem. Abstr.*, **74**, 32530s (1971).

43. K. Nützel, F. Haas and G. Marwede, **Ger.Offen.** 1,919,048, 15.03.1969; *Chem. Abstr.*, **74**, 64592c (1971).

44. K. Nützel and F. Haas, **Fr.** 2.008.180, 9.05.1968; *Chem. Abstr.*, **73**, 4623 (1970).

45. J. Witte, G. Pampus, N. Schön and G. Marwede, **Ger.Offen.** 1,957,026, 19.05.1971; *Chem. Abstr.*, **75**, 65075 (1971).

46. N. Schön, G. Pampus, J. Witte and D. Theisen, **Ger.Offen.** 2,006,776, 19.08.1971; *Chem. Abstr.*, **76**, 46705f (1972).

47. N. Schön, G. Pampus and J. Witte, **Ger.Offen.** 1,770,844, 10.07.1968 ; *Chem. Abstr.*, **73**, 110697 (1970).
48. W. Oberkirch, P. Günther and G. Pampus, **U.S.** 3,787,381, 22.01.1974 ; *Chem. Abstr.*, **81**, 38655 (1974).
49. W. Oberkirch, P. Günther and G. Pampus, **Ger.Offen.** 1,811,653, 17.09.1970 ; *Chem. Abstr.*, **73**, 110694 (1970).
50. R. Streck and H. Weber, **Ger.Offen.** 2,028,716, 16.12.1971 ; *Chem. Abstr.*, **76**, 113903 (1972).
51. R. Streck and H. Weber, **Ger.Offen.** 2,027,905, 20.01.1972 ; *Chem. Abstr.*, **76**, 127817 (1972).
52. R. Streck and H. Weber, **Ger. Offen.** 2.028,935, 16.12.1971 ; *Chem. Abstr.*, **76**, 113891 (1972).
53. C. A. Uraneck and W. J. Trepka, **Fr.** 1,542,040, 11.10.1968 ; *Chem. Abstr.*, **71**, 22762 (1969).
54. J. E. Burgleigh and C. A. Uraneck, **U.S.** 3,640,986, 8.02.1972 ; *Chem. Abstr.*, **76**, 141537 (1972).
55. A. J. Amass, T. A. McGourtey and C. N. Tuck, *European Polymer J.*, **12**, 93 (1976).
56. Uniroyal Inc., **Brit.** 1,131,160, 23.10.1968 ; *Chem. Abstr.*, **70**, 20488 (1969).
57. G. Natta, G. Dall'Asta and G. Mazzanti, **Fr.** 1,425,601, 21.01.1966 ; *Chem. Abstr.*, **65**, 10692 (1966).
58. G. Natta, G. Dall'Asta and G. Mazzanti, **Belg.** 667,392, 15.11.1965 ; *Chem. Abstr.*, **65**, 3996e (1966).
59. K. W. Scott *et al.*, *Adv. Chem. Ser.*, **91**, 399 (1969).
60. a) N. Calderon and M. C. Morris, *J. Polymer Sci.*, *A—2*, **5**, 1283 (1967) ; b) N. Calderon and W. A. Judy, **Brit.** 1,124,416, 21.08.1968 ; *Chem. Abstr.*, **69**, 77970 (1968).
61. C. P. C. Bradshaw, **Brit.** 1,252,799, 10.11.1971 ; *Chem. Abstr.*, **76**, 60726 (1972).
62. P. R. Marshall and B. J. Ridgewell, *European Polymer J.*, **5**, 29 (1969).
63. F. W. Kuepper, R. Streck and K. Hummel, **Ger.Offen.** 2,051,798, 24.04.1972 ; *Chem. Abstr.*, **77**, 62529 (1972).
64. F. W. Kuepper, R. Streck, H. T. Heims and H. Weber, **Ger.Offen.** 2,051,799, 27.04.1972 ; *Chem. Abstr.*, **77**, 62528 (1972).
65. N. Calderon, **U.S.** 3,932,373, 13.01.1976 ; *Chem. Abstr.*, **84**, 106843 (1976).
66. N. Calderon, **U.S.** 3,597,406, 3.08.1971 ; *Chem. Abstr.*, **75**, 141573 (1971).
67. K. W. Scott *et al.*, *Rubber Chem. Technol.*, **44**, 1341 (1971).
68. R. Rossi and R. Giorgi, *Tetrahedron Letters*, **1975**, 1721.
69. G. Dall'Asta and R. Manetti, *European Polymer J.*, **4**, 154 (1968).
70. F. W. Kuepper and R. Streck, *Makromol. Chem.*, **175**, 2055 (1974).
71. G. Natta, G. Dall'Asta and G. Mazzanti, **Ital.** 733,857, 7.01.1964 ; *Chem. Abstr.*, **65**, 3996c (1966).
72. L. M. Vardanyan *et al.*, *Doklady Akad. Nauk S.S.S.R.*, **207**, 345 (1972).
73. E. A. Ofstead and N. Calderon, *Makromol. Chem.*, **154**, 21 (1972).
74. G. Sartori, F. Ciampelli and W. Camelli, *Chimica e Industria*, **45**, 1473 (1962).
75. F. W. Michelotti and W. P. Keaveney, *J. Polymer Sci.*, *A*, **3**, 895 (1965).
76. R. E. Rinehart and H. P. Smith, *J. Polymer Sci.*, *B*, **3**, 1049 (1965).
77. K. Hiraki, A. Kuroiwa and H. Hirai, *J. Polymer Sci.*, *A—1*, **9**, 2323 (1971).
78. T. Oshika, S. Murai and Y. Koga, *Chem. High Polymer Japan*, **22**, 211 (1968).
79. T. Oshika and H. Tabuchi, *Bull. Chem. Soc. Japan*, **41**, 211 (1968).
80. H. G. G. Dekking, *J. Polymer Sci.*, **55**, 525 (1961).
81. J. Takada, T. Otsu and M. Imoto, *Kogyo Kogaku Zasshi*, **69**, 715 (1966).
82. P. Hepworth, **Ger.Offen.** 2,231,955, 18.01.1973 ; *Chem. Abstr.*, **78**, 98287 (1973).
83. T. Ueshima, S. Kobayashi and M. Matsuoka, **Ger.Offen.** 2,316,087, 4.10. 1973 ; *Chem. Abstr.*, **80**, 83932 (1974).
84. T. Ueshima, Y. Tanaka and S. Kobayashi, **Japan. Kokai,** 75 128,799, 11.10.1975 ; *Chem. Abstr.*, **84**, 90770 (1976).
85. J. Nakamura et al., **Japan.Kokai,** 76 107,351, 19.03.1975 ; *Chem. Abstr.*, **86**, 17577 (1977).
86. J. Nakamura et al., **Japan.Kokai,** 75 142,661, 15.11.1975 ; *Chem. Abstr.*, **84**, 90959 (1976).
87. J. Nakamura et al., **Japan.Kokai,** 76 112,866, 29.03.1976 ; *Chem. Abstr.*, **86**, 56162 (1977).
88. J. Nakamura et al., **Japan.Kokai,** 76 12,865, 5.10.1976 ; *Chem. Abstr.*, **86**, 90838 (1977).
89. Y. Kobayashi, T. Ueshima and S. Kobayashi, **Japan. Kokai,** 77 36, 200, 19.03.1977 ; *Chem. Abstr.*, **87**, 53856 (1977).
90. Y. Kobayashi, T. Ueshima and S. Kobayashi, **Japan.Kokai,** 77 26,599, 28.02.1977 ; *Chem. Abstr*, **87**, 40076 (1977).

91. Y. Kobayashi, T. Ueshima and S. Kobayashi, Japan. Kokai, 77 42,698, 02.04.1977; Chem. Abstr., 87, 53860 (1977).
92. Y. Kobayashi, T. Ueshima and S. Kobayashi, Japan. Kokai, 77 45,698, 11.04.1977; Chem. Abstr., 87, 53875 (1977).
93. Y. Kobayashi, T. Ueshima and S. Kobayashi, Japan. Kokai, 77 44,899, 8.04.1977; Chem. Abstr., 87, 53857 (1977).
94. Y. Kobayashi, T. Ueshima and S. Kobayashi, Japan. Kokai, 77 37,998, 24.03.1977; Chem. Abstr., 87, 40075 (1977).
95. Y. Kobayashi et al., Ger.Offen. 2,540,553, 17.03.1977; Chem. Abstr., 86, 156245 (1977).
96. S. Kurosawa et al., Ger.Offen. 2,529,963, 15.01.1976; Chem. Abstr., 84, 122565 (1976).
97. S. Kurosawa et al., Japan.Kokai, 75 61,500, 27.05.1975; Chem. Abstr., 83, 179939 (1975).
98. S. Kurosawa et al., Japan. Kokai, 75 110,000, 29.08.1975; Chem. Abstr., 84, 31737 (1976).
99. F. Kanega et al., Japan. Kokai, 76 111,300, 1.10.1976; Chem. Abstr., 86, 44222 (1977).
100. F. Imaizumi, K. Suzuki and M. Kawakami, Japan. Kokai, 76 136,800, 26.11.1976; Chem. Abstr., 87, 23950 (1977).
101. F. Arai et al., Japan. Kokai, 76 101,060, 7.09.1976.
102. H. Ikeda, S. Matsumoto and K. Makiuo, Ger.Offen. 2,447,687, 17.04.1975; Chem. Abstr., 83, 44046 (1975).
103. G. Sartori, A. Valvassori and S. Faina, Atti Acad. naz. Lincei, Rend., Classe Sci. fis. mat. nat., 34, 565 (1963).
104. N. Calderon, cited after reference 13.
105. R. G. Schulz, J. Polymer Sci., B, 4, 541 (1966).
106. G. Dall'Asta and G. Motroni, J. Polymer Sci., A—1, 6, 2405 (1968).
107. E. A. Ofstead, cited after reference 73.
108. G. Dall'Asta and G. Motroni, Chimica e Industria, 50, 972 (1968).
109. G. Dall'Asta, G. Motroni, R. Manetti and C. Tossi, Makromol. Chem., 130, 153 (1969).
110. G. Dall'Asta and G. Motroni, Ger.Offen. 2,237,353, 15.02.1973; Chem. Abstr., 79, 6517b (1973).
111. G. Dall'Asta, G. Motroni and G. Carella, Fr. 1,545,643, 15.11.1968; Chem. Abstr., 71, 13733 (1969).
112. K. W. Scott and N. Calderon, Ger.Offen. 2,058,198, 9.07.1971; Chem. Abstr., 75, 118801 (1971).
113. K. W. Scott and N. Calderon, U.S. 4,010,224, 1.03.1977; Chem. Abstr., 86, 140943 (1977).
114. G. Pampus, N. Schön, J. Witte and G. Marwede, Ger.Offen. 1,961,865, 10.12.1969; Chem. Abstr., 75, 77954 (1971).
115. J. Witte, N. Schön and G. Pampus, Fr. 2,014,139, 17.04.1970; Chem. Abstr., 74, 4463 (1961).
116. J. D. Donald, C. A. Uraneck and J. E. Burgleigh, Ger.Offen. 2,137,524, 3.02.1972; Chem. Abstr., 76, 142106 (1972).
117. C. A. Uraneck and J. A. Burgleigh, U.S. 3,681,300, 1.08.1972; Chem. Abstr., 77, 127236 (1972).
118. Y. Tanaka, T. Ueshima and S. Kurosawa, Japan. Kokai, 75 160,400, 25.12.1975; Chem. Abstr., 84, 151261 (1976).
119. Y. Tanaka, T. Ueshima and S. Kobayashi, Japan. Kokai, 75 159,598, 24.12.1975; Chem. Abstr., 84, 165412 (1976).
120. K. Komatsu, S. Matsumoto and S. Aotani, Japan. Kokai, 76 11,900, 30.01.1976; Chem. Abstr., 86, 56201 (1977).
121. T. Kotani et al., Japan.Kokai, 77 32,100, 10.03.1977; Chem. Abstr., 87, 53832 (1977).
122. G. Dall'Asta and G. Motroni, European Polymer J., 7, 707 (1971).
123. G. Dall'Asta, G. Motroni and L. Motta, J. Polymer Sci., A—1, 10, 1601 (1972).

Chapter 5

THERMODYNAMIC ASPECTS

The metathesis of olefins and the ring-opening polymerization of cyclo-olefins exhibit a series of characteristic thermodynamic features related to thermodynamic parameters such as the reaction enthalpy and entropy, and the equilibrium of the process. This chapter will briefly survey several aspects of these thermodynamic properties.

5.1. Olefin Metathesis

5.1.1. Enthalpy and Entropy of Metathesis

Metathesis is not accompanied by significant structural changes between the reactants and the reaction products; the total number of chemical bonds and their type are the same before and after the reaction. Consequently, the enthalpy difference between the products and reactants, (ΔH_R), is very small or even zero. The free energy of the reaction

$$\Delta G_R = \Delta H_R - T\Delta S_R \qquad (5.1)$$

is mainly determined by the reaction entropy, ΔS_R.

The metathesis of acyclic olefins is clearly controlled by this thermodynamic feature. The metathesis of simple linear olefins, in the homogeneous phase, is a thermally neutral process which proceeds to the thermodynamic equilibrium determined by the reaction entropy.[1] Variation of temperature has a very weak influence on the process.

In the metathesis of branched olefins, the reaction enthalpy is non-vanishing. The increase in reaction enthalpy for the metathesis of branched olefins is accompanied by a corresponding rise in the free energy of formation for a series of linear and branched acyclic olefins[2] (Table 5.1).

When the enthalpy is significantly different from zero, e.g. for branched olefins such as isobutene and 2-methyl-2-butene, temperature variation considerably influences the distribution of the reaction products. Thus,

TABLE 5.1. Reaction enthalpy, reaction entropy, and free energy of formation for the metathesis of acyclic olefins[a]

Olefin	ΔH, kcal/mol		ΔS, cal/mol.deg		ΔG, cal/mol	
	cis-	trans-	cis-	trans-	cis-	trans-
Propene	1.13	1.64	−3.25	−4.29	2042	1352
1-Butene	0.93	1.49	−2.9	−4.04	1762	1142
Isobutene	2.59		1.22		5042	
cis-2-Pentene	0.98	1.49	−2.69	−3.93	2860	370
trans-2-Pentene	0.20	0.31	−0.09	−1.13	3680	1190

[a] Data from ref. [2].

in the metathesis of isobutene a low concentration (1.4 %) of the reaction products is reached at low temperatures (20°C) whereas at high temperatures (over 100°C) the limiting value of the product concentration is obtained, which is 25 %. Similarly, for the reaction of 2-methyl-2-butene, the product concentration rises from 7.45 % to the limiting value of 25 %, when the temperature varies from 20°C to over 100°C.

5.1.2. Equilibrium Conversion in Metathesis

The metathesis reaction is a reversible process: the equilibrium distribution between products and reactants is readily reached. This fact is readily established for reaction in the homogeneous phase. Thus, in the metathesis of 2-pentene with the catalytic system $WCl_6 \cdot EtAlCl_2 \cdot EtOH$,[3] an equilibrium molar ratio of 2 : 1 : 1 between 2-pentene : 2-butene : 3-hexene is obtained starting either from 2-pentene or from an equimolar mixture of 2-butene and 3-hexene. This equilibrium ratio corresponds to a statistical distribution of alkylidene groups between the reactant and the products, the position of equation mainly being controlled by the reaction entropy.

Atlas and co-workers [4,5] calculated the equilibrium constants, the equilibrium compositions and the degrees of conversion in the metathesis of propene for the temperature range from 50 to 300°C. These authors obtained a variation of the equilibrium constant from 24×10^3 to 53.7×10^3 for a rise in temperature from 50 to 300°C at low pressure. The variation of the equilibrium constant brings about an important variation of the degree of conversion, from 31.1 to 47.7 %, and a smaller variation of the equilibrium concentration, from 11.8 to 15.9 % (Table 5.2). Increase of pressure to 30 atm leads to minor modifications of the equilibrium constant and of the degree of conversion in the reaction of propene.

TABLE 5.2. Equilibrium constants and compositions, as well as conversion degree for the metathesis of propene[a]

Temperature, °C	Equilibrium constant $K_j \times 10^{-3}$	$C_2(n-C_4)$ concentration (mole %)	Conversion (mole %)
50	24.0	11.8	31.1
75	28.2	12.6	35.7
100	30.2	12.9	35.0
125	35.4	13.6	37.6
150	40.0	14.3	40.4
175	44.7	14.9	42.4
200	46.8	15.0	44.9
300	53.7	15.9	47.7

[a] Data from ref. [4].

The values calculated for the equilibrium conversion and the distribution of the reaction products in the metathesis of propene are very close to those found experimentally. Such results were obtained by Heckelsberg, Banks and Bailey [6] in the metathesis of propene with a silica-supported tungsten oxide catalyst (Table 5.3).

TABLE 5.3. Conversion and product distribution for the metathesis of propene at 427°C[a,b]

	Experimental values	Calculated values
Conversion, %	44.8	49.8
Product composition (mole %)		
Ethene	9.5	10.0
Propane	40.0	40.0
Propene	33.1	30.1
1-Butene	5.4	5.2
trans-2-Butene	6.7	8.5
cis-2-Butene	4.7	6.2

[a] Catalyst: WO_3/SiO_2; [b] Data from ref. [5].

The degree of conversion reached experimentally was slightly below the calculated value. Consequently, the concentration of propene found in the reaction mixture was higher than the calculated one and the concentrations of the products, ethene and 2-butene, lower than the calculated values. The presence of 1-butene in the reaction product indicates that the catalytic system which was employed promotes the isomerization of 2-butene to 1-butene.

5.2. Ring-Opening Polymerization of Cyclo-Olefins

The ring-opening polymerization of cyclo-olefins presents a series of specific thermodynamic charcteristics, distinct from those of the metathesis of acyclic olefins. This behaviour is determined, primarily, by the thermodynamic stability of cyclo-olefins which has an important contribution in the reaction enthalpy. Furthermore, the nature of cyclo-olefins strongly influences the entropy contribution to the free energy.

The thermodynamic parameters of ring-opening polymerization are completely different from those of conventional addition polymerization. This difference can be readily inferred from the nature of the chemical bonds involved in the two processes. In addition polymerizations, a double bond of the monomer is converted into two single bonds of the polymer, whereas in ring-opening polymerizations the overall number and type of bonds in the monomer are the same as in the polymer. As a consequence, addition polymerizations are normally favoured strongly by enthalpy and weakly opposed by entropy, while ring-opening polymerizations are mainly favoured by enthalpy and, in some cases, also by entropy. In ring-opening polymerizations the enthalpy part is determined by the release of ring strain. This value is particularly large in four-membered and five-membered rings with endocyclic double bonds, and small in higher-membered rings; cyclohexene, which is an almost strain-free ring, is an exception because the equilibrium favours the monomer.

Up to the present time, studies carried out on the thermodynamics of cyclo-olefin polymerization are relatively few and consequently the present chapter will present the main features of the thermodynamics of ring-opening polymerization of cyclo-olefins such as the thermodynamic stability of the rings, the thermodynamic parameters, and the equilibrium of the reaction.

5.2.1. Thermodynamic Stability of Various Rings

The thermodynamic stability of cyclo-olefins is a factor which determines to a great extent the enthalpy and the free energy of the reaction and even the polymerizability of these monomers. Taking into account the fact that the double bond makes a minor contribution to the stability of the ring, at least for medium and long rings, the thermodynamic stability can easily be followed along the entire range of cyclo-olefins, by comparison with that of cyclo-alkanes. The thermodynamic stability of cyclo-alkanes was described in detail by Eliel [7] on the basis of the heats of combustion of these compounds (Table 5.4).

TABLE 5.4. Heats of combustion for n-membered cyclo-alkanes[a]

n	3	4	5	6	7	8	9	10	11	12	13	14	15	16
Heat of [a,b] combustion, kcal/mol	9.2	6.55	1.3	0.0	0.9	1.2	1.4	1.2	1.0	0.3	0.4	0.0	0.1	0.2
Ring strain, [c] kcal/mol	27.5	26.1	6.1	0.1	6.1	9.7	12.5	—	—	—	—	—	—	—

[a] Increment per CH_2 group to the value of 157.4 kcal/mol for the CH_2 group of alkanes or cyclohexane; [b] Data from ref. [7]; [c] Data from ref. [8].

Small rings with three or four carbon atoms, with high heats of combustion, have the lowest stability because of the considerable deviation of the ring angles from the normal tetrahedral angle of sp^3-hybridized carbon atoms, 109°28' (Table 5.5). Cyclopentanes, medium and large rings (but not cyclohexane) present slight deviations from the normal value, because of eclipsed conformations and of interactions between the methylene groups; however, cyclohexane, which has a normal heat of combustion, has a strain-free chair conformation and possesses the highest stability.

TABLE 5.5. Deviation from tetrahedral angle in cyclo-alkanes

n	3	4	5
	24°44'	9°44'	0°44'

Owing to their structural relationship, cyclo-olefins exibit behaviour which closely relates to that of cyclo-alkanes.

Since the two sp^2-hybridized olefinic carbon atoms have the normal valency angle of 120°, the steric strain of n-membered cyclo-olefins is slightly higher than that of n-membered cyclo-alkanes. In the polymerization reaction, the cyclo-olefins proved to have peculiarities closely related to the stability of the ring. Small strained rings, such as those of cyclobutene and cyclopentene, are easily polymerizable [9,10] while cyclohexene has not been polymerized by ring-openning, [11] and the polymerizability of the larger rings decreases with increase of the ring size.[12] The non-polymerizability of cyclohexene is attributed to the high stability of this ring. Cyclohexene exists generally in the chair-form. According to literature data,[13] this conformation is with 29.2 kJ/mol more stable than the boat-form. The stability of the chair conformer considerably reduces its mobility and its possibility of conversion into other conformers (boat or twist forms), which would allow the access to the reaction centre. Determinations of the barrier of pseudo-rotation from the chair form into another such form, involving a boat form intermediate give values of 22.2 [14] or 26.7 [15] kJ/mol. Attempts to polymerize cyclohexene with metathesis catalytic systems[16] actually led to vinylic polymers which contain even more stable cyclohexane rings in the chain. Cyclohexene adopts the less stable boat conformation in a series of bridged, strained cyclic compounds such as norbornene, di(cyclopentadiene), tricyclodecene and [2.2.2]bicyclo-octene. The lower stability of the boat form of these compounds favours their ring-opening polymerization. The polymerization of these structures is possible because it has as driving force the release of the ring strain.

Natta and Dall'Asta [17] established the dependence of the polymerizability of cyclo-olefins on the ring strain of small and medium rings. They showed that polymerizability, for small-ring cyclo-olefins, is directly correlated to ring strain, while for medium rings, where the strain energy is low, polymerizability is favoured by other thermodynamic factors, such as the entropy change of the reaction.

5.2.2. Thermodynamic Parameters for the Ring-Opening Polymerization of Cyclo-Olefins

Dainton and co-workers [18,19] employed a semi-empirical method to calculate the enthalpy, entropy and free energy changes in the hypothetical polymerization of cyclo-alkanes to the corresponding polymers (polyethene or polymethylene) (Table 5.6).

TABLE 5.6. Reaction enthalpy, entropy and free energy for the theoretical polymerization of cyclo-alkanes[a]

Cyclo-alkanes $(CH_2)_n$	$\Delta H°$, kcal/mol	$\Delta S°$, cal/mol·deg	$\Delta G°$, kcal/mol
$n = 3$	-27.0	-16.5	-22.1
4	-25.1	-13.2	-21.2
5	-5.2	-10.2	-2.2
6	$+0.7$	-2.5	$+1.4$
7	-5.1	-0.7	-4.9
8	-8.3	$+8.9$	-11.0

[a] Data from ref. [18].

The lowest (highest negative) values of the free energy of polymerization were found for cyclopropane and cyclobutane, whereas the highest value (a positive one) was found for cyclohexane. For small rings, although the reaction entropy is negative, the reaction enthalpy determines the value of the free energy change and makes the polymerization of these rings possible. Cyclohexane, with positive reaction enthalpy possesses a positive free energy of reaction as well. The polymerization of this ring is thermodynamically impossible. Large rings possess negative free energies of reaction, which make them polymerizable. Along with the reaction enthalpy for these rings, the entropy makes a significant contribution to the free energy. In some cases, i.e. for cyclo-octane, the reaction entropy is even positive.

Cyclo-olefins show a similar trend to that calculated for cyclo-alkanes. Estimates of the thermodynamic behaviour of cyclo-olefins come from reports of Natta and Dall'Asta[17], Calderon,[20] Teyssié,[21] Höcker[22] and other authors.[23,24] On the basis of an empirical method analogous to that used by Dainton for the hypothetical polymerization of cycloalkanes, Teyssié and co-workers [21] calculated the thermodynamic parameters for a series of cyclo-olefins with the monomers in the liquid state and polymers in condensed state. For small and highly strained rings, polymerization is allowed because of the high value of the reaction enthalpy. Generally, for these rings, the enthalpy change is negative and it determines, to a great extent, the free energy of polymerization. Natta and Dall'Asta [17] showed that the reaction enthalpy for small rings depends on ring strain, the latter being the driving force of the polymerization of these cyclo-olefins. For example cyclobutene, a cyclo-olefin with large ring strain, shows a high negative enthalpy value favouring ring-opening polymerization while the negative entropy makes a positive contribution to the free energy.

Cyclopentene polymerizes readily, regardless of the fact that the reaction enthalpy is lower than that for cyclobutene, and the reaction entropy is negative. Experimental determinations indicate for this reaction, enthalpy values of -18.4, -18.8, or -20.5 kJ/mol [20,23,24] and entropy value of -60.25 J/mol·degree. [17] As a result of these tests, the average value of the reaction enthalpy is very close to the known strain energy of cyclopentene (20.5 kJ/mol).[25] The high polymerizability of cyclopentene is favoured by (i) the release of ring strain and (ii) by the higher-energy eclipsed conformation of the hydrogen atoms in the ring, while in the polypentenamer they are free to adopt the lower-energy staggered conformation. Although the entropy is negative and has a relatively high absolute value at temperatures between 0 and 40°C, it does not sensibly influence the course of the reaction.

As has been shown above, cyclohexene does not undergo ring-opening polymerization because of the high stability of this cyclo-olefin in the chair conformation. It is also known that the polymerization of cyclohexane is thermodynamically impossible, this reaction having positive values for free energy and enthalpy changes (Table 5.6). The presence of the double bond brings about only a minor modification in the free energy situation for cyclohexene. Although exact thermodynamic data for the polymerization of cyclohexene are not available, it is possible to make quantitative estimates by comparison with cyclohexane. The reaction enthalpy

will have a value close to zero and the reaction entropy, very likely negative, will be low. These facts will lead to a low free energy of reaction.
Even if the reaction is thermodynamically possible, the equilibrium will
favour the reverse reaction as a consequence of the very high stability of
cyclohexene. This characteristic explains why no ring-opening polymerization of cyclohexene has been possible so far.

As pointed out by Dainton and Ivin [19] for monomers possessing low
ring-strain energy, the entropy is often the determining factor of the free
energy of polymerization. Indeed, for cyclo-olefins with $n = 5 - 7$ carbon atoms, the reaction enthalpy for polymerization is lower than that of
small rings and contributes only slightly to the free energy of polymerization while the reaction entropy is the major determinant factor. As the
ring size increases, the polymerizability of the ring increases too. For still
larger cyclo-olefins ($n = 8-12$) strain factors again are involved. Ofstead
and Calderon [20] confirmed the contribution of entropy factors to the nature and extent of the free energy change through their studies of the polymerizability of 1,9,17-cyclotetraeicosatriene.

There is a significantly different contribution of transational or
of torsional and vibrational entropies in ring-opening polymerization
depending on the ring size. For small rings the negative translational entropy involved in polymerization is very high and becomes less negative
for larger rings. In contrast, the torsional and vibrational entropies which
are always positive, decrease to a much smaller extent with increasing
ring size. As a result, the negative translational entropy prevails over
the other two types of entropy in small rings up to cyclohexane and the
positive torsional and vibrational entropies prevail over the translational
one in larger rings.

5.2.3. Thermodynamic Equilibrium of Ring-Opening Polymerizations

Owing to the fact that the ring-opening polymerization of cyclo-olefins
proceeds through a metathetic mechanism, this reaction presents all the
thermodynamic characteristics of the equilibrium reaction, already outlined for the metathesis of cyclic olefins. The ring-opening polymerization
is a reversible process : under given working conditions, equilibrium conversion is reached. This equilibrium is determined by the stability of the
monomer and polymer, by the characteristic thermodynamic parameters
such as enthalpy and entropy, and by the working temperature. The
equilibrium concentration of the monomer is given by the relationship

$$\ln [M]_e = \frac{\Delta H_p}{RT} - \frac{\Delta S_p^0}{R} \tag{5.2}$$

where $[M]_e$ represents the equilibrium concentration of the monomer,
ΔH_p the enthalpy variation during the polymerization, ΔS_p^0 the standard entropy variation during the polymerization, T the working temperature, and R the ideal gas constant.

The existence of a thermodynamic equilibrium in the ring-opening polymerization reaction was first noticed by Calderon and co-workers.[26] They demonstrated both the first equilibrium which was established between the growing rings and the initial monomer, as well as the second ring-chain cometathetical equilibrium between macrocycles and polyalkenamer.

In the reaction of cyclopentene carried out at 0°C for 40 minutes, 78% conversion of cyclopentene was obtained.[20] By carrying out the reaction at 25°C for 30 minutes, the conversion was 66%, and after 80 minutes 64%. The conversion of the polypentenamer into cyclopentene with a 76.9% reverse conversion was accomplished by carrying out the depolymerization of the polyalkenamer in the presence of metathesis catalytic system. Calderon and co-workers [26] inferred on the basis of the monomer-polymer equilibrium, the possibility of forming macrocycles of a certain concentration, starting either from the monomer or the polyalkenamer.

The monomer concentration at equilibrium is a measure of the polymerizability of cyclo-olefins. Depending on the nature of the ring, the terms of equation (5.2) make different contribution to the value of the concentration of the monomer at equilibrium, i.e. of the polymerizability. Thus, Natta and Dall'Asta [17] showed that for small-ring cyclo-olefins the polymerizability is determined, principally, by the strain energy in the ring. On the other hand, the polymerizability of larger rings is mainly determined by the entropy of the reaction. The concentration of the monomer at equilibrium decreases with increase in the ring size.

Ofstead and Calderon [20] quantitatively determined the concentration of the monomer at equilibrium in the reaction of cyclopentene at various temperatures in the presence of the ternary catalytic system $WCl_6 \cdot EtOH \cdot EtAlCl_2$. By carrying out the reaction at temperatures of 0, 10, 20 and 30°C, they found significant dependence of the cyclopentene concentration on temperature (Table 5.7). The minimum equilibrium concentration, 0.51 mol $\cdot$ L^{-1}, was obtained at 0°C, which corresponds to a maximum cyclopentene concentration of 78%. The maximum equilibrium concentration, 1.19 mol $\cdot$ L^{-1},

TABLE 5.7. Equilibrium concentration of cyclopentene at various temperatures [a, b]

Temperature, °C	Equilibrium concentration $[M]_e$, mol $\cdot$ L^{-1}
0	0.51 ± 0.01
10	0.70 ± 0.01
20	0.88 ± 0.01
30	1.19 ± 0.04

[a] Catalyst : $WCl_6 \cdot EtOH \cdot EtAlCl_2$; $[M]_0 = 2.2 [M]$; monomer : $WCl_6 = 4.5 \times 10^3$; [b] Data from ref. [20].

was reached at 30°C, corresponding to a conversion below 50%. From the equilibrium concentration of the mononer as a function of temperature, the thermodynamic parameters can be estimated.

Detailed studies on ring-chain equilibrium for the polymerization of cyclo-octene have been reported by Scott [27] and Höcker.[28] Scott found that the polyoctenamer obtained in the absence of deliberately added acyclic olefin consisted of 15—20% ring fraction and of 80—85% linear chains.[27] Höcker *et al.*[28] correlated the product distribution as a function of the initial cyclo-octene concentration. They showed that below a threshold monomer concentration of 0.1M practically no polymer was present; at higher concentrations the presence of cyclic oligomers and linear polymer was evidenced.

The ring-chain equilibrium for cyclo-octene as well as for cyclododecene and cyclopentadecene has been treated by Höcker *et al.*[28] in the limits of Jacobson and Stockmayer theory [29]. According to Jacobson and Stockmayer, the equilibrium constant, K_n, of the resulting oligomer, estimated to be equal to the equilibrium concentration of the corresponding oligomer, is inversely proportional to the power 2.5 of the degree of polymerization, n,

$$K_n = \left(\frac{3}{2\pi}\right)^{\frac{3}{2}} \frac{1}{N} \left(\frac{1}{\nu l^2 C_n}\right)^{\frac{3}{2}} n^{-\frac{5}{2}} \tag{5.4}$$

where N_A is Avogadro's number, ν is the number of bonds in the monomer unit, 1 is the average bond length, C_n is the constant characteristic ratio. When plotting the logarithm of the equilibrium constant, K_n, of the oligomers of cyclo-octene, cyclododecene and cyclopetadecene *versus* the logarithm of the degree of polymerization, straight lines with a slope of —5/2 are obtained (Fig. 5.1).

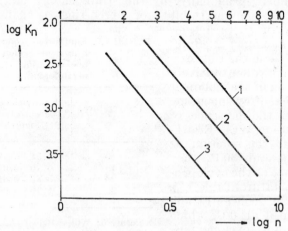

Fig. 5.1. — Plot of log K_n for different oligomers *versus* log n. Cyclo-olefins: cyclo-octene (1), cyclododecene (2) and cyclopentadecene (3).[28]

Deviations were observed for the oligomers with $n > 4$. The larger the ring, the lower is the position of the straight line.

An important thermodynamic feature of cyclo-olefin ring-opening polymerization is the constant oligomer distribution which can be obtained starting from monomer, from a fraction of higher oligomers, from the polymer, or from a single oligomer.[28] The absolute concentrations of cyclic oligomer were found to be constant, above a minimum initial monomer concentration, and independent of catalyst concentration or temperature in a certain range between 0 and 100°C. Up to a minimum monomer concentration, which is called the out-off point (threshold value), the monomer is generally converted into cyclic oligomers, while above this concentration linear polymer is also formed. It is noteworthy that higher ratios polymer/oligomer result from the use of higher monomer concentrations.

The equilibrium oligomer concentration $[M]_e$, was related directly to the entropy of formation, $\Delta S°$, by equation 5.5.

$$R \ln [M]_e = \Delta S^0 \qquad (5.5)$$

Significantly, the plot of the entropy per carbon atom *versus* the number of carbon atoms for cyclobutene, cyclo-octene, cyclododecene and cyclopentadecene resulted in a common curve.[28] Consequently, the number of double bonds has no effect on the entropy of formation or polymerization.

The oligomer-polymer equilibrium in the reaction of cyclo-octadiene observed by Calderon and co-workers [26, 30] has been thoroughly studied by Chauvin and co-workers [31] under various conditions. The latter authors found that the molecular weight distribution in both oligomer and polymer is constant whatever the starting material (1,5-cyclo-octadiene, polymer, oligomer or a mixture oligomer-monomer). They inferred that a true thermodynamic equilibrium between cyclic oligomers and high molecular weight polymer without participation of the monomer is attained (5.6)

$$1,5\text{-cyclo-octadiene} \longrightarrow [\text{oligomer} \rightleftarrows \text{polymer}] \qquad (5.6)$$

The existence of this equilibrium, which accounts for the "back biting" process in the ring-opening polymerization, was ascribed, in part, to conformational effects. In addition, no significant variations in the composition of the equilibrium or in the molecular weight of the polymer with temperature were observed at high conversion level. The absence of such a variation of equilibrium oligomer concentration at low monomer content was attributed to the strain-free nature of the cyclo-octadiene oligomers. Consequently, the enthalpic term is negligible and the entropy governs the thermodynamic equilibrium. Furthermore, Chauvin et al.[31] found that the equilibrium oligomer concentration increased with increasing concentration of the reactants when chlorobenzene or heptane were used as solvents instead of benzene. This finding was ascribed to interactions between solute and solvent.

References

1. N. Calderon, E. A. Ofstead, J. P. Ward, W. A. Judy and K. W. Scott, *J. Amer. Chem. Soc.*, **90**, 4137 (1968)
2. V. Drăguțan, *Studii Cercetări Chim.*, **20**, 1171 (1972).
3. N. Calderon, N. Y. Chen and K. W. Scott, *Tetrahedron Letters*, **1967**, 3327.
4. V. V. Atlas, I. I. Pis'man and A. M. Bakhshi-Zadé, *Neftepererab. Neftekhim.* **5**, 30 (1969).
5. V. V. Atlas, I. I. Pis'man and A. M. Bakhshi-Zadé, *Khim. Prom.*, **10**, 17 (1969).
6. L. F. Heckelsberg, R. L. Banks and G. C. Bailey, *Ind. Eng. Chem., Prod. Res. Develop.*, **7**, 29 (1968).
7. E. Eliel, "Stereochemistry of Carbon Compounds", McGraw-Hill, New York, 1962, pp. 188.
8. J. D. Cox, *Tetrahedron*, **19**, 1175 (1963).
9. G. Natta, G. Dall'Asta, G. Mazzanti and G. Motroni, *Makromol. Chem.*, **69**, 163 (1963).
10. G. Natta, G. Dall'Asta and G. Mazzanti, *Angew. Chem.*, **76**, 765 (1964).
11. N. Calderon, *J. Macromol. Sci.-Revs. Macromol. Chem.*, **C7**, 111 (1972).
12. G. Natta, G. Dall'Asta, I. W. Bassi and G. Carella, *Makromol. Chem.*, **91**, 87 (1966).
13. R. Bucourt and D. Hainaut, *Bull. Soc. chim. France*, **1965**, 1366.
14. F. A. L. Anet and M. Z. Haq, *J. Amer. Chem. Soc.*, **87**, 3147 (1965).
15. N. L. Allinger and J. T. Sprague, *J. Amer. Chem. Soc.*, **94**, 5734 (1972).
16. M. F. Farona and C. Tsonis, *J. Chem. Soc., Chem. Commun.*, **1977**, 363.
17. G. Natta and G. Dall'Asta, in "Polymer Chemistry of Synthetic Elastomers", J. P. Kennedy and E. Thornquist (eds.), Interscience Publ., New York, 1969, pp. 703.
18. F. S. Dainton, T. R. E. Delvin and P. P. Small, *Trans. Faraday Soc.*, **51**, 1710 (1955).
19. F. S. Dainton and K. J. Ivin, *Quart. Revs.*, **12**, 61 (1958).
20. E. A. Ofstead and N. Calderon, *Makromol. Chem.*, **154**, 21 (1972).
21. L. Höcks, D. Berck, A. J. Hubert and Ph. Teyssié, *J. Polymer Sci., Polymer Letters*, **13**, 391 (1975).
22. H. Höcker, W. Reimann, L. Reif and K. Riebel, *J. Mol. Catalysis*, **8**, 191 (1980).
23. D. Kranz and M. Beck, *Angew. Makromol. Chem.*, **27**, 29 (1972).
24. K. Nützel and V. Henze, cited after reference 23.
25. F. D. Rossini *et al.*, "Selected Values of Physical and Thermodynamic Properties of Hydrocarbons and Related Compounds", Carnegie Press, Pittsburg, 1953.
26. K. W. Scott, N. Calderon, E. A. Ofstead, W. A. Judy and J. P. Ward, *Adv. Chem. Ser.*, **91**, 399 (1969).
27. K. W. Scott, N. Calderon, E. A. Ofstead, W. A. Judy and J. P. Ward, *Rubber Chem. Technol.*, **44**, 1341 (1971).
28. H. Höcker, L. Reif, W. Reimann and K. Riebel, *Rec. Trav. chim.*, **96**, M47 (1977).
29. H. Jacobson and W. H. Stockmayer, *J. Chem. Phys.*, **18**, 1600 (1950).
30. N. Calderon, E. A. Ofstead and W. A. Judy, *J. Polymer Sci.*, A—1, **5**, 2209 (1967).
31. Y. Chauvin, D. Commereuc and G. Zaborowski, *Rec. Trav. chim.*, **96**, M131 (1977).

Chapter 6

KINETICS AND MECHANISM

The kinetics and mechanism of metathesis and ring-opening polymerization have not been fully elucidated at the present time. Although studies on the reaction kinetics and mechanism have been very numerous recently, the complexity of the process has been responsible for a variety of interpretations rather than a unanimous point of view.

6.1. Reaction Kinetics

The first investigations on the kinetics of metathesis were carried out for the reaction in the heterogeneous phase. The results obtained in these studies do not agree among themselves, neither do they fully agree with the results obtained in the homogeneous phase. Since interpretation of the reaction kinetics is the first step in the elucidation of the mechanism, this chapter will present the main data obtained in the kinetic studies.

6.1.1. Metathesis in the Heterogeneous Phase

Most of the kinetic studies on the reaction in heterogeneous phase were carried out on the disproportionation of propene with various catalytic systems based on transition metal oxides. The results of these investigations are divergent : some of them correspond to a unimolecular kinetic model while others to a bimolecular one.

6.1.1.1. Kinetic Models for Heterogeneous-Phase Reaction

Several kinetic models were examined for the heterogeneous metathesis reaction. Of these the most thoroughly studied are the Langmuir-Hinshelwood and Rideal models.

Langmuir-Hinshelwood Model. The Langmuir-Hinshelwood kinetic model is based on the hypothesis that the interaction occurs between two molecules adsorbed on the surface of the catalyst. The fraction of the active centres of the catalytic surface which adsorb from the gas phase of the system can be determined by employing a simple Langmuir adsorption relationship. With propene, the fraction of the active centres of the catalytic surface which adsorb propene is given by the relation :

$$\theta_p = \frac{K_p C_p}{1 + K_p C_p + K_e C_e + K_b C_b + \Sigma K_i C_i} \tag{6.1}$$

where K_e, K_p, K_b and K_i are the adsorption coefficients, while C_e, C_p, C_b and C_i are the concentrations of ethene, propene, butene and of the impurities, respectively.

If we suppose that the direct reaction between two adsorbed propene molecules is second-order, we shall have for the equation of the direct reaction rate the following relationship

$$-\frac{dC_p}{dt} = 2k_1\theta_p^2 \tag{6.2}$$

where k_1 represents the rate constant for the direct reaction, d_t is the time variation, and θ_p is the fraction of active centres with adsorbed propene. In this case it is very likely that the reverse reaction also occurs through second-order reaction between a molecule of ethene and one of 2-butene. The rate of disproportionation is therefore given by the relationship

$$-\frac{1}{2} dC_p/dt = k_1\theta_p^2 - k_2\theta_e\theta_b \tag{6.3}$$

where θ_e and θ_b represent the fractions of active centres which adsorb ethene and butene, and k_2 is the rate constant for the reverse reaction.

By substituting for time in relationship (6.3) with the volume of the catalyst and the volumetric flow, we obtain the following equation for the rate of propene disproportionation

$$-\frac{\frac{1}{2} V dC_p}{dV_c} = \frac{k_1 K_p^2 C_p^2 - k_2 K_e K_b C_e C_b}{(1 + K_p C_p + K_e C_e + K_b C_b + \Sigma K_i C_i)^2} \tag{6.4}$$

where V represents the volumetric flow and V_c the volume of the catalyst.

By considering the relationship between the concentrations of the products and reactants

$$K_e K_b = K_p^2 \tag{6.5}$$

$$K_p(C_p)_0 + K_e C_e + K_b C_b = K_p(C_p)_0 \tag{6.6}$$

where $(C_p)_0$ is the initial concentration of propene, the rate equation for propene disproportionation becomes

$$-\frac{\frac{1}{2} V d C_p}{d V_0} = \frac{K_0^2 (K_1 C_p^2 - k_2 C_e C_b)}{[1 + K_p (C_p)_0 + K_i C_i]^2} \tag{6.7}$$

One can convert equation (6.7) into a form which can be integrated, if the following relationships are used

$$X = \frac{C_p}{(C_p)_0} \tag{6.8}$$

$$\frac{1 - X}{2} = \frac{(C_p)_0 - C_p}{2(C_p)_0} = \frac{C_e}{(C_p)_0} = \frac{C_b}{(C_p)_0} \tag{6.9}$$

where X represents the fraction of unreacted propene.
The following rate equation is obtained

$$-\frac{\frac{1}{2} V d X}{d V_c} = \frac{k_1 X^2 - \frac{1}{4} k_2 (1 - X)^2}{\{[1 + K_p (C_p)_0 + \Sigma K_i C_i]/K_p (C_p)_0\}^2} \times$$

$$\times \int_1^X \frac{d X}{\left(k_1 - \frac{1}{4} k_2\right) X^2 + \frac{1}{2} k_2 X - 2k_2} =$$

$$= -\int_0^{V_c} \frac{d V_c}{\{V(C_p)_0 [1 + K_p (C_p)_0 + K_i C_i]/K_p (C_p)_0\}^2} \tag{6.10}$$

which, in a final integrated form, becomes

$$\tanh^{-1} \frac{2\left(K_{eq} - \frac{1}{4}\right) X + \frac{1}{2}}{K_{eq}^{1/2}} - \tanh^{-1} \frac{2\left(K_{eq} - \frac{1}{4}\right) + \frac{1}{2}}{K_{eq}^{1/2}} =$$

$$= \frac{V_c k_2 K_{eq}^{1/2}}{V(C_p)_a \{[1 + K_p (C_p)_0 + \Sigma K_i C_i]/K_p (C_p)_0\}^2} \tag{6.11}$$

where K_{eq} is the equilibrium constant for propene disproportionation.

Rideal Model. The kinetic model of Rideal is based on the hypothesis that the interaction occurs between an adsorbed molecule and a species in the gas phase. In the case of propene disproportionation, the expression for the rate of the direct reaction is

$$-\frac{dC_p}{dt} = 2k_1\theta_p C_p \tag{6.12}$$

and the expression for the rate of the reverse reaction is

$$-\frac{dC_p}{dt} = 2[k'\theta_e C_b + k''C_e\theta_b] \tag{6.13}$$

The rate of propene disproportionation will be

$$\frac{-\dfrac{1}{2} dC_p}{dt} = k_1\theta_p C_p - k'\theta_e C_b - k''C_e\theta_b \tag{6.14}$$

By using relationship (6.1) for the fraction of adsorbed propene, and the volume of the catalyst and the volumetric flow instead of time, and supposing that there the relationship $k' = k'' = k_2$ holds, the expression for the disproportionation rate becomes

$$\frac{-\dfrac{1}{2} VdC_p}{dV_c} = \frac{k_1 K_p C_p^2 - k_2 C_e C_b(K_e + K_b)}{1 + K_e C_e + K_p C_p + K_b C_b + \Sigma K_i C_i} \tag{6.15}$$

By approximating $K_e + K_b = 2K_p$ the rate equation results as follows

$$\frac{-\dfrac{1}{2} VdC_p}{dV_c} = \frac{k_1 K_p C_p^2 - 2k_2 K_p C_e C_b}{1 + K_e C_e + K_p C_p + K_b C_b + \Sigma K_i C_i} \tag{6.16}$$

From equations (6.8) and (6.9) giving concentrations as functions of unreacted propene, the following relation is obtained for the rate of propene disproportionation

$$\frac{-\dfrac{1}{2} VdX}{dV_c} = \frac{k_1 X^2 - 2k_2 \dfrac{1}{4}(1 - X)^2}{[1 + K_p(C_p)_0 + \Sigma K_i C_i]/K_p(C_p)_0} \tag{6.17}$$

By integrating this expression, the final form of the integrated rate equation results

$$\tanh^{-1}\frac{2\left(K_{eq}-\dfrac{1}{4}\right)X+\dfrac{1}{2}}{K_{eq}^{1/2}}-\tanh^{-1}\frac{2\left(K_{eq}-\dfrac{1}{4}\right)+\dfrac{1}{2}}{K_{eq}^{1/2}}=$$

$$=\frac{V_c k_2 K_{eq}^{1/2}}{V\{[1+K_p(C_p)_0+\Sigma K_i C_i]/K_p(C_p)_0\}} \tag{6.18}$$

6.1.1.2. Kinetic Results

Begley and Wilson,[1] by using a catalytic system based on tungsten oxide, carried out the first kinetic studies on the disproportionation of propene in an integral reactor. These authors supposed that this reaction can be based on one of the two kinetic models mentioned above, namely the Langmuir-Hinshelwood and Rideal models.

If the kinetics were defined by the Langmuir-Hinshelwood model, the data obtained for the rate constant for disproportionation should correspond to an equation of the following form

$$k_{LH}=\frac{k_2}{[1+K_p(C_p)_0+\Sigma K_i C_i(C_i/K_p(C_p)_0]^2} \tag{6.19}$$

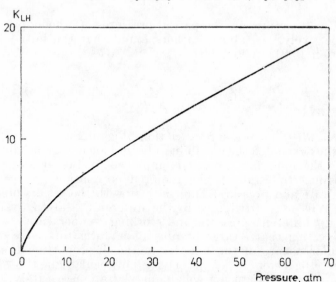

Fig. 6.1. — Effect of pressure on rate constant k_{LH}.[1]

The values obtained by Begley and Wilson for k_{LH} do not correlate as expected with the pressure at higher values (see Fig. 6.1).

For high pressures, the calculated Langmuir-Hinshelwood "rate constant" increases steadily and does not reach a limiting value.

In contrast to these results, the rate constant obtained using the Rideal model is usually independent of pressure even at higher pressures. Although there appear to be some slight variations in the rate constant, they can be attributed to experimental error.

Consequently, the results obtained by Begley and Wilson with the Rideal model indicate first-order kinetics for the disproportionation of propene, which corresponds to an expression of the following form

$$k_R = \frac{k_2}{1 + [K_i C_i / K_p (C_p)_0]} \tag{6.20}$$

Later, Lewis and Wills,[2] studying the initial rates in propene disproportionation with a catalytic system based on cobalt molybdate, in a differential reactor, concluded that their results are better correlated by a bimolecular model of the Langmuir-Hinshelwood type. By starting from a Hougen-Watson type reaction rate expression for the Langmuir-Hinshelwood mechanism

$$r = \frac{k(P_p^2 - P_e P_b / K_{eq})}{(1 + K_e P_e + K_p P_p + K_b P_b)^2} \tag{6.21}$$

where P_e, P_p, P_b represent the partial pressures of ethene, propene, and butene, respectively, the above authors found that the initial rates, for which equation (6.21) reduces to

$$r_0 = \frac{k P_p^2}{(1 + K_p P_p)^2} \tag{6.22}$$

correlate well with the pressure variation (Fig. 6.2).

Thus, they excluded the possibility of a unimolecular mechanism showing that, in the experimental range employed by Begley and Wilson, it was not possible to establish the required correlation between disproportionation rate and pressure. The contrast with the results obtained by Begley and Wilson is attributed to the different catalytic systems employed and to different pressure and working temperature.

It is interesting that a large number of investigators have confirmed the bimolecular nature of the kinetics in the disproportionation of propene with heterogeneous catalysts. Thus, bimolecular kinetics were found by Moffat and Clark,[3] by using cobalt molybdtate as a catalytic system, by Lin, Aldag and Clark[4] or by Pletneva, Usov and Skvortsova[5] with a rhenium oxide system, by Davie, Whan and Kemball[6] with alumina-supported tungsten hexacarbonyl, by Luckner, McConchie and Wills[7] or by Hattikudur and Thodos[8] with silica-supported tungsten oxide.

More recently Mol *et al.*[9] derived a larger number of possible expressions for the reaction rate using more realistic models for the metathesis of
propene. In their derivation they used the steady-state approximation
and considered one or two elementary processes to be rate determining.

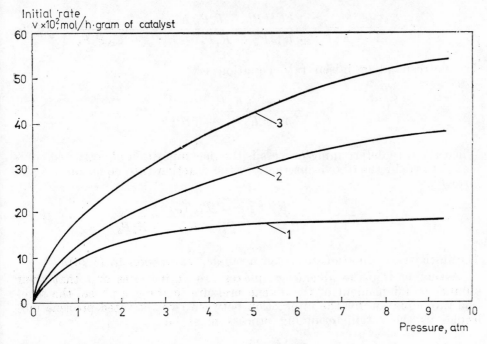

Fig. 6.2. — Initial rates of propene disproportionation as a function of pressure : 1 — 394°K ;
2 — 436°K ; 3 — 478°K.[1]

In a Langmuir-Hinshelwood treatment, Mol *et al.* assumed that the
rate-determining process is the interaction between two propene molecules
adsorbed on two active sites. In this case, for the reaction rate they
derived the following equation

$$r = \frac{kK_p^2(P_p^2 - P_eP_b/K_{eq})}{(1 + K_eP_e + K_pP_p + K_bP_b)^2}$$

(6.23)

and the corresponding initial rate expression

$$r_0 = \frac{kK^2P_p^2}{(1 + KP_p)^2}$$

(6.24)

where K is the equilibrium constant for propene adsorption.

In a second kinetic model they assumed that both propene molecules are chemisorbed, *via* two successive steps on the same active centre. If the rate-determining process is the interaction between two activated propene molecules, the resulting reaction rate equation is

$$r = \frac{kK_p^2(P_p^2 - P_eP_b/K_{eq})}{1 + K_pP_p + K_p2P_p^2 + K_eP_e + K_bP_b + K_{eb}P_eP_b} \qquad (6.25)$$

The corresponding initial rate equation is

$$r_0 = \frac{kK_2P_p^2}{1 + K_1P_p + K_2P_p^2} \qquad (6.26)$$

When the rate-determining process is the decomposition of activated complex formed in the above interaction, the reaction rate equation has the form

$$r = \frac{kK_p^2(P_p^2 - P_eP_b/K_{eq})}{1 + K_pP_p + K_p^2 2P_p^2 + K_eP_e + K_bP_b} \qquad (6.27)$$

The initial rate equation takes on a similar expression to (6.26).

Assuming that the alkene complexes are active sites and that their number is independent of the alkene pressure in the gas phase, the reaction rate equation for the interaction between two activated propene molecules as the rate-determining process must be

$$r = \frac{kK_p^2(P_p^2 - P_eP_b/K_{eq})}{P_p + K_p2P_p^2 + K_eP_e + K_bP_b + K_{eb}P_eP_b} \qquad (6.28)$$

and the initial rate equation is

$$r_0 = \frac{kKP_p}{1 + KP_p} \qquad (6.29)$$

If the rate-determining process is the decomposition of the resulted activated complex the reaction rate equation would be slightly changed to the form

$$r = \frac{kK_p2(P_p^2 - P_eP_b/K_{eq})}{P_p + K_p2P_p^2 + K_eP_e + K_bP_b} \qquad (6.30)$$

but the initial rate equation would remain the same as (6.29).

More complex rate equations were derived assuming that two elementary processes are rate-determining for alkylidene complexes as active sites

$$r = \frac{kK_p^2(P_p^2 - P_e P_b/K_{eq})}{P_p + K_p2P_p^2 + K_e P_e + K_b P_b + K_{eb}P_e P_b + K_{pe}P_p P_e + K_{pb}P_p P_b}$$

(6.31)

$$r = \frac{kK_p2(P_p^2 - P_d P_b/K_{eq})}{P_p + K_p2P_p^2 + K_e P_e + K_b P_b + K_{pe}P_p P_e + K_{pb}P_p P_b}.$$

(6.32)

These lead to the same initial rate expression as equation (6.29) for the above mentioned model.

By comparing these kinetic models, Mol *et al.* found that the available experimental kinetic data cannot be reconciled well with the Langmuir-Hinshelwood mechanism. They concluded that there must exist a preference for kinetic models which are different from the classical Langmuir-type ones, and which are in agreement with all known data.

Kinetic data on the "break-in" period in propene metathesis over tungsten oxide on silica are reported by Wills *et al.*[10][11] They found that the variation in the rate of propene metathesis with time during "break--in" period may be represented by a first-order expression such as

$$\frac{r_{ss} - r}{r_{ss} - r_0} = e^{-kt}$$

(6.33)

where r_{ss} is the steady-state rate, r_0 is the rate at $t = 0$, and k is the "break-in" rate constant. The "break-in" rate constant was determined to be first order with respect to propene pressure and to have an Arrhenius temperature dependence with an activation energy of 197.4 kJ/g.mole.

$$k = 1.993 \times 10^{12} \ e^{-(47.2)/RT)}P_{C_3H_6} \ \text{min}^{-1}$$

(6.34)

Numerous factors are involved in the heterogeneous-phase reaction, every one of which may be rate-determining. Among them, the most important are : diffusion of the reactants on the catalyst surface, adsorption of the reactant by the active centres, reaction of the adsorbed molecules on the catalyst surface, desorption of the reaction products from the active centres, and diffusion of reaction product from the catalyst surface. Some of these factors have been evaluated quantitatively by Begley and Wilson for the disproportionation of propene with cobalt molybdate catalyst. They showed that the diffusion of reactant propene and of the disproportionation products does not play a substantial role in determining the rate. Estimates of the rate of adsorption of propene on cobalt molybdate indicated very high values of this quantity as compared with that of the actual disproportionation reaction. The desorption of the reaction products, ethene and butene, from the active centres of the catalyst could not be quantified because the relevant adsorption isotherms are not known. The role of rate-determining process is thus attributed to the reaction on the catalyst surface. However, it is not impossible that, under different conditions (catalytic systems, pressures, temperatures, etc.), the other factors could contribute to the rate of the process to a greater extent.

The disproportionation of propene homologues in the heterogeneous phase is an even more complex process, so that their kinetics have been little studied up to now. In the disproportionation of 2-pentene with an alumina-supported molybdenum oxide catalyst, Ismayel-Milanovic *et al.*[12] have found a variable kinetics approximating to first-order when the reaction temperature varies.

In conclusion, the vast amount of experimental kinetic data have not so far been fully reconciled with one or another kinetic model for the heterogeneous metathesis reaction. The early results obtained by Begley and Wilson suggested a unimolecular mechanism of the Rideal type; later on, many reports supported a dual-site mechanism of the well-known Langmuir-Hinshelwood type with the surface reaction as the rate-controlling process. More recently, the detailed kinetic studies of Mol *et al.* have pointed out the correlation of available experimental data with kinetic models based on mechanisms in which only one metal atom is involved in the actual reaction step ; the latter results afford a better fit to the accepted carbene mechanism. Anyway, the published kinetic data for heterogeneous metathesis reactions demonstrate a complexity, parallel to that of other well-known related heterogeneous catalytic processes.

6.1.2. Homogeneous-Phase Metathesis

The first detailed studies on homogeneous-phase kinetics of metathesis were carried out by Hughes[13] in the reaction of 2-pentene with catalytic systems consisting of molybdenum nitrosyl complexes and alkylaluminium halides. Hughes observed a significant correlation of certain reaction parameters such as catalyst or olefin concentration with the reaction rate and order, and postulated a reaction mechanism on the basis of these results. Thus, by studying the influence of the catalyst concentration on the reaction rate in the metathesis of 2-pentene with the above mentioned catalyst, variable reaction rates and orders were obtained. As seen from Table 6.1, the reaction rates were in the range $0.73-2.10 \times 10^2$ mol/L·min corresponding to reaction orders between 0.84 and 1.22. These data indicate that 2-pentene metathesis follows first-order kinetics with respect to the catalyst concentration.

TABLE 6.1. Variation of rate and order with catalyst concentration in 2-pentene metathesis[a,b]

Catalyst type[c]	Catalyst concentration $\times 10^3$, mol $\times$ L^{-1}	$v \times 10^{-2}$ mol $\times$ L$^{-1} \times$ min^{-1}	Reaction order
I	2.08	0.73	0.84
	4.16	1.07	1.22
	8.32	2.50	1.21
II	4.16	1.30	0.84
	6.24	1.76	1.21
	8.32	2.50	1.22

[a] Data from ref.[13]; [b]All reactions were carried out in chlorobenzene at 0°C with initial 2-pentene concentration of 0.58 mol $\times$ L^{-1}; [c] Catalyst: $Py_2Mo(NO)_2Cl_2 \cdot EtAlCl_2$, I — formed for 1 hr at room temperature, II — formed for 0.5 hr at 0°C.

In similar studies concerning the influence of 2-pentene concentration on the reaction rate, Hughes obtained orders between 0.7 and 1.7 (Table 6.2).

TABLE 6.2. Variation of rate and order with olefin concentration in 2-pentene metathesis[a,b]

Catalyst	$[C_5]$, mol $\times$ L^{-1}	$[C_5]:[Mo]$	$v \times 10^{-2}$ mol $\times$ L$^{-1} \times$ min^{-1}	Reaction order
$Py_2Mo(NO)_2Cl_2 \cdot Me_3Al_2Cl_3$[c,d]	0.26	39	0.8	1.7
,,	2.06	307	5.7	0.7
$Py_2Mo(NO)_2Cl_2 \cdot EtAlCl_2$[c,e]	1.28	178	3.4	1.4
,,	2.56	356	9.1	1.0—1.4
,,	5.12	712	18.3	1.0

[a]Data from ref. [13]; [b]All reactions were carried out in chlorobenzene at 0°C; [c]Catalyst ratio Al:Mo = 3.8:1; [d][Mo] = 6.7 $\times$ 10^{-3} mol $\times$ L^{-1}; [e][Mo] = 7.2 $\times$ 10^{-3} mol $\times$ L^{-1}.

Hughes interpreted his kinetic results in the light of a bimolecular mechanism considering that the transformation of the diolefin complex is the rate-controlling step. For the initial reaction rate, r_0, expression (6.35) was used

$$r_0 = \frac{kK_1K_2[\text{Cat}]_0[C_5]_0^2}{1 + K_1[\text{Cat}]_0 + K_1K_2[C_5]_0^2} \tag{6.35}$$

where k is the rate constant of the rate-controlling step, K_1 and K_2 represent equilibrium constants, $[\text{Cat}]_0$ is the initial catalyst concentration and $[C_5]_0$ is the initial 2-pentene concentration. Hughes found that the experimental data agreed with that predicted. At constant olefin concentration, the reaction rate follows first-order kinetics with respect to catalyst concentration. The values for the reaction order are around 1.0. The order with respect to olefin concentration, at constant catalyst concentration, is more complex. At very low olefin concentration, the reaction rate approaches a second-order dependence on olefin concentration, whereas at very high olefin concentration the order changes between one and zero.

Detailed kinetic studies on 2-pentene metathesis were performed by Basset and co-workers [14] using various catalytic systems mainly based on $W(CO)_5L[L = PPh_3, PBu_3, P(OPh)_3]$. These authors found that the initial rates of *cis-trans* isomerization and of the metathesis of 2-pentene are in agreement with the proposed mechanism of coordination of the internal olefin on the two different metallacarbene moieties. A kinetic model was developed which accounts for the linear relationship observed between *trans/cis*-C_4 and *trans/cis*-C_5, and also between *trans/cis*-C_6 and *trans/cis*-C_5. The final mathematical expressions derived from the model conceived were

$$\frac{trans\text{-}C_4}{cis\text{-}C_4} = \frac{0.73 - 0.27\alpha}{1 - 2.97\alpha} \qquad (6.36)$$

$$\frac{trans\text{-}C_6}{cis\text{-}C_6} = \frac{0.81 - 0.069\alpha}{0.92 - 2.63\alpha} \qquad (6.37)$$

$$\frac{trans\text{-}C_5}{cis\text{-}C_5} = \frac{1.54 - 2.27\alpha^2}{1 - 2.97\alpha} \qquad (6.38)$$

in which α, equal to *cis*-C_4/*cis*-C_5, is a parameter depending on the conversion. Values for *trans/cis*-C_4, *trans/cis*-C_6 and *trans/cis*-C_5 were calculated by varying α between 0 and 0.1. The calculated data plotted in Fig. 6.3 agreed with the experimental ones.

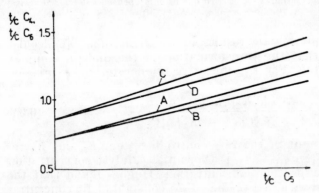

Fig. 6.3. — Ratios *trans*-C_4-*cis*-C_4 and *trans*-C_6/*cis*-C_6 *ver/sus* the *trans*-C_5/*cis*-C_5 ratio in the metathesis of *cis*-2-pentene : A — *trans*-C_4/*cis*-C_4 calculated ; B — *trans*-C_4/*cis*--C_4 experimental ; C — *trans*--C_6/*cis*-C_6 calculated ; D — *trans*-C_6 /*cis*-C_6 experimental (data from reference [11a])

Other kinetic studies on the metathesis of 2-pentene have been carried out by Marin and Labunskaya [15] by employing catalytic systems of tungsten hexachloride and organomagnesium compounds. These authors concluded that the reaction rate is first-order with respect to olefin and catalytic system. They showed that the rate-determining step is the formation of the initial complex between olefin and catalyst; these authors also discussed possible structures of the proposed complex.

The kinetics of *cis, cis*-2,8-decadiene metathesis was examined by Grubbs and Hoppin [16] in connection with the stereochemical and mechanistic aspects of this reaction. Based on a non-pairwise metallacarbene mechanism, they devised a kinetic model analogous to that of Basset and co-workers, mentioned above for 2-pentene. The mathematical equation they obtained was

$$\frac{trans\text{-}C_4}{cis\text{-}C_4} = \frac{0.25 + 0.5\alpha}{1 - 0.34\alpha} \tag{6.39}$$

$$\frac{trans\text{-}C_{10}}{cis\text{-}C_{10}} = \frac{0.25\alpha - 0.21\alpha^2}{1 - 0.34\alpha} \tag{6.40}$$

where α, equal to $cis\text{-}C_4/cis\text{-}C_{10}$, is a measure of the amount of productive metathesis. By varying α between 0 and 0.4, the values for $trans\text{-}C_4/cis\text{-}C_4$ and $trans\text{-}C_{10}/cis\text{-}C_{10}$ were calculated. A linear relationship was obtained by plotting $trans\text{-}C_4/cis\text{-}C_4$ *versus* $trans\text{-}C_{10}/cis\text{-}C_{10}$ (Fig. 6.4).

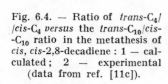

Fig. 6.4. — Ratio of *trans*-C$_4$/ /*cis*-C$_4$ *versus* the *trans*-C$_{10}$/*cis*- -C$_{10}$ ratio in the metathesis of *cis, cis*-2,8-decadiene : 1 — calculated ; 2 — experimental (data from ref. [11c]).

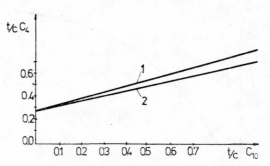

The slope of the calculated curve was in good agreement with that obtained experimentally.

6.1.3. Kinetics of Polymerization of Cyclo-Olefins

It is reasonable to adopt the kinetic treatment of metathesis to ring-opening polymerization. However, the complex nature of the latter process has to accomodate the well-known difficulties of the related Ziegler-Natta systems. So far, few data have been published concerning polymerization rates, the nature of the main kinetically important species, the determination of the concentration of the active centres, and the catalytic activity. The main kinetic data reported come from homogeneous catalytic systems involving the ring-opening polymerization of cyclopentene, cyclo-octene, cyclo-octadiene and norbornene.

Early studies of Hodjemirov et al.[17] on the ring-opening polymerization of cyclopentene with WCl_6-based catalytic systems showed that the kinetics correspond to a first-order rate equation, the following expression being found for the rate of disappearance of the monomer:

$$\ln \frac{M - M_0}{M - M_e} = k_{cr} \times t \tag{6.41}$$

where M, M_0, M_e represent the concentrations of the monomer at any moment, at the initial moment and at equilibrium, respectively, k_{cr} is the rate constant and t the time.

Amass and Tuck [18] investigated the polymerization kinetics of cyclopentene with $WCl_6 \cdot (i-C_4H_9)_3Al$ (molar ratio $1:2$) in toluene in more detail; they observed complex kinetic behaviour. A first-order dependence of the rate of the catalyst concentration was found, whereas the order with respect to the monomer concentration was more complicated. The initial rate of polymerization was of first order with respect to cyclopentene concentration but the rate rapidly decreased as reaction proceeded (Fig. 6.5).

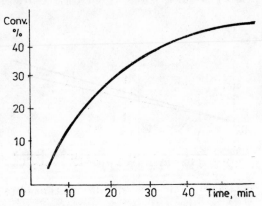

Fig. 6.5. — Variation of conversion with time in the polymerization of cyclopentene (catalyst WCl_6 $(i-C_4H_9)_3Al)$. [18]

Their results indicate the presence of two species active in the polymerization. One is formed by the reaction of a complex between cyclopentene and WCl_6 with two moles of $(i-C_4H_9)_3Al$. This species is partially converted by reaction with cyclopentene into a new species with much lower activity. The authors suggested a simple kinetic scheme illustrating the generation of the two distinct active centres

$$WCl_6 + C_5H_8 \xrightarrow{k_1} \underset{(I)}{W^*} \xrightarrow{Al(iBu)_3} W^*/Al \tag{6.42}$$

$$\underset{(I)}{W^*} + C_5H_8 \xrightarrow{k_2} \underset{(II)}{W^{**}} \xrightarrow{Al(iBu)_3} W^{**}/Al \tag{6.43}$$

Species (II) has an unknown activity and exhibits complex behaviour. A significant finding was that the rate of polymerization under a given set of conditions depended upon the time that elapses between the addition of the two catalyst components (WCl_6 and i-Bu_3Al) to the cyclopentene; the pre-mixing time for WCl_6 before adding i-Bu_3Al that gave the highest polymerization rate can be defined by the equation

$$t_{max} = \frac{1}{(k_1 - k_2)\,[M]_0} \ln \frac{k_1}{k_2} \qquad (6.44)$$

were $[M]_0$ is the initial cyclopentene concentration.

The optimum premixing time, t_{max}, was inversely proportional to the monomer concentration. Accordingly, in the above scheme the initial rate of polymerization, R_{p0}, was represented by the kinetic equation

$$R_{p0} = kp[W^*/Al]_0\,[M]_0 \qquad (6.45)$$

where $[W^*/Al]_0$ was estimated to be equal to $[I]$. Thus, the initial rate of polymerization could be used to measure the concentration of the first active species, $[W^*/Al]$, at any time.

Recent investigations of Amass and Zurimendi [19], by dilatometric measurements, indicate that during the polymerization of cyclopentene the rate of propagation decreases faster than can be accounted for by consumption of monomer alone. These authors inferred that the decrease in the rate of polymerization is a second-order reaction with respect to the concentration of active species. In agreement with such a statement is the fact that the reciprocal rate of polymerization, equivalent to the concentration of the propagating species, is a linear function of time. Furthermore, these authors found that the catalyst produced from WCl_6 and i-Bu_3Al in cyclopentene polymerization is not a living system. They demonstrated that deactivation of the system occurs during the polymerization. The kinetic behaviour of this system suggested the existence of a termination reaction of the second order (6.46) with respect to the active species, similar to the bimolecular termination reaction that occurs in the Ziegler-Natta polymerization of linear olefins

$$W^* + W^* \rightarrow Polymer \qquad (6.46)$$

A refined kinetic treatment was developed by Ivin et al.[20, 21] in the interpretation of the propagation mechanism for cyclopentene, norbornene and norbornadiene and their derivatives on the basis of the structure and stereochemistry of the resulting polymers. The variation with temperature of the fractions of cis and trans double bonds σ_c and σ_t, respectively, in the polymers were correlated with the rate constant for interconversion or relaxation of the intermediate metallacarbenes, k, and those of propagation to the corresponding metallacyclobutanes, k_1, k_{2c} and k_{2t}

Scheme 6.1.

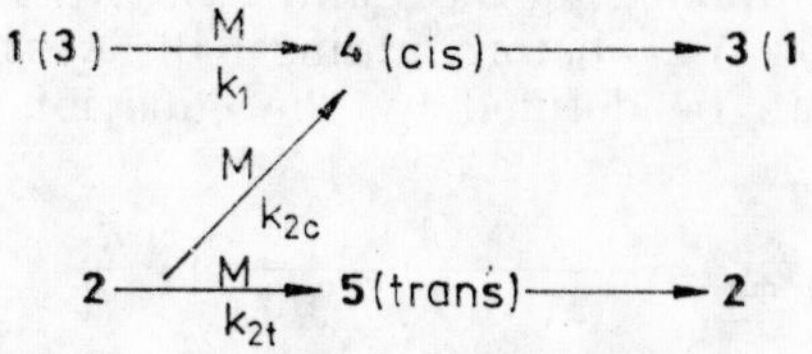

1, 2 and **3** represent intermediate metallacarbenes, **4** and **5** the corresponding metallacyclobutanes and M represents cyclopentene monomer. A steady-state treatment of scheme 6.1 leads to equation (6.47)

$$\frac{\sigma_c}{\sigma_t} - \frac{k_{2c}}{k_{2t}} = \frac{k_{2c}}{k_{2t}} \cdot \frac{k_1}{k} \, [M] \qquad (6.47)$$

If at high temperature $k_1[M] \ll k$, then from equation 6.45 one obtains

$$\frac{\sigma_c}{\sigma_t} = \frac{k_{2c}}{k_{2t}} \qquad (6.48)$$

The fall in σ_c with increasing temperature will thus be governed by the rate of conversion of **2** to **5** trans. At low temperature $k_1[M] \ll k$, and equation (6.47) will become

$$\frac{\sigma_c}{\sigma_t} = \frac{k_{2c}}{k_{2t}} \cdot \frac{k_1}{k} \, [M] \qquad (6.49)$$

In this case σ_c is high but should decrease as $[M]$ is reduced, as is observed experimentally. Ivin and co-workers correlated the rates of relaxation and propagation with the extent of d-orbital occupancy around the central metal ion of the catalyst.

Significant contributions to the kinetics of cyclopentene polymerization were made by Ceauşescu, Bittman, Dimonie et al.[22-24] in studies with WCl_6-based catalysts. From the experimental data, a termination reaction second-order with respect to the concentration of active centres was inferred

$$\ln \frac{M_0 - M_e}{M_0 - M} = \frac{k_{cr}}{k_t} \, (1 + k_t C_0' t) \qquad (6.50)$$

where k_t is the rate constant for the termination reaction, M_0, M and M_e represent the monomer concentration at the initial moment, at a given time, and at equilibrium, and C_0 is the concentration of active species. The kinetics thus found for the termination reaction show similarities to the known bimolecular termination reaction which occurs in the Ziegler-Natta polymerization of linear olefins.[24]

The kinetics of polymerization of higher cyclo-olefins have received little attention and are not well defined. Höcker[25] and Reimann[26] investigated the kinetics of cyclo-octene polymerization in chlorobenzene as solvent, using three types of catalysts: $WCl_6 \cdot EtOH \cdot EtAlCl_2$, $(Ph_3P)_2(NO)_2MoCl_2 \cdot EtAlCl_2$ and $WCl_6 \cdot SnMe_4$. They obtained typical conversion curves for different reaction times (Figs 6.6. and 6.7).

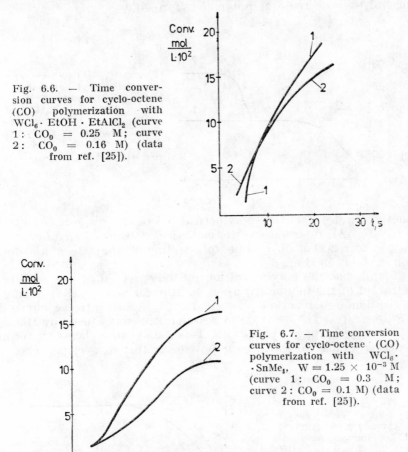

Fig. 6.6. — Time conversion curves for cyclo-octene (CO) polymerization with $WCl_6 \cdot EtOH \cdot EtAlCl_2$ (curve 1: $CO_0 = 0.25$ M; curve 2: $CO_0 = 0.16$ M) (data from ref. [25]).

Fig. 6.7. — Time conversion curves for cyclo-octene (CO) polymerization with $WCl_6 \cdot SnMe_4$, $W = 1.25 \times 10^{-3}$ M (curve 1: $CO_0 = 0.3$ M; curve 2: $CO_0 = 0.1$ M) (data from ref. [25]).

The reaction proved to be of first-order with respect to the catalyst concentration and second-order with respect to the monomer when $WCl_6 \cdot EtOH \cdot EtAlCl_2$ was used in the range $1 - 5 \times 10^{-4}$M. However, in contrast to this system, for $WCl_6 \cdot SnMe_4$ and $(Ph_3P)_2(NO)_2MoCl_2 \cdot EtAlCl_2$ first-order kinetics were found with respect to monomer concentration. The activation energy for the cyclo-octene reaction was determined to be 21.7 kJ/mol and the activation entropy was — 55 e.u. The above authors observed pronounced kinetic control of reaction products in the early stage of the reaction : at high initial monomer concentration, immediate formation of polymer occurs, while at low concentration the homologous series of oligomers is first formed and eventually the polymer. The concentrations of oligomers increase as the reaction progresses, reach a maximum and then decrease as polymer is formed.

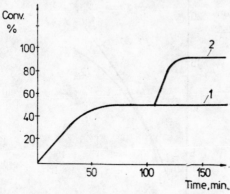

Fig. 6.8. — Time conversion curve of 1,5-cyclo-octadiene (COD) polymerization with $WCl_6 \cdot (i - C_4H_9)_3Al$, COD = 1 mol L^{-1} (curve 1: initial addition of catalyst; curve 2: new addition of catalyst) (data from ref. [27]).

Kinetic studies for the polymerization of 1,5-cyclo-octadiene with $WCl_6 \cdot (i-C_4H_9)_3Al$, reported by Korshak et al.,[27] showed significant similarities to the reaction of cyclopentene, although the rate of the former reaction was lower (Fig. 6.8).

It was found that the polymerization did not proceed at the equilibrium concentration of the monomer ; upon addition of new catalyst during the polymerization, the reaction proceeded at a higher rate, as curve 3 in Fig. 6.8 illustrates. The decrease in reaction rate was caused by the significant consumption of active species. This behaviour indicates a termination reaction similar to that in the polymerization of cyclopentene.

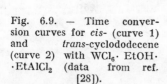

Fig. 6.9. — Time conversion curves for cis- (curve 1) and trans-cyclododecene (curve 2) with $WCl_6 \cdot EtOH \cdot EtAlCl_2$ (data from ref. [28]).

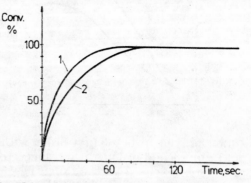

The form of polymerization kinetics strongly depends on the nature of the stereoisomer of the cyclo-olefin substrate and on the type of catalyst used. Such data are known for cyclododecene and cyclopentadecene polymerization. By using $WCl_6 \cdot EtOH \cdot EtAlCl_2$ as catalyst, Riebel[28] found that cis-cyclododecene reacts at a higher rate than the trans-isomer (Fig.6.9).

The same finding has been made by Reif [29] in the reaction of cyclopenta-diene (Fig. 6.10).

In contrast, when using $WCl_6 \cdot SnMe_4$ as catalyst, the reactivity of the two stereo isomers is reversed, i.e. the *trans*-isomer reacts faster than the *cis*-isomer.

Fig. 6.10. — Time conversion curves for the *cis*- (curve 1) and *trans*-cyclopentadecene (curve 2) with $WCl_6 \cdot EtOH \cdot EtAlCl_2$ (data from ref. [29])

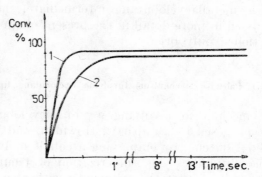

6.2. Reaction Mechanism

Elucidating the mechanisms of olefin metathesis and the ring-opening polymerization of cyclo-olefins has provided one of the most challenging problems in modern heterogeneous and homogeneous catalysis, in coordination, organometallic and polymer chemistry. From the various models conceived since the discovery of these reactions, only within recent years have models been designed which begin to reveal the important mechanistic features of these reactions. Most of the investigations have been directed towards deciding whether the reaction proceeds by a "pairwise" interchange of the carbons between two olefins or through a one-carbon chain mechanism implying a carbenic intermediate. Although there is a growing consensus which favours the carbene mechanism, many of the experimental observations are hard to explain on this basis. It is the aim of this section to gather most of the available information concerning the known mechanisms of these reactions, and to discuss the most relevant features defining the mechanism.

6.2.1. Survey of Proposed Mechanisms for Metathesis

The first mecanism proposed for metathesis, which seemed to have an experimental basis, was the pairwise or bimolecular mechanism involving a four-centred cyclobutane intermediate. Later, other mechanisms were advanced, e.g. the mechanism involving a metallacyclopentane intermediate, the carbenic mechanism involving a metallacyclobutane intermediate and the carbene mechanism involving a multi-centred intermediate. Since the experimental data could be reconciled only with two of the possible mechanisms, namely the bimolecular mechanism involving a cyclobutane four-centred intermediate and the carbene mechanism involving a metallacyclobutane intermediate, these two mechanisms will be discussed in more detail in the present chapter, whereas the others will be mentioned only briefly.

6.2.1.1. Pairwise Mechanisms Involving Cyclobutane Intermediates

The mechanism involving a "quasicyclobutanic" transition state was first proposed by Bradshaw, Howman, and Turner.[30] They disproportionated 1-butene, by employing a cobalt molybdate catalyst, which simultaneously causes the isomerization of 1-butene to 2-butene. These authors noticed that the selectivity with respect to ethene and hexene varies inversely proportional to the tendency for isomerization to 2-butene; when the isomerization was inhibited by treating the catalyst with sodium bicarbonate, only ethene and hexene were obtained. In order to explain the formation of these reaction products, the authors proposed the following mechanism involving a "quasicyclobutane" four-centred intermediate.

$$\tag{6.51}$$

The structure and distribution of the reaction products, both in the first studies carried out by Banks and Bailey[31] and in the subsequent research of other investigators apparently agreed with the reaction mechanism involving the four-centred intermediate. Thus, Adams and Brandenberger,[32] while studying the metathesis of pentene and hexene, noticed a distribution of reaction products which corresponded to the mechanism proposed by Bradshaw, Howman and Turner; likewise, Crain[33] obtained data in agreement with this mechanism for the metatheses of heptene and octene.

Support for this mechanism has been further supplied by numerous studies on the metathesis of isotopically-labelled propene carried out by several groups of investigators. Mol, Moulijn and Boelhouwer,[34] by employing 1-^{14}C-propene in the reaction, found the whole of the radioactivity in the ethene, while use of 2-^{14}C-propene and 3-^{14}C-propene led to radioactive 2-butene but non-radioactive ethene. These results were explained by a four-centred mechanism, as in the following equations

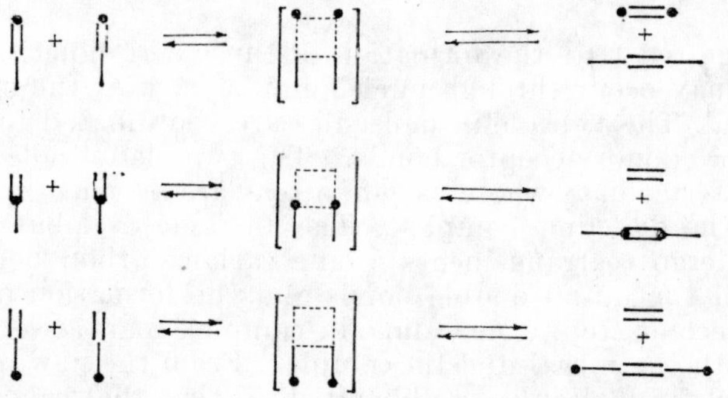

(6.52)

(6.53)

(6.54)

Similar results were recorded by Woody, Lewis and Wills [35] in the disproportionation of 1-^{14}C-propene, and by Clark and Cook [36] in the disproportionation of 1-^{14}C-propene and of 2-^{14}C-propene. Furthermore, kinetic studies in the heterogeneous phase lend support to a pairwise mechanism involving a four-centred cyclobutane intermediate.

For the homogeneous-phase reaction, Calderon and co-workers [37] conceived a pairwise mechanism based on results obtained in the reaction of deuterated 2-butene. These authors showed that the cometathesis of 2-butene with 2-butene-d_8, in the presence of the ternary catalytic system $WCl_6 \cdot EtAlCl_2 \cdot EtOH$ yields only a transalkylidenation product, namely $C_4H_4D_4$. Also, the reaction of 3-hexene with 2-butene-d_8 yields $C_5H_6D_4$ as the unique reaction product. The formation of these products was interpreted by Calderon through a transalkylidenation mechanism consisting of the following steps:

a. the formation of a diolefin-metal complex

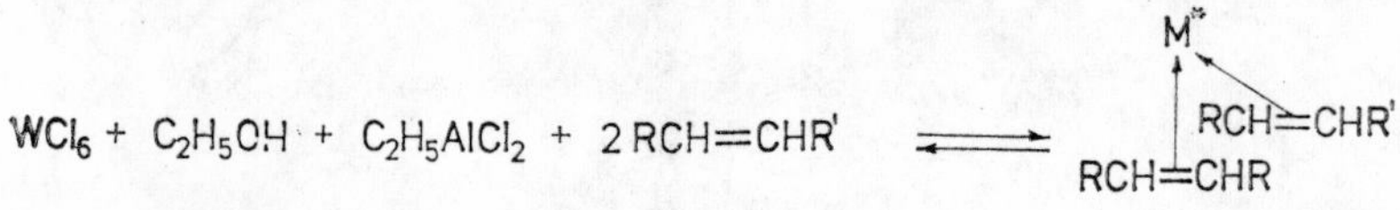
WCl6 + C2H5OH + C2H5AlCl2 + 2 RCH=CHR'
M*
RCH=CHR'
RCH=CHR

(6.55)

b. transalkylidenation

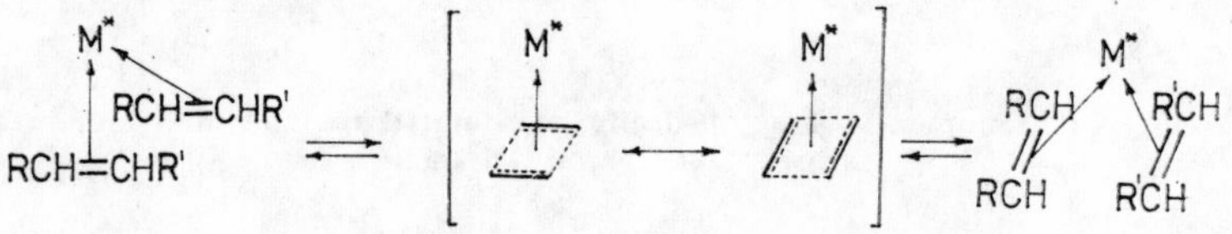
M*
RCH=CHR'
RCH=CHR'
M*
M*
M*
RCH
RCH
RCH
RCH

(6.56)

c. olefin exchange

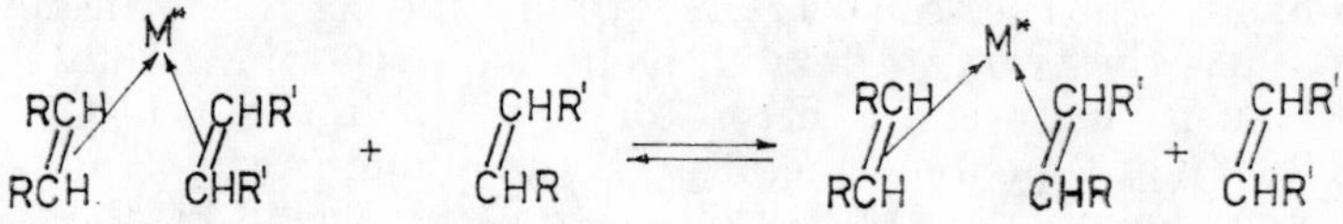
M*
RCH
CHR'
RCH
CHR'
CHR'
CHR
M*
RCH
CHR'
RCH
CHR
CHR'
CHR'

(6.57)

where M* represents the transition metal including any ligands from the catalytic system, while R and R′ represent hydrogen atoms or alkyl groups.

It was suggested that the formation of the metal-diolefin complex in the first step may occur through nucleophilic attack of the olefin on the transition metal. The two olefin molecules are coordinated to the transition metal by π donor-acceptor bonds.[38] The two olefin molecules form a cyclobutane intermediate whose carbon atoms are attached to the transition metal in the diolefinic complex. The unstable cyclobutane intermediate may undergo rearrangements of the carbon-carbon bonds, namely the cleavage of the initial double bonds and the formation of new ones. Thus, in the second step, a new diolefin complex may arise, which is in equilibrium with the initial diolefin complex. From the new diolefin complex, two new olefin molecules are liberated, so that the reaction products are obtained through decomplexation, or through substitution with other olefins present in the system. Analogous olefin-exchange reactions are known in such complexes.[39,40] They generally proceed at a high rate, and are consistent with the stereochemistry of metathesis.

Hughes [13] proposed a pairwise mechanism involving a cyclobutane intermediate for the metathesis of 2-pentene with a catalytic system based on molybdenum. According to this mechanism, a diolefin complex of molybdenum with 2-pentene in the *cis* position is formed in a first step. In this complex, through a cyclic transition state, the cleavage of π and σ bonds of the initial 2-pentene occurs, and new π and σ bonds are formed leading to the reaction products (Fig. 6.11).

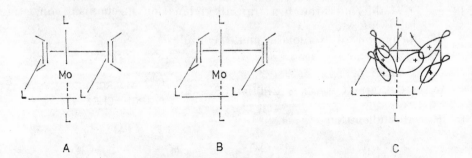

Fig. 6.11. — *cis*-Complexes of octahedrally coordinated molybdenum and rotation pathway for complexed olefins.

In order to have positive overlap between the π orbital of the olefin and the d-orbitals of the transition metal, facilitating the formation of the transition state, the two coordinated olefin molecules must rotate through the cyclic transition state in a disrotatory way, as in Fig. 6.11 C. The high stereospecificity of 2-pentene metathesis was explained by Hughes on the basis of a four-membered transition state as in Fig. 6.11 above. In the case of two molecules of *cis*-2-pentene, the cyclobutane intermediate leads to *cis*-2-butene and *cis*-3-hexene, while in the case of *trans*-2-pentene it leads to *trans*-2-butene and *trans*-3-hexene. By employing the concept of orbital symmetry, Hughes [13] advanced a theoretical interpretation of the proposed mechanism, specifying the role of the transition metal in the cleavage of the old bonds from the olefin and the formation of new ones.

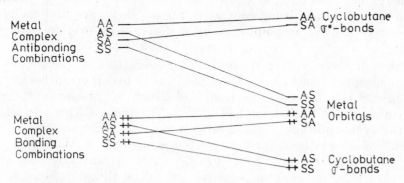

Fig. 6.12. — Correlation diagram for cyclobutanation of two complexed olefin molecules (after Mango and Schachtschneider [41]); S is symmetric A is antisymmetric with respect to the symmetry plane.

The pairwise mechanism involving a cyclobutane-type transition state was given quantitative interpretation through the molecular-orbital calculations effected by Mango and Schachtschneider.[41-43] Extending the Woodward and Hoffmann [44] concept of conservation of orbital symmetry,

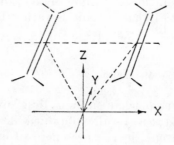

Fig. 6.13. — Symmetry elements for complexed olefins; X, Y, Z represent the symmetry axes.

on the basis of EHT calculations to the cyclo-addition reaction catalyzed by transition metals, these authors obtained correlation diagrams for the formation of the cyclobutane transition state in the presence of the transition metal (Figs 6.12 and 6.13).

It was concluded that in this process, the electron density of AS and SA orbitals in the complex may become redistributed according to the correlation diagram : in the AS orbital, the electron density increases in the *forming* σ bond of cyclobutane and in the SA orbital the electron density increases in the metallic d orbital. Consequently, on forming a cyclobutane transition state, the SA combination of olefin orbitals increases in energy, thus diminishing the overlap of the ligand-metal orbitals and increasing the electron density in the d orbital. The AS combination of olefin orbitals, being converted into a σ bond, acquires an increased electron density accomplished by a decrease of overlap with the d orbital.

Although several other theoretical considerations of the cyclobutane transition state in metathesis have been made by other authors,[45-48] we shall briefly mention only the results obtained by Armstrong [49] through the semiempirical CNDO method. The calculations of this author show that the cyclo-addition reaction between two molecules of ethene, in the presence of a metathesis catalytic system consisting of WCl_6 and a co-catalyst, is a thermically allowed process. In this transformation, the $6p$ orbitals of tungsten play an essential part. The tungsten atom supplies the vacant and occupied $6p$ orbitals which, with suitable energy and symmetry, can promote the cyclo-addition process $(C_2H_4)_2 \cdot WCl_4 \rightarrow \rightarrow C_4H_8 \cdot WCl_4$.

The complex $C_4H_8 \cdot WCl_4$ possesses a doubly degenerate orbital occupied by a single electron, so that this complex can become distorted to interact with a pair of coordinated ethene molecules. Armstrong's results which refine to a higher accuracy those of Mango and Schachtschneider, confirm the possibility of forming a four-centred cyclobutane transition state in metathesis. Unlike Mango and Schachtschneider, Armstrong has demonstrated that p-orbitals rather than d-orbitals of tungsten play the essential role in the process.

6.2.1.2. Carbene Mechanisms

The carbene mechanism was first proposed by Hérisson and Chauvin [50] to interpret the results obtained in the telomerization of cyclopentene with 2-pentene, in the presence of catalytic systems based on tungsten. This mechanism involves insertion of the olefin into the carbene complex of the transition metal resulting in a metallacyclobutane-type transition state (Scheme 6.2). The coordination of the olefin to the transition metal occurs only in the processes of formation and cleavage of the metalla-cyclobutane transition state.

Scheme 6.2

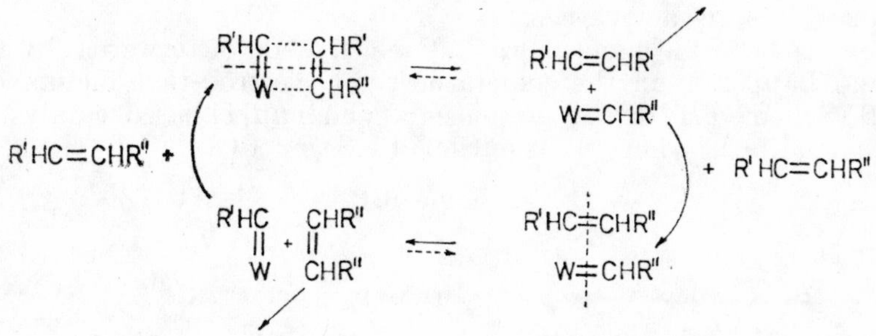

Hérisson and Chauvin proposed, for the telomerization reaction, the following scheme which involves the insertion of acyclic and cyclic olefins into the carbene complex.

Scheme 6.3

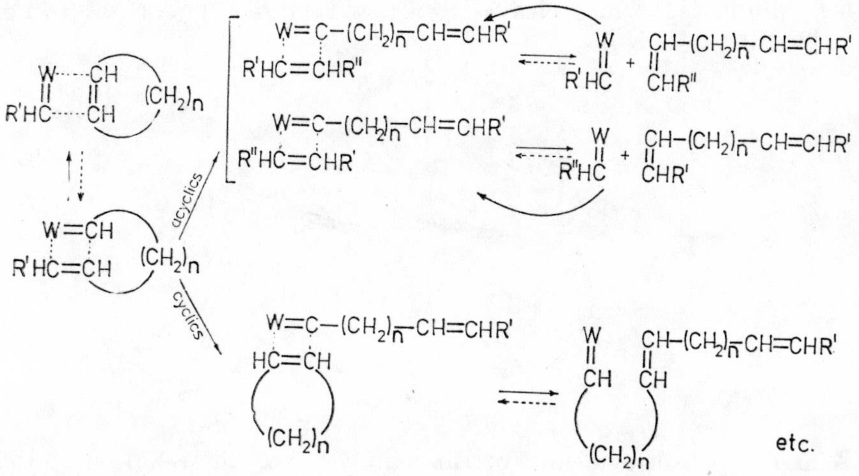

On the basis of this mechanism, these authors explained the formation of the product "triads" in the telomerization of cyclopentene with 2-pentene. They carried out a statistical calculation for the distribution of the reaction products, obtaining results that fully agree with the experimental data. However, the initial proposal of Hérrison and Chauvin did not explain the nature of the initiation reaction yielding the carbene complex, the product distribution in the cometathesis of α-olefins with cyclo-ole-

fins, or the nature of the fragmentation reaction of cyclic dienes and trienes, which involves the incorporation of at least two unsaturated fragments on the same active centres.

A new carbene mechanism was advanced soon afterwards by Cardin,
Doyle and Lappert [51] for the cometathesis of electron-rich olefins of type
A and B (Scheme 6.4) in the presence of rhodium[I]-based catalysts such
as $L(Ph_3P)_2RhCl$, where L is either Ph_3P or CO.

Scheme 6.4

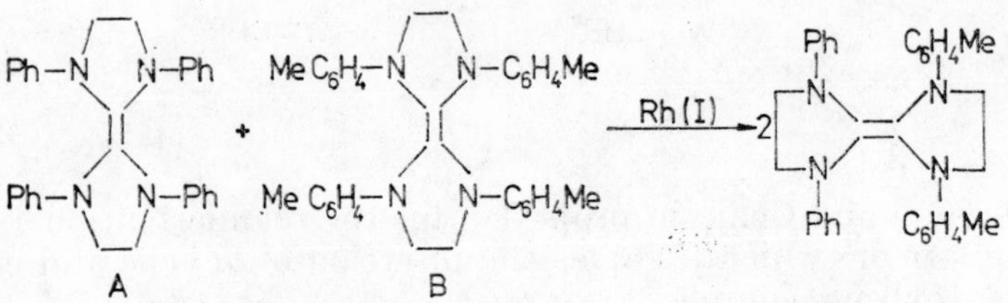

According to this mechanism, in a first stage a carbene complex of
rhodium (I) is formed through interaction with the electron-rich olefin.
This complex and the olefin lead to a metallacyclobutane intermediate
with rhodium (III) which cleaves to a new rhodium (I) carbene complex
and the product olefin.

Scheme 6.5

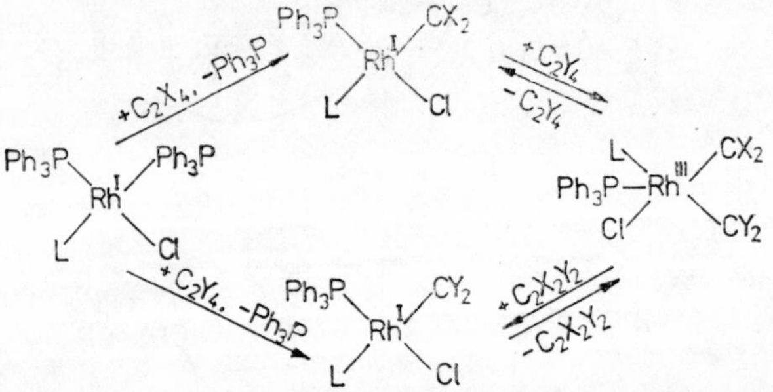

A major item of evidence for this pathway was the isolation and characterization of the carbene complexes $L(Ph_3P)Rh(CX_2)Cl$ and
$L(Ph_3P)Rh(CY_2)Cl$, where $L = Ph_3P$ or CO, which possess catalytic activity. Furthermore, these authors succeeded in isolating dicarbene complexes $L(Cl)Rh(CX_2)(CY_2)$, with $L=CO$, which were believed to play a
part in the reaction.

For the formation of the metallacyclobutane intermediate from metallacarbene and olefin, Cardin et al.[51] suggested the intervention of one
of the two transition state (C) and (D) (Fig. 6.14), with transition state
(C) being more likely. The highly nucleophilic character of electron-rich

olefins and $Rh^{(I)}$, and the electrophilic character of a carbenic carbon atom, are in good agreement with this suggestion. The electronic processes leading to the formation of transition states (C) or (D) need not be synchronous; the electron character of the olefin and rhodium(I) complex suggests that process α probably precedes β.

A carbene mechanism was described by Casey and Burkhardt [52] for the transalkylidenation and cyclopropanation occurring in the reaction of olefins with (diphenylcarbene)-pentacarbonyltungsten. When isobutene reacts with the tungsten carbene complex, 1,1-diphenylethene, 1,1-dimethyl-2,2-diphenylcyclopropane and $W(CO)_6$ are formed

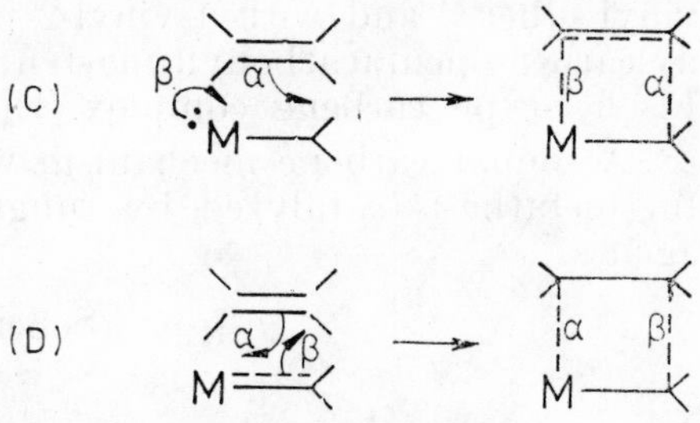

Fig. 6.14. — Transition states for the olefin—metallacarbene reaction.

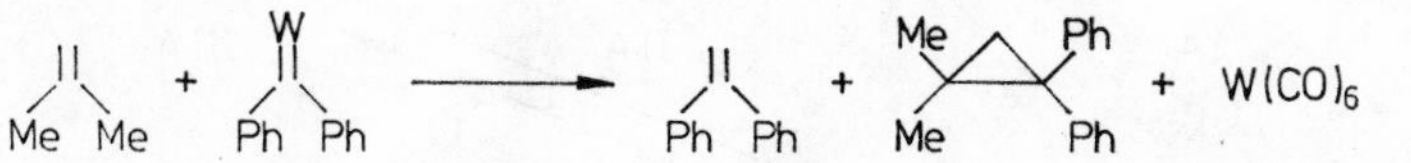

$$\text{(6.58)}$$

Casey and Burkhardt [52] proposed the formation of a metal complex containing both a carbene and an alkene ligand which rearranges to a metallacyclobutane, the key intermediate in the reaction. The metallacyclobutane, can undergo a reductive elimination to give a cyclopropane or can undergo cleavage to give a new metal complex containing both the coordinated alkene and the new carbene ligand.

Scheme 6.6

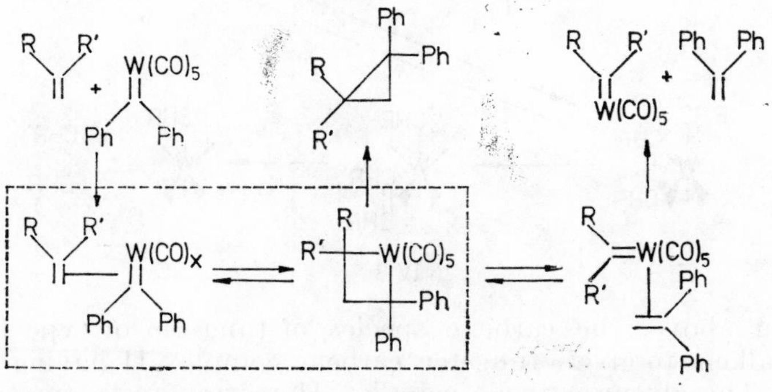

The main feature of this mechanism is the formation and interconversion of metallacyclobutane and a metal complex containing both an alkene and a carbene ligand. Although the metallacyclobutane is the more likely key intermediate, it remains to be proved which of the two structures within the dashed frame has the predominant role. As we shall see later, in subsequent work, Casey et al.[53] focussed on the elucidation of this objective but the results are not yet conclusive. It is remarkable that this mechanism explains Fischer's observation of 1-methoxy-1-phenyl-ethene as a product in the reaction of $(CO)_5Cr(OCH_3)C_6H_5$ with ethyl vinyl ether [51] and with 1-vinyl-2-pyrrolidone.[55] The behaviour of (diphenylcarbene)pentacarbonyltungsten in the above reaction reveals, in this Fischer-type carbene complex, the electrophilic character of C_{carb}.

A similar carbene mechanism was advanced by Muetterties[56] for olefin metathesis catalyzed by tungsten hexachloride and alkylating reagents.

Scheme 6.7

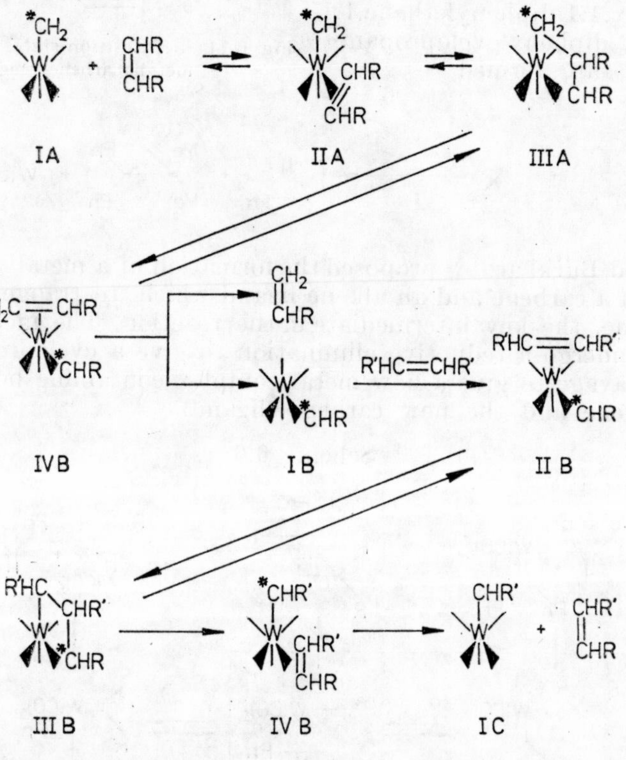

IA   IIA   IIIA

$H_2C = CHR$

IVB   IB   IIB

IIIB   IVB   IC

As seen above, the carbene species of tungsten of type I can coordinate an alkene to give a tungsten carbene complex II having the alkene as ligand. The alkene carbene complex II rearranges to tungstacyclobutane III which equilibrates to a new alkene carbene complex IV. The alkene carbene complex IV gives rise to the product alkene and regenerates the active carbene species I for the reaction cycle. Muetterties suggested the formation of carbene species I as a result of the alkylation reaction of tungsten hexachloride with an alkyl-metal compound, followed by α-hydrogen elimination from the tungsten-alkyl group. When dimethyl-zinc reacted with tungsten hexachloride in benzene, methane was produced and probably a methylidene tungsten species is formed.

Scheme 6.8

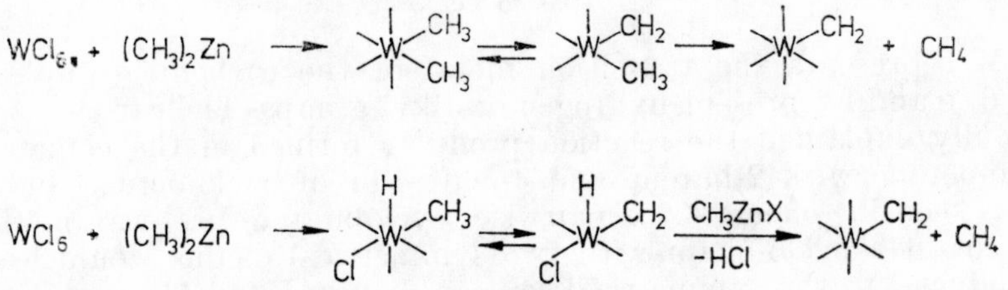

$$WCl_6 + (CH_3)_2Zn \longrightarrow \overset{|}{\underset{|}{W}}\overset{CH_3}{\underset{CH_3}{}} \rightleftharpoons \overset{|}{\underset{|}{W}}\overset{CH_2}{\underset{CH_3}{}} \longrightarrow \overset{|}{\underset{|}{W}}CH_2 + CH_4$$

$$WCl_5 + (CH_3)_2Zn \longrightarrow \overset{H}{\underset{Cl}{\overset{|}{W}}}\overset{CH_3}{} \rightleftharpoons \overset{H}{\underset{Cl}{\overset{|}{W}}}CH_2 \xrightarrow[-HCl]{CH_3ZnX} \overset{|}{\underset{|}{W}}CH_2 + CH_4$$

The nature of the tungsten methylidene species WCl_xCH_2 is probably more complex. Muetterties proposed its interaction with metal compounds formed from the alkylating agent through carbene, alkyl, and halogen bridges (Fig. 6.15A). A similar complex can also arise in the initial interaction of tungsten hexachloride with metal-alkyl (Fig. 6.15 B). In such a complex α-hydrogen elimination may become a more facile process explaining the generation of the carbene species in the system. The finding that the reactivity of the binary systems WC_6-MR varies drastically with the nature of M suggests that the metal M must be intimately involved in the relevant bonding of the active species. Admitting the formation of a tungstacyclobutane intermediate of type C (Fig. 6.15) in the above mechanistic scheme, Muetterties rationalized the prevalence of degenerate metathesis *versus* productive metathesis in the reaction of terminal olefin.[57] In this case, assuming that an RCH carbene carrier

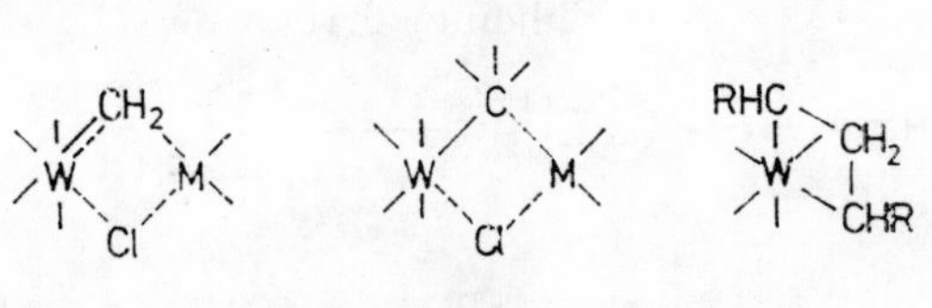

A   B   C

Fig. 6.15. — Bridged complex and metallacy-
clobutane as intermediates in metathesis.

is preferred over a CH_2 carbene species, it is evident that terminal olefins would preferentially metathesize to regenerate the original terminal olefins.

Strong support for the carbene mechanism is supplied by the detailed studies of Katz et al.[58-60] and Grubbs et al.[61-63]

Katz and McGinnis [58] interpreted the metathesis mechanism by a chain reaction propagated by the union of an olefin with a metallacarbene (equation 6.59).

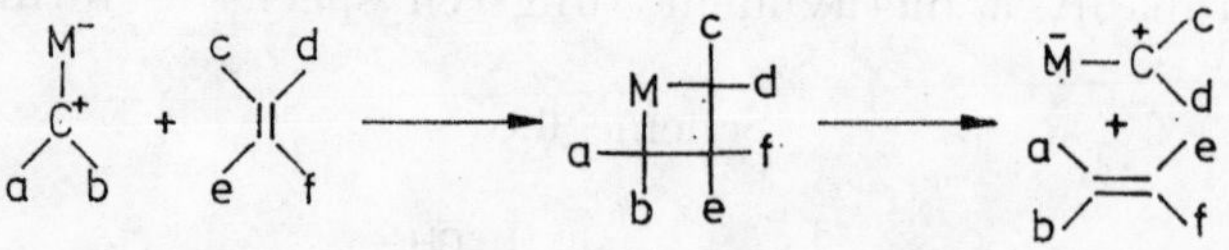

$$\tag{6.59}$$

where M represents the transition metal of the carbenic complex and a, b, c, d, e and f represent hydrogen or alkyl groups. The carbene scheme successfully explained the reaction products formed in the cometathesis of cyclo-octene with 2-butene and 4-octene or of cyclo-octene and 2-hexene. As seen in scheme 6.9 a pairwise (cyclobutane) pathway in the cometathesis of a cyclo-olefin with two symmetrical olefins would lead, as first products, to the symmetrical scission dienes, I and II.

Scheme 6.9

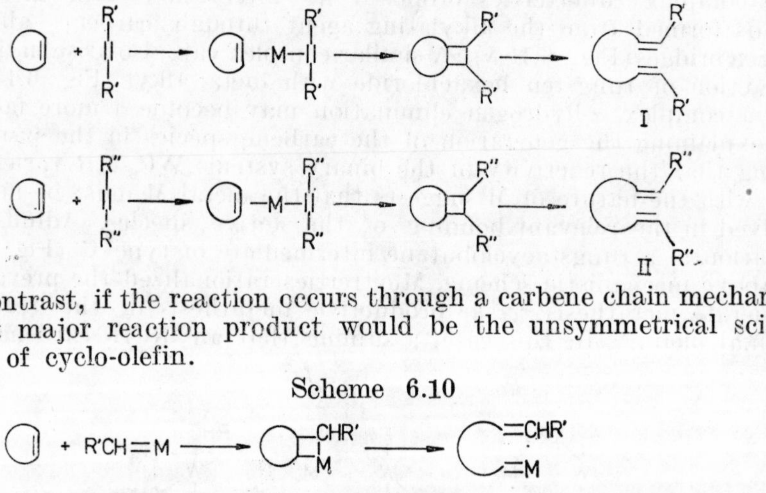

In contrast, if the reaction occurs through a carbene chain mechanism, the first major reaction product would be the unsymmetrical scission product of cyclo-olefin.

Scheme 6.10

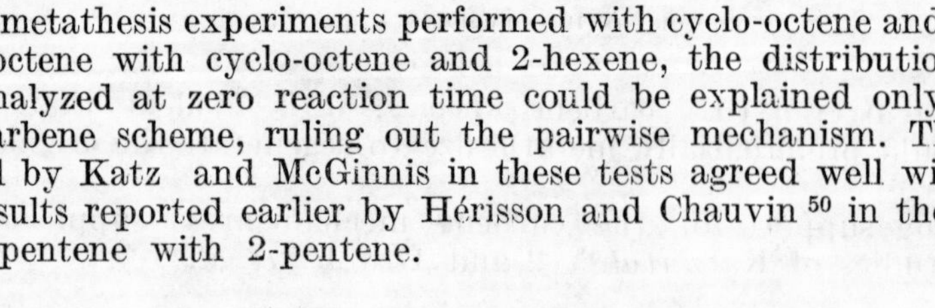

In cometathesis experiments performed with cyclo-octene and 2-butene or of 4-octene with cyclo-octene and 2-hexene, the distribution of products analyzed at zero reaction time could be explained only by the chain carbene scheme, ruling out the pairwise mechanism. The results obtained by Katz and McGinnis in these tests agreed well with the similar results reported earlier by Hérisson and Chauvin [50] in the reaction of cyclopentene with 2-pentene.

By means of the above scheme Katz and McGinnis interpreted a series of known facts such as the acetylene metathesis of acetylene by a metal-lacarbyne mechanism, the formation of large rings and catenanes from cyclo-olefins by cyclization of the terminal carbene at an internal double bond of the chain, and the formation of products containing $(C_4H_6)_n$ fragments in the oligomers from 1,5-cyclo-octadiene or 1,4,7-cyclododecatriene or from their adducts with 1-pentene, the high molecular weight of polymers formed early in the reaction of cyclo-olefins and the generation of cyclopropane and methylcyclopropane from ethylene through metallacarbene involvement.

Further evidence for the carbene mechanism was presented by Katz and Rothchild [60] on studying the tungsten- and molybdenum-catalyzed metathesis of 2,2'-divinylbiphenyl-d_4 and unlabelled 2,2'-divinylbiphenyl. The pairwise mechanism predicts the formation of ethene-d_4 and unlabelled ethene as major products (besides phenanthrene), at short reaction time assuming that the diene reacts more readily bound than unbound.

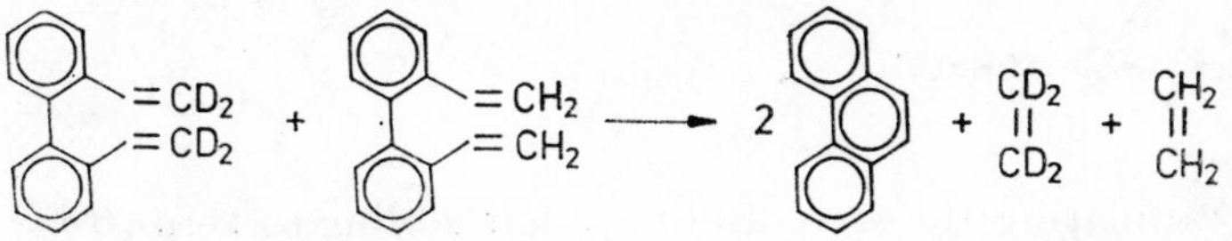

$$(6.60)$$

The molar ratio of $1:2:1$ found for the mixture of ethene-d_4 and ethene-d_6, respectively, formed besides phenanthrene in the above reaction, was interpreted in terms of the carbene chain mechanism.

Scheme 6.11

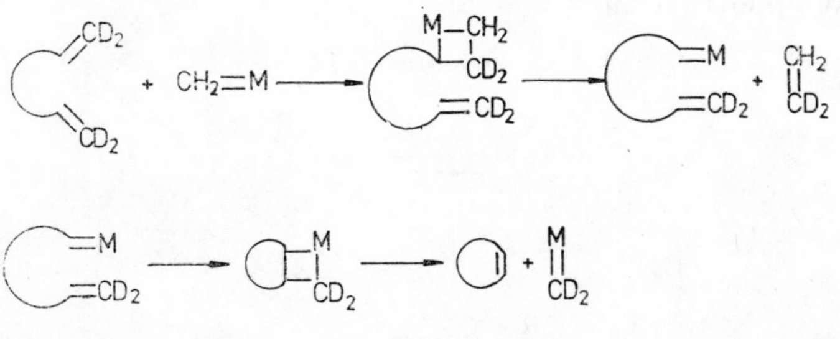

Elegant studies on the carbene mechanism were reported by Grubbs and co-workers [61-63] by using deuterated linear dienes as starting materials. Examination of the reaction products obtained from the mixture of 1,7-octadiene-1,1,8,8-d_4 [61, 62] allowed a distinction to be made between the pairwise and chain carbene mechanism. The formation of a completely randomized ethene-d_4:-d_2:-d_6 equilibrium mixture in a $1:2:1$ ratio (in addition to cyclohexene) supported the operation of a carbene mechanism and not a "pairwise" carbon interchange.

Scheme 6.12

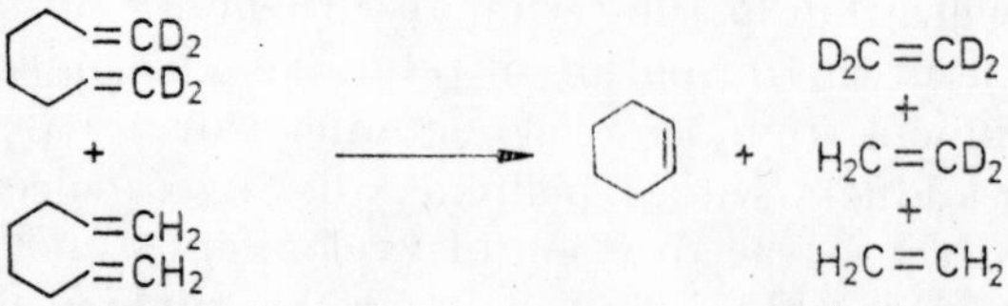

A similar conclusion was drawn by analyzing the product distribution in the metathesis of cis-cis-2,8-decadiene and cis-cis-2,8-decadiene-1,1,1, 10,10,10-d_6[63], taking into account the isotopic scrambling over a wide range of conversions.

Scheme 6.13

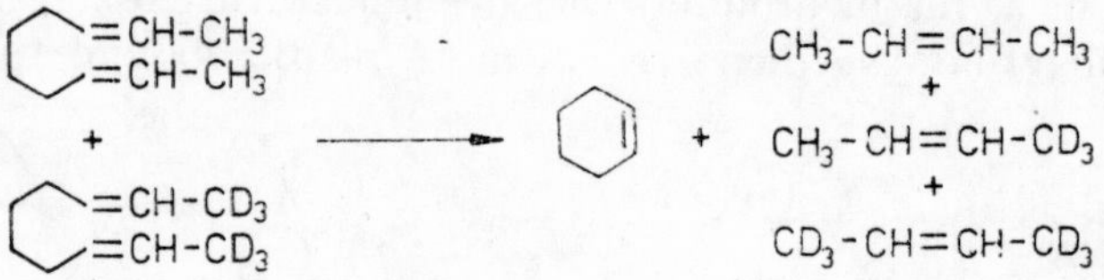

By extrapolating the stereoisomeric scrambling to zero (trans :-cis = 0), the plot of cis-: trans-2-butene versus label ratio, the values of the isotopic label ratio corresponding to zero scrambling were obtained. The extrapolated value was that expected according to the carbene chain mechanism.

An attractive carbene mechanism for olefin metathesis without involving the intermediacy of a metallacarbene-olefin complex was proposed by Puddephatt et al.[64]

Scheme 6.14

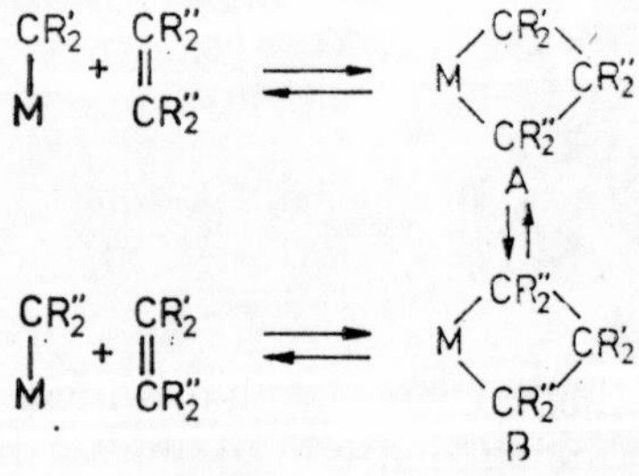

The direct metallacyclobutane isomerization A⇌B has experimental support from the chemistry of substituted platinacyclobutanes. Nevertheless, it was unclear whether the isomerization A⇌B might occur through a concerted pathway without complete carbon-carbon bond cleavage (sequence 1—2, Scheme 6.14a) or through a stepwise pathway (sequence 3—5, Scheme 6.14a) involving carbon-carbon bond cleavage with the intermediacy of a metallacarbene-olefin complex.

Scheme 6.14a

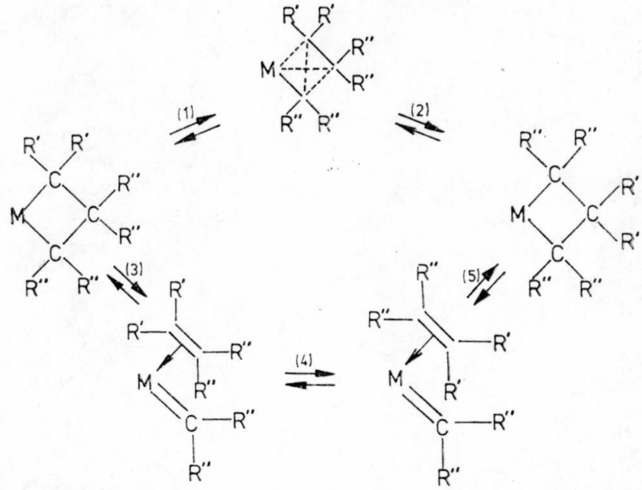

Based on Chauvin's mechanism, Basset and co-workers [14, 65, 66] proposed a stereochemical process for the coordination of the olefin to the metallacarbene moiety, which accounted for the nature of the metathesis product. In this model they assumed that the C_{carb} had an sp^2 character with a bond order equal to 2, and that the product distribution is determined by the geometry of approach of the olefin (Fig. 6.16).

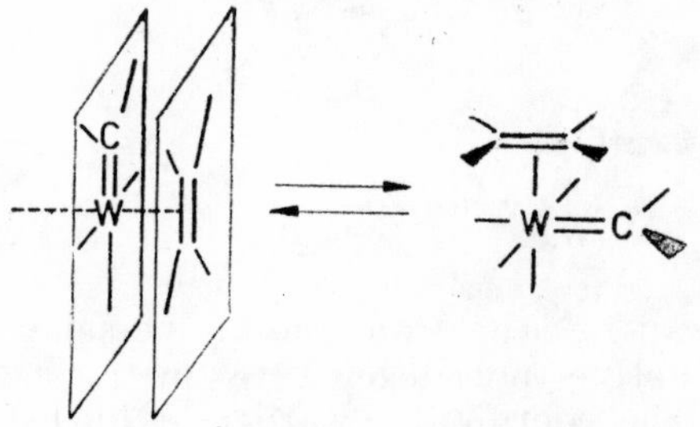

Fig. 6.16. — Steric model for olefin coordination to the metallacarbene.

This mechanism of approach rationalized the simultaneous occurrence of productive metathesis, degenerate metathesis, and cis-trans isomerization. Applied to the reaction of 2-pentene, the possible modes of olefin approach to the two tungsten-carbene moieties, $W = CHCH_3$ and $W = CHC_2H_5$, accounted for the formation of cis and trans butenes and hexenes, cis and trans isomerization of 2-pentene as well as the degenerate transformation of *cis* or *trans*-2-pentene.

Scheme 6.15

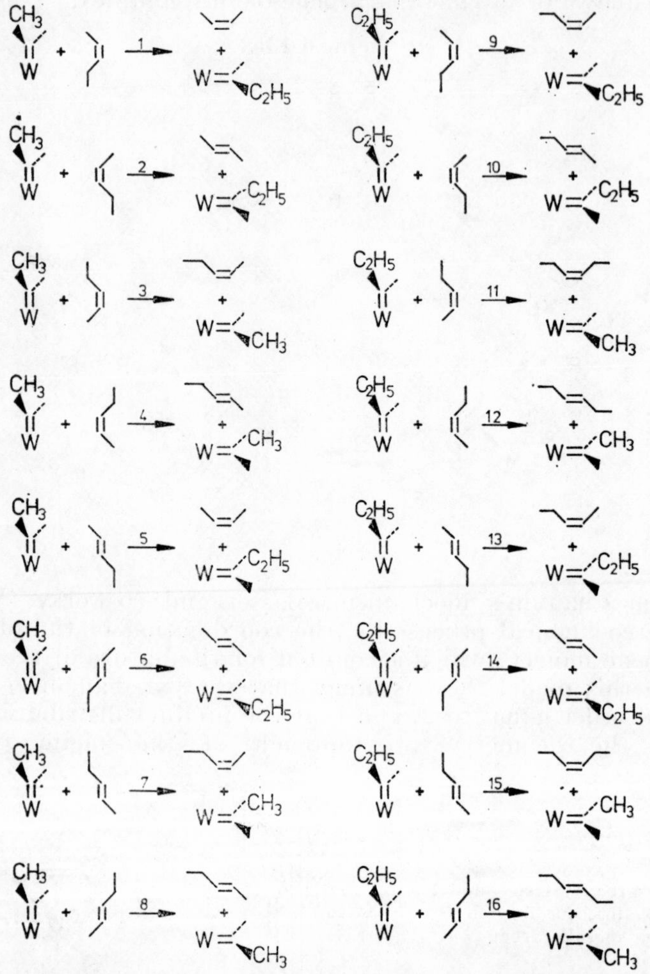

As can be seen from the above scheme, the elementary steps leading to productive (non-degenerate) metathesis result in a change of the nature of the coordinated carbene while the steps leading to *cis-trans* isomerization of the starting olefin do not change the metallacarbene moiety. Furthermore, the "regenerative" (degenerate) reactions occur with retention or inversion of the steric configurations. Kinetic data obtained at low conversion with 2-pentene seem to confirm the validity of this mechanism adopted by Basset and co-workers.

By studying the nature of the olefins first formed in the metathesis of 1,7-octadiene or 4-octene catalyzed by rhenium- or molybdenum-based systems, Farona *et al.*[67][68] were able to collect considerable evidence in support of a coordinated carbene mechanism. The detection of 1-butene and 1,7-decadiene in certain amounts from 1,7-octadiene in the presence of the system $Re(CO)_5Cl \cdot C_2H_5AlCl_2$, besides the normal metathesis products, ethene and cyclohexene, was rationalized by a mechanistic scheme involving the coordination of the olefin substrate to the carbene initiator followed by cleavage of the intermediate metallacyclobutane and subsequent decomplexation of the olefin products.

Scheme 6.16

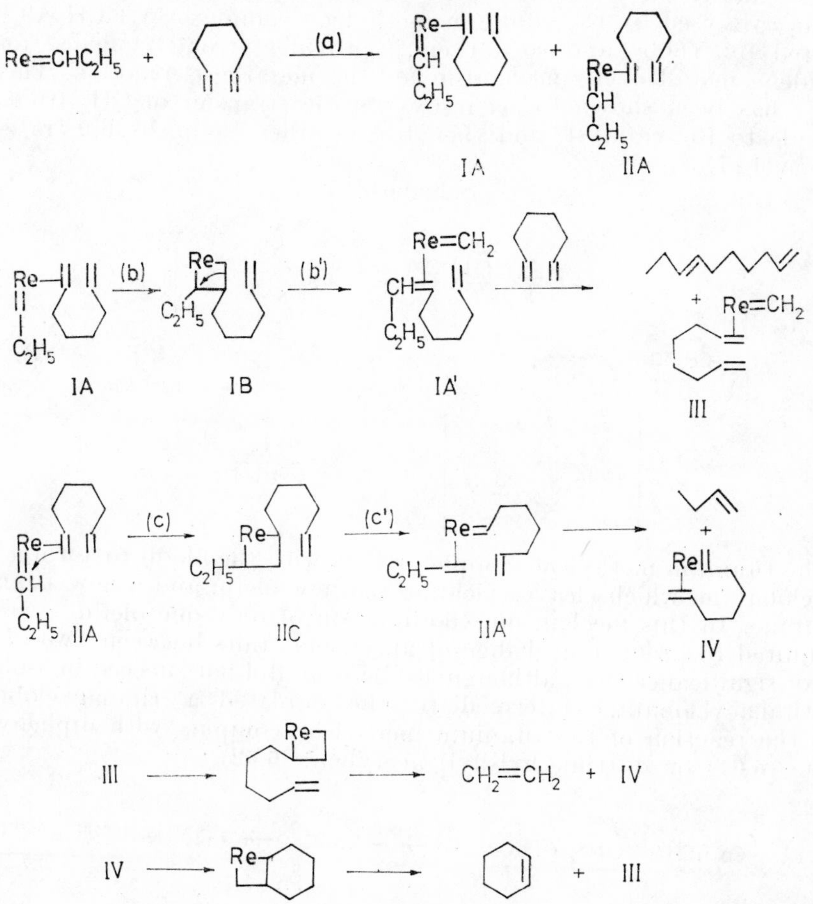

$Re{=}CHC_2H_5$ +

(a)

IA   IIA

(b)   (b')

IA   IB   IA'   III

(c)   (c')

IIA   IIC   IIA'   IV

III $\longrightarrow$ $\longrightarrow$ $CH_2{=}CH_2$ + IV

IV $\longrightarrow$ $\longrightarrow$ + III

The first olefins formed in this reaction were produced in minor quantities approximately equal to the amount of the catalyst. The propagating carbenes alternate between methylene and the carbene formed from the

C_7 fragment, the major products being ethylene and cyclohexene. The same conclusion was drawn from the metathesis of 4-octene with the above catalyst. When CH_3AlCl_2 was used instead of $C_2H_5AlCl_2$ as cocatalyst, the first olefins formed were identified as propene and 2-hexene, for the metatheses of 1,7-octadiene and 4-octene, respectively. In these studies evidence was brought forward for the distinct nature of the carbene species in the initiation and propagation steps. In addition, Farona *et al.* pointed out the existence of alternate pathways for the formation of the initial carbene moiety from the catalytic systems based on rhenium and molybdenum; this feature will be discussed in a later section.

The methylene exchange between isobutene and methylene-cyclohexane catalysed by the titanium methylene complex $Cp_2TiCH_2AlClMe_2$, reported by Tebbe and co-workers,[69] provides strong evidence for an alkylidene metallacycle mechanism for the metathesis reaction. The exchange has been shown to occur by stepwise transfer of CH_2 from one molecule to the catalyst, and then to the other olefin, as illustrated in Scheme 6.17.

Scheme 6.17

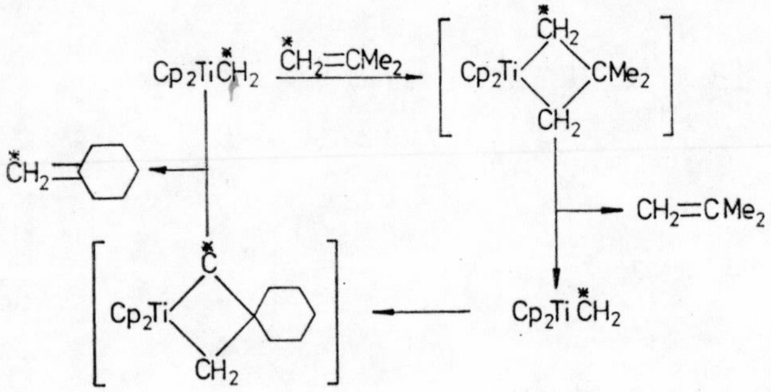

The titanium methylene complex reacts with the olefin to form a titanacyclobutane which cleaves yielding the new olefin and a new titanacyclobutane. In this mechanism, coordination of only one olefin at a time is required to produce exchange of alkylidene units between two olefins. It is of significance that, although Tebbe *et al.* did not succeed in isolating the titanacyclobutane intermediate, they isolated a titanacyclobutene from the reaction of the titanium methylene complex with diphenylacetylene (6.61) or di(trimethylsilyl) acetylene (6.62).

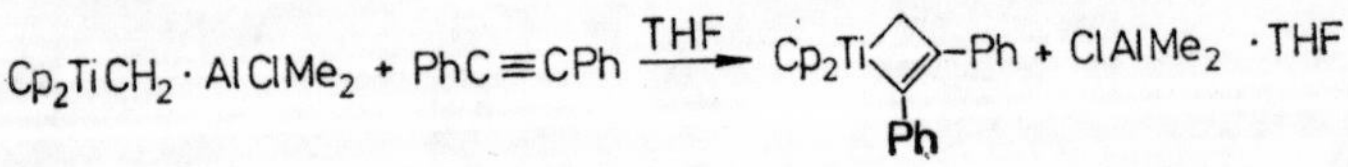

$$Cp_2TiCH_2 \cdot AlClMe_2 + PhC \equiv CPh \xrightarrow{THF} Cp_2Ti\!\!<\!\!>\!\!-Ph + ClAlMe_2 \cdot THF \qquad (6.61)$$

$$Cp_2TiCH_2 \cdot AlClMe_2 + Me_3SiC \equiv CSiMe_3 \xrightarrow{THF} Cp_2Ti\!\!<\!\!>\!\!-SiMe_3 + ClAlMe_2 \cdot THF \qquad (6.62)$$

Although the chain carbene mechanism seems to be correct, the metallacyclobutane intermediates proposed do not satisfactorily explain stereoselectivity of the reaction. For example, as will be outlined in a subsequent chapter, the interactions which have arisen in connection with metallacyclobutane structure do not correlate with the high *trans* content of the cyclobutane and low *cis* stereoselectivity of 2-alkene metathesis. In order to account for the high steric constraints observed in metathesis, Garnier *et al.*[70] proposed the intermediate participation of a new type of a dinuclear species I (Fig. 6.17) where M = tungsten and X = halogen. This proposal is based on the experimentally observed tungsten metathesis complexes stabilized by halide bridges,[71] the possible stabilization of the chlorinated metallacarbene precursor $(CO)_4ClW=CCl_2$ through a μ-dichloro-ditungsten complex, as well as on the essential role of the halogen ligands of the catalytic intermediates derived from WCl_6.[70]

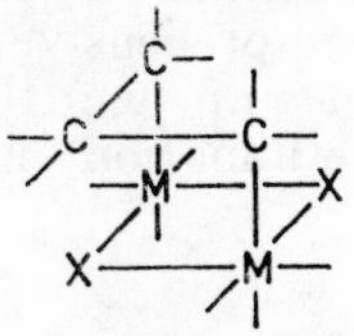

Fig. 6.17. — Proposed dimetallic intermediate for olefin metathesis.[70]

The new mechanistic scheme involves, as in the chain carbene mechanism, a metal carbene as initiator, a metal olefin π-complex and a trans-alkylidenation step. These all are associated with a dinuclear halogen bridged species, as shown is Scheme 6.18.

Scheme 6.18

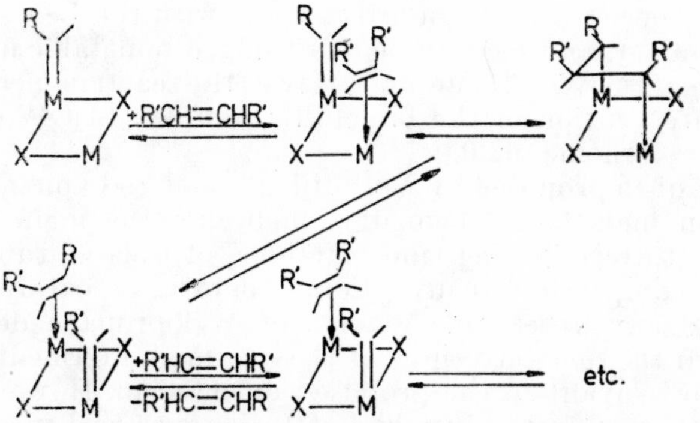

Owing to the larger steric constraints associated with the dinuclear intermediate, the mechanistic scheme proposed by Garnier et al. correlates better with the stereoselectivity observed experimentally in metathesis influenced by substituent interactions.

Important objections against the simple carbene mechanism were set forth by Verkuijlen[72] based on experimental data obtained in metathesis reactions carried out with the catalyst $WOCl_4 \cdot Me_4Sn$ and partially on results reported in the literature. This author observed that the Me_3SnCl derivative of the cocatalyst is essential in maintaining the activity of the system in metathesis of methyl 10-undecenoate or methyl oleate; on completely removing Me_3SnCl, the catalyst becomes inactive. The co-catalyst seems not only to be a source for generating the carbene species in the initiation step, but governs the level of activity in propagation and termination steps as well. In consequence, Verkuijlen presented a more comprehensive mechanistic scheme for metathesis reactions induced by bicomponent catalyst systems comprising initiation, propagation and termination steps.

Scheme 6.19

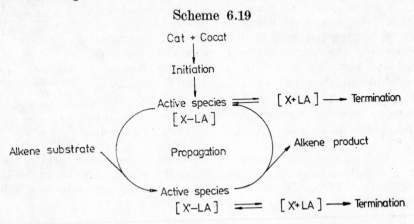

The active species arise from interactions with the Lewis acids derived from the cocatalyst, and seem to be of a bridged bimetallic nature. For the particular case of the $WOCl_4 \cdot Me_4Sn$ catalyst, the reaction mechanism would involve the intervention of the bimetallic complex $Cl_3OW-CH_2-SnMe_3$ as depicted in Scheme 6.20.

The mechanism proposed by Verkuijlen seems to explain several facts encountered in metathesis chemistry, including the ineffectiveness, and the sensitivity to terminating side reactions, of isolable tungsten carbene complexes as compared to conventional metathesis catalysts. The non-generation, to any observable extent, of cyclopropane derivatives, the instability, and the ineffectiveness of the assumed intermediates in meta-thesis, are also relevant to the proposed mechanism. However, in the new mechanistic scheme, deeper insight has to be exercised in relation to the correlation of reaction stereoselectivity with the applied cocatalyst, as well as to the nature and structure of the suggested bimetallic interme-diates. For the generation of the active species $Cl_3OW-CH_2-SnMe_3$, an interaction scheme between $WOCl_4$ and Me_4Sn was advanced which will be dealt with in a subsequent section (p. 248).

An interesting mechanism for olefin metathesis emphasizing the role of the cocatalyst was discussed recently by Grubbs et al.[73] By using the catalyst $Cp_2TiCH_2 \cdot AlClMe_2$, reported earlier by Tebbe, [69] Grubbs succeed-ed in isolating stable titanacyclobutanes presumed to be intermediates

Scheme 6.20

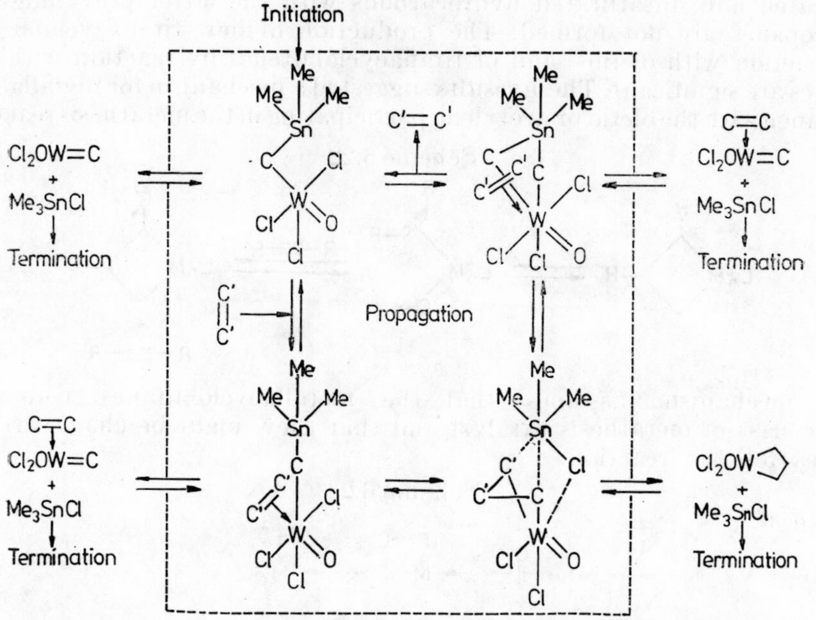

in metathesis reaction. The main reactions of formation and destruction of these titanacyclobutanes are outlined in Scheme 6.21.

Scheme 6.21

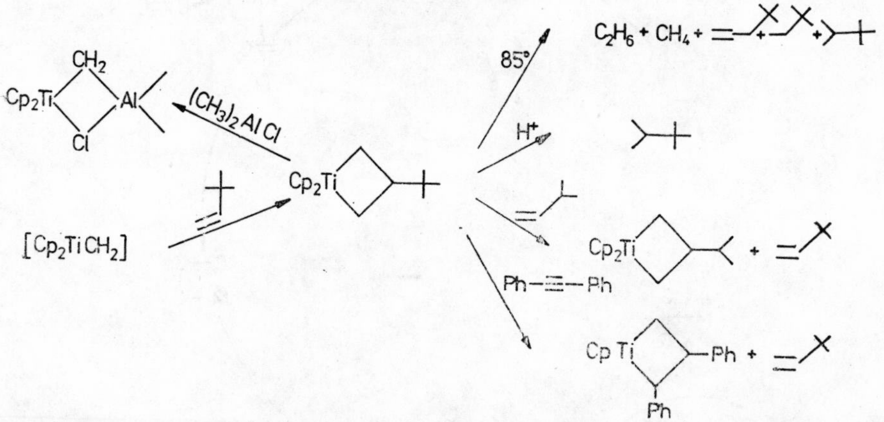

Several important features deserve special attention in the above scheme. The Lewis acid cocatalyst facilitates the decomposition of tita-nacyclobutane favouring the regeneration of the catalyst while pyridine or one of its derivatives removes the cocatalyst from the titanium com-pelx, liberating the reactive Cp_2TiCH_2 fragment responsible for olefin reaction. The thermal decomposition of the titanacyclobutane provides saturated and unsaturated hydrocarbons with the latter prevailing. Cyclopropanes are not formed. The production of new titanacyclobutanes by reaction with olefins, and of titanacyclobutenes by reaction with acetylenes are significant. These results suggested a mechanism for metallacycle exchange with the olefin or acetylene participating in the metathesis reaction.

Scheme 6.22

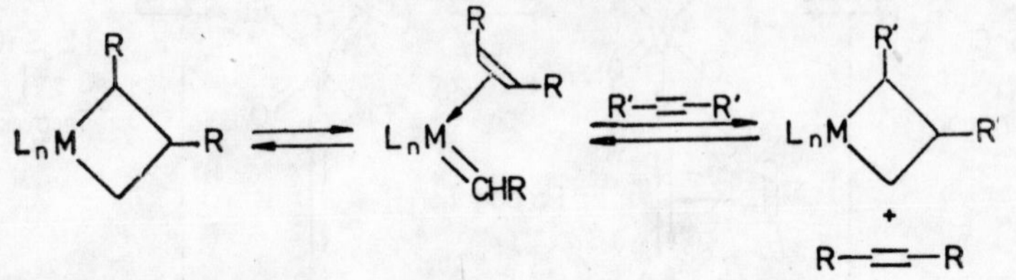

This mechanism assumes that the metallacyclobutanes constitute a new class of metathesis catalyst and that they might be chain-carrying species for the reaction.

Scheme 6.23

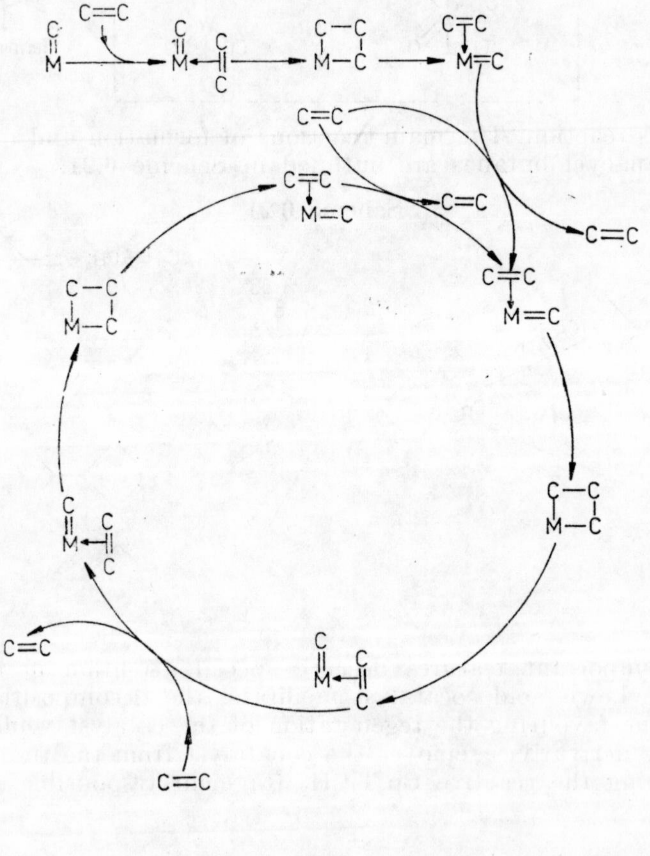

Based on the available experimental data on carbene mechanism, Masters [74] discussed a generalized catalysis cycle for transition metal-catalyzed alkene metathesis. The proposed cycle involved an initiating metallacarbene, the complexation of the olefin to the transition metal, the formation of a metallacyclobutane intermediate, the cleavage of this intermediate to yield a new olefin metal complex, the decomplexation of the product olefin, and further complexation of the substrate olefin.

6.2.1.3. Other Proposed Mechanisms

In the initial stage of the study on metathesis, for the disproportionation of propene to ethene and 2-butene, [75] a mechanism was proposed consisting of an addition reaction between two molecules of propene followed by an elimination reaction. This mechanism was soon ruled out on the basis of the first results obtained when radioactive propene was reacted (6.63)

$$\text{(6.63)}$$

According to this proposal, 1-^{14}C-propene would yield ethene and butene both radioactive; however, the metathesis of 1-^{14}C-propene yields radioactive ethene and non-radioactive butene.

Calderon [37] discussed at an early stage the possibility of a mechanism of olefin metathesis through transalkylation.

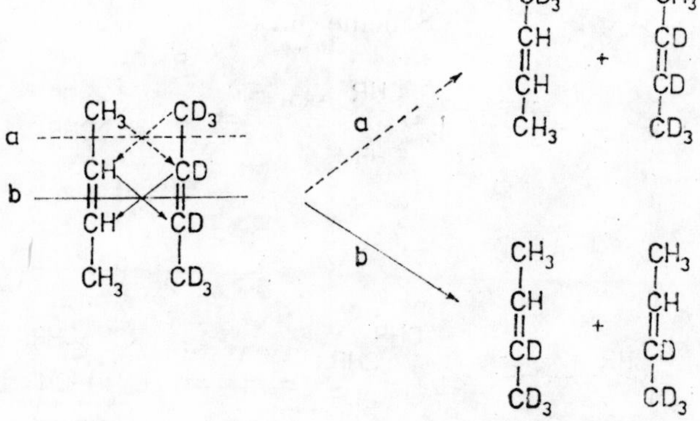

$$\text{(6.64)}$$

If the reaction were to proceed through transalkylation with the alkyl groups undergoing reciprocal substitution, the reaction of 2-butene with 2-butene-d_8 should yield 2-butene-d_3 and 2-butene-d_5 as reaction products. However, this reaction yields only 2-butene-d_4. So it is clear that the mechanism involves transalkylidenation, i.e. the alkylidene groups change their position reciprocally (6.64b).

Another mechanism, which suggested, for the first time, the possibility of involving carbene structures (namely methylene) during the metathesis, was put forward by Lewandos and Pettit [76] postulating the formation of a four-centred organometallic complex in the reaction

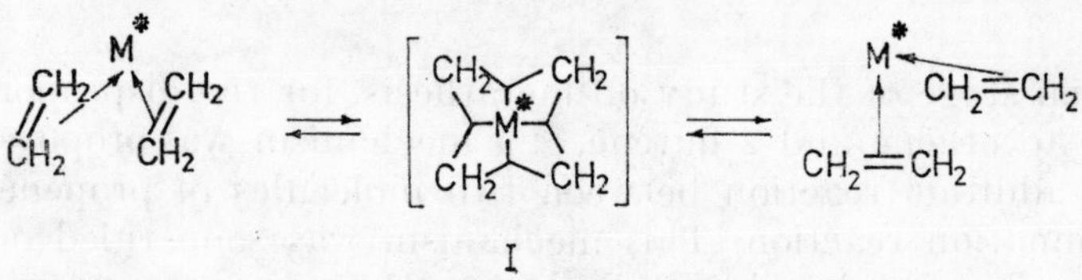

$$
\begin{array}{ccc}
\underset{\substack{CH_2 \\ \| \\ CH_2}}{\overset{\substack{M^* \\ CH_2}}{}} \quad\rightleftharpoons\quad \left[\underset{\substack{CH_2 \\ \| \\ CH_2}}{\overset{\substack{CH_2 \\ M^* \\ CH_2}}{}}\right] \quad\rightleftharpoons\quad \underset{CH_2=CH_2}{\overset{\substack{M^* \\ CH_2=CH_2}}{}} & & (6.65)
\end{array}
$$

I

According to this mechanism, in the first step, a diolefin complex of the transition metal arises, which forms a four-centred organometallic intermediate (I) rather than a cyclobutane transition state. The intermediate I can give rise to reaction products or can lead to an initial diolefinic complex, through a reversible process. Evidence supporting this mechanism was derived from an attempt to cleave cyclobutane under the conditions employed for the metathesis of monodeuterated ethene. The cleavage reaction occurred in a very low yield (3%). Under similar conditions, ethene yielded cyclobutane, also in very small amounts (below 0.1%). Calculations effected by Lewandos and Pettit, using the molecular orbital method, showed that the formation of the intermediate I from the diolefinic complex is symmetry-allowed.

Taube and Seyferth [77] supposed the intermediacy of a dialkylidene or a dicarbene complex which *via* oxidative addition of a coordinated olefin could give rise to a metallacyclobutane species.

Scheme 6.24

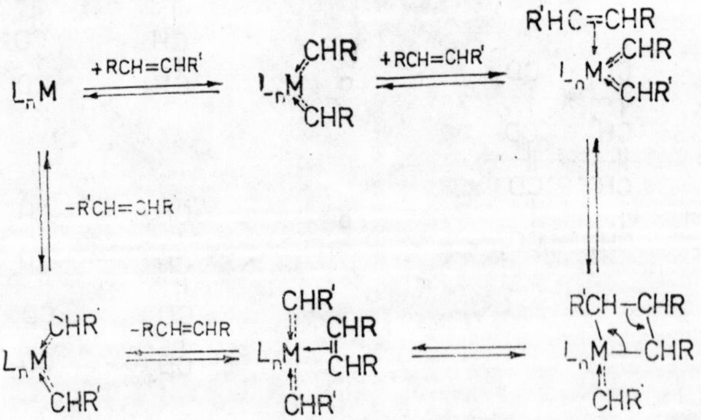

This metallacyclobutane species may generate a new dicarbene complex and release the new olefin. By the proposed mechanism the above authors explained the formation of the asymmetrical products in cometathesis reaction of α-olefins with cyclo-olefins catalysed by tungsten complexes.

Haines and Leigh [78] suggested the possibility of forming tetramethylene and trimethylene intermediates in metathesis, in the presence of heterogeneous catalytic systems.

Scheme 6.25

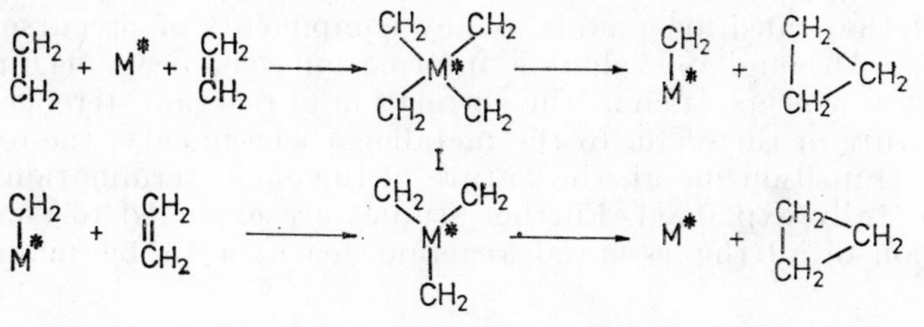

$$
\begin{array}{c}
CH_2 \\
\| \\
CH_2
\end{array}
+ M^* +
\begin{array}{c}
CH_2 \\
\| \\
CH_2
\end{array}
\longrightarrow
\begin{array}{c}
CH_2 \quad CH_2 \\
M^* \\
CH_2 \quad CH_2
\end{array}
\longrightarrow
\begin{array}{c}
CH_2 \\
| \\
M^*
\end{array}
+
\begin{array}{c}
CH_2 \\
CH_2
\end{array} CH_2
$$

I

$$
\begin{array}{c}
CH_2 \\
| \\
M^*
\end{array}
+
\begin{array}{c}
CH_2 \\
\| \\
CH_2
\end{array}
\longrightarrow
\begin{array}{c}
CH_2 \quad CH_2 \\
M^* \\
CH_2
\end{array}
\longleftarrow
M^* +
\begin{array}{c}
CH_2 \\
CH_2
\end{array} CH_2
$$

II

These authors explained, on the basis of this mechanism, the production of cyclopropane in the metathesis of ethene. The tetramethylene intermediate I can form cyclopropane and a carbene species; the latter may give rise to a trimethylene intermediate II which can lead to a new cyclopropane regenerating the catalytic system.

A pairwise mechanism involving a metallacyclopentanic intermediate was proposed by Grubbs and Brunk.[79] According to this suggestion, a diolefinic complex is first formed, which leads to a metallacyclopentane intermediate III; this rearranges to a new metallacyclopentane intermediate IV which yields the reaction products (6.66)

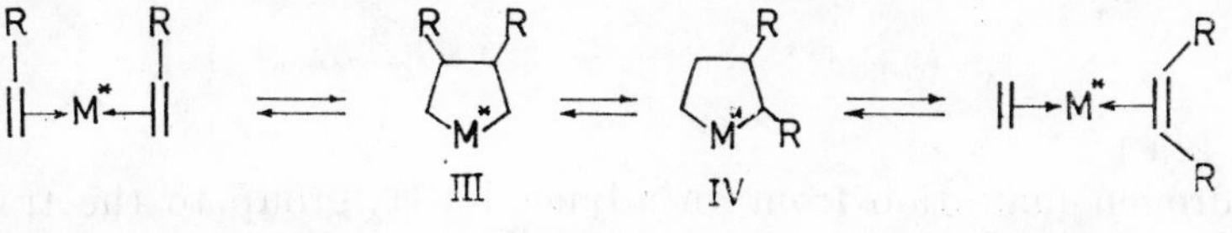

$$
\begin{array}{c}
R \quad R \\
\| \rightarrow M^* \leftarrow \|
\end{array}
\rightleftharpoons
\underset{III}{\overset{R \quad R}{M^*}}
\rightleftharpoons
\underset{IV}{\overset{R}{M^*}} R
\rightleftharpoons
\begin{array}{c}
R \\
\| \rightarrow M^* \leftarrow \\
R
\end{array}
\tag{6.66}
$$

This mechanism, based on a series of reactions involving metallacyclic complexes [80] is consistent with the conversion of 1,4-dilithiumbutane (I, M = Li; R = H) into ethene in the presence of a catalytic system based on WCl_6.

6.2.2. Features of Carbene Mechanism

The chain mechanism involving carbenes for olefin metathesis has received firm experimental support from kinetic studies, examination of the first-formed reaction products, and studies on other aspects of metallacarbene chemistry related to metathesis. Considerable effort has been devoted to understanding the nature of the initiation, propagation and termination steps. The main studies aimed at elucidation of reaction mechanism have concerned the mode of generation of the metallacarbene active species, the reaction of metallacarbene with olefins, the role of the alkylidene complex or metallacyclobutane as the dominant active intermediate, the nature of the chain-carrying metallacarbene, the characterization of the metallacarbene active species, the role of ligands on the transition metal, the role of the cocatalyst in metathesis activity, and the influence of the transition metals as key components of precursor of active species. Although a valuable information has been accumulated, many essential aspects, such as the correlation of reaction stereochemistry with the nature of the olefin to the metallacarbene moiety, the oxidation state of the transition metal, the nature of the chain termination step(s), are still not fully explained. Further studies are expected to enable full interpretation of all the essential reaction features to be made.

6.2.2.1. Formation of the Initiating Metallacarbene

Depending on the nature of the catalytic system and on the presence or absence of cocatalyst, several pathways have been suggested for the formation of the initiating metallacarbene species.

The majority of homogeneous metathesis catalysts are obtained from tungsten-, molybdenum- or rhenium halides or complexes and a metal alkyl. Typical examples were given in a previous chapter. It is well known that in such systems alkyl transfer readily occurs from the organometal compound to the transition metal giving rise to alkylated transition metal complexes. For example, the ethyl aluminium compound can alkylate a transition metal halide, for instance

$$C_2H_5AlCl_2 \; + \; MX_n \longrightarrow C_2H_5MX_m \; + \; AlCl_3 \tag{6.67}$$

where $X = Cl$.

α-Hydrogen migration from an adjacent CH_2 group to the transition metal (whose coordination and oxidation numbers increase accordingly) would give a carbene species which could then initiate the metathesis process

$$CH_3-\overset{\overset{\textstyle H}{|}}{\underset{\underset{\textstyle H}{|}}{C}}-MX_m \longrightarrow CH_3-CH\overset{\overset{\textstyle H}{|}}{=}MX_m \tag{6.68}$$

α-Hydrogen eliminations of this type have been shown by Green et al.[81] to occur in tungsten-methyl complexes formed on treating $[W(n^5-C_5H_5)_2CH_3(C_2H_4)]PF_6$ with PMe_2Ph.

The formation of methane in the reaction of WCl_6 with $(CH_3)_2Zn$ was rationalized by Muetterties[56] through α-hydrogen elimination from a methyl group attached to tungsten.

Scheme 6.26

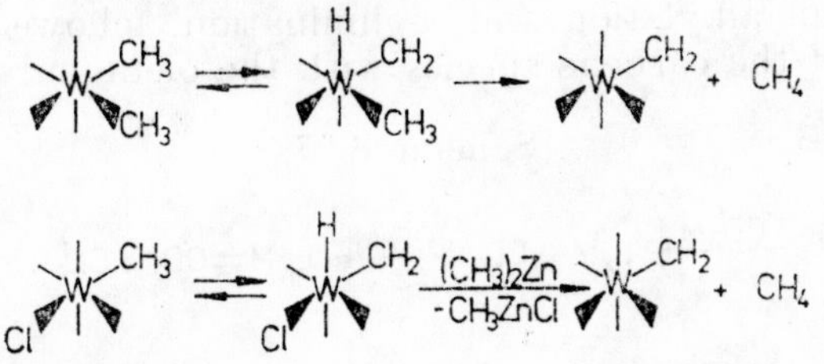

The resulting tungsten methylene complex would then initiate the metathesis. The existence of a tungsten methylene complex in a simple form is very unlikely. It is probable that the carbene complex interacts with the organometallic compounds or the resulting Lewis acid to give bridged dinuclear compounds, which are more stable than the initial carbene. Furthermore, the initial interaction of the transition metal halide with the metal alkyl probably leads to alkyl bridged complex wherein α-hydrogen elimination may occur more easily. The strong effect of the nature of metal alkyl on the reactivity of WCl_6-MR systems seems to support this interaction model.

α-Hydrogen abstraction from alkyl ligands of transition metals of groups V_b and VI_b has been well documented by Schrock and co-workers who synthesized and isolated such alkylidene compounds.[82,83]

Good evidence for the generation of carbene species in alkylaluminium-molybdenum and alkyltin-tungsten complexes was reported by Grubbs and Hoppin.[84] On mixing the transition metal complex with the appropiate methylaluminium or methyltin cocatalyst, methane and ethylene were formed (equations 6.69 and 6.70).

$$Mo(Ph_3P)_2Cl_2(NO)_2 + Me_3Al_2Cl_3 \xrightarrow{PhCl} CH_4 + C_2H_4 + Mo-cat. \qquad (6.69)$$

$$WCl_6 + Me_4Sn \xrightarrow{PhCl} CH_4 + C_2H_4 + W-cat. \qquad (6.70)$$

When $(CD_3)_4Sn$ was used, the hydrocarbon products were perdeuterated. The reaction of 2,8-decadiene in the presence of these catalysts gave propene along with the usual metathesis products 2-butene and cyclohexene.

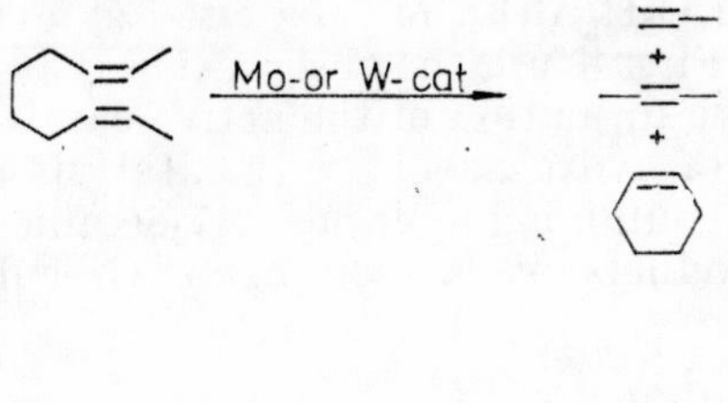

$$(6.71)$$

When $(CD_3)_4Sn \cdot WCl_6$ was used, propene-1,1-d_2 resulted. When the reactions were carried out with 2,8-decadiene-1,1,1,10,10,10-d_6 the initial product was propene-3,3,3,-d_3 while 2,8-decadiene-1, 1-d_2 in the presence of $(CD_3)_4Sn \cdot WCl_6$ gave propene-1,1,3,3,3,-d_5.

Grubbs and Hoppin rationalized their results by an initiating mechanism involving an alkylation and α-elimination, followed by the subsequent reaction of the carbene species with the olefin substrate.

Scheme 6.27

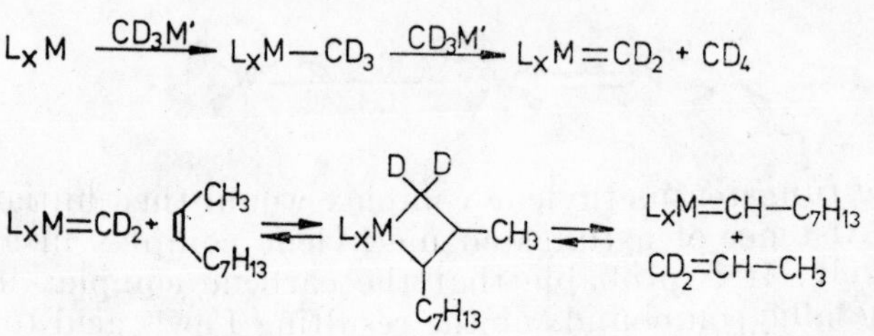
$L_xM \xrightarrow{CD_3M'} L_xM{-}CD_3 \xrightarrow{CD_3M'} L_xM{=}CD_2 + CD_4$
$L_xM{=}CD_2 +$ CH₃ / C₇H₁₃ $\rightleftharpoons$ L_xM D D CH₃ C₇H₁₃ $\rightleftharpoons$ $L_xM{=}CH{-}C_7H_{13}$ + $CD_2{=}CH{-}CH_3$

The yield of propene relative to the amount of transition metal was correlated with the fraction of active catalyst formed in the initiation step from the total amount of catalyst. These authors observed that the molybdenum system is much more active for metathesis than the tungsten system and that it produces a much higher yield of propene (90% compared to 8%). The formation of ethylene was attributed to dimerization of the metallacarbene during initiation

$$2L_xM{=}CH_2 \longrightarrow CH_2{=}CH_2 \qquad (6.72)$$

This reaction indicates one of the chain termination mechanisms.

Farona *et al.* [67,68] pointed out different ways of generating the carbene species in binary systems of rhenium or molybdenum carbonyl complexes and organo-aluminium compounds. In the case of a catalyst consisting of rhenium chloropentacarbonyl and ethylaluminium dichloride, the initial carbene was shown to be formed by attack of an ethyl group on coordinated CO, followed by exchange of oxygen for hydrogen to create a coordinated propylidene molecule; the second site for olefin coordination results from transfer of Cl from Re to Al.

The following observation on the metathesis of 1,7-octadiene catalyzed by $Re(CO)_5Cl$ and $C_2H_5AlCl_2$ may be cited as evidence for the above process (Scheme 6.28). First of all, the observation that no gas (hydrocarbon, CO) was evolved in the formation of the active catalyst, is in contrast to the known formation of hydrocarbons with catalysts based on WCl_6. The presence of an oxygen-containing molecule, 1,7-octadien-3-one, was detected among metathesis products of 1,7-octadiene. Greenlee and Farona sug-

Scheme 6.28

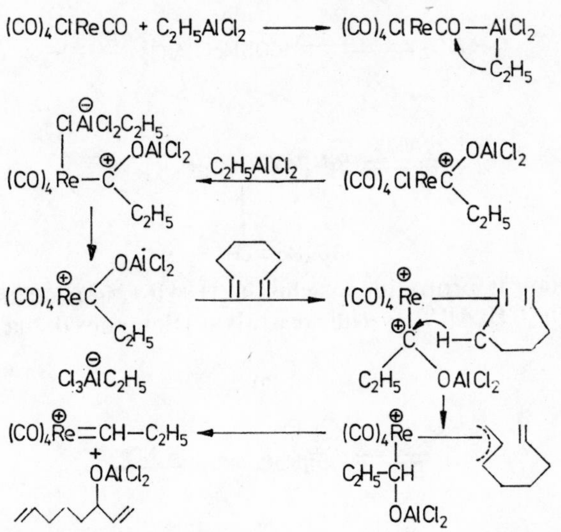

gested that 1,7-octadien-3-one was formed by decomposition of the $ROAlCl_2$ complex to the carbonyl compound and $HAlCl_2$

(6.73)

Furthermore, the nature of first-formed olefins, 1-butene and 1 7-de-cadiene, which occur in catalytic amounts beside the normal metathesis products, as well as the detection of cyclopropane derivatives formed by insertion of the propylidene moiety into the double bond of ethylene, 1-butene or cyclohexene, supports the proposed mode of generation of the initiating propylidene species in the catalytic system.

For the binary catalyst consisting of $Mo(CO)_5Py$ and $C_2H_5AlCl_2$ or CH_3AlCl_2, Motz and Farona [68] found that the formation of the initial co-ordinated carbene is dependent on the nature of the solvent. In chloro-benzene, the initial carbenes are ethylidene and methylene when em-ploying $C_2H_5AlCl_2$ or CH_3AlCl_2 as cocatalysts, respectively; however, in heptane as solvent, coordinated propylidene and ethylidene arise from the corresponding cocatalysts. In the first case, the formation of the car-bene species, which occurred with gas evolution, was interpreted, accord-ing to the mechanism of Muetterties as alkylation of Mo followed by subsequent α-hydrogen elimination. Scheme 6.29 illustrates the formation of ethylidene compounds for the $C_2H_5AlCl_2$ cocatalyst. By a similar pathway the methylene species can arise from CH_3AlCl_2. In heptane, the mechanism of generation of the carbene species is analogous to that found by Greenlee and Farona for $Re(CO)_5Cl$ catalyst.

Scheme 6.29

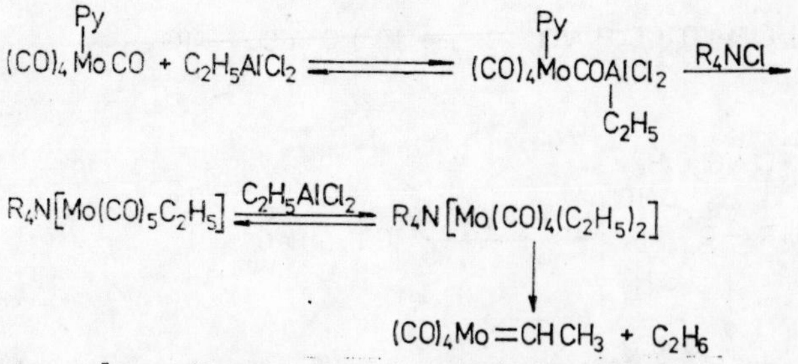

$$(CO)_4\overset{Py}{Mo}CO + C_2H_5AlCl_2 \rightleftharpoons (CO)_4\overset{Py}{\underset{C_2H_5}{Mo}}COAlCl_2 \xrightarrow{R_4NCl}$$

$$R_4N[Mo(CO)_5C_2H_5] \xrightarrow{C_2H_5AlCl_2} R_4N[Mo(CO)_4(C_2H_5)_2]$$

$$\downarrow$$

$$(CO)_4Mo\!=\!CHCH_3 + C_2H_6$$

The initiating carbene is propylidene when $C_2H_5AlCl_2$ is used as cocatalyst and ethylidene when CH_3AlCl_2 is the cocatalyst (Schemes 6.30a and 6.30b).

Scheme 6.30

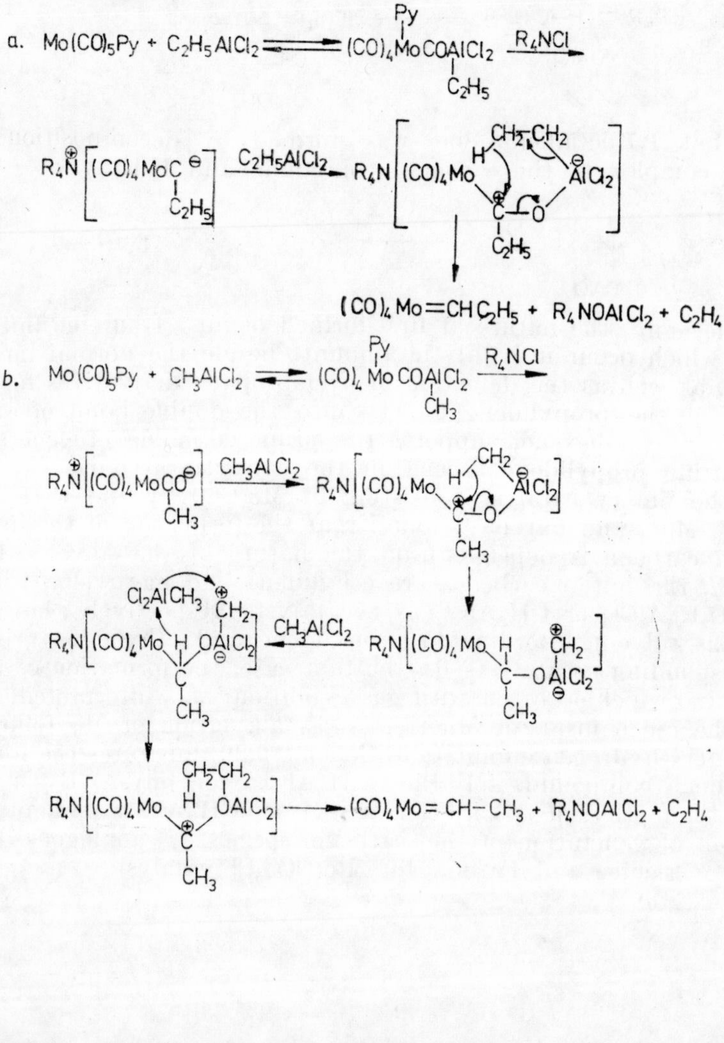

a. $\quad Mo(CO)_5Py + C_2H_5AlCl_2 \rightleftharpoons (CO)_4\overset{Py}{\underset{C_2H_5}{Mo}}COAlCl_2 \xrightarrow{R_4NCl}$

$$R_4\overset{\oplus}{N}\left[(CO)_4Mo\overset{\ominus}{\underset{C_2H_5}{C}}\right] \xrightarrow{C_2H_5AlCl_2} R_4N\left[(CO)_4Mo\cdots\right]$$

$$\downarrow$$

$$(CO)_4Mo\!=\!CHC_2H_5 + R_4NOAlCl_2 + C_2H_4$$

b. $\quad Mo(CO)_5Py + CH_3AlCl_2 \rightleftharpoons (CO)_4\overset{Py}{\underset{CH_3}{Mo}}COAlCl_2 \xrightarrow{R_4NCl}$

$$R_4\overset{\oplus}{N}\left[(CO)_4Mo\overset{\ominus}{\underset{CH_3}{C}O}\right] \xrightarrow{CH_3AlCl_2} R_4N\left[(CO)_4Mo\cdots\right]$$

$$R_4N\left[(CO)_4Mo\cdots\right] \xleftarrow{CH_3AlCl_2} R_4N\left[(CO)_4Mo\cdots\right]$$

$$\downarrow$$

$$R_4N\left[(CO)_4Mo\cdots\right] \longrightarrow (CO)_4Mo\!=\!CH\!-\!CH_3 + R_4NOAlCl_2 + C_2H_4$$

The nature of the evolved gases and metathesis products in the reaction of 1,7-octadiene confirmed the above proposed mechanisms.

Thorn-Csányi [85,86] demonstrated the generation of metallacarbene as initiating species from the binary system $WCl_6 \cdot (CH_3)_4Sn$ in kinetic studies on *trans*-2-pentene. The formation of propene was observed in two stages of the reaction. On varying the amount of 1-pentene in the feed, the percentage of propene changes only in the second reaction stage. It is thought that the carbene complex which is formed in this stage is responsible for the main metathesis activity; the rate of propene evolution and its quantitative yield were correlated with the kinetic course of metathesis activity. Analogously, in these studies, it was found that the formation of ethene occurs in two separate stages in the absence of olefin.

Recent studies carried out by Muetterties and Band [87] on the $CH_3WOCl_3 \cdot O(C_2H_5)_2$ complex, obtained from the reaction of $WOCl_4$ with $(CH_3)_2Mg$, suggested the intervention of a CH_2WOCl_4 species as metathesis intermediate. Thermal decomposition of CH_3WOCl_3 appeared to be very complex, with three major reductive elimination products, methyl chloride, polymethylene and methane, as well as minor amounts of ethene and ethane. In the presence of an olefin, the $CH_3WOCl_3 \cdot O(C_2H_5)_2$ complex initiated the metathesis reaction of the olefin. From 2-pentene, in addition to the normal products (2-butene and 3-hexene), propene and 1-butene were formed in molar amounts comparable to the moles of CH_3WOCl_3 used. When the methyltungsten complex was $CD_4WOCl_3 \cdot O(C_2H_5)_2$ the abnormal metathesis products consisted solely of the $C_3H_4D_2$ and $C_4H_6D_2$ isotopic species. The CH_2 fragment of the CH_2WOCl_2 intermediate is thus shown to be incorporated into the olefin metathesis products. For the generation of CH_2WOCl_2 species, Muetterties and Band suggested a plausible scheme where methyl chloride formation is the initiating step of a chain reaction involving $WOCl_2$ as the chain-propagating species.

Scheme 6.31

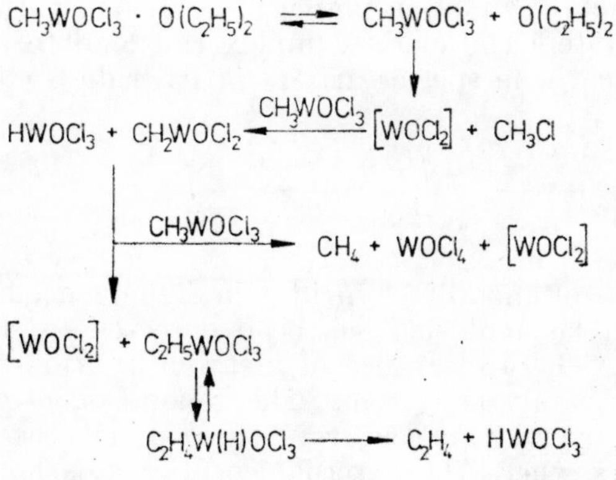

$$CH_3WOCl_3 \cdot O(C_2H_5)_2 \rightleftharpoons CH_3WOCl_3 + O(C_2H_5)_2$$

$$HWOCl_3 + CH_2WOCl_2 \xleftarrow{CH_3WOCl_3} [WOCl_2] + CH_3Cl$$

$$\xrightarrow{CH_3WOCl_3} CH_4 + WOCl_4 + [WOCl_2]$$

$$[WOCl_2] + C_2H_5WOCl_3$$

$$C_2H_4W(H)OCl_3 \longrightarrow C_2H_4 + HWOCl_3$$

These authors noted that the formation of ethene from the thermal decomposition of CH_3WOCl_3 was completely suppressed by the presence of the olefin.

A somewhat different carbene species was proposed by Verkuijlen for the initiation and propagation of metathesis catalyzed by the $WOCl_4$·-Me_4Sn system.[72] The active carbene species is supposed to involve dimetallic coordination, the organometallic compound playing a more active part in these steps than just that of alkylating agent.

Scheme 6.32

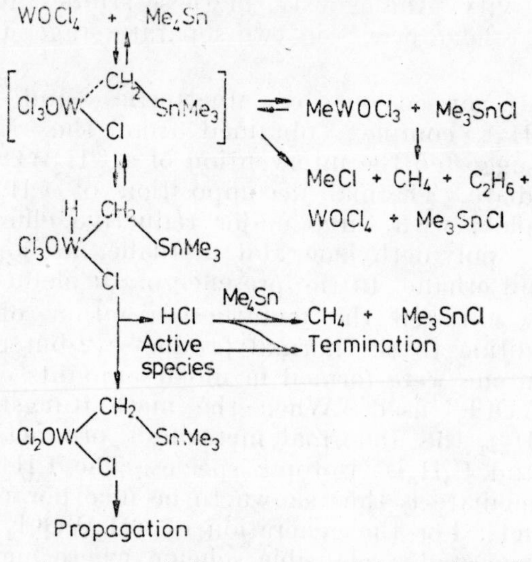

The bimetallic complexes explain more satisfactorily the role of the cocatalyst during the initiation and propagation steps of the metathesis reaction.

In catalytic systems devoid of cocatalyst the olefin complex with the transition metal can be a source of carbene species. If a hydride is present in the system the olefin complex can lead to an alkyl compound from which the carbene species arises via hydride α-elimination.

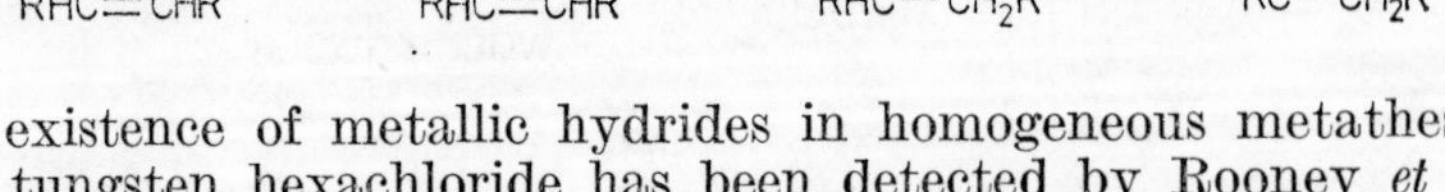

$$(6.74)$$

The existence of metallic hydrides in homogeneous metathesis catalysts of tungsten hexachloride has been detected by Rooney $et\ al.$[88] by IR spectroscopy. The appearance of metallic hydrides is highly probable in heterogeneous catalytic systems. The generation of carbene species via a metal hydride could thus occur in these widely used and extremely efficient catalysts where the external cocatalyst is not necessary.

A new method for the generation of carbene species in the absence of cocatalyst may start from π-allylic olefin complexes *via* a metallacyclobutane intermediate

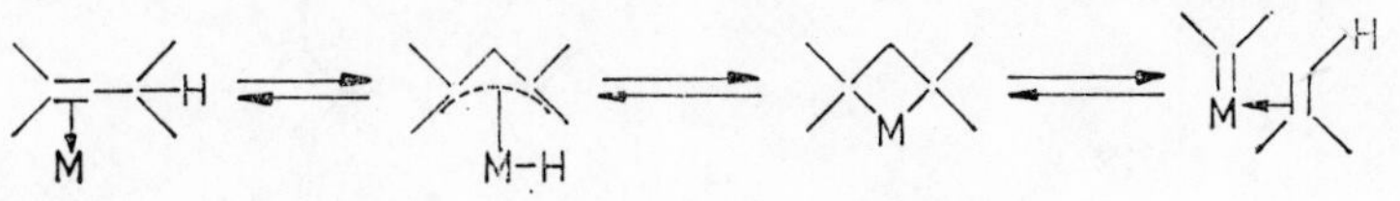

$$(6.75)$$

In detailed studies Green *et al.* [89,90] confirmed the formation of metallacyclobutanes in reactions of π-allyl tungsten complexes with nucleophilic reagents. Surprisingly, the metallacyclobutanes of tungsten decomposed thermally to cyclopropanes and saturated hydrocarbons while photochemical decomposition led to olefins with fewer carbon atoms than the initial metallacycle. The hydrogen source in the thermal decompositions remains unknown. The appearance of olefins in photochemical decompositions was interpreted by a photoinduced $\eta^5 - \eta^3$ shift of an η-cyclopentadienyl ring to provide a suitable vacant site on the metal centre.

Scheme 6.33

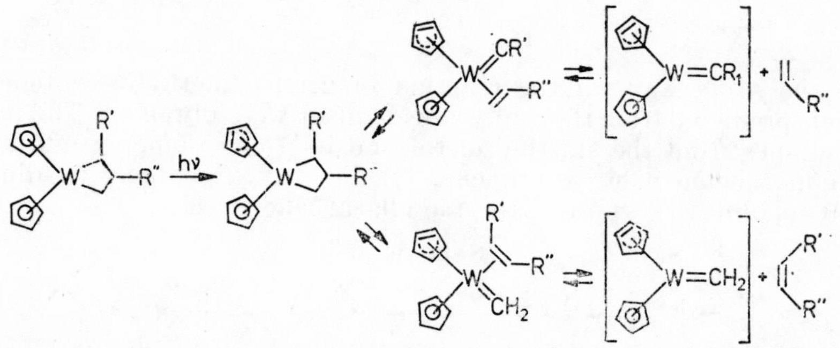

The π-allyl mechanism based on Green's studies received a significant support from work reported by Farona and Tucker [91] on reactions of 2-pentene and 2,4,4-trimethyl-2-pentene with molybdenum hexacarbonyl on alumina. The formation of nonbranched olefins (ethene, propene and 1-butene) in the reaction of 2-pentene was rationalized by the intervention of two π-allyl complexes and the corresponding metallacyclobutanes. The absence of branched olefins in the reaction products seems to rule out direct formation of the carbene species by the coordination of the

Scheme 6.34

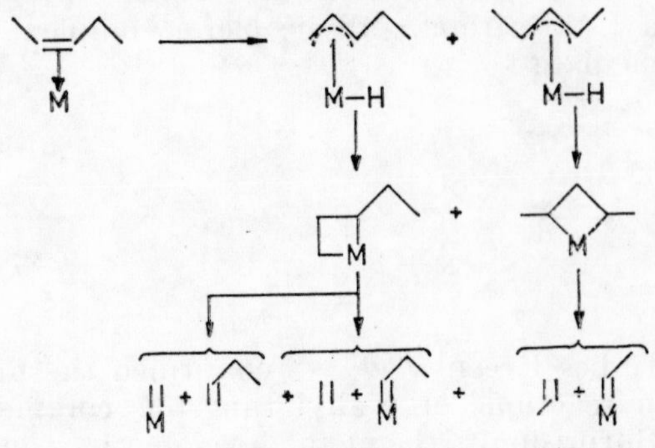

olefin and subsequent 2,1-hydrogen shift or by hydride migration from the catalyst and subsequent α-elimination (See scheme 6.35).

Scheme 6.35

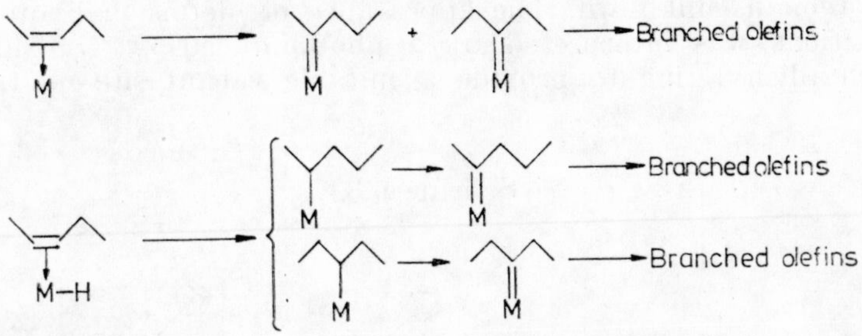

In the same way, in the reaction of 2,4,4-trimethyl-2-pentene, only products predicted from the π-allyl mechanism were obtained. The metallacyclobutanes from the starting olefin and its 1-ene isomer would lead, as shown in scheme 6.36, to propene, 4,4-dimethyl-2-pentene, 4,4-dimethyl-1-pentene and the appropriate metallacarbene species.

Scheme 6.36

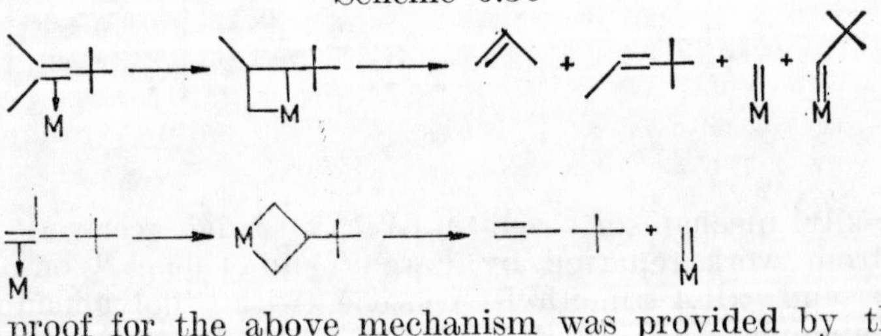

Additional proof for the above mechanism was provided by the metathesis of isotopically-labelled 4,4-dimethyl-1-pentene and 4,4-dimethyl-2-pentene which ought to lead, according to the carbene mechanism, to identical metallacyclobutanes; the predicted firts-formed olefins, ethylene and 3,3-dimethyl-1-butene, were observed in both cases and differences were observed only in their rates of formation.

Initiation of the starting olefin by a π-allyl complex seems to be in agreement with the isotope effect observed by Grubbs and Swetnick[92] for the induction or "break-in" period registered in the reaction of 2,8-decadiene and 2,8-decadiene-1,1,1,10,10,10-d_6 over $MoO_3 \cdot CoO \cdot Al_2O_3$. Significantly, the isotope effect is of the same order as in other reactions which involve allylic intermediates. Furthermore, the insignificant isotope effect found in the propagation reaction indicates that formation of carbene species during propagation occurs by a different mechanism from that operating during initiation.

6.2.2.2. The Nature of Propagation Reaction

In their original proposal of the carbene mechanism, Hérisson and Chauvin[50] suggested an insertion type reaction involving a metallacyclobutanic transition state and a metallacarbene as the propagating species.

Scheme 6.37

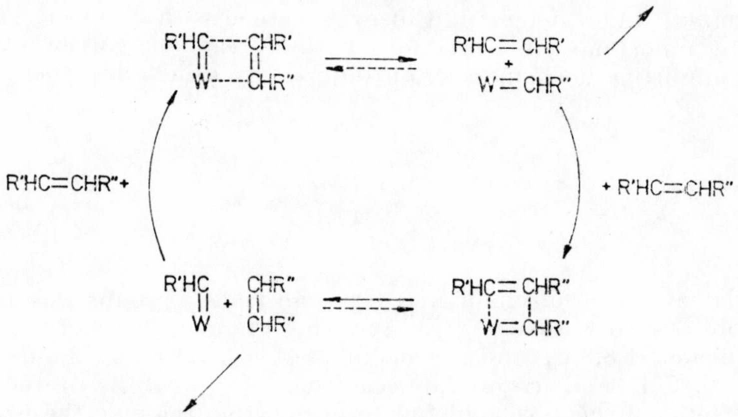

```
        R'HC-----CHR'           R'HC == CHR'
         ‖      ‖        ⇌          +
        W-----CHR"                W == CHR"

R'HC==CHR"+                                    + R'HC==CHR"

        R'HC   CHR"              R'HC == CHR'
         ‖  +  ‖        ⇌          │      │
        W      CHR"              W == CHR"
```

Thereafter, Cardin, Doyle and Lappert[51] provided strong evidence for the occurrence of a metallacarbene intermediate in the metathesis of electron-rich alkenes catalysed by rhodium(I) complexes. They postulated an equilibrium between rhodium-carbene complexes and metallacycles. The formation of metallacycles was suggested to occur by a stepwise transition state such as A or B in Fig. 6.18, where the (α) covalency change probably precedes (β). In state A, electron-rich alkenes and Rh(I) are good nucleophiles and C_{carb} is an electrophilic centre. Although for this particular case state A seems to be the more probable, it is difficult to generalize to other substrates and metallic systems.

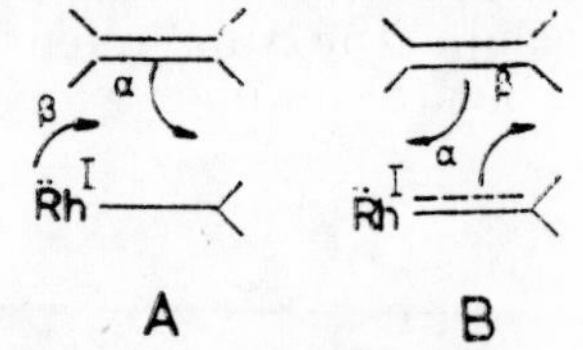

Fig. 6.18.— Proposed transition states for the formation of rhodium (I)-cyclobutane.

The equilibrium between a metallacyclobutane and a metal complex containing both an alkene and a carbene ligand was invoked by Casey and Burkhardt [52] in alkene scission reactions initiated by tungstacarbene complexes (6.76).

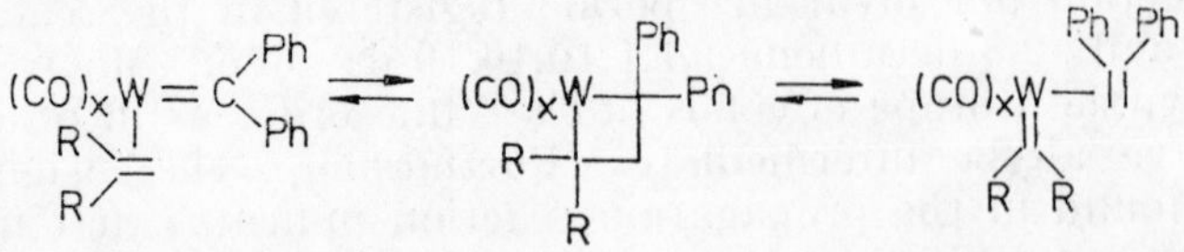

$$(CO)_x W = C \overset{Ph}{\underset{Ph}{\diagdown}} \quad \rightleftharpoons \quad (CO)_x W \overset{Ph}{\underset{R}{\square}} Pn \quad \rightleftharpoons \quad (CO)_x W - \overset{Ph \; Ph}{\underset{R \; R}{\parallel}} \tag{6.76}$$

This transformation involves a formal oxidation at the metal centre and the creation of one additional coordination site. Consequently the above equilibrium depends strongly on the oxidation state and coordination number of the metal. Recently Casey and Shusterman [53] succeeded in synthesizing such metallacarbene olefin complexes with a tungsten system and Schrock and co-workers, [93] with a tantalum system.

Katz and McGinnis [58,59] performed relevant experiments concerning the nature of propagation reaction in the cross-metathesis of cyclo-octene with 2-butene and 4-octene and of cyclo-octene with 2-hexene. The formation of anomalous cross-products in these reactions can be explained only by admitting a chain reaction induced by a metallacarbene species (6.77).

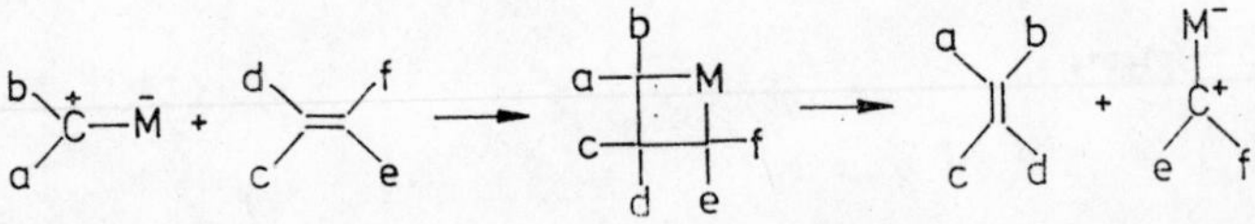

$$\overset{b}{\underset{a}{\diagdown}} \overset{+}{C} - \bar{M} \; + \; \overset{d}{\underset{c}{\diagdown}} \overset{f}{\underset{e}{=}} \quad \longrightarrow \quad \overset{b}{\underset{c}{\underset{d}{\square}}} \overset{M}{\underset{e}{\square}} \quad \longrightarrow \quad \overset{a}{\underset{c}{\diagdown}} \overset{b}{\underset{d}{=}} \; + \; \overset{M^-}{\underset{e}{\underset{f}{\square}}} \overset{}{C^+} \tag{6.77}$$

Similar evidence for chain transfer of alkylidene groups was provided by Grubbs and co-workers [61,63] by studying the metathesis of mixtures of 1,7-octadiene-1,1,8,8-d_4 and 1,7-octadiene, or of cis,cis-2,8-decadiene-1,1,1,10,10,10-d_6 and cis,cis-2,8-decadiene. The analysis of the scrambling patterns of the acyclic olefins resulting from the metathesis of these 1,7-diene systems favored a "non pair-wise" interchange of carbons between two olefins.

There is at present a general consensus that olefin metathesis proceeds by coordination of the olefin to the transition metal, and that the coordinated complex then rearranges to a metallacyclobutane. The metallacyclobutane subsequently cleaves to furnish a new coordinated olefin carbene complex which liberates the product olefin (6.78).

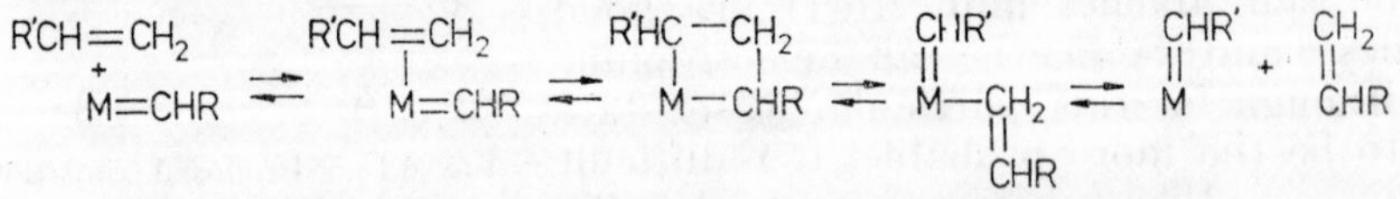

$$\begin{array}{ccccc} R'CH = CH_2 & R'CH = CH_2 & R'HC - CH_2 & CHR' & CHR' \;\;\; CH_2 \\ + & | & | \;\;\;\; | & \| & \| \;\;\;\;\;\; \| \\ M = CHR & M = CHR & M - CHR & M - CH_2 & M \;\;\;\;\; CHR \\ & & & \| & \\ & & & CHR & \end{array} \tag{6.78}$$

In the above scheme it is not yet firmly established whether or not the olefin coordinates to transition metal. The stereochemical results obtained by Calderon and co-workers [94] in studies with cis- and non-specific catalysts, and their rationalization in terms of the bidentate chelating concept, seem to favour a precoordination of the olefin to the transition metal prior to cyclobutane formation and rearrangement.

Numerous data seem to support the intermediacy of metallacyclobutanes as the key step in the propagation reaction. Green and co-workers[90,95] prepared a wide range of tungsta- and molybda-cyclobutanes and studied their chemistry. Thermal and photochemical decomposition evidenced that metathesis-like olefins were formed by the latter process. The mechanism of olefin formation was interpreted through a photoinduced $\mu^5 - \mu^3$ shift of a cyclopentadienyl ring pertaining to the metallic complex.

Scheme 6.38

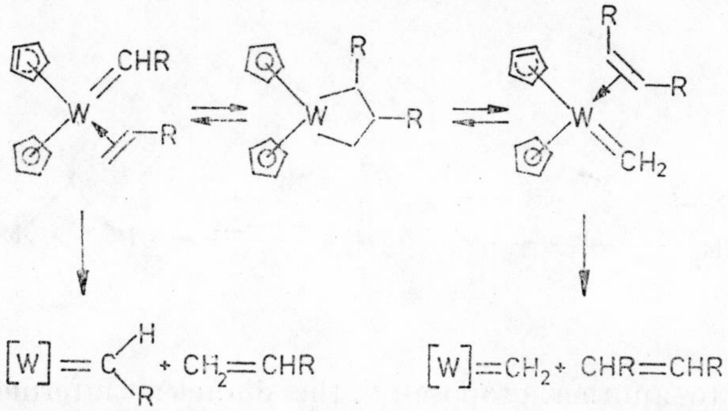

The possibility of direct rearrangement of substituted metallacyclobutanes without the intermediacy of a metallacarbene olefin complex was suggested by Puddephatt et al.[64] when the metal is platinum.

Scheme 6.39

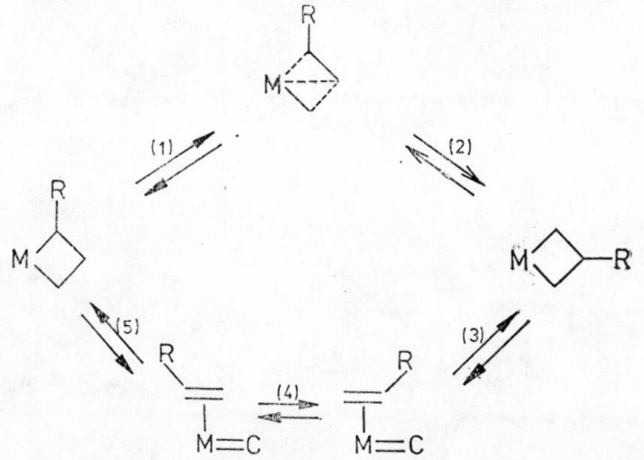

Based on stereochemical considerations, these authors prefer a concerted pathway for metallacyclobutane isomerization (sequence 1—2, Scheme 6.39) and rule out stepwise dissociation by cleavage of a carbon-carbon bond (sequence 3—5, Scheme 6.39).

Interesting proposals suggest the appearence of dimetallic species as important intermediates in propagation reactions [96,97]. In such a proposal, one considers the possibility of a dinuclear intermediate in which one of the metal centres carries the carbene and the other the olefin.

Scheme 6.40

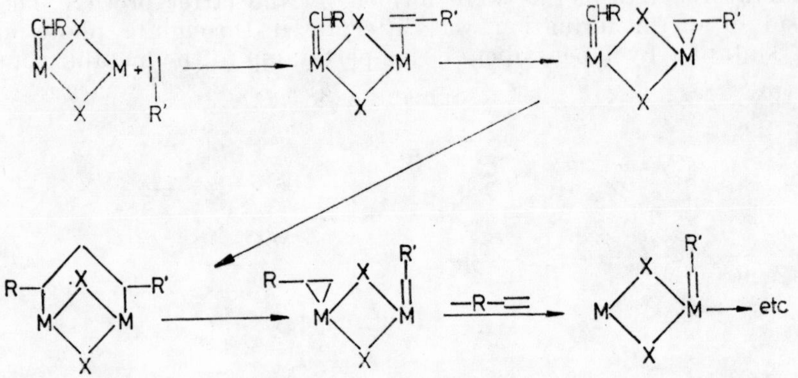

According to another proposal, [97] the dinuclear intermediate carries the carbene species bound to both metal centres and the olefin to one of the metal centres.

Scheme 6.41

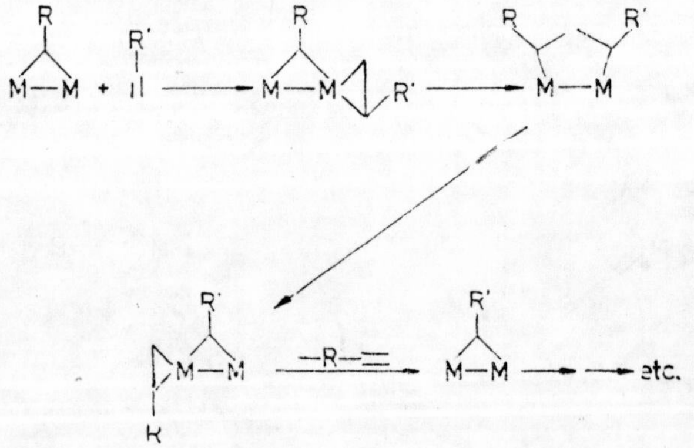

The possible appearance of such dimetallic species in metathesis finds important support in the recent isolation and characterization by Levi-salles and co-workers [98] of ditungsten carbene compounds.

Alternative dimetallic intermediates involving the participation of the metal from the cocatalyst were proposed by Verkuijlen [72].

Scheme 6.42

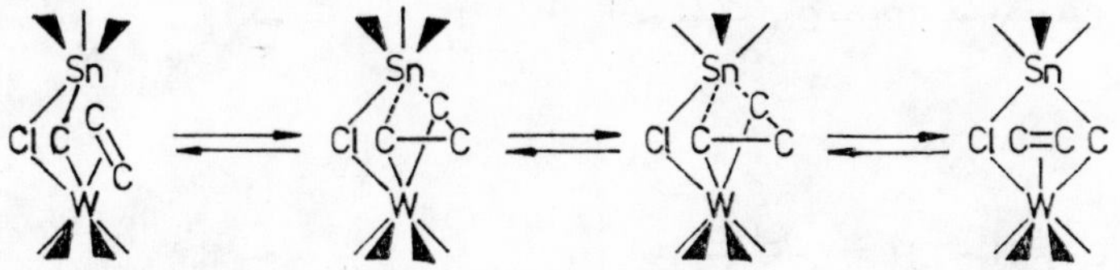

Further work is required to provide conclusive evidence for participation of dimetallic intermediates in olefin metathesis.

The intervention of metallacyclobutane compounds in metathesis was convincingly pointed out by Grubbs and co-workers [99] by synthesising, and studying the metathesis behaviour of, titanacyclobutanes. The formation of titanacyclobutanes occurs readily by reaction of titanocene—methylene complexes with olefins in the presence of a base (pyridine or one of its derivatives) (6.79).

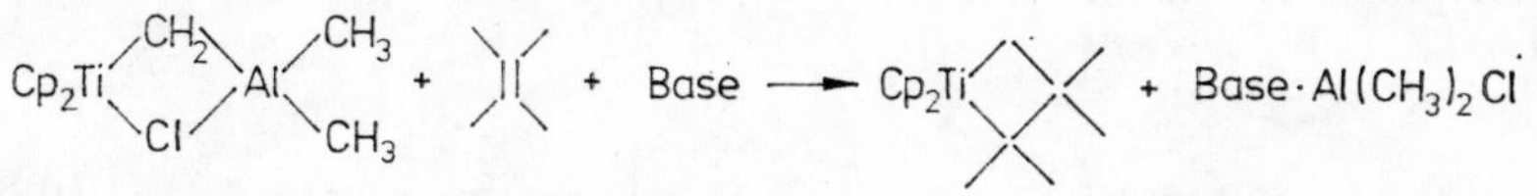

$$\hspace{12cm} (6.79)$$

The main pathway for decomposition of such titanacyclobutanes is C-C bond cleavage to produce a carbene—olefin complex which facilitates olefin exchange in metathesis reactions (6.80).

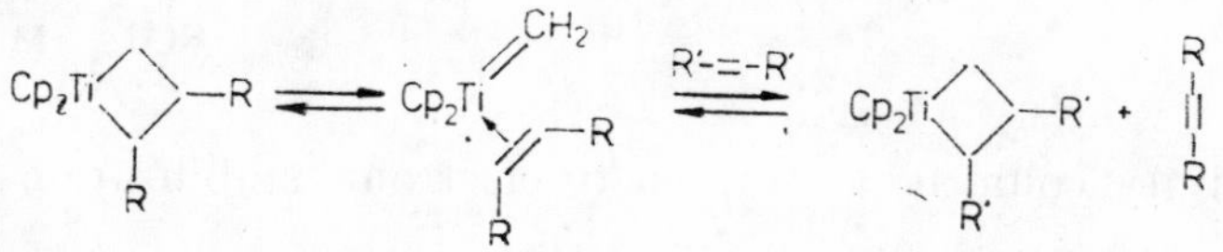

$$\hspace{12cm} (6.80)$$

This reaction pathway indicates the mode by which metallacyclobutanes play the role of chain-carrying species.

One of the most intriguing problems in the chemistry of metathesis was the nature of the chain-carrying metallacarbene intermediate. This problem was extensively examined in degenerate metathesis of terminal alkenes.

Some authors [100] demonstrated, by deuterium labelling, the occurrence of degenerate metathesis of terminal alkenes at rates more rapid than productive metathesis which yields ethene and an internal alkene. Two explanations were advanced to account for the degenerate metathesis of terminal alkenes, assuming a metallacarbene as chain carrier. If the reaction proceeds through an $M=CH_2$-carbene complex, this will react selectively with a terminal alkene transfering the most highly substituted alkylidene unit and regenerating an $M{=}CH_2$-carbene complex; reaction (6.81) will be preferred relative to reaction (6.82).

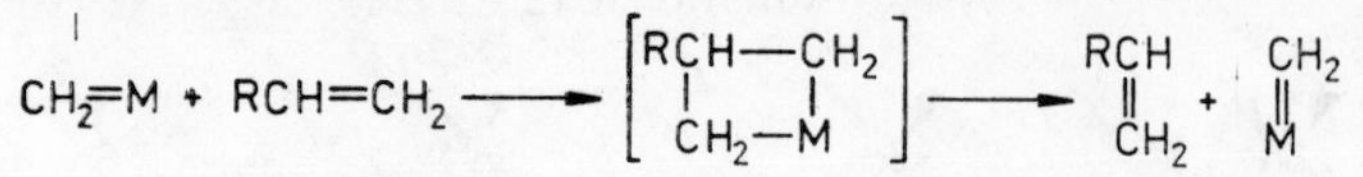

$$CH_2{=}M + RCH{=}CH_2 \longrightarrow \begin{bmatrix} RCH{-}CH_2 \\ | \qquad | \\ CH_2{-}M \end{bmatrix} \longrightarrow \begin{matrix} RCH \\ \| \\ CH_2 \end{matrix} + \begin{matrix} CH_2 \\ \| \\ M \end{matrix} \qquad (6.81)$$

$$CH_2{=}M + CH_2{=}CHR \longrightarrow \begin{bmatrix} CH_2{-}CHR \\ | \qquad | \\ CH_2{-}M \end{bmatrix} \longrightarrow \begin{matrix} CH_2 \\ \| \\ CH_2 \end{matrix} + \begin{matrix} CHR \\ \| \\ M \end{matrix} \qquad (6.82)$$

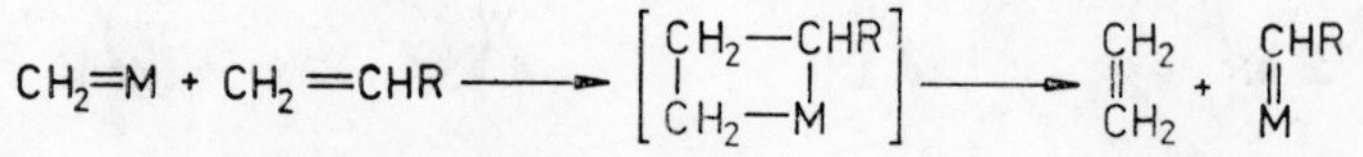

This selectivity could be attributed either to steric preferences for olefin coordination or to electronic stabilization of a negatively polarized carbene ligand, $\overset{\delta+}{M}{-}\overset{\delta-}{CH_2}$.

Alternatively, if the reaction proceeds through an $M{=}CHR$—carbene complex, this will react selectively with a terminal alkene transferring the least substituted alkylidene unit and regenerating an $M{=}CHR$-carbene complex. In this case, reaction (6.83) will be preferred to reaction (6.84).

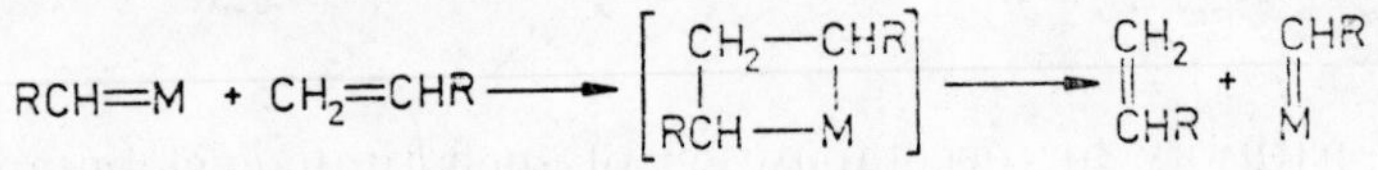

$$RCH{=}M + CH_2{=}CHR \longrightarrow \begin{bmatrix} CH_2{-}CHR \\ | \qquad | \\ RCH{-}M \end{bmatrix} \longrightarrow \begin{matrix} CH_2 \\ \| \\ CHR \end{matrix} + \begin{matrix} CHR \\ \| \\ M \end{matrix} \qquad (6.83)$$

$$RCH{=}M + RCH{=}CH_2 \longrightarrow \begin{bmatrix} RCH{-}CH_2 \\ | \qquad | \\ RCH{-}M \end{bmatrix} \longrightarrow \begin{matrix} RCH \\ \| \\ RCH \end{matrix} + \begin{matrix} CH_2 \\ \| \\ M \end{matrix} \qquad (6.84)$$

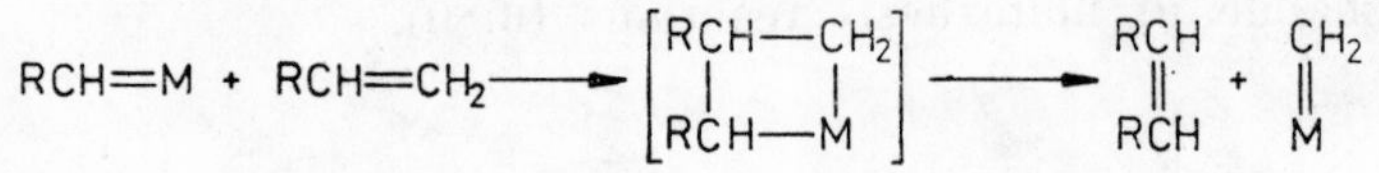

This selectivity could be attributed to electronic stabilization of a positively polarized carbene ligand, $\overset{\delta-}{M}{-}\overset{\delta+}{CHR}$.

There are several contrasting experimental facts concerning the polarization of carbene complexes which are active in metathesis. Thus, the reaction of the nucleophilic carbene complexes of tantalum with olefins leads to major products which could be explained in terms of equation 6.81.[101] Likewise, the metathesis of terminal olefins $R{-}CH{=}CH_2$ ($R{=}Et$, Pr, Bu) in the presence of the catalyst $C_6H_5WCl_3 \cdot AlCl_3$ gave significant amounts of $R{-}CH{=}CH{-}C_2H_5$, which was interpreted in favour of a nucleophilic carbenoid species.[102] Furthermore, a titanocene—methylene catalyst of the form $Cp_2Ti\underset{Cl}{\overset{CH_2}{\diagup\diagdown}}AlMe_2$, active in the metathesis of terminal olefins,[69] exhibited pronounced nucleophilicity at the carbene site.

In contrast to these results, the reaction of the electrophilic carbene complex $(CO)_5W=C(p\text{-Tol})_2$ with alkenes [103] led to the selective transfer of the least substituted alkylidene unit of a terminal alkene to the diarylcarbene, providing a good model for reaction sequence 6.83. The intervention of an $RCH=M$-carbene as chain-carrier having an electrophilic carbene carbon has been established by results recorded in the cross-metathesis of cyclopentene with 1-pentene,[104] of 1-octene-1,1-d_2 with (Z)-1-decene-1-d_1,[105] of norbornadiene with 1-hexene,[106] of cyclopentene with 1,7-octadiene [106] and of 1-pentene with ethene-d_1.[107]

The non-symmetrical structure of the dominant products obtained by Kelly and Calderon [104] in the cross-metathesis of cyclopentene with 1-pentene, $CH_2=[CH(CH_2)_3CH=]_nCHC_3H_7$ has been explained by postulating chain-carrier of the form $C_3H_7 CH=W$, $C_3H_7 CH=(CH_2)_7CH=W$ and $RCH=W$.

Scheme 6.43

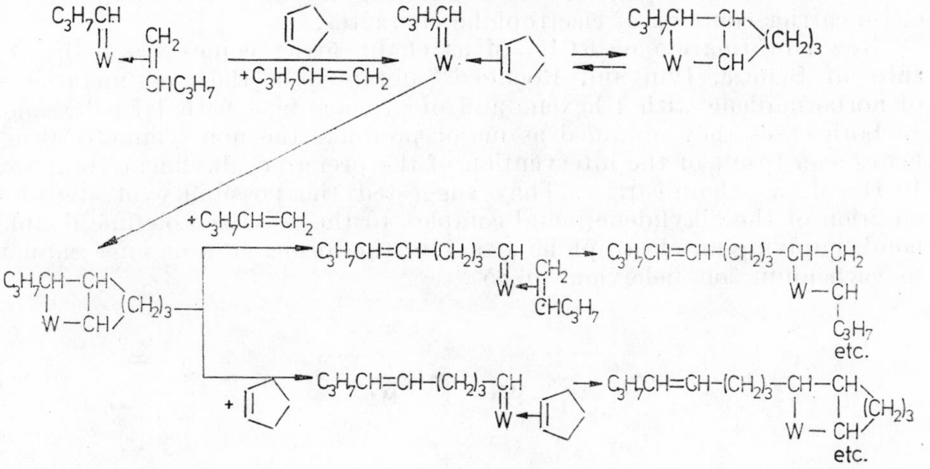

The results obtained by Casey and Tuinstra [105] in the stereochemical study of the degenerate metathesis of 1-octene-1,1-d_2 with (Z)-1-decene-1-d_1 favour an $M=CHR$ intermediate of electrophilic character in this process. Thus, according to the above alternative, the cross-metathesis of 1-octene-1,1-d_2 with (Z)-1-decene-1-d_1 would proceed via either pathway (A) or (B) as illustrated in Scheme 6.44. Pathway (A) would lead to a steric ratio of (Z)- to (E)-1-octene-1-d_1 of 1 : 1 since neither the metallacarbene — alkene complex (I) nor the metallacyclobutane (II) is capable of retaining the stereochemistry. Alternatively, if the reaction proceeds via pathway (B), for both the metallacarbene—alkene complex (III) and the metallacyclobutane (IV), a preference for retention of stereochemistry would be expected, and consequently the steric ratio Z : E would be greater than 1 : 1. The results recorded for ratios of (Z)- to

Scheme 6.44

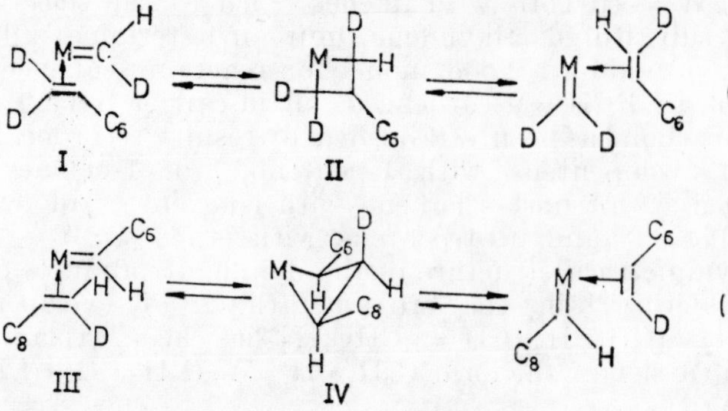

(E)-1-octene-1-d_1 ranged between 1.3 and 2.4 depending on the catalyst used. They were explained by postulating an M=CHR species as a chain-carrier complex of electrophilic character.

New confirmation of RCH=M as chain-carrier comes from the results of Bencze, Ivin and Rooney [106] obtained in the cross-metathesis of norbornadiene with 1-hexene and of cyclopentene with 1,7-octadiene. In both cases they obtained as major products the non-symmetrical alkenes as a result of the intervention of the preferred alkylidene complex, RCH=M, as chain-carrier. They suggested the possibility of stepwise addition of the alkylidene metal complex to the polarized olefinic double bond, implying an electrophilic attack of the former in a manner similar to carbenium ion behaviour (6.85).

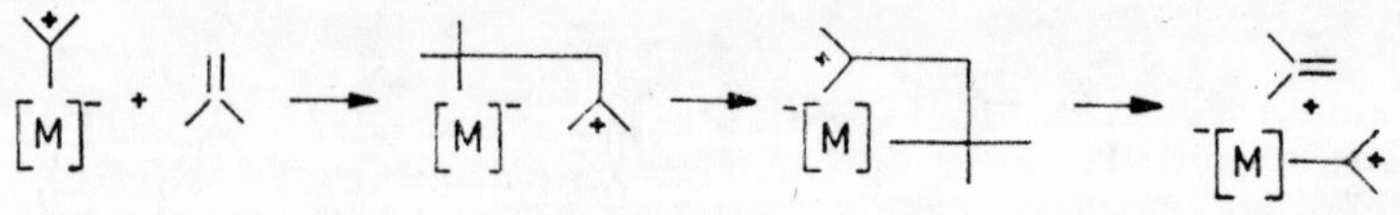

$$\text{(6.85)}$$

Moreover, they found that a methylene complex, CH_2=M, when formed, is more reactive than substituted alkylidene complexes and the strongly electrophilic methylene ligand selectively adds to the terminal olefins at the C-1 position.

On studying the degenerate metathesis of terminal alkenes and of ethene Casey and Brondsema [107] found varying rates as a function of the nature of the starting material. Thus, the degenerate metathesis of 1-pentene is faster than the degenerate metathesis of ethene-d_1 while the reaction of 1-pentene is greatly inhibited in the presence of ethene. In an internal competition the degenerate metathesis of 1-pentene is slower than that of ethene-d_1. The results were explained by the concurrent competition of the CH_2=M and RCH=M carbene complexes with preference for the latter as the chain-carrying species.

6.2.2.3. The Nature of Termination Step

There are some indications that the termination step in olefin metathesis proceeds by recombination of carbenes from metallacarbene complexes (6.86).

$$2 \text{ L}_x\text{M} = \overset{*}{\text{C}}\text{H}_2 \longrightarrow \overset{*}{\text{C}}\text{H}_2 = \overset{*}{\text{C}}\text{H}_2 \qquad (6.86)$$

Experimental support for such a reaction comes from Schrock [108] for the tantalum—methylene complex and from Casey [109] for tungsten—carbene complexes.

Grubbs and Hoppin [84] correlate the production of ethene in the metathesis of 2,8-decadiene in the presence of $WCl_6 \cdot Me_4Sn$ or $(Ph_3P)_2(NO)_2Cl_2Mo \cdot Me_3Al_2Cl_3$ with metallacarbene dimerization during initiation. These dimerization reactions proceed probably *via* either 1,3-dimetallacyclobutanes or bis(carbenemetal)intermediates.[99b] A bimolecular mechanism was suggested also by Ephritikhine and Green [89] for the formation of ethene through thermal or photochemical decomposition of the carbene intermediate $[W(\mu^5\text{-}C_5H_5)_2L=CHR]$ where R=H.

More possibilities for termination reactions arise in the metathesis of functionalized olefins. In this case, inhibition of the reaction may occur by formation of stable adducts between a functional group and a metallacarbene carrier or the Lewis acid used as cocatalyst.

For the metathesis of ethyl 3-pentenoate in the presence of $WCl_6 \cdot Me_4Sn$, Otton *et al.*[110] suggested the formation of a stable allylic complex between the functionalized olefin and the transition metal by abstraction of a hydrogen atom from an allylic position (6.87).

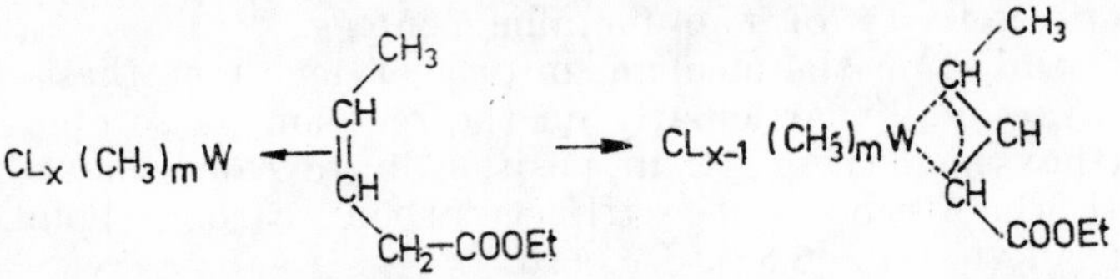

$$CL_x (CH_3)_m W \longleftarrow \overset{CH_3}{\underset{\underset{CH_2-COOEt}{CH}}{\overset{CH}{\underset{\|}{}}}} \longrightarrow CL_{x-1} (CH_3)_m W \overset{CH\diagdown CH_3}{\underset{CH\diagdown COOEt}{\diagup}} CH \qquad (6.87)$$

This termination pathway could explain why the tungsten catalytic system is less active in the case of ethyl 3-pentenoate than for methyl oleate where hydrogen abstraction is less probable.

6.2.3. Mechanism of Alkyne Metathesis

Pennella, Banks and Bailey [111] suggested a mechanism involving a four-centred intermediate for the metathesis of acetylenes with heterogeneous catalytic systems based on tungsten. According to these authors, a diacetylenic complex could be formed during a first step, giving rise to a cyclobutadiene complex in which the rearrengement of the initial π and σ bonds occurs with formation of a new diacetylenic complex, and which would finally give the reaction products.

Scheme 6.45

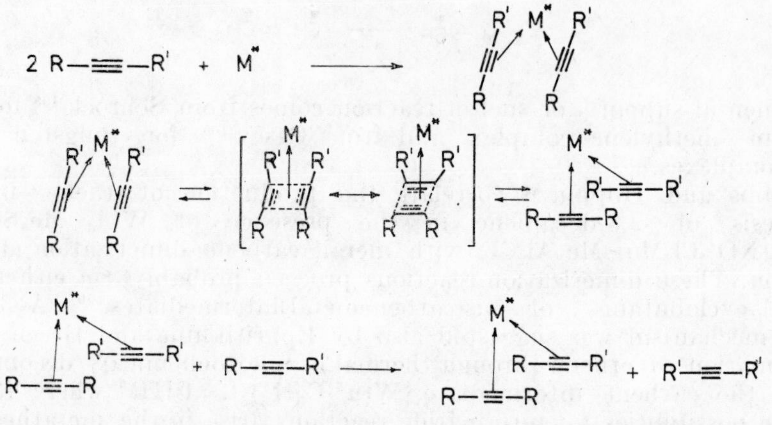

The formation of diacetylenic complexes of the transition metals is well known.[112] Calculations by means of the molecular orbital method carried out by Mango and Schachtschneider [113] on the transformation of bis(acetylene)—iron to cyclobutadiene—iron, indicate that concerted cyclo-addition of the two molecules of acetylene in the complex to cyclo-butadiene is a high energy process as compared to the similar transformation of olefins. These authors showed that when the cyclo-addition occurs in the ground state, it can be a stepwise process or it may involve the catalytic activity of two metallic centres.

Recent studies on the mechanism of acetylene metathesis carried out by Mortreux et al.,[114] particularly on the reaction of isotopically-labelled 1-phenyl-1-hexyne having ^{13}C in position 1, revealed that the reaction occurs with the fission of the triple carbon—carbon bond, involving direct transalkynation (6.88).

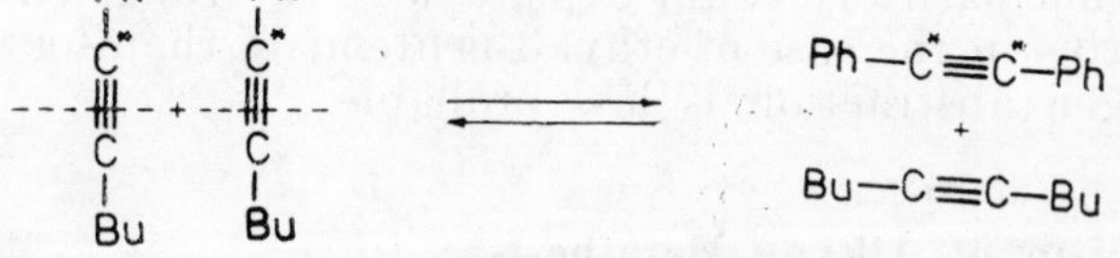

$$\text{(6.88)}$$

These authors proposed the possibility of the involvement of the phenolic group of the catalyst, $Mo(CO)_6 \cdot PhOH$, along with the transition metal during transalkynation. Similar results obtained by Mortreux et al.[115] in metathesis of disubstituted alkynes seem to support the transalkynation mechanism.

As a matter of fact, the intermediacy of metallacarbynes in acetylene metathesis was proposed earlier by Katz and McGinnis.[58] They considered that acetylenes add to metallacarbyne to give a metallacyclobutadiene which rearranges to give the new metallacarbyne as chain-carrier (6.89).

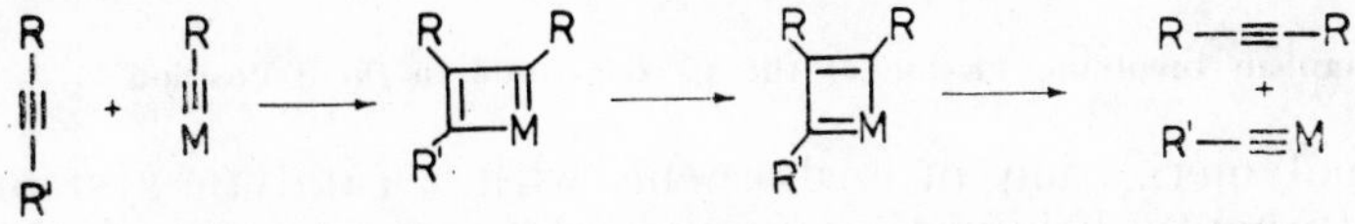

$$(6.89)$$

The formation of such carbyne complexes of transition metals is well known from the extensive work of Fischer and co-workers[116] and is now under investigation.

More recently, the intermediacy of metallacarbyne species in acetylene metathesis was demonstrated by Schrock and co-workers[117] by preparing and characterizing several carbyne complexes which metathesise acetylenes. The finding that these complexes exhibit much greater activity than that of the known heterogeneous oxide or homogeneous phenolic systems is of great interest.

A dimetallic intermediate similar to those proposed for the olefin reaction was also considered to be present in acetylene metathesis.[118] According to this suggestion, a dimetallic carbyne species would lead to a 1,2-dimetallacyclopentene intermediate which generates the new dimetallacarbyne as chain-carrier (6.90).

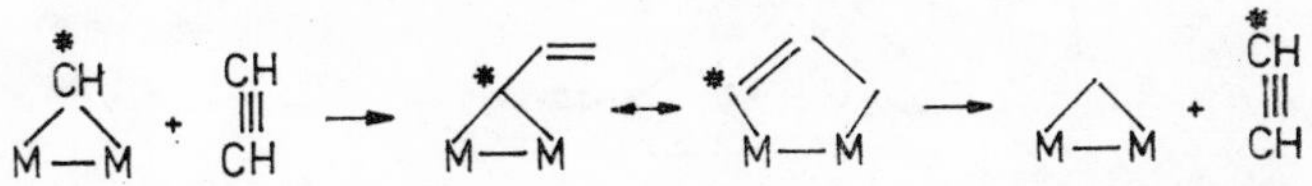

$$(6.90)$$

The possibility of an alternate mechanism involving a metallacycles of type I A and I B was suggested by Haines and Leigh[78] taking into account the fact that such complexes are known[119] (6.91).

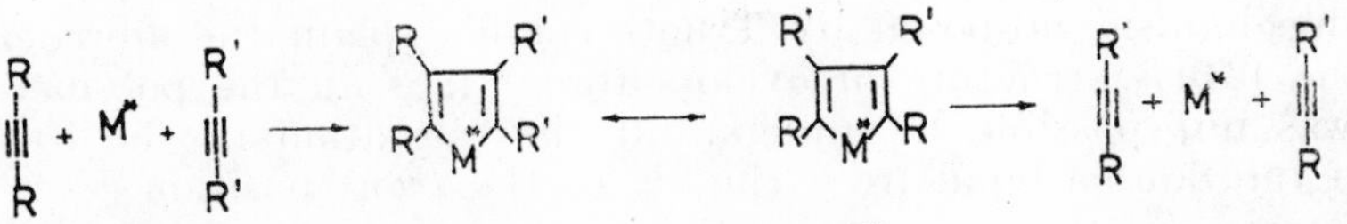

$$(6.91)$$

6.2.4. Mechanism of the Ring-Opening Polymerization of Cyclo-Olefins

Initially, it was believed that the ring-opening polymerization of cyclo-olefins occurs by the fission of the carbon—carbon σ-bond situated in the α or β position relative to the double bond. Later, after having accumulated more data concerning the mechanism of acyclic alkene metathesis and after having identified the relation of this reaction to the ring-opening polymerization of cyclo-olefins, it was recognized that the latter reaction occurs through a metathetic pathway. This section will survey the main mechanisms proposed for ring-opening polymerization with emphasis on the metathetic mechanism.

6.2.4.1. Mechanism Involving Fission of the C—C σ-Bond in the α-Position

For the polymerization of norbornene with a catalytic system consisting of tetraheptyl-lithiumaluminium and titanium tetrachloride, Truett and co-workers[120] proposed a mechanism involving fission of the C—C σ-bond situated in the α-position relative to the double bond. They showed that the active catalyst is titanium in a valence state lower than IV, bearing an alkyl group as a ligand. This valence state is electrophilic enough to coordinate the double bond of norbornene. In the complex thus formed electron rearrangement with fission of a single C—C bond in the α-position may occur with the formation of two new bonds, one between titanium and the double bond, and the other between a titanium alkyl group and the carbon atom in α-position (6.92).

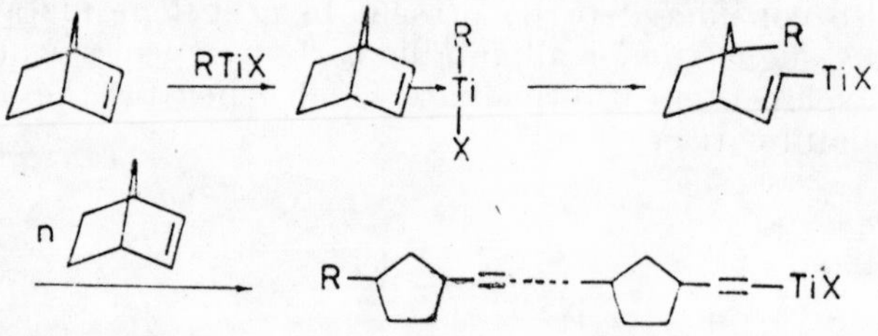

$$(6.92)$$

In this process, titanium regains its initial valency and is ready for further coordination. The newly formed alkyl group contains the monomeric unit of norbornene cleaved at the double bond. By repeating the same process with new monomer molecules, growth of the macromolecular chain ensues.

The mechanism proposed by Truett could explain the stereochemistry of the 1,3-*cis* structure of cyclopentane rings in the polymer chain but it was not possible to understand the migration of the hydrogen atoms of the double bond from the *cis* to the *trans* position.

Michelotti and Keavney[121] proposed a similar mechanism for the polymerization of norbornene with transition metal salts in polar media. These authors showed that the solvent takes a direct part in the initial step, its nucleophilic character weakening the C—C bond in the α-position.

For the polymerization of small strained cyclo-olefins such as cyclobutene and cyclopentene, Natta[122] and Dall'Asta[123] formulated a new mechanism involving fission of a C—C bond in the α-position. Later, this mechanism was generalized [124] to the polymerization of non-strained cy-clo-olefins such as cycloheptene, cyclo-octene and cyclododecene. The main experimental evidence for such a mechanism was the presence of vinyl group at the end of the polymer chain.

Subsequently, the same polymerization mechanism *via* fission of a C—C bond in the α-position was also adopted by Oshika and Tabuchi,[125] for the interpretation of the structure of the polymers obtained from nor-bornene, *endo*-di(cyclopentadiene) and *exo*-trimethylenenorbornene. Now-adays this mechanism has only historical significance.

6.2.4.2. Allylic Mechanism

Kormer *et al.*[126] considered that the ring-opening polymerization of cyclo-olefins could be promoted by π-allyl complexes which are formed by cyclo-olefins through fission of the C—C bond in the β-position. According to these authors, the polymerization may also proceed through a chelate intermediates of the π-allylic type, formed by dimerization of two cyclo-olefin molecules in the presence of the catalytic system (6.93).

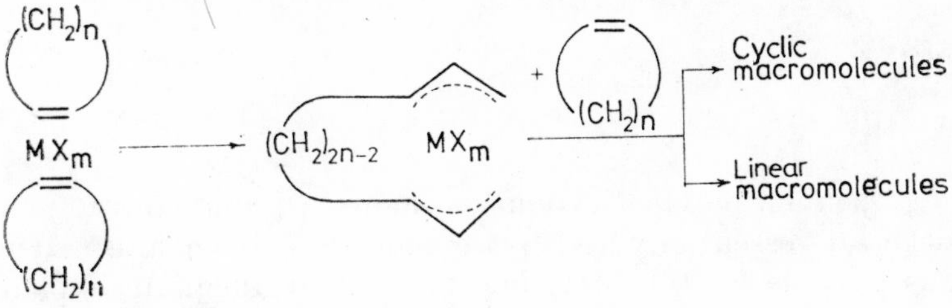

$$(6.93)$$

This mechanism also is obsolete.

6.2.4.3. Metathetic Mechanism

Calderon and co-workers [127] proposed a mechanism for the polymeriza-tion of cyclo-olefins which is totally different from those presented above, a mechanism where the cyclo-olefin ring-opening occurs through cleavage of the C=C double bond, as in the metathesis of acyclic olefins. Accord-ing to this mechanism, in the initiation step, the cyclo-olefin molecules react at the double bond, in the presence of the catalytic system, passing through a four-centred intermediate similar to that proposed by Brad-shaw, Howman and Turner [30] for the metathesis of acyclic olefins and yielding a dimeric cyclic diolefin. The dimer thus formed reacts further with monomer or dimer molecules forming progressively larger macro-cyclic polyalkenamers. These macrocyclics finally lead to linear polymers either by a cleavage reaction at a double bond in the chain with an acyclic olefin from the reaction medium or by another termination reaction.

Scheme 6.46

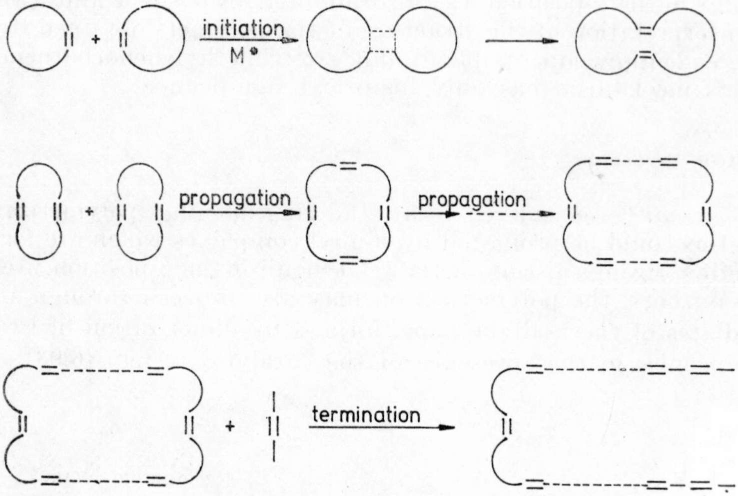

The ring-opening polymerization mechanism proposed by Calderon and co-workers is essentially a stepwise metathesis mechanism. On this basis, it was possible for the first time to offer an adequate explanation of the formation of polyalkenamers and macrocyclic compounds in the ring-opening polymerization of cyclo-olefins.

Dall'Asta and co-workers [128] provide overwhelming evidence in favour of the metathetic mechanism as a result of their studies on the polymerization of isotopically labelled cyclo-olefins. Thus, by using cyclopentene labelled at the carbon atoms of the double bond in the copolymerization reaction with cyclo-octene, they proved that the structure of the copolyalkenamer coresponds to fission of cyclopentene at the double bond and not at a C—C single bond. The reaction products after ozonation and reductive degradation will be differently labelled for the two pathways of monomer ring cleavage. This fact is easily understandable from the scheme proposed by Dall'Asta for the nature of the reaction products formed in the two modes of cyclo-olefin ring cleavage (see equations 6.94 and 6.95).

a) In the case of ring-opening at the double bond the following glycols would be obtained after oxidative degradation at the double bond of the polymer, as was experimentally found (6.94):

$$
\begin{cases}
2 \ HOH_2C-(CH_2)_6-CH_2OH \\
+ \\
HOH_2\overset{*}{C}-(CH_2)_3-\overset{*}{C}H_2OH
\end{cases}
$$

(6.94)

b) In the case of ring-opening at the single C—C bond in the α-position, the following glycols would have been (but were not) obtained after oxidative degradation of the polymer (6.95):

$$
\begin{cases}
HOH_2C-(CH_2)_6-\overset{*}{C}H_2OH \\
+ \\
HOH_2\overset{*}{C}-(CH_2)_3-CH_2OH \\
+ \\
HOH_2C-(CH_2)_6-CH_2OH
\end{cases}
$$

(6.95)

Through isotopic labelling techniques Dall'Asta and co-workers [129] proved that the copolymerization of cyclobutene with 3-methylcyclobutene proceeds via metathesis. After oxidative degradation of the copolymer obtained from cyclobutene, which had been ^{14}C-labelled at the double bond together with unlabelled methylcyclo-octene, they obtained diols whose distribution of radioactivity corresponded only to ring—opening at the double bond. After studies carried out on the polymerization and copolymerization reactions of a large number of cyclo-olefins with catalytic systems based on molybdenum, tungsten, titanium or ruthenium, Dall'Asta was able to generalize the metathetic mechanism for cyclo-olefin polymerization to polyalkenamers.

As early as 1970, Hérisson and Chauvin [50] proposed a metathetic mechanism for cyclo-olefin polymerization involving an insertion type of reaction through carbene species.

Scheme 6.47

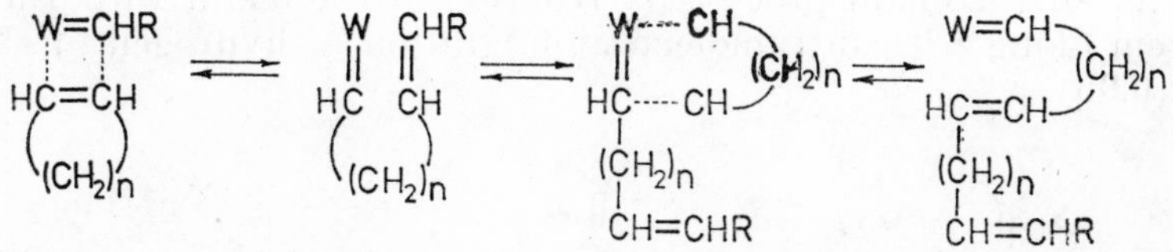

The propagation reaction assumes a metallacyclobutane transition state formed by insertion from the cyclo-olefin and the metallacarbene. This proposal was based on kinetic results obtained in the telomerization of cyclopentene with 2-pentene in the presence of tungsten catalytic systems.

For the telomerization reaction, a similar mechanism was proposed; it explained the distribution of telomers obtained in the cross-metathesis

see also p. 222

of cyclo-olefins with internal non-symmetrical olefins.

Scheme 6.48

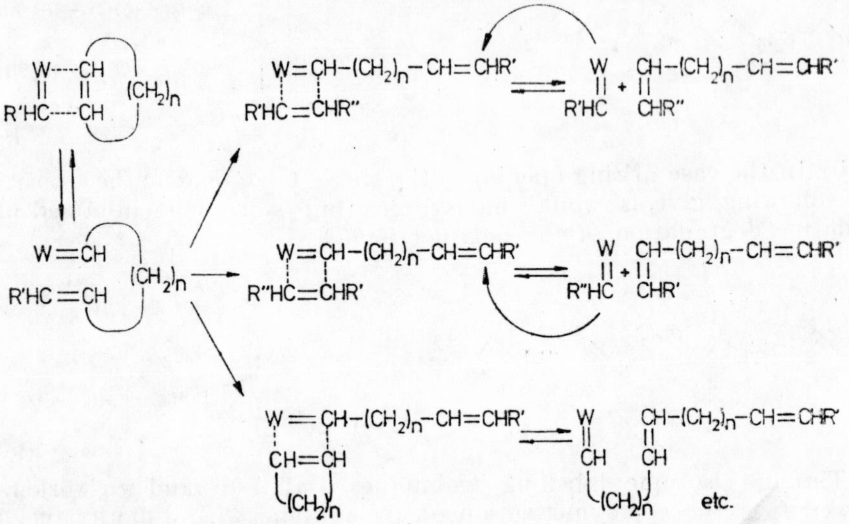

Although the mechanism of Hérisson and Chauvin afforded a correct interpretation of polyalkenamer formation and their degradation to macrocycles, at that time the mode of generation of the carbene species, the mode of cyclodiene or cyclotriene fragmentation, and the product distribution in the telomerization of cyclo-olefins with α-olefins were still unclear.

A chain carbene mechanism for the ring-opening polymerization of cyclo-olefins was suggested by Dolgoplosk and co-workers [130–132] based on extensive studies on cyclo-olefin polymerization initiated by carbene precursors. Dolgoplosk and co-workers [130] considered that the chain process of cyclo-olefin ring-opening involves the participation of active centres which are complexes of carbenes with transition metals. These authors set forth a formulation for the generation of carbene species in the initial step of the chain process by reaction of the olefin with the catalytic system along with intramolecular migration of hydrogen in the olefin group (6.96).

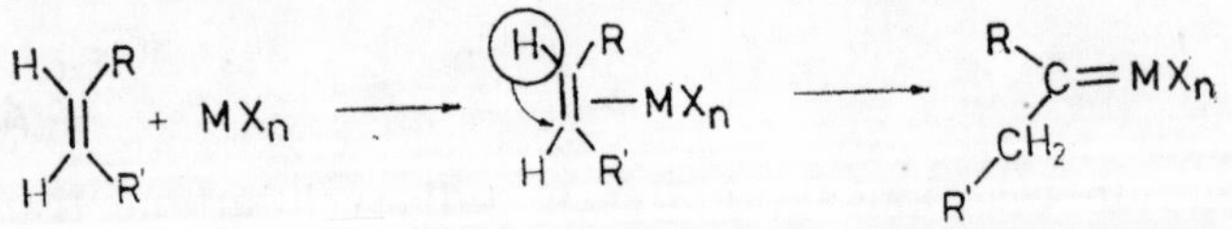

(6.96)

Relevant experimental results obtained by Dolgoplosk and co-workers provide firm evidence for carbene initiation of the chain process of ring-opening polymerization. For instance, they found that the decomposition of phenyldiazomethane, ethyldiazoacetate and trimethylsilyl-diazomethane under the influence of tungsten chlorides initiates the reaction of cyclopentene and 1,5-cyclo-octadiene with the formation of very high molecular weight polymers. With phenyldiazomethane the reaction proceeds practically instantaneously at room temperature. Furthermore, the formation of high molecular weight products with a low yield of oligomers even in the earliest stages of the reaction indicated a chain process and the generation of some highly reactive centres in the system. The chain propagation was represented by reversible monomer coordination to metal followed by monomer insertion into the chain.

Scheme 6.49

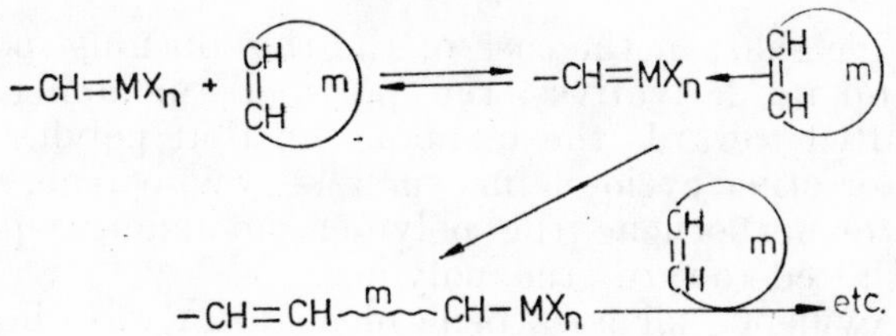

$$-CH{=}MX_n + \begin{pmatrix} CH \\ \| \quad m \\ CH \end{pmatrix} \rightleftharpoons -CH{=}MX_n \leftarrow \begin{pmatrix} CH \\ \| \quad m \\ CH \end{pmatrix}$$

$$-CH{=}CH\overset{m}{\sim\!\sim\!\sim} CH{-}MX_n \longrightarrow \begin{pmatrix} CH \\ \| \quad m \\ CH \end{pmatrix} \longrightarrow \text{etc.}$$

Investigations carried out with cyclopentene, cyclo-octadiene and norbornene showed that polymers of very high molecular weight have been formed in the early stages of monomer conversion. The polymer formed initially is characterized by a narrow molecular weight distribution, gradually broadening at high conversions. The amount of oligomeric compounds at low conversion of monomer was found to be small and their formation was associated with polymer cyclodestruction and not with the polymerization process. The cyclodestruction was interpreted in terms of a reversible polymer—monomer equilibrium promoted by the reaction of the active centre with the nearest double bond of its own chain (6.97).

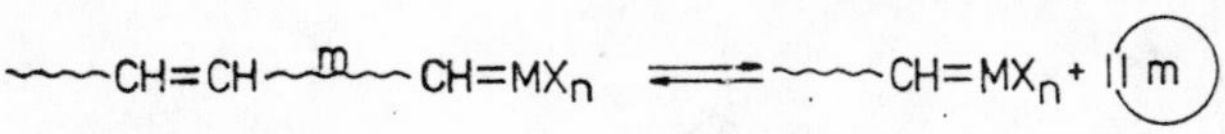

$$\sim\!\sim\!\sim CH{=}CH\overset{m}{\sim\!\sim\!\sim} CH{=}MX_n \rightleftharpoons \sim\!\sim\!\sim CH{=}MX_n + \begin{pmatrix} \| \, m \end{pmatrix} \qquad (6.97)$$

Naturally, if the reaction of the active centre occurs with other double bonds of its own chain, a set of cyclo-oligomers could be formed by cyclodestruction (6.98).

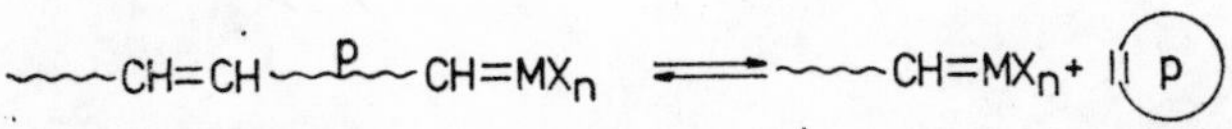

$$\sim\!\sim\!\sim CH{=}CH\overset{p}{\sim\!\sim\!\sim} CH{=}MX_n \rightleftharpoons \sim\!\sim\!\sim CH{=}MX_n + \begin{pmatrix} \| \, p \end{pmatrix} \qquad (6.98)$$

This polymer—cyclo-oligomer equilibrium will be governed by entropy factors and by the stability of cyclo-oligomers. The cyclo-oligomer concentration will consequently increase with monomer conversion. When an active centre reacts with the double bond of a different chain, redistribution of the molecular weights by chain transfer results (6.99).

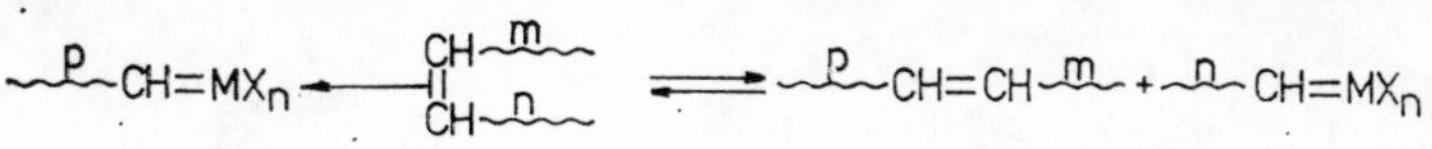

$$\sim\!\!\overset{p}{\sim\!\sim} CH{=}MX_n \leftarrow \begin{matrix} CH\overset{m}{\sim\!\sim} \\ \| \\ CH\overset{n}{\sim\!\sim} \end{matrix} \rightleftharpoons \sim\!\!\overset{p}{\sim\!\sim} CH{=}CH\overset{m}{\sim\!\sim} + \overset{n}{\sim\!\sim} CH{=}MX_n \qquad (6.99)$$

The polymer—monomer equilibrium has been thoroughly investigated for a wide range of cyclo-olefins. It was found that for cyclopentene the polymer—monomer equilibrium depends considerably on the *cis* or *trans* configuration of the terminal unit of the growing chain. The equilibrium is related to the steric probability of coordination of the nearest double bond of the polymer chain to the metallacarbene centre (6.100).

$$\tag{6.100}$$

It is probable that, in the case of the ring-opening polymerization of cyclohexene and its derivatives, the polymer—monomer equilibrium is completely shifted towards the monomer so that polyhexenamer cannot be obtained. For other cyclo-olefins such as cyclo-octene, cyclo-octadiene, cyclododecene or norbornene, the polymer—monomer equilibrium seems to be wholly shifted towards the polymer.

Additional evidence for a carbene mechanism was obtained by cyclodegradation of linear polymers with metathesis catalysts [133] as well as by degradation of polymers with acyclic olefins in the presence of metathesis catalysts.[134]

Rooney *et al.*[135] found that metal-hydride complexes (whose concentration is increased for WCl_6 by a little water and diminished by LiC_4H_9) could be largely responsible for dimerization of norbornene and are essential for polymerization. Such hydrides, which could arise from reaction of HCl and/or H_2O with W^{IV} subsequent to reduction of W^{VI} by norbornene, seem to generate metallacarbene species involved in the polymerization process.

Scheme 6.50

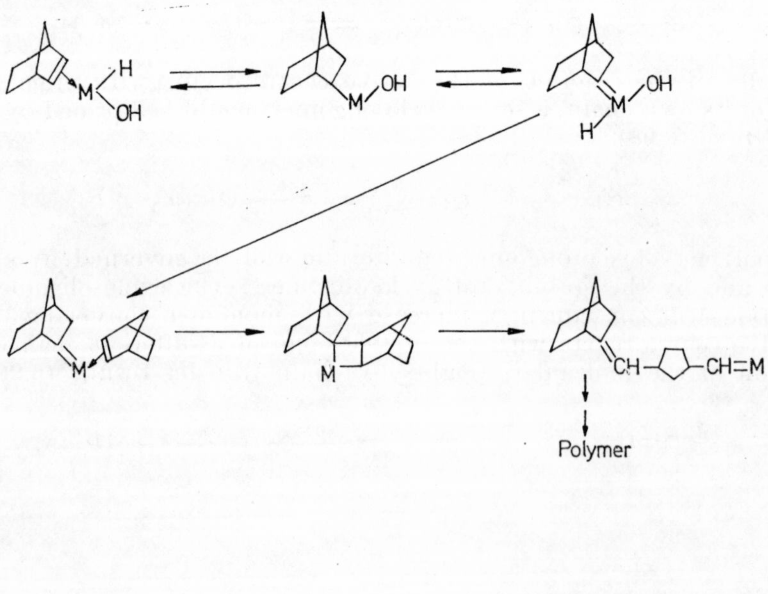

Polymer

Hydrides of transition metal complexes, for instance a hydrated 1,5-cyclo-octadiene complex of $IrCl_3$, active in polymerizing norbornene, have been detected since they possess an M—H bond active in IR spectra.

Katz and Acton [136] interpreted the mechanism of the polyalkenamer formation from cyclo-olefins in the presence of carbene systems $(C_6H_5)_2C=W(CO)_5$ and $C_6H_5(CH_3O)C=W(CO)_5$ through a chain process where the initiation and propagation are attributed to carbene complexes (6.101)

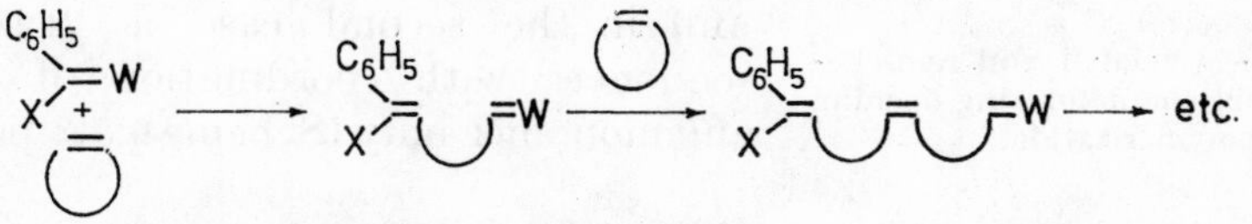

$$(6.101)$$

where $X = C_6H_5$ or CH_3O.

In addition, Katz and co-workers [137] pointed out how alkynes can induce cyclo-alkene polymerization with stabilized metallacarbenes. Thus, they supposed that alkynes combine readily with stabilized metallacarbenes of the type $C_6H_5(CH_3O)C=W(CO)_5$ producing by an insertion process more reactive metallacarbenes that initiate cyclo-alkene polymerization like common metallacarbenes.

Scheme 6.51

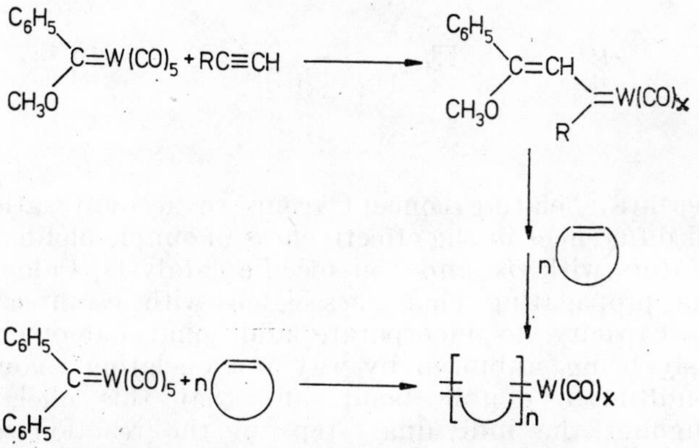

They assumed that alkynes could quench the polymerization of cyclo-alkenes by the same insertion-type reaction (6.102).

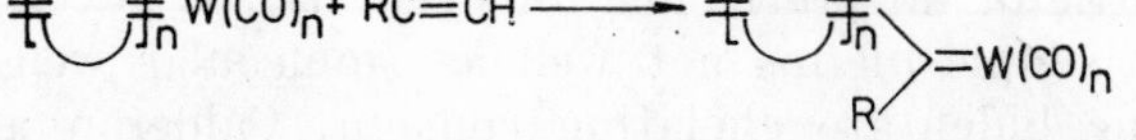

$$(6.102)$$

For ring-opening polymerization of cyclopentene with cis- and non-specific metathesis catalysts, Calderon and co-workers [138] advanced a bidentate chelating concept involving a multi-dentate growing chain for *cis*-specific catalysts, A, and a non-chelated growing chain for non-specific catalysts, B (Fig. 6.19).

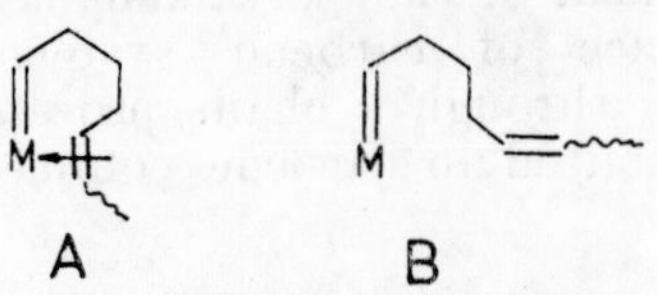

A   **B**

Fig. 6.19. — Chelated and non-chelated growing chain for ring-opening polymerization.

In the first case the propagation reaction involves a "three-ligand sequence" implying simultaneous coordination of the carbene, monomer and polymer chain to the metal centre (Scheme 6.52, path (1)) and in the second case a "two-ligand sequence" with coordination of carbene and monomer only (Scheme 6.52, path (2)).

Scheme 6.52

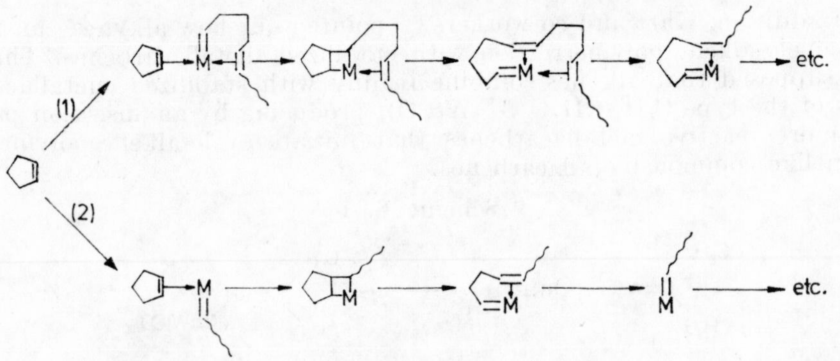

The bidentate chelating concept seems to account satisfactorily for the observed differences in the effectiveness of simple olefins as molecular weight regulators with *cis*- and non-specific catalysts. Calderon considered that the propagating chain associated with *cis*-directing catalysts maintains its capacity to incorporate additional monomer units while simultaneously being stabilized by way of a chelating π-complex involving the penultimate double bond, and that this chelation remains intact throughout the individual steps of the reaction sequence. For this reason *cis*-directing catalysts exhibit a preference for monomer cyclo-olefin rather than acyclic olefins, as found experimentally. By contrast, non-selective catalysts do not require stabilization *via* chelation with the penultimate double bond of the growing chain and therefore they exhibit little or no preference for cyclic *versus* acyclic olefins. With these catalysts simple olefins act well as molecular weight regulators. Using the same bidentate chelating concept, Calderon and co-workers rationalized the effectiveness of α,ω-olefins, namely 1,6-heptadiene, 1,5-hexadiene and 1,7-octadiene to act as molecular weight regulators.[94]

Based on the steric structure of polyalkenamers, Ivin and co-workers [20,21,139] postulated the presence of two kinetically distinct species, not in equilibrium with one another, to explain the mechanism of ring-opening polymerization. This mechanism was interpreted within the limits of the carbene concept as involving metallacarbenes which first coordinate and then add olefin to give a transient metallacyclobutane which ruptures to yield a new metallacarbene and a newly formed double bond. It was assumed that metallacarbenes have approximately octahedral symmetry and one vacant position □ for coordination of olefin. In this case the following four possible orientations of the carbene ligand with respect to the vacancy were considered (Fig. 6.20):

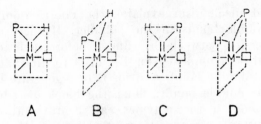

A   B   C   D

Fig. 6.20. — Orientations of the carbene ligand
in octahedral complexes with respect to the
vacancy.

where P denotes the polymer chain and M the metal centre.

In the polymerization of norbornene only two types of metallacyclobutane structures are possible, E and F, which were termed *cis* and *trans* (Fig. 6.21)

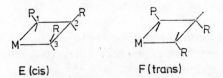

E (cis)   F (trans)

Fig. 6.21. — Types of possible metallacyclobutanes in ring-opening polymerization.

where the norbornene ring system is represented as R, R at C_2 and C_3. By detailed analysis of all possible modes of rupture of transient metallacyclobutanes E and F, Ivin and co-workers concluded that the most likely pathways are those shown by the full lines in Scheme 6.53, together with some contribution from the pathways with broken lines.

Scheme 6.53

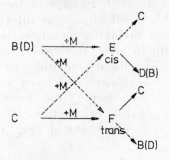

The proposed mechanism explains the non-random distribution of *cis* and *trans* double bonds in the polymer. If there is not too much direct interconversion between B(C) and D before each addition of monomer, there would be a tendency for B and D to alternate giving rise to a sequence of *cis* double bonds and for C to regenerate C giving rise to a sequence of *trans* double bonds. Thus, the block distribution of *cis* and *trans* double bonds in the polymer can be rationalized. In the same way the tacticity of *cis* and *trans* structures in the polymer was interpreted.

The interconversion of B, C, D (and A) which accounts for the structure of the polymer may proceed by rotation about $M=C\langle$, by ligand migration or by relaxation of the geometry from octahedral to square pyramidal or trigonal bipyramidal. Interconversion by rotation about the $M=C\langle$ double bond is known to involve a relatively small energy barrier. The π-component of this bond formed by combination of a metal d_π orbital with a carbene p_π orbital is considered to be relatively weak in an active metallacarbene species, corresponding to a rotational barrier of 70—90 kJ/mol. Ligand migration could also be a favorable relaxation pathway of low energy barrier, closely correlated with the steric environment of the metallacarbene.

New possibilities arise for interconversion of B, C and D by relaxation of the geometry from octahedral to other structures, such as square pyramidal or trigonal bipyramidal, with a plane of symmetry through the metal—carbene bond. For the octahedral carbene complex C the new relaxed forms are C′ and C″ (Fig. 6.22). This relaxation would involve only a change in bond angles consequent upon valence-bond electrons repulsion and activation energy would be expected to be small unless reaction occurs within a cluster complex or at the surface of a crystal.

To account for the oxygen activation of cyclopentene polymerization in the presence of WCl_6 catalyst, Amass and McGourtey [140] pointed to the

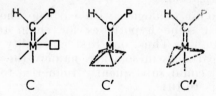

Fig. 6.22. — Possible relaxed forms for
octahedral carbene complexes.

formation of an oxygen ligand to the active species in the initial step responsible for the generation of the carbene chain-carrier (6.103).

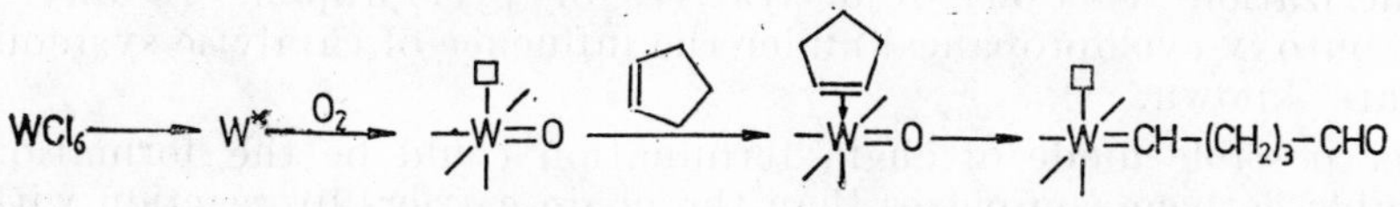

$$WCl_6 \longrightarrow W^* \xrightarrow{O_2} \quad -\overset{\square}{\underset{|}{W}}=0 \longrightarrow \quad -\overset{\square}{\underset{|}{W}}=0 \longrightarrow \quad -\overset{\square}{\underset{|}{W}}=CH-(CH_2)_3-CHO \qquad (6.103)$$

Infra-red spectral analysis of the polymer indicated the presence of a carbonyl group in the chain. The oxygen ligand seems to be essential in the initiation step with this catalyst but it is not directly implied in the propagation step.

More recently, Levisalles and co-workers [141] proposed a new insertion-type mechanism for cyclo-olefin polymerization involving a bimetallic propagation centre.

Scheme 6.54

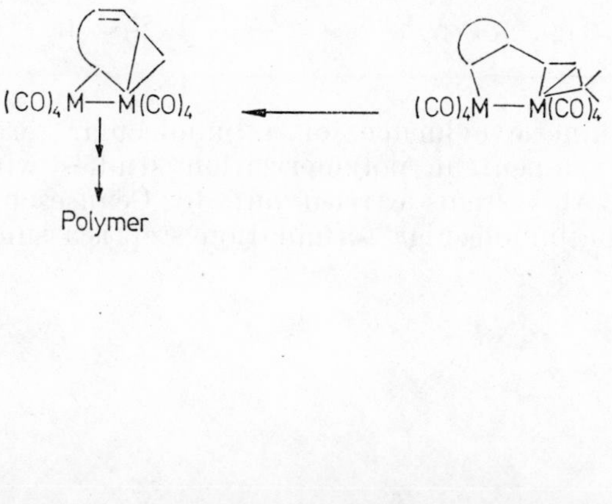

$$(CO)_5M \overset{\triangle}{-} M(CO)_4 \quad \xrightarrow{+\,\square} \quad (CO)_4M - M(CO)_4$$

$$(CO)_4M - M(CO)_4 \quad \longleftarrow \quad (CO)_4M - M(CO)_4$$

Polymer

Evidence for this reaction sequence is based on the isolation and characterization of the bimetallic complex (I) (where $M=W$) active in alkene and alkyne polymerization.

Taking into account analogies with carbene chemistry in solution, Dolgoplosk advanced some hypotheses for chain termination in ring-opening polymerization.[132] Thus, one first possibility would be the most specific reaction of carbenes in solution, namely the formation of cyclopropane groups that could subsequently isomerize to the corresponding unsaturated products (6.104).

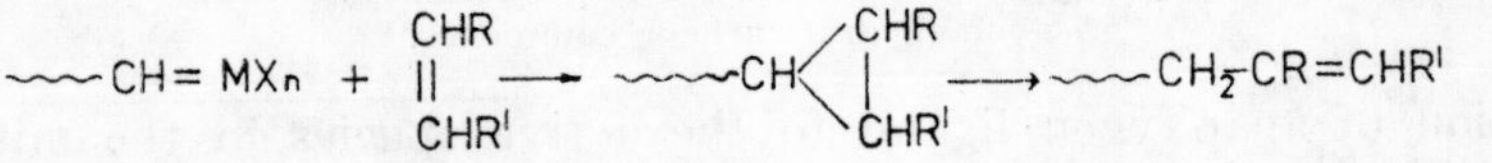

$$\sim\sim CH=MX_n + \begin{matrix} CHR \\ || \\ CHR^I \end{matrix} \longrightarrow \sim\sim CH \begin{matrix} CHR \\ | \\ CHR^I \end{matrix} \longrightarrow \sim\sim CH_2 CR=CHR^I \qquad (6.104)$$

Such isomerization reactions of derivatives of cyclopropane (n-amyl-, phenyl-, n-butoxy-cyclopropanes) under the influence of catalytic systems of WCl_6 are known.

Another possible mode of chain termination could be the formation of more stable carbene complexes than the chain-carriers by reaction with impurities of unsaturated character like olefins, dienes, acetylenes or similar compounds. Thus, it was found that in the polymerization of cyclo-olefins with $(CH_3)_3SiCH_2Li \cdot WCl_6$ and $C_2H_5AlCl_2 \cdot WCl_6$, small amounts of trimethylvinylsilane completely inhibit the process.[142] This fact was attributed to the formation of the more stable carbene complex $(CH_3)_3SiCH=WX_n$.

The reduction of the transition metal and the formation of a free radical could be involved in chain termination taking into account that such transformations are energetically favourable (6.105).

$$\sim\sim CH=MX_n \longrightarrow \sim\sim\sim CHX^\bullet + MX_{n-1} \qquad (6.105)$$

Further, another hypothesis assumes the migration of the hydrogen atom in the carbene complex with the formation of unsaturated chain ends (6.106).

$$\sim\sim CH_2CH=MX_n \longrightarrow \sim\sim\sim CH=CH_2 + MX_n \qquad (6.106)$$

Substantial kinetic evidence for a bimolecular termination reaction was found in cyclopentene polymerization studies with $WCl_6 \cdot Et_2AlCl$ and $WCl_6 \cdot i\text{-}Bu_3Al$ systems carried out by Ceauşescu, Bittman and Dimonie.[22-24] The bimolecular termination step is assumed to occur by a disproportionation pathway common to transition metal organometallics (6.107).

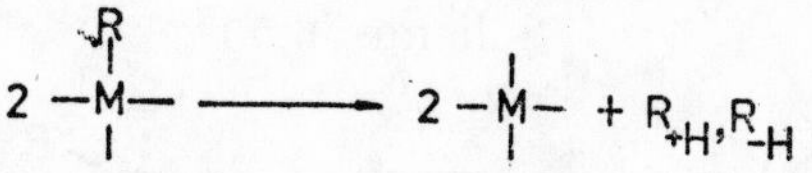

$$2 \; -\overset{\overset{\displaystyle R}{|}}{\underset{|}{M}}- \quad \longrightarrow \quad 2 \; -\overset{|}{\underset{|}{M}}- \; + \; R_{+H}, R_{-H} \tag{6.107}$$

This assumption has been corroborated with hydrocarbon formation from organotungsten compounds and supported by data obtained concerning molecular weight variation during polymerization for both catalytic systems. It seems highly probable that deactivation by disproportionation reactions would be correlated with the reduction of valence state of the tungsten atom.

Based on second order kinetics for the termination reaction in cyclopentene polymerization, Amass and Zurimendi [143] proposed a bimolecular mechanism for the termination step involving intermolecular hydrogen migration (6.108).

$$\begin{array}{c} W \vdots CH_2-CH_2 \sim \\ \nwarrow \; + \\ \sim CH_2-CH_2 \vdots W \end{array} \quad \longrightarrow \quad \begin{array}{c} CH_3-CH_2 \sim \\ + \\ CH_2 = CH \sim \end{array} \tag{6.108}$$

Together with such a second order process these authors suggested another termination reaction which is again second order overall, but first order with respect to the active species and first order with respect to the concentration of destroyed species. Since these termination reactions may lead to the production of a species that can be regenerated upon the addition of oxygen, these authors inferred the probability that these termination reactions would involve a lower oxidation state of the transition metal compound.

It is now evident that the chain carbene mechanism, in contrast to earlier proposed schemes, offers a better interpretation for the formation of high molecular weight polyalkenamers at the beginning of the polymerization and gives a satisfactory explanation for the formation of macrocycles as well as for the occurrence of polymer—monomer equilibrium in the system.

6.2.5. Mechanism of Metathesis Polymerization of Acetylenes

As with olefin metathesis and cyclo-olefin polymerization the polymerization of acetylenes induced by carbene metathesis catalysts was proposed to proceed by a chain reaction involving a metallacyclobutene intermediate as a key step for acetylene insertion in the metallacarbene bond.[144]

Scheme 6.55

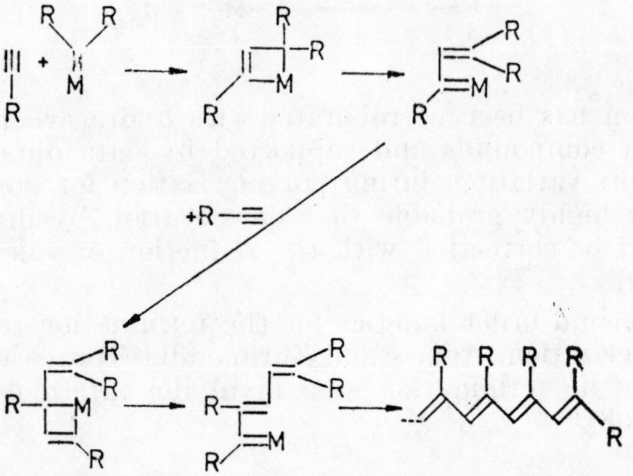

This mechanism could rationalize the results obtained in the polymerization of various acetylenes with (phenylmethoxycarbene)-pentacarbonyltungsten and (diphenylcarbene)-pentacarbonyltungsten as catalysts.[136] By studying the kinetics of polymerization of phenylacetylene in $CDCl_3$ solution, initiated by (phenylmethoxycarbene)-pentacarbonyltungsten, Thoi, Ivin and Rooney [145] observed kinetics of zero order with respect to monomer and of fractional order (0.7) with respect to initiator, the variation of rate with temperature corresponding to an apparent activation energy of 116 kJ/mol. These authors discussed the mechanism of initiation and propagation in terms of elementary steps involving formation and reactions of metallacarbenes, metallacarbene—monomer complexes, and metallacyclobutenes. It is interesting to note the retardation of the reaction by the presence of carbon monoxide, oxygen or air, but not by nitrogen. This fact was attributed to coordination of CO or O_2 to the 16-electron metallacyclobutene intermediate in the presence of monomer (and not to the metallacarbenes). It was inferred that coordination of these gases results in chain termination.

Metallacyclobutenes, presumed intermediates in the metathesis polymerization of acetylenes, have been obtained by reaction of titanocene carbene complex $Cp_2TiCH_2 \cdot AlClMe_2$ with diphenylacetylene [69] or of titanacyclobutanes such as

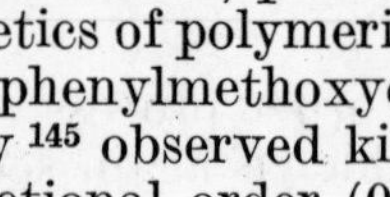

Cp₂Ti◇—⟨

with diphenylacetylene.[99]

An alternate possible mechanism for acetylene polymerization was suggested by Levisalles and co-workers [141] involving a dimetallic carbene intermediate as the chain-carrier.

Scheme 6.56

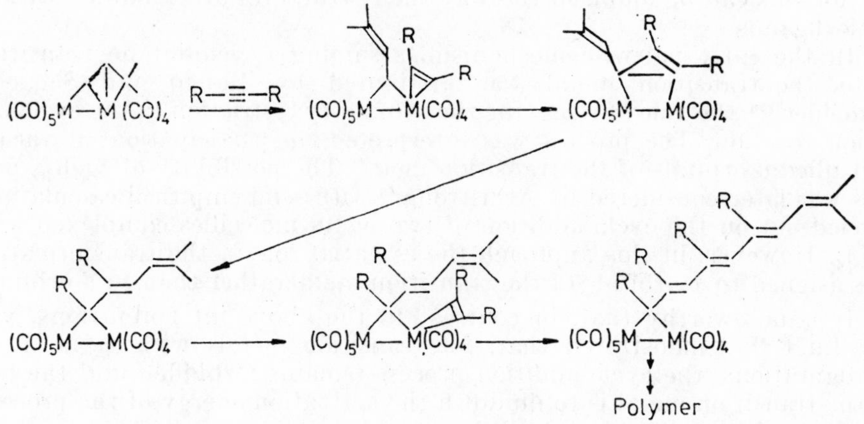

Successive insertions of monomer could lead to a polymer which grows on two metal centres. This mechanism seems to explain well the structure of the polyacetylenes.

6.2.6. The Role of Transition Metal

It is now evident that the metathesis reaction proceeds in the presence of active complexes of a limited number of transition metals, mainly belonging to groups V—VIII of the Periodic Table, of which molybdenum, tungsten and rhenium are the most effective. It is natural to consider that these transition metals exert their general behaviour because of their properties common to transition metal chemistry as well as their specific property of generating metathesis species. This specific role for metathesis activity may be attributed to both electronic and geometric features of these metals. Although some theoretical interpretations have been advanced and abundant experimental data are available, the exact role of the transition metal for metathesis activity is still unknown.

The outstanding ability of transition metals to coordinate donor ligands including olefins, acetylenes and even carbenes or carbynes, to exchange ligands or to favour their rearrangement, readily to change their oxidation state and coordination number, and to pass from one configuration to another, to generate vacant sites and to liberate ligands are widespread in the organometallic chemistry of transition metals. More-

over, in the metathesis process the transition metals seem especially to
generate carbene or carbyne ligands and to arrange them and the reactant molecules (i.e. alkenes or alkynes) in a favourable position for exchanging alkylidene or alkylidyne groups. Finally, the metathesis transition metals themselves behave as carbene or carbyne species. The reason for this can be found in the way their structures are modified by suitable ligands.

In the early pairwise mechanism, assuming a cyclobutane transition
state, the transition metal was attributed by Mango and Schachtschneider [146] the role of removing the symmetry restriction in the cyclo-addition reaction. The process was interpreted *via* participation of vacant
and filled d orbitals of the transition metal. The possibility of such a process was later considered by Armstrong [49] with semi-empirical calculations
carried out on the cyclo-addition of two olefin molecules complexed with
WCl_4. However, in this approach the essential role in the transformation
was asigned to p orbitals of the transition metal rather than to d orbitals.

It is noteworthy that, in contrast to the above interpretations, van
der Lugt [147] demonstrated that, for transition metals with d^8 and d^{10}
configurations, the cyclo-addition process remains forbidden and the role
of the transition metal is to diminish the activation energy of the process.
Similarly, Dewar [148] referred to electrocyclic processes catalysed by transition metals as forbidden processes. Dewar showed that an electrocyclic
transformation is catalysed by transition metals only if the latter lead to
an anti-aromatic transition state. The tendency to form an anti-aromatic
transition state is due to the fact that the transition metal can coordinate
this state more strongly, and this is the cause of its higher stability.

Recent data confirm that the active species in metathesis are of carbene and carbyne type. The transition metals must be able to generate
such active metallacarbene or metallacarbyne species. The mode of
generation which has gained particular experimental attention was presented in section 6.2.2.1. However, it is important to remark that not
all transition metallacarbenes or metallacarbynes are active in metathesis.
The reason for this, which is attributed to the nature of the transition metal, is still unclear.

The essential role of the transition metal for metathesis proceeding
by a carbene or carbyne mechanism seems to be that of bringing the reactive partners in a position favourable for exchanging carbene or carbyne
groups and for retaining a bound carbene or carbyne so as not to produce cyclopropane or another cyclic structure. This transformation may
occur *via* a coordinate alkene or alkyne metallacarbene or metallacarbyne complex and a subsequent metallacyclobutane or metallacyclobutadiene intermediate, or some other non-classical cyclic structure. This
role can be correlated with the d electron configuration of transient species formed from metathesis transition metals.

Rappé and Goddard [149] have shown theoretically how the nature of
the transition metal may crucially affect the mode of metallacyclobutane

evolution in the metathesis process. They found by *ab initio* calculations that the energetics of C—C bond cleavage in metallacyclobutane could be changed dramatically when passing from Mo and W to Cr. While for Mo and W the C—C bond cleavage (pathway **a**, Scheme 6.57) may be favoured relative to reductive elimination (pathway **b**, Scheme 6.57), for Cr reductive elimination will be the predominant mode of transformation.

Scheme 6.57

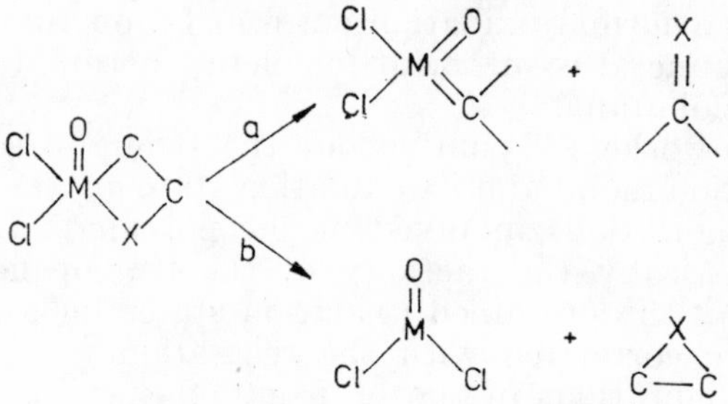

This fact was attributed to the increased σ bond strengths for coordinated ligands in Mo and W compounds.

For a carbene mechanism acting in alkene metathesis, Hoffmann and co-workers [150] pointed out that a "collinear" conformation of the alkene metallacarbene complex is essential for metathesis activity. From EHT calculations they found that the more positively charged the metal, the lower the activation energy required to reach the collinear geometry. Furthermore, for second and third transition series metals there is a small barrier to attaining the collinear geometry for d^4 and d^6 configurations but the collinear arrangement is strongly favoured for d^2 configurations. The carbene—alkene interaction is considered to be strong and to depend on the nature of transition metal. The CC overlap populations generated along the linear transit from a metal (carbene)-(olefin) complex to a C_{2v} metallacyclobutane indicate that the metal seems to hold two reactive components together in a geometry favorable for their interaction and to prevent, for d^0-d^4 electron configurations, cyclopropane elimination by direct carbene—alkene cyclo-addition; at the same time, cyclopropane elimination is allowed [151] for d^6 states.

There is an optimum oxidation state of the transition metal that affords metathesis active species. So far, from physico-chemical investigations on homogeneous and heterogeneous catalysts the precise oxidation state at a given centre is generally uncertain.

For homogeneous catalytic systems based on tungsten, it seems that oxidation state IV is optimal for producing active species, but other species may appear in which tungsten would be in a lower oxidation state. For the W(IV) oxidation state a minimum of two d electrons are necessary for the formation of the metal—carbene bond. Similar oxidation states may be considered also for molybdenum catalysts. Other transition metals could lead to active species of higher (Re(V)) or lower (Ru(III), Os(III), Rh(I) and Ir(III)) oxidation states.

The high oxidation states involve geometrical restrictions around the central atom while the lower oxidation states favour easier ligand rearrangement and alkene or alkyne complexation. Thus, it is known that group VIb metals can form more stable olefin complexes in lower valence states, even down to oxidation state zero.

For a carbene system generated from WCl_6 and $(CH_3)_2Zn$, Muetterties[56] suggested a formal oxidation state of IV on tungsten. This oxidation state on the metal is sufficient for olefin coordination and transition to a metallacyclobutane.

Ivin and co-workers[21] pointed out the strong effect exerted by the nature of transition metal and its oxidation state on the rate of relaxation of possible configurations involved in the polymerization reaction. This can change drastically the tacticity of the ring-opened polymers. The electron counts of the presumed oxidation states for a number of transition metals were correlated with the relaxation rates between possible geometrical configurations of the propagation species. Thus, in d^2 states of Re(V), W(IV) and Mo(IV) where two of the three t_{2g} (d_{xy}, d_{yz}, d_{xz}) orbitals will be empty and available to attract electrons from Cl ligands, the relaxation of the quasi-octahedral configuration to other related configurations will presumably be restrained. This will result in predominant participation of the former configuration in a propagation reaction with enhancement of polymer tacticity. On the other hand, Ru(III), Os(III) or Ir(III), having either five or six d electrons, will not impose the same restriction on relaxation of the configuration, so that atactic polymers with a random cis/trans distribution will be produced; Mo- and W-based catalysts involving variable oxidation states e.g. M(III) or M(II) (M=Mo or W) will therefore form atactic polymers with random cis/trans distribution.

Investigations concerning the valence state of heterogeneous catalytic systems based on molybdenum have indicated different and still inconclusive results.

Davie, Whan and Kemball,[152] through infra-red spectroscopic studies on a heterogeneous alumina-supported $Mo(CO)_6$ catalyst, noticed that the active Mo species usually has valence III, but may also have valence IV. On the other hand, Mortreux et al.,[153] through electron paramagnetic resonance studies on the catalytic systems $MoO_3 \cdot SiO_2$ and $MoO_3 \cdot Al_2O_3$, showed that the active species is rich in molybdenum ions of valence V. These authors found a correlation between the activity of the catalyst and its content of Mo^{5+} ions in the disproportionation of propene. Similar results were also obtained by Nakamura and co-workers,[154] through electron paramagnetic resonance studies on the above catalytic systems. They observed that the catalytic system $MoO_3 \cdot SiO_2$ has low disproportionation activity after treatment with nitrogen, and that this activity is correlated with the intensity of the ESR signals associated with

Mo $^{5+}$ ions in the catalyst and with double bond isomerization. On reducing the catalyst with hydrogen, the disproportionation activity increases and, at the same time both the intensity of the Mo $^{5+}$ signal in the ESR spectrum and the degree of dispersion of the $\overline{MoO_3}$ on the inert carrier are increased.

Investigations on allyl and alkoxide Mo, W and Re compounds anchored on silica or alumina, carried out by Ermakov and co-workers,[155,156] revealed the role of coordinatively unsaturated metal ions with oxidation number +4 on the catalyst surface. They found, in the metathesis of propene, that surface compounds containing Mo, W and Re ions with oxidation number +4 have maximum activity. In addition, the coordination environment of the metal (in particular, the nature and the number of the nearest ligands) determines the activity even in this favorable oxidation state.

Coordinatively unsaturated tetravalent molybdenum species, chemically bonded to a support through surface oxygen atoms, are found by Iwasawa et al.[157] to be the actual active species in propene and 1-butene metathesis. Such species were identified by means of XPS and volumetric analysis in highly active catalysts with well-defined surface structures prepared through the reaction of $Mo(\pi\text{-}C_3H_5)_4$ with surface acidic OH groups of alumina, silica and silica-alumina. Furthermore, no correlation between the amount of pentavalent molybdenum and metathesis activity was found by ESR spectroscopy. Similarly, from XPS measurements, hexavalent and divalent molybdenum species were shown to have no catalytic activity under the experimental conditions.

The nature of the ligand coordinated to the Mo $^{4+}$ ions in this class of supported catalysts, as well as the support employed, had a marked influence on the catalytic activity and selectivity. The carrier also acted as a supplier of reactants for the active sites. For a wide range of oxidation states of the molybdenum ions (namely, 2+, 3+, 4+, 5+ and 6+) which are encountered in a sequence of surface structures (monomeric and dimeric molybdenum compounds) a correlation with metathesis activity was later reported by Iwasawa and Kubo.[157] Tetravalent molybdenum species were also considered to be important active component of this series, although other species such as Mo^{3+} may also play a significant role.

By contrast, experimental studies of reflectance spectroscopy reported by Andreev et al.[48,158] with the $Re_2O_7 \cdot Al_2O_3$ system indicated that Re(VII) and the partially reduced form Re(VI) are responsible for catalytic activity. Furthermore, they established that reduction to Re(IV) results in an almost complete loss of activity in metathesis.[159] Surprisingly, however, the $Re_2O_7 \cdot MgO$ system proved to be active in 1-butene metathesis at high temperatures, when reduction to metallic rhenium occurs.[160] In this case, Edreva-Kardjieva and Andreev supposed that metallic rhenium is responsible for the metathesis activity whereas partially reduced Re(IV) promotes double bond isomerization.

6.2.7. The Nature and Role of Ligands

The nature of the ligands bound to the transition metal is critical for catalytic activity and for the selectivity of metathesis catalysts. The ligands act upon the reaction centre through both electronic and steric effects.

The ligands commonly employed for metathesis catalysts can be divided into four main groups : (i) ligands which do not influence the catalytic activity and selectivity, such as chlorine, carbonyl and similar groups; (ii) ligands which do influence the catalytic activity and selectivity such as oxo and similar weak electron donor groups; (iii) ligands which influence the coordination environment of transition metal, such as phosphine, pyridine or other strong electron donor groups and (iv) ligands responsible for the catalytic activity of the reaction centre, such as carbene and carbyne groups.

Ligands of the first category are essential for maintaining the coordination number and oxidation state necessary for metathetic activity. This category includes a limited number of anionic or neutral ligands; among them the most important are chlorine and carbonyl groups. For a more detailed bonding description, the reader should consult the special literature.[161]

Chlorine ligands are generally attached to transition metals through σ bonds formed by overlapping chlorine $3s$ orbitals with metal orbitals having suitable symmetry and energy. Electron transfer from the transition metal to the chlorine atoms maintains the high oxidation state of the transition metal and stabilizes the metal coordination state. π-Bonding is also possible between the metal and the chlorine *via* the empty chlorine $3d$ orbitals but the energy of the latter orbitals is too high relative to those of the metal for any significant overlap to occur; the chlorine—metal bond is therefore basically σ in character so π-donation is practically absent.

Carbonyl (carbon monoxide) ligands form σ/π bonds with transition metals in unidentate complexes involving both σ and π components. The σ component is formed by interaction between an empty metal σ-orbital and a filled sp carbon orbital, whereas the π component is formed by interaction between a filled metal $d\pi$-, or hybrid $dp\pi$-orbital and an empty antibonding $p\pi$-orbital of the carbon monoxide. The σ component results in a net transfer of electron density from the ligand to the metal and the π component in a net transfer from the metal to the ligand. The bonding is synergistic and results in a decrease in the $C-O$ bond order. In bidentate and tridentate complexes, the carbonyl ligand is bridging two or three metal centres; this kind of bonding occurs *via* the carbonyl carbon as in unidentate complexes.[162] The net result is a substantial decrease in the $C-O$ bond order from A to C mode of coordination, as illustrated in Fig. 6.23.

Carbonyl complexes are known in which the π-system of the carbonyl group can be bonded to one or two other metal centres [163] (Fig. 6.24) but this type of bonding seems to be of no consequence for metathesis activity.

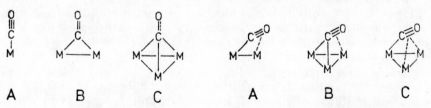

Fig. 6.23. — Symmetrical modes of carbon monoxide coordination.

Fig. 6.24. — Asymmetrical modes of carbon monoxide coordination.

Nitrosyl groups make use of their filled sp nitrogen orbitals for interaction with the empty metal σ orbitals, and their unfilled and partially filled $p\pi$ orbitals for interaction with filled and partially filled metal $d\pi$ or hybrid $dp\pi$ orbitals, respectively. Thus, the NO ligand contributes three electrons to a σ-π bond with the metal while the CO ligand contributes only two electrons.

Remarkably, the NO ligand can coordinate in both linear and bent fashion [164] owing to its ambiguous Lewis base and Lewis acid character, respectively. The tautomerism between base and acid behaviour can convert a coordinatively saturated metal to a more reactive unsaturated species without the normal requirement of dissociation of other ligands.

Oxygen ligands generally form σ and π bonds with transition metals by using s and p orbitals of suitable symmetry and energy. One π-bond arises by transfer of electron density from a filled metal $d\pi$- or hybrid $dp\pi$-orbital to an empty oxygen $p\pi$-orbital while a second π-bond can be formed by transfer of electron density from a filled oxygen $d\pi$-orbital to an empty metal $d\pi$- or hybrid $dp\pi$-orbital. Thus, the metal—oxygen bond could gain quasi-triple bond character with the transfer of more than two electrons from oxygen to metal.[165] The same bonding picture can arise in other oxo ligands having one additional substituent on oxygen like alkoxy, phenoxy, etc.

Bonding pictures similar to those above can be considered for various donor ligands containing phosphorus, arsenic, nitrogen, sulphur, etc. For instance, with phosphine ligands the σ component is formed by overlap of an empty metal σ-orbital and a filled sp phosphorus orbital while the π component is formed by overlap of a filled metal (d or dp hybrid) orbital with an empty $3d$, or hybrid $3d3p$, phosphorus orbital.

Some of the above mentioned groups, like oxo or phosphine ligands can exert a pronounced $trans$ effect generally labilizing the group situated in the *trans* position of a square planar or an octahedral structure. This behaviour may be relevant for the possible access of a reactant molecule to a *trans* position when the active species has favourable geometry, such as square planar.

There is now convincing evidence that carbene and carbyne ligands are responsible for the catalytic activity of metathesis catalysts. These ligands may arise in catalyst species by several ways discussed in an earlier section (p. 242).

Carbene and carbyne ligands are generally alkylidene and alkylidyne groups coordinated to the transition metal. There are several excellent reviews on metallacarbene and metallacarbyne complexes to which the reader is referred.[166-168] Some relevant features related to metathesis chemistry will be dealt with in the following paragraphs.

A variety of metallacarbenes of transition metals from groups IV—VIII of the Periodic Table have been synthesised and characterized, but only few of them are active in metathesis. Moreover, the activity of these metallacarbenes is low as compared to the most active binary or ternary metathesis catalysts.

Generally, the stable metallacarbenes carry the carbene ligand as a terminal unidentate group but bridged structures through metal—carbon—metal bonds are possible for reactive intermediates. In the first case the carbenic carbon C_{carb} may be considered to be a trigonal sp^2C bound to the metal by electron acceptor—donor bonds. The bonding arises as a result of donation of a pair of electrons from an sp^2 set of the singlet C_{carb} atom to the deficient metal d orbital and back—donation of electrons from the metal into the vacant p orbital of C_{carb}. The π acceptor : σ-donor ratio for carbene ligands, which depends largely on the transition metal, its ligands and C_{carb} substituents, determines the structure of the metal—carbene bond, the electrophilic or nucleophilic character of C_{carb}, and the stability and reactivity of the metallacarbene.

The metal—carbon bond in terminal carbene complexes has partial double bond character and is shorter than the corresponding single bonds of an sp^2 hybridized carbon atom. The barriers to rotation about metal—carbene bonds are generally low but significant barriers may also be encountered.[169] The C_{carb} atom is known to have electrophilic character for Fischer-type complexes[170] and nucleophilic character relative to Ta and Nb complexes synthesised by Schrock.[171] It seems that the charge character of C_{carb} varies widely within the limits I and II with the nature of the transition metal and of the ligands (Fig. 6.25).

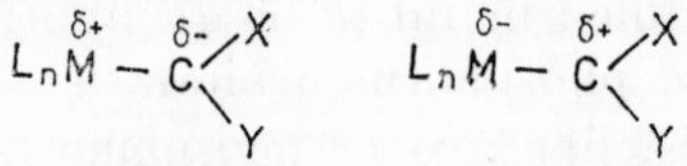

$$L_nM \overset{\delta+}{-} \overset{\delta-}{C} \overset{X}{\underset{Y}{\diagdown}} \qquad L_nM \overset{\delta-}{-} \overset{\delta+}{C} \overset{X}{\underset{Y}{\diagdown}}$$

I   II

Fig. 6.25. — Possible polarization in metallacarbene complexes.

A carbene ligand may exert a significant *trans* effect upon the existing ligand on the transition metal when the geometry is favourable. The *trans* effect is more frequently found in square planar complexes and may easily labilize the ligands for the creation of a coordinatively unsaturated site.

Carbene complexes of electron—deficient transition metals display characteristic structural deformation in which the carbene appears to pivot while the C_{carb}—H bond weakens. A molecular orbital analysis of the carbene complexes traces the deformation to intramolecular electrophilic interaction of acceptor orbitals of the metal with the carbene lone pair.[172] Hydride transfer from C_{carb}-H to the metal was found to be a forbidden reaction, at least for five-coordinate, 14-electron complexes. The metal—hydrogen bonding interaction guides the hydride to a neighbouring alkyl group and facilitates an α-elimination mode, characteristic of the reactions of these complexes.

The bonds in bridged metallacarbene complexes appear to be somewhat different in that the C_{carb} is no longer trigonal.[16]

Metallacarbyne complexes of transition metals of groups V—VIII of the Periodic Table are now synthesised in large and increasing numbers.[173] The carbyne ligand may be bound to the transition metal terminally or y a bridged structure to more than one metal atom. For a terminal carbyne ligand sp hybridization of C_{carb} may be assumed, but a different mode will be encountered in bridged carbynes.

The metal-carbyne bond is generally described in terms of the donor—acceptor principle involving the metal and filled, partially filled, and unfilled orbitals on carbon.[174] For the terminal metallacarbyne complexes, the bond between the metal and the alkylidyne ligand lies in proximity of a triple M≡C bond.[117] The C_{carb} may be of electrophilic or nucleophilic character. Many representatives of group VI transition metals exhibit definite electrophilic character.[175]

Experimental data confirm that the ligands bonded to the transition metal may vary widely and that they strongly influences the activity of the metathesis catalyst. Thus, earlier observations showed that controlled treatment of WCl_6 with protic compounds such as ethanol, phenol, water, acetic acid or with ether, epichlorohydrin, chlorinated unsaturated compounds, etc., prior to the addition of the organoaluminium compound, significantly increases the catalytic activity of systems $WCl_6 \cdot R_xAlCl_{3-x}$.[176,177] The reaction of WCl_6 with ethanol forms a stoichiometric amount of HCl together with species such as WCl_5OEt. The bonding of such an oxygenated ligand to tungsten seems to be essential for catalysts based on WCl_6 and organoaluminium compounds. It is therefore not surprising that tungsten compounds containing oxygenated ligand, e.g. as $WOCl_4$, $W(OC_6H_5)_6$, $W(OC_6H_5)_4Cl_2$, $W(OC_6H_4R)_4Cl_2$, $WO(OC_6H_5)_4$, $WCl_5[(MeO_2C)C_6H_4O]$, etc. are active metathesis catalysts in the presence of cocatalysts.[50]

An important class of tungsten carbonyl derivatives, $[W(CO)_5L]$ or $[W(CO)_4L_2]$, in the presence of $AlCl_3$ or $EtAlCl_2$[178] as well as WCl_6 or

$ReCl_5$ with Et_3Al [179], promote the metathesis reaction provided that trace amounts of oxygen are present. Although some explanations have been advanced, the role of oxygen as a promoting ligand has not been fully elucidated.[180-182]

Ramain and Trambouze [178] suggested that, in the presence of tungsten carbonyl complexes, oxygen inserts into a carbon—aluminium bond in $[EtAlCl_2]_2$ and the resultant dimer then rearranges to give an oxo-bridged dialuminium compound. This compound was believed to bind to the tungsten at a site originally occupied by a carbonyl group. Likewise, Basset and co-workers[180] suggested that the main effect of O_2 on the $W(CO)_5PPh_3 \cdot EtAlCl_2$ system is to enhance the Lewis acidity of $EtAlCl_2$ so as to favour an electrophilic attack of the Lewis acid on the tungsten through the carbonyl group. Infra-red spectroscopic studies indicated a complexing of the carbonyl group *trans* to PPh_3 with the Lewis acid resulting in a large decrease of the back bonding of tungsten d electrons to the anti-bonding orbitals of the CO groups in the square plane. This would produce mobility of the carbonyl ligands in the square plane favouring their release and alkene coordination. For the $W(PMe_2Ph)_4L_2 \cdot EtAlCl_2$ system activated by oxygen, Haines [78] explained the catalytic activity by assuming the formation of vacant or potentially vacant sites through the removal of dimethylphenylphosphine as its oxide.

It is, however, highly probable that the oxygen ligand has to maintain the electron demand at the bonding orbitals of the transition metal necessary for carbene generation and its activation for subsequent metallacyclobutane formation. For WCl_6, activated by treatment with oxygen, Amass and McGourtey [140] found that oxygen inserts as a carbonyl group in the polypentenamer chain formed by polymerization. Infra-red studies on oxygen-treated $Re_2O_7 \cdot Al_2O_3$ catalysts carried out by Nakamura et al.[183] indicated that O_2^- radicals are formed and stabilized on the aluminium ions adjacent to the active sites. These authors suggested that the active sites are formed by electron transfer from low-valent rhenium ions to the adsorbing oxygen molecules.

The catalytic activity of metathesis catalysts consisting of group VI metal halide complexes and organoaluminium compounds is increased by treatment with carbon monoxide. For instance, the interaction of CO with $WCl_4Py_2 \cdot EtAlCl_2$ gives tungsten carbonyl species which are highly active in converting 2-pentene into 2-butene and 3-hexene. Likewise, the activity of a number of molybdenum systems can be stimulated by treatment with nitric oxide prior to addition of an organoaluminium compound.[185] This has been found in particular for $[MoCl_4L_2]$ (L=PPh_3, Py, n-PrCN), μ-$C_5H_5Mo(CO)_3$, $MoCl_3(PhCO_2)_2$, $MoO_2(acac)$, $MoOCl_3$ and $MoCl_5$ in the presence of ligands such as Py, Ph_3PO, n-BuPO and n-OctPO. This treatment should yield molybdenum nitrosyl species which are highly effective catalysts in combination with organoaluminium compounds.

It is known that certain donors reduce the activity of catalysts derived from WCl_6 while other donors enhance it. [186] Thus, treatment of $WCl_6 \cdot EtAlCl_2$ with pyridine or triphenyl-phosphine inhibits metathesis. These donors probably reduce WCl_6 to WCl_4L_2 (L = pyridine or triphenylphosphine) and the new ligands block the possible coordination sites for the olefin. By contact with the cocatalyst the ligands could be removed more or less easily by complexation and the effect of the ligand should be diminished.

The catalytic activity of systems consisting of compounds such as $Mo(NO)_2L_2X_2$ was found by Hughes [13] to be dependent on the nature of L and of the anion X. An increase in activity has been observed along the L series $AsPh_3 < 4\text{-}EtC_5H_4N \simeq Ph_3PO$ and along the X series $I < < Br < Cl$. These results suggest that the neutral and anionic ligands are not removed from the metal in the formation of the actual catalytic intermediate. Infra-red spectroscopic measurements carried out by Leconte et al.[187] on $W(NO)_2L_2X_2$ (L = Ph_3P and X = Cl) in combination with $EtAlCl_2$ revealed that nitrosyl ligands change their initial cis position to trans in the active intermediate while phosphines probably change from trans to cis. These ligands seem to be unaffected by cocatalyst complexation; in particular, phosphines do not seem to interfere with the metallacyclobutane moiety.

The basicity of ligands in precursor complexes of the type $M(CO)_5L[M=Mo$ or W, L=CO, $P(OPh)_3$, PPh_3, $PBu_3]$ or $M(CO)_{6-x}L_x[x \leqslant 3$, $L_x=PPh_3$, PMe_3, ethylenediamine, diethylenetriamine, bis-(1,2-diphenylphosphine) ethane, bispyridine, toluene, p-xylene, or mesitylene] was found strongly to influence the activity of the catalyst.[188] For $M(CO)_5L$ complexes, the lower the basicity of the ligand L, the higher the initial rate of metathesis. The final conversion is the highest for the smallest initial activity. In the case of $M(CO)_{6-x}L_x$ complexes, arene ligands drastically increased both the initial rates of metathesis and the final conversions. These experimental observations were explained by Basset and co-workers on the basis of changes induced by the basicity of the ligands on acid—base equilibrium in adduct formation from the catalyst complex and cocatalyst.

Dodd and Rutt [189] reported the activity of $W(OPh)_4Cl_2$ to be enhanced on substituting two phenoxy groups of $W(OPh)_6$ by chlorine. Electron withdrawing groups in the 4-position of the phenoxide caused very considerable increase in reactivity. A linear Hammett relationship was observed for a range of substituents (Cl, Br, H, Me, MeO) on phenyl.

6.2.8. Coordination of Alkenes and Alkynes

It has not been possible to establish firmly whether the alkene or alkyne coordinates to the transition metal of the metathesis catalyst, or whether it reacts without prior coordination to give the intermediate metallacycle. It is a general assumption that the alkene or alkyne coordinates to the transition metal, although similarities to the related complexation or polymerization reactions of these unsaturated substrates are not evident.

The first experimental indication that the alkene could coordinate to the metathesis transition metal was afforded by the isolation of stable metallacarbene alkene complexes of tungsten [53] and tantalum.[93] Further work is necessary, however, to demonstrate the participation of these complexes in alkene metathesis.

If the coordination of an unsaturated substrate occurs, the stability of the coordinative bond is a factor which can determine the conversion rate of the alkene or alkyne complex, and the reaction rate for decomplexation and ligand exchange. The rate therefore ultimately depends mainly on the nature of the coordination bond of the coordinated substrate, on the nature of the ligands in the catalytic system and on the nature of the transition metal itself.

The nature of the bond in olefin coordination to transition metals is well known from the molecular orbital formalism of Dewar [190] and Chatt and Duncanson [38] for platinum complexes. Coordination of the donor — acceptor type may occur through the formation of σ and π bonds between the olefin orbitals and those of the transition metal. The σ bonds are formed through overlap between a π bonding orbital of the olefin and vacant d orbitals of the transition metal, while π bonds result from overlap of the antibonding π orbital of the olefin with the occupied d orbitals of the transition metal (Fig. 6.26).

The extent of the overlap between the highest occupied or lowest unoccupied molecular orbitals of the olefin (π and π^* respectively) with the orbitals of metal atom determines, to a first approximation, the stability of the metal—olefin complex. Any factor which increases this overlap will strengthen the metal—olefin bond. For the metal the most important variables are charge or oxidation state and configuration, and for the olefin the effect of substituents and strain. It is inferred that removal of charge from the metal will cause the energy of the metal orbitals to decrease, making the extent of overlap less significant while, conversely, supply of charge will cause the energy of the metal orbitals to increase so that overlap becomes more significant.

Calculated energy differences between various possible configurations indicate that the barrier to rotation is generally low and that rotation of the olefin molecule may occur in this type of complex.

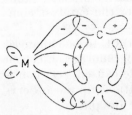

Fig. 6.26. — Coordinative bonding for complexes of alkenes with transition metals.

The stability of the π bond depends, at the same time, on the hybridization of the filled d orbital of the transition metal. The stability can be increased as a consequence of hybridization of the filled d orbital with the vacant p orbitals of the transition metal and also by forming dp hybrid orbitals more suitable for overlapping with the bonding π orbital of the olefin. The stability can be diminished by hindering this hybridization when the p orbitals of the metal are involved in bonds with other ligands. In this last case both the stability of the π bond and the coordination capacity of the olefin are decreased.

Electron-withdrawing groups on the olefin will lower the energy of both the π and π^* orbitals, which may cause the energy of the π^* orbital to fall below that of the highest occupied metal orbital so that direct transfer of an electron from the metal to the olefin may occur with resulting oxidation of the metal. In contrast, electron-releasing groups on the olefin will increase the energy of both the π and π^* orbitals and in the extreme situation overlap with the π^* orbital will be so small that no back-bonding from the metal to the olefin can occur.

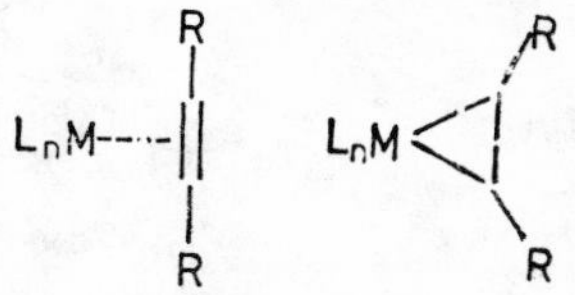

Fig. 6.27. — Valence bond description for alkene transition metal complex.

The introduction of strain into the molecule of the olefin will lead to a decrease in energy of the π^* orbital and an increase in energy of the π orbital, resulting in more favourable overlap for bonding.

The valence bond formalism has also been used to describe the bonding in transition metal—olefin complexes (Fig. 6.27). The limiting structures are I in which the metal is in a lower oxidation state and the olefin is undistorted, and II in which the metal is in a higher oxidation state and rehybridization of the olefin carbon atoms from sp^2 to sp^3 occurs to give a bis-σ-alkyl metallacyclopropane ring. When R is an electron-releasing group, the real structure may be expected to be closer to I while when R is an electron-withdrawing group the structure will be closer to II.

Distortion of the olefin on complex formation may cause the C=C bond to weaken and increase in length. The carbon—carbon bond length in the complexed olefin generally lies between that of the free olefin (C=C) and saturated hydrocarbons (C—C). This fact results from the decrease of electron density in the olefin pπ orbital and the increase of electron density in the pπ^* orbital. The complexed olefin molecule loses its planarity and the angle between substituents on carbon atoms will decrease. Furthermore, in the case of planar configurations, the complexed olefin molecule is twisted out of the plane formed by the metal and the corresponding ligands. As a result of complexation to transition metal, the olefin molecule is activated towards nucleophilic or electrophilic atack in varying degree depending upon the metal oxidation state.

The same acceptor—donor model of Dewar-Chatt-Duncanson [38,190] could be used for the description of the metal—alkyne bond for alkyne complexation. By extension of olefin model to alkyne complexes, the use of dp^2 trigonal hybrid orbitals instead of sp^2 hybrids becomes possible.

The bonding scheme [191,192] involves overlap of a dp^2 hybrid ($d_{xy} + p_x + p_y$) orbital with the alkyne π^-_{xy} orbital and overlap of a d^2 hybrid ($d_{xz} + d_{yz}$) orbital with the alkyne π_z orbital (Fig. 6.28)

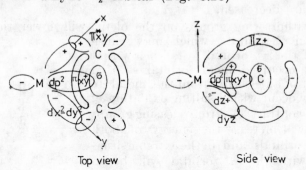

Top view   Side view

Fig.6.28. — Coordinative bonding for complexes of alkynes
with transition metals.

These two resulting bonds are synergistically opposed by overlap of $d_{x^2-y^2}$ orbital with the π_{xy}-antibonding orbital and overlap of the $d^2(d_{xz} + d_{yz})$ orbital with the π_z^*-antibonding orbital. The appropriate molecular orbitals in the alkyne molecule are lower in energy than those in an alkene molecule. This apparently leads to a more favourable bonding situation.

The valence bond approach has been used as for alkene complexation to describe bonding in the transition metal—alkyne complexes. The limiting structures are I in which the metal is in a lower oxidation state and the alkyne is left undistorted, and II in which the metal is in a higher oxidation state and rehybridization of the alkyne carbon atoms from sp to sp^2 has occurred to give a bis-σ-alkenyl metallacyclopropene ring (Fig. 6.29). When R is an electron-releasing group, the complex structure may be more closely related to I, and when R is an electron-withdrawing group the structure may be closer to II.

The distortions of complexed alkyne molecules are mainly similar to those encountered in the alkene complexes. Upon complex formation the carbon—carbon distance increases and becomes generally intermediate in length between that for normal carbon-to-carbon double and triple bonds. This increase in bond length is manifested in the infra-red spectrum as a shift in $\nu_{C\equiv C}$ to longer wavelengths.

Rotation about the coordinative bond is also possible for complexed alkyne molecules but the rotation barrier is somewhat higher than that for the alkene complexes.

Fig. 6.29. — Valence bond description for alkyne transition metal complex.

The substituents on alkyne can strongly influence on the metal—alkyne bond. Electron-releasing substituents will stabilize the coordinative bond by enhancing the σ-donor capacity of the triple bond while electron-withdrawing substituents will strengthen the metal—alkyne bond by lowering the energy of the antibonding orbitals of the organic moiety, which enhances back-bonding. The substituents attached to the alkyne group mainly adopt a *cis* configuration upon complex formation. This arrangement was observed in the majority of the transition metal alkyne complexes investigated by X-ray techniques and is inferred from the infra-red spectra for complexes of symmetrically substituted alkynes.

6.2.9. The Role of the Cocatalyst

It is now evident that the cocatalyst used in most homogeneous systems plays an essential part in the whole process of metathesis. Besides its first action as alkylating and reducing agent for the transition metal compound, the cocatalyst may interact with active centres during propagation and may act upon metal ligands, labilizing them and creating vacant sites for coordination.

The metathesis cocatalysts generally have Lewis acid properties. They interact as such with the transition metal complex or are converted into more active species which are involved in the process. This interaction depends upon the nature of the transition metal and its ligands, the nature of cocatalyst metal and its ligands, the solvent and the reaction conditions.

Numerous Lewis salt type adducts are known in which Lewis acids attach to halogen, carbonyl or dinitrogen ligands of transition metal complexes.[193-195] Such adducts were thought by Bencze [184,196] to appear in the reductive nitrosation of molybdenum and tungsten chlorides in the presence of $EtAlCl_2$ or in homogeneous catalytic systems based on tungsten and molybdenum halocarbonyl and halonitrosyl, as well as by Hughes [13,197] in molybdenum halonitrosyl compounds ((6.109) and (6.110)).

$$M(NO)_2L_2Cl_2 + 2EtAlCl_2 \rightleftarrows M(NO)_2L_2Cl_2 \cdot 2EtAlCl_2 \qquad (6.109)$$

$$M(CO)_3L_2Cl_2 + 2AlCl_3 \rightleftarrows M(CO)_3L_2Cl_2 \cdot 2AlCl_3 \qquad (6.110)$$

For one class of tungsten and molybdenum halocarbonyl and halonitrosyl compounds Bencze and co-workers [198,199] found that adduct formation is followed by rapid equilibration to ion pairs and partial solvation ((6.111) and (6.112)).

$$W(CO)_3Cl_2L_2 \cdot 2AlCl_3 \rightleftarrows [W(CO)_3L_2S]^{2+}[AlCl_4]_2^- \qquad (6.111)$$

$$[W(CO)_3L_2S]^{2+}[AlCl_4]_2^- \rightleftarrows [W(CO)_3L_2S]^{2+} + 2[AlCl_4]^- \qquad (6.112)$$

It seems that in these ion pairs the Lewis acid (which provides the negatively charged counter ion) is situated in the outer coordination sphere of the transition metal. From infra-red and conductivity measurements it was observed that adduct formation and the conductivity of the system depend both on the molar ratio and the total concentration. Molar ratios Al/M necessary for complete adduct formation decreased with increasing concentrations.[199]

From these results Bencze concluded that one of the most important changes in the transition metal complexes effected by the addition of Lewis acids would be the enhanced mobility of their ligands. The order of this ligand mobility induced by the cocatalyst would be L $<$CO $<$Cl. Thus, on addition of the cocatalyst to the transition metal complexes, two free coordination sites may form and the mobility of other ligands would then be enhanced. Consequently, the adduct could provide the necessary condition for alkene coordination.

Numerous reports pointed out the strong effect of $AlCl_3$ and other Lewis acids on the metathesis activity of catalytic systems. Thus, $C_6H_5WCl_3$ is active without any cocatalyst in metathesis of acyclic and cyclic olefins, but in combination with $AlCl_3$ a remarkable increase in activity is observed.[200] Similar effects were obtained for benzyltungsten as catalyst and for other unicomponent and bicomponent systems.[201] The unicomponent system WCl_6, which is inactive without addition of other compounds, becomes an active catalyst in combination with $AlCl_3$ or $AlBr_3$.[202] Similar behaviour was observed with WCl_4 obtained by reduction of WCl_6 with H_2 at high temperatures.[203] The activity of the binary system $C_4H_9Li \cdot WCl_6$ is increased 100-fold when an equimolar amount of $AlCl_3$ is added to the reaction.[204] Many similar examples are encountered in the patent literature.[205-207]

Organoaluminium halides can readily exchange their organic groups with transition metal halides (6.113).

$$WCl_6 + C_2H_5AlCl_3 \rightleftarrows C_2H_5WCl_5 + AlCl_3 \qquad (6.113)$$

The organocompound of the transition metal thus formed can be further reduced and serves as a source of metallacarbene complex. The presence of $AlCl_3$, formed during the reduction of W(VI), may lead to associations *via* μ-chloride bonding (6.114) and acid—base equilibria (6.115).

$$WCl_n + AlCl_3 \rightleftarrows \begin{array}{c} Cl \\ \diagdown \\ Cl \end{array} Al \begin{array}{c} Cl \\ \diagup \diagdown \\ Cl \end{array} WCl_{n-1} \qquad (6.114)$$

$$WCl_n + AlCl_3 \rightleftarrows [WCl_{n-1}]^+ [AlCl_4]^- \qquad (6.115)$$

Recent studies on zero-valent complexes of tungsten carried out by Basset and co-workers[208,209] established the formation of a 1 : 1 adduct by acid—base interaction between $W(CO)_5L[L=CO, PPh_3, P(OPh)_3,$ $PBu_3]$ and $EtAlCl_2$ or $AlX_3[X=Cl, Br]$ (6.116).

$$W(CO)_5L + AlX_3 \rightleftharpoons LW(CO)_4CO \mapsto AlX_3 \qquad (6.116)$$

The adduct identified by infra-red spectroscopy is relatively unstable and gives rise to dissociation of the ligand L with the formation of the unsaturated zero-valent tungsten species $W(CO)_5$ complexed with the Lewis acid (6.117).

$$LW(CO)_4 \mapsto AlX_3 \rightleftharpoons L + \square - W(CO)_4CO \mapsto AlX_3 \qquad (6.117)$$

The new coordinatively unsaturated tungsten complex is more stable in solution and may favour ligand interchange (formation of $W(CO)_6$) as well as oxidation of zero-valent tungsten to higher oxidation states (for instance to $W(CO)_4Cl_2$). There are some indication of further formation of non-carbonyl complexes of tungsten, which might be highly unsaturated zero-valent tungsten or highly oxidized tungsten such as $WOCl_2$ or WCl_4.

The free ligand competes with the tungsten catalyst in complexing the Lewis acid (6.118).

$$L + AlX_3 \rightleftharpoons L \mapsto AlX_3 \qquad (6.118)$$

This acid—base equilibrium, which depends on ligand basicity and cocatalyst acidity governs the initial catalyst—cocatalyst interaction and ultimately the stability and activity of the catalyst.

The identities of the species formed in the reaction of WCl_6 with $C_2H_5AlCl_2$ have been proved by Singleton and Eatough [210] by titration calorimetry. They observed that the first two molar equivalents of $C_2H_5AlCl_2$ react with WCl_6 to form $WCl_4 \cdot 2AlCl_3$ (6.119).

$$WCl_6 + 2C_2H_5AlCl_2 \rightleftharpoons C_2H_4 + C_2H_6 + WCl_4 \cdot 2AlCl_3 \qquad (6.119)$$

Further reaction of $C_2H_5AlCl_2$ gives a second species which is active in olefin coordination (6.120).

$$WCl_4 \cdot 2AlCl_3 + 2C_2H_5AlCl_2 \rightleftharpoons WCl_4 \cdot 2AlCl_3 \cdot 2C_2H_5AlCl_2 \qquad (6.120)$$

These authors proposed possible structures for $WCl_4 \cdot 2AlCl_3$ and $WCl_4 \cdot 2AlCl_3 \cdot 2C_2H_5AlCl_2$ with a μ-chloride bridge bonding the W and Al atoms. Some of the structures for the second species include carbene bridges, metal—carbene double bonds or coordinated solvent sited at the transition metal. In the latter situation the function of $C_2H_5AlCl_2$ would be to free bonding sites for coordination with the incoming olefin.

It is assumed that ethanol added to WCl_6-based catalyst leads to $WCl_5OC_2H_5$ formation. Höcker and Jones [211] suggested a two-stage reaction through the intermediacy of $WCl_4(OC_2H_5)_2$ ((6.121) and (6.122)).

$$WCl_6 + 2C_2H_5OH \rightleftharpoons WCl_4(OC_2H_5)_2 \qquad (6.121)$$

$$WCl_4(OC_2H_5)_2 \rightleftharpoons 2WCl_5OC_2H_5 \qquad (6.122)$$

In the reaction of $C_2H_5AlCl_2$ with the $(WCl_6 + C_2H_5OH)$ system, Singleton and Eatough [210] observed the participation of five moles of $C_2H_5AlCl_2$ in three distinct steps. They proposed the following compounds to be involved in this process $((6.123)-(6.125))$:

$$WCl_5(OC_2H_5) + 3C_2H_5AlCl_2 + HCl \rightarrow WCl_3(OC_2H_5) \cdot 3AlCl_3 +$$

$$+ 2C_2H_6 + C_2H_4 \qquad (6.123)$$

$$WCl_3(OC_2H_5) \cdot 3AlCl_3 + C_2H_5AlCl_2 \rightarrow WCl_3(OC_2H_5) \cdot 3AlCl_3 \cdot C_2H_5AlCl_2$$
$$(6.124)$$

$$WCl_3(OC_2H_5) \cdot 3AlCl_3 \cdot C_2H_5AlCl_2 + C_2H_5AlCl_2 \rightarrow$$

$$\rightarrow WCl_3(OC_2H_5) \cdot 3AlCl_3 \cdot 2C_2H_5AlCl_2 \qquad (6.125)$$

Further on, in the reaction of $WOCl_4$ with $C_2H_5AlCl_2$ only two moles of aluminium compound entered the reaction (6.126).

$$WOCl_4 + 2C_2H_5AlCl_2 \rightarrow WOCl_2 \cdot 2AlCl_3 + C_2H_6 + C_2H_4 \qquad (6.126)$$

Stoichiometric formation of complexes does not occur upon addition of further $C_2H_5AlCl_2$ to the system. More data and information are necessary for structure assignments for these compounds.

Leconte et al.[187] studied the stereospecificity of the nitrosyl ligands in the metathesis of acyclic olefins and inferred from spectroscopic results that $EtAlCl_2$ gives with $W(NO)_2Cl_2(PPh_3)_2$ first a 1 : 1 adduct and then a 2 : 1 adduct. It seems that the nitrosyl and phosphine ligands remain coordinated to the tungsten in both adducts since it was possible to regenerate unchanged $W(NO)_2Cl_2(PPh_3)_2$. However, the spectroscopic data indicate that in these adducts the nitrosyl ligands are in the *trans* position while the phosphine ligands are *cis*, the exact reverse of their initial positions. The 2 : 1 adduct is believed to be the precursor compound of the active catalyst, being the only detectable nitrosyl complex in the catalytic reaction. In these adducts the Lewis acid is probably coordinated to the halogen by acid—base complexation and the catalytic activity results from slow intramolecular alkylation in the 2 : 1 adduct. It seems that the 2 : 1 molar ratio of cocatalyst to catalyst is essential for the catalytic activity of this system, and this fact favours intramolecular rearrangement of organic ligands.

Muetterties pointed out that the activity of the $WCl_6 \cdot MR_n$ systems varies drastically with the nature of M. [56] He suggested that the organometallic cocatalyst gives rise to Lewis acids which interact with the tungsten complex intermediates through halogen, alkyl, and carbene bridges. As a consequence of this interaction, active species having M—Cl—W bridge bonding are probably operative in the propagation reaction. For carbene bridges three-centre two-electron bonds between W, C and the second metal would be involved in the active species. These bridged structures may act as chain-carriers. Furthermore, the initial interaction of the cocatalyst with the transition metal compound could be of a bridged nature wherein α-hydrogen elimination may become a more facile process for the generation of carbene.

A noteworthy activating effect is exerted by $AlCl_3$ or organoaluminium compounds on metallacarbene complexes active in metathesis. The oxoalkylidene tungsten complex, obtained recently by Schrock and co-workers,[212] showed low activity in metathesis; however, when $AlCl_3$ was added, the activity of the complex increased considerably. This effect was attributed to the formation of an empty coordination site and/or to the formation of a more soluble complex in the system.

The extensive studies of Tebbe [213-215] on organotransition metal complexes convincingly demonstrated the specific mode of aluminium interaction with carbene complexes. In the titanocene methylene complex, found to be active in metathesis, the titanacarbene moiety is trapped by the aluminium compound in a bridged cyclic structure **1** (6.127).

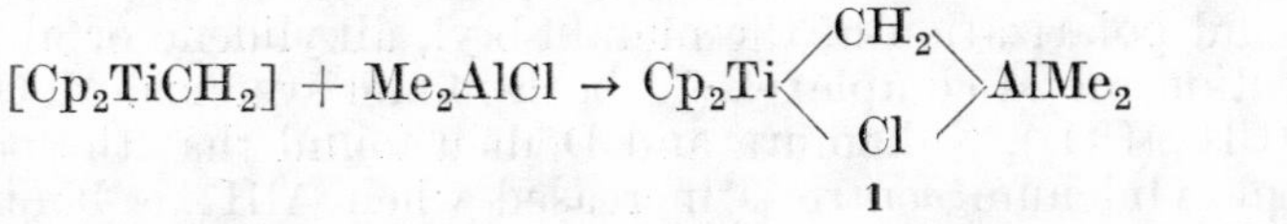

$$[Cp_2TiCH_2] + Me_2AlCl \rightarrow Cp_2Ti \underset{Cl}{\overset{CH_2}{<\quad>}} AlMe_2 \qquad (6.127)$$

1

A competition between the organoaluminium compound and the unsaturated olefinic or acetylenic substrate for titanocene carbene species seems to exist. The relative stability of complexed bridged cycles formed from metallacarbene and olefin or acetylene and the organoaluminium compound will govern the equilibrium of these interactions. In titanacyclobutane formation and decomposition, Grubbs and co-workers[73] identified the significant role of Me_2AlCl for such an equilibrium (6.128).

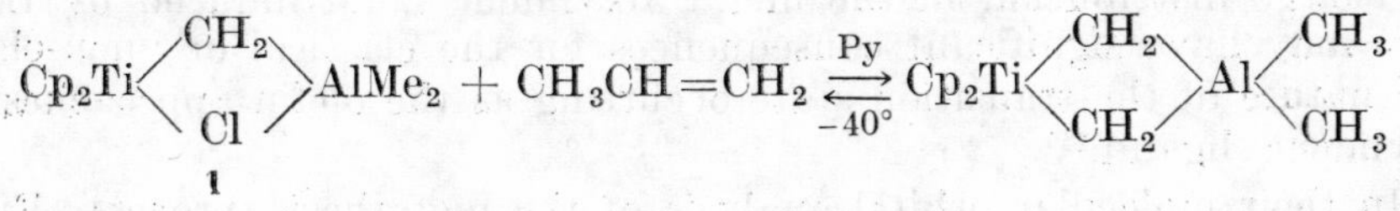

$$Cp_2Ti \underset{Cl}{\overset{CH_2}{<\quad>}} AlMe_2 + CH_3CH = CH_2 \underset{-40°}{\overset{Py}{\rightleftharpoons}} Cp_2Ti \underset{CH_2}{\overset{CH_2}{<\quad>}} Al \underset{CH_3}{\overset{CH_3}{<\quad>}}$$

1

$$(6.128)$$

Pyridine or one of its derivatives remove Me_2AlCl from compound **1** and the released $[Cp_2TiCH_2]$ fragment reacts immediately with the olefin, while Me_2AlCl facilitates the metallacyclobutane → carbene transformation.

The fact that the Lewis acid binds to metallacarbene or metallacarbyne complexes of the Schrock-type "to protect" the carbene or carbyne moiety has been proved by structure characterization of several such adducts.[216,217] This "protection" is readily removed through base treatment and the free metallacarbene or -carbyne may be released.

The nature of the binding of Lewis acid to alkyl, alkylidene or alkylidyne complexes may be greatly changed by the ligands existing on the transition metal. From LCAO SCF calculations for the system $MoOCl(CH_3)_3 \cdot AlH_3$ and $MoOCl(CH_3)(CH_2) \cdot AlH_3$, Nakamura and Dedieu [218] found that the Lewis acid binds preferentially to the oxo end of a molybdenum compound when the complex stabilization is at a maximum. This result has been rationalized in terms of interactions between the molecular orbitals of the molybdenum compound and AlH_3 fragments. No stabilization was found through interaction of the Lewis acid with the methylene ligand either directly or by producing a bridged species in the case of $MoOCl(CH_3)(CH_2)$.

In addition, a rather important rotational barrier for the CH_2 unit is found although AlH_3 diminishes the barrier when bound to the oxo ligand (in this case the barrier decreases from 128 to 96.6 kJ/mol). Similar results concerning the site of preferential binding and the extent of stabilization were found for AlH_2Cl as the Lewis acid.

A possible role of the Lewis acid would be to modify the charge distribution and polarization of the metal alkyl, alkylidene or alkylidyne bond in transition metal complexes. In both model systems, $MoOCl(CH_3)_3$ and $MoOCl(CH_3)(CH_2)$, Nakamura and Dedieu found that the positive charge at the molybdenum centre is increased when AlH_3 is bound to the oxo ligand. The greater electron deficiency at the metal centre in the alkyl molybdenum compound may thus favour the formation of the alkylidene complex while the greater charge separation of the $Mo=C$ bond in the alkylidene compound will lower the rotational barrier and consequently increase the configurational population favorable for interaction with an incoming olefin. This may have some influence on the formation of the metallacyclobutane transition state which is dependent on the relative orientation of the carbene ligand and the reactant olefin. Furthermore, the charge distribution in the metal alkylidene bond induced by Lewis acids may have significant consequences for the classical or "non-classical" nature of the transition state occurring as the olefin approaches the alkylidene ligand.[150]

In their molecular orbital analysis of the metathesis process, Eisenstein and Hoffmann [151] pointed out the role of the low-lying empty orbital substantially localized on the carbene and available for attack by nucleophiles. In the metallacycle transition state, this low-lying orbital favours metathesis transformation through its bonding counterparts. The coordination of a base on either side of the metallacycle seems to hamper the ease of metathesis. For this transition state, and consequently for the propagation step, the role of the Lewis acid may be rationalized as sequestering any base which would obstruct metathesis route by coordination.

For the system $WOCl_4 \cdot Me_4Sn$, Verkuijlen [72] has found that the presence of the cocatalyst derivative Me_3SnCl is essential to maintain the activity of this catalytic system. Consequently, this author suggested a reaction scheme involving dimetallic alkylidene bridged complexes with the active participation of the cocatalyst metal atom during the initiation, propagation and termination steps. In the initiation step Me_4Sn can be a methyl source for the bridged methylene dimetallic complex $Cl_2OWCH_2 \cdot Me_3SnCl$, considered to be an active species for propagation.

Scheme 6.58

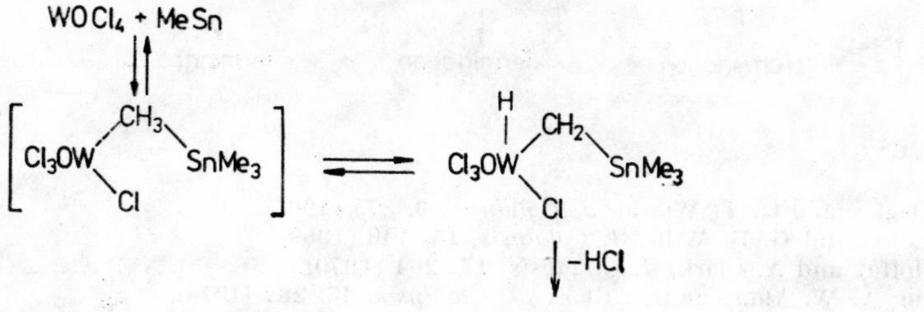

During propagation the Me_3SnCl takes an active part in the dimetallic bridged carbene complex through its assistance in the metallacyclobutane transition state of the catalytic cycle.

Scheme 6.59

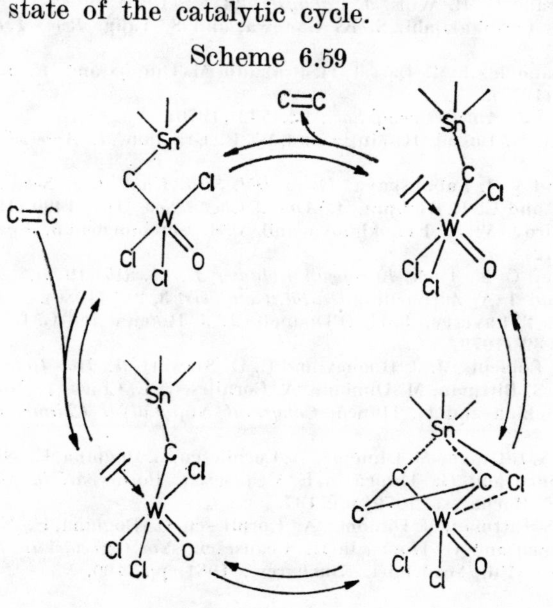

As to the various modes of Me_3SnCl participation in termination reactions, it was suggested that Me_3SnCl may react with every transient species in metathesis.

Scheme 6.60

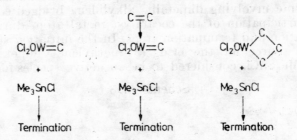

References

1. J. W. Begley and R. T. Wilson, *J. Catalysis*, **9**, 375 (1967).
2. M. J. Lewis and G. B. Wills, *J. Catalysis*, **15**, 140 (1969).
3. A. J. Moffat and A. Clark, *J. Catalysis*, **17**, 264 (1970).
4. C. J. Lin, A. W. Aldag and A. Clark, *J. Catalysis*, **45**, 287 (1976).
5. E. V. Pletneva, Yu. N. Usov and E. V. Skvortsova, *Neftekhimiya*, **15**, 347 (1975).
6. E. S. Davie, D. A. Whan and C. Kemball, *J. Catalysis*, **24**, 272 (1972).
7. R. C. Luckner, G. E. McConchie and G. B. Wills, *J. Catalysis*, **28**, 63 (1973).
8. U. R. Hattikudur and G. Thodos, *Adv. Chem. Ser.*, **133**, 80 (1974).
9. a. F. H. M. van Rijn and J. C. Mol, *Rec. Trav. Chim.*, **96**, M96 (1977); b. F. Kapteijn, H. L. G. Bredt, E. Homburg and J. C. Mol, *Ind. Eng. Chem.*, *Prod. Res. Develop.*, **20**, 457 (1981).
10. R. C. Luckner and G. B. Wills, *J. Catalysis*, **28**, 83 (1973).
11. G. B. Wills, J. Fathikalajahi, S. K. Gangwal and S. Tang, *Rec. Trav. Chim.*, **96**, M110 (1977).
12. A. Ismayel-Milanovic, J. M. Basset, H. Praliaud, M. Dufaux and L. de Mourgues, *J. Catalysis*, **31**, 408 (1973).
13. W. B. Hughes, *J. Amer. Chem. Soc.*, **92**, 532 (1970).
14. J. L. Bilhou, J. M. Basset, R. Mutin and W. F. Graydon, *J. Amer. Chem. Soc.*, **99**, 4083 (1977).
15. V. I. Mar'in and V. I. Labunskaya, *Issled. Obl. Sint. Katal. Org. Soed.*, **1975**, 46.
16. R. H. Grubbs and C. R. Hoppin, *J. Amer. Chem. Soc.*, **101**, 1499 (1979).
17. V. A. Hodjemirov, V. A. Evdokimova and V.M. Cerebnichenko, *Vysokomol. Soedinenya*, B **14**, 727 (1972).
18. A. J. Amass and C. N. Tuck, *European Polymer J.*, **14**, 817 (1978).
19. A. J. Amass and J. A. Zurimendi, *J. Mol. Catalysis*, **8**, 243 (1980).
20. K. J. Ivin, D. T. Laverty, J. H. O'Donnell, J. J. Rooney and C. D. Stewart, *Makromol. Chem.*, **180**, 1989 (1979).
21. K. J. Ivin, G. Lapienis, J. J. Rooney and C. D. Stewart, *J. Mol. Catalysis*, **8**, 203 (1980).
22. E. Ceaușescu, S. Bittman, M. Dimonie, A. Cornilescu, C. Cincu, E. Nicolescu, M. Popescu, S. Coca, C. Boghină and G. Hupcă, *Congresul Național de Chimie*, 11 September 1978, Bucharest.
23. E. Ceaușescu, S. Bittman, M. Dimonie, A. Cornilescu, C. Boghină, E. Nicolescu, M. Popescu, S. Coca, C. Cincu and G. Hupcă in E. Ceaușescu, *Participări la Manifestări Științifice*, Ed. Academiei, Bucharest, 1979, pp. 137.
24. E. Ceaușescu, S. Bittman, M. Dimonie, A. Cornilescu, C. Boghină, E. Nicolescu, M. Popescu, S. Coca, C. Cincu and G. Hupcă in E. Ceaușescu, *Noi Cercetări în Domeniul Compușilor Macromoleculari*, Ed. Academiei, Bucharest, 1981, pp. 100.

25. H. Höcker, W. Reimann, L. Reif and K. Riebel, *J. Mol. Catalysis*, **8**, 191 (1980).
26. W. Reimann, *Ph. D. Thesis*, Mainz, 1977.
27. Yu. V. Korshak, B. A. Dolgoplosk and M. A. Tlenkopachev, *Rec. Trav. Chim.*, **96**, M64 (1977).
28. K. Riebel, *Ph. D. Thesis*, Mainz, 1976.
29. L. Reif, cited after reference 25.
30. C. P. C. Bradshaw, E. J. Howman and L. Turner, *J. Catalysis*, **7**, 269 (1967).
31. R. L. Banks and G. C. Bailey, *Ind. Eng. Chem., Prod. Res. Develop.*, **3**, 170 (1964).
32. C. T. Adams and S. G. Brandenberger, *J. Catalysis*, **13**, 360 (1969).
33. D. L. Crain, *J. Catalysis*, **13**, 110 (1969).
34. J. C. Mol, J. A. Moulijn and C. Boelhouwer, *J. Catalysis*, **11**, 87 (1968).
35. F. L. Woody, M. J. Lewis and G. B. Wills, *J. Catalysis*, **14**, 389 (1969).
36. A. Clark, C. Cook, *J. Catalysis*, **15**, 420 (1969).
37. N. Calderon, E. A. Ofstead, J. P. Ward, W. A. Judy and K. W. Scott, *J. Amer. Chem. Soc.*, **90**, 4133 (1969).
38. J. Chatt and R. Duncanson, *J. Chem. Soc.*, **1953**, 2939.
39. R. Cramer, *J. Amer. Chem. Soc.*, **86**, 217 (1964).
40. R. Cramer, *J. Amer. Chem. Soc.*, **89**, 4621 (1967).
41. F. D. Mango and J. H. Schachtschneider, *J. Amer. Chem. Soc.*, **89**, 2484 (1967).
42. F. D. Mango, *Top. Curr. Chem.*, **45**, 39 (1974).
43. F. D. Mango, *Coord. Chem. Rev.*, **15**, 109 (1975).
44. R. B. Woodward and R. Hoffmann, *Angew. Chem.*, **81**, 797 (1969); *Internat. Edition*, **8**, 781 (1969).
45. G. L. Caldow and R. A. McGregor, *J. Chem. Soc., A*, **1971**, 1654.
46. R. G. Pearson, *Top. Curr. Chem.*, **41**, 75 (1973).
47. K. Fukui and S. Inagaki, *J. Amer. Chem. Soc.*, **97**, 4445 (1975).
48. A. A. Andreev, R. M. Edreva-Kardjieva and N. M. Neshev, *Rec. Trav. Chim.*, **96**, M23 (1977).
49. D. R. Armstrong, *Rec. Trav. Chim.*, **96**, M17 (1977).
50. J. L. Hérisson and Y. Chauvin, *Makromol. Chem.*, **141**, 161 (1970).
51. D. J. Cardin, M. J. Doyle and M. F. Lappert, *J. Chem. Soc., Chem. Commun.*, **1972**, 927.
52. C. P. Casey and T. J. Burkhardt, *J. Amer. Chem. Soc.*, **96**, 7808 (1974).
53. C. P. Casey and A. J. Shusterman, *J. Mol. Catalysis*, **8**, 1 (1980).
54. E. O. Fischer and K. H. Dötz, *Chem. Ber.*, **105**, 3966 (1972).
55. E. O. Fischer and B. Dorrer, *Chem. Ber.*, **107**, 1156 (1974).
56. E. L. Muetterties, *Inorg. Chem.*, **14**, 951 (1975).
57. T. Mocella, M. A. Busch and E. L. Muetterties, *J. Amer. Chem. Soc.*, **98**, 1283 (1976).
58. T. J. Katz and J. McGinnis, *J. Amer. Chem. Soc.*, **97**, 1592 (1975).
59. T. J. Katz and J. McGinnis, *J. Amer. Chem. Soc.*, **99**, 1903 (1977).
60. T. J. Katz and R. Rothchild, *J. Amer. Chem. Soc.*, **98**, 2519 (1976).
61. R. H. Grubbs, P. L. Burk and D. D. Carr, *J. Amer. Chem. Soc.*, **97**, 3265 (1975).
62. R. H. Grubbs, D. D. Carr, C. R. Hoppin and P. L. Burk, *J. Amer. Chem. Soc.*, **98**, 3478 (1976).
63. R. H. Grubbs and C. R. Hoppin, *J. Amer. Chem. Soc.*, **101**, 1499 (1979).
64. a. R. J. Puddephatt, M. A. Quyser and C.F.H. Tipper, *J. Chem. Soc., Chem. Commun.*, **1976**, 626; b. R. J. Al-Essa, R. J. Puddephatt, M. A. Quyser and C. F. H. Tipper, *J. Amer. Chem. Soc.*, **101**, 364 (1979).
65. M. Leconte and J. M. Basset, *J. Amer. Chem. Soc.*, **101**, 7296 (1979).
66. M. Leconte and J. M. Basset, *Ann. N. Y. Acad. Sci.*, **333**, 165 (1980).
67. W. S. Greenlee and M. F. Farona, *Inorg. Chem.*, **15**, 2129 (1976).
68. V. W. Motz and M. F. Farona, *Inorg. Chem.*, **16**, 2545 (1977).
69. a. F. N. Tebbe, G. W. Parshall and D. W. Ovenall, *J. Amer. Chem. Soc.*, **101**, 5074 (1979); b. U. Klabunde, F. N. Tebbe, G. W. Parshall and R. L. Harlow, *J. Mol. Catalysis*, **8**, 37 (1980).
70. a. F. Garnier and P. Krausz, *J. Mol. Catalysis*, **8**, 91 (1980); b. F. Garnier, P. Krausz and H. Rudler, *J. Organomet. Chem.*, **186**, 77 (1980).
71. H. Funk and H. Naumann, *Z. Anorg. Allgem. Chem.*, **343**, 294 (1966).
72. E. Verkuijlen, *J. Mol. Catalysis*, **8**, 107 (1980).
73. T. R. Howard, J. B. Lee and R. H. Grubbs, *J. Amer. Chem. Soc.*, **102**, 6876 (1980).

74. C. Masters, "Catalysis, a gentle art", Methuen, London, 1980, pp. 210.
75. D. A. Dowden, *Anales real. Soc. Espan. Fis. Quim.*, **61**, 326 (1965).
76. a. G. S. Lewandos and R. Pettit, *Tetrahedron Letters*, **1971**, 789; b. G. S. Lewandos and R. Pettit, *J. Amer. Chem. Soc.*, **93**, 7087 (1971).
77. R. Taube and F. Seyferth, Wiss Z. TH Leuna-Merseburg, **17**, 350 (1975).
78. R. J. Haines and G. J. Leigh, *Chem. Soc., Revs.*, **4**, 155 (1975).
79. R. H. Grubbs and T. K. Brunk, *J. Amer. Chem. Soc.*, **94**, 2538 (1972).
80. G. C. Biefeld, H. A. Eick and R. H. Grubbs, *Inorg. Chem.*, **12**, 2166 (1973).
81. M. Ephritikhine, M. L. H. Green and R. E. MacKenzie, *J. Chem. Soc., Chem. Commun.*, **1976**, 619.
82. R. R. Schrock, *J. Amer. Chem. Soc*, **96**, 6796 (1974).
83. D. N. Clark and R. R. Schrock, *J. Amer. Chem. Soc.*, **100**, 6774 (1978).
84. R. H. Grubbs and C. R. Hoppin, *J. Chem. Soc., Chem. Commun.*, **1977**, 634.
85. E. Thorn-Csányi, *Nach. Chem. Tech. Lab.*, **29**, 700 (1981).
86. E. Thorn-Csányi, *Angew. Makromol. Chem.*, **94**, 181 (1981).
87. E. L. Muetterties and E. Band, *J. Amer. Chem. Soc.*, **101**, 6572 (1980).
88. D. T. Laverty, J. J. Rooney and A. Stewart, *J. Catalysis*, **45**, 110 (1976).
89. M. Ephritikhine and M. L. H. Green, *J. Chem. Soc., Chem. Commun.*, **1976**, 926.
90. G. J. A. Adam, S. G. Davies, K. A. Ford, M. Ephritikhine, P. F. Todd and M. L. H. Green, *J. Mol. Catalysis*, **8**, 15 (1980).
91. M. F. Farona and R. L. Tucker, *J. Mol. Catalysis*, **8**, 85 (1980).
92. R. H. Grubbs and S. J. Swetnick, *J. Mol. Catalysis*, **8**, 25 (1980).
93. A. J. Shultz, R. K. Brown, J. M. Williams and R. R. Schrock, *J. Amer. Chem. Soc.*, **103**, 169 (1981).
94. E. A. Ofstead, J. P. Lawrence, M. L. Senyek and N. Calderon, *J. Mol. Catalysis*, **8**, 227 (1980).
95. M. L. H. Green, *Ann. N. Y. Acad. Sci.*, **333**, 229 (1980).
96. a. R. H. Grubbs, *Progr. Inorg. Chem.*, **24**, 1 (1978); b. K. C. Ott and R. H. Grubbs, *J. Amer. Chem. Soc.*, **103**, 5922 (1981).
97. F. Garnier, P. Krausz and J. E. Dubois, *J. Organomet. Chem.*, **170**, 195 (1979).
98. a. J. Levisalles, H. Rudler, F. Dahan and J. Jeannin, *J. Organomet. Chem.*, **178**, C8 (1979); b. J. Levisalles, H. Rudler, F. Dahan and J. Jeannin, *J. Organomet. Chem.*, **187**, 233 (1980); c. J. Levisalles, H. Rudler, F. Dahan and J. Jeannin, *J. Organomet. Chem.*, **188**, 193 (1980).
99. R. H. Grubbs, *Internat. Symp. Metathesis*, Belfast, 1981.
100. a. J. McGinnis, T. J. Katz and C. Hurwitz, *J. Amer. Chem. Soc.*, **98**, 605 (1976); b. M. T. Mocella, M. A. Busch and E. L. Muetterties, *J. Amer. Chem. Soc.*, **98**, 1238 (1976).
101. S. J. McLain, C. D. Wood and R. R. Schrock, *J. Amer. Chem. Soc.*, **99**, 3519 (1977).
102. P. G. Gassman and T. H. Johnson, *J. Amer. Chem. Soc.*, **99**, 622 (1977).
103. C. P. Casey, H. E. Tuinstra and M. C. Saeman, *J. Amer. Chem. Soc.*, **98**, 608 (1976).
104. W. J. Kelly and N. Calderon, *J. Macromol. Sci., Chem.*, **A9**, 911 (1975).
105. C.P. Casey and H. E. Tuinstra, *J. Amer. Chem. Soc.*, **100**, 2270 (1978).
106. L. Bencze, K. J. Ivin and J. J. Rooney, *J. Chem. Soc., Chem. Commun.*, **1980**, 834.
107. C. P. Casey and P. J. Brondsema, *Internat. Symp. Metathesis*, Belfast, 1981.
108. a. R. R. Schrock, *J. Amer. Chem. Soc.*, **97**, 6577 (1975); b. R. R. Schrock and P. R. Sharp, *J. Amer. Chem. Soc.*, **100**, 2389 (1978).
109. C. P. Casey, T. J. Burkhardt, C. A. Bunnell and J. C. Calabase, *J. Amer. Chem. Soc.*, **99**, 2127 (1977).
110. J. Otton, Y. Colleuille and J. Varagnat, *J. Mol. Catalysis*, **8**, 313 (1980).
111. F. Pennella, R. L. Banks and G. C. Bailey, *J. Chem. Soc., Chem. Commun.*, **1968**, 1548.
112. G. N. Schrauzer, *Angew. Chem.*, **76**, 28 (1964).
113. F. D. Mango and J. H. Schachtschneider, *J. Amer. Chem. Soc.*, **91**, 1030 (1969).
114. A. Mortreux, F. Petit and M. Blanchard, *Tetrahedron Letters*, **1978**, 4967.
115. A. Ben-Cheick, M. Petit, A. Mortreux and F. Petit, *Internat. Symp. Metathesis*, Belfast, 1981.
116. E. O. Fischer and U. Schubert, *J. Organomet. Chem.*, **100**, 59 (1975).
117. a. J. H. Wengrovius, J. Sancho and R. R. Schrock, *J. Amer. Chem. Soc.*, **103**, 3932 (1981); b. J. H. Wengrovius, J. Sancho and R. R. Schrock, *Internat. Symp. Metathesis*, Belfast, 1981.

118. H. Rudler, *J. Mol. Catalysis*, **8**, 53 (1980).
119. R. P. Dodge and V. Schomaker, *J. Organomet. Chem.*, **3**, 274 (1965).
120. W. L. Truett, D. R. Johnson, I. M. Robinson and B. A. Montague, *J. Amer. Chem. Soc.*, **82**, 2337 (1960).
121. F. E. Michelotti and W. P. Keavney, *J. Polymer Sci.*, *A*, **3**, 895 (1965).
122. G. Natta, G. Dall'Asta and L. Porri, *Makromol. Chem.*, **81**, 253 (1965).
123. a. G. Dall'Asta, *J. Polymer Sci.*, *A—1*, **6**, 2397 (1968); b. G. Dall'Asta and G. Motroni, *J. Polymer Sci.*, *A—1*, **6**, 2405 (1968).
124. G. Natta, G. Dall'Asta, I. W. Bassi and G. Caralla, *Makromol. Chem.*, **91**, 87 (1966).
125. T. Oshika and H. Tabuchi, *Bull. Chem. Soc. Japan*, **41**, 211 (1968).
126. V. A. Kormer, I. A. Poletayeva and T. L. Yufa, *J. Polymer Sci.*, *A—1*, **10**, 251 (1972).
127. K. W. Scott, N. Calderon, E. A. Ofstead, W. A. Judy and J. P. Ward, *Adv. Chem. Ser.*, **91**, 399 (1969).
128. a. G. Dall'Asta, *Makromol Chem.*, **154**, 1 (1972); b. G. Dall'Asta and G. Motroni, *European Polymer J.*, **7**, 707 (1971); c. G. Dall'Asta and G. Motroni, *Angew. Makromol. Chem.*, **16/17**, 51 (1971).
129. G. Dall'Asta, G. Motroni and L. Motta, *J. Polymer Sci.*, **10**, 1601 (1972).
130. B. A. Dolgoplosk, K. L. Makovetsky, T. G. Golenko, Yu. V. Korshak and E. I. Tinyakova, *European Polymer J.*, **10**, 905 (1974).
131. B. A. Dolgoplosk, I. A. Oreshkin and S. A. Smirnov, *European Polymer J.*, **15**, 237 (1980).
132. B. A. Dolgoplosk, *J. Polymer Sci.*, *Polymer Symposium*, **67**, 99 (1980).
133. E. N. Kropacheva, B. A. Dolgoplosk, D. E. Sterenzat and Yu. A. Patrushin, *Dokl. Akad. Nauk SSSR*, **195**, 1383 (1970).
134. a. K. Hummel and W. Ast, *Makromol. Chem.*, **160**, 39 (1973); b. K. Hummel, D. Wewerka, F. Lorber and G. Zeplichal, *Makromol. Chem.*, **166**, 45 (1973).
135. D. T. Laverty, M. A. McKervey, J. J. Rooney and A. Stewart, *J. Chem. Soc., Chem. Commun.*, **1976**, 193.
136. a. T. J. Katz and N. Acton, *Tetrahedron Letters*, **1976**, 4251; b. T. J. Katz, S. J. Lee and M. A. Shippey, *J. Mol. Catalysis*, **8**, 219 (1980).
137. a. T. J. Katz, S. J. Lee, M. Nair and E. B. Savage, *J. Amer. Chem. Soc.*, **102**, 7940 (1980); b. T. J. Katz, E. B. Savage, S. J. Lee and M. Nair, *J. Amer. Chem. Soc.*, **102**, 7942 (1980).
138. N. Calderon, J. P. Lawrence and E. A. Ofstead, *Adv. Organomet. Chem.*, **17**, 449 (1979).
139. K. J. Ivin, J. J. Rooney and H. H. Thoi, *Internat. Symp. Metathesis*, Belfast, 1981.
140. A. J. Amass and T. A. McGourtey, *European Polymer J.*, **16**, 235 (1980).
141. J. Levisalles, F. Rose-Munch, H. Rudler, J.-C. Daran, Y. Dromzée and Y. Jeannin, *J. Chem. Soc., Chem. Commun.*, **1981**, 152.
142. S. A. Smirnov, I. A. Oreshkin, I. A. Kopieva, B. A. Dolgoplosk and E. I. Tinyakova, *Dokl. Akad. Nauk SSSR*, **239**, 392 (1978).
143. A. J. Amass and J. A. Zurimendi, *European Polymer J.*, **17**, 1 (1981).
144. T. J. Katz and S. J. Lee, *J. Amer. Chem. Soc.*, **102**, 422 (1980).
145. H. H. Thoi, K. J. Ivin and J. J. Rooney, *Internat. Symp. Metathesis*, Belfast, 1981.
146. F. D. Mango and J. H. Schachtschneider, *J. Amer. Chem. Soc.*, **93**, 1123 (1971).
147. W. Th. A. M. van der Lugt, *Tetrahedron Letters*, **1970**, 2281.
148. M. J. S. Dewar, *Angew. Chem.*, **83**, 859 (1971).
149. A. K. Rappé and W. A. Goddard III, *J. Amer. Chem. Soc.*, **102**, 5115 (1980).
150. O. Eisenstein, R. Hoffmann and A. R. Rossi, *J. Amer. Chem. Soc.*, **103**, 5582 (1981).
151. O. Eisenstein and R. Hoffmann, *Internat. Symp. Metathesis*, Belfast, 1981.
152. E. S. Davie, D. A. Whan and C. Kemball, *J. Chem. Soc., Chem. Commun.*, **1969**, 1430.
153. A. Mortreux, E. Digny, D. Nyranth and M. Blanchard, *Bull. Soc. Chim. France*, **1974**, 241.
154. a. R. Nakamura and E. Echigoya, *Bull. Jap. Petrol Inst.*, **14**, 187 (1972); b. R. Nakamura, Y. Morita and E. Echigoya, *Nippon Kagaku Kaishi*, **1973**, 244.
155. a. A. N. Startsev, B. N. Kuznetsov and Yu. I. Ermakov, *React. Kinet. Catal. Lett.*, **3**, 321 (1975); b. Yu. I. Ermakov, B. N. Kuznetsov and Yu. A. Ryndin, *J. Catalysis*, **42**, 73 (1976); c. B. N. Kuznetsov, A. N. Startsev and Yu. I. Ermakov, *J. Mol. Catalysis*, **8**, 135 (1930).

156. a. Yu. I. Ermakov, *Catalysis Rev.*, **13**, 77 (1976); b. Yu. I. Ermakov, "Supported Complexes as Catalysts", Novosibirsk, 1977; c. Yu. I. Ermakov, B. N. Kuznetsov and V. A. Zaharov, "Studies in Surface Science and Catalysis. 8. Catalysis by Supported Complexes", Elsevier, Amsterdam, 1981.

157. a. Y. Iwasawa, S. Ogasawara and M. Soma, *Chem. Letters*, **1978**, 1039; b. Y. Iwasawa, H. Ichinose and S. Ogasawara and M. Soma, *J. Chem. Soc., Faraday Trans. I*, **77**, 1763 (1981); c. Y. Iwasawa and H. Kubo, *J. Chem. Soc., Faraday Trans. I*, **78** (1982).

158. R. M. Edreva-Kardjieva, A. A. Andreev and N. M. Neshev, *Internat. Symp. Metathesis*, Lyons, 1979.

159. R. M. Edreva-Kardjieva and A. A. Andreev, *React. Kinet. Catal. Lett.*, **5**, 465 (1976).

160. R. M. Edreva-Kardjieva and A. A. Andreev, *Rec. Trav. Chim.*, **96**, M141 (1977).

161. See for instance F. A. Cotton and G. Wilkinson, "Advanced Inorganic Chemistry", 4th Edition, John Wiley, New York, 1980.

162. E. L. Muetterties and J. Stein, *Chem. Rev.*, **79**, 479 (1979).

163. W. A. Herrmann, M. L. Ziegler, K. Wiedenhammer and H. Biersack, *Angew. Chem.*, **91**, 1026 (1979).

164. M. O. Visscher and K. G. Caulton, *J. Amer. Chem. Soc.*, **94**, 5923 (1972).

165. W. E. Griffith, *Coord. Chem. Rev.*, **5**, 459 (1970).

166. a. D. J. Cardin, B. Cetinkaya and M. F. Lappert, *Chem. Rev.*, **72**, 541 (1972); b. D. J. Cardin, B. Cetinkaya, M. J. Doyle and M. F. Lappert, *Chem. Soc. Rev.*, **2**, 99 (1973).

167. F. J. Brown, *Progr. Inorg. Chem.*, **27**, 1 (1980).

168. E. O. Fischer, *Adv. Organomet. Chem.*, **14**, 1 (1976).

169. L. J. Guggenberger and R. R. Schrock, *J. Amer. Chem. Soc.*, **97**, 6578 (1975).

170. E. O. Fischer, C. Kreis, G. C. Kreiter, J. Müller, G. Hutner and H. Lorenz, *Angew. Chem.*, **85**, 618 (1973).

171. R. R. Schrock, *Accounts Chem. Res.*, **12**, 98 (1979).

172. R. J. Goddard, R. Hoffmann and E. D. Jemmis, *J. Amer. Chem. Soc.*, **102**, 7667 (1980).

173. U. Schubert, in "Chemistry of the Metal-Carbon Bond", S. Patai, Ed., John Wiley, New York, in press.

174. N. M. Kostić and R. F. Fenske, *J. Amer. Chem. Soc.*, **103**, 4677 (1981).

175. E. O. Fischer, U. Schubert and H. Fischer, *Pure Appl. Chem.*, **50**, 857 (1978).

176. a. N. Calderon, H. Y. Chen and K. W. Scott, *Tetrahedron Letters*, **1967**, 3327; b. N. Calderon, *Accounts Chem. Res.*, **5**, 127 (1972); c. N. Calderon and H. Y. Chen, **Brit.** 1125529, 28.08.1968, *Chem. Abstr.*, **69**, 105851 (1968).

177. D. Medema, W. Brunmayer-Schilt and R. van Helden, **Brit.** 11939—43, 3.06.1970, *Chem. Abstr.*, **73**, 76637 (1970).

178. L. Ramain and Y. Trambouze, *Compt. Rend.*, **273C**, 1409 (1971).

179. a. A. Uchida, M. Hidai and T. Tatsumi, *Bull. Chem. Soc. Japan*, **45**, 1158 (1972); b. A. Uchida, K. Kobayashi and S. Matsuda, *Ind. Eng. Chem., Prod. Res. Develop.*, **11**, 4, 389 (1972).

180. J. M. Basset, G. Coudurier, R. Mutin, H. Praliaud and Y. Trambouze, *J. Catalysis*, **34**, 196 (1974).

181. a. J. Kress, M. J. M. Russell, M. G. Wesolek and J. A. Osborn, *J. Chem. Soc., Chem. Commun.*, **1980**, 431; b. J. Kress, M. J. Wesolek, J. P. Le Ny and J. A. Osborn, *J. Chem. Soc., Chem. Commun.*, **1981**, 1039.

182. K. J. Ivin, B. S. R. Reddy and J. J. Rooney, *J. Chem. Soc., Chem. Commun.*, **1981**, 1062.

183. a. R. Nakamura, F. Abe and E. Echigoya, *Chem. Letters*, **1981**, 51; b. R. Nakamura and E. Echigoya, *Internat. Symp. Metathesis*, Belfast, 1981.

184. a. L. Bencze and L. Markó, *J. Organomet. Chem.*, **28**, 271 (1971); b. L. Bencze, A. Rédey and L. Markó, *Hung. J. Ind. Chem.*, **1**, 453 (1973).

185. E. A. Zuech, W. B. Hughes, D. H. Kubicek and E. T. Kittleman, *J. Amer. Chem. Soc.*, **92**, 528 (1970).

186. M. Kothari and J. J. Tazuma, *J. Org. Chem.*, **36**, 2951 (1971).

187. M. Leconte, Y. Ben Taarit, J. L. Bilhou and J. M. Basset, *J. Mol. Catalysis*, **8**, 263 (1980).

188. J. L. Bilhou, R. Mutin, M. Leconte and J. M. Basset, *Rec. Trav. Chim.*, **96**, M5 (1977).

189. H. D. Dodd and K. J. Rutt, *Internat. Symp. Metathesis*, Belfast, 1981.

190. M. J. S. Dewar, *Bull. Soc. Chim. France*, **18**, C79 (1951).

191. J. H. Nelson, K. S. Wheelock, L. C. Cusachs and H. B. Jonassen, *J. Chem. Soc., Chem. Commun.*, **1969**, 1019.
192. U. Bellucco, "Organometallic and Coordination Chemistry of Platinum", Academic Press, London, 1974.
193. J. Chatt, J. L. Dilworth, G. J. Leigh and R. L. Richards, *J. Chem. Soc., Chem. Commun.*, **1970**, 955.
194. A. E. Crease and P. Legzdins, *J. Chem. Soc., Chem. Commun.*, **1972**, 268.
195. M. Pankowski, B. Demerseman, G. Bouquet and M. Bigorgne, *J. Organomet. Chem.*, **35**, 155 (1972).
196. L. Bencze, *J. Organomet. Chem.*, **56**, 303 (1973).
197. W. B. Hughes, *Organomet. Chem. Synth.*, **1**, 341 (1972).
198. L. Bencze, J. Pallos and L. Markó, *J. Mol. Catalysis*, **2**, 139 (1977).
199. E. K. Tarcsi, F. Ratkovics and L. Bencze, *J. Mol. Catalysis*, **1**, 435 (1975/76).
200. R. Opitz, L. Bencze, L. Markó and K.-H. Thiele, *J. Organomet. Chem.*, **71**, C3 (1974).
201. L. Bencze, K.-H. Thiele and V. Marquardt, *Rec. Trav. Chim.*, **96**, M8 (1977).
202. P. R. Marshall and B. J. Ridgewell, *European Polymer J.*, **5**, 29 (1969).
203. W. A. Judy, cited after 176b.
204. J. L. Wang and H. R. Menapace. *J. Org. Chem.*, **33**, 3794 (1968).
205. G. Dall'Asta and G. Carella. **Ital**. 784,307, 2.08.1966.
206. D. M. Singleton, **Fr.** 2,013,672, 25.08.1968.
207. D. M. Singleton, **U.S.** 3,530.196, 29.01.1970.
208. J. L. Bilhou and J. M. Basset, *J. Organomet. Chem.*, **132**, 395 (1977).
209. J. L. Bilhou, A. K. Smith and J. M. Basset, *J. Organomet. Chem.*, **148**, 53 (1978).
210. D. M. Singleton and D. J. Eatough, *J. Mol. Catalysis*, **8**, 175 (1980).
211. H. Höcker and F. R. Jones, *Makromol. Chem.*, **161**, 251 (1972).
212. R. R. Schrock, S. Rocklage, J. Wengrovius, G. Rupprecht and J. Fellmann, *J. Mol. Catalysis*, **8**, 73 (1980).
213. F. N. Tebbe, *J. Amer. Chem. Soc.*, **95**, 5412 (1973).
214. F. N. Tebbe and L. J. Guggenberger, *J. Chem. Soc., Chem. Commun.*, **1973**, 227.
215. F. N. Tebbe, G. W. Parshall and G. S. Reddy, *J. Amer. Chem. Soc.*, **100**, 3611 (1978).
216. R. R. Schrock and P. R. Sharp, *J. Amer. Chem. Soc.*, **100**, 2389 (1978).
217. P. R. Sharp, S. J. Holmes, R. R. Schrock, M. R. Churchill and H. J. Wasserman, *J. Amer. Chem. Soc.*, **103**, 965 (1981).
218. S. Nakamura and A. Dedieu, *Nouveau J. Chim.*, **6**, 23 (1982).

Chapter 7

STEREOCHEMISTRY OF OLEFIN METATHESIS AND RING-OPENING POLYMERIZATION

The metathesis of acyclic olefins and the ring opening polymerization of cyclo-olefins are stereoselective. The degree of stereoselectivity varies) with the nature of the catalytic system and the olefin (or cyclo-olefin substrate employed in the reaction. Furthermore, the stereoselectivity of the reaction products depends on the reaction conditions, especially the reaction temperature and the conversion. As the stereochemistry of the reaction may provide further information enhancing understanding of the process and helping to clarify the reaction mechanism, this chapter will present the main aspects of the stereochemistry of the metathesis reaction.

7.1. Metathesis of Acyclic Olefins

The metathesis of acyclic olefins exhibits a characteristic stereospecificity. This is manifested through definite stereoselectivity of the reaction products, which depends on the nature of the catalytic system and on the steric interactions which occur during the reaction.

7.1.1. Experimental Results Relating to the Stereoselectivity of Metathesis Products

The first results concerning the stereoselectivity of the reaction products to be mentioned are those obtained by Banks and Bailey [1] in the metathesis of acyclic olefins in the presence of heterogeneous catalysts consisting of alumina-supported cobalt molybdate. On studying the relative quantities of the two geometric isomers of 2-butene generated as metathesis reaction products from propene, 1-butene, 2-butene, 1-pentene and 2-pentene within a temperature range between 95° and 300°C, Banks and Bailey observed that, in most cases, the *trans-* : *cis*-2-butene ratio in the product reached values close to that corresponding to thermodynamic equilibrium (Table 7.1).

In the reactions of propene and 2-pentene, it was found that the *trans-* : *cis*-2-butene ratio was very close to the thermodynamic equilibrium ratio.

TABLE 7.1. *trans-* :*cis*-2-Butene ratio in the metathesis product of acyclic olefins [a, b]

Olefin	Temperature, °C			
	95	160	200	300
Propene	2.15	1.95	1.78	1.80
1-Butene	0.46	1.24	1.49	1.49
2-Butene	2.31	2.21	2.21	—
1-Pentene	1.0	2.0	—	1.5
2-Pentene	2.0	1.8	—	1.8
Equilibrium ratio	2.4	2.0	1.9	1.7

[a] Catalyst : $CoO-MoO_3/Al_2O_3$; [b] Data from ref. [1].

In the reactions of 1-pentene and 1-butene, the ratio was smaller, i.e. the stereoselectivity of the *cis* isomer formation was higher. The very small value of the *trans-* :*cis*-2-butene ratio for 1-butene suggests that the mechanism of the isomerization of 1-butene to 2-butene in the presence of the metathesis catalytic system favours the *cis* stereoisomer.

It is important to note that the *trans-* :*cis*-2-pentene ratio in the above reactions had values more remote from the corresponding thermodynamic equilibrium (Table 7.2).

TABLE 7.2. *trans-* : *cis*-2-Pentene ratio in the metathesis product of acyclic olefins [a, b]

Olefin	Temperature, °C			
	95	160	200	300
Propene	—	3.0	3.4	1.8
1-Butene	2.0	2.4	2.4	2.1
2-Butene	3.0	2.6	2.4	—
1-Pentene	1.2	2.2	—	2.1
2-Pentene	3.8	2.7	—	2.2
Equilibrium ratio	1.6	1.5	1.4	1.3

[a] Catalyst : $CoO-MoO_3/Al_2O_3$; [b] Data from ref. [1].

As it can be seen from Table 7.2, in all the reactions reported, the *trans-* : *cis*-2-pentene ratio had values which were considerably higher than those for thermodynamic equilibrium. It is interesting to note that the *trans-* : *cis*-2-pentene ratio is smaller than the equilibrium values at lower temperatures and higher for higher temperatures in the reaction of 1-pentene.

Similar results concerning the stereoselectivity of the propene metathesis with heterogeneous catalytic systems were obtained by Furukawa [2] and Davie, Whan and Kemball.[3] The ratio between *trans-* and *cis*-2-butene for high conversions of propene had values very close to the thermodynamic equilibrium value. For low conversions, Davie, Whan and Kemball observed that this ratio differs substantially from the equilibrium value, favouring the *cis*-isomer (Table 7.3). These authors

TABLE 7.3. Product distribution in the metathesis of propene [a, b]

Time, minutes	Product distribution (mole %)		
	1-Butene	*trans*-2-Butene	*cis*-2-Butene
1	0.0	70.1	29.9
8	0.2	69.3	30.3
15	0.5	69.9	29.6
23	1.1	70.6	28.3
73	1.4	72.0	26.6
Equilibrium composition	2.7	73.6	23.7

[a] Catalyst : $Mo(CO)_6/Al_2O_3$; [b] Data from ref. [3].

interpreted this result as involving a steric effect, namely postulating that the *cis*-stereoisomer can be formed more easily at the surface of the catalytic system as a consequence of lower steric compression.

On studying the stereochemistry of *cis*-2-pentene metathesis over a catalytic system consisting of cobalt molybdate, Ismayel-Milanovic *et al.*[4] found very high selectivity in favour of the *cis* isomers, especially at low temperatures. Thus, when the reaction was carried out at temperatures up to 50°C, the *trans*- :*cis*-2 butene ratio was considerably lower than 1, i.e. *cis*-2-butene predominated in the product mixture. In contrast at temperatures above 100°C, the *trans*- : *cis*-2-butene ratio increased to between 1 and 2, tending towards thermodynamic equilibrium (Fig. 7.1).

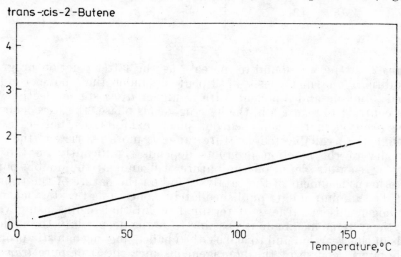

Fig. 7.1. — Effect of reaction temperature of the ratio *trans*- :*cis*-2-pentene.

Moreover, the kinetic data showed that the rate of *cis*-2-pentene metathesis was ten times higher than that of *trans*-2-pentene.

Relevant stereochemical results were obtained recently by Tanaka and co-workers [5,6] on studying the stereoselectivity of propene, 2-butene and 2-pentene metathesis over molybdenum oxide supported on titanium oxide. They observed that in propene metathesis on this catalyst, the formation of *trans*-2-butene was preferred in a ratio of *trans* : *cis* $\geqslant$ 6. Further, degenerate cometathesis of *trans*-2-butene-d_0 and d_8 proved to yield practically 100% *trans*-2-butene-d_4 (7.1) while the reaction of

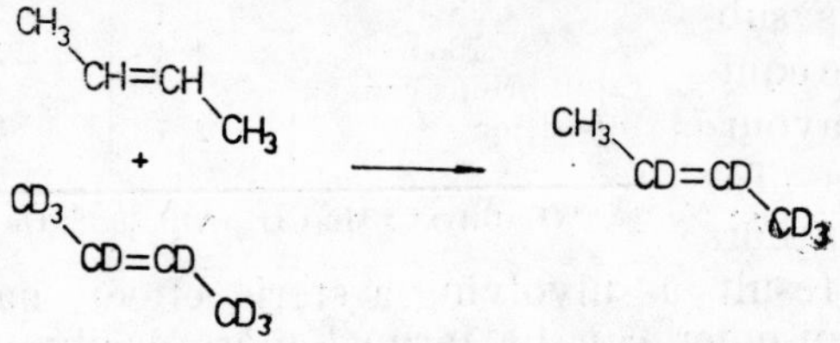

$$(7.1)$$

cis-2-butene-d_0 and d_8 afforded *cis*-2-butene-d_4 in more than 60% yield. The *trans*-2-butene formed in the reaction of *cis*-2-butene-d_0 and d_8 was composed of d_0-, d_4- and d_8-components (7.2). The ratio of $d_4/(d_0 + d_8)$

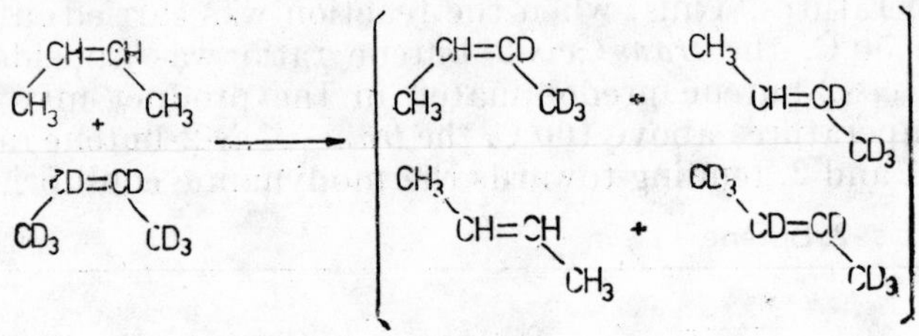

$$(7.2)$$

in *trans*-2-butene was found to increase as the latter reaction progressed. In addition, the metathesis of 2-pentene under the above conditions yielded 2-butene and 3-hexene with a higher *trans* stereoselectivity.

In contrast to reaction in the heterogeneous phase, the stereochemistry of the reaction in the homogeneous phase exhibits distinctive features. Although the available temperature range is more restricted, the stereoselectivity of the reaction products depends significantly on the steric purity of reactant olefin, on the nature of ligands at the double bond, on the steric environment of the catalyst, and on the nature of the transition metal. The amount of data published lately, concerning the stereochemistry of homogeneous metathesis of terminal or internal olefins using various catalytic systems has made it possible to probe deeper into the molecular details of this reaction and to clarify the challenging mechanistic problems.

The steric course of the homogeneous metathesis of pure *trans*- and *cis*-2-pentene was initially examined by Calderon et al.[7] in their earlier work, with the catalytic system $WCl_6 \cdot EtOH \cdot EtAlCl_2$. The isomer composition of the reaction product illustrated in Table 7.4 is particularly interesting.

TABLE 7.4. Metathesis of 2-pentene with the ternary catalyst $WCl_6 \cdot EtOH \cdot EtAlCl_2$ [a]

2-Pentene	Al : W	Conversion, %	*trans* :*cis* Ratio		
			2-Butene	2-Pentene	3-Hexene
96% *cis* — 4% *trans*	4	9	0.7	0.1	—
,,	2	46	2.3	2.1	—
,,	1	53	1.9	2.8	—
97% *cis* — 3% *trans*	1	40	1.1	0.9	—
,,	4	47	1.5	1.6	2.5
,,	4	50	2.5	4.5	6.7
100% *trans*	4	15	1.5	30.0	2.1
,,	4	41	2.2	11.5	3.0
,,	4	45	2.3	4.5	4.2

[a] Data from ref. [7].

The results show that substantial amounts of both *cis* and *trans* isomers of 2-butene and 3-hexene are formed at the outset, whereas the composition of the residual 2-pentene very gradually approached its equilibrium *cis* content. It was also found that the initially formed butene and hexene were somewhat higher in *cis* content than the equilibrium values for these olefins, regardless of the steric structure of the initial pentene. In addition, these authors observed that the *cis* :*trans* isomer ratio for 2-butene was always different from that for 3-hexene over the entire range of conversions. Further, the steric course of 2-pentene metathesis with Calderon's catalyst proved to be independent of the Al : W ratio within the range $1 \leqslant$ Al : W $\leqslant 4$.

With the binary catalytic system $WCl_6 \cdot BuLi$, the reaction of 2-pentene isomers has a somewhat different stereochemistry than that depicted above. Under these conditions, Wang and Menapace [8] noticed that the *trans* : *cis* ratio of the reaction products depends strongly on the steric structure of the initial olefin (Table 7.5).

TABLE 7.5. Stereochemistry of the metathesis products of 2-pentene with the catalyst $WCl_6 \cdot BuLi$ [a]

2-Pentene	Reaction time, minutes	% Conversion	*trans* :*cis* Ratio	
			2-Butene	2-Pentene
96% *cis* — 4% *trans*	120	40	0.8	0.8
100% *trans*	120	38	3.3	7.4
54% *cis* — 46% *trans*	120	43	1.6	3.0
96% *cis* — 4% *trans*	240	50	1.8	2.9
100% *trans*	240	40	3.1	6.8
54% *cis* — 48% *trans*	240	51	2.5	4.2
96% *cis* — 4% *trans*	1025	50	1.8	2.9
100% *trans*	1025	40	3.1	7.2
54% *cis* — 46% *trans*	1025	51	2.6	4.2

[a] Data from ref. [8].

The *trans* : *cis* ratio of a mixture of geometric isomers of 2-pentene increases non-linearly when the conversion increases and reaches a maximum of 4.2 corresponding to 81% *trans* at 50% conversion. The ratio of 2-pentene isomers and the conversion maintain steady values even after a prolonged reaction time; the *trans*-content in butene was 72%.

For *cis*-2-pentene, at 50% conversion, the ratio of *trans-/cis*-2-pentene was 2.9, corresponding to 75% *trans*. This value was significantly lower than the 81% found above for 2-pentene. The conversion of pure *trans*-2-pentene levelled out at only 40%; at this value the *trans* :*cis* ratio was 6.8 (87% *trans*).

These differences in behaviour between *trans* and *cis* isomers of 2-pentene were attributed to the higher steric hindrance in coordinating to tungsten atoms for the *trans* isomer.

Higher stereoselectivity in 2-pentene metathesis, as compared to that reported for tungsten catalysts, was obtained by Hughes [9] with homogeneous catalytic systems based on molybdenum complexes. In this case, the reaction proved to be kinetically controlled. The ratio of the stereoisomers in the product favours the steric isomer existing in the starting material (Table 7.6).

TABLE 7.6. Stereoisomer ratio in the metathesis product of 2-pentene with $L_2MoCl_2(NO)_2$ [a]

Starting material	Reaction time, minutes	*cis* : *trans* Ratio		
		2-Butene	2-Pentene	3-Hexane
cis-2-Pentene	2	4.4	28.8	2.3
	4	2.7	7.2	2.0
trans-2-Pentene	2	0.1	0.04	0.0
	4	0.2	0.08	0.1

[a] Data from ref. [9].

From these results it can be easily deduced that *cis*-2-pentene disproportionates preferentially to *cis*-2-butene and *cis*-3-hexene, while *trans*-2-pentene disproportionates to *trans*-2-butene and *trans*-3-hexene.

Abundant experimental data concerning the stereochemistry of 2-pentene metathesis (as well as the reaction of other substituted olefins) in the presence of various tungsten, molybdenum or chromium catalysts were reported by Basset and co-workers.[10-19] These authors have undertaken the systematic kinetic study of both the metathesis reaction and the *cis-trans* isomerization of 2-pentene. They correlated the stereoselectivity of 2-pentene metathesis with the structure of the coordination sphere of the catalyst. They studied (i) the influence of the substitutents in the α-position to the double bond on the stereoselectivity and (ii) the effect of the nature of transition metal on the steric course of 2-pentene metathesis. The results were interpreted taking into account the main steps in the one-carbene exchange mechanism proposed by Chauvin.

The metathesis of *cis*-2-pentene with a wide variety of tungsten and molybdenum catalysts displayed a varying degree of selectivity but always showed a preference for retention of the *cis* configuration (Table 7.7).

TABLE 7.7. Stereoselectivity of *cis*-2-pentene metathesis in homogeneous phase at zero conversion [a]

Catalytic system	*trans*- : *cis* — C_4	*trans*- : *cis* — C_6
$W(CO)_5PPh_3 + EtAlCl_2 + O_2$	0.73	0.88
$W(CO)_5(n\text{-}C_4H_9)_3 + EtAlCl_2 + O_2$	0.76	0.83
$W(CO)_5P(OPh)_3 + EtAlCl_2 + O_2$	0.76	0.89
$W(CO)_3(Mes) + EtAlCl_2$	0.72	0.88
$WCl_6 + (CH_3)_4Sn$	0.73	—
$W(NO)_2Cl_2(PPh_3)_2 + EtAlCl_2$	0.44	0.74
$W(NO)_2Cl_2(PPh_3)_2 + Me_3Al_2Cl_3$	0.40	0.60
$Mo(NO)_2Cl_2(PPh_3)_2 + EtAlCl_2$	0.200	0.45
$Mo(NO)_2Br_2(PPh_3)_2 + EtAlCl_2$	0.20	0.43
$Mo(NO)_2Cl_2Py_2 + EtAlCl_2$	0.23	0.45
$Mo(NO)_2Cl_2Py_2 + Me_3Al_2Cl_3$	0.24	0.41
$Mo(NO)_2Cl_2 + Me_3Al_2Cl_3$	0.22	0.40

[a] Data from ref. [11, 15, 17].

The stereoselectivity was determined from the *trans* : *cis* ratio of 2-butenes and 3-hexenes extrapolated to 0% conversion. Basset *et al.*[13,17] observed a linear relationship for *trans*- : *cis*-C_5(C_5 = 2-pentene) over a wide range of conversions (Fig. 7.2). The intercept of this line at a *trans*- : *cis*-C_5

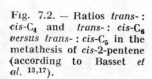

Fig. 7.2. — Ratios *trans*- : *cis*-C_4 and *trans*- : *cis*-C_6 *versus* *trans*- : *cis*-C_5 in the metathesis of *cis*-2-pentene (according to Basset *et al.* [13,17]).

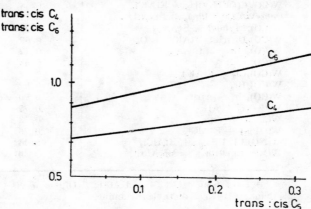

ratio equal to zero appears to give a precise value for the stereoselectivity of the propagation step. It is worth mentioning that the degree of retention of the *cis* configuration of the starting olefin is rather low with

TABLE 7.8. Stereoselectivity of cis-2-pentene metathesis over η-alumina-supported tungsten complexes [a]

Tungsten compound	$trans-$: $cis-C_4$ at 0% conversion
$W(CO)_6$	0.37
$W(CO)_5P(n-C_4H_9)_3$	0.38
$W(CO)_5P(C_6H_5)_3$	0.38
$W(CO)_5P(OC_6H_5)_3$	0.40

[a] Data from ref. [11].

many tungsten-based catalysts but higher with molybdenum. It is also noteworthy that the $trans$: cis ratio of the resulting 3-hexenes was always higher than the $trans$: cis ratio of 2-butenes, with catalysts derived from either molybdenum or tungsten.

Interestingly, in the presence of supported tungsten catalysts, cis-2-pentene exhibited a high stereoselectivity with retention of configuration [11] (Table 7.8). These results indicate that a significant steric effect is exerted by the catalyst support upon the course of the reaction.

Valuable data concerning the stereoselectivity of cis-2-pentene were also reported [17] by changing the ligands of the precursor complex (Table 7.9).

TABLE 7.9. Effect of the ligand of the tungsten complex on the stereoselectivity of cis-2-pentene metathesis [a]

Catalytic system	$trans-$: $cis-C_4$	$trans-$: $cis-C_6$
$W(CO)_6 + EtAlCl_2$	0.69	—
$W(CO)_5(C_6H_5NH_2) + EtAlCl_2$	0.78	—
$W(CO)_5P(n-Bu)_3 + EtAlCl_2$	0.76	0.83
$W(CO)_5PPh_3 + EtAlCl_2$	0.73	0.88
$W(CO)_5P(OPh)_3 + EtAlCl_2$	0.76	0.89
$W(CO)_5C(OCH_3)CH_3 + EtAlCl_2$	0.68	—
$trans$-$W(CO)_4(PPh_3)_2 + EtAlCl_2 + O_2$	0.77	—
$W(CO)_4Diphos^b + EtAlCl_2$	0.72	
$W(CO)_3Dien^c + EtAlCl_2 + O_2$	0.69	—
$W(CO)_3(Mes) + EtAlCl_2$	0.72	0.88
$W(CO)_4Cl_2$	0.68	—
$W(CO)_4Cl_2 + EtAlCl_2$	0.74	—
$W(CO)_4Br_2$ —	0.57	—
$W(CO)_4Br_2 + AlBr_3$	0.70	—
$W(CO)_4I_2 + AlI_3$	0.77	—
$WCl_6 + Me_4Sn$	0.73	—
$W(OPh)_6 + EtAlCl_2$	0.73	—
$W(NO)_2Cl_2(PPh_3)_2 + EtAlCl_2$	0.44	0.74
$W(NO)_2Cl_2(PPh_3)_2 + Me_3Al_2Cl_3$	0.40	0.60

[a] Data from ref. [17] and [18]; [b] Diphos = bis (1,2-diphenyl-phosphino)-ethane; [c] Dien = diethylen-triamine.

Monosubstituted complexes of the type $W(CO)_5L$ where L is carbonyl, phosphine, phosphite, amine or carbene exhibited the same stereoselectivity within experimental error. Similar steric behaviour was observed for zero-valent disubstituted complexes $W(CO)_4L_2$ (where L_2 represents either a dichelating diphosphine or two triphenylphosphine groups in *trans* positions) and for zero-valent trisubstituted complexes $W(CO)_3L_3$ (L_3 is a diene or arene ligand). Tungsten (II) and tungsten (IV) derivatives exhibited the same stereoselectivity in *cis*-2-pentene methathesis regardless of the nature of the ligand in the precursor complex or of the cocatalyst. By contrast, nitrosyl complexes of tungsten or molybdenum $M(NO)_2X_2L_2$, associated with $EtAlCl_2$ or $Me_3Al_2Cl_3$ show particularly high stereoselectivity in *cis*-2-pentene metathesis as compared with any other two component catalysts (Table 7.10).

TABLE 7.10 Effect of ligand of the molybdenum complex on the stereoselectivity of *cis*-2-pentene metathesis [a]

Catalytic system	*trans*- : *cis*-C_4	*trans*- : *cis*-C_6
$Mo(CO)_5PPh_3 + EtAlCl_2 + O_2$	0.60	—
$Mo(CO)_3(Mes) + EtAlCl_2$	0.60	—
$Mo(NO)_2Cl_2(PPh_3)_2 + EtAlCl_2$	0.20	0.45
$Mo(NO)_2Br_2(PPh_3)_2 + EtAlCl_2$	0.205	0.43
$Mo(NO)_2I_2(PPh_3)_2 + EtAlCl_2$	0.17	—
$Mo(NO)_2Cl_2Py_2 + EtAlCl_2$	0.23	0.45
$Mo(NO)_2Cl_2Py_2 + Me_3Al_2Cl_3$	0.24	0.41
$Mo(NO)_2Cl_2 + Me_3Al_2Cl_3$	0.22	0.40

[a] Data from ref. [18].

trans-2-Pentene with zero-valent tungsten systems ($W(CO)_5PPh_3 \cdot EtAlCl_2 + O_2$ or $W(CO)_3(Mes) \cdot EtAlCl_2 + O_2$) gives no reaction when the starting olefin was pure.[13] However, if the *trans*-2-pentene is admixed with a *cis* or terminal olefin (even in less than 0.2% molar ratio) the metathesis reaction takes place after an induction period of varying length; thus the reaction is autocatalytic for the first minutes. Similar behaviour had already been observed in the metathesis of *trans*-2-butene[20] with the catalytic system $WCl_6 \cdot MeAlCl_2$ and in the cross-metathesis between allyl-benzene and *trans*-1-phenyl-2-butene [21] with the catalytic system $Mo(NO)_2Cl_2(PPh_3)_2 \cdot EtAlCl_2$.

Basset and co-workers [13] explained this unexpected behaviour of *trans*-2-pentene by considering the initiation step corresponding to the first metallacarbene separately from the propagation step. Steric factors related to the nature of the geometric isomers seem to be a key factor during the initiation period. Thus, *trans*-2-pentene alone is unable to produce the first metallacarbene while *cis*-2-pentene can, and traces of *cis* isomer promote the metathesis reaction of the *trans* isomer. The new *cis* isomers formed will continue to produce the metallacarbene responsible for the propagation step.

High stereoselectivity was obtained in *trans*-2-pentene metathesis
with tungsten and molybdenum nitrosyl complexes (Table 7.11).

TABLE 7.11. Stereoselectivity of *trans*-2-pentene metathesis in
homogeneous phase at zero conversion [a]

Catalytic complex	*trans*- : *cis*-C_4	*trans*- : *cis*-C_6
$W(NO)_2Cl_2(PPh_3)_2$	6.2 ± 1	11.4 ± 1.0
$Mo(NO)_2Cl_2(PPh_3)_2$	12.5 ± 0.5	25.0 ± 1.0
$Mo(NO_2)Br_2(PPh_3)_2$	12.5 ± 0.5	25.0 ± 1.0

[a] Data from ref. [17].

The *trans* : *cis* ratio for 3-hexenes was higher than the *trans* : *cis* ratio for
2-butene when the catalyst was either molybdenum or tungsten. A similar observation was mentioned for *cis*-2-pentene metathesis but for *trans*-
2-pentene the stereoselectivity was higher.

By using various Cr, Mo and W catalysts in the metathesis of *cis*-
and *trans*-2-pentene, a wide range of catalytic activities was reported.
Most chromium complexes were found to be almost inactive except for
the arene complexes $(ArH)Cr(CO)_3(ArH = $ benzene or mesitylene) in
the presencec of $EtAlCl_2$; on going to molybdenum and tungsten, the
activity increased. At the same time, the stereoselectivity observed with
cis- or *trans*-2-pentene decreased in the order $Cr > Mo \gg W$, chromium
being the most, and tungsten the least stereoselective (Table 7.12).

TABLE 7.12. Effect of transition metal on the stereoselectivity of metathesis of *cis*- or
trans-2-pentene [a]

cis-2-pentene	Catalyst [b]					
	I		II		III	
	$t-/c-C_4$	$t-/c-C_6$	$t-/c-C_4$	$t-/c-C_6$	$t-/c-C_4$	$t-/c-C_6$
W	0.80	1.1	0.73	0.88	0.44	0.74
Mo	0.60		0.60		0.20	0.45
Cr	0.45					
$Cr(CO)_3$ (Benzene)	0.42					

trans-2-pentene	Catalyst [b]	
	III	
	$t-/c-C_4$	$t-/c-C_6$
W	6.2	11.4
Mo	12.5	25

[a] Data from ref. [15]; [b] The catalysts used were: I = $M(CO)_3(Mes)$; II = $M(CO)_5PPh_3$
and III = $M(NO)_2Cl_2(PPh_3)_2$;

These drastic changes in metathesis stereoselectivities encountered with group VI transition metals were attributed to the nature of the transition metal itself.

Moderate stereoselectivities in *cis*- and *trans*-2-pentene metathesis were obtained by Bosma, Xu and Mol [22] with the catalyst precursor $WCl_6 \cdot SnR_4$ and $WCl_6 \cdot Sn_2R_6$. The *cis* : *trans* ratios of the 2-butene and 3-hexene products were studied by changing the nature of the ligands in the cocatalyst. No significant differences in stereoselectivity were observed when the length of the alkyl group in SnR_4 was varied from CH_3 to C_4H_9 or C_8H_{17}. When, however, the R group was C_6H_5, the product stereoselectivities changed in favour of the *cis* isomers. These observations were rationalized in terms of the relative stabilities of the metallacyclobutane intermediates of the reaction.

The highest stereospecificity in 2-pentene metathesis was reported by Katz and Hersh [23] with the carbenic system $Ph_2C = W(CO)_5$. In this case, by conducting the reaction at a high degree of conversion ($\sim 87\%$), *cis*-2-pentene yielded a product with a *cis*-stereoisomer content of about 90%. Extrapolated to zero conversion, the product had a very high steric purity (Table 7.13).

TABLE 7.13. 2-Pentene metathesis with the carbenic system $Ph_2C = W(CO)_5$ [a]

Starting 2-pentene	% *cis* at zero conversion		
	2-Butene	2-Pentene	3-Hexene
cis-2-Pentene	94.3	96.0	92.5
trans-2-Pentene	27.2	0.2	16.6

[a] Data from ref. [23].

It is of interest that in this reaction, *trans*-2-pentene yielded a mixture of *trans* and *cis* isomers in ratios close to the thermodynamic equilibrium values. As will be shown below, the above results were interpreted on the basis of steric interactions in the metallacyclobutane intermediate, assuming in some cases the cleavage of the metal-carbon bond in the metallacyclobutane with the formation of a metallopropyl cation in which the remaining bonds may rotate.

Important stereochemical data were reported by Basset *et al.* [14,15] for the reaction of α-olefins and internal olefins of the type *cis*- or *trans*-RCH = CHCH$_3$ where R is an alkyl, aryl or aralkyl group. These data provide a sound basis for the proper interpretation of steric effects during metathesis.

The metathesis of propene, 1-butene and 1-pentene with $Mo(NO)_2Cl_2(PPh_3)_2$ in the presence of $EtAlCl_2$ afforded various stereoselectivities, as functions of the substituents at the double bond (Table 7.14).

TABLE 7.14. *trans-* : *cis-* Product in metathesis of
α-olefins [a]

Olefin	*trans-* : *cis* ratio of internal olefinic product [b]	
	Determn.	Equilibrium
Propene	1.1	3.0
1-Butene	1.42	6.0
1-Pentene	2.0	4.7

  [a] Data from ref. [15];
  [b] Catalyst : $Mo(NO)_2 Cl_2(PPh_3)_2$ + $EtAlCl_2$.

The stereoselectivity increased in the order: propene > 1-butene > > 1-pentene.

Internal olefins of the type *cis-* or *trans-*$RCH=CHCH_3$, in which R is ethyl, n-propyl, n-butyl, n-pentyl, isopropyl, *tert*-butyl, phenyl or benzyl, displayed various stereoselectivities, depending on the nature of the R group and the catalyst. With zero-valent complexes of tungsten, $W(CO)_5PPh_3$ or $(ArH)W(CO)_3$ (ArH = p-xylene), in the presence of organoaluminium compounds and sometimes of oxygen, Basset observed low stereoselectivity regardless of the *cis* or *trans* nature of the initial olefin (Table 7.15).

TABLE 7.15. Effect of the substituents at the double
bond on stereoselectivity in olefin
metathesis with the catalytic system
$W(CO)_5PPh_3 + EtAlCl_2 + O_2$ [a]

Substituted olefin	*trans-* : *cis* — C_4 at 0% conversion
cis-2-Pentene	0.73
cis-2-Hexene	0.79
cis-4-Methyl-2-pentene	1.02
cis-4,4-Dimethyl-2-pentene	1.00
trans-2-Pentene	1.25
trans-2-Hexene	1.25
trans-2-Heptene	1.25
trans-2-Octene	1.25
trans-1-Phenyl-2-butene	1.21
trans-4-Methyl-2-pentene	1.16
trans-4,4-Dimethyl-2-pentene	1.10

  [a] Data from ref. [15].

It is worth mentioning that the length of the unsubstituted R group ($R \neq H$) has no significant effect on the reaction stereoselectivity; the *trans-* : *cis*-2-butene ratio remains almost unchanged going from ethyl to n-propyl or n-pentyl. When R is isopropyl or *tert*-butyl, the *trans-* : *cis*-ratio of 2-butene tends toward unity, regardless of the *cis* or *trans* nature of the olefin; simultaneously, the rate of reaction decreases.

With molybdenum catalysts, Basset observed higher stereoselectivity (Table 7.16).

TABLE 7.16. Effect of the substituents at the double bond on stereoselectivity in olefin metathesis with molybdenum cataysts [a]

Substituted olefin	*trans-* : *cis*-C_4 at 0% conversion		
	Molybdenum catalyst [b]		
	I	II	III
cis-2-Pentene	0.20	0.205	0.17
cis-2-Hexene	0.22	—	—
cis-4-Methyl-2-pentene	0.42	0.38	—
cis-4,4-Dimethyl-2-pentene	0.44	0.41	—
trans-2-Pentene	12.5	12.5	10±2
trans-2-Hexene	11.1	—	—
trans-2-Heptene	12.5	—	—
trans-2-Octene	10.9	—	—
trans-4,4-Dimethyl-2-pentene	8.3	—	—
trans-1-Phenyl-1-propene	8.0	—	—
trans-1-Phenyl-2-butene	8.2	7.9	—

[a] Data from ref.[15];

[b] The catalysts used were: I = $Mo(NO)_2Cl_2(PPh_3)_2$, II = = $Mo(NO)_2Br_2(PPh_3)_2$ and III = $Mo(NO)_2I_2(PPh_3)_2$.

It is interesting to note that *cis* and *trans* olefins exhibited, respectively, high *cis* or *trans* stereoselectivity and that the stereoselectivity was much more pronounced for *trans* olefins than for *cis* olefins. With tungsten catalysts, when the size of the R group increased the reaction stereoselectivity decreased; with a *cis* olefin, the *trans* : *cis* ratio for 2-butene increased and with a *trans* olefin, this *trans* : *cis* ratio decreased. Simultaneously, on increasing the size of the R group, the reactivity of the olefin decreased considerably. This effect was quite noticeable for R isopropyl or *tert*-butyl, indicating that the steric effect of R induces different stereoselectivities and reactivities.

Remarkably high stereoselectivities were obtained by Garnier and Krausz [24] in the metathesis of internal *cis* and *trans* RCH = CHMe olefins with the system $W(CO)_6 \cdot CCl_4 - h_\nu$ (Table 7.17). *cis*-Olefins exhibit pronounced retention of *cis* configuration whereas *trans*-olefins retain *trans* configuration. By increasing the size of the R group the stereoselectivity of *cis*-olefin metathesis decreases while the stereoselectivity of *trans*-olefins is almost unchanged. These results were interpreted by involving a new type of intermediate in the metathesis reaction.

TABLE 7.17. Effect of the substituents at the double bond on stereoselectivity in olefin metathesis with the catalytic system: $W(CO)_6 - CCl_4 - h\nu$ [a]

Olefin	*cis*- : *trans*-C_4	*trans*- : *cis*-C_4
Propene	2	
cis-2-Pentene	3.7	
cis-2-Hexene	3.3	
cis-4-Methyl-2-pentene	3.3	
cis-2-Heptene	3.0	
cis-5-Methyl-2-hexene	3.8	
trans-2-Pentene		20
trans-2-Hexene		20
trans-4-Methyl-2-pentene		20
trans-2-Heptene		20
trans-5-Methyl-2-hexene		20

[a] Data from ref. [24].

The metathesis of *cis,cis*-2,8-decadiene in the presence of molybdenum- or tungsten-based catalysts shows moderate stereospecificity at low conversions, giving predominantly *cis*-2-butene.[25] As the reaction progressed, *trans*-2-butene increased until a *trans* : *cis*-2-butene ratio of 2.9 was reached (close to the thermodynamic equilibrium). The *trans*- : *cis*- C_4 ratio plotted against time gave a linear dependence similar to that obtained by Basset and co-workers for the reaction of 2-pentene. Extrapolated to $t = 0$, a value of about 0.25 was found for the *trans* : *cis* C_4 ratio representing the reaction stereoselectivity at zero conversion. Other olefins, such as *cis*-2-decene, *trans*-2-decene, *trans,trans*-2,8-decadiene were metathesised under the same conditions, leading to extrapolated values for *trans*- : *cis*- C_4 presented in Table 7.18. These results indicate that *cis*-2-decene and *cis, cis*-2,8-decadiene exhibit roughly the same stereoselectivity in the early part of the reaction as *cis*-2-pentene, and analogously *trans*-2-decene and *trans,trans*-2,8-decadiene as *trans*-2-pentene. Therefore, the length of an alkyl group on one side of a methyl-substituted olefin does not markedly affect the stereospecificity.

The metathesis of *trans, trans*-2,8-decadiene proceeded at approximately half the rate of that of the *cis, cis*-isomer and displayed induction periods which were eliminated by addition of a trace of 1-hexene. These results confirm the finding that *trans*-olefins react more strongly than *cis*-olefins and that *trans*-olefins do not participate in the initiating steps of the metathesis.

The stereochemistry of the degenerate cometathesis of (Z)-1-decene-1-d_1 and 1-octene-1,1-d_2 in the presence of molybdenum-, tungsten- and

TABLE 7.18. Stereoselectivity in metathesis of 2,8-decadiene and 2-decene with molybdenum and tungsten catalysts[a]

Olefin	trans- :cis-C_4	
	Catalyst [b]	
	I	II
cis, cis-2,8-Decadiene	0.25	0.61
cis-2-Decene	0.21	0.62
trans, trans-2,8-Decadiene	2.0	1.9
trans-2-Decene	2.1	—

Data from ref. [25]; [b] Catalysts: I = Mo(NO)$_2$Cl$_2$(PPh$_3$)$_2$ + Me$_3$Al$_2$Cl$_3$, II = WCL$_6$ + Me$_4$Sn.

rhenium-based catalysts has been studied by Casey and Tuinstra [26] in order to obtain information concerning the nature of the chain-carrying metallacarbene intermediate (7.3)

$$C_6H_{13}CH{=}CD_2 + (Z){-}C_8H_{17}CH{=}CHD \longrightarrow (Z/E){-}C_6H_{13}CH{=}CHD + C_8H_{17}CH{=}CD_2 \qquad (7.3)$$

They found a preference for retention of the (Z)-configuration in the cross-product 1-octene-1-d$_1$ for all catalytic systems employed (Table 7.19).

TABLE 7.19. Stereochemistry of degenerate metathesis of 1-octene-1,1-d$_2$ and (Z)-1-decene-1-d$_1$ [a]

Catalytic system	% 1-octene-1-d$_1$	(Z) : (E)-1-octene-1-d$_1$
MoCl$_2$(NO)$_2$(PPh$_3$)$_2$ + Me$_3$Al$_2$Cl$_3$	6.3	2.3
PhWCl$_3$ + AlCl$_3$	9.5	1.7
WCl$_6$ + Me$_4$Sn	6.3	1.5
W(CO)$_5$(PPh$_3$) + EtAlCl$_2$+O$_2$	10.9	1.3
Re(CO)$_5$Cl + EtAlCl$_2$	5.1	2.4

[a] Data from ref. [26].

The stereochemical results permit identification of the nature of the transient intermediates (metallacyclobutane and chain-carrying metallacarbene) occurring in degenerate metathesis reaction.

The steric course of the metathesis of 4-methyl-2-pentene has been thoroughly examined by Calderon and co-workers, [27,28] providing new information on the reaction stereoselectivity. By reacting 4-methyl-cis-2-pentene in the presence of $WCl_6 \cdot Et_2O \cdot Me_4Sn$, 2,5-dimethyl-*trans*-3-hexene was formed with a stereospecificity $\geqslant 99\%$ (higher than the thermodynamically predicted value of $\sim 93\%$) (7.4)

(7.4)

The 2-butene which was formed in this reaction had an initial *cis* content of $\sim 63\%$, close to that usually observed in the metathesis of *cis*-2-pentene, but this gradually approached the equilibrium content of $\sim 28\%$. Under the same conditions 4-methyl-*trans*-2-pentene led to 2,5-dimethyl-*trans*-3-hexene of higher than 99% *trans* content throughout the course of the reaction (7.5)

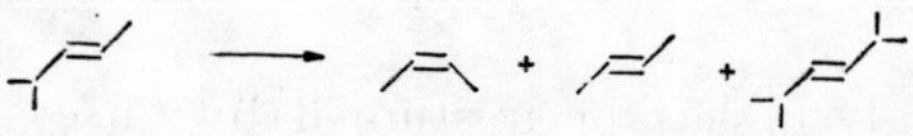

(7.5)

The 2-butene formed at early stages of the reaction was richer in *trans*-isomer than the equilibrium value, but it also tended to approach the thermodynamically predicted value.

The striking feature of the data obtained by Calderon and co-workers [28] in the reaction of 4-methyl-2-pentene is the highly specific formation of 2,5-dimethyl-*trans*-3-hexene, regardless of the steric structure of the starting olefin. It is obvious that the greater steric requirement of the isopropyl group (in comparison with an ethyl or methyl group) results in a marked change in the steric course of the reaction. These data, however, are somewhat different from those obtained by Basset [13] and Garnier [24] for the metathesis of 4-methyl-2-pentene or other highly crowded olefins with tungsten- and molybdenum-based catalysts.

The steric course of the metathesis of functionalized olefins is of great interest for the reaction mechanism and stereochemistry. The functional group is expected to interfere considerably with the central atom during the formation of transition state, resulting in a change of stereochemistry. Recently, Edwige *et al.* [29] undertook a comparative stereochemical study of the metathesis of the following unsaturated substituted amine and olefin (7.6)

$$trans\text{-}CH_3\text{—}CH = CH\text{—}(CH_2)_3\text{—}NH\text{—}CH_3 \text{ and}$$

(7.6)

$$trans\text{-}CH_3\text{—}CH\text{—} = CH\text{—}(CH_2)_3\text{—}CH_2\text{—}CH_3$$

in the presence of tungsten and molybdenum complexes associated with $EtAlCl_2$ and sometimes with molecular oxygen. The stereoselectivity for both reactions was determined by plotting the *trans* : *cis* ratios of the 2-butenes produced *versus* the overall conversion. The results indicated that the sterically similar long-chain olefins 2-hexene-6-(N-methyl)amine and 2-octene had the same steric behaviour ; the ratio *trans* : *cis*-2-butene equals 1.25 at zero conversion for both olefinic compounds. As a consequence, these authors suggested that the amine group was not coordinated to the tungsten catalyst during the formation of the transition state.

7.1.2. The Nature of Steric Interactions in Olefin Metathesis

The steric interactions in olefin metathesis will be considered mainly in the light of the Chauvin carbene mechanism [30] involving a metallacyclo-butane transition state and briefly of other proposed transition states, such as the diolefinic metallic complex, the non-classical four centre structure and the dimetallic intermediate. It is generally observed that the overal stereoselectivity of the reaction products depends significantly on the stereochemistry of the propagation step, although the initial stereo-selectivity at low conversions depends on the stereochemistry of the initiation step, and the final stereoselectivity may be drastically changed by the stereochemistry of the termination step.

Steric interactions are important in determining the stereochemistry of the reaction ; this was demonstrated in earlier reports on reaction stereo-chemistry for both heterogeneous [1-4] and homogeneous catalytic systems.[7-9] It is worth mentioning that Calderon and co-workers [7] first suggested steric effects in the transition state due to differences in steric require-ments of the methyl and ethyl group in the metathesis of 2-pentene. Thereafter, Hughes [9] interpreted the stereochemistry observed in the reaction of 2-pentene in terms of the pairwise mechanism by considering the most stable configurations of the metal-bisolefin complexes : the com-plexes containing two *cis*- and two *trans*-pentenes in the octahedral en-vironment of the transition metal are shown in Figs 7.3A and 7.3B, respectively.

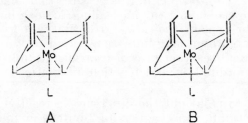

Fig. 7.3. — Coordination of olefins in octahe-dral molybdenum complexes in the *cis* mode.

In this manner *cis*-2-pentene would lead preferentially, *via* complex A to *cis*-2-butene and *cis*-3-hexene ; similarly, *trans*-2-pentene, *via* complex B, would transform to *trans*-2-butene and *trans*-3-hexene. The two proposed diolefin complexes were also able to rationalize the automerization and isomerization reactions of 2-pentene.

Most stereochemical data obtained in heterogeneuos metathesis reaction are difficult to interpret owing to competitive secondary reactions promoted by the heterogeneous catalyst. So far, only a few heterogeneuos reactions are stereochemically well-defined : e.g. metathesis over molybdenum oxide supported on titanium oxide [5,6] or over supported tungsten zero-valent complexes.[11] Heterogeneous metathesis reactions lead in most cases to a steric composition close to thermodynamic equilibrium. However, under kinetic control, *cis*-stereoisometric products result predominantly in reactions of lower α- and internal olefins over conventional molybdenum oxide on alumina.[1] In addition, significant dependence on the steric nature of the starting olefin was also observed : *cis* isomers favour *cis* product formation while *trans* isomers favour *trans* products. The latter stereoisomers also have a lower level of reactivity.

Stereoselectivity in heterogeneous metathesis reactions could be correlated with steric interactions assuming, as the rate determining step, either the complexation of the incoming olefin on the catalyst surface or metallacyclobutane formation. Owing to the multifunctional activity of such heterogeneous systems the stereochemistry of metathesis might easily be changed by other competitive physico-chemical phenomena such as physical adsorption, chemisorption, bond cleavage and bond rotation. These phenomena could significantly determine the reaction stereochemistry by the steric interactions involved. Furthermore, the initiation step (which corresponds to the formation of the first metallacarbene species) is probably very complex and may give rise to metallacarbene species different from those produced during propagation. This would result in different stereoselectivity during the initial stages of reaction.

If the initiating metallacarbene species is generated by direct interaction of the starting olefin with the heterogeneous metallic site, significant differences in steric interaction may arise in coordination and reaction for *cis*- and *trans*-olefins. Along with smaller steric compression for the *cis*-olefin *versus* the *trans*-olefin, it is possible that smaller steric effects are involved in the *cis*-olefin reaction leading to metallacyclobutane formation. This would explain the lower reactivity of *trans*-olefins in the initial stages indicated by a long induction period or "break-in". In addition, the lower overall reaction rate of *trans*-olefins indicated that similar steric effects for coordination and reaction may also be involved during propagation.

It is important to point out the difference in reactivity that may appear between the *cis* and *trans* isomers of 2-butene in metallacarbene generation *via* a metallacyclobutane intermediate. From *cis-trans* isomerization studies of d_0- and d_8-*cis*- or *trans*-2-butene, Tanaka and co-work-

ers[6] suggested that in the initial stages the formation of metallacyclobutane from *cis*-2-butene might be reversible, unlike that of *trans*-2-butene.

Scheme 7.1.

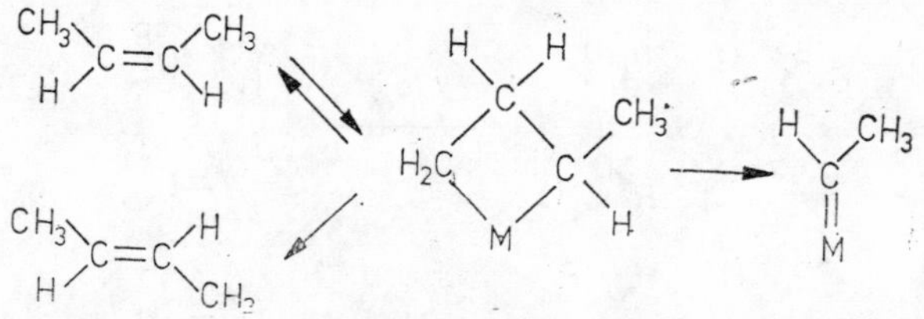

Assuming that metallacyclobutane formation is the rate-determining
step in propagation, the above mentioned authors correlated the conservation of the *cis* or *trans* configuration of the starting olefins (reacted over
molybdenum oxide supported on titanium oxide) with steric interactions
involved in the cyclic intermediates I and II.

Scheme 7.2

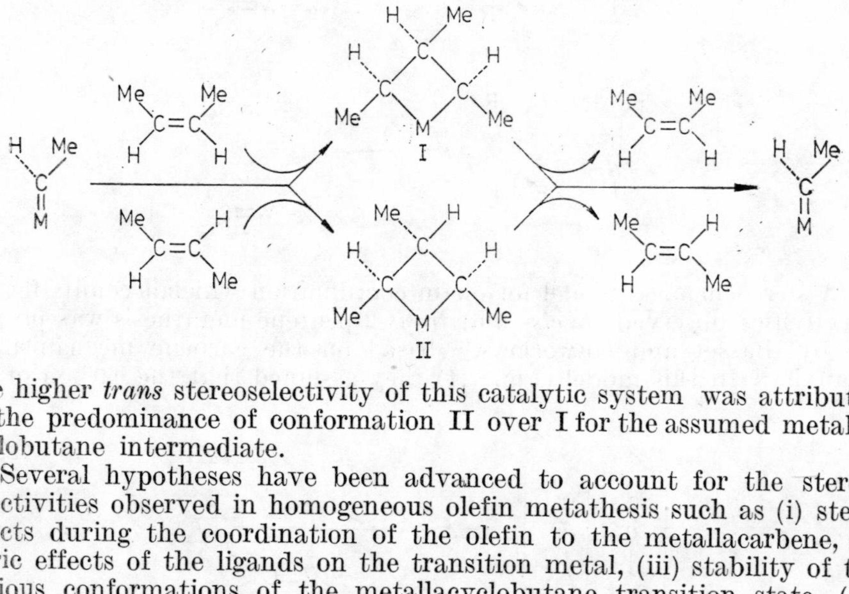

The higher *trans* stereoselectivity of this catalytic system was attributed
to the predominance of conformation II over I for the assumed metallacyclobutane intermediate.

Several hypotheses have been advanced to account for the stereoselectivities observed in homogeneous olefin metathesis such as (i) steric
effects during the coordination of the olefin to the metallacarbene, (ii)
steric effects of the ligands on the transition metal, (iii) stability of the
various conformations of the metallacyclobutane transition state, (iv)
scission of the carbon-metal bond in the metallacyclobutane intermediate
leading to a substituted metallopropyl cation in which the remaining
bonds rotate. Each of these hypotheses has been used to explain certain aspects of experimental data concerning stereoselectivity.

Assuming that complexation is the rate-determining step, two types
of steric interaction may occur during the coordination of the olefin to
the metallacarbene moiety (Scheme 7.3) : (i) interaction between the alkyl

group of the coordinated carbene and that of the approaching olefin (reactions 1 and 2) and (ii) interaction between the metal and/or its ligands X and the alkyl group of the coordinating olefin (reactions 3 and 4).

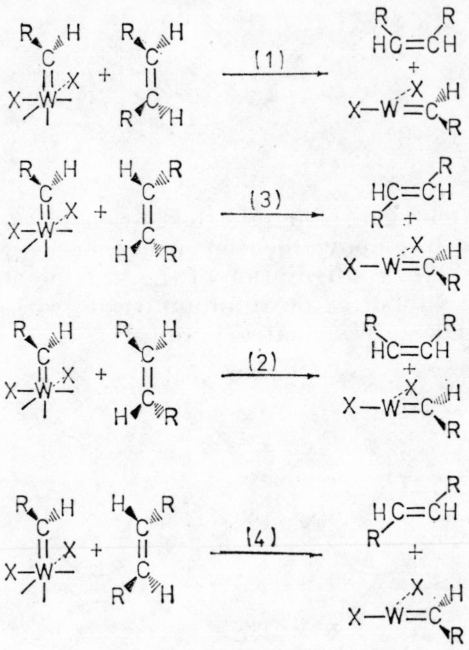

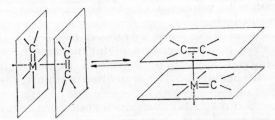

Scheme 7.3

A stereochemical model for olefin coordination which accounts for the selectivities observed in *cis-* and *trans-*2-pentene metathesis was proposed by Basset and co-workers,[13] based on the carbene mechanism of Chauvin.[30] In this model (Fig. 7.4) they assumed that the carbon of the

Fig. 7.4. — Steric model of olefin coordination to metallacarbene (according to Bilhou *et al.*).[13]

coordinated carbene had sp² character with a bond order greater than unity and probably equal to two ; the rotation barrier in the coordinated complex containing carbene and olefin is high, and the stereoche-

mistry is determined by the geometry of the approach of the olefin. This
model accounted for degenerate metathesis and *cis-trans* isomerization
in olefin metathesis. The reaction of *cis*-2-pentene in the presence of
tungsten catalysts is illustrated in Scheme 7.4.

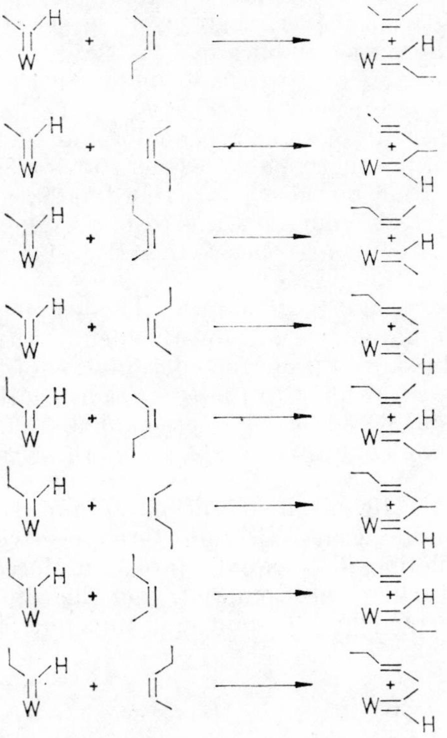

Scheme 7.4

As can be seen from the above scheme, depending on the type of coordination of *cis*-2-pentene to one or the other carbene moiety, one may
obtain *cis* or *trans* isomers of 2-butene and 3-hexene for the pairs of productive reactions and *cis* and *trans* isomers of the starting olefin, 2-pen-
tene for the pairs of regenerative processes. One may observe that in the
regenerative reactions, there are reactions with retention of the *cis* configuration (degenerate metathesis) and reactions with inversion of configuration (*cis-trans* isomerization).

Although it is not certain if complexation is indeed the rate-determining step, the kinetic data obtained by Basset and co-workers.[11-13] in the
metathesis of 2-pentene seem to confirm the validity of this stereochemical model. The stereoselectivities reported in *cis*-2-pentene metathesis
with a variety of zero-valent tungsten catalysts revealed that the ratios
trans- : *cis*-2-butene (*trans-* : *cis*-C$_4$) and *trans-* : *cis*-3-hexene (*trans-* :
cis-C$_6$) plotted *versus* the ratios *trans-* : *cis*-2-pentene (*trans-* : *cis*-C$_5$)

gave linear relationships from which *trans-/cis*-C_4 and *trans-/cis* C_6 values at zero conversion of 0.72—0.86 and 0.88—0.89, respectively, were found. The fact that with these catalysts *trans : cis* ratios close to unity for C_4 and C_6 were found indicated that steric factors were not very important in determining the *cis* or *trans* approach of the olefin with respect to the metallacarbene moiety; however, there exists a slight preference for the *cis* approach. Furthermore, values of 2.5—2.6 for *trans*-C_5 : *trans*-C_4 ratios instead of the value of 2 expected indicated that certain modes of olefin coordination were slightly preferred; nonetheless, upside down orientation with respect to metallacarbene does not result in a substantial change of the rates of productive and regenerative reactions.

The steric effects due to the ligands coordinated to the metal may by relevant for the steric course of the reaction. The ligand distribution around the metallacarbene may be asymmetric (Fig. 7.5) or symmetric (Fig. 7.6).

With asymmetric ligand distribution, the ligands on the transition metal repel the coordinated carbene and the olefin away from the ligands. This interaction will result in preferred coordination of the carbene moiety and reactant olefin with respect to the transition metal. Such stereoselective effects are considered to be predominant factors, when one increases the size of one ligand X, for example, in the case of bulky groups or a surface.

With a symmetric ligand distribution around the metallacarbene, the repulsion of olefinic groups is non-stereoselective and *cis* or *trans* coordination of the olefin will be equally probable. Moreover, the pronounced repulsion of the carbenic group may favour the rotation of the carbene around the metal-carbon double-bond resulting in a significant decrease of stereoselectivity.

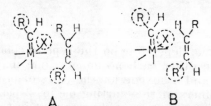

A   B

Fig. 7.5. — Steric interaction with asymmetric ligand arrangement around transition metal atoms for a *cis* olefin (left) and a *trans* olefin (right).

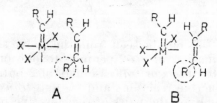

A   B

Fig. 7.6. — Steric interaction with symmetric ligand arrangement around transition metal atoms for a *cis* olefin (left) and a *trans* olefin (right).

Steric effects during the coordination seem to be important, along with inductive effects, in the degenerate metathesis of α-olefins. However, steric interaction of the environment may easily change the regioselectivity in α-olefin reaction. Detailed studies by Calderon,[31] Katz[32] and Casey[33] indicated that with α-olefins the "head to tail" type of coordination leading to regenerative metathesis is much preferred to the "head to head" type leading to productive metathesis.

The retention of steric configuration of the starting olefin observed with most tungsten, molybdenum and chromium homogeneous catalysts has been rationalized by Katz,[23] Casey [34,35] and Basset [15-17] in terms of the stability of the possible conformations of the metallacyclobutanes, if such metallacyclobutanes are real intermediates.

Katz and Hersh [23] accounted for the stereochemistry of 2-pentene metathesis by minimizing the steric interactions in the intermediate four-membered ring between juxtaposed axial groups and ring atoms, between 1,2-diequatorial groups and between 1,3-diaxial groups (Schemes 7.5 and 7.6)

Scheme 7.5

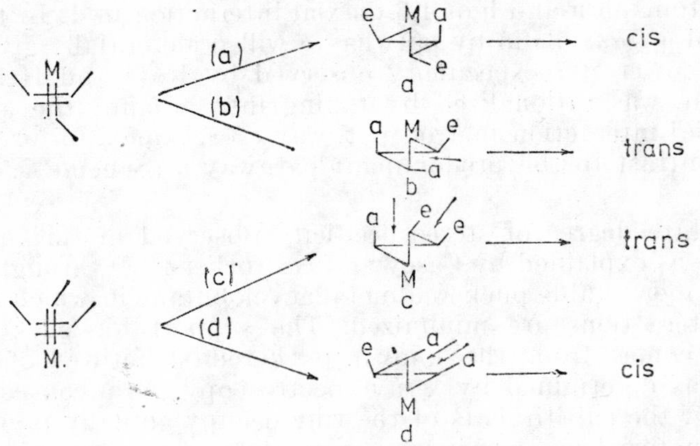

Scheme 7.6

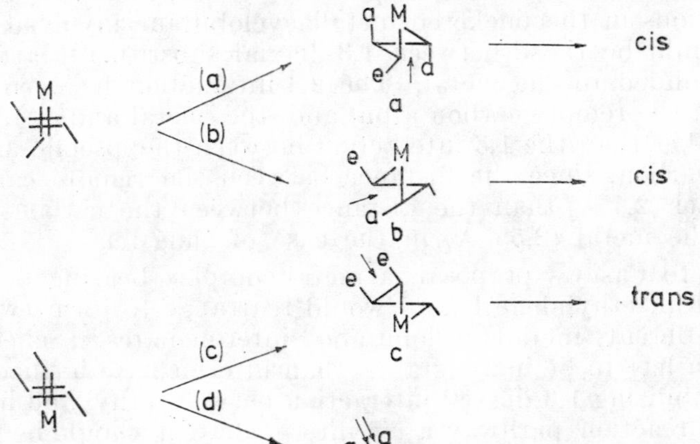

Thus, in *cis* olefin metathesis the cyclobutane conformation **a** (Scheme 7.5) has the lowest energy as compared to the conformations **b**, **c** and **d** where stronger steric interactions between juxtaposed axial groups and carbon ring atoms (**b** and **c**), 1,2-diequatorial groups (**c**), and 1,3-diaxial groups (**d**) occur. If we consider the geometry of metallacyclobutane to be puckered, the distance across the ring from carbon to metal should be much greater than from carbon to carbon ; the minimal steric interaction occur in the most stable metallacyclobutane conformation **a**. Accordingly, reaction of *cis* olefins by pathway **a** will be the fastest and the *cis* product will predominate.

Analogously, in the metathesis of *trans* olefins, the conformation **c** (Scheme 7.6) has the lowest energy. The 1,2-diequatorial interaction in **c** is weaker than the interactions between juxtaposed axial groups and carbon ring atoms in **a** and **b** or 1,3-diaxial interaction in **d**. In this case, the reaction of a *trans* olefin by patwhay **c** will preferentially give a *trans* product. The lower stereospecificity observed by Katz and Hersh [23] for *trans*-2-pentene was rationalized by taking into account the additional 1,2-diequatorial interaction met on pathway **c** (Scheme 7.6) for the *trans* reaction in contrast to the predominant pathway **a** (Scheme 7.5) for the *cis* reaction.

The modetate degree of stereospecificity observed in the metathesis of 2-pentene was explained by Casey and co-workers [34, 35] through the formation of the more stable puckered metallacyclobutane in which repulsive 1,3-diaxial interactions are minimized. The support for puckered ring configuration comes from the known puckered structures of platinacyclobutanes as determined by x-ray spectroscopy. As a consequence of ring puckering, the substituents of the ring occupy actually pseudo-axial and pseudo-equatorial positions in which the interactions can be enhanced or weakened. Casey and co-workers proposed that the most important steric interactions in the puckered metallacyclobutane involved in olefin metathesis would be those between 1,3-diaxial substituents attached to the carbons bonded to the metal. The 2,4-interaction between an axial substituent on the remote carbon atom and the metal and/or its ligands should be weaker than the 1,3-interaction between the pseudo-axial substituents on carbon, since the distance between the remote carbon and metal is greater (2.7 Å) than the distance between the carbons directly attached to the metal (2.55 Å) in the case of tungsten.

According to Casey's proposal, a metal complex bearing both a *cis*-2-pentene and an ethylidene ligand would rearrange to form two stereochemically different metallacyclobutane intermediates (Scheme 7.7). Intermediate **a** has to be more stable than intermediate **b** since the latter has a destabilizing 1,3-diaxial interaction between ethyl and hydrogen. Consequently, reaction pathway **a** *via* intermediate **a** whould be preferred and *cis*-2-butene would be the kinetically favoured product. Likewise, the greater stability of intermediate **c** relative to intermediate **d** would explain the preferred formation of *trans*-2-butene from *trans*-2-pentene.

Scheme 7.7

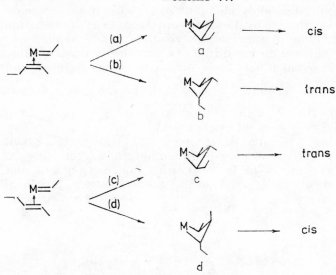

One way to test the stereochemical implications of a puckered metalla-cyclobutane intermediate in metathesis, as proposed by Casey and co-workers,[34] would be the study of the reactions of *cis* and *trans* disubstituted alkenes with a metallacarbene in which the carbene ligand posses-ses one sterically large (L) and one sterically small (S) substituent. By acting thus on the steric interactions occurring in the puckered metalla-cyclobutane formed as intermediate, the "large-small" metallacarbene would react with *cis* olefins to give new *cis* olefins, with *trans* olefins to give new *trans* olefins and with α-olefins to give preferentially *cis* substituted cyclopropanes (Scheme 7.8).

Scheme 7.8

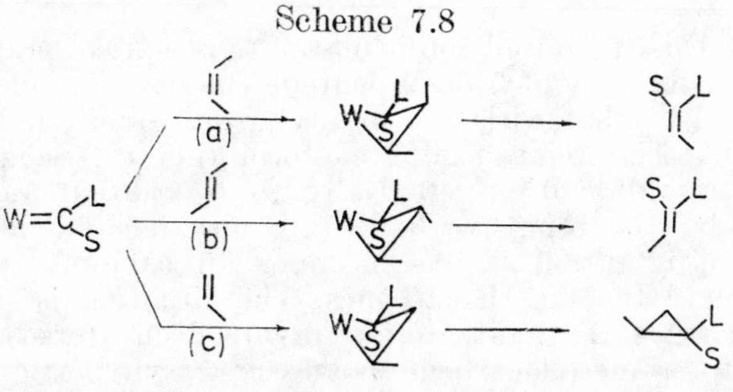

Although the retention of the steric configuration in the reaction of *cis* and *trans* olefins in the presence of substituted carbenes seems to confirm the above stereochemical hypothesis (pathways **a** and **b**), the formation of *cis* substituted cyclopropane *via* a puckered metallacyclo-butane is still unclear (pathway **c**).

The retention of configuration of *cis*- and *trans*-2-pentene largely encountered with tungsten, molybdenum and chromium homogeneous catalysts, has been attributed by Leconte and Basset [15-17] to the stability of possible conformations of the metallacyclobutane intermediates by analogy with the stability of disubstituted cyclobutane derivatives. Among the various interactions which may arise in disubstituted metallacyclobutane namely, $1-2, 1-3, 2-4$, and $3-4$ $(\overset{4}{M}-\overset{1}{C}-\overset{2}{C}-\overset{3}{C})$, the above authors have shown that only the $1-3$ and $3-4$ interactions are able to explain satisfactorily the retention of configuration.

By considering $1-3$ diaxial interaction in the metallacyclobutane transition state occurring in *cis*-2-pentene metathesis, the most favoured metallacyclobutane ought to be that with one methyl and one ethyl group in $1-3$ diequatorial relationship (conformation A Scheme 7.9).

Scheme 7.9

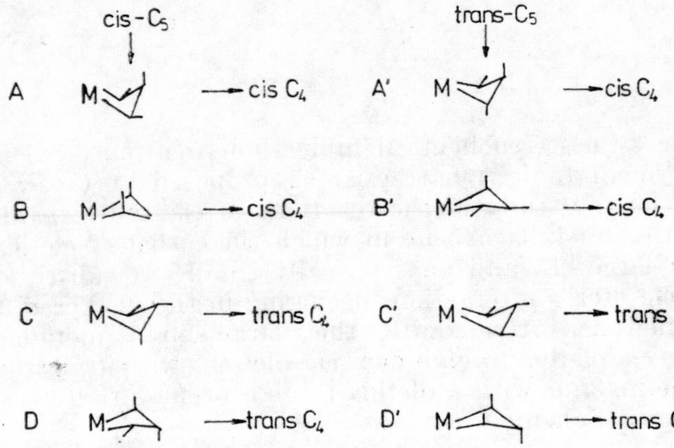

Consequently, this favoured conformation gives rise preferentially to *cis*-2-butene. In the case of *trans*-2-pentene the most favoured metallacyclobutane would be that with two methyl groups and one ethyl group in $1-2-3$ triequatorial relationship (conformation C′, Scheme 7.9). This conformation would lead preferentially to *trans*-2-butene.

Alternatively, the repulsive $3-4$ interaction could account for the retention of configuration if one assumes an asymmetric distribution of ligands around the metallacarbenes. The situation is similar to that considered for the steric interaction involved in stereoselective olefin coordination to a metallacarbene possessing asymmetric ligand distribution discussed earlier (see Fig. 7.4). In the present case the bulky ligands on the transition metal would interact more strongly with the substituents of the olefins as well as with the substituent on the coordinated carbene. As will be seen later on, the possible $3-4$ interaction has not been verified in the reaction of substituted olefins bearing bulky groups.

The more pronounced retention of configuration for 3-hexene as compared to 2-butene, obtained in *cis*- and *trans*-2-pentene metathesis with molybdenum and tungsten catalysts has been attributed by Leconte and Basset [15,17] to repulsive 1—2 interactions in the transient metallacyclobutane. As an illustration, the transition states leading to 2-butene (C_4) and 3-hexene (C_6) in the metathesis of *cis*-2-pentene are represented in Scheme 7.10.

Scheme 7.10

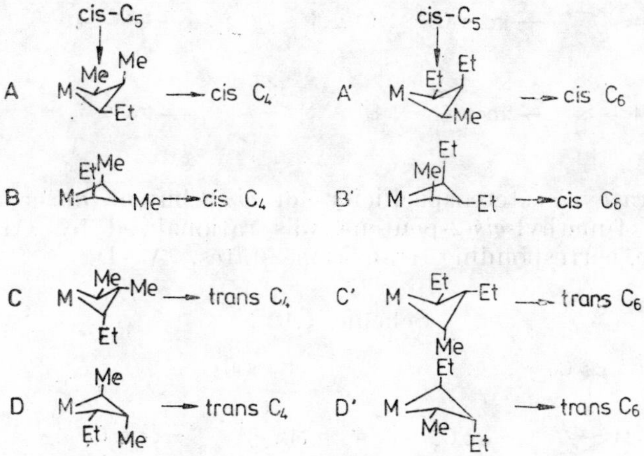

A   M—Me(Me)—Et cis-C_5 → cis C_4 A' M—Et(Et)—Me cis-C_5 → cis C_6

B   M—Et(Me)—Me → cis C_4 B' M—Me(Et)—Et → cis C_6

C   M—Me(Me)—Et → trans C_4 C' M—Et(Et)—Me → trans C_6

D   M—Me(Et Me) → trans C_4 D' M—Me(Et)(Me Et) → trans C_6

As can easily be seen, each transition state leading to *cis* and *trans* isomers of C_4 and C_6 has the same substituent in the 1 and 3 positions. Consequently, the difference of stereoselectivity between C_4 and C_6 can be attributed to the presence of a larger substituent in position 2. This substituent may increase the repulsive 1—2 and 2—4 interactions in metallacyclobutane. However, it was inferred from the steric behaviour of α-olefins (see below) that the repulsive 1—2 interactions is more stereoselective than the 2—4 interaction.

The predominance of 1—2 interaction between substituents on adjacent carbons of the metallacyclobutane was demonstrated by Calderon and co-workers [27,28] in the metathesis of 4-methyl-2-pentene. Regardless of the *cis* or *trans* nature of the initial olefin, Calderon obtained *trans*-2,5-dimethyl-3-hexene with stereospecificity higher than 99 % throughout the reaction. On considering the various transition states leading to *cis*- or *trans*-2,3-dimethyl-3-hexene starting from *cis*- or *trans*-4-methyl-2-pentene (Scheme 7.11) one observes that the 1—2 interaction prevails over the 1—3 interaction resulting predominantly in the *trans* product *via* C and C' metallacyclobutanes.

Scheme 7.11

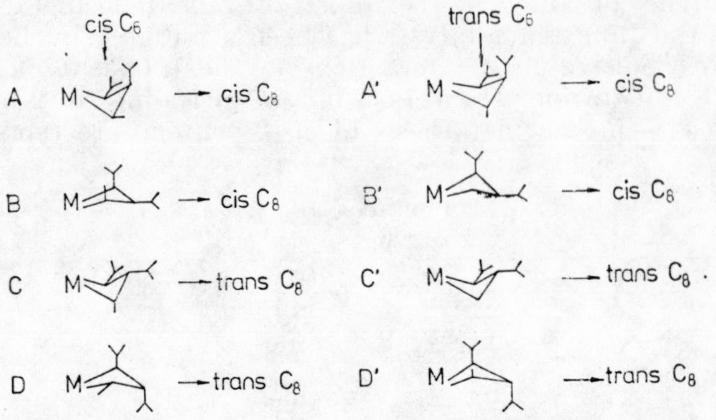

The moderate *cis* stereospecificity for *cis*-2-butene observed in the metathesis of 4-methyl-*cis*-2-pentene was rationalized by Calderon by examining the corresponding transition states, A—D.

Scheme 7.12

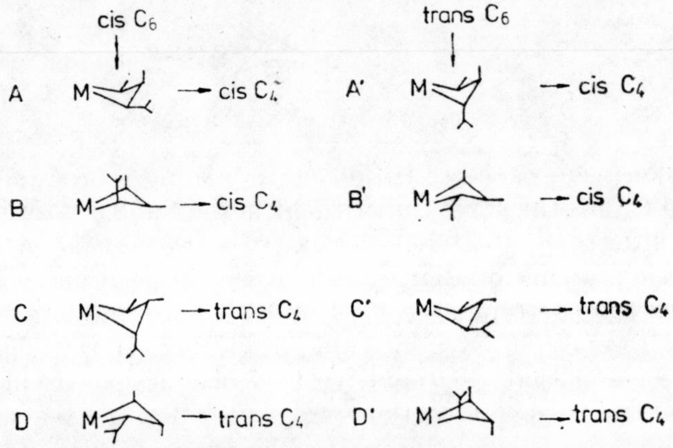

In this case the tendency to minimize the interaction in the 1—3 positions between the methyl and isopropyl groups seems to be offset by weak repulsions caused by 1,2-dimethyl substitution slightly favouring transition state A over transition state D. By contrast, in the case of 4-methyl-*trans*-2-pentene, transition state C′ is evidently preferred over the other transition states since all substituents occupy equatorial positions resulting in high *trans* stereospecificity for *trans*-2-butene.

In addition to significant 1—2 repulsive interactions, Calderon and co-workers highlighted the interaction occurring between ligands about the transition metal and the bulky group of the olefin in the regenerative metathesis of 4-methyl-2-pentene. If one considers the regenerative reaction of 4-methyl-*cis*-2-pentene, assuming that isopropyl groups occupy equatorial positions, transition states A and B (Fig. 7.7) would be preferred.

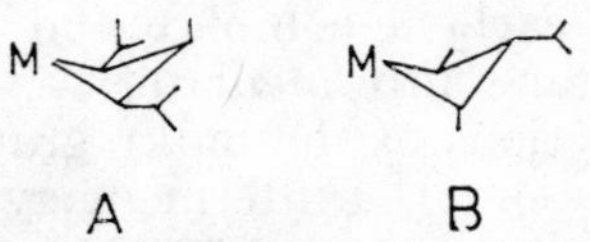

Fig. 7.7. — Preferred transition states for regenerative metathesis of 4-methyl-*cis*-2-pentene.

Transition state A leads to retention of steric configuration (automerization) whereas transition state B leads to isomerization of the starting olefin. Experimental results obtained in the metathesis of 4-methyl-*cis*-2-pentene indicated that isomerization prevails over automerization as a consequence of the ligand field about the transition metal, which directs the bulky group in state B away from the metal.

The repulsive 1—2 interactions are relevant to the stereochemistry of α-olefin metathesis. The increase of the *trans* : *cis* ratios of the products in the order 2-butene < 3-hexene < 4-octene when going from propene to 1-pentene as starting materials was rationalized only by enhanced repulsive 1—2 interaction in the metallacyclobutane intermediate.

Scheme 7.13

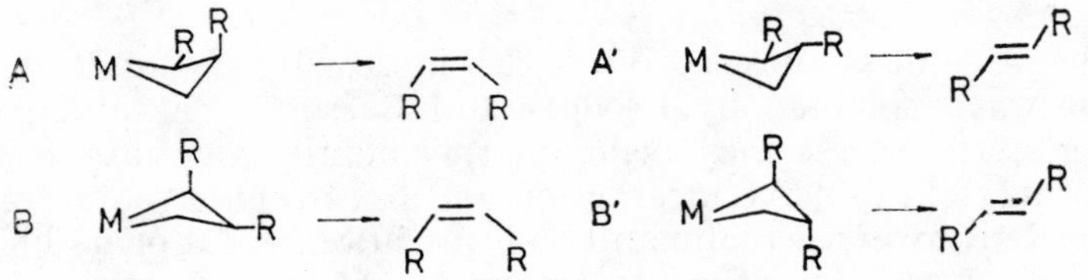

As may be inferred from the four possible transition states involved in the metathesis of α-olefins, *viz.*, A, B, A′ and B′, the 1—3 diaxial interaction may be negligible (the substituent in position 3 is H) while the 1—2 interaction may be significant. Upon increasing the size of group R, transition states A′ and B′, in which the substituents have diequatorial or diaxial positions, will be progressively favoured relative to transition states A and B, which have axial-equatorial substituents. This may explain the appearance of higher amounts of *trans* isomers in the products with increasing R size on going from propene to 1-pentene. Propene, however, is a limiting case (*trans*- : *cis*-C$_4$ ratio equal to 1) since the 1—2 effect is very small for two methyl groups in positions 1 and 2 of the corresponding metallacyclobutane.

As mentioned earlier (see Table 7.18), by using (Z)-1-decene-1-d$_1$ in cometathesis with 1-octene-1,1-d$_2$, Casey and Tuinstra [26] found a preference for the retention of configuration in degenerate metathesis promoted by catalytic systems based on molybdenum, tungsten and rhenium. The retention of configuration was explained by considering M=CHR as being the chain-carrying carbene complex and by assuming a favoured puckered metallacyclobutane intermediate, A, in which the 1—3 positions minimize the interactions of the bulky groups (Scheme 7.14). These stereochemical results rule out the intermediacy of a metallacyclobutane B which would not be able to retain stereochemical information.

Scheme 7.14

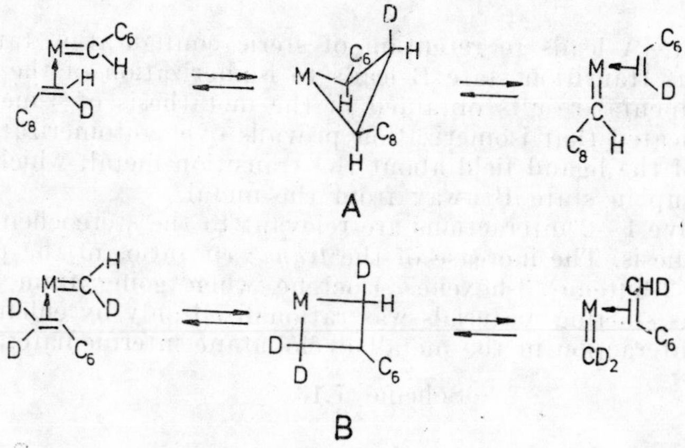

The decrease of stereoselectivity with increasing size of R above a certain value was explained by Leconte and Basset [14,15] assuming that two competing steric effects may occur in the metallacyclobutane intermediate : 1—3 diaxial repulsive interaction and 3—4 repulsive interaction between the R group of the olefin and the transition metal or its ligands. The 1—3 diaxial interaction leads to retention of configuration and prevails for rather small substituents. The 3—4 repulsive interaction is "non-stereoselective" with a symmetric ligand arrangement around the metallacarbene and starts to compete over a certain size during both coordination and reaction. In this situation, it will be equally probable whether a cis or trans olefin coordinates and reacts in a cis or trans position with respect to metallacarbene and the stereoselectivity will decrease. In contrast, the 3—4 repulsive interaction with asymmetric ligand arrangement around the transition metal would result in an increase in stereoselectivity by increasing the bulkiness of R. Thus, the decrease of stereoselectivity seems to rule out the asymmetric model which earlier was able to explain the retention of configuration in cis- and trans-2-pentene metathesis.

In addition to the geometric ligand arrangement (symmetric and asymmetric) around the metallacarbene, the nature of the ligand attached to the transition metal may be relevant to the stereochemistry of the reaction. The ligands on the transition metal may be classified into two main types : non-stereoselective and stereoselective ligands.

The first type of ligand includes those whose size and position induce small changes in reaction stereochemistry, e.g. the halogens, by weak 3—4 interactions during coordination and reaction. The low stereoselectivity changes observed for Cl, Br and I may be assigned to small variations of covalent radii between these ligands on metal or to the absence of the living catalyst on the coordination sphere, owing to the double alkylation of the metal by the cocatalyst.

The stereoselective ligands, e.g. nitrosyl groups, may induce substantial variations in reaction stereochemistry (Table 7.9) owing to their particular geometry (bulky linear or bent structure[36]) or to their active presence on the living catalyst during the metathesis process.[18] These ligands may direct the bulky groups of the olefin away from the metal centre minimizing both 3—4 and 2—4 repulsive interactions during olefin coordination or metallacyclobutane formation.

The decrease of metathesis stereoselectivity in the series $Cr > Mo \gg W$ has been attributed to changes in geometry which may be induced by the transition metal in the metallacyclobutane intermediate. First, on going from Cr to W, in the transition state $[M-C_1-C_2-C_3]$ the $M-C_1$

and $M-C_3$ bond lengths would increase owing to the increase of the corresponding covalent radius of the metal. This would increase, therefore, the C_1-C_3 distance and decrease the 1—3 repulsive interaction in the metallacycle. The lowest 1—3 repulsive interaction in tungstacyclobutane would lead to the lowest metathesis stereoselectivity characteristic for tungsten. Second, the polarity or puckering of the metallacyclobutane intermediate may be very different for these transition metal series. For tungsten planar or slightly puckered geometries of the tungstacyclobutane intermediate would induce negligible 1—2 and 1—3 repulsive interactions which would result in low stereoselectivity. By contrast, the strongly puckered metallacyclobutane intermediates, which would be involved with molybdenum and chromium, may significantly increase the 1—2 and 1—3 repulsive interactions resulting in high stereoselectivity. So far, there is no X-ray structure available for metallacyclobutanes of group VI transition metals but some comparative data for platinacyclobutane derivatives reveal the change induced on metallacycle planarity. Thus, for platinacyclobutane structures A, B and C (Fig. 7.8) a vary-

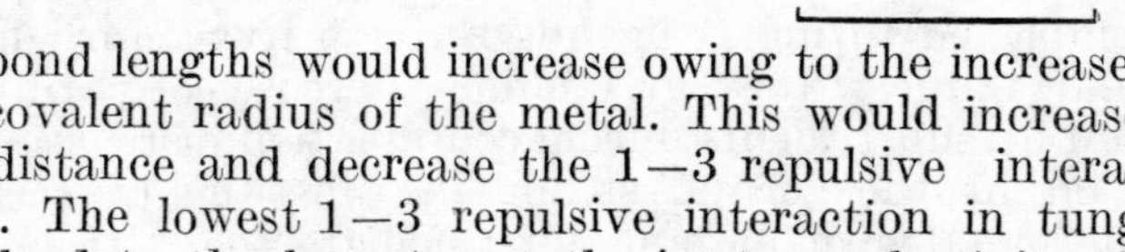

Fig. 7.8.— Platinacyclobutane derivatives of known geometries.

ing degree of puckering is encountered. The puckering angle between the $C_1C_2C_3$ plane and the MC_1C_3 plane [37] (α, Fig. 7.9) varies from 12° for the unsubstituted metallacycle structure A, to 26° and 28°, respectively, for the substituted structures B and C. For the Cr, Mo, and W series of metallacycles, the puckering angle is expected to vary over a wide range, from highly puckered to planar structures, as a function of the nature of the metal and the geometry of its ligands as well as of the bulk of substituents on the carbene and olefinic carbon atoms.

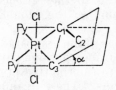

Fig. 7.9.— X-Ray structure for $Py_2Cl_2PtC_3H_6$ (according to Gillard *et al.*). [37]

The cocatalyst may significantly influence the stereochemistry of the metathesis by complexing with the metallacarbene ligand or with ligands on the transition metal. When the cocatalyst attaches to the carbene ligand the rotation barrier about the $M-C_{carbene}$ bond may be diminished resulting in a change of metallacarbene geometry. In addition, the substituents on the cocatalyst may interact with the olefin during coordination thus favouring a non-stereoselective reaction. Studies carried out by Mol *et al.*[22] showed that the alkyl groups R on the cocatalyst had no pronounced effect on *cis*-2-pentene metathesis with the system $WCl_6 \cdot SnR_4$. By contrast, when the cocatalysts are complexed to metal ligands through oxygen or chlorine bridges in bimolecular adducts, they may induce stereoselective interaction with the olefin and its substituents during coordination or reaction. In this case the interaction of the cocatalyst and its substituents with the substituents of the metallacyclobutane intermediate may strongly change the puckered geometry of the metallacycle ring. The pronounced stereoselective interaction reported for phenyl groups of the system $WCl_6 \cdot SnPh_4$ in 2-pentene metathesis as favouring *cis* products, may be attributed to both steric and inductive effects exerted by the complexed cocatalyst on the metallacyclobutane transition state.

In summary, assuming that a metallacyclobutane is indeed an intermediate in olefin metathesis, the reaction stereochemistry would be determined by competition among the various repulsive steric interactions in the corresponding metallacyclobutane intermediate such as 1—2 (*e-a*, *a-e* or *e-e*), 1—3 (*a-a*), 2—4 and 3—4 interactions. It seems that there is often a retention of the steric configuration of the starting olefin : with *cis* olefins the 1—3 (*a-a*) interaction in the metallacyclobutane intermediate would be more severe than the 1—2 (*e-a*, *a-e*) interaction, thus favouring the formation of *cis* products, while with *trans* olefins the two simultaneous, 1—2 (*e-a* or *a-e*) and 1—3 (*a-a*) interactions would compete favouring the formation of *trans* products. For bulky olefins, 1—2 (*e-a* or *a-e*), 2—4 or 3—4 interactions may be comparable with the 1—3 (*a-a*) interaction or even dominate, resulting in a drastic change of stereoselectivity. In α-olefins bearing bulky substituents, 1—2 interactions are predominant.

Recently, Basset and co-workers [19,38] related the stereoselectivity observed in the metathesis of acyclic and cyclic olefins with the respective energy levels of the coordinated olefin and of the metallacyclobutane. They showed that when the coordinated olefin has an energy level which is smaller than that of the two possible metallacyclobutanes leading to *cis* or *trans* products, the reaction stereoselectivity will be governed by the energy levels of these two metallacycles. Alternatively, when the coordinated olefin has an energy level which is higher than that of the *cis* or *trans* directing metallacyclobutanes, the reaction stereoselectivity will not be governed by the energy levels of these two metallcycles and the reaction will lose its stereoselectivity, so that the *trans* : *cis* ratio will tend towards unity. Most acyclic olefins belong to the former category while highly strained olefins such as norbornene, are in the latter one.

The slight changes of reaction stereoselectivity induced by the ligands of the precursor complexes were related to electronic effects on the energy levels of the metallacycle intermediate. Likewise, the Lewis acids as co-catalysts may decrease the energy levels of the metallacyclobutane intermediate by their ability to complex with the metal and its ligands resulting in a decrease of stereoselectivity. In addition, the increase of metathesis stereoselectivity with group VI transition metals, on going from W to Cr, may be assigned to the corresponding increase in metallacyclobutane energy levels, in agreement with the *ab initio* calculations of Rappé and Goddard.[39,40]

Based on a theoretical analysis of the approach of ethylene for coordination to metallacarbene, Eisenstein and Hoffmann [41] have shown that there should be strong electron-count and transition-series dependences of the relative orientation of the carbene and ethylene. Among the possible modes of olefin coordination to the metallacarbene moiety (Fig. 7.10), only the "collinear" geometry A is favourable for metathesis reaction.

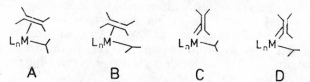

Fig. 7.10. — Conformations for the approach of an olefin
to a metallacarbene.

Attainment of this "collinear" geometry seems to be the essential step. Stereoselectivity would then be determined by the steric interactions occurring in attaining this favoured "collinear" geometry and that of the non-classical minima which may contain highly polarized olefin structures (Fig. 7.11). Although from calculations puckering in the metallacycle appears to be unimportant, other deformations which would play a part in reaction stereochemistry, for instance a change in bite angle φ (Fig. 7.12), may tend to stabilize the d^0-d^4 metallacyclobutanes.

The level of lower stereoselectivity in the metathesis of internal olefins obtained with tungsten catalysts or in the presence of excess metal

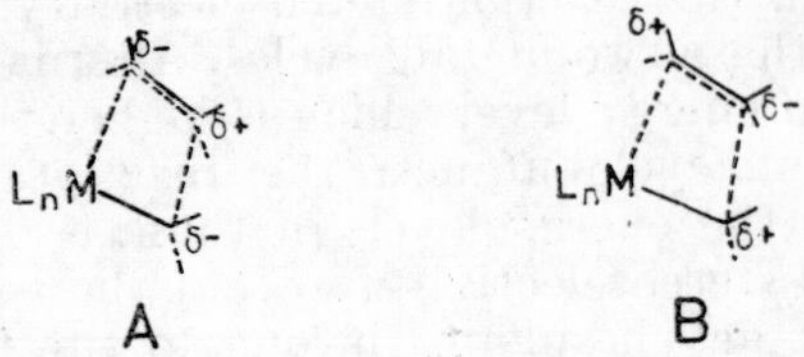

A   **B**

Fig. 7.11. — Possible non-classical structures in olefin metathesis.

halides (e.g. Lewis acids) was explained by Katz and Hersh [23] *via* cleavage of the carbon-metal bond in metallacyclobutane intermediate and formation of a metallopropyl cation in which the remaining bonds rotate. They related metallopropyl cation formation to the sensitivity of the system towards the effects of substituents that determine the selectivity. Support for this proposal comes from the stereoselectivity observed in the metathesis of cyclopentene with tungsten catalysts in the presence of varying amounts of Lewis acids [42] and at various temperatures.[43,44]

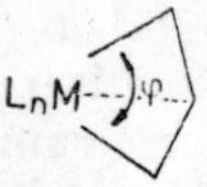

Fig. 7.12. — Possible deformation in a metallacyclobutane.

In a stereochemical study of internal 2-alkene metathesis Garnier and Krausz [24] argued that the metallacyclobutane hypothesis for the transition state could not explain several experimental data concerning reaction stereoselectivity. A first observation is that the relative stability of the metallacyclobutane intermediate in the metathesis of propene should lead to a *trans*- : *cis*- 2-butene ratio higher than 1, whereas experimental results give ratios lower than 1. Secondly, for the reaction of *cis* olefins through a metallacyclobutane intermediate, increasing the bulk of the substituent would result in a higher *cis* stereoselectivity, while in fact a decrease in stereoselectivity is obtained. And thirdly, for *trans* olefins a higher content of *trans* product ought to be obtained along with a substantial amount of *cis* product but stereoselectivity for the *trans* product is often higher than 95 %.

Consequently, they proposed a new type of dimetallic intermediate which involves larger steric strains in the transition state than those expected with a metallacyclobutane. This intermediate would arise *via* interaction between a carbene coordinated to one metallic centre and an olefin coordinated to another. For *cis* olefins, the preferred intermediate would be of the form **a** or **b** (Scheme 7.15) bearing the R groups outside the cycle. By minimizing the 1—2 repulsive interactions between R′ groups, the intermediate **a** would be favoured relative to **b**, resulting subsequently in a higher proportion of *cis* product. Increasing the

Scheme 7.15

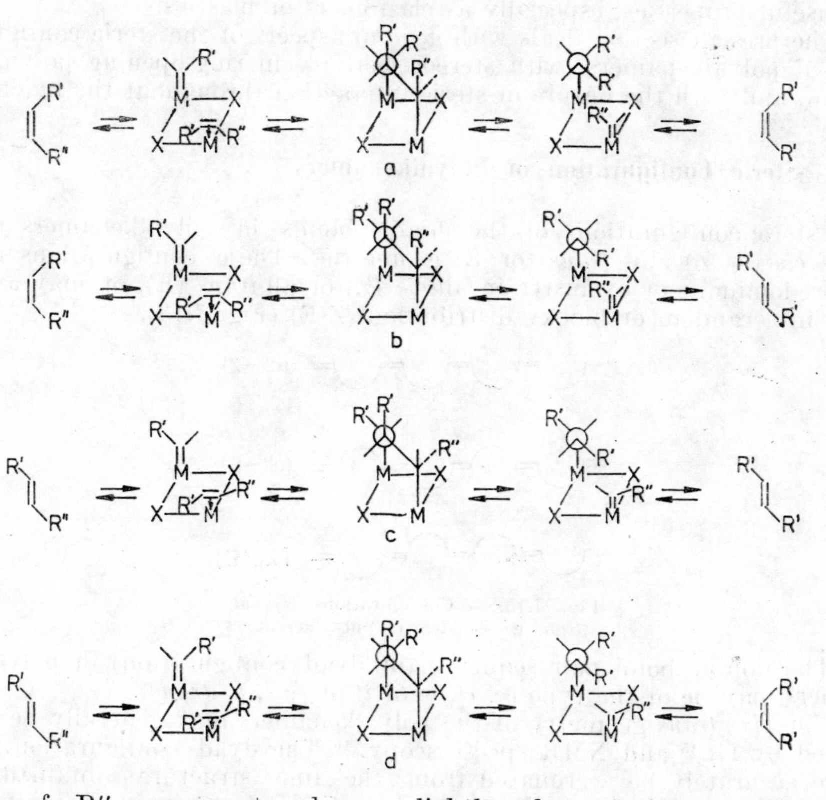

bulk of R'' appears to change slightly the relative stabilities of intermediates **a** and **b** and, consequently, the *trans* : *cis* ratio of the 2-butene product. For a *trans* olefin, the preferred intermediate would be **c** and **d** with the R' group outside the cycle. By minimizing the 1—2 and 1—3 repulsive interactions between the R groups, the intermediate **c** should be much more stable than **d**, leading thus to a high *trans* content. In the particular case of propene (R'=CH$_3$, R''=H), the intermediates **b** and **d** would be more stable and, consequently, would lead to a higher *cis*-2-butene ratio.

7.2. Ring-Opening Polymerization of Cyclo-Olefins

As with the metathesis of acyclic olefins, the ring-opening polymerization of cyclo-olefins very often exhibits high stereoselectivity, depending on the catalyst system. The reaction thereby affords a means for the synthesis of polyalkenamers with high *cis* and *trans* content or of highly stereoregular polyalkenamers. Some of these polymers have outstanding useful properties, especially as elastomers or plastics.

The present section deals with general aspects of the steric configuration of polyalkenamers, with stereoselectivity in ring-opening polymerization, and with the nature of steric interaction throughout the reaction.

7.2.1. Steric Configuration of Polyalkenamers

The steric configurations of the double bonds in polyalkenamers may have *cis* (or Z) and *trans* (or E) geometries. These configurations may be predominant or exclusive in all-*cis* (Z) or all-*trans* (E), or may alternate in a random or blocky distribution (Z/E) (Fig. 7.13).

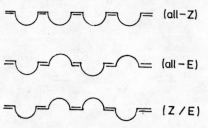

Fig. 7.13. — Configurations (*cis* and *trans* geometries) in polyalkenamers

The double bond pair sequence (or dyad configuration) in polyalkenamers may be of the type *cc*, *ct*, *tc* or *tt* (*c=cis*, *t=trans*).

The *cis* : *trans* geometry of the polyalkenamers may generally be evaluated by IR [45] and NMR spectroscopy. [46] The dyad configuration may more accurately be estimated from the fine structure obtained by ^{13}C-NMR spectroscopy. [47] This dyad configuration provides significant information about the polymerization mechanism.[48]

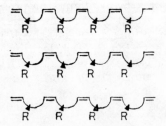

Fig. 7.14. — Isotactic, syndiotactic and atactic structures in (*all*-Z)-polyalkenamers.

If the polyalkenamers have chiral centres, the monomer units may have an isotactic, syndiotactic or atactic relationship (Fig. 7.14). In fully isotactic and syndiotactic structures the dyad relationship can be of the

HT or TH type (H = head, T = tail), while in atactic structures the dyad sequence will be of the HH, TH, HT, and TT type (Fig. 7.15). The tacticity of chiral polyalkenamers may also be determined by ^{13}C NMR spectroscopy.[49]

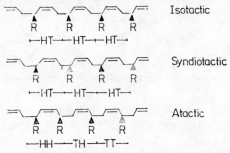

Fig. 7.15. — Dyad sequences in polyalkenamers.

When the substituent on the cyclo-olefin is situated at an olefinic carbon, the all-*cis* polyalkenamers formed may have a translationally invariant structure.[50] In this polymer, the unit sequence will be of the HT or TH type, analogous to the isotactic or syndiotactic structures (Fig. 7.16). In polyalkenamers derived from bicyclic or polycyclic olefins, ad-

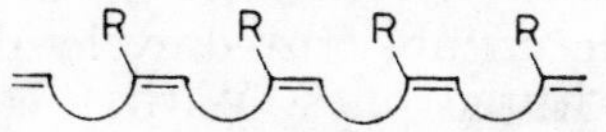

Fig. 7.16. — Translationally invariant structures for substituted polyalkenamers.

jacent rings may have an isotactic (*m*), syndiotactic (*r*) or atactic (*m/r*) relationship. For polynorbornenamer these structures are represented in Fig. 7.17. Polyalkenamers of non-symmetrically substituted bicyclo-

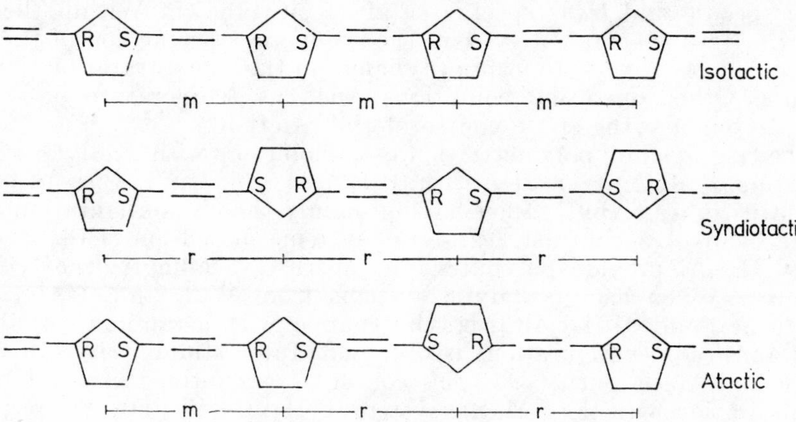

Fig. 7.17. — Tacticity of ring sequence in polynorbornene.

olefins, such as 5-alkylnorbornene, may have the monomer dyad in an HH, TT or TH sequence (Fig. 7.18).

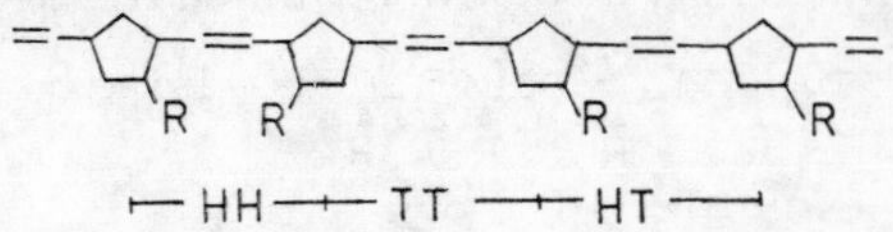

Fig. 7.18. — Dyad sequence in poly (5-alkyl-norbornene) ignoring steric configurations.

Thus, regioselectivity problems arise in addition to stereochemical problems connected with the orientation *(cis/trans) (trans/cis)* of the double bonds.

7.2.2. Stereoselectivity in the Ring-Opening Polymerization of Cyclo-Olefins

There is now growing evidence that the ring-opening polymerization of cyclo-olefins provides an interesting approach for the synthesis of polyalkenamers with high *trans-* or *cis*-content. The reaction has been effected with nearly all available unsubstituted cyclo-olefins, except cyclohexene, and with various substituted ones. With the exception of polyhexenamers, polyalkenamers of predominantly *trans* structure have thus been prepared from cyclobutene to cyclododecene, and polyalkenamers of predominantly *cis* structure from cyclobutene to cyclo-octene. In addition, polyalkenamers of high steric purity have been obtained from norbornene and its derivatives and other bicyclic and polycyclic olefins.

The nature of the catalyst is crucial in directing the polymerization stereoselectivity towards *cis* or *trans* polyalkenamers; definite classes of *cis* specific and *trans* specific catalysts are known. Within the same binary or ternary catalyst system, the *cis* or *trans* proportion of the polymer can be varied by appropriate change in the molar ratio of the components. Other operating conditions, such as temperature or solvent, may also change the steric course of the reaction.

The ring-opening polymerization of cyclobutene with catalytic systems consisting of $MoCl_5$ associated with $(C_2H_5)_3Al$ or of $RuCl_3$ without cocatalyst, leads to a polybutenamer having mainly the *cis* configuration at the double bond. By contrast, catalytic systems based on $TiCl_3$ or $TiCl_4$ and $(C_2H_5)_3Al$ provide polybutenamer with prevailingly *trans* configuration.[51,52] When using catalytic systems from WCl_6 and $(C_2H_5)_3Al$, or $MoO_2(acac)_2$ and $(C_2H_5)_2AlCl$, polybutenamer with a random distribution of *cis* and *trans* configurations is obtained. Interestingly, *cis*-polybutenamer of high steric purity (93 % *cis* content), was formed *via* cyclobutene polymerization with the carbenic system $(C_6H_5)_2C{=}W(CO)_5$ without using cocatalyst.[53]

High stereoselectivity has been reported for the polymerization of 1-methylcyclobutene with some catalysts. For instance, the carbenic system $(C_6H_5)C=W(CO)_5$ gave a polymer having essentially a polyisoprene structure of 84—87% Z and 13—16% E configuration whereas $WCl_4 \cdot nBuLi$ afforded a polyisoprene of about 83% Z configuration having additional 2-butene and 2,3-dimethyl-2-butene units as a result of methyl group migration in the polymer chain.

Stereospecific polymerization of cyclopentene to *trans*- or *cis*-polypentenamer may readily be achieved with many of the common binary and ternary catalysts. Thus, a *trans*-polypentenamer with up to 95% *trans* configuration is obtained with a wide range of binary and ternary catalytic systems derived from WCl_6, $TiCl_4$, BCl_3 and alkylaluminium halides with or without the addition of a ternary component, such as alcohol, peroxide, hydroperoxide, chlorinated compound, nitro compound, etc.[42] A *cis*-polypentenamer with a *cis* structure over 90% is commonly synthesized with catalytic systems consisting of WF_6, $MoCl_5$ or $ReCl_5$ and alkylaluminium compounds [28,46] or with the carbenic system $(C_6H_5)_2C=W(CO)_5$ mentioned earlier.[53] Remarkably, polypentenamers of more than 99% *cis* configuration have been prepared by using the catalyst combination $MoCl_5 \cdot (C_2H_5)_3Al$ and $WCl_6 \cdot (CH_2=CHCH_2)_4Si$.[44]

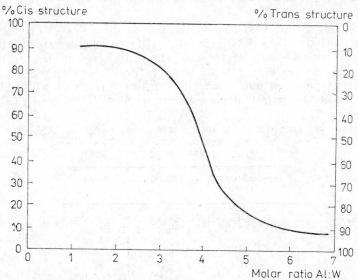

Fig. 7.19.—Effect of molar ratio Al:W on the configuration of polypentenamer.

The steric configuration of the polypentenamer is readily modified by changing the composition of the catalyst or the reaction temperature. Thus, for instance, in the reaction of cyclopentene with the catalytic system WCl_6 and $(C_2H_5)_3Al_2Cl_3$ a range of polypentenamers having from 85% *cis* to 95% *trans* configuration were obtained by varying the molar ratio Al:W between 1 and 7 (Fig. 7.19).

However, with the catalyst derived from $MoCl_5$ and $(C_2H_5)_3Al$ mentioned earlier, the variation of molar ratio Al : Mo between 0.25 and 10 led to only small changes of the steric configuration in the polypentenamer [55] (Table 7.20).

TABLE 7.20. Effect of Al:Mo molar ratio on the *cis*-content and yield of polypentenamer [a]

Al : Mo, molar ratio	*cis*-Content %	Yield %
0.25	90.1	1.0
0.5	98.0	10.7
1.0	99.4	41.4
1.5	99.1	49.2
2.0	99.4	45.3
2.5	99.3	47.2
3.5	99.4	7.4
5.0	98.9	3.0
10.0	99.0	2.3

[a] Data from ref. [55].

It was found that variation of temperature below $-10°C$ slightly modifies the content of *cis* configuration in the polypentenamer [55] but above this temperature a sharp decrease in the initial *cis* content occurs. With $MoCl_5 \cdot (C_2H_5)_3Al$ as catalyst for temperatures between $-80°$ and $-10°$, a *cis* content between 99.4% and 97.7% was observed while above this temperature a sharp fall in *cis* content to 85.4% is found (Table 7.21).

TABLE 7.21. Effect of temperature on the *cis*-content and yield of polypentenamer [a]

Temperature, °C	*cis*-Content, %	Yield, %
−80	99.4	3.3
−55	98.1	14.6
−40	99.3	40.8
−30	98.8	27.0
−10	97.7	13.3
+10	91.9	2.0
+30	85.4	0.3

[a] Data from ref. [55].

Similarly, with $WCl_6 \cdot (CH_2{=}CHCH_2)_4$ Si as catalyst for temperatures between $-80°$ and $-30°$, the *cis* content of polypentenemer is close to 100% while at 30° the content of *cis* double bonds is no more than 30%.

A polypentenamer having prevailingly *trans* configuration at the double bond has been obtained in the presence of most binary and ternary catalytic systems based on WCl_6 and $MoCl_5$ associated with organometallic compounds.[54] It is surprising that even the catalytic system consisting of $MoCl_5$ and $(C_2H_5)_3Al$ (which polymerizes cyclopentene to *cis*-polypentenamer) leads to *trans*-polyheptenamer. By contrast, a *cis*-polyheptenamer was obtained with the carbenic complex $(C_6H_5)_2C=W(CO)_5$, which is known to be a highly *cis* stereospecific system.[53] The proportion of *cis* configuration in this polymer was very high, about 98%.

The polymerization of cyclo-octene induced by catalytic systems consisting of WCl_6 in combination with $(C_2H_5)_3Al$, $C_2H_5AlCl_2$ or other organoaluminium compound provides mainly *trans*-polyoctenamer.[54] By modifying the reaction time in the cyclo-octene polymerization with WCl_6 and $(C_4H_9)_3Al$, Calderon and Morris[56] were able to vary the content of *trans*-configuration in the product. A polyoctenamer with *cis* configuration up to 93% was obtained using the binary catalytic system $WF_6 \cdot C_2H_5AlCl_2$ while high *cis* polyoctenamer ($\sim 97\%$ *cis*) was produced by the *cis* specific carbenic system $(C_6H_5)_2C=W(CO)_5$.[53] It is interesting to note that in these polymerizations only *cis* cyclo-octene is reactive while *trans* cyclo-octene remains inert. As will be seen later, the reverse was found for 1-methylcyclo-octene.

A *trans*-polydecenamer is readily formed through ring-opening polymerization of *cis*-cyclodecene with various binary and ternary systems based on tungsten such as $WCl_6 \cdot C_2H_5AlCl_2$, $WCl_6 \cdot C_2H_5OH \cdot C_2H_5AlCl_2$, $WOCl_4 \cdot Cum_2O_2 \cdot (C_2H_5)_2AlCl$.[57] When the reaction of a *cis*-/*trans*-cyclodecene mixture is carried out with $WCl_6 \cdot (C_2H_5)_2AlCl$, a polydecenamer with predominantly *trans* configuration results, but the residual monomer contains less *cis* isomer than the initial ratio. This result indicated that of the two cyclodecene stereoisomers, the *cis*-cyclodecene manifests a higher reactivity than *trans*-cyclodecene.

Among the reactions of the monomers bearing several double bonds, the polymerizations of *cis, cis*-1,5-cyclo-octadiene, *cis, trans, trans*- and *trans, trans, trans*-1,5,9-cyclododecatriene are briefly mentioned. These monomers yield, in the presence of the catalytic systems $WCl_6 \cdot C_2H_5AlCl_2$, $WCl_6 \cdot C_2H_5OH \cdot C_2H_5AlCl_2$ and $WCl_6 \cdot (C_2H_5)_2AlCl$ polybutenamers in which the *trans* structure prevails.[58]

High stereospecificity is recorded in ring-opening polymerization of bicyclic and polycyclic olefins using various catalytic systems.[47,53,59-62] Thus norbornene yields *trans*-polynorbornenamer[59,62] in the presence of $MoCl_5$ or $RuCl_3$, *cis*-polynorbornenamer[59,60] with $ReCl_5$ or $(C_6H_5)_2C=W(CO)_5$ while mixtures of *cis* and *trans* configurations are obtained[61,62] when binary and ternary systems of WCl_6 or other transition metal (Ir, Rh, Os) catalysts are used. In certain products a non-random distribution of *cis* and *trans* double bonds in the chain has been observed. Polynorbornenamer with a moderately high *cis* double bond content manifests generally a blocky distribution of *cis* and *trans* double bonds. As the *cis* content decreased from 100% to 35% a trend from fully syndiotactic to atactic ring dyads has been observed.[62] A comparison of the stereoselectivity in norbornene ring-opening polymerization with the stereoselectivity in metathesis of an acyclic olefin, *cis*-2-pentene, in the presence of arene or nitrosyl complexes of Group VI transition metals, revealed that the stereochemistry of norbornene reaction does not depend on the transition metal and its ligands, whereas that of *cis*-2-pentene does depend on these parameters.[19] As will be seen later, this result was attributed to the cyclic character of norbornene.[38]

exo-Trimethylnorbornene and the Diels-Alder dimer of cyclopentadiene behave similarly to norbornene in ring-opening polymerization.[59] In the presence of the above *cis* and *trans* directing catalysts, *cis* or *trans* polyalkenamers of these tricyclic olefins may be obtained. It is important to mention at this point that the reaction stereoselectivity with bicyclic and tricyclic systems is, in several cases, the reverse of that encountered with monocyclic olefins.

With few exceptions, the polymerization of substituted cyclo-olefins is less stereospecific than that of unsubstituted ones.[63,64] Thus, *cis*-2-methylcyclo-octene, *trans*-3-methylcyclododecene or mixtures of *cis*- : *trans*-3-methylcyclododecene, in the presence of the catalytic system WCl_6 · $C_2H_5AlCl_2$ · peroxide, lead mainly to atactic poly-3-methylalkenamers, with no stereospecificity with respect to double bond configuration or the arrangement of tertiary carbon atoms.[65] Of particular significance is the observation that, in the reaction of 1,2-dimethyl-1,5-cyclo-octadiene, only the double bond between unsubstituted carbons underwent metathesis and the polymer had an equal amount of *cis* and *trans* configurations at the unsubstituted double bonds.[65]

High stereoselectivity was often reported with particular substituted cyclo-olefins. When the substituent is located at the double bond translationally invariant polymers are readily formed in the presence of the carbenic system $(C_6H_5)_2C=W(CO)_5$, for example high stereoselectivity in 1-methylcyclobutene polymerization, leading to essentially a polyisoprene of Z stereostructure, was mentioned earlier.[50] Likewise, 1-trimethylsilyl-cyclobutene gave the most perfectly translationally invariant polymer made by metathesis till now : it has, within the limits of spectroscopic measurements, practically 100% *cis* structure at the double bonds.[66] 1-Methylcyclopentene failed to polymerize with the above carbenic system or with other tungsten catalysts [67] while 3-methylcyclopentene, like cyclopentene, with catalysts based on WCl_6, gave polyalkenamers of over 90% *trans* configuration.[42] The polymerization of 1-methyl-*trans*-cyclo-octene induced by the carbenic system $(C_6H_5)_2C=W(CO)_5$ gave a perfectly alternating polyalkenamer with predominantly E stereostructure, while 1-methyl-*cis*-cyclo-octene failed to polymerize with the same catalytic system.[68] In another example, 2-methylnorbornene gave a translationally invariant substituted polyalkenamer having prevailingly E stereostructure.[66]

It is of interest for the mechanism of ring-opening polymerization that highly tactic polyalkenamers have been prepared from 5-substituted norbornene such as *endo-* and *exo-*5-methylnorbornene [48,49] [69-72] or 5,5-dimethylnorbornene [73,74] in the presence of a wide range of transition metal catalysts. The content as well as the blocky or random distribution of *cis* and *trans* double bond configurations in these polyalkenamers changed significantly with the nature of the transition metal, the presence or absence of electron acceptors or donors, and the reaction temperature. A direct comparison of the fraction of *cis* double bonds obtained for polymers of norbornene and dimethylnorbornene with a range of catalysts is represented in Table 7.22.

TABLE 7.22. Content of *cis* double bonds in polymers of norbornene and 5,5-dimethylnorbornene with various catalysts[a]

Catalytic system	Fraction of *cis* double bonds	
	Poly(norbornene)	Poly(5,5-dimethyl-norbornene)
$RuCl_3$	0.05	0.05
$RuCl_3$ — COD [b]	0.2	0.35—0.65
$IrCl_3$	0.45	0.30
$OsCl_3$	0.50	0.41
$WCl_6 \cdot EtAlCl_2$	0.55	0.55
$WCl_6 \cdot Bu_4Sn$	0.70	0.60
$WCl_6 \cdot Me_4Sn$	0.55	0.69
$WCl_6 \cdot Ph_4Sn$	0.55	0.69
$Ph(CH_3O)C=W(CO)_5$	0.75	0.70
$Ph_2C=W(CO)_5$	1.0	0.92
$ReCl_5$	1.0	1.0

[a] Data from ref. [74];   [b] COD = 1,5-cyclo-octadiene.

As can be easily observed from the above data, substituents at position 5 of norbornene may increase (with $WCl_6 \cdot Me_4Sn$, $WCl_6 \cdot Ph_4Sn$ and $RuCl_3 \cdot COD$) or decrease (with $IrCl_3$, $OsCl_3$, $WCl_6 \cdot Bu_4Sn$, $Ph(MeO)C=W(CO)_5$ and $Ph_2C=W(CO)_5$) the *cis* content, or may leave it unaffected (with $RuCl_3$, $WCl_6 \cdot EtAlCl_2$, $ReCl_5$). It is noteworthy that with $ReCl_5$ catalyst all-*cis* polymers have been obtained from norbornene derivatives characterized by high syndiotacticity of structural units while $RuCl_3$-based catalysts provided polymers having predominantly *trans* double bonds but with a varying degree of isotacticity.

Generally, copolymerizations of cyclo-olefins are distinct and less stereospecific than homopolymerizations of individual monomers. In copolymerizations between cyclopentene-cycloheptene, cyclopentene-cyclooctene, cyclo-octene-cyclododecene in the presence of $WOCl_6 \cdot$ benzyl peroxide $\cdot Et_2AlCl$, random distribution of the monomer units and prevailingly *trans* configurations at the double bonds have been obtained

(Table 7.23). The *trans* contents of the above copolymers are generally less high than those of homopolymers (which are usually over 80% *trans* tactic) prepared with the same catalysts.

TABLE 7.23. Configuration of *cis* and *trans* double bonds in copolymers : cyclopentene — cycloheptene, cyclopentene —
cyclo-octene and cyclo-octene — cyclododecene [a]

Monomer pair	Monomer composition (molar ratio)	Copolymer composition (molar ratio)	*trans* : *cis* Ratio of the double bonds
Cyclopentene-cycloheptene	50 :50	72 :28	75 :25
	25 :75	44 :56	80 :20
	10 :90	18 :82	80 :20
Cyclopentene-cyclo-octene	50 :50	97 :3	70 :30
	25 :75	53 :47	65 :35
	10 :90	13 :87	75 :25
Cyclo-octene-cyclododecene	80 :20	85 :15	80 :20
	50 :50	69 :31	82 :20
	5 :95	14 :86	70 :30

[a] Data from ref. [75].

When using various transition metal salts as catalysts in the copoly-merization of cyclobutene and 3-methylcyclobutene, a correlation of *cis* and *trans* stereospecificity with the nature of transition metal was observ-ed.[76] The alkenamer units in the copolymer were prevailingly of the *trans* type with the metals titanium and ruthenium and of the *cis* type with tungsten and molybdenum (Table 7.24).

TABLE 7.24. Microstructure of copolymers of 3-methylcyclobutene-cyclobutene using various
ring-opening catalysts [a]

Catalyst (molar ratio)	Cyclobutene : 3-methylcy-clobutene, molar ratio	Cyclobutene : 3-methylcy-clobutene units in copolymer, molar ratio	*trans* Alkenamer units wt. %	*cis* Alkenamer units wt. %	Saturated units wt. %
WCl$_6$AlEt$_3$ (1 :3)	30 :70	50 :50	25	60	15
MoO$_2$(acac)$_2$·AlEt$_2$Cl (1 :5)	35 :65	40 :60	25	70	5
TiCl$_4$·AlEt$_3$ (1 :2)	26 :74	25 :60	55	30	15
RuCl$_3$·xH$_2$O·EtOH	35 :65	45 :55	~80	traces	20

[a] Data from ref. [76].

A varying degree of stereoselectivity has been reported for cyclopentene-norbornene copolymerization with a wide range of catalyst systems.[48,77] The fractions of *cis* double bonds for norbornene-norbornene dyads of the copolymer are illustrated in Table 7.25.

TABLE 7.25. Content of *cis* double bonds for compositional dyads in cyclopentene (M_1)—norbornene(M_2) copolymers [a]

Catalytic system	Monomer feed composition (mol fraction of M_1)	Polymer composition (mol fraction of M_1)	Fraction of *cis* double bonds for M_2M_2 dyads
$RuCl_3 \cdot COD$ [b]	0.68	0.04	0.11
$IrCl_3 \cdot COD$ [b]	0.68	0.23	0.33
$WCl_6 \cdot Ph_4Sn$	0.66	0.46	0.53
$WCl_6 \cdot nBu_4Sn$	0.61	0.13	0.60
$WCl_6 \cdot (allyl)_4Sn$	0.73	0.04	0.68
$WCl_6 Ph_4Sn \cdot EA$ [c]	0.66	0.51	0.71
$ReCl_5$	0.67	0.04	0.86

[a] Data from ref. [48]; [b] COD = 1,5-cyclo-octadiene; [c] EA = ethyl acrylate.

A slight trend towards lowering the *cis* content as compared to homopolymer was observed.[48]

7.2.3. The Nature of Steric Interactions

The stereoselectivities encountered in the ring-opening polymerization of cyclo-olefins indicate that significant steric interactions occur throughout the reaction course. Although several hypotheses have been advanced, they do not fully explain all the features of reaction stereoselectivity.

Assuming a metallacarbene mechanism for ring-opening polymerization, the following main factors may be relevant to the steric course of reaction : (i) mode of cyclo-olefin approach to the coordination site, (ii) nature of coordination of cyclo-olefin, (iii) geometric configuration of the intermediate metallacarbene, (iv) steric environment around the transition metal, (v) rotation about metal-carbene bond, (vi) mode of addition or insertion of coordinated cyclo-olefin to metallacarbene, (vii) steric interactions arising in transition state of reaction intermediate, (viii) mode of decoordination of the newly formed double bond of the polymer chain, (ix) mode of rupture of the supposed cyclic intermediate and (x) nature of the transition metal.

The modes of cyclo-olefin approach to the coordination site, e.g. parallel (A), orthogonal (B) or intermediate (C) (Fig. 7.20) are significantly determined by two main factors : the geometry of the cyclo-olefin and the steric environment around the transition metal. The *cis* or *trans* geometry of the cyclo-olefin determines the orientation of cyclo-olefin approach for larger rings, for substituted cyclo-olefins and for bicyclic or polycyclic olefins. For these particular cyclo-olefins strong interactions with ligands around the metal arise, which may govern the parallel or orthogonal mode of approach. It is probable that, initially, the orthogonal mode prevails owing to minimized steric interactions with ligands and polymer chain during coordination but subsequent rotation around the coordinative bond provides the parallel orientation required for positive orbital overlap to occur.

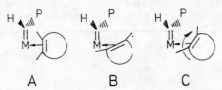

Fig. 7.20. — Modes of approach of a cyclo-olefin to a metallacarbene.

The possible modes of approach of cyclo-olefin to metallacarbene to form the intermediate metallacycle have been fully illustrated by Ivin and co-workers [47,48] for the ring-opening polymerization of norbornene and its derivatives. In the polymerization of norbornene, monomeric molecules may approach so as to bring the polymeric unit either to the left or to the right-hand side.

Scheme 7.16

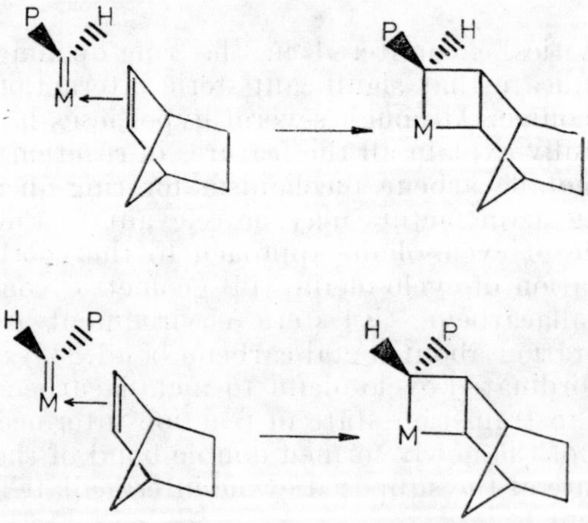

There are four possibilities, two of which are shown in Scheme 7.16, where both P and the bridging CH_2 groups may be on the same or opposite sides, leading to cis or trans double bond formation in the polymer chain, respectively.

Rotation of cyclo-olefin around the carbon-carbon double bond during the coordination process is another requisite for positive overlap interaction (Fig. 7.21).

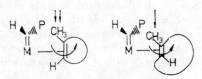

Fig. 7.21. — Steric interactions in rotating coordinated cyclo-olefins.

The restriction imposed by steric interaction of cyclo-olefin with metal ligands on the barrier to rotation around the carbon-carbon double bond may render substituted cyclo-olefins, such as 1-methyl-*cis*-cyclo-octene, unpolymerizable. For this cyclo-olefin, the rotation around the carbon-carbon double bond, necessary for the positive orbital overlap, seems to be restricted on grounds of strong steric interactions with ligands. On the contrary, for the isomeric *trans* cyclo-olefin, the distorted geometry obviates the necessity of further rotation around the carbon-carbon double bond for effective overlap.

The nature of coordinative bonds depends greatly on both the nature of the transition metal and its oxidation state. The more electronegative transition metals will coordinate more strongly to the cyclo-olefin and will restrict rotation around the coordinative bond, while the more electropositive ones will weaken the coordination bond and thus will allow rotation of cyclo-olefin. A similar phenomenon occurs when considering variable oxidation states where different electron densities exist at the metal site. For higher oxidation states, lower electron density will favour stronger coordination and restricted rotation of cyclo-olefins, while higher electron density will loosen the coordination and favour rotation of the coordinated olefin. This feature may account for the varying degree of stereoselectivity in polymerizations using various catalysts such as $MeCl_5$, $ReCl_5$, WCl_6, or $RuCl_3$, $IrCl_3$, $OsCl_3$, where the transition metals differ substantially in electron density by adopting variable oxidation states.[62,74]

The geometry of the coordinated double bond will directly affect the strength of the coordinative bond. Generally, *cis* double bonds in the incoming olefin or polymer terminal coordinate more strongly than *trans* double bonds owing to weaker steric interactions with the ligands of the transition metal. This fact will change the equilibrium in favour of coordination of the *cis* monomer followed by rapid reaction with the metallacarbene and polymer formation. When the newly formed double bond is *cis*, this bond will be preferred in coordination, giving high *cis* stereospecificity. If the newly formed double bond is *trans*, this bond will decoordinate more easily, generating a *trans* polymer.

Two proposals for the interpretation of *cis* and *trans* stereospecificity of polyalkenamers have been advanced. First, Calderon and co-workers[27,28] proposed a chelating concept to explain the high stereospecificity in cyclopentene polymerization. According to this concept a "three ligand" sequence involving simultaneous coordination of carbene, cyclo-olefin and polymer chain will lead in the presence of *cis* specific catalysts, to highly *cis* polymer, whereas a "two-ligand" sequence, involving simultaneous coordination of carbene and cyclo-olefin in the presence of *trans* specific catalysts, will lead to *trans* polymer.

Scheme 7.17

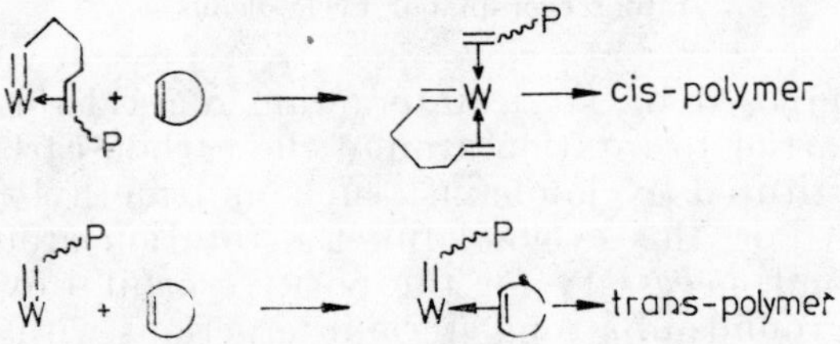

The ligand chelating concept has firmly been supported by results obtained by Calderon and co-workers in regulating the molecular weight of polypentenamer with acyclic olefins.[28]

The second concept was proposed by Ivin and co-workers[62] and involves steric crowding at the metal site. Originally, Ivin and co-workers postulated that :(i) a general order of complexing strength exists at the metal site : norbornene > cyclopentene > 1-alkene > 2-alkene ; (ii) steric crowding at the metal centre by ligands other than the olefin favours that orientation of complexed olefin which leads to *cis* double bond formation ; and (iii) steric crowding at the metal centre widens the scale of relative complexing abilities of olefins. These postulates adequately explain the stereoselectivities obtained in the ring-opening polymerization of norbornene, with various catalysts based mainly on WCl_6 and $MoCl_5$, with $EtAlCl_2$, BuLi, Ph_4Sn and $(allyl)_4Sn$ as cocatalysts, and particularly the effect of additives, monomer concentration, and temperature on polymer geometry. Essentially, the steric crowding at the metal centre exerts a direct effect on monomer and on the additive coordination resulting in preferential formation of the metallacycle in the *cis* or *trans* orientation. More recently, Ivin and co-workers[74] postulated that when the coordinated double bond of the polymer terminal is *cis*, the mode of entry of the monomer is such as to lead to the formation of another *cis* double bond. This fourth postulate explains the high *cis*-stereospecificity of ring-opening polymerization. It is worth mentioning that the above two concepts have close similarities to the well-known "back-biting" concept advanced earlier by Furukawa [78] to explain *cis*- and *trans*-stereospecificity in diene polymerization.

At present, it is considered that the configuration around the metallacarbene centre for some active catalyst species such as tungsten is octahedral, having one vacant position *cis* to the carbene ligand available for olefin coordination [48] (Fig. 7.22 A). This type of configuration is possible in higher oxidation states of transition metals and imposes a particular steric course on the reaction. However, other configurations, such as trigonal bipyramidal (Fig. 7.22 B) or square pyramidal (Fig. 7.22 C),

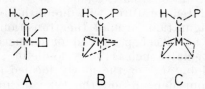

A   B   C

Fig. 7.22. — Possible configurations of metallacarbene species.

in which the cyclo-olefin may attach to the metallacarbene directly without requiring a vacancy, are also possible. These configurations may appear in lower oxidation states or in more electropositive transition metals and particularly in homogeneous systems.

The octahedral metallacarbene may coordinate and add the cyclo-olefin in one of the following four conformations, depending on the relative orientation of the polymer chain and of the hydrogen attached to the carbenic carbon (Fig. 7.23).

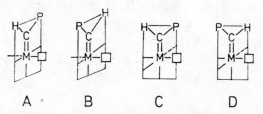

A   B   C   D

Fig. 7.23. — Octahedral metallacarbene species.

The first two non-symmetrical conformations (A and B) are mirror images to each other when the four ligands are identical and are considered to be of a chiral type. They have the same energy and reactivity, and are kinetically indistinguishable. The alternate formation of these species leads mainly to the production of *cis* polymers. The last two symmetrical conformations (C and D) of achiral type are kinetically distinct species. Species C may add monomer at a different rate to yield mainly *trans* polymer while species D seems unlikely to appear.

Interconversion between the above octahedral conformations either by rotation about the metal-carbene bond or by rapid migration from one ligand position to the other is possible. The substantial rotational barrier of the metal-carbene bond reported in the literature for tantalum carbenic species [79] suggests restriction of interconversion between successive additions of monomer, resulting in high stereospecificity. The rate of interconversion by rotation will certainly depend on the strength of the coordinative donor-acceptor metal-carbene bond and consequently on the electro-negativity and oxidation state of the transition metal. Factors lowering the rotational barrier (such as reduced state, presence of ligands, and elevated temperature) will diminish stereoselectivity.

The original configuration around the transition metal atom in the metallacarbene may be either preserved or changed in subsequent reaction steps or intermediates such as the cyclo-olefin—metallacarbene complex, the metallacyclobutane, or the newly formed coordinated metallacarbene. The behaviour depends on the type of addition of monomer, rupture of metallacycle and decoordination of the newly formed double bond in the polymer chain.

It is generally accepted that the addition of monomer is of the *cis* type to fulfill the maximum overlap requirements. So far there is no definite correlation between the steric course of reaction and orbital interactions in terms of the carbenic mechanism. Obviously, in addition to direct interaction between the appropriate relevant orbitals of metallacarbene and cyclo-olefin, important non-bonded interactions have to arise between the cyclo-olefin and metal ligands.

The formation of *cis* and *trans* double bonds in the polymer was suggested to occur *via* intermediate *cis* and *trans* metallacyclobutanes as represented in Scheme 7.18 in planar form as I and II and in puckered form as Ia, Ib, IIa and IIb, where R, R represent the monomer unit.[62]

Scheme 7.18

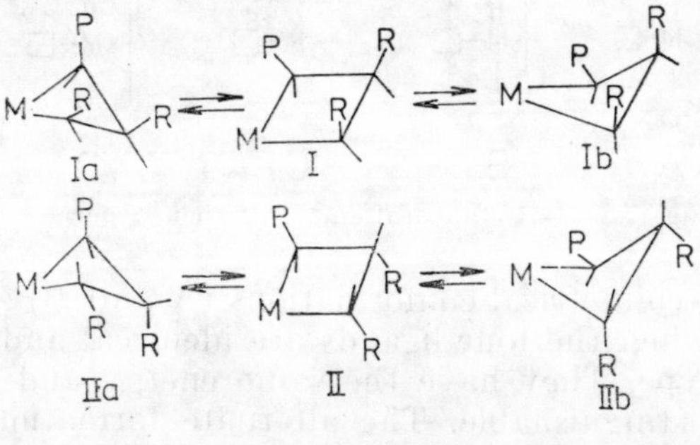

Ia   I   Ib

IIa   II   IIb

Ivin *et al.* argued that, under conditions where steric crowding is severe, norbornene adds to metallacarbene leading mainly to *cis*-metallacycle I.[62] This metallacycle adopts the puckered conformation Ib rather than Ia, which has unfavourable 1,3-diaxial interactions. On the contrary, under conditions of less severe steric crowding, norbornene will add to the metallacarbene with the formation of a *trans*-metallacycle II. The two puckered forms IIa and IIb will be energetically equivalent having similar steric interaction.

The fact that steric interactions arising in the transient metallacyclo-butane are determinant for stereoselectivity has been advocated by Katz in rationalizing the product stereoselectivity in 1-methylcyclo-octene polymerization.[68] Thus, the predominant formation of the E stereoiso-mer in 1-methyl-*trans*-cyclo-octene polymerization has been assigned to lesser 1,3-diaxial interaction between the methyl groups in the interme-diate structure A as compared to structure B where the methyl groups interact more strongly with the polymer chain.

Scheme 7.19

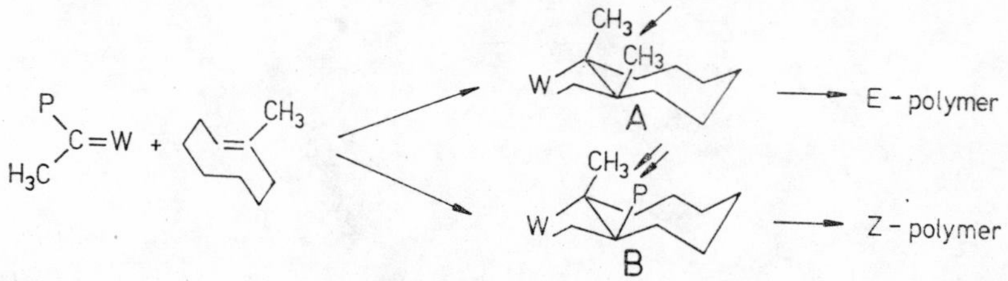

In an alternative explanation advanced recently by Basset and co-work-ers[19] the energy levels of *cis*- and *trans*-directing metallacyclobutanes derived from strained cyclo-olefins such as norbornene are not relevant for reaction stereochemistry. Based on stereochemical comparison of the reactions of norbornene and *cis*-2-pentene in the presence of catalysts such as $M(CO)_3(ArH)(M=Cr, Mo, W)$ or $M(NO)_2Cl_2(PPh_3)_2(M=Mo,W)$ associated with $C_2H_5AlCl_2$, the above authors deduced that the energy level of metallacyclobutanes arising from cyclic olefins is lower than that with acyclic olefins, and also lower than that of coordinated cyclo-olefin, owing to a pronounced strain release in the metallacyclobutane inter-mediate. Accordingly, the differences of energy between the *cis* and *trans* directing metallacyclobutanes for norbornene will not be a discriminating factor for stereoselectivity. Furthermore, the small energy differences of metallacyclobutane conformations caused by ligands interactions and transition metal variation will not affect reaction stereoselectivity. In contrast to cyclic olefins, stereoselectivity in the metathesis of acyclic olefins will be governed by the energy levels of the *cis* and *trans* directing metallacyclobutanes. The change in stereoselectivity observed with acy-clic olefins, on varying the ligands or the transition metal, were assigned to the change of energy levels for the corresponding metallacyclobutanes.

The modes of rupture of intermediate metallacyclobutane have been discussed by Ivin *et al.*[74] for the polymerization of norbornene and its deri-vatives. Two main modes of metallacyclobutane rupture have been pos-tulated to explain the stereoselectivity of reaction products: (i) conrota-tory or parallel and (ii) disrotatory or orthogonal (Scheme 7.20).

According to the first mode of rupture, (a), the ring breaks such that the $p\pi$ orbitals on the unsaturated carbons in the product separate in parallel allignment. This rupture leads to the formation of chiral conformations of the intermediate metallacarbene and seems to be characteristic of more electronegative transition metals. Racemization of the resulting chiral conformations may occur only after the propagation step. This mode of rupture leads preferentially to highly tactic polymers.

Scheme 7.20

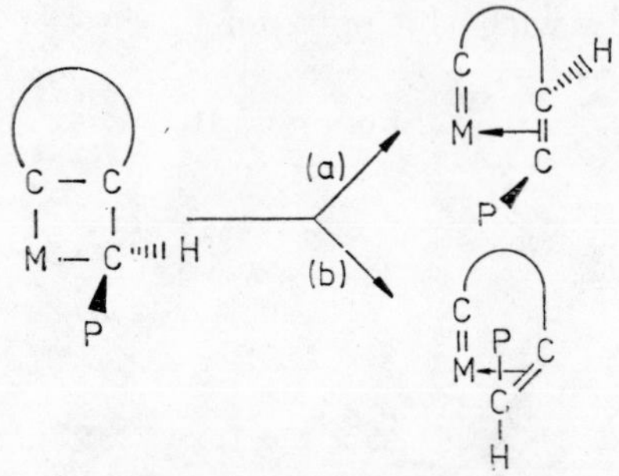

C — C   (a)   C=M—C(H), C   P

M — C''''H   (b)

P   C=M—P—C, C—H

In the second mode of rupture, (b), the ring breaks so that the pπ orbitals rotate into orthogonal planes as they separate. This rupture leads to the formation of achiral "symmetrical" conformations of intermediate metallacarbene and seems to prevail for less electronegative metals. Racemization may occur during the propagation process. By this type of rupture, Ivin *et al.* explained the independence of tacticity on dilution of monomer and the fall in tacticity with increasing temperature.[74]

The nature of the transition metal atom in the catalyst seems to be significant for the steric course of ring-opening polymerization. As discussed earlier, the transition metal will directly affect the coordination of cyclo-olefin and carbenic ligand, the geometry and stability of the intermediate metallacyclobutane, the mode of rupture of the metallacycle, and the (de)coordination of the unsaturated polymer chain. The more electronegative transition metals and/or higher oxidation states will prefer to coordinate more strongly the cyclo-olefin, the carbenic ligand and the unsaturated polymer chain. Such metals and oxidation states will imply more severe steric interactions with metal ligands resulting in higher barriers to rotation or to interconversion between the conformations of the metallacarbene species. On the contrary, the more electropositive transition metals and/or lower oxidation states tend to weaken the coordination of cyclo-olefin or polymer chain, and to facilitate conformational isomerizations. The electron density at the metal site is related to the metal electronegativity and oxidation state. Low electron densities will strengthen coordination to the double bond resulting in higher stereoselectivity whereas higher electron densities favour decoordination, permitting thereby rotation of the approaching monomer and interconversion between the corformations of the intermediate metallacarbenes with a corresponding lowering of stereoselectivity.

References

1. R. L. Banks and G. C. Bailey, *Ind. Eng. Chem., Prod. Res. Develop.*, **3**, 170 (1964).
2. S. Furukawa, *Kogyo Kagaku Zasshi*, **74**, 2471 (1971).
3. E. S. Davie, D. A. Whan and C. Kemball, *J. Catalysis*, **24**, 272 (1972).
4. A. Ismayel-Milanovic, J. M. Basset, H. Praliaud, M. Dufaux and L. de Mourgues, *J. Catalysis*, **31**, 408 (1973).
5. K.Tanaka, K.-I. Tanaka and K. Miyahara, *Chem. Letters*, **1970**, 623.
6. K. Tanaka, K. Miyahara and K.-I. Tanaka, *J. Mol. Catalysis*, **15**, 133 (1982).
7. N. Calderon, E. A. Ofstead, J. P. Ward, W. A. Judy and K. W. Scott, *J. Amer. Chem. Soc.*, **90**, 4133 (1968).
8. J. L. Wang and H. R. Menapace, *J. Org. Chem.*, **33**, 3794 (1968).
9. W. B. Hughes, *J. Chem. Soc., Chem. Commun.*, **1969**, 431.
10. J. M. Basset, G. Coudurier, R. Mutin and H. Praliaud, *J. Catalysis*, **34**, 152 (1974).
11. J. M. Basset, J. L. Bilhou, R. Mutin and A. Theolier, *J. Amer. Chem. Soc.*, **97**, 7376 (1975).
12. J. L. Bilhou, J. M. Basset, R. Mutin and W. F. Graydon, *J. Chem. Soc., Chem. Commun.*, **1976**, 970.
13. J. L. Bilhou, J. M. Basset, R. Mutin and W. F. Graydon, *J. Amer. Chem. Soc.*, **99**, 4083 (1977).
14. M. Leconte, J. L. Bilhou, W. Reimann and J. M. Basset, *J. Chem. Soc., Chem. Commun*, **1978**, 341.
15. M. Leconte and J. M. Basset, *J. Amer. Chem. Soc.*, **101**, 7296 (1979).
16. M. Leconte and J. M. Basset, *Nouveau J. Chim.*, **3**, 429 (1979).
17. M. Leconte and J. M. Basset, *Ann. N. Y. Acad. Sci.*, **333**, 165 (1980).
18. M. Leconte, Y. Ben Taarit, J. L. Bilhou and J. M. Basset, *J. Mol. Catalysis*, **8**, 263 (1980).
19. C. Larroche, J. P. Laval, A. Lattes, M. Leconte, F. Quignard and J. M. Basset, *Nouveau J. Chim*, **6**, 61 (1982).
20. G. Pampus, "Catalyse par Coordination", Guyot (Ed.), Paris, 1974, p. 172.
21. P. Chevalier, D. Sinou, G. Descotes, R. Mutin and J. M. Basset, *J. Organomet. Chem.*, **113**, 1 (1976).
22. R. H. A. Bosma, X. D. Xu and J. C. Mol, *J. Mol. Catalysis*, **15**, 187 (1982).
23. T. J. Katz and W. H. Hersh, *Tetrahedron Letters*, **1977**, 585.
24. F. Garnier and P. Krausz, *J. Mol. Catalysis*, **8**, 91 (1980).
25. R. H. Grubbs and C. P. Hoppin, *J. Amer. Chem. Soc.*, **101**, 1499 (1979).
26. C. P. Casey and H. E. Tuinstra, *J. Amer. Chem. Soc.*, **100**, 2270 (1978).
27. N. Calderon, J. P. Lawrence and E. A. Ofstead, *Adv. Organomet. Chem.*, **17**, 449 (1979).
28. E. A. Ofstead, J. P. Lawrence, M. L. Senyek and N. Calderon, *J. Mol. Catalysis*, **8**, 227 (1980).
29. C. Edwige, A. Lattes, J. P. Laval, R. Mutin, J. M. Basset and R. Nouguier, *J. Mol. Catalysis*, **8**, 297 (1980).
30. J. L. Hérisson and Y. Chauvin, *Makromol. Chem.*, **141**, 161 (1970).
31. W. J. Kelley and N. Calderon, *J. Macromol. Sci., Chem.*, **9**, 911 (1975).
32. J. McGinnis, T. J. Katz and S. Hurwitz, *J. Amer. Chem. Soc.*, **98**, 605 (1976).
33. C. P. Casey, H. E. Tuinstra and M. C. Saeman, *J. Amer. Chem. Soc.*, **98**, 608 (1976).
34. C. P. Casey, L. D. Albin and T. J. Burkhardt, *J. Amer. Chem. Soc.*, **99**, 2533 (1977).
35. C. P. Casey, S. W. Polichnowski, A. J. Shusterman and C. R. Jones, *J. Amer. Chem. Soc.*, **101**, 7282 (1979).
36. M. O. Visscher and K. G. Coulton, *J. Amer. Chem. Soc.*, **94**, 5923 (1972).
37. R. G. Gillard, M. Keeton, R. Mason, M. F. Pilbrow and D. Russel, *J. Organomet. Chem.*, **33**, 247 (1971).
38. N. Tagizadeh, F. Quignard, M. Leconte, J. M. Basset, C. Larroche, J. P. Laval and A. Lattes, *J. Mol. Catalysis*, **15**, 219 (1982).
39. A. K. Rappé and W. A. Goddard III, *J. Amer. Chem. Soc.*, **102**, 5115 (1980).
40. A. K. Rappé and W. A. Goddard III, *J. Amer. Chem. Soc.*, **104**, 448 (1982).
41. O. Eisenstein, R. Hoffmann and A. R. Rossi, *J. Amer. Chem. Soc.*, **103**, 5582 (1981).

42. P. Günther, F. Haas, G. Marwede, K. Nützel, W. Oberkirch, G. Pampus, N. Schön and H. Witte, *Angew. Makromol. Chem.*, **16/17**, 27 (1971); *ibid.* **14**, 87 (1970).

43. R. J. Minchak and H. Tucker, *Polymer Prepr., Amer. Chem. Soc. Div., Polymer Chem.*, **13**, 885 (1970).

44. I. A. Oreshkin, L. I. Rednina, T. L. Kershenbaum, G. M. Chernenko, K. L. Makovetsky, E. I. Tinyakova and B. A. Dolgoplosk, *European Polymer J.*, **13**, 447 (1977).

45. M. A. Golub, *J. Polymer Sci., Polymer Letters Ed.*, **12**, 295 (1971).

46. M. A. Golub, S. A. Fuqua and N. S. Bhacca, *J. Amer. Chem. Soc.*, **84**, 4981 (1962).

47. K. J. Ivin, D. T. Laverty and J. J. Rooney, *Makromol. Chem.*, **178**, 1545 (1977).

48. K. J. Ivin, G. Lapienis, J. J. Rooney and C. D. Stewart, *J. Mol. Catalysis*, **8**, 203 (1980).

49. K. J. Ivin, G. Lapienis and J. J. Rooney, *Polymer*, **21**, 436 (1980).

50. T. J. Katz, J. McGinnis and C. Altus, *J. Amer. Chem. Soc.*, **98**, 606 (1976).

51. G. Natta, G. Dall'Asta, G. Mazzanti and G. Motroni, *Makromol. Chem.*, **69**, 163 (1963).

52. G. Natta, G. Dall'Asta and L. Porri, *Makromol. Chem.*, **81**, 253 (1965).

53. T. J. Katz, J. McGinnis, S. J. Lee and N. Acton, *Tetrahedron Letters*, **1976**, 4246.

54. G. Natta, G. Dall'Asta and G. Mazzanti, *Angew. Chem.*, **76**, 765 (1964).

55. G. Dall'Asta and G. Motroni, *Angew Makromol. Chem.*, **16/17**, 51 (1971).

56. N. Calderon and M. G. Morris, *J. Polymer Sci.*, A−2, **5**, 1283 (1967).

57. G. Dall'Asta and R. Manetti, *European Polymer J.*, **4**, 145 (1968).

58. N. Calderon, E. A. Ofstead and W. A. Judy, *J. Polymer Sci.*, A−1, **5**, 2209 (1967).

59. T. Oshika and H. Tabuchi, *Bull. Chem. Soc. Japan*, **411**, 211 (1968).

60. T. J. Katz and N. Acton, *Tetrahedron Letters*, **1976**, 4251.

61. K. J. Ivin, D. T. Laverty and J. J. Rooney, *Makromol. Chem.*, **179**, 253 (1978).

62. K. J. Ivin, D. T. Laverty, J. H. O'Donnell, J. J. Rooney and C. D. Stewart, *Makromol. Chem.*, **180**, 1989 (1979).

63. G. Gianotti, G. Dall'Asta, A. Valvassori and V. Zamboni, *Makromol. Chem.*, **149**, 117 (1971).

64. G. Dall'Asta, *Makromol. Chem.*, **154**, 1 (1972).

65. G. Dall'Asta, *Rubber Chem. Technol.*, **47**. 511 (1974).

66. T. J. Katz, S. J. Lee and M. A. Shippey, *J. Mol. Catalysis*, **8**, 219 (1980).

67. J. McGinnis, cited after reference 68.

68. S. J. Lee, J. McGinnis and T. J. Katz, *J. Amer. Chem. Soc.*, **98**, 7818 (1976).

69. K. J. Ivin, D. T. Laverty, J. J. Rooney and P. Watt, *Rec. Trav. Chim.*, **96**, M54 (1977).

70. K. J. Ivin, G. Lapienis and J. J. Rooney, *J. Chem. Soc., Chem. Commun.* **1979**, 1068.

71. K. J. Ivin, L.-M. Lam and J. J. Rooney, *Makromol Chem.*, **182**, 1847 (1981).

72. K. J. Ivin, *Pure Appl. Chem.*, **52**, 1907 (1980).

73. H. H. Thoi, K. J. Ivin and J. J. Rooney, *Makromol. Chem.*, to be published.

74. H. H. Thoi, K. J. Ivin and J. J. Rooney, *J. Mol. Catalysis*, **15**, 245 (1982).

75. G. Motroni, G. Dall'Asta and I. W. Bassi, *European Polymer J.*, **9**, 257 (1973).

76. G. Dall'Asta, G. Motroni and L. Motta, *J. Polymer Sci.*, A−1, **10**, 160 (1972).

77. K. J. Ivin, J. H. O'Donnell, J. J. Rooney and C. D Stewart, *Makromol. Chem.*, **180**, 1975 (1979).

78. J. Furukawa, *Pure Appl. Chem.*, **42**, 415 (1975).

79. R. R. Schrock, L. M. Messerle and C. D. Wood, *J. Amer. Chem. Soc.*, **100**, 3393 (1978).

Chapter 8

PRACTICAL APPLICATIONS OF METATHESIS AND RING-OPENING POLYMERIZATION

In less than three years after its discovery, the metathesis in the hetero-geneous phase was applied on an industrial scale to prepare olefins, thus widening the range of methods for obtaining some of the hydrocarbons employed in petrochemistry, organic chemistry, the polymer industry and in other branches of the chemical industry. Subsequently, new processes were devised to apply the olefin metathesis for obtaining high octane gasoline. When extended to cyclo-olefins, the reaction made possible the synthesis of polyalkenamers, as well as macrocycles, cate-nanes and other compounds with interlocked structures, which are difficult to obtain by conventional methods. More recently, olefin metathesis has been used to synthesize natural compounds, to prepare olefins and acetylenes substituted with functional groups, and for other types of organic compounds.

8.1. Industrial Processes for the Synthesis of Olefins

The first metathesis process applied on an industrial scale (The Triolefin Process) comprised the disproportionation of propene to ethene and 2-butene in the presence of alumina-supported molybdenum or tungsten oxide cata-lysts.[1] According to this process (Fig. 8.1). the, reaction is carried out in a tu-bular reactor with a fixed bed of catalyst. Separation of the reaction products was effected by fractionating columns and subsequent purification by normal distillation. A de-ethanized mixture of propane and propene is fed into the disproportionation reactor A, which contains cobalt molybdate as catalyst. The disproportionation products are separated in column C into a light fraction consisting mainly of ethene, a medium fraction containing unreacted propane and propene, and a heavy fraction formed from 2-butene and higher unsaturated hydrocarbons. The ethene can be employed directly as a monomer for manufacturing polyethylene, without any special purification. Pure propene is separated from the medium fraction in column B and the residual propene-propane mixture is recirculated to the disproportiona-tion reactor. From the heavy fraction, stripper D separates a fraction of butene and another fraction containing pentene, hexene and higher homologues. The butene fraction contains over 90 % 2-butene : it can be used as a starting material for the synthesis of gasoline or it can be isomerized in an additional isomerization unit to high purity 1-butene which is in demand as a co-monomer in polyethylene production. This fraction

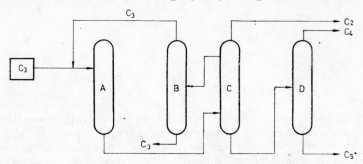

Fig. 8.1. — Triolefin Process for the disproportionation of propene into ethene and butene.

can also constitute a source for the manufacture of high-purity butadiene *via* dehydrogenation, since it contains branched hydrocarbons only in traces. The industrial process of propene disproportionation presents economic advantages, because it is simple and facilitates the inexpensive production of pure olefins ; the process does not require the difficult separations of ethane from ethene and of butane from butene which are involved in other processes.

A variation of the Triolefin Process is a three-stage reaction giving a mixture of hexenes, heptenes and octenes, as well as ethene from propene. The hexene-heptene-octene product has 95 % linearity and can be used in the oxo alcohol synthesis for production of PVC plasticizers.

A new process (SHOP Process), recently applied on the industrial scale by Shell,[2] disproportionates the olefins obtained through a complex process of oligomerization and isomerization to yield internal olefins which are used as starting materials for detergent productions (Fig. 8.2). In the SHOP Process, ethene is oligomerized in a first stage, A, to a mixture of olefins C_4-C_{20}. The olefins from this mixture are then partly separated in a fractionating column, B, and isomerized to α-olefins in an isomerization reactor, D. They are then disproportionated by metathesis to internal olefins in the disproportionation reactor, E, and the product is fractionated in distillation columns, F, into fractions of higher olefins.

In a very short time, numerous processes have been developed for producing olefins or olefin mixtures useful as starting materials in industrial organic chemistry, petrochemistry, the polymer industry, etc. An example is the process for the simultaneous production of ethene and butadiene through propene disproportionation [3] (Fig. 8.3).

The metathesis of propene is directly integrated in the cracker and the separation unit for the petroleum fractions. The ethene resulting from disproportionation enriches the ethene content of the C_2 fraction from

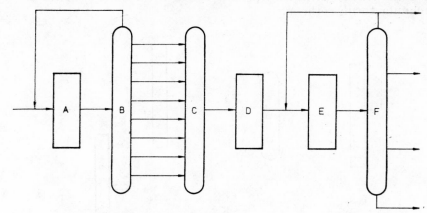

Fig. 8.2. — SHOP Process for the disproportionation of olefins.

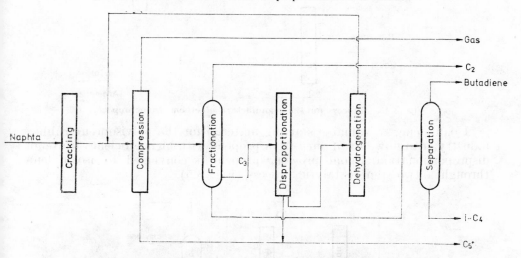

Fig. 8.3. — Synthesis of ethene and butadiene by disproportionation.

cracking. Butadiene is formed by the subsequent dehydrogenation of the butene fraction resulting from disproportionation after being recirculated and fractionated.

Another example is an improved process for manufacturing ethene from propene through a series of successive disproportionations[4] (Fig. 8.4). In this process also the disproportionation stages are directly connected to the petroleum cracking unit which supplies the propene for disproportionation. In the first stage, propene is converted to ethene and

2-butene, then the butene, after enrichment in the 1-butene isomer through isomerization, is converted to ethene, pentene, and hexene. Pentene and hexene are then hydrogenated to pentane and hexane which are recirculated in the cracker unit. This process is one of the most economical ways for manufacturing ethene.

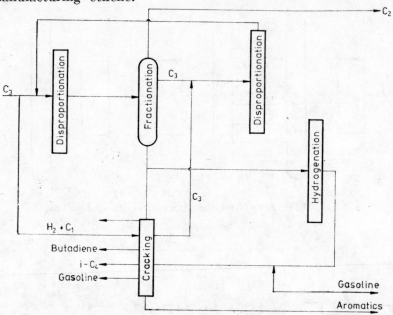

Fig. 8.4. — Process for the manufacture of ethene from propene.

Isoamylene, a valuable starting material for the polyisoprene rubber industry, can now be obtained from propene with high efficiency through disproportionation. In one process,[5] propene is converted to isoamylene through a two-step catalytic process (Fig. 8.5).

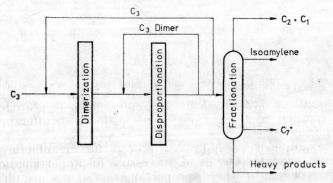

Fig. 8.5. — Two-stage process for the manufacture of isoamylene from propene.

In the first stage propene is dimerized, and in the second stage the dimer of propene is codisproportionated with propene. After separation of the reaction products, a fraction rich in 2-methylbutenes is obtained, which can be employed for manufacturing isoprene through dehydrogenation. In another three-step process,[6] propene is first disproportionated to ethene and 2-butene, then the butene is dimerized to a branched dimer and finally this dimer is codisproportionated with ethene (Fig. 8.6).

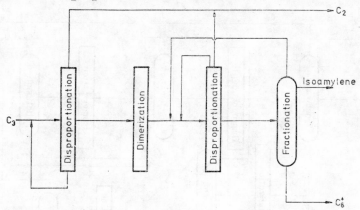

Fig. 8.6. — Three-stage process for the manufacture of isoamylene from propene.

Other economic processes for synthesizing isoamylene have been reported employing the codisproportionation of isobutene with propene,[7] or of isobutene with 2-butene,[7] etc.

A proven Phillips process [7] thus converts isobutene with n-butene or propene in a disproportionation reactor and obtains isoamylene after a separation unit (Fig. 8.7).

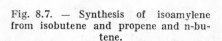

Fig. 8.7. — Synthesis of isoamylene from isobutene and propene and n-butene.

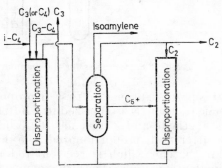

The higher fraction is cleaved further to a C_3-C_4 fraction with ethene formed in the first stage, and recirculated to the first stage for isobutene disproportionation. The resulting isoamylene may be used for isoprene synthesis by dehydrogenation.

Propene continues to be converted into a large number of higher acyclic olefins by employing several variants of codisproportionation between the reaction products formed at successive stages of disproportionation. Thus, a process for producing 5-decene from propene [8] was developed, which consists of three stages of disproportionation combined with isomerization (Fig. 8.8).

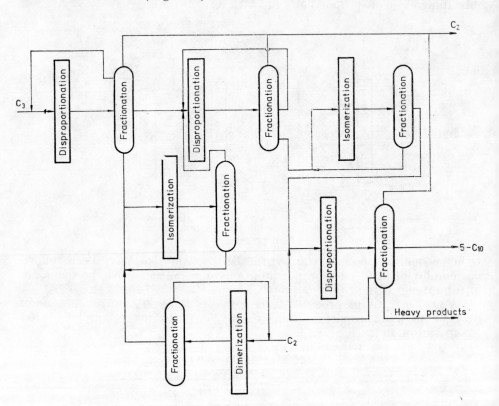

Fig. 8.8. — Process for the manufacture of 5-decene from propene.

In this process, propene is first disproportionated to ethene and 2-butene, the butene fraction is isomerized to 1-butene, then 1-butene is disproportionated to ethene and 3-hexene; the 3-hexene is isomerized to 1-hexene and the latter is disproportionated to 5-decene and ethene. Ethene is dimerized to butene which is recirculated to the butene fraction prior to the isomerization stage. The 3-hexene can also be used as such directly for various purposes after separation from the disproportionation products of 1-butene.

In another process (Fig. 8.9),[9] propene is converted into higher acyclic olefins ranging from undecene to pentadecene, by employing a combination of the disproportionation and separation stages of the reaction products. Propene is disproportionated in a first reactor, which is also fed with the light product obtained in the separation stage after the reaction ; the products of the first disproportionation are disproportionated, in a second reactor, in combination with the light products from the later separation stage ; similarly, the reaction products are disproportionated in

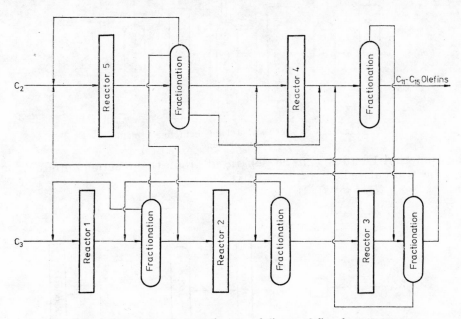

Fig. 8.9. — Process for the manufacture of linear olefins from propene.

third and fourth reactors ; one finally obtains a mixture of linear olefins with 11 to 15 carbon atoms. The heavier compounds with branched structures are removed from the process at each separation stage. Ethene, formed during the disproportionation stage, is oligomerized to olefins of various lengths which, after separation, are introduced into the disproportionation flow to the corresponding olefinic fraction.

A similar disproportionation process has been developed by Phillips for the production of $C_{12}-C_{16}$ olefins from propene or butenes (Fig. 8.10).[7] The medium fractions are recirculated and codisproportionated with propene or lower fractions. Ethene formed in the first stage is separated and used in an ethene flow. The technology may economically be integrated with a Triolefin unit using a propene or butene line as source of starting material.

Branched acyclic olefins, which have many applications in the petrochemical industry, can easily be obtained from ethene and isobutene through a series of disproportionation reactions. By a simple process [10] "neohexene" i.e. 3,3-dimethylbutene, is obtained together with ethene by the disproportionation of isobutene using a catalytic system formed from silica-supported WO_3 and MgO (Fig. 8.11).

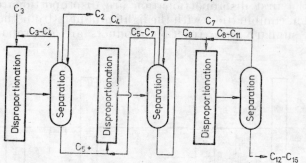

Fig. 8.10. — Synthesis of long-chain linear olefins from propene.

3,3-Dimethyl-1-butene can be separated from the reaction product or can be recirculated.

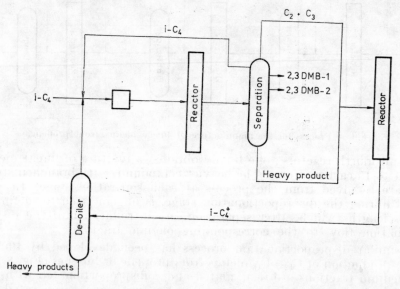

Fig. 8.11. — Process for the synthesis of neohexene.

Another neohexene process, now operated on a commercial scale (Fig. 8.12), produces neohexene starting either from commercial di-isobutene (a mixture of β- and α- or 3- and 4-[2,2,4-trimethyl] pentenes) or from iso-butene with subsequent dimerization to a di-isobutene fraction.[7] The disproportionation of di-isobutene by ethene cleavage of the β-isomer gives the desired neohexene product. The process uses a separate dimerization unit for di-isobutene synthesis and a selective isomerization unit continuosly to shift the double bond of the α isomer to the β as the latter is being consumed.

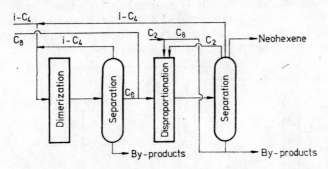

Fig. 8.12. — Synthesis of neohexene from isobutene or di-isobutene.

1,2-Di-t-butylethene, a component which markedly enhances synthetic gasolines, can be obtained starting from isobutene and ethene by using a disproportionation catalyst having activity for concurrent double bond isomerization; neohexene (3,3-dimethyl-1-butene), is formed as a by-product [11] (Fig. 8.13).

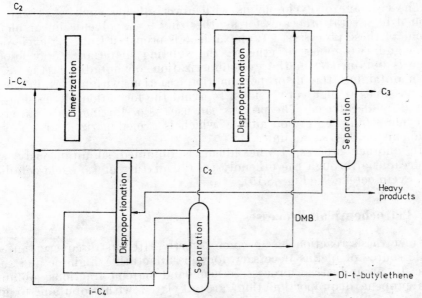

Fig. 8.13. — Process for the manufacture of di-tert-butylethene.

Isobutene is dimerized in a first reactor to di-isobutene (2,4,4-tri-methyl-pentene) which is simultaneously isomerized and disproportionated to di-t-butylethene in a second reactor. 2,3-Dimethyl-2-butene formed as by-product is introduced into a disproportionation unit and cleaved with ethene to give isobutene, which is recirculated to the initial feedstock. It is also possible to use neohexene (3,3-dimethyl-1-butene) as the starting material. In this case the process contains only the disproportionation and separation stages of the reaction products.

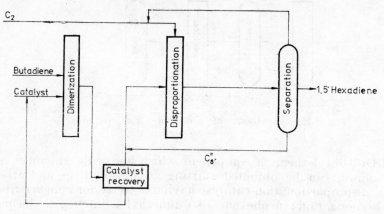

Fig. 8.14. — Process for the manufacture of 1,5-hexadiene.

The cleavage of cyclo-olefins with lower alkenes such as ethene was applied in several processes for synthesizing α, ω-polyenic linear olefins. In one of these processes,[12] 1,5-hexadiene is prepared from 1,5-cyclo-octadiene and two moles of ethene; the starting material, 1,5-cyclo-octadiene is obtained from the cyclodimerization of butadiene (Fig. 8.14). By maintaining the disproportionation reaction within certain limits, it is possible, at high conversions, to avoid the formation of the side product, 1,5,9-decatriene. The process can use, as a starting material, either 1,5-cyclo-octadiene, or butadiene which is cyclodimerized to 1,5-cyclo-octadiene in the same unit.

In another process, 1,5-hexadiene is obtained simultaneously with 1,9-decadiene through the ethenolysis of a mixture of 1,5-cyclo-octadiene and cyclo-octene [13] (Fig. 8.15).

8.2. Petrochemical Processes

The metathesis reaction has become, shortly after its discovery, an economic source of olefins necessary for manufacturing high octane gasolines. In an industrial process, ethene and butene fractions, obtained from propene disproportionation, are alkylated with isobutane to give saturated hydrocarbons with a high octane number [14] (Fig. 8.16).

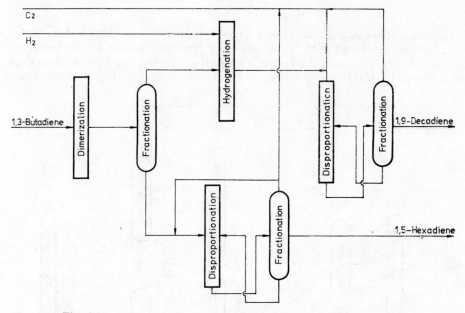

Fig. 8.15. — Process for the simultaneous synthesis of 1,5-hexadiene and 1,9-decadiene.

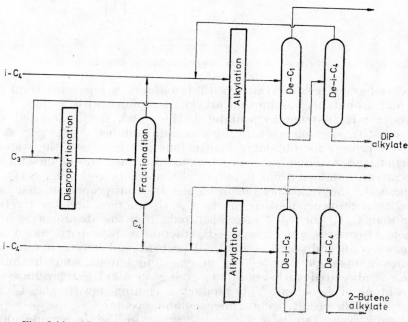

Fig. 8.16. — Process for the manufacture of saturated hydrocarbons *via* disproportionation and alkylation.

The ethene fraction, after having been separated from the butene fraction, is first alkylated with isobutene; the product is then separated from methane and isobutane, the result being an alkylated product with a high "di-isopropyl" (DIP i.e. 2,3-dimethylbutane) content. The butene frac-

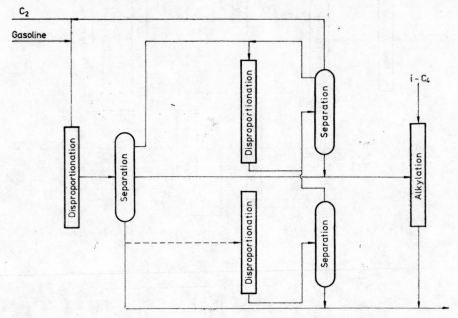

Fig. 8.17. — Process for the disproportionation of gasoline fractions.

tion, after having been alkylated with isobutane, is separated from propane and isobutane, yielding an alkylated product with a high isooctane content (2,2,3-trimethylpentane in this case). Both alkylated products constitute gasolines with high octane number.

A new process for obtaining high octane gasolines uses the step-disproportionation reaction of olefin fractions from mixtures of unsaturated hydrocarbons obtained from light petroleum fractions [15] (Fig. 8.17). In this process, the light petroleum olefins are codisproportionated with ethene, then propene, pentene or the C_5-C_6 fraction, and the fractions higher than $C_6(C_6^+)$ or $C_7(C_7^+)$ are separated from the disproportionation product. Afterwards, the C_5 or C_5-C_6 fraction is disproportionated with ethene when an additional amount of propene and butene is obtained. Propene is disproportionated to ethene and butene, and the butene fraction is alkylated with isobutane, giving an alkylated product with high content of isooctane. This product is combined with the C_5^+ and C_6^+ fractions, the result being a high octane gasoline.

Alkenolysis has an important application in the fragmentation of unsaturated petroleum fractions obtained through the catalytic cracking of benzines. High octane gasoline is obtained from the ethenolysis of the pentene fraction combined with disproportionation of the propene-rich fraction, followed by dimerization of the isobutene fraction and by alkylation of butene and ethene fractions [16] (Fig. 8.18).

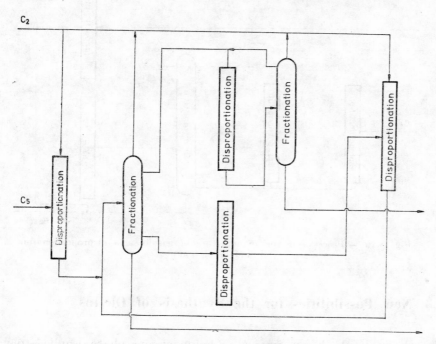

Fig. 8.18. — Process for the manufacture of high octane gasoline *via* disproportionation and dimerization.

In another process(Fig. 8.19), the disproportionation of mixtures of linear and branched olefins in the presence or absence of ethene is improved by adjusting the conversion so as to increase selectivity with respect to the disproportionation of linear olefins as compared to that of branched olefins.[17] Thus, the C_5 fraction (containing pentene and isopentane) is selectively disproportionated with or without ethene with transformation of mainly n-pentene to butenes; the alkylation of the resulting butenes leads to high octane gasoline as a by-product. A modification of this process permits the disproportionation of the propene formed as a by-product, to butenes, which are used for alkylation, and to ethene, which is used for the disproportionation of the C_6 fraction.

Several processes for the manufacture *via* disproportionation of linear or branched olefins have been patented,[18-99] the range of catalysts used for these processes has been extended [100-147] and complex procedures-have been elaborated to transform the unsaturated hydrocarbons from petroleum,[148-155] thus diversifying the applications of this type of reaction.

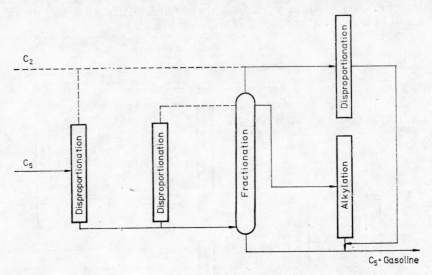

Fig. 8.19. — Process for the manufacture of gasoline *via* disproportionation and alkylation.

8.3. New Possibilities for the Synthesis of Olefins

Metathesis, in the heterogeneous or homogeneous phase, applied to various types of acyclic olefins, constitutes a general method for the synthesis of olefins. By varying the nature of the olefins, it is possible to obtain a large variety of olefinic structures.

The metathesis of α-olefins, whose simplest representative is propene, provides a route to the synthesis of symmetrical internal olefins, along with ethene as by-product where $R = CH_3$, C_2H_5, C_3H_7, etc. This reac-

$$\underset{R}{\overset{\|}{|}} + \underset{R}{\overset{\|}{|}} \; \rightleftharpoons \; \overset{=}{} + R-\!\!=\!\!-R \tag{8.1}$$

tion was extensively discussed in chapter 3 and it is commonly used, in the homogeneous or heterogeneous phase, for the synthesis of a large number of symmetrical olefins, such as 2-butene, 3-hexene, 4-octene, 5-decene, etc.

Internal symmetrical olefins also result from the reaction of α-olefins

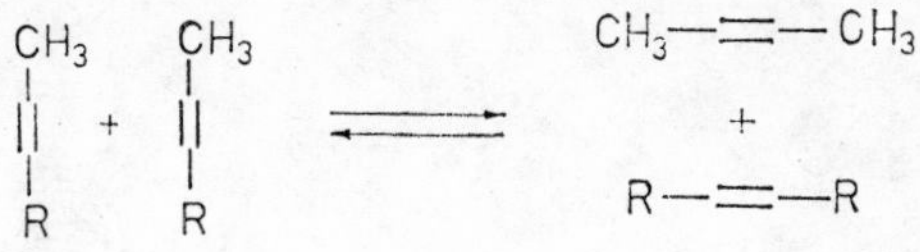

$$
\begin{array}{c}
\text{CH}_3 \quad \text{CH}_3 \\
\| \quad + \quad \| \\
\text{R} \qquad \text{R}
\end{array}
\quad \rightleftharpoons \quad
\begin{array}{c}
\text{CH}_3\text{—}\!\!=\!\!\text{—CH}_3 \\
+ \\
\text{R—}\!\!=\!\!\text{—R}
\end{array}
\qquad (8.2)
$$

where R $=$ CH$_3$, C$_2$H$_5$, C$_3$H$_7$, etc. In this case, 2-butene is the by-product, but usually it can easily be separated from the higher internal olefin.

For branched olefins, metathesis offers a simple and elegant method for the synthesis of some internal olefins with more complex structures

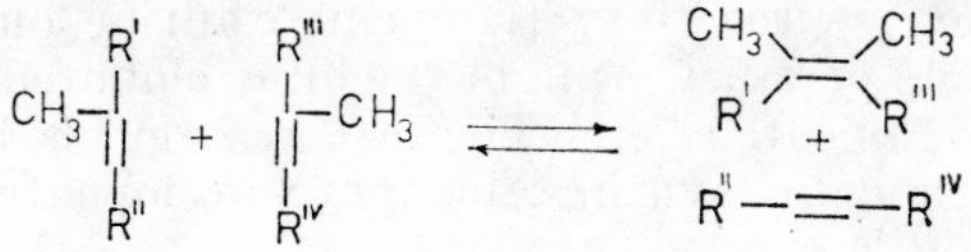

$$
\begin{array}{c}
\text{R}' \qquad \text{R}''' \\
\text{CH}_3\text{—}\| \quad + \quad \|\text{—CH}_3 \\
\text{R}'' \qquad \text{R}^{IV}
\end{array}
\quad \rightleftharpoons \quad
\begin{array}{c}
\text{CH}_3 \diagdown \quad \diagup \text{CH}_3 \\
\text{R}' \diagup \quad \diagdown \text{R}''' \\
+ \\
\text{R}''\text{—}\!\!=\!\!\text{—R}^{IV}
\end{array}
\qquad (8.3)
$$

where R^I, R^{II}, R^{III} or R^{IV} $=$ CH$_3$, C$_2$H$_5$, C$_3$H$_7$ etc. Unsaturated hydrocarbons, such as 2,3-dimethyl-2-butene or 3,4-dimethyl-3-hexene thus become easily accessible through direct syntheses.

Cometathesis reactions allow, in many variants, to obtain a large range of unsymmetrical internal olefins

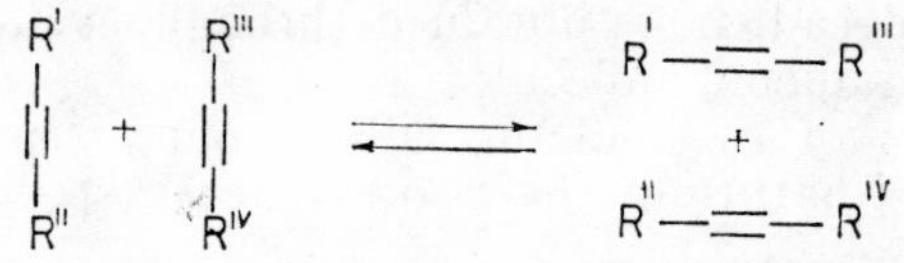

$$
\begin{array}{c}
\text{R}' \qquad \text{R}''' \\
\| \quad + \quad \| \\
\text{R}'' \qquad \text{R}^{IV}
\end{array}
\quad \rightleftharpoons \quad
\begin{array}{c}
\text{R}'\text{—}\!\!=\!\!\text{—R}''' \\
+ \\
\text{R}''\text{—}\!\!=\!\!\text{—R}^{IV}
\end{array}
\qquad (8.4)
$$

where R^I, R^{II}, R^{III} and R^{IV} are different alkyl groups.

In the series of cyclo-olefins, it is possible to obtain, *via* metathesis, cyclic dienes or cyclic polyenes, macrocycles, or compounds with interlocked cyclic structure, polyalkenamers or macrocyclic structures; all these compounds can be obtained only with great difficulty by conventional methods, or are otherwise inaccessible. Here we shall only give examples of cyclo-olefin syntheses, because the other compounds were discussed in other chapters.

The metathesis of a cyclo-olefin may yield, under mild conditions, a cyclic diene,

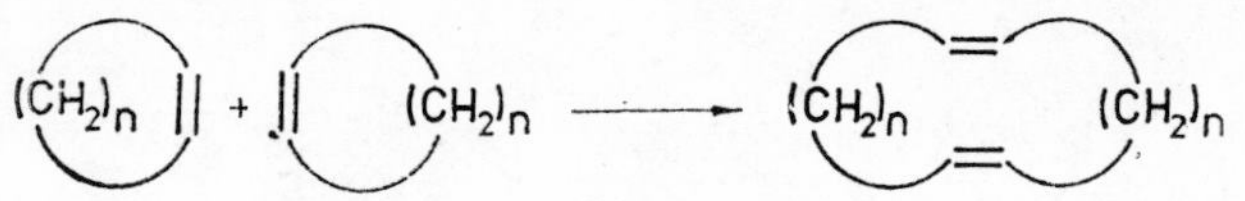

$$
\text{(CH}_2)_n \, \| \quad + \quad \| \, \text{(CH}_2)_n \quad \longrightarrow \quad \text{(CH}_2)_n \overset{=}{\underset{=}{}} \text{(CH}_2)_n
\qquad (8.5)
$$

where n = 1, 2, 3, etc. Thus, cyclopropene yields 1,4-cyclohexadiene; cyclobutene yields 1,5-cyclo-octadiene while cyclopentene forms 1,6-cyclodecadiene etc.

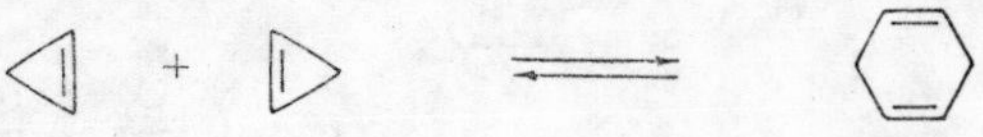

(8.6)

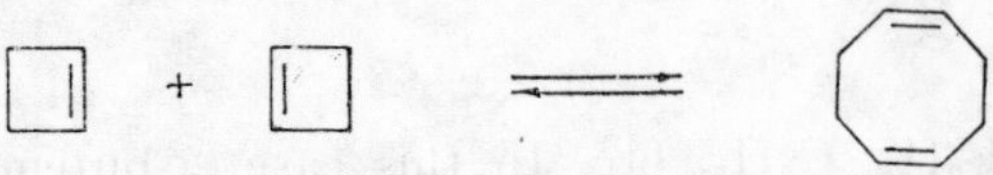

(8.7)

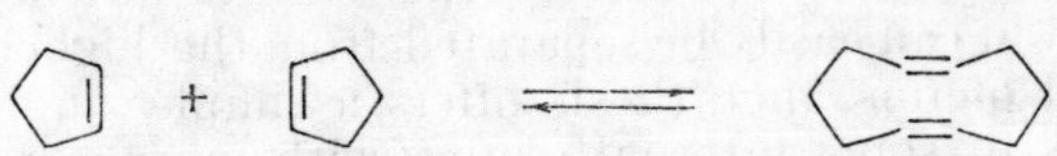

(8.8)

Through the reverse reaction, 1,4-cyclohexadiene ought to yield cyclopropene, 1,5-cyclo-octadiene might yield cyclobutene and 1,6-cyclodecadiene yields cyclopentene etc. The last reaction is known and constitutes a good method for synthesizing pure cyclopentene starting from 1,6-cyclodecadiene [156,157]

(8.9)

thus extending the range of starting materials for the synthesis of this monomer. 1,6-Cyclodecadiene is obtained through cyclodimerization from ethene and two butadiene molecules.

The cometathesis of cyclo-olefins allows one to prepare any cyclic diene by varying the nature of the two coreactants

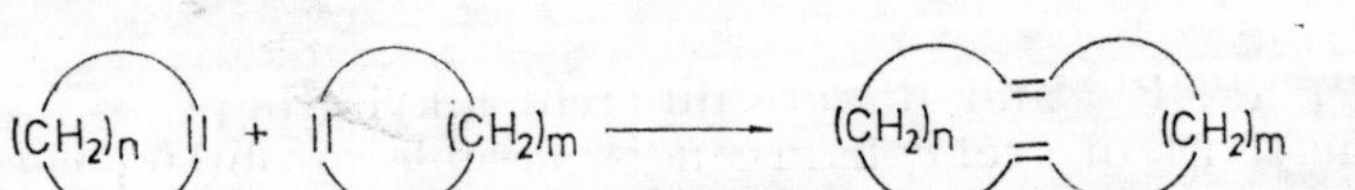

(8.10)

where m ≠ n = 1, 2, 3, 4, etc. Through partial or total hydrogenation, it is possible to prepare, *via* this route, higher cyclo-olefins (cyclo-alkenes) or cyclo-alkanes, starting from lower cyclo-olefins.

The cleavage or alkenolysis reaction of cyclo-olefins also constitutes an elegant method for synthesizing certain simple or substituted linear dienes or polyenes,

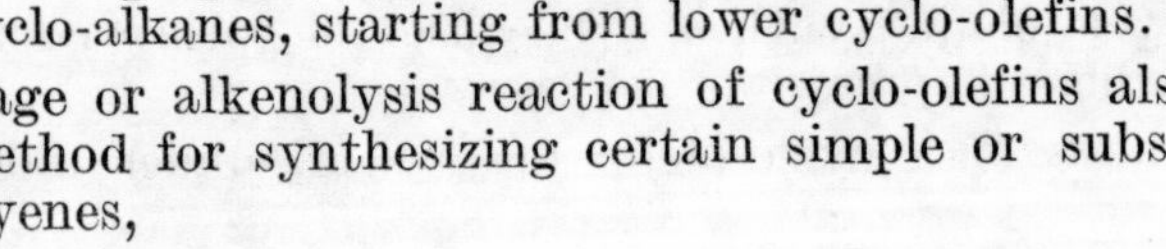

(8.11)

where n = 1, 2, 3, etc. and $R^I \neq R^{IJ} = H$, CH_3, C_2H_5, C_3H_7 etc. For example, cyclopentene yielded, through ethenolysis, 1,6-heptadiene [158]

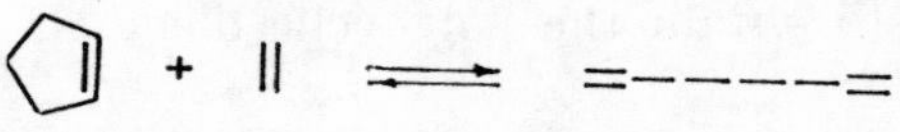

$$\text{⬠} + \| \rightleftharpoons = \text{-----} = \qquad (8.12)$$

and cyclo-octene afforded 1,9-decadiene.[158,159]

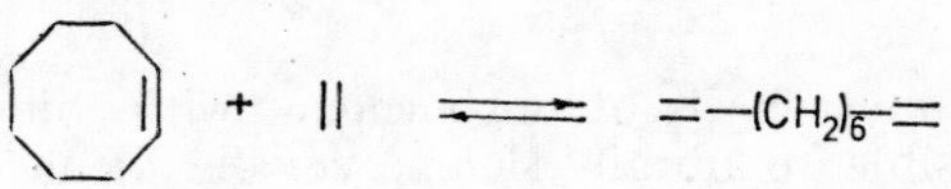

$$\text{⯃} \| + \| \rightleftharpoons = -(CH_2)_6 = \qquad (8.13)$$

Cyclic dienes or polyenes yield under mild conditions, linear trienes or polyenes; under more drastic conditions they can be converted to lower acyclic dienes

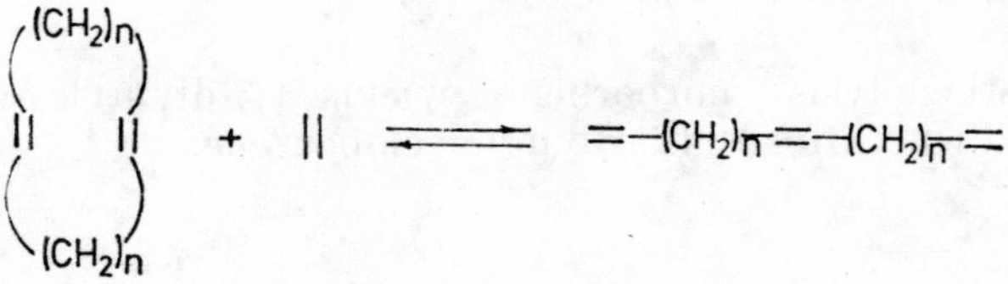

$$\begin{pmatrix} (CH_2)_n \\ \| \quad \| \\ (CH_2)_n \end{pmatrix} + \| \rightleftharpoons = -(CH_2)_{\overline{n}} = -(CH_2)_{\overline{n}} = \qquad (8.14)$$

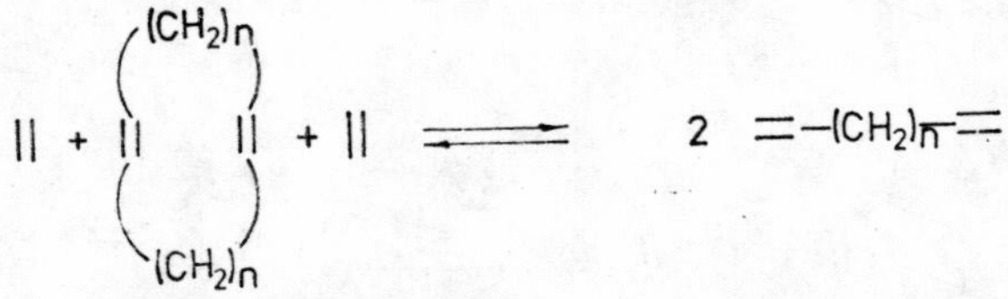

$$\| + \begin{pmatrix} (CH_2)_n \\ \| \quad \| \\ (CH_2)_n \end{pmatrix} + \| \rightleftharpoons 2 = -(CH_2)_{\overline{n}} = \qquad (8.15)$$

where n = 1, 2, 3, etc. For example, 1,5-cyclo-octadiene, through cleavage at the double bond with a molecule of ethene, yields 1,5,9-decatriene [74] while, through a reaction with two molecules of ethene at both double bonds, it yields 1,5-hexadiene.[160]

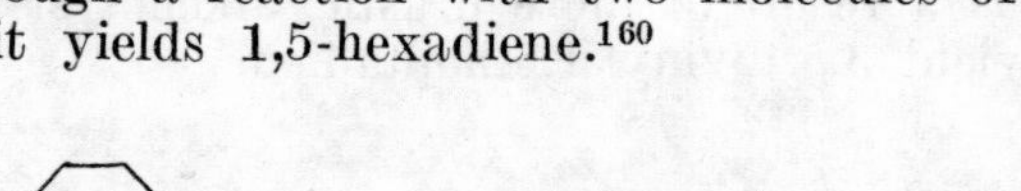

$$\text{⯃} + \| \rightleftharpoons \wedge\wedge\wedge\vee \qquad (8.16)$$

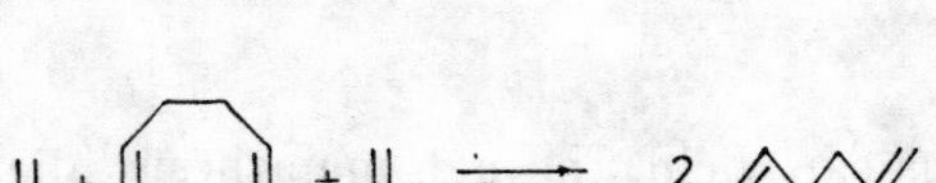

$$\| + \text{⯃} + \| \rightleftharpoons 2 \wedge\wedge\vee \qquad (8.17)$$

The metathesis or cometathesis reactions of polycyclic olefins are simple syntheses of unsaturated polycyclic hydrocarbons which are otherwise difficultly accessible by more familiar methods. Thus, from norbornene, it is possible to obtain the polycyclic diene (I)

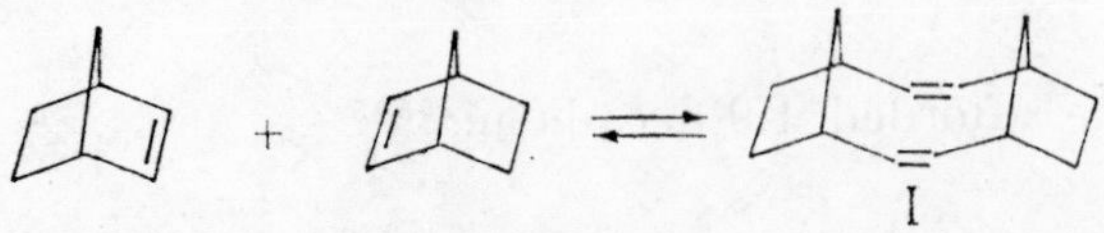

$$(8.18)$$

I

Through the cometathesis of norbornene with bicyclo-[2,2,2]octene it is similarly possible to obtain the polycyclic diene (II)

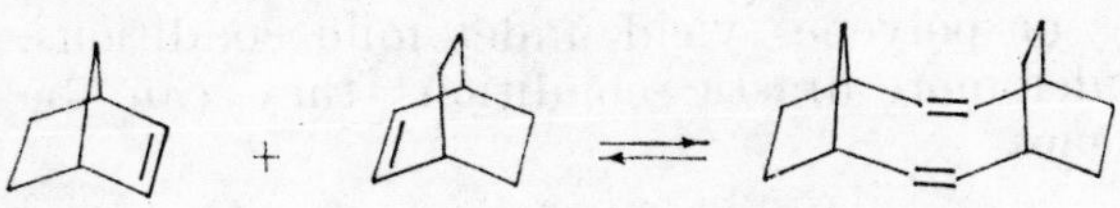

$$(8.19)$$

II

Through ethenolysis, norbornene yields 1,3-divinylcyclopentane,[73] while norbornadiene affords 1,3-divinylcyclopentene.

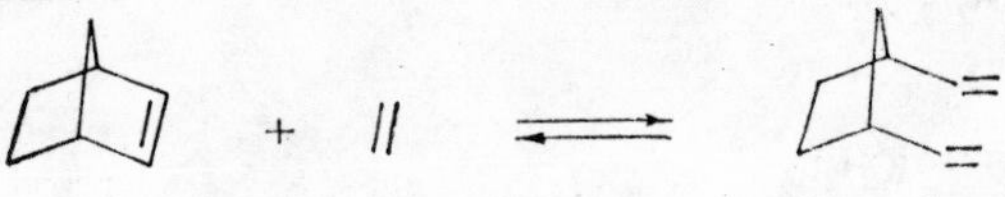

$$(8.20)$$

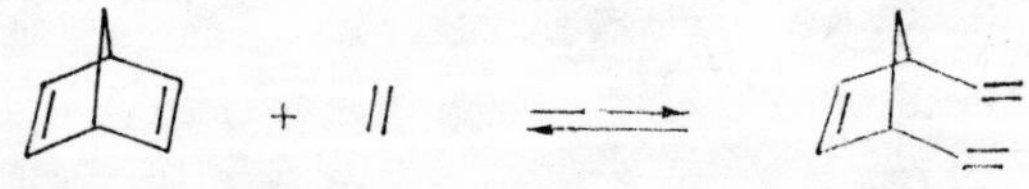

$$(8.21)$$

Polycyclic olefins with several double bonds can yield branched polyenic compounds as a result of more drastic alkenolysis. Norbornadiene and ethene can yield 3,5-divinyl-1,6-heptadiene.

$$(8.22)$$

Exocyclic olefins can yield, through metathesis, alicyclic olefins which are difficult to synthesize by known methods. By metathesis, it is possible to prepare di(cyclobutylidene) from methylenecyclobutane, and di(cyclohexylidene) from methylenecyclohexane [161,162] :

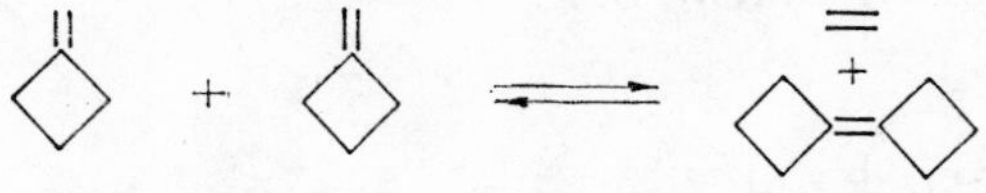

$$(8.23)$$

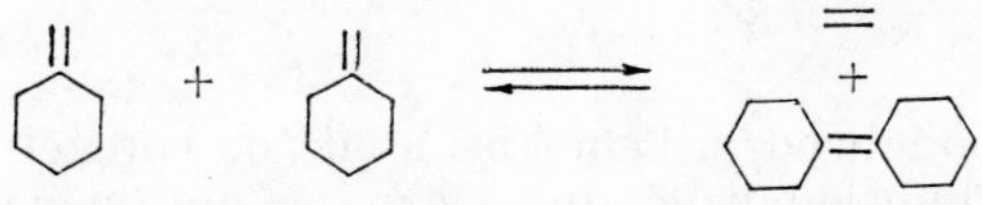

$$(8.24)$$

Extended to alicyclic hydrocarbons with more complicated structures, the method may provide other unsaturated hydrocarbons, such as di(bicyclo-octylidene) (III), di(adamantylidene) (IV), etc.

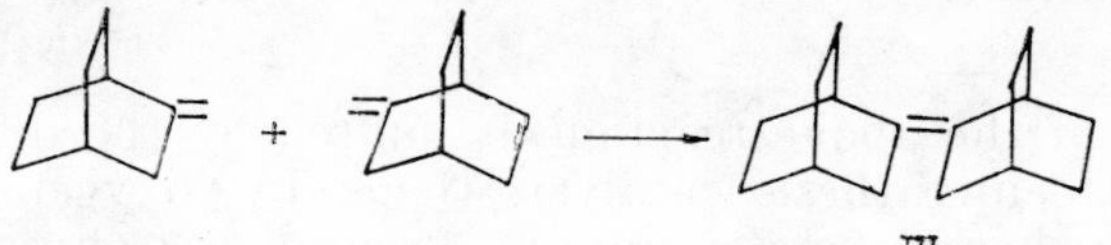

III

$$(8.25)$$

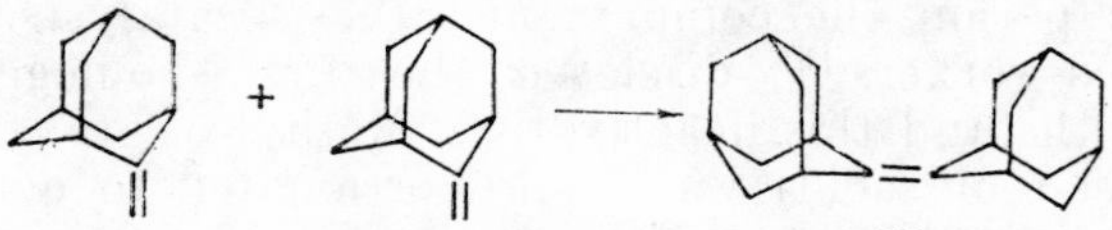

IV

$$(8.26)$$

8.4. Unsaturated Compounds with Functional Groups

The application of the metathesis reaction to alkenes bearing functional groups opens a new route for the synthesis of various mono- and di-functional derivatives of hydrocarbons. By homometathesis, monosubstituted alkenes can form disubstituted compounds which are obtained with difficulty by conventional methods.

$$(8.27)$$

Cometathesis with acyclic olefins offers the possibility of synthesizing a wide range of homologues of functionalized alkenes

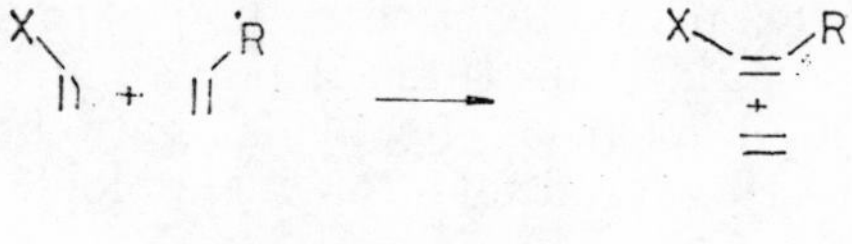

$$(8.28)$$

whereas the reaction with cyclo-olefins will lead to linear long-chain unsaturated functionalized alkenes.

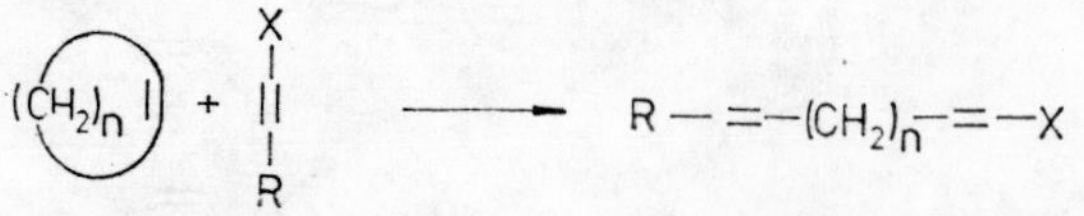

$$(CH_2)_n + \underset{R}{\overset{X}{\underset{|}{\overset{|}{||}}}} \longrightarrow R-=-(CH_2)_n-=-X \qquad (8.29)$$

When a difunctionalized alkene is used in coreaction with a cyclo-olefin, an α,ω-difunctionalized diene can be obtained.

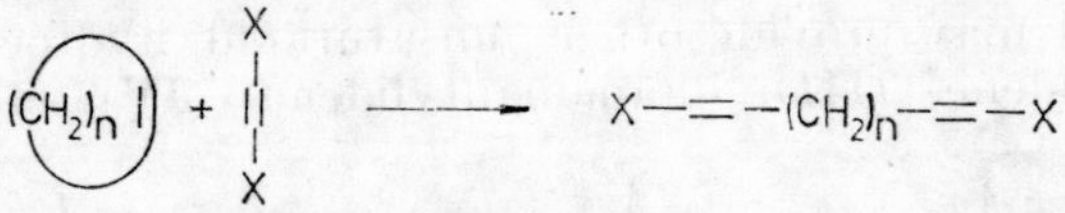

$$(CH_2)_n + \underset{X}{\overset{X}{\underset{|}{\overset{|}{||}}}} \longrightarrow X-=-(CH_2)_n-=-X \qquad (8.30)$$

Application of the above reactions is limited by the difficulty of finding an adequate metathesis catalyst. Generally, oxygen-, halogen-, nitrogen- or other heteroatom-containing functional groups strongly deactivate or even poison the common metathesis catalysts. In 1972 Boelhouwer and co-workers[163] disclosed the first homogeneous catalyst which could withstand this deactivation. Methyl oleate (the methyl ester of *trans*-octadec-9-oic acid) was disproportionated to octadec-9-ene and to the dimethyl ester of octadec-9-enedionic acid by using $WCl_6 \cdot ZnMe_4$ catalyst.

The use of Boelhouwer's catalyst thereafter permitted a large number of unsaturated fatty esters to be metathesized for synthetic purposes.[164-169]

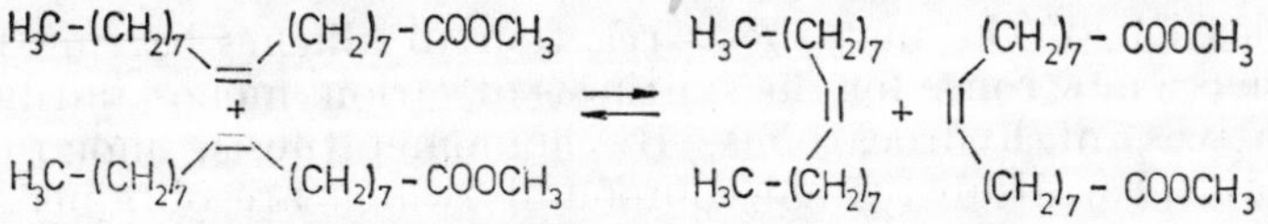

$$\begin{array}{c} H_3C-(CH_2)_7 \diagdown \diagup (CH_2)_7-COOCH_3 \\ + \\ H_3C-(CH_2)_7 \diagup \diagdown (CH_2)_7-COOCH_3 \end{array} \rightleftharpoons \begin{array}{c} H_3C-(CH_2)_7 \qquad (CH_2)_7-COOCH_3 \\ || + || \\ H_3C-(CH_2)_7 \qquad (CH_2)_7-COOCH_3 \end{array} \qquad (8.31)$$

Other homogeneous and heterogeneous catalytic systems have been reported.[170-180] The reaction has been applied to erucic, linoleic, linolenic, undecenoic esters as well as to low-molecular weight alkenoic esters.

Unsaturated dicarboxylic esters may thus find a use in the production of unsaturated polyesters and polyamides of a new type. Thus, unsaturated polyesters, formed by transesterification of dimethyl octadec-9-ene-dioate with 1,2-propanediol, possess outstanding properties as a stationary phase in gas-liquid chromatography[166]; such esters, after vulcanization with sulphur or peroxide agents, may provide elastomers with good adhesive properties.[166] Unsaturated polyamides, which can be prepared from unsaturated dicarboxylic acids, may be vulcanizable, thus providing new interesting products for technological application.

Unsaturated diacids may lead, *via* ketodecarboxylation to ketones or, *via* intramolecular condensations, to other cyclic compounds which are difficult to obtain by conventional methods. For instance, the ketodecarboxylation of octadec-9-enedioic acid obtained from the corresponding ester offers a new route to the production of civetone, a valuable compound for the perfume industry (8.32).

$$
\begin{array}{c}
\text{C} \begin{array}{l} (CH_2)_7 - COOH \\ (CH_2)_7 - COOH \end{array} \quad \longrightarrow \quad \text{C} \begin{array}{l} (CH_2)_7 \\ (CH_2)_7 \end{array} \hspace{-6pt}\diagdown\hspace{-4pt} CO + H_2O + CO_2
\end{array}
\qquad (8.32)
$$

The cometathesis of unsaturated esters with linear or branched olefins offers a new possibility of preparing homologues of unsaturated esters with well-defined structures. These may be employed for the synthesis of saturated and unsaturated fatty acids or esters, and polyunsaturated acids or esters with high molecular weight.[166,175]

When applied to polyunsaturated fatty oils, metathesis facilitates the enlargement of molecules without decrease of unsaturation. For instance, the reaction of olive oil results in the formation of 9-octadecene and an oil of increased molecular weight and viscosity, which includes dimeric and higher molecular weight glycerides. Likewise, metathesis of soyabean oil and linseed oil leads to the formation of stand oils with high viscosity index and improved drying properties.

Low-molecular unsaturated esters of the general formula $RCH=CH-(CH_2)_n-COOR'$, where $n \geqslant 1$, were found to metathesize to the corresponding unsaturated dicarboxylic esters.[168] Thus, esters of 1-butenoic, 1-, 2-, and 3-pentenoic, and 3- and 4-hexenoic acids gave the expected products with high conversion and selectivity. When the double bond is in α,β-position (i.e. with propenoic esters and its homologues), the reaction takes place only if an additional methyl substituent is present, as in 2-butenoic and 2-methyl-2-butenoic esters. Higher activities were reported in the cometathesis of these esters with symmetrical alkenes, e.g. 3-hexene. The reaction may find synthetic use for esters such as acrylic, methacrylic, crotonic, maleic, fumaric, and hydromuconic esters.

The potential for reaction of low-molecular weight saturated ester is considerably enhanced by using cyclo-olefins as coreactants. Cometathesis of cyclo-octene with unsaturated mono- or diesters leads to the synthesis of long-chain linear mono- and diesters.[176] Similar compounds may be obtained using other cyclo-olefins. Furthermore, cyclodienes, e.g. 1,5-cyclo-octadiene, with unsaturated diesters afford polyenes which may be used as prepolymers in condensation or reticulation processes.[177]

As the carboxyl group of unsaturated acids has a high affinity for binding with metal catalysts, the metathesis of free acids is strongly inhibited. Hummel and Imamoglu [178] carried out the cometathesis of erucic acid with 7-tetradecene in the presence of excess of 4-octene but the selectivity was low and the acid reacted preferentially with the tungsten catalyst. Further work may lead to the discovery of more refined catalysts for the direct metathesis of free unsaturated acids.

The metathesis of unsaturated nitriles affords unsaturated dinitriles and the corresponding olefin. The reaction of nitriles $CH_2=CH(CH_2)_n$ CN in the presence of the homogeneous system $WCl_6 \cdot Me_4Sn$ occurs for $n = 2, 3$ in low conversion.[181] Higher yields are obtained in cometathesis with linear olefins for $n = 1, 2, 3, 4$ when higher homologues of nitriles are formed. Acrylonitrile, however, can be cometathesized with propene on a heterogeneous catalyst of tungsten oxide or molybdenum oxide giving crotononitrile and ethylene.[182] Higher unsaturated nitriles such as 9-octadecenenitrile have been transformed to linear dinitriles with homogeneous tungsten and molybdenum catalysts but the selectivity was low.[172,183]

The metathesis of unsaturated amines of general formula $CH_2=CH(CH_2)_nNRR'$ or of related hydrochlorides gives diamines or their hydrochlorides in the presence of homogeneous tungsten or molybdenum catalysts.[181-186] The reaction has maximum activity for $R = R' = H$ and $n = 3$. Substitution of R and R' for alkyl groups drastically lowers the conversion.

By reacting unsaturated ethers, Ast and co-workers [179] found that vinyl butyl ether and allyl butyl ether do not transform metathetically, while higher homologues do. Thus, the metathesis of 3-butenyl-, 4-pentenyl-, 3-hexenyl- and 10-undecenyl butyl ethers with $WCl_6 \cdot SnMe_4$ leads to the corresponding diethers. The cometathesis of unsaturated ethers with cyclo-olefins may be of use in providing telechelic oligomers. Telechelic polymers or oligomers have been defined as polymeric (oligomeric) molecules having two functional terminal groups. Such products were detected in the coreaction of 4-pentenyl butyl ether with cyclo-octene or ambrettolide (7-hexadecene-(16)-olide) when a statistical equilibrium was attained. By contrast, Mol and Woerlee [180] succeeded in metathesizing allyl ethyl ether on the heterogeneous catalysic system $Re_2O_7(+Me_4Sn) \cdot Al_2O_3$. The selectivity to diether proved to be high and free from double bond isomerization.

The synthesis of unsaturated diketones appears to be readily available by metathesis of unsaturated ketones. In the presence of the abovementioned heterogeneous catalytic system, Mol and Woerlee transformed 3-butenyl methyl ketone into the expected diketone with high selectivity.

When carried out with unsaturated haloderivatives, the metathesis reaction opens a new route for the preparation of compounds of this class as potential intermediates for organic synthesis. Thus, metathesis of monohaloalkenes forms dihaloalkenes, while cometathesis with terminal or internal alkenes may lead to homologues of haloalkenes.

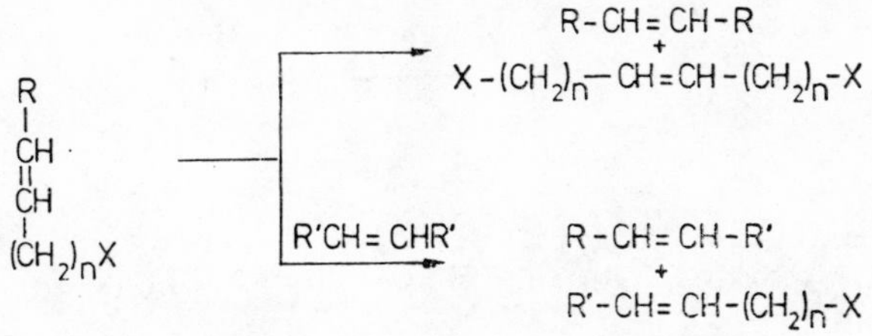

$$
\begin{array}{c}
R-CH=CH-R \\
+ \\
X-(CH_2)_n-CH=CH-(CH_2)_n-X
\end{array}
\qquad (8.33)
$$

Schema:
$$
\begin{array}{c}
R \\
| \\
CH \\
\| \\
CH \\
| \\
(CH_2)_n X
\end{array}
\longrightarrow
\begin{array}{c}
R-CH=CH-R \\
+ \\
X-(CH_2)_n-CH=CH-(CH_2)_n-X
\end{array}
$$

$$
R'CH=CHR' \qquad
\begin{array}{c}
R-CH=CH-R' \\
+ \\
R'-CH=CH-(CH_2)_n-X
\end{array}
$$

The reaction of vinyl bromide or chloride with propene, isobutene, 1-butene or cyclohexene on the heterogeneous catalytic system $Re_2O_7 \cdot Al_2O_3(+Bu_4Sn)$ yielded the corresponding homologues of bromo- or chloro-alkenes.[187] Allyl bromide forms, by heterogeneous self-metathesis, 1,4-dibromo-2-butene whereas the coreaction with propene gives 1-bromo-2-butene. Similar reactions of synthetic utility lead to higher haloalkenes such as 5-bromo-1-pentene, 5-bromo-1-hexene, 6-bromo- and 6-chloro-2-hexene and 1-chloro-9-octadecene in heterogeneous or homogeneous media.[187,188]

A silyl group in alkyl silyl unsaturated derivatives does not essentially affect the metathesis of this class of compounds. Trimethylsilylethene metathesizes to bis(trimethylsilyl)ethene on the Et_4Sn- or Bu_4Sn-promoted heterogeneous catalysts $MoO_3 \cdot Al_2O_3$ and $Re_2O_7 \cdot Al_2O_3$.[189] Trichlorosilyl-, alkylchlorosilyl- or oxysilyl-alkenes react in the same manner.[190] The coreaction of silyl derivatives such as vinyltrichlorosilane or methylvinyldichlorosilane with cyclo-olefins or polyalkenamers has proved to be of practical use.[191,192] In this case they may act as molecular weight regulators or cleaving agents to telechelic polymers. Similar results have been reported recently by using stannyl derivatives in cyclo-olefin metathesis polymerization.[190,193]

8.5. Synthesis of Acetylenes

The metathesis reaction, discovered by Banks and Bailey for the acyclic olefins, was later successfully applied to acetylenes.[194] More recently, the reaction has been extended to a larger number of substituted acetylenes and acetylenic derivatives, thus becoming a useful synthetic method for this class of compounds.[195-201]

Through the metathesis of acetylenes substituted with alkyl groups, it was possible to obtain, along with acetylene, internal acetylenes symmetrically substituted with alkyl groups e.g.

$$
\underset{R}{\overset{|}{\underset{|}{|||}}} + \underset{R}{\overset{|}{\underset{|}{|||}}} \longrightarrow \quad \overset{\equiv\equiv}{} + \quad R-\equiv-R \tag{8.34}
$$

where $R = CH_3$, C_2H_5, C_3H_7 and C_6H_5. The reaction was carried out with propyne, butyne, pentyne,[195] hexyne,[196] phenyl acetylene,[196] and 1-phenylhexyne.[197]

The metathesis of propargyl alcohol [198] constitutes the direct synthesis of butyne diol.

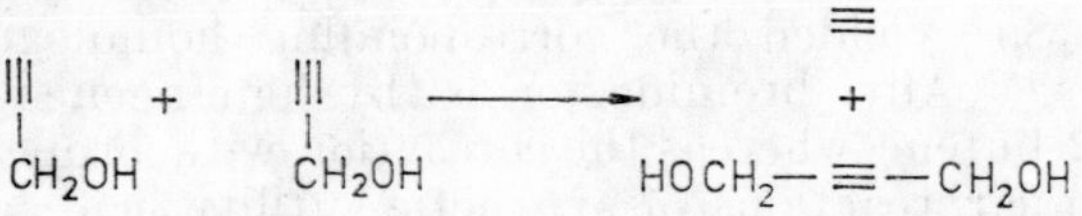

$$
\underset{CH_2OH}{\overset{|}{\underset{|}{|||}}} + \underset{CH_2OH}{\overset{|}{\underset{|}{|||}}} \longrightarrow \quad \overset{\equiv}{} + \quad HOCH_2-\equiv-CH_2OH \tag{8.35}
$$

The metathesis of certain acetylene derivatives possessing functional groups can yield compounds which are difficult to obtain through common methods. One example is the metathesis of functional phenyl-1-but-1-yne derivatives with molybdenum carbonyls or homogeneous catalysts giving tolane and disubstituted acetylenes[201a]

$$
2C_6H_5C \equiv C(CH_2)_2X \rightleftharpoons C_6H_5C \equiv CC_6H_5 + X(CH_2)_2C \equiv C(CH_2)_2X \tag{8.35a}
$$

where $X = OH$, $OCOCH_3$, Br, COOH, COOCH_3 and CN.

Extended to cyclic acetylenes, in the variants known from cyclo-olefins, metathesis can constitute a new method for synthesizing linear or cyclic diynes and polyynes.

$$
(CH_2)_n\,||| + \underset{R}{\overset{|}{\underset{|}{|||}}} \longrightarrow \quad \equiv-(CH_2)_n-\equiv-R \tag{8.36}
$$

$$
(CH_2)_n\,||| + |||\,(CH_2)_n \longrightarrow (CH_2)_n \overset{\equiv}{\underset{\equiv}{}} (CH_2)_n \tag{8.37}
$$

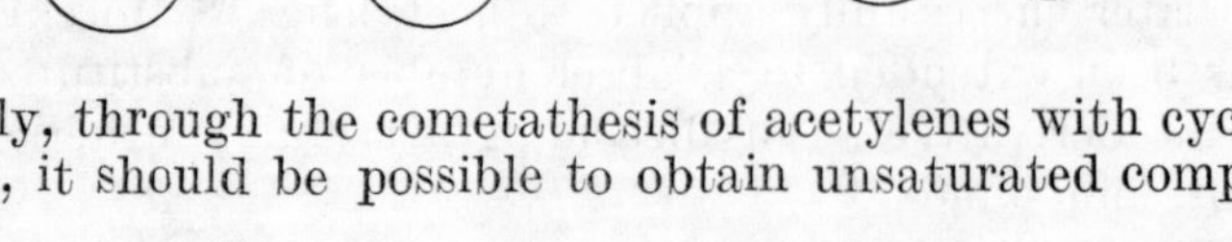

Similarly, through the cometathesis of acetylenes with cyclic and acyclic olefins, it should be possible to obtain unsaturated compounds with double and triple bonds in the chain :

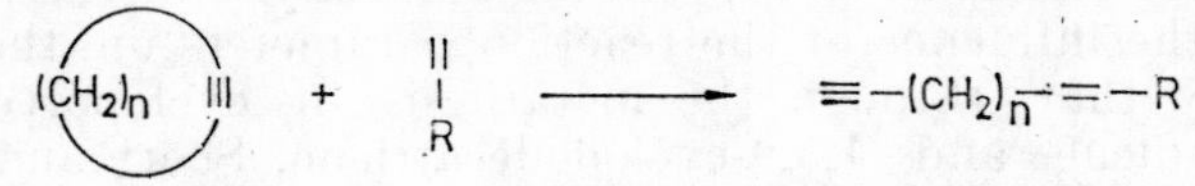

$$\text{(8.38)}$$

$$\text{(8.39)}$$

where R is an alkyl or aryl group.

8.6. Macrocycles, Catenanes, Knots and Rotaxanes

8.6.1. Synthesis of Macrocycles

Metathesis provides an elegant and simple method for preparing unsaturated carbocyclic macrocycles. This method is more advantageous than the conventional routes as it comprises only a one-step process. In the presence of a metathesis catalytic system, cyclo-olefins yield, through cyclo-oligomerization, carbocycles whose content of carbon atoms is a multiple of that in the initial cyclo-alkene.

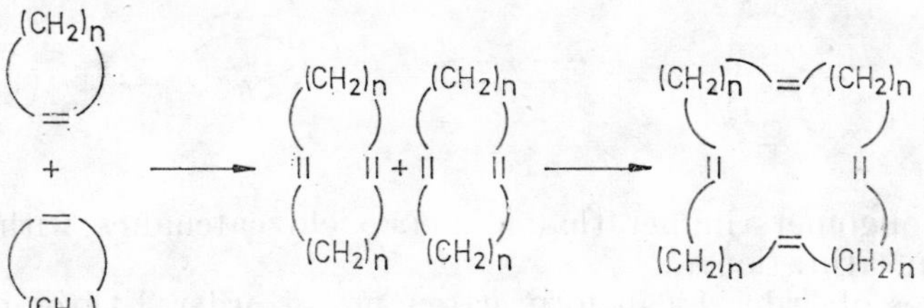

$$\text{(8.40)}$$

Starting from cyclo-octene, Wasserman, Ben-Efraim and Wolovsky [202] obtained unsaturated carbocycles with up to 120 carbon atoms (oligomerization degree up to 15). The reaction proceeded under mild metathesis conditions, with the homogeneous catalyst $WCl_6 \cdot EtAlCl_2 \cdot EtOH$, at $5-20°C$ in benzene as solvent. The separation of the reaction products was accomplished through gas-chromatography and the analysis of the fractions obtained was performed by mass-spectrometry. The corresponding saturated carbocycles were also obtained by catalytic reduction.

The oligomerization products of cyclo-octene obtained through the metathesis of cyclo-octene, were also studied by Scott,[203] Calderon,[204,205] and Höcker and co-workers.[206-209] Scott and Calderon [203] were the first to define the nature of oligomeric macrocycles formed by the metathesis of cyclo-octene. Höcker and Musch [206] elaborated in more detail the methods of chromatographic and spectrometric analysis of these compounds, and studied the influence of the reaction parameters on their nature. It is noteworthy that, through the metathesis of cyclopolyenes such as 1,5-cyclo-octadiene and 1,5,9-cyclododecatriene, Scott and Calderon [203] obtained carbocycles with similar structures. The higher degree of unsaturation led, in this case, to greater fragmentation of the reaction products.

A wide range of unsaturated carbocycles was synthesized by Wolovsky and Nir [210] by the metathesis of cyclododecene. The unsaturated oligomers, obtained in the first stage, were reduced to the corresponding mono-olefins which were further on subjected to oligomerization, in the presence of the catalytic system. By this method, these authors easily synthesized carbocycles with 23, 36 or 48 carbon atoms.

8.6.2. Syntheses of Catenanes, Knots and Rotaxanes

Metathesis is an elegant way for preparing compounds with interlocked cyclic structures such as catenanes, knots, rotaxanes and other analogous compounds.[211-214]

Syntheses of catenanes. The dimers and cyclo-oligomers which result from larger cyclo-olefins can readily form catenanes through intramolecular metatheses. A catenane with an interlocking topological number $\alpha = 1$ is formed when the dimeric diene, prior to intramolecular metathesis, undergoes half-twists (n=2).

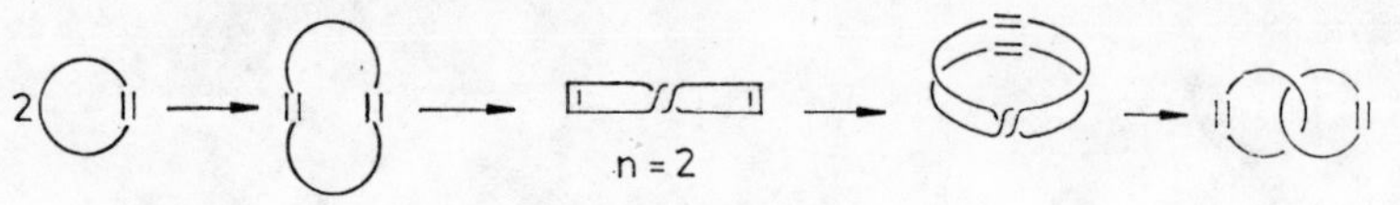

$$(8.41)$$

Cyclo-oligomers higher than dimers yield catenanes with a higher degree of unsaturation.

Catenanes of order 1 can form catenanes of order 2 through intermolecular metathesis with a cyclo-olefin, followed by intramolecular metathesis accompanied by the same number of half-twists.

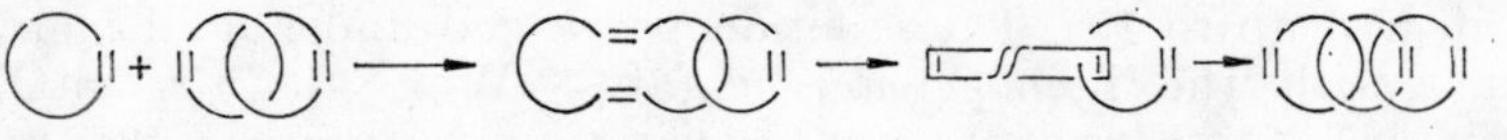

$$(8.42)$$

Metathesis between two catenanes of order 1 can yield catenanes of order 3.

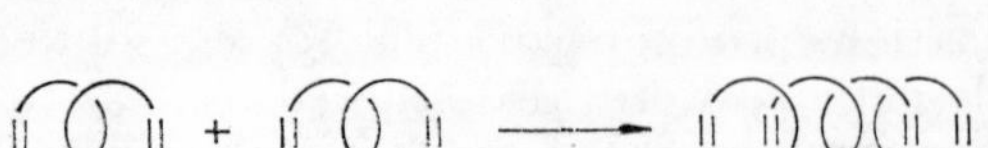

$$(8.43)$$

Through similar reactions between catenanes or between the latter and cyclo-olefins, still higher-order catenanes can be formed.

Synthesis of knots. Molecular knots can be formed from cyclo-olefins, through metathesis, in a reaction in which the dimeric diolefin undergoes three or more half-twists, prior to metathesis. The reaction in which the diolefin undergoes three half-twists gives rise to a knot with $N = 3$ interlockings.

$$(8.44)$$

Reactions accompanied by 4, 5, or 6 half-twists give rise to knots with interlocking number $N = 4$, 5, or 6, respectively.

$$(8.45)$$

$$(8.46)$$

$$(8.47)$$

Metathesis reactions between olefinic knots and catenanes or between two unsaturated knots can lead to more complex structures called knot-catenanes or polyknots.

Syntheses of rotaxanes. The alkenolysis reaction applied to unsaturated catenanes or to other interlocked rings makes possible the synthesis of rotaxanes. Thus, a catenane with interlocking order 1 can give rise to a rotaxane of order 1

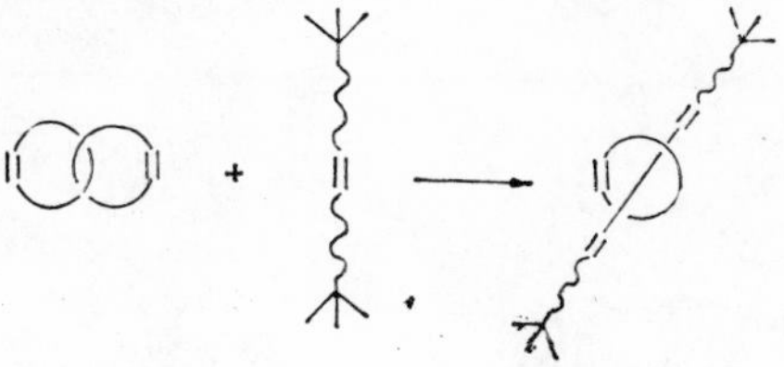

$$(8.48)$$

Catenanes with a higher interlocking number can lead to higher-order rotaxanes. For example, a catenane with interlocking number 3 forms a rotaxane of order 2.

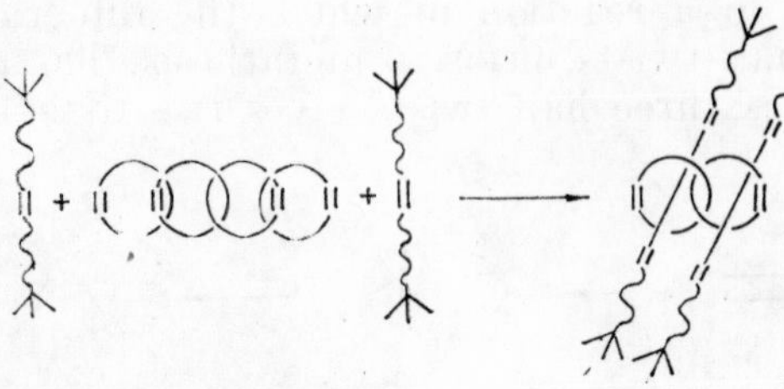

$$(8.49)$$

Rotaxanes having complex structures can be formed by alkenolysis of knots or of other interlocked ring structures.

8.7. Polyalkenamers

Recently, metathesis has become an important method for manufacturing polyalkenamers from cyclo-olefins, thus extending the range of application of these polymers in the field of elastomers, plastics, synthetic resins, etc. This method provides a means of direct synthesis of polyalkenamers from cyclo-olefins, as well as the chemical conversion of polyalkenamers through insertion or grafting reactions at the double bond in the polymer. The present chapter will demonstrate the importance for the elastomer industry of the metathesis synthesis of polyalkenamers. The properties of the main polyalkenamers obtained from cyclo-olefins will also be presented.

8.7.1. Syntheses of Polyalkenamers

Polybutenamer, generally known as 1,4-polybutadiene, is at present obtained on a large scale by the polymerization of butadiene.[215] The metathesis reaction affords a new way of synthesizing this polyalkenamer through the ring-opening polymerization of cyclobutene.[216-219] The same polymer can be also obtained from 1,5-cyclo-octadiene or 1,5,9-cyclodecatriene. [203, 220]

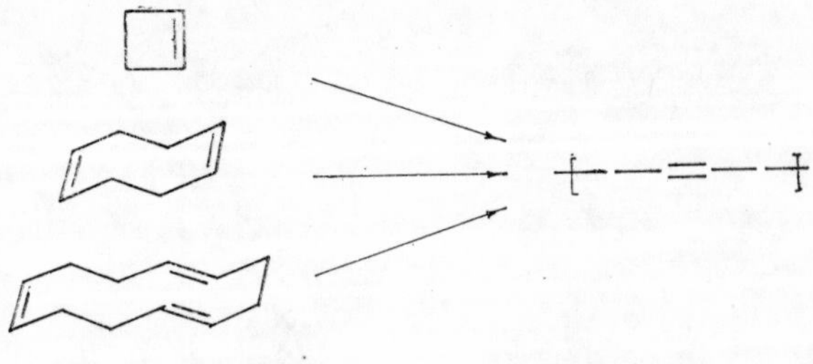

$$(8.50)$$

The polybutenamer thus obtained possesses a high degree of stereospecificity [219] the nature of which depends on the catalytic system employed; it exhibits properties similar to those of the conventional polymer. The new method can utilize any of the three cyclo-olefins, the choice depending on the availability of the possible starting materials. 1,5-Cyclooctadiene and 1,5,9-cyclododecatriene are readily obtained from butadiene through cyclodimerization [221] and cyclotrimerization,[222] respectively. However, up to the present date, this method for the preparation of polybutenamer has not been economically competitive.

The synthesis of poly-1-methylbutenamer (polyisoprene), a polymer which has stimulated intensive investigations for a long period of time and which is now obtained synthetically by the polymerization of isoprene, can be achieved by metathesis of 1-methylbutenamer or 1,5-dimethyl-1,5-cyclo-octadiene.

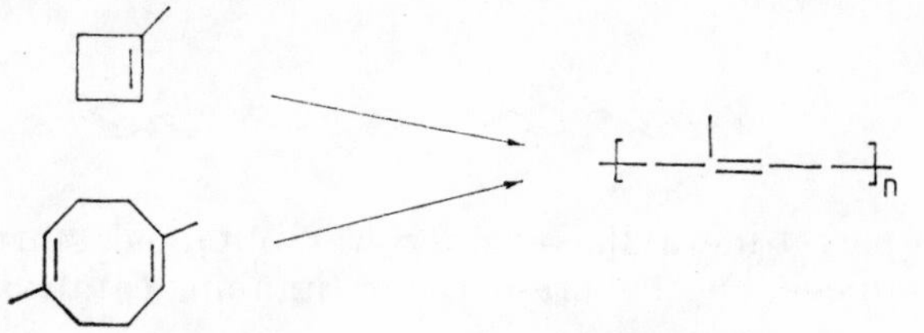

$$\text{(8.51)}$$

The reaction of 1-methylcyclobutene with carbene metathesis systems yields a polyisoprene with a very high content of the *cis* isomer ($\sim 90\%$).[223] This high content of *cis* isomer imparts elastomeric properties to the polymer very similar to those of natural rubber. Unfortunately, the synthesis of polyisoprene through this method is limited by the poor availability of the starting material, 1-methyl-cyclobutene and 1,5-dimethyl-1,5-cyclo-octadiene.

The polymerization of cyclopentene to polypentenamer found application immediately after the discovery of olefin metathesis.[224-228] In the presence of metathesis catalytic systems, cyclopentene yields *trans*- or *cis*-polypentenamer, depending on the catalytic system employed.[229-243]

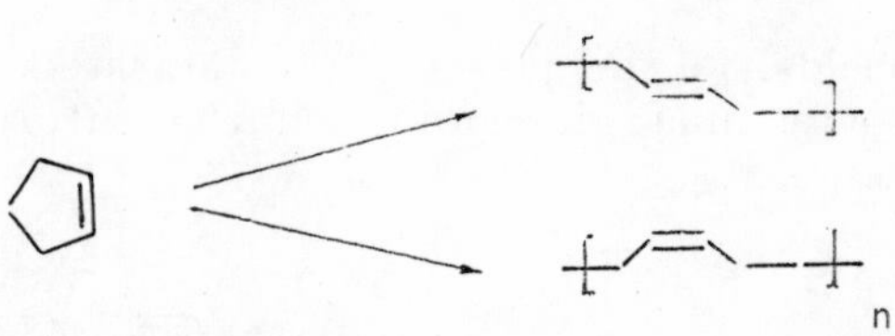

$$\text{(8.52)}$$

The properties of the polypentenamer obtained by this method are particularly interesting; *trans*-polypentenamer has some properties close to those of natural rubber while *cis*-polypentenamer is similar to some special rubbers. This fact may explain the interest that cyclopentene polymerization has aroused in the rubber industry.[224,227]

In a very short time, the Bayer A. G. Company developed an industrial process for polypentenamer production for general use in tyre fabrication.[226,244]

Polyhexenamer can not be obtained through the metathesis of cyclohexene. The polymerization of cyclohexene in the presence of metathesis catalysts yields only vinyl polymers with the polycyclohexane structure.[245] The synthesis of polyhexenamer was achieved through copolymerization of ethene with butadiene in the presence of ternary catalytic systems of the Ziegler-Natta type, based on vanadium or titanium.[246]

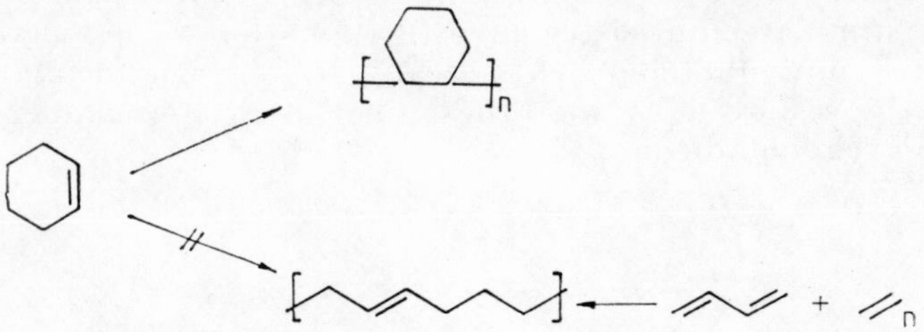

$$(8.53)$$

Similarly, poly-4-methylhexenamer was obtained starting from propene and butadiene, in the presence of titanium catalysts.[247,248]

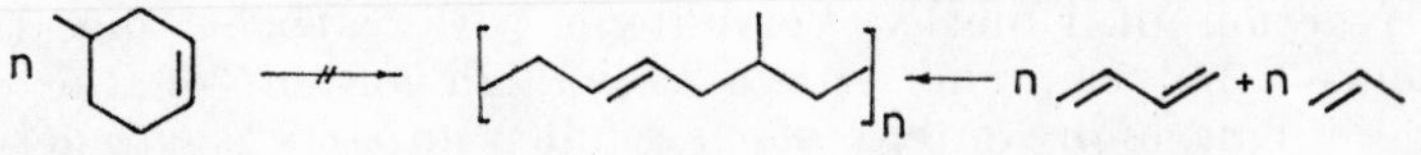

$$(8.54)$$

Polyheptenamer is obtained through metathesis of cycloheptene in the presence of binary or ternary catalytic systems based on tungsten or molybdenum.[249-251] The polymer obtained has the *trans* configuration at the double bond.

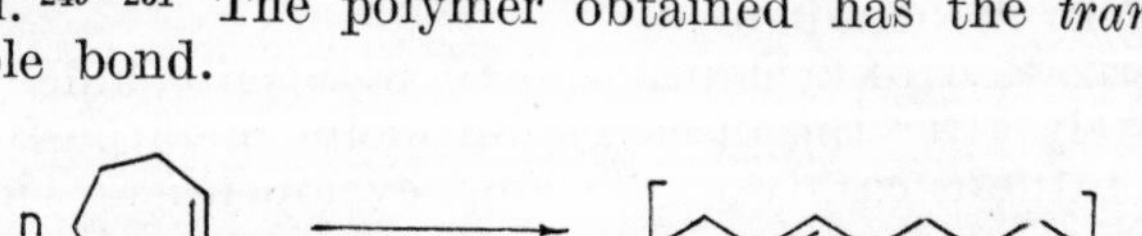

$$(8.55)$$

Cyclo-octene yields, in the presence of metathesis catalytic systems, *cis*- or *trans*-polyoctenamer, depending on the nature of the catalytic system.[251]

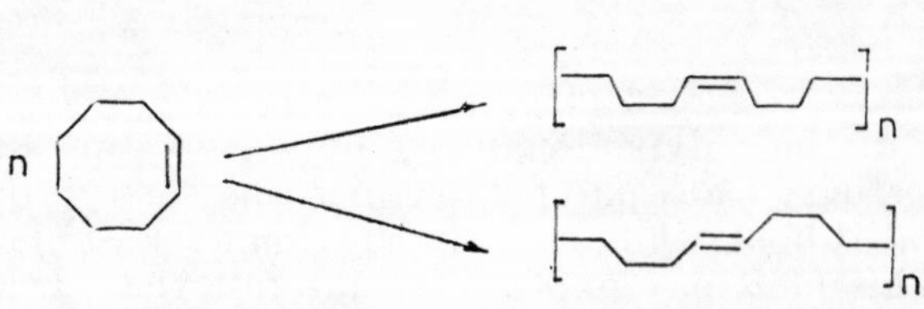

$$(8.56)$$

Binary catalytic systems based on WCl_6 lead to polyoctenamer with the *trans* configuration at the double bond [252] and those based on WF_6 to the corresponding *cis* configuration.[253] The polymer with high *trans* content has excellent processing properties and valuable physical characteristics after vulcanization in blends with other rubbers. *Trans*-polyoctenamer is now obtained on commercial scale by Chemische Werke Hüls ("Veste-namer 8012") for the rubber industry [254] in a process based on cyclo-octene production *via* butadiene cyclodimerization and subsequent par-tial hydrogenation.

The polymerization of cyclodecene can also yield either conformer of the polydecenamer, *cis* or *trans*, depending on the catalytic system.[251]

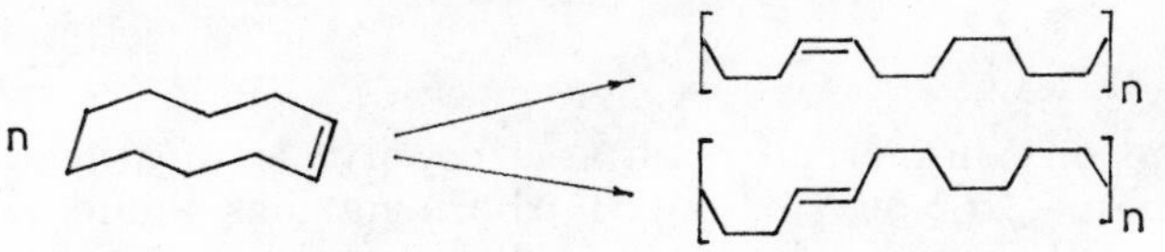

$$\text{(8.57)}$$

trans-Polydecenamer has been obtained with a catalytic system based on $WOCl_4$ and it possesses elastomeric properties. [255]

A polydodecenamer, with high *trans* configuration content, is obtain-ed from cyclododecene by using catalytic systems based on WCl_6. [251, 252]

$$\text{(8.58)}$$

Among the substituted cyclo-olefins, the metathesis reaction can be extended to a large number of substituted polyalkenamers. Among them, mention will be made only of those that are of interest for elastomer in-dustry.

The polyalkenamers which are likely to be of practical interest are those obtained from 3-methylcyclo-octene, 3-phenylcyclo-octene, 3-methyl-cyclodecene. [252, 256, 257] Poly(3-methyloctenamer) and poly(3-methylde-cenamer), both atactic unsaturated polymers, yield, through hydrogena-tion, poly(methylalkenamers) with elastomer properties. They possess better properties than the corresponding copolymers of ethene and pro-pene [258]

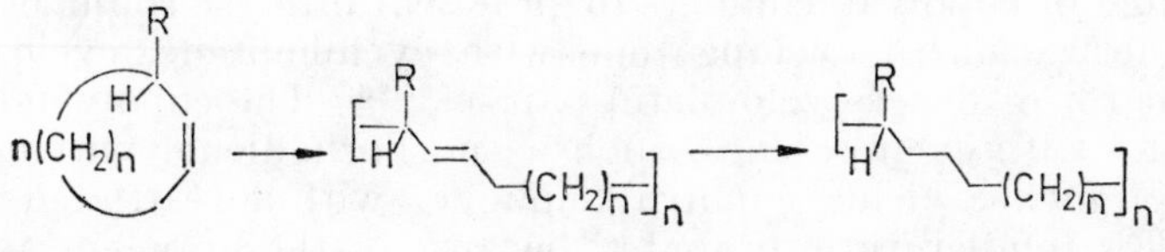

$$\text{(8.59)}$$

where $n = 4$; $R = CH_3$ or C_6H_5 or $n = 6$ and $R = CH_3$.

Poly(norbornene) and its alkyl-substituted derivatives offer a large class of polyalkenamers with potential utility as elastomers. Norbornene, owing to its high reactivity, leads in the presence of a whole range of catalysts based on tungsten, molybdenum, rhenium as well as ruthenium, osmium and iridium, to a high yield of *trans-* or *cis-*polynorbornene.

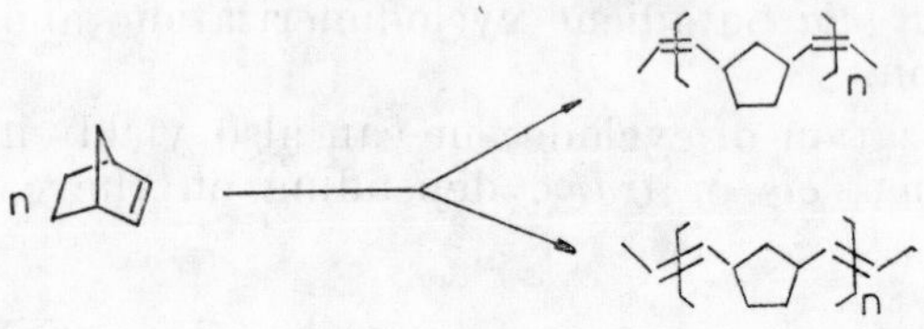

$$\tag{8.60}$$

Produced on commercial scale as a copolymer of ethene and propene under the name Norsorex [259] poly(norbornene) has found applications as a novel low hardness vulcanizate. The reaction might successfully be extended to various alkyl-substituted norbornenes of potential practical interest.

Copolymers with a structure which alternates butadiene and substituted butadiene are readily obtainable through the metathesis of 2-substituted 1,5-cyclo-octadiene, in the presence of ternary catalytic systems based on WCl_6 [260]

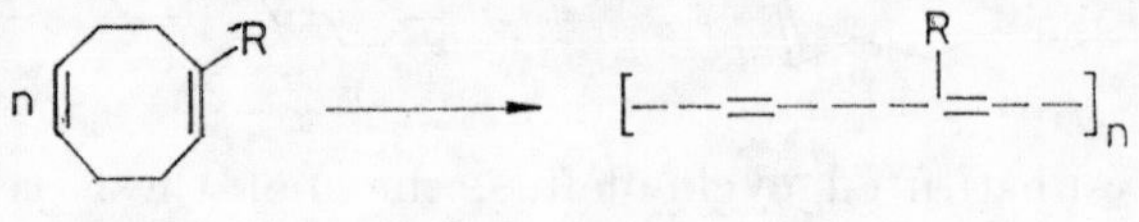

$$\tag{8.61}$$

where $R = CH_3$, C_2H_5, Cl or of 1,2-disubstituted 1,5-cyclo-octadiene

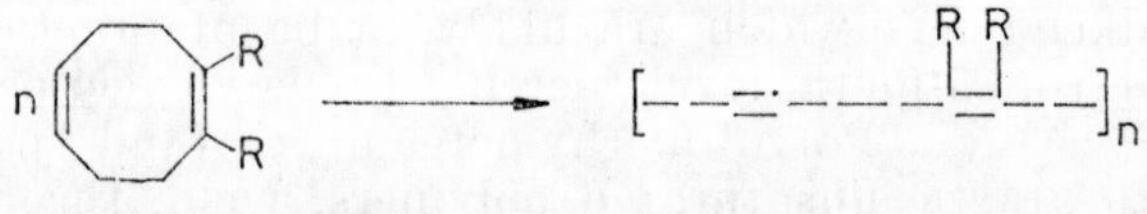

$$\tag{8.62}$$

where $R = CH_3$. The reaction proceeds in this manner because the substituted double bond is less reactive than its unsubstituted counterpart, so that ring-opening polymerization takes place only at the latter site. A wide range of copolyalkenamers of practical interest could be obtained by copolymerization of cyclopentene with cycloheptene, cyclo-octene or cyclodecene (or of other cyclo-olefin pairs).[261,262] The copolymerization of cyclopentene with cyclic dienes, such as norbornadiene, vinylnorbornene or cyclo-octadiene, yields copolyalkenamers with a restricted flow properties at low temperature, probably because of the branches and bridges existing between chains.[263,264] The copolyalkenamers obtained from a

simple cyclo-olefin such as cyclopentene or cyclo-octene, and a polychlorinated cyclo-olefin, such as hexachlorocyclopentadiene exhibit high resistance to lubricating oils, comparable to that of polychloroprene rubber [265]
because of the high content of chlorine.

Through metathesis, it is possible to obtain polyalkenamers with microstructures of block polymers, graft polymers or branched polymers.
The method has lately been applied for obtaining polymers with improved elastomeric or plastic properties.

The block polymers are formed by the cometathesis of two polyalkenamer chains, or of other polymers containing double bonds in the molecule.

$$
\begin{array}{l}
AAAAAHC{=}CHAAA \\
\quad\quad\quad + \\
BBBBC{=}CHBBB
\end{array}
\;\rightleftarrows\;
\begin{array}{l}
AAAAACH \quad CHAAA \\
\quad\;\| \;+\; \| \\
BBBCH \quad CHBBB
\end{array}
\qquad (8.63)
$$

This reaction was applied by Scott and Calderon [266,267] to obtain AB-type
block polymers from polyoctenamer and polydodecenamer, in the presence
of the catalytic system $WCl_6 \cdot EtOH \cdot EtAlCl_2$. Similarly, Pampus and
co-workers [268] obtained block polymers in the reaction between polypentenamer and polybutenamer or butadiene-styrene copolymer, a ter-
nary catalytic system $WCl_6 \cdot$ epichlorohydrin $\cdot Et_2AlCl$. This type of reaction does not modify the average molar mass of the two polymers but
changes the configuration of the chain.

Graft polymers are easily obtainable through cometathesis between
a polymer with an unsaturated side-chain, and a cyclo-olefin or any other
polymer containing a double bond in the chain.

$$
\begin{array}{l}
AAAAAAHC{-}CH_2AAA \\
\quad\quad | \\
\quad\quad CH{=}CH_2 \\
\quad\quad\quad + \\
\quad\bigcirc\!(CH_2)_n
\end{array}
\;\rightleftarrows\;
\begin{array}{l}
AAAAAAHC{-}CH_2AAA \\
\quad\quad | \\
\quad\quad CH \quad CH_2 \\
\quad\quad\; \| \quad\;\; \| \\
\quad\quad (CH_2)_n
\end{array}
\qquad (8.64)
$$

Thus, Medema, Alkema and Van Helden [269] grafted cyclo-octene on the
unsaturated branches of natural rubber in the presence of $ReCl_5 \cdot Et_3Al$.
Pampus and co-workers [270-273] obtained graft copolymers from the reaction of cyclopentene with 1,2-polybutadiene, with butadiene-styrene copolymer or with ethene-propene-di(cyclopentadiene) terpolymer. Similarly, Scott and Calderon [266,267] obtained graft polymer from ethene-propene-diene terpolymer and 1,5-cyclo-octadiene. The molecular mass of
the polymer usually increases as a result of the grafting reaction.

Polymerization of cyclo-olefins combined with insertion of a polyalkenamer can be carried out with all types of polyalkenamers or polymers containing a double bond in the chain.

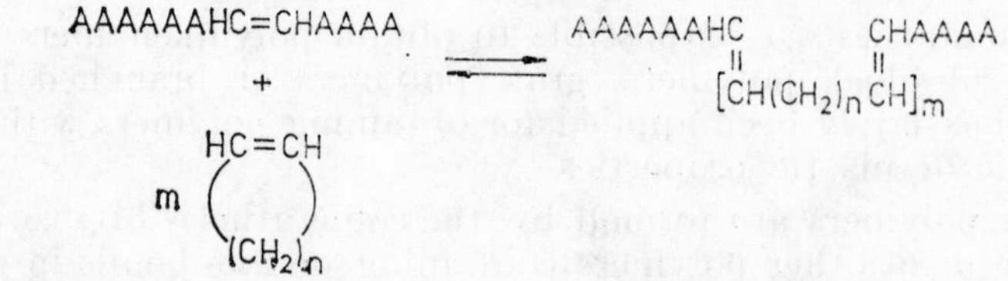

$$\begin{array}{c} \text{AAAAAAHC=CHAAAA} \\ + \\ \text{HC=CH} \\ m \left(\begin{array}{c} \\ \\ \end{array} \right) \\ (\text{CH}_2)_n \end{array} \quad \rightleftharpoons \quad \begin{array}{c} \text{AAAAAAHC} \quad \text{CHAAAA} \\ \overset{\|}{[\text{CH}(\text{CH}_2)_n\text{CH}]_m} \end{array} \quad (8.65)$$

This reaction leads to an increase of the molar mass and to substantial modification of the initial polyalkenamer properties. In several cases, the cyclo-olefin units are randomly distributed in the product, for example, in the polymerization of cyclo-octene with $WCl_6 \cdot EtOH \cdot EtAlCl_2$ in the presence of 1,4-polybutadiene.[266] However, in other insertion reactions of cyclo-olefins on polyalkenamers, copolymers with a microstructures close to that of block polymers were obtained. Thus, the copolymer obtained by Dall'Asta and Motroni[274] in the polymerization of cyclo-octene with insertion of *trans*-polypentenamer possessed (after partial ozonolysis) a block distribution of the two types of monomer unit. The insertion of cyclo-olefin molecules in long sequence is attributed to the high stability of the catalyst.

If, in the above reaction, an acyclic olefin is used instead of cyclo-olefin, the polyalkenamer is cleaved to lower polymers or fragments of low molar mass.

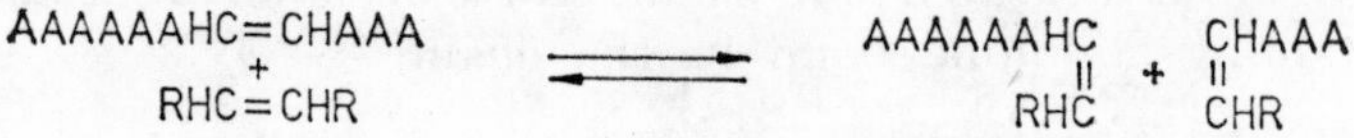

$$\begin{array}{c} \text{AAAAAAHC=CHAAA} \\ + \\ \text{RHC=CHR} \end{array} \quad \rightleftharpoons \quad \begin{array}{c} \text{AAAAAAHC} \quad \text{CHAAA} \\ \overset{\|}{\text{RHC}} + \overset{\|}{\text{CHR}} \end{array} \quad (8.66)$$

$$\begin{array}{c} \text{AAAHC=CHAAHC=CHAAHC=CHAA} \\ + \\ \text{RHC=CHRRHC=CHRRHC=CHR} \end{array} \quad \rightleftharpoons \quad \begin{array}{c} \text{AAAHC} \quad \text{CHAAHC} \quad \text{CHAA} \\ \overset{\|}{\text{RHC}} + 2 \overset{\|}{\text{CHR}} \quad \overset{\|}{\text{CHR}} + \overset{\|}{\text{CHR}} \end{array} \quad (8.67)$$

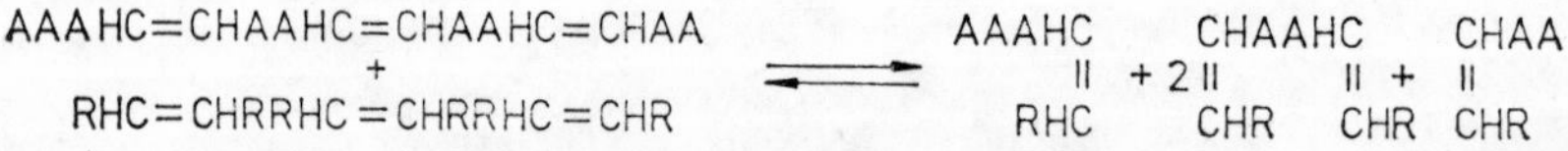

This type of reaction is increasingly used for adjusting the molar mass of polyalkenamers. Through the metathesis of dienic polymers (such as 1,4-polybutadiene) or of polypentenamer with small proportions of 4-octene or 6-dodecene, polymers with the desired lower molar masses result.[275, 276] The reaction is applied also directly in the polymerization of cyclo-olefins by employing a small amount of acyclic alkene.[277-281] Cleavage of the polyalkenamers with acyclic olefins is an elegant method for analyzing the structure of unsaturated polymers.[282-288] The chemical or physico-chemical identification of small fragments with low molar mass obtained through this degradation of the polyalkenamers, constitutes a highly accurate method for the determination of the structure of the polymers.

Polyalkenamers can also be obtained by the metathesis of dienes or acyclic polymers, so that the presence of a cyclo-olefin is not necessary.

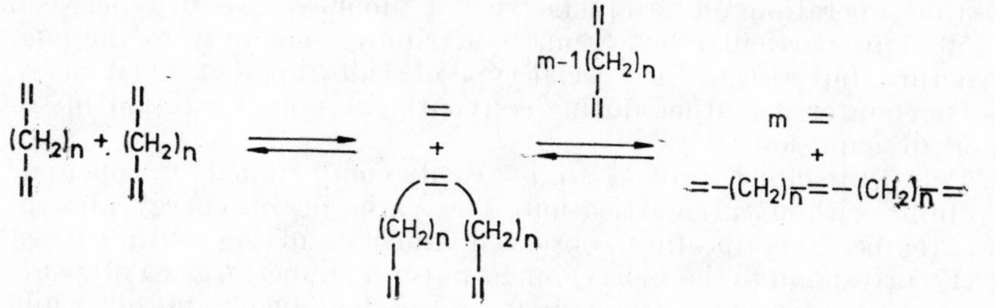

$$(8.68)$$

This reaction is known under the name "metathetic polycondensation".[289] The method, which extends the range of starting materials employed for the synthesis of polyalkenamers, is readily applicable to α,ω-dienes. For example, 1,5-hexadiene in the presence of certain metathesis catalytic systems, leads to oligomers and polymers with the polybutenamer structure.[289] Similarly, 1,4-pentadiene led to linear C_{28}-oligomers of the type $H_2C=CH-(CH_2-CH=CH)_8-CH=CH_2$. Conjugated dienes, however, cannot undergo metathetic polycondensation. 1,7-Octadiene yields cyclohexene under the above conditions through intramolecular metathesis, whereas butadiene yields vinylcyclohexene through cyclodimerization.

8.7.2. Elastomeric Properties of Polyalkenamers

Polyalkenamers are polymers with a number of important elastomeric properties. They have low glass transition temperatures and melting points and show high rates of crystallization. The high molar mass permits easy processing. Their chains constituted of methylenic and double bond sequences, possess a high flexibility and the ability to be vulcanized with sulphur. Owing to these and other properties, polyalkenamers have acquired importance in the elastomer industry.[225,227,228,254,259,290]

Studies have so far been reported on the properties of trans-polypentenamer, cis-polypentenamer, trans-polyoctenamer and cis-polyoctenamer. Some of the most important elastomeric properties of these polymers will be dealt with in this chapter.

trans-*Polypentenamer*. This polymer has a fairly low melting point (18°C), very close to that of natural rubber and has very good processing properties. The low glass transition temperature (-97°C) imparts also easy processability and good elastomeric qualities. trans-Polypentenamer practically does not crystallize at room temperature in a finite period of time. However, under strain it crystallizes readily so that the polymer has good processing qualities.[228]

Investigations on the compounding and processing properties of trans-polypentenamer revealed that it behaves well on mixing, on dispersing ingredients and fillers, on extrusion, calendering, tyre assembling or any other operation on products with a Mooney viscosity between 30 and 150. This particular behaviour is attributed not only to the chemical structure, but also to the special type of wide range of molar mass, which is encountered neither during conventional polymerization in solution nor in emulsion.

trans-Polypentenamer proved to be easily compounded in open or Banbury mills without premastication. The variation of energy absorption and temperature profiles observed during mixing with carbon black HAF correspond to the behaviour of natural rubber. trans-Polypentenamer can be loaded with large amounts of carbon black and oil, while maintaining its good mechanical properties; this reduces the price of the product below that of butadiene-styrene rubber admixed with natural rubber. The dispersion of carbon black in trans-polypentenamer is comparable with that of natural rubber and is obviously superior to that of butadiene or butadiene-styrene rubbers.

An important property of trans-polypentenamer is its high stress-crystallization in the unvulcanized state, which leads to an increase in resistance during processing as a consequence of high green strength. This quality, superior to that of natural rubber, permits easy processability, especially during the operation of building up a tyre carcass.

trans-Polypentenamer can easily be vulcanized with sulphur. It is possible to obtain vulcanized products with high crosslinking yields requiring only low consumption of sulphur, vulcanizing agent and accelerator. The vulcanized products thus obtained possess high tensile strength and modulus. They also have high abrasion resistance, superior to that of natural rubber, of polyisoprene or butadiene-styrene rubbers and similar to that of butadiene styrene-butadiene (1 : 1) mixes.

The very low permeabilities of trans-polypentenamer to air is also remarkable. Studies of this parameter showed that air-permeability decreases with increasing filler content. The permeability is lower than those of butadiene and isoprene rubbers and approaches that of butyl rubber.

Another property of trans-polypentenamer is that of good aging resistance, superior to that of butadiene rubber and similar to that of butadiene-styrene rubber. This quality is primarily due to the trans-configuration of the double bond, as well as to the lower content of double bond per monomer unit. trans-Polypentenamer also possesses good stability to ultra-violet (UV) irradiation. Unlike butadiene rubber, which under UV radiations, undergoes trans/cis isomerization and chain scission, trans-polypentenamer exhibits only trans/cis isomerization and does not suffer appreciable chain scission. This difference in behaviour may be attributed to the fact that, in the case of polybutadiene, chain scission between two methylene groups leads to formation of two stabilized allylic radicals, while in the case of polypentenamer only one such radical is formed.

trans-Polypentenamer is extremely resistant to ozone and weathering. The degradation of this polymer under the action of ozone in the presence or absence of antiozonants or antioxidants is almost unnoticeable. According to some studies [291] this degradation occurs much more slowly with this polymer than with natural rubber, butadiene, or butadiene-styrene rubbers, and it is similar in this respect to chloroprene rubber.

trans-Polypentenamer can be utilized in compositions with various diene rubbers, exhibiting good compatibility and covulcanizability with the latter.[292,293] It is thus compatible with natural, isoprene, butadiene and butadiene-styrene rubbers, or with ethene-propene-diene terpolymers, yielding compositions which are largely used in industry for tyres and rubber articles. *trans*-Polypentenamer confers on these blends high abrasion and aging resistance, good elasticity and low air permeability.

cis-*Polypentenamer*. This polymer has a very low melting point (−41°C), much lower than that of *trans*-polypentenamer, *cis*-poly(dienes) or other known *cis*-polyalkenamers. The glass transition temperature is also very low (−114°C). Consequently, *cis*-polypentenamer has poor elastomeric properties.[294]

cis-Polypentenamer is difficultly compounded with ingredients or fillers at room temperature. Its compounding ability improves at increased temperatures, e.g. at 80−100°C, but it remains much inferior to that of *trans*-polypentenamer.

The physico-chemical properties of *cis*-polypentenamer are generally inferior to those of *trans*-polypentenamer. It is only at low temperatures that *cis*-polypentenamer exhibits improved qualities, and this limits its utilization to such conditions. In this respect, *cis*-polypentenamer can compete with rubbers, such as polypropylene oxide or *cis*-polybutadiene, known for their utilization at low temperatures.

trans-*Polyoctenamer*. *trans*-Polyoctenamer has a fairly high melting point (73°C) (as determined by extrapolation of the plot of melting point versus *trans* content[295]), even higher than that of natural or *trans*-polypentenamer rubber. The melting point decreases markedly on increasing the proportion of *cis* double bonds in the chain. Thus, for 90 and 80% *trans* double bonds, the melting point is 62 and 55°C, respectively. The glass transition temperature for 80% *trans* content is −65°C. The *trans*-polymer has a lower crystallization rate than the *cis*-polymer, but it has high viscosity at room temperature and low viscosity at processing temperature.

trans-Polyoctenamer has good processing properties and possesses good physical characteristics after vulcanization.[254,296,297] It is suitable for use in blends with other rubbers such as natural, butadiene, isoprene, styrene-butadiene rubbers, etc. As a blend component, it leads to easier filler incorporation, better filler distribution, higher green strength and rigidity of the uncured compound at room temperature, lower viscosity at processing temperatures, and improved flow behaviour during calendering, extrusion, injection, moulding, and pressing. The vulcanization rate and accelerator requirement are comparable to those of a slowly vulcanizing styrene-butadiene rubber. It can be covulcanized with a wide range of rubbers using sulphur, peroxides or other cross-linking systems. After vulcanization, it has good tensile strength and elasticity, high modulus and hardness, excellent abrasion resistance and low temperature properties, and high dynamic behaviour, as well as good resistance to aging or ozone and to swelling in organic media or water.

As a result, *trans*-polyoctenamer has many applications in rubber processing for injection moulding, calendered rubber grades, fabric coatings, extrusion of rubber compounds, components of radial tyres, rubber coatings for rolls, production of plasticizer-free compounds etc.[298]

cis-*Polyoctenamer*. This has the same melting temperature as *trans*-polypentenamer (18°C), but a higher crystallization rate at room temperature. In order to diminish the crystallization rate, it is preferable to decrease the proportion of *cis* configuration content at the double bond. The glass transition temperature (−108°C) is also lower than that of *trans*-polypentenamer.[228]

cis-Polyoctenamer has poor processing properties, especially at temperatures below 100°C. Between 100 and 120°C it can be compounded with fillers, ingredients and oil, but it fails to form good quality mixes. Its toleration for sulphur is also lower than that of polypentenamers.[251]

cis-Polyoctenamer has better aging resistance than diene rubbers, but inferior to that of *trans*-polypentenamer. The higher aging resistance of polybutenamer is attributed to the smaller number of double bonds in the macromolecule; the lower resistance than that of *trans*-polypentenamer is attributed to the *cis* configuration at the double bond.

Polynorbornene imparts to rubber compositions special qualities leading to easier processability and better mechanical characteristics.[299, 301] Thus, Norsorex rubber is a white, freely-flowing powder which, compounded with other rubbers, can absorb large amounts of plasticizers and fillers at a uniform level of distribution. These products have excellent dynamic and mechanical properties. The vulcanizates can be rendered ozone resistant by adding suitable antiozonants in combination with a wax to the compound. The cold resistance varies with the plasticizer and reaches a high level when using paraffin oils. The vulcanized rubber has remarkable properties for vibration dampering, impact absorption and sound deadening. Other special characteristics are achieved by using suitable compositions and additives.

8.7.3. Syntheses of Polymers with Functional Groups

Several processes exist which apply metathesis to cyclo-olefins bearing functional groups; the functional group is rarely adjacent to the double bond, more generally it is placed in a remote position. An interesting example which may have good prospects in polymer synthesis is the polymerization of 1-trimethylsilylcyclobutene by tungsten carbene catalysts [302] leading to a perfectly regular invariant *cis* polyalkenamer.

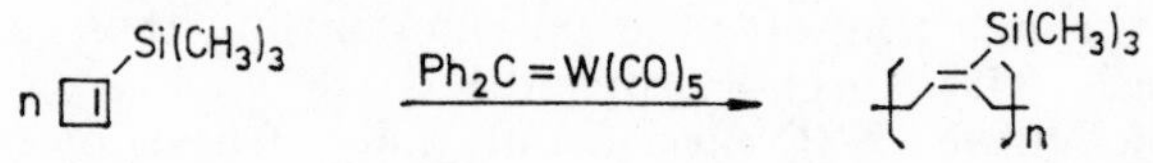

$$n \quad \square^{Si(CH_3)_3} \quad \xrightarrow{\quad Ph_2C=W(CO)_5 \quad} \quad \left\{ = \left\langle {}^{Si(CH_3)_3} \right\}_n \quad (8.69)$$

The trimethylsilyl groups are uniformely arranged on the each monomer units and are unaffected by the catalyst. After polymerization, they can easily be replaced by other functional groups which otherwise would react with the catalyst during polymerization. For instance, a sulphur-containing polymer can be obtained *via* substitution of the silyl groups in the polymer.

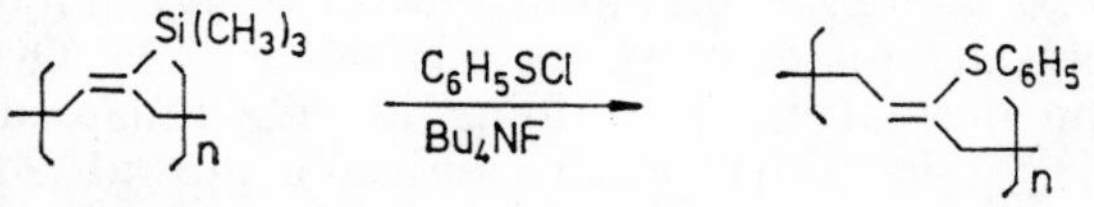

$$\left\{ = \left\langle {}^{Si(CH_3)_3} \right\}_n \quad \xrightarrow{\quad C_6H_5SCl \quad}_{\quad Bu_4NF \quad} \quad \left\{ = \left\langle {}^{SC_6H_5} \right\}_n \quad (8.70)$$

The monomers containing silyl groups are preferred in copolymerization with unsubstituted cyclo-olefins since they confer desirable properties on the polymer.[190] Suitable substituted comonomers are, for instance, 10-trichlorosilyl-1,5-cyclododecadiene, 5-trichlorosilylnorbornene, 5-trimethyloxysilylnorbornene, etc. Thus, the copolymer obtained from *cis,cis*-1,5-cyclo-octadiene with 10 mole % of 5-trichlorosilylnorbornene in the presence of 15 mole % of 4-vinylcyclohexene is an efficient adhesion promoter for coupling rubber to siliceous fillers.

When a cyclo-olefin bearing a stannyl group in a remote position is copolymerized with unsubstituted cyclo-olefins, polymers with biocidal activity (with possible use as marine anti-fouling coatings, in wood preservation, etc.) may be obtained.[190] Tributylstannyl groups substituted on cyclo-octene, tricyclo[8.2.1.$^{2.9}$]-tridecene or norbornene are found to be suitable for such purposes.

Polyalkenamers containing halogen are of practical importance owing to the special elastomeric properties they possess as rubbers. When substituted at the double bond, halogens strongly deactivate the catalyst, and the metathesis is suppressed. Thus, 1-chlorocyclo-alkenes are found to be inert. However, metathesis reactions have been reported for halo-alkenes having remote halogen substituents. 1-Chloro-1,5-cyclo-octadiene polymerizes to give poly-5-chloro-1,5-octadienamer which is the perfectly alternating copolymer of butadiene and chloroprene.[303, 304] Likewise, the Diels-Alder adduct of hexachlorocyclopentadiene was found to give a chlorinated polyalkenamer.[305]

The ring-opening polymerization of cyclo-alkenes carrying ester and ether groups may extend the application of metathesis to the synthesis of polyfunctional polymers. For instance, Ast and co-workers [179] reported a polyoctenamer substituted with both acetate and ether groups. Other examples are numerous, particularly in the norbornene and norbornadiene class; through ring-opening polymerization, these compounds yield polyalkenamers with special properties.

Feast and Wilson [306,307] carried out the synthesis of fluorinated polyalkenamers which are not available through conventional polymerization methods. Thus, polynorbornene substituted in the cyclopentane ring with trifluoromethyl and fluorine or chlorine, or polynorbornadiene substituted on the cyclopentene ring with trifluoromethyl groups have been obtained. Such polymers are of major importance since they display high thermal stability and good solvent resistance, associated with good mechanical properties. Other polyfluorinated polymers obtainable by metathesis were recently reported by Feast and co-workers.[308]

Through the polymerization of 5-cyanonorbornene, it is possible to prepare polymers which, compounded with vinyl polymers, generate resins with increased impact or shock resistance.[309-313] Other norbornene or norbornadiene derivatives containing a wide range of polar groups, such as halogen, nitrile, ester, ether, amide, acid, anhydride, or amide, are described in the patent literature. [314-328]

The polyalkenamers obtained by Nakamura et al. through the polymerization of 5,6-bis-(chloromethyl)norbornene are suitable for preparing compositions with improved mechanical properties and higher resistance to heat.[329,330] Other polymers with low flammability and high glass transition temperature were synthesized by Imaizumi et al.[331] through the polymerization of norbornene derivatives containing halogenated ester or ether groups. Owing to the attractive special properties of the above mentioned substituted polymers, studies concerning the synthesis and processing of such compounds are presently being carried out by several research groups.

8.7.4. Metathetic Ring-Opening Polymerization of Heterocycles

Unsaturated hetero-atom containing cycles may undergo ring-opening polymerization in the presence of metathesis catalysts; through cleavage at the double bond, these heterocycles lead to unsaturated linear polymers containing the hetero-atom in the chain.

$$n \left[\parallel \ X \right] \longrightarrow \left[- X - \right]_n \tag{8.71}$$

Depending on the nature of the hetero-atom of the initial heterocycle, metathesis may form polymers pertaining to polyethers, polyesters or other poly(hetero-substituted) compounds. The polymers thus formed may have high structural purity and display distinct physico-chemical properties as compared to similar conventional polymers.

As a first example, we mention the ring-opening polymerization of 1,1-dimethyl-1-silacyclopent-3-ene carried out by Lammens et al.[332]

$$n \; \underset{\text{Si}}{\bigcirc} \quad \longrightarrow \quad \left\{ \wedge \text{Si} \wedge \right\}_n \tag{8.72}$$

The structure of the polymer corresponds to opening at the carbon-carbon double bond of the initial heterocycle.

Metathesis polymerization of unsaturated lactones affords unsaturated polyesters. Ast et al.[333] synthesized a rubber-like polyester by metathesizing the lactone of 16-hydroxy-6-hexadecenoic acid in the presence of $WCl_6 \cdot Me_4Sn$ catalyst.

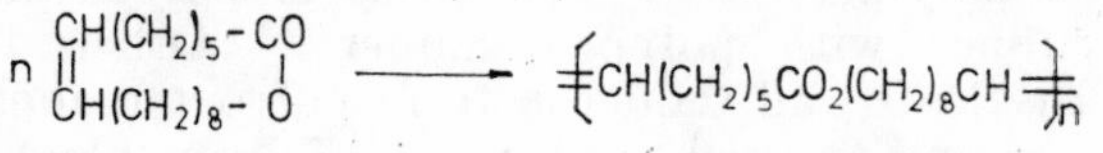

$$n \; \begin{array}{l} CH(CH_2)_5 - CO \\ \parallel \qquad\qquad | \\ CH(CH_2)_8 - O \end{array} \quad \longrightarrow \quad \left\{ CH(CH_2)_5 CO_2(CH_2)_8 CH \right\}_n \tag{8.73}$$

The product is a fibrous, non-tacky polymer having an average molar mass of 95,000 g × mol⁻¹.

Recently, Höcker and co-workers [324] reported the polymerization of 2,3-dihydrofuran with tungsten and chromium carbenes. The polymer obtained had a polyether structure but with a cis : trans ratio close to 1.

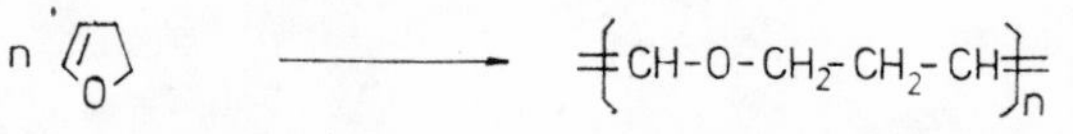

$$n \; \underset{O}{\bigcirc} \quad \longrightarrow \quad \left\{ CH-O-CH_2-CH_2-CH \right\}_n \tag{8.74}$$

The application of metathesis polymerization to unsaturated heterocycles seems to be a new promising synthetic tool for organic chemistry.

8.7.5. Conversion of Unsaturated Polymers by Metathesis

Unsaturated polymers, mainly of the polydiene, polyalkenamer or poly-(acetylene) type, may undergo metathesis or cometathesis in the presence of certain metathesis catalysts, affording new uses for these compounds. Of particular practical significance are the preparation of graft, block and crosslinked unsaturated polymers and copolymers by metathesis, unsaturated polymer metathetic degradation, and telechelic conversion.

The metathetic reactivity of unsaturated polymers facilitates their conversion to a wide range of new polymers and copolymers. Thus, block copolymers with a large variety of structures can be obtained by reaction of two or more polydiene or polyalkenamers in the presence of metathesis catalysts.

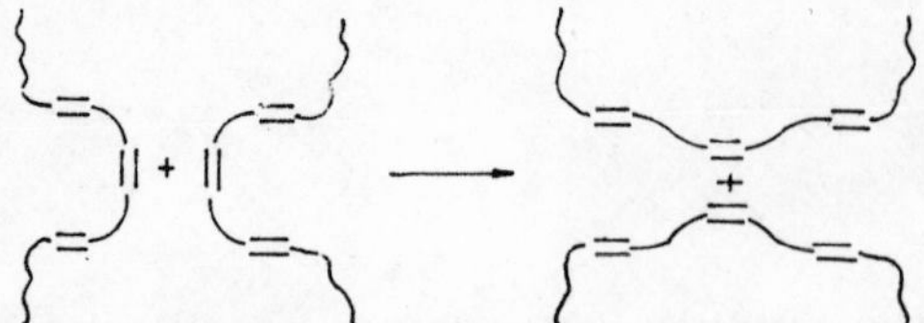

$$(8.75)$$

Polybutadiene, or the butadiene-styrene copolymer, or the ethylene-propylene-diene terpolymer can cometathesize among themselves or with polypentenamer, polyoctenamer, etc. leading to block polymers with interesting elastomeric properties. Pampus and co-workers synthesized such block polymers by cometathesis of polybutadiene and butadiene-styrene rubber with polypentenamer.[268] Other block polymers were obtained by Scott and Calderon from polyoctenamer and polydodecenamer or other unsaturated rubbers.[266, 267] New block polymers may become available by direct ring-opening polymerization of cyclo-olefins in the presence of diene polymers.

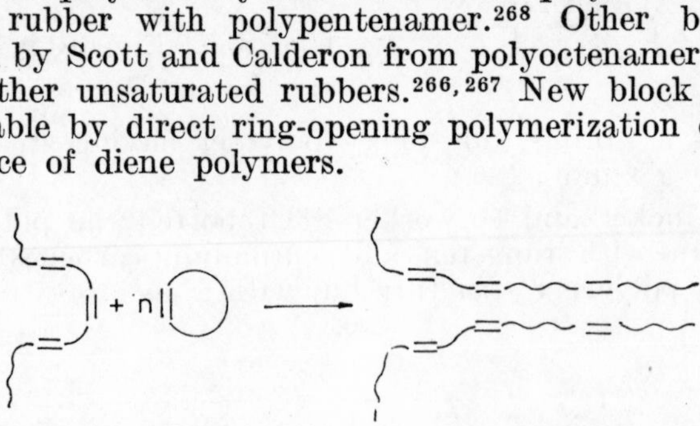

$$(8.76)$$

Graft copolymers are easily obtained by cometathesis of dienes (or other polymers bearing unsaturated branches) with polydienes or polyalkenamers.

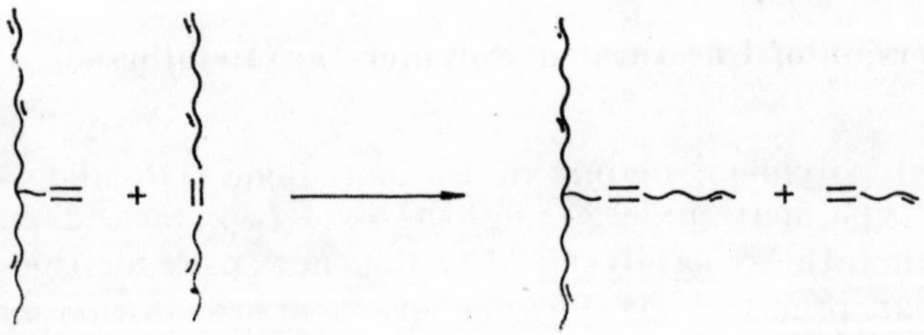

$$(8.77)$$

Thus, ethene-propene-diene rubbers were grafted with 1,4-polybutadiene, poly-1,5-octadienenamer or other polyalkenamers.[266] Likewise, 1,2-polybutadiene was grafted with polypentenamer.[335] Grafting may also be achieved by the direct ring-opening polymerization of cyclo-olefins in the presence of unsaturated branched polymers.

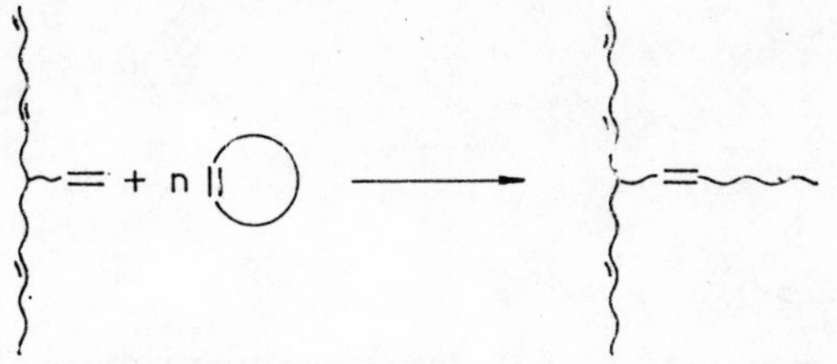

$$
\text{(8.78)}
$$

By polymerizing cyclopentene in the presence of 1,2-polybutadiene, polypentenamer grafted onto 1,2-polybutadiene was obtained.[273]

Crosslinked polymers may be prepared by metathetic conversion of unsaturated branched polymers.

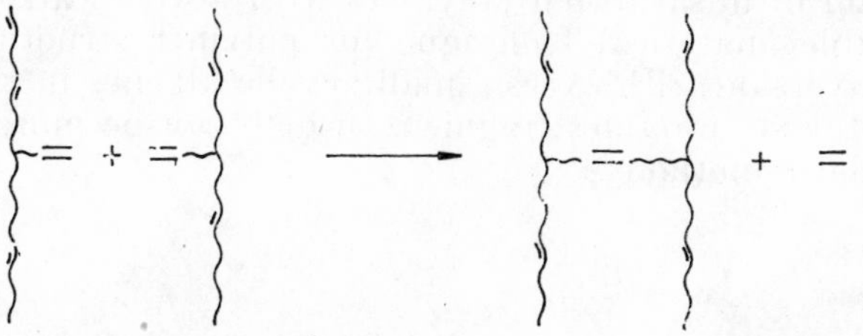

$$
\text{(8.79)}
$$

Thus, unsaturated bridges between chains are formed when metathesizing or cometathesizing 1,2-polybutadiene, ethene-propene-dicyclopentadiene copolymer or other similar polymers.[266]

Metathetic degradation of unsaturated polymers may proceed in the absence or presence of an acyclic alkene. Degradations in the absence of acyclic alkenes lead to either linear destruction or cyclodestruction of the polymer chain.

Linear destruction results in polymer fragmentation by successive intermolecular metathesis reactions

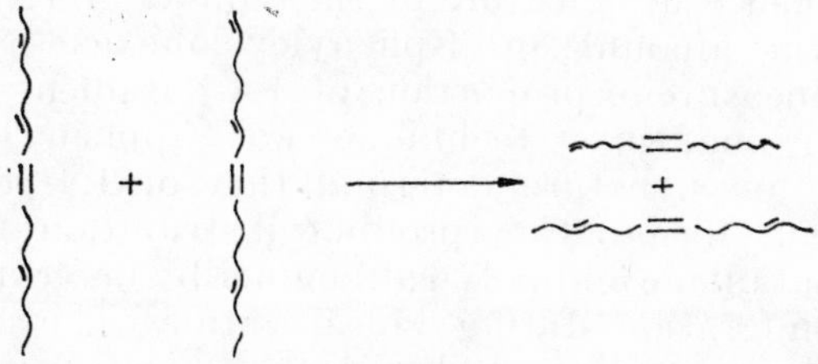

$$
\text{(8.80)}
$$

with broadening of the molar mass distribution. Polymers of high molecular weight may be converted to low and medium molar mass fractions, suitable as additives or special oils.

Cyclodestruction of polymers by metathesis occurs intramolecularly and forms cyclic oligomers along with linear low molar mass fragments.

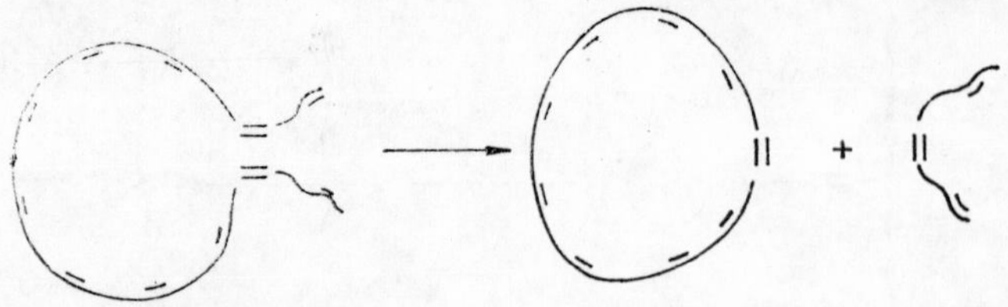

$$\text{(8.81)}$$

The net result is a reduction of the average molar mass and the formation of unsaturated carbocycles. This phenomenon was thoroughly investigated on polypentenamer, polyoctenamer, poly-1,5-octadienamer and other diene polymers.[336-338] In some cases, simpler cyclic compounds are formed by advanced cyclodegradaiton, such as 4-methylcyclohexene from propene-butadiene copolymer.[336]

Degradation of unsaturated polymers with acyclic alkenes proved to be a remarkable analytical technique for polymer structure determination.[339,340] Conversion with excess small acyclic alkenes in the presence of metathesis catalysts provides fragments which can be more easily identified by standard methods.

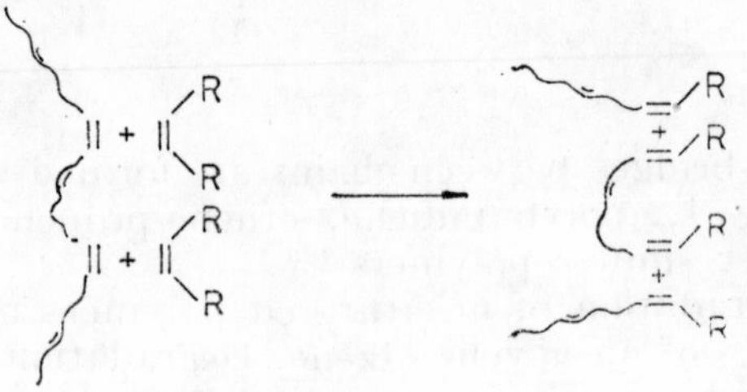

$$\text{(8.82)}$$

Mikhailov and Harwood [341] applied this procedure to 1,4-polybutadiene and butadiene-styrene copolymers which were fragmented with excess 2-butene in the presence of $WCl_6 \cdot EtOH \cdot EtAlCl_2$. The amount of 2,6-octadiene formed was a measure of the number of 1,4-butadiene homolinkages, while the amount of 4-phenylcyclohexene and 5-phenyl-2,8-decadiene was a measure of proportion of 1,4-butadiene-styrene-1,4-butadiene triads. An analogous technique was applied by Hummel and co-workers [342-344] for structure determination of 1,4-polybutadiene and derivatives thereof, formulating product distribution relations and for vulcanized polybutadiene and polypentenamer by determining the amount of double bond migration during vulcanization. The 1,7-polybutadiene structure was shown to occur by metathetical degradation, in the first case by Abendroth and Canji.[286] By further investigating the degradation of polybutadiene and butadiene-styrene copolymers, Csanyi et al.[345-247] determined 1,2- and 1,4-sequences in these polymers. 1,2-Butadiene sequences obtained by metathetical degradation of polybutadiene may constitute a useful resource for the synthesis of substituted cyclopentene and cyclohexene derivatives.[348] Microstructural characterization of partially hydrogenated or crosslinked polybutadiene reported by Ast and co-workers to give information on polymer constitution or on vulcanization and other crosslinking reactions.[284]

Cometathesis of unsaturated polymers with functionalized alkenes provides a facile route for functionalizing polymers.

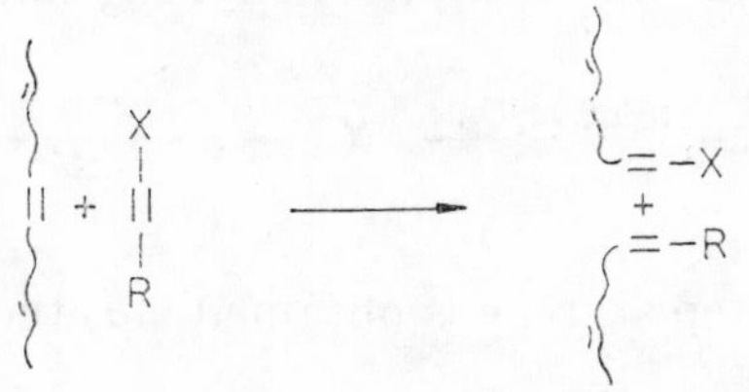

$$
\begin{array}{c}
\text{(8.83)}
\end{array}
$$

Likewise, cyclic polymers form telechelic compounds *via* ring cleavage with suitably substituted alkenes.

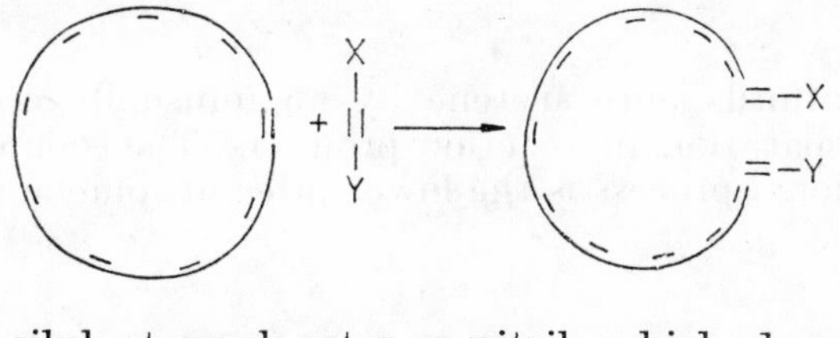

$$
\begin{array}{c}
\text{(8.84)}
\end{array}
$$

Groups such as silyl, stannyl, ester or nitrile which do not inhibit metathesis catalysts are preferably employed.[190-193]

8.8. Other Applications

Recently, metathesis has become a useful method for synthesis of compounds of various classes in organic chemistry. Olefin metathesis can be applied to the preparation of natural compounds with uncommon structures, high purity monomers, and various unsaturated compounds of low or high molar mass.

8.8.1. Synthesis of Monomers

The metathesis of cyclo-alkenes, carried out with a suitable catalytic system, leads to cyclic dienes from mono-olefins or to cyclo-alkenes from cyclic dienes; in both cases, the product has a high degree of purity. It is noteworthy that, through this method, it is possible to obtain monomers with a higher purity than is furnished by conventional methods. An example is the synthesis of pure cyclopentene from 1,6-cyclodecadiene, in the presence of catalytic systems based on tungsten hexachloride carried out by Küpper and Streck.[156,157] The method can be extended also to other simple or substituted cyclo-olefins, accessible only with difficulty through common methods.

The ethenolysis of stilbene could become a convenient method for the production of styrene. Thus, Montgomery et al.[349] conceived a two-step process for the synthesis of styrene from toluene. In a first step, stilbene is formed from toluene by catalytic dehydrogenative coupling

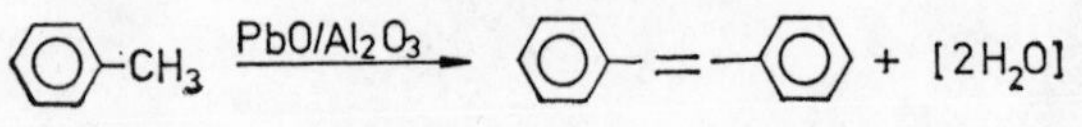

$$\langle\!\!\bigcirc\!\!\rangle\text{-CH}_3 \xrightarrow{\text{PbO/Al}_2\text{O}_3} \langle\!\!\bigcirc\!\!\rangle\text{-}=\text{-}\langle\!\!\bigcirc\!\!\rangle + [2\,\text{H}_2\text{O}] \qquad (8.85)$$

and then in a second step styrene is obtained via ethenolysis over a disproportionation catalyst.

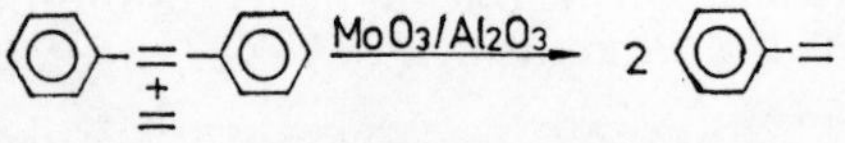

$$\langle\!\!\bigcirc\!\!\rangle\text{-}\overset{=}{\underset{+}{=}}\text{-}\langle\!\!\bigcirc\!\!\rangle \xrightarrow{\text{MoO}_3/\text{Al}_2\text{O}_3} 2\,\langle\!\!\bigcirc\!\!\rangle\text{-}= \qquad (8.86)$$

This process affords pure styrene by an unusually economical route involving easy separation of reaction products. The economic advantage over the conventional process is the lower price of toluene relative to benzene.

8.8.2. Synthesis of Conjugated Polyenes (Poly(acetylenes))

Metathesis affords a new promising route for the synthesis of poly(acetylenes), polymers with outstanding electrochemical properties. The unsaturated polymers thus obtained seem to have high structural purity in comparison to most poly(acetylenes) available through conventional methods. The reaction was performed mainly in the presence of carbene complexes including both the Casey carbene and Fischer-type carbenes but extension to other metathesis catalysts is expected.

Katz and co-workers [302,350] synthesized a variety of poly(acetylenes) starting from monosubstituted acetylenes (propyne, t-butylacetylene, 1-hexyne, phenylacetylene), disubstituted acetylenes (2-butyne), or mixtures of monosubstituted and disubstituted acetylenes (4-octyne and phenylacetylene). The polymers obtained from monosubstituted acetylenes are generally perfectly regular and of high steric purity.

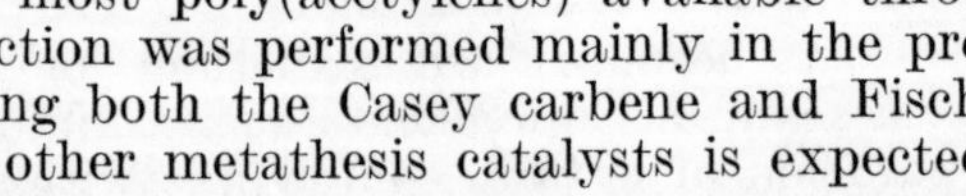

$$3n\,\text{R}\text{—}\equiv \qquad\longrightarrow\qquad \left(\!\!\!\begin{array}{c}\text{R}\quad\text{R}\quad\text{R}\\ \bigwedge\!\!\bigvee\!\!\bigwedge\end{array}\!\!\!\right)_n \qquad (8.87)$$

An alternative route for poly(acetylene) synthesis was reported by Edwards and Feast [351] using highly strained polycyclic olefins. According to this procedure, in the first stage, a polyalkenamer is formed by ring-opening polymerization of the strained cyclo-alkene and, in the second stage, the poly(acetylene) is released by an elimination reaction. The driving force for the latter reaction is the formation of the aromatic 1,2-bis-trifluoromethylbenzene. The procedure was performed with 7,8-bis(trifluoromethyl)tricyclo[4.2.2.0 $^{2.5}$]deca-3,7,9-triene leading to polyacetylene *via* the corresponding intermediate polyalkenamer.

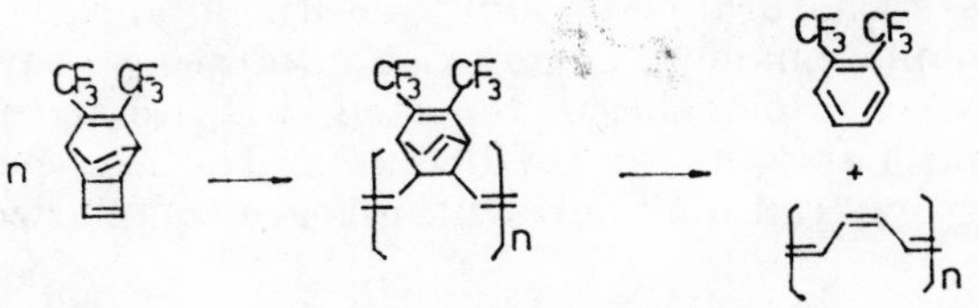

$$(8.88)$$

In practice, the first stage of this reaction sequence should yield a polyalkenamer which can conveniently be fabricated into a desired form (fibre, film, surface coating, etc.) and the second should then be achieved quantitatively and without disruption of the physical integrity of the material. Further examples given by Feast *et al.*[352] seem to extend this approach to other cyclo-olefin substrates. In addition, a new route to substituted polyconjugated systems in which the substituent contains a hetero-atom adjacent to the polyconjugated system is inherent in this reaction.

8.8.3. Synthesis of Natural Compounds

The first application of metathesis for obtaining natural compounds was the synthesis of poly(isoprene) from 1-methylcyclobutene.

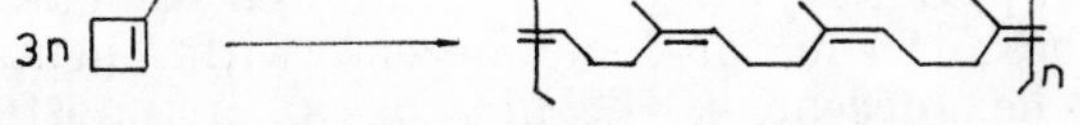

$$(8.89)$$

The polyalkenamer obtained by Katz [223] through the ring-opening polymerization of 1-methylcyclobutene with the catalytic system diphenylcarbene-pentacarbonyltungsten contains the *cis* stereoisomer of 1,4-polyisoprene in a proportion of 84—87%. In this product, the monomer unit adopts a structure very close to that of the natural compound.

The formation of 1,4-poly(isoprene) by the above reaction suggested the idea of applying the metathesis reaction to the synthesis of other compounds with isoprenic structures. Thus, the synthesis of farnesyl acetate, a compound of the terpene class has been described by Wilson and Schalk[353] employing the metathesis of geranyl acetate with 1-methyl-cyclobutene in the presence of the catalytic system $WCl_6 \cdot (CH_3)_4Sn$.

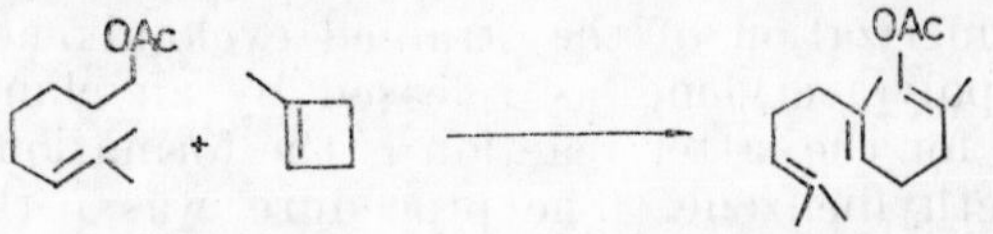

$$(8.90)$$

Although the yield is very low (1—2 %), the method may have prospects for terpene syntheses.

Another class of natural compounds, easily obtainable through metathesis is that of pheromones. There exist interesting examples of pheromone synthesis, *via* metathesis reported by Rossi, [354] Küpper and Streck [355-358] and Levisalles and Villemin.[359] The classes of pheromones and the starting material for their synthesis are summarized in Table 8.1.

TABLE 8.1. Syntheses of pheromones *via* cometathesis reaction

Pheromone	Starting olefins	References
9-Tricosene	1-Decene + 1-Pentadecene	354, 355
	1-Decene + 2-Hexadecene	355, 356
	9-Octadecene + 1-Pentadecene	355, 356
	9-Octadecene + 2-Hexadecene	355, 356
13-Heptacosene	1-Tetradecene + 2-Hexadecene	355
13-Nonacosene	1-Tetradecene + 2-Octadecene	355, 357, 358
14-Nonacosene	1-Hexadecene + 2-Hexadecene	355, 357, 358
2-Methylheptadecane	4-Methyl-1-pentene + 1-Tetradecene	354, 355
2-Methyl-7-octadecene	7-Methyl-1-octene + 1-Dodecene	355, 356
7,8-Epoxy-2-methylocta-decane	7-Methyl-1-octene + 1-Dodecene	356

Thus, 9-tricosene (muscalure) has been prepared by Rossi [354] through the cometathesis of an equimolecular mixture of 1-decene and 1-pentadecene in the presence of a catalytic system consisting of molybdenum complexes and by Küpper and Streck [355, 356] through cometathesis of 9-octadecene with 2-hexadecene, of 9-octadecene with 1-pentadecene or of 1-decene with 2-hexadecene in the presence of a tungsten hexachloride catalytic system. The latter authors also succeeded in synthesizing tricosene from 9-octadecene and 2-hexadecene with heterogeneous catalysts based on molybdenum or rhenium oxides.[356]

The same general route has been applied to the synthesis of other representatives of this class; 13-heptacosene was prepared from 1-tetradecene and 2-hexadecene, 13-nonacosene from 1-tetradecene and 2-octadecene, and 14-monacosene from 1-hexadecene and 2-hexadecene. Methyl-substituted pheromones (2-methylheptadecene and 7,8-epoxy-2-methyloctadecane), were separated by combining the above disproportionation procedure with corresponding hydrogenation or epoxidation reactions.

Thus, 2-methylheptadecane was obtained through cometathesis of 4-methyl-1-pentene with 1-tetradecene, followed by the hydrogenation of the 2-methyl-4-heptadecene formed as an intermediate,[354] whereas 7,8-epoxy-2-methyloctadiene was formed through cometathesis of 1-dodecene with 7-methyl-1-octene and subsequent epoxidation.[356] The synthesis of other compounds belonging to the same class is presently under investigation.

Another series of insect pheromones is the class of esters of alkenoic acids which were synthesized by Levisalles and Villemin [359] using cross-metathesis of alkenes with ω-unsaturated esters. The reaction was carried out with terminal or internal alkenes and the corresponding ω-unsaturated acetate, leading to the corresponding pheromone structure.

$$R-CH=CH_2 + CH_2=CH-(CH_2)_n-OAc \longrightarrow R-CH=CH-(CH_2)_n-OAc + CH_2=CH_2 \quad (8.91)$$

$$R-CH=CH-R + CH_2=CH-(CH_2)_n-OAc \longrightarrow R-CH=CH-(CH_2)_n-OAc + R-CH=CH_2 \quad (8.92)$$

Thus, when $R = C_2H_5$ and $n = 8$ the pheromone $C_2H_5-CH=CH-(CH_2)_8OAc$ was synthesized with 24.4% selectivity *via* reaction (8.92). Other pheromones with similar structures may likewise be obtained. Although the method is very simple, the stereoselectivity with respect to the alkenoic acid is low.

The metathesis reaction has been useful in synthesizing natural compounds of the macrolide type such as exaltolide or higher homologues. The synthesis of the macrocyclic lactone exaltolide was carried out by Villemin [360] starting either from methyl 11-undecenoate and hexenyl acetate, through metathesis and subsequent hydrogenation and cyclization

Scheme 8.1

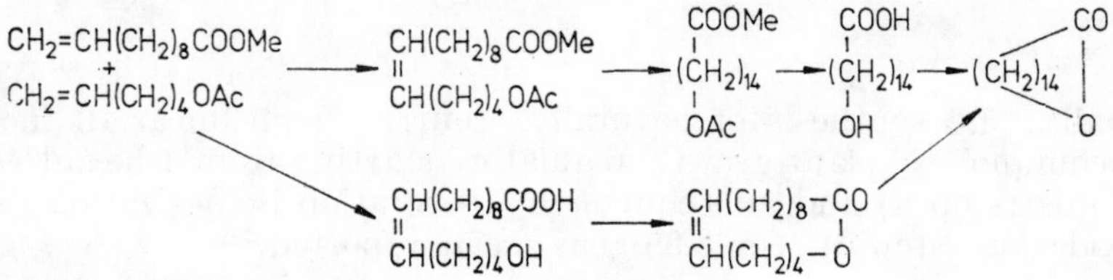

or from unsaturated linear lactone (I), through metathesis and subsequent hydrogenation.

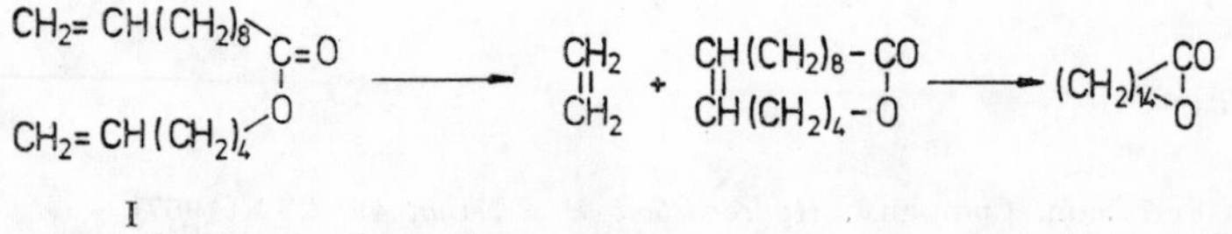

Higher macrolides such as 9-octadecen-18-lide and 10-eicosen-20-lide have been obtained by Tsuji and Hashiguchi [361] by metathesis cyclization of the corresponding unsaturated linear lactones which are easily available by lactonization from the corresponding acid and alcohol.

Scheme 8.2

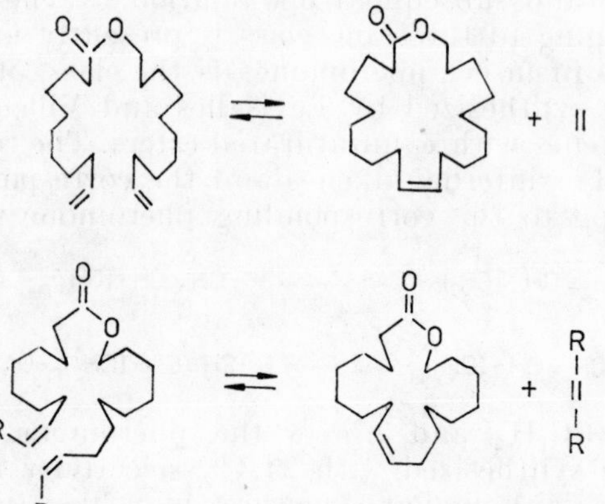

Civetone (cyclo-9-heptadecenone) used in the perfume industry, has been obtained by the same authors by ketocyclization of 9-octadecene-1,18-dioate, which was obtained in one step by the metathesis of ethyl oleate.

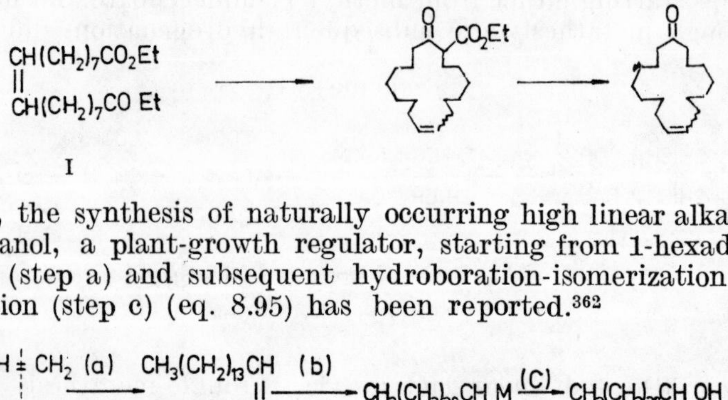

$$
\begin{array}{c}
CH(CH_2)_7CO_2Et \\
\| \\
CH(CH_2)_7CO\ Et
\end{array}
\qquad (8.94)
$$

I

Finally, the synthesis of naturally occurring high linear alkanols such as triacontanol, a plant-growth regulator, starting from 1-hexadecene *via* metathesis (step a) and subsequent hydroboration-isomerization (step b) and oxidation (step c) (eq. 8.95) has been reported.[362]

$$
\begin{array}{c}
CH_3(CH_2)_{13}CH \doteq CH_2 \\
| \\
CH_3(CH_2)_{13}CH \doteq CH_2
\end{array}
\xrightarrow{(a)}
\begin{array}{c}
CH_3(CH_2)_{13}CH \\
\| \\
CH_3(CH_2)_{13}CH
\end{array}
\xrightarrow{(b)}
CH_3(CH_2)_{28}CH\ M
\xrightarrow{(c)}
CH_3(CH_2)_{28}CH\ OH
\qquad (8.95)
$$

References

1. Phillips Petroleum Company, *Hydrocarbon Processing*, **46**, 232 (1967).
2. E. R. Freitas and C. R. Gum, *Chem. Eng. Progr.*, **75**, 73 (1979).
3. P. H. Johnson, *Hydrocarbon Processing*, **46**, 149 (1967).
4. R. E. Dixon, **U.S.** 3, 485, 890, 23.12.1969; *Chem. Abstr.*, **72**, 42719h (1970).

5. R. L. Banks, **U.S.** 3, 538, 181, 3.11.1970.
6. R. E. Dixon and J. E. Hutto, **U.S.** 3, 544, 649, 1.12.1970.
7. R. L. Banks and R. B. Regier, *Ind. Eng. Chem., Prod. Res. Develop.*, **10**, 46 (1971).
8. L. F. Heckelsberg, **U.S.** 3, 485, 891, 18.05.1967.
9. J. R. Kenton, D. L. Crain and R. F. Kleinschmidt, **U.S.** 3, 491, 163, 20.10.1970.
10. R. E. Reusser, **U.S.** 3, 729, 524, 24.04.1973.
11. R. E. Reusser and S. D. Turk, **U.S.** 3, 760,026, 15.01.1974.
12. Van C. Vives, **U.S.** 3,786,111, 15.01.1974.
13. Van C. Vives and R. E. Reusser, **U.S.** 3,792,102, 30.12.1971.
14. R. S. Logan and R. L. Banks, *Hydrocarbon Processing*, **47**, 135 (1968).
15. R. L. Banks, **U.S.** 3,785,957, 15.01.1974; *Chem. Abstr.*, **80**, 110630 (1974).
16. R. L. Banks, **U.S.** 3,767,565, 20.10.1973; *Chem. Abstr.*, **80**, 50190 (1974).
17. R. L. Banks, **U.S.** 3,785,956, 15.01.1974; *Chem. Abstr.*, **80**, 110622 (1974).
18. M. A. Albright, **U.S.** 3,330,882, 24.01.1963; *Chem. Abstr.*, **62**, 2706 (1965).
19. H. J. Alkema and R. van Helden, **Brit.** 1,117,968, 26.06.1968; *Chem. Abstr.*, **69**, 95906 (1968).
20. H. J. Alkema and R. van Helden, **Neth.** 6,814,835, 22.04.1969; *Chem. Abstr.*, **71**, 49211 (1969).
21. H. J. Alkema and R. van Helden, **U.S.** 3,536,777; 27.10.1970.
22. H. J. Alkema, D. Medema and F. Wattimena, **Ger. Offen.** 1,929,136, 11.12.1969; *Chem. Abstr.*, **72**, 81244 (1970).
23. J. K. Allen, B. P. McGrath and B. M. Palmer, **S. African** 69,06,673, 19.03.1970; *Chem. Abstr.*, **73**, 87395 (1973).
24. J. K. Allen, B. M. Palmer and B. P. McGrath, **Ger. Offen.** 2,230,490, 11.01.1973; *Chem. Abstr.*, **78**, 98245 (1973).
25. R. P. Arganbright, **U.S.** 3,702,827, 14.11.1972; *Chem. Abstr.*, **78**, 110518 (1973).
26. R. P. Arganbright, **U.S.** 3,792,108, 12.02.1974; *Chem. Abstr.*, **80**, 95212 (1974).
27. R. L. Banks, **Neth.** 6,400,549, 27.07.1964.
28. R. L. Banks, **U.S.** 3,261,876, 19.07.1966; *Chem. Abstr.*, **65**, 12105e (1966).
29. R. L. Banks, **U.S.** 3,431,316, 4.03.1969; *Chem. Abstr.*, **70**, 96142s (1969).
30. R. L. Banks and R. B. Regier, **Ger. Offen.** 2,217,675, 9.11.1972; *Chem. Abstr.*, **78**, 42839 (1973).
31. R. L. Banks and F. Pennella, **U.S.** 3,729,525, 24.04.1973; *Chem. Abstr.*, **78**, 158879 (1973).
32. R. L. Banks, **U.S.** 3,883,606, 13.05.1975; *Chem. Abstr.*, **83**, 58066 (1975).
33. a. A. N. Bashkirov, Yu. B. Kryukov, R. A. Fridman, S. M. Nosakova, S. I. Beilin, **U.S.S.R.** 445,259, 21.02.1973; *Chem. Abstr.*, **83**, 205746 (1975); b. A. N. Bashkirov, R. A. Fridman, S. M. Nosakova, L. G. Liberov, **U.S.S.R.** 517,575, 15.06.1976; *Chem. Abstr.*, **85**, 159377 (1976).
34. a. A. N. Bashkirov, R. A. Fridman, L. G. Liberov, S. M. Nosakova, **U.S.S.R.** 564,301, 5.07.1977; *Chem. Abstr.*, **87**, 117556 (1977); b. A. N. Bashkirov, L. G. Liberov, R. A. Fridman, S. M. Nosakova, **U.S.S.R.** 565,028, 15.06.1977; *Chem. Abstr.*, **87**, 153189 (1977).
35. D. A. Bol'shakov, A. I. Yablonskaya, V. Sh. Fel'dblyum, A. M. Kut'in, G. A. Stepanov, T. I. Baranova, E. A. Shmulevich, O. P. Karpov, **U.S.S.R.** 566,814, 30.07.1977; *Chem. Abstr.*, **87**, 167518 (1977).
36. C. P. C. Bradshaw, **Ger. Offen.** 1,900,490, 31.07.1969; *Chem. Abstr.*, **71**, 90763 (1969).
37. C. P. C. Bradshaw, **Brit.** 1,282,592, 19.07.1972; *Chem. Abstr.*, **77**, 100700 (1972).
38. British Petroleum Company, **Brit.** 1,064,829, 19.11.1964; *Chem. Abstr.*, **65**, 13539a (1966).
39. British Petroleum Company, **Fr.** 1,554,287, 17.01.1969; *Chem. Abstr.*, **71**, 60655 (1969).
40. N. Calderon and Y. Chen, **U.S.** 3,535,401, 20.10.1970.
41. D. L. Crain, **Belg.** 691,711, 23.06.1967.
42. D. L. Crain, **Fr.** 1,516,853, 15.03.1968; *Chem. Abstr.*, **71**, 12498 (1969).
43. J. W. Davison, **U.S.** 3,442,970, 6.05.1969; *Chem. Abstr.*, **71**, 38316 (1969).
44. A. J. De Jong, H. J. Heijmen and R. van Helden, **Ger. Offen.** 2,604,261, 19.08.1976; *Chem. Abstr.*, **85**, 176800 (1976).
45. R. E. Dixon, **Belg.** 633,482, 11.12.1963; *Chem. Abstr.*, **62**, 4446 (1965).
46. E. D. M. Eades and P. R. Hugnes, **U.S.** 3,652,704, 1.08.1969.
47. E. D. M. Eades and P. R. Hugnes, **Ger. Offen.** 1,942,635, 26.02.1970; *Chem. Abstr.*, **72**, 110756 (1970).

48. A. F. Ellis and E. T. Sabourin, **Ger. Offen.** 2,306,813, 30.08.1973 ; *Chem. Abstr.*, **79**, 125796 (1973).
49. V. Fattore, S. D. Milanese, M. Mazzei and B. Notari, **U.S.** 3,792,107, 21.07.1971.
50. V. Fattore, M. Mazzei and B. Notari, **Fr.** 2,089,461, 11.02.1972 ; *Chem. Abstr.*, **77**, 125924 (1972).
51. L. F. Heckelsberg, **U.S.** 3,340,322, 5.09.1967 ; *Chem. Abstr.*, **67**, 116529 (1967).
52. L. F. Heckelsberg, **S. African** 68 01,707, 15.08.1968 ; *Chem. Abstr.*, **70**, 67557 (1969).
53. L. F. Heckelsberg, **S. African** 68 01,620, 15.08.1968 ; *Chem. Abstr.*, **70**, 67560 (1969).
54. L. F. Heckelsberg, **S. African** 68 01,877, 20.08.1968 ; *Chem. Abstr.*, **70**, 67558b (1969).
55. L. F. Heckelsberg, **U.S.** 3,395,196, 30.07.1968 ; *Chem. Abstr.*, **69**, 60530g (1968).
56. L. F. Heckelsberg, **S. African** 68 01,878, 20.08.1968 ; *Chem. Abstr.*, **70**, 67559 (1969).
57. L. F. Heckelsberg, **Fr.** 1,562,396, 4.04.1969 ; *Chem. Abstr.*, **71**, 93357 (1969).
58. L. F. Heckelsberg, **U.S.** 3,444,262, 13.05.1969 ; *Chem. Abstr.*, **71**, 38317 (1969).
59. L. F. Heckelsberg, **S. African** 68 01,408, 31.08.1968 ; *Chem. Abstr.*, **70**, 69924 (1969).
60. L. F. Heckelsberg, **U.S.** 3,418,390, 24.12.1968 ; *Chem. Abstr.*, **70**, 46802 (1969).
61. L. F. Heckelsberg, **U.S.** 3,546,311, 8.12.1970.
62. L. F. Heckelsberg, **U.S.** 3,365,513, 23.01.1968 ; *Chem. Abstr.*, **68**, 61424s (1968).
63. L. F. Heckelsberg, **Brit.** 1,282,592, 19.07.1972 ; *Chem. Abstr.*, **77**, 100701 (1972).
64. L. F. Heckelsberg, **U.S.** 3,485,891, 23.12.1969.
65. L. F. Heckelsberg, **U.S.** 3,723,562, 27.03.1973 ; *Chem. Abstr.*, **79**, 18040 (1973).
66. E. J. Howman and B. P. McGrath, **Brit.** 1,123,500, 14.08.1968.
67. E. J. Howman and L. Turner, **U.S.** 3,448,163, 3.06.1969.
68. Y. Kamiya and E. Ogata, **Japan. Kokai** 75 69,002, 9.06.1975.
69. E. T. Kittleman and E. A. Zuech, **Fr.** 1,561,025, 21.03.1969.
70. H. Knoche, **Ger. Offen.** 2,024,835, 10.12.1970 ; *Chem. Abstr.*, **74**, 44118b (1971).
71. C. A. Kubicek, **U.S.** 3,974,100, 10.08.1976 ; *Chem. Abstr.*, **86**, 43287 (1977).
72. L. G. Larson, **U.S.** 3,564,314, 8.12.1970.
73. F. D. Mango, **U.S.** 3,424,811, 28.01.1969 ; *Chem. Abstr.*, **70**, 106042 (1969).
74. R. M. Marsheck, **Belg.** 678,016, 19.09.1966.
75. R. M. Marsheck, **Brit.** 1,128,091, 25.09.1968 ; *Chem. Abstr.*, **70**, 19540t (1969).
76. B. P. McGrath, **Ger. Offen.** 1,940,433, 19.02.1970 ; *Chem. Abstr.*, **72**, 110755 (1970).
77. B. P. McGrath, **Brit.** 1,216,278, 19.07.1971 ; *Chem. Abstr.*, **75**, 151335 (1971).
78. B. P. McGrath and K. V. Williams, **Brit.** 1,159,056, 23.06.1969 ; *Chem. Abstr.*, **72**, 12044 (1970).
79. D. Medema, W. Brunmayer-Schilt and R. van Heldaen, **Brit.** 1,193,943, 3.06.1970 ; *Chem. Abstr.*, **73**, 76637 (1970).
80. D. P. Montgomery, **Ger. Offen.** 2,160,151, 20.07.1972 ; *Chem. Abstr.*, **77**, 125928 (1972).
81. P. H. Phung and G. Lefebvre, **Fr.** 1,594,582, 17.07.1970 ; *Chem. Abstr.*, **74**, 87316 (1971).
82. G. C. Ray and D. L. Crain, **Belg.** 694,420, 22.08.1967.
83. G. C. Ray and D. L. Crain, **Fr.** 1,511,381, 26.01.1968 ; *Chem. Abstr.*, **70**, 114570 (1969).
84. R. B. Regier, **U.S.** 3,792,106, 12.02.1974 ; *Chem. Abstr.*, **80**, 95211 (1974).
85. R. E. Reusser, **S. African** 68 01,879, 20.08.1968 ; *Chem. Abstr.*, **70**, 77280 (1969).
86. R. E. Reusser, **U.S.** 3,836,480, 12.09.1974 ; *Chem. Abstr.*, **81**, 151498 (1974).
87. H. W. Ruhle, **Ger. Offen.** 2,062,448, 16.09.1971 ; *Chem. Abstr.*, **75**, 51341 (1971).
88. Shell Internat. Res. Maatschappij, **Neth.** 7,111,882, 25.05.1972 ; *Chem. Abstr.*, **77**, 125930 (1972).
89. L. Turner, E. J. Howman and C. P. C. Bradshaw, **Brit.** 1,056,980, 1.02.1967 ; *Chem. Abstr.*, **66**, 75651 (1967).
90. L. Turner and K. V. Williams, **Brit.** 1,096,200, 20.12.1967 ; *Chem. Abstr.*, **68**, 77691 (1968).
91. L. Turner and K. V. Williams, **Brit.** 1,096,250, 20.12.1967 ; *Chem. Abstr.*, **68**, 77693 (1968).
92. L. Turner and C. P. C. Bradshaw, **Brit.** 1,103,976, 21.02.1968 ; *Chem. Abstr.*, **68**, 86834 (1968).
93. L. Turner, C. P. C. Bradshaw and E. J. Howman, **U.S.** 3,526,676, 1.09.1970.
94. British Petroleum Company, **Fr.** 1,554,287, 17.01.1969 ; *Chem. Abstr.*, **71**, 60655 (1969).
95. Y. Uchida and M. Hidai, **Japan. Kokai** 72 31,903, 14.11.1972 ; *Chem. Abstr.*, **78**, 83784 (1973).
96. Y. Uchida, M. Hidai and T. Tatsumi, **Japan. Kokai** 72 31,904, 14.11.1972 ; *Chem. Abstr.*, **78**, 79100 (1973).

97. Y. Uchida, M. Hidai and T. Tatsumi, **Japan. Kokai** 74 85,004, 15.04.1974; *Chem. Abstr.*, **82**, 3738 (1975).
98. R. T. Wilson and L. G. Larson, **U.S.** 3,346,661, 10.10.1967; *Chem. Abstr.*, **68**, 43639m (1968).
99. R. L. Banks, **U.S.** 3,538,181, 3.10.1970.
100. R. L. Banks, **U.S.** 3,442,969, 6.05.1969; *Chem. Abstr.*, **71**, 38315 (1969).
101. R. L. Banks, **U.S.** 3,546,313, 8.12.1970; *Chem. Abstr.*, **74**, 78085 (1971).
102. J. M. Basset, J. Bousquet and R. Mutin, **Ger. Offen.** 2,511,359, 9.10.1975; *Chem. Abstr.*, **84**, 30375 (1976).
103. D. A. Bol'shakov, A. I. Yablonskaya, V. Sh. Fel'dblyum, T. I. Baranova, A. M. Kut'in, G. A. Stepanov, **U.S.S.R.** 536,837, 30.11.1976; *Chem. Abstr.*, **86**, 79436 (1977).
104. D. A. Bol'shakov, A. I. Yablonskaya, G. S. Paramanova, V. Sh. Fel'dblyum, A. M. Kut'in, G. A. Stepanov, T. I. Baranova, **U.S.S.R.** 525,467, 25.08.1976; *Chem. Abstr.*, **85**, 182916 (1976).
105. D. A. Bol'shakov, A. I. Yablonskaya, G. A. Stepanov, L. A. Morozova, N. F. Chernoknizhnaya, **U.S.S.R.** 480,442, 15.08.1975; *Chem. Abstr.*, **83**, 153213 (1975).
106. D. A. Bol'shakov, A. I. Yablonskaya, G. A. Stepanov, L. A. Morozova, E. K. Lopatina, N. F. Chernoknizhnaya, E. A. Shmulevich, E. V. Orlov, **U.S.S.R.** 514,626, 25.05.1976; *Chem. Abstr.*, **85**, 83669 (1976).
107. C. P. C. Bradshaw and L. Turner, **Brit.** 1,121,806, 31.07.1968; *Chem. Abstr.*, **69**, 76595 (1968).
108. C. P. C. Bradshaw, **Ger. Offen.** 1,909,951, 16.10.1969; *Chem. Abstr.*, **72**, 12047 (1970).
109. C. P. C. Bradshaw and J. I. Blake, **Brit.** 1,266,340, 8.03.1972; *Chem. Abstr.*, **76**, 126352 (1972).
110. British Petroleum Company, **Brit.** 1,054,864, 8.09.1964; *Chem. Abstr.*, **64**, 19408 (1966).
111. British Petroleum Company, **Brit.** 1,089,956, 23.07.1965; *Chem. Abstr.*, **67**, 53613 (1967).
112. British Petroleum Company, **Brit.** 1,093,784, 25.08.1965; *Chem. Abstr.*, **67**, 53624 (1967).
113. British Petroleum Company, **Brit.** 1,105,563, 23.04.1965; *Chem. Abstr.*, **66**, 48080 (1967).
114. British Petroleum Company, **Brit.** 1,105,564, 23.04.1965; *Chem. Abstr.*, **66**, 48081 (1967).
115. British Petroleum Company, **Fr.** 1,485,859, 23.06.1967; *Chem. Abstr.*, **68**, 61420 (1968).
116. G. Doyle, **Ger. Offen.** 2,047,270, 1.04.1971; *Chem. Abstr.*, **75**, 5202a (1971).
117. A. Greco, F. Pirinoli and G. Dall'Asta, **U.S.** 3,956,178, 5.11.1976.
118. L. F. Heckelsberg, **Fr.** 1,562,397, 4.04.1969; *Chem. Abstr.*, **71**, 93358 (1969).
119. L. F. Heckelsberg, **U.S.** 3,546,312, 8.12.1970.
120. E. J. Howman, C. P. C. Bradshaw and L. Turner, **U.S.** 3,424,812, 28.01.1969.
121. W. B. Hughes and E. A. Zuech, **Fr.** 1,575,779, 25.07.1969; *Chem. Abstr.*, **72**, 99985 (1970).
122. M. L. Khidekel', A. D. Shebaldova, V. I. Mar'in, S. M. Kursanov, I. V. Kalechits, **U.S.S.R.** 491,397, 15.11.1975; *Chem. Abstr.*, **84**, 65758 (1976).
123. E. T. Kittleman and E. A. Zuech, **U.S.** 3,708,551, 2.01.1973; *Chem. Abstr.*, **78**, 71382 (1973).
124. E. T. Kittleman and E. A. Zuech, **U.S.** 3,554,924, 12.01.1971; *Chem. Abstr.*, **74**, 99440 (1971).
125. W. R. Kroll, G. Doyle and W. H. Ruhle, **U.S.** 3,691,095, 12.09.1972; *Chem. Abstr.*, **78**, 110511 (1973).
126. W. R. Kroll and G. Doyle, **U.S.** 3,689,433, 5.09.1972; *Chem. Abstr.*, **77**, 151431 (1972).
127. D. H. Kubicek and E. A. Zuech, **U.S.** 3,703,561, 21.11.1972; *Chem. Abstr.*, **78**, 71378 (1973).
128. M. J. Lawrenson, **Ger. Offen.** 2,340,081, 28.02.1974; *Chem. Abstr.*, **81**, 54837 (1974).
129. B. P. McGrath and K. V. Williams, **Brit.** 1,159,055, 23.06.1969; *Chem. Abstr.*, **71**, 70066 (1969).
130. K. M. Minachev, E. S. Mortikov, A. S. Leont'ev, T. S. Papko, A. A. Masloboev-Shvedov, N. F. Kononov, **U.S.S.R.** 521,007, 15.07.1976; *Chem. Abstr.*, **85**, 113139 (1976).
131. R. B. Regier, **U.S.** 3,652,703, 28.03.1972; *Chem. Abstr.*, **76**, 139910j (1972).
132. R. B. Regier, **U.S.** 3,923,920, 2.12.1975; *Chem. Abstr.*, **84**, 58594 (1976).
133. R. B. Regier, **U.S.** 3,761,427, 25.09.1973; *Chem. Abstr.*, **80**, 41318 (1974).
134. R. E. Reusser, **U.S.** 3,511,890, 12.05.1970.
135. R. E. Reusser, **U.S.** 3,538,180, 3.11.1970.
136. R. E. Reusser and S. D. Turk, **U.S.** 3,786,112, 15.01.1974; *Chem. Abstr.*, **80**, 95210 (1974).

137. R. E. Reusser, **U.S.** 3,915,897, 28.10.1975 ; *Chem. Abstr.*, **84,** 16737 (1976).
138. L. Turner, E. J. Howman and K. V. Williams, **Brit.** 1,106,015, 23.07.1975 ; *Chem. Abstr.*, **67,** 53607t (1967).
139. L. Turner, E. J. Howman and K. V. Williams, **Brit.** 1,106,016, 23.07.1965 ; *Chem. Abstr.*, **67,** 53607t (1967).
140. L. Turner, E. J. Howman and C. P. C. Bradshaw, **Brit.** 1,054,864, 11.12.1970 ; *Chem. Abstr.*, **75,** 140216 (1971).
141. R. van Helden, C. P. Kohll and P. A. Verbrugge, **Ger. Offen.** 2,109,011, 16.09.1971 ; *Chem. Abstr.*, **75,** 140218 (1971).
142. R. van Helden, W. F. Engel, A. J. Alkema and P. Krijger, **Neth.** 6,608,427, 18.12.1967 ; *Chem. Abstr.*, **69,** 43389a (1969).
143. E. A. Zuech, **Ger. Offen.** 2,017,841, 29.10.1971 ; *Chem. Abstr.*, **74,** 5300 (1971).
144. E. A. Zuech, W. B. Hughes, D. H. Kubicek and E. T. Kittleman, *J. Amer. Chem. Soc.*, **92,** 528 (1970).
145. E. A. Zuech, **U.S.** 3,865,892, 11.02.1975 ; *Chem. Abstr.*, **82,** 149367 (1975).
146. E. A. Zuech, **U.S.** 3,981,940, 21.09.1976 ; *Chem. Abstr.*, **86,** 4920 (1977).
147. E. A. Zuech, **U.S.** 4,010,217, 1.03.1977 ; *Chem. Abstr.*, **86,** 189167 (1977).
148. R. L. Banks, **U.S.** 3,696,163, 3.10.1972 ; *Chem. Abstr.*, **78,** 44200 (1973).
149. R. L. Banks, **Ger. Offen.** 1,280,869, 24.10.1968 ; *Chem. Abstr.*, **70,** 19539 (1969).
150. a. P. A. Gautier, F. J. F. Van der Plas, P. A. Verbrugge, **Ger. Offen.** 2,208,058. 7.10.1972 ; *Chem. Abstr.*, **77,** 151421 (1972) ; b. P. A. Gautier, **U.S.** 3,776,974, 4.12.1973
151. R.S. Logan and R. L. Banks, *Oil Gas J.*, **66,** 131 (1968).
152. J. E. Phillips, **U.S.** 3,236,921, 22.02.1966 ; *Chem. Abstr.*, **64,** 12440f (1966).
153. R. E. Reusser, **U.S.** 3,396,165, 3.10.1972 ; *Chem. Abstr.*, **78,** 44199 (1973).
154. R. E. Reusser, **U.S.** 3,689,589, 5.09.1972 ; *Chem. Abstr.*, **78,** 15473 (1973).
155. E. A. Zuech, **U.S.** 3,590,095, 29.06.1971 ; *Chem. Abstr.*, **75,** 76133g (1971).
156. F. W. Küpper and R. Streck, *Makromol. Chem.*, **175,** 2055 (1974).
157. F. W. Küpper and R. Streck, **Ger. Offen.** 2,326,196, 12.12.1974 ; *Chem. Abstr.*, **82,** 155528 (1975).
158. Van C. Vives and R. E. Reusser, **U.S.** 3,792,102, 12.02.1974 ; *Chem. Abstr.*, **80,** 95214 (1974).
159. Van C. Vives and R. E. Reusser, **U.S.** 3,878,262, 15.04.1975 ; *Chem. Abstr.*, **83,** 27518 (1975).
160. Van C. Vives, **U.S.** 3,786,111, 15.01.1974 ; *Chem. Abstr.*, **80,** 82034 (1974).
161. A. M. Popov, R. A. Fridman, E. Sh. Finkel'stein, N. S. Nametkin, V. M. Vdovin, N. A. Bashkirov, Yu. B. Kryukov, L. G. Liberov, *Izv. Akad. Nauk SSSR, Ser. Khim.*, **1973,** 1429.
162. a. E. S. Finkel'stein, B. S. Strel'chik, V. G. Zaikin, V. M. Vdovin, *Neftekhimiya*, **15,** 667 (1975) ; b. E. S. Finkel'stein, B. S. Strel'chik, E. B. Postnykh, V. M. Vdovin, N. S. Nametkin, *Dokl. Akad. Nauk SSSR*, **232,** 1322 (1977).
163. P. B. van Dam, M. C. Mittelmeijer and C. Boelhouwer, *J. Chem. Soc., Chem. Commun.*, **1972,** 1221.
164. P. B. van Dam, M. C. Mittelmeijer and C. Boelhouwer, *J. Amer. Oil Chem. Soc.*, **51,** 389 (1974).
165. P. B. van Dam, M. C. Mittelmeijer and C. Boelhouwer, *Fette, Seifen, Anstrichsmittel*, **67,** 264 (1974).
166. C. Boelhouwer and E. Verkuijlen, *Riv. Ital. Sost. Grasse*, **53,** 237 (1976).
167. E. Verkuijlen and C. Boelhouwer, *Riv. Ital. Sost. Grasse*, **53,** 234 (1976).
168. E. Verkuijlen, R. J. Dirkes and C. Boelhouwer, *Rec. Trav. Chim.*, **96,** M86 (1977).
169. E. Verkuijlen and C. Boelhouwer, *Chem. Phys. Lipids*, **24,** 305 (1979).
170. R. Nakamura, S. Fukuhara, S. Matsumoto and K. Komatsu, **Japan. Kokai** 76 125,317, 1.11.1976 ; *Chem. Abstr.*, **86,** 170899 (1977).
171. R. Nakamura, S. Fukuhara and S. Matsumoto, *Chem. Letters*, **1976,** 253.
172. a. R. Nakamura, S. Matsumoto and E. Echigoya, *Chem. Letters*, **1976,** 1019 ; b. R. Nakamura and E. Echigoya, *Chem. Letters*, **1977,** 1227.
173. R. Baker and M. J. Grimmin, *Tetrahedron Letters*, **1977,** 441.
174. E. Verkuijlen, F. Kapteijn, J. C. Mol and C. Boelhouwer, *J. Chem. Soc., Chem. Commun.*, **1977,** 198.

175. R. H. A. Bosma, F. van den Aardweg and J. C. Mol, *J. Chem. Soc., Chem. Commun.*, **1981**, 1130.
176. J. Otton, Y. Colleuille and J. Varagnat, *J. Mol. Catalysis*, **8**, 313 (1980).
177. C. P. Pinazzi, I. Campistron, M. C. Croissandeau and D. Reyx, *J. Mol. Catalysis*, **8**, 325 (1980).
178. K. Hummel and Y. Imamoglu, *Colloid. Polymer Sci.*, **253**, 225 (1975).
179. W. Ast, G. Rheinwald and R. Kerber, *Rec. Trav. Chim.*, **96**, M127 (1977).
180. J. C. Mol and E. F. G. Woerlee, *J. Chem. Soc., Chem. Commun.*, **1979**, 330.
181. a. J. O. Apatu, D. C. Chapman and H. Heaney, *J. Chem. Soc., Chem. Commun.*, **1981**, 1079; b. R. H. A. Bosma, A. P. Kouwenhoven and J. C. Mol, *J. Chem. Soc., Chem. Commun.*, **1981**, 1081.
182. G. Foster, **Ger. Offen.** 2,063,150, 2.01.1970; *Chem. Abstr.*, **75**, 63172 (1971).
183. R. Nakamura and B. Fukuhara, **Ger. Offen.** 2,456,112, 23.10.1975.
184. J. P. Laval, R. Mutin, A. Lattes and J. M. Basset, *J. Chem. Soc., Chem. Commun.*, **1977**, 502.
185. R. Nouguier, R. Mutin, J. P. Laval, G. Chapelet, J. M. Basset and A. Lattes, *Rec. Trav. Chim.*, **96**, M91 (1977).
186. C. Edwige, A. Lattes, J. P. Laval, R. Mutin, J. M. Basset and R. Nouguier, *J. Mol. Catalysis*, **8**, 297 (1980).
187. R. A. Fridman, A. N. Bashkirov, L. G. Liberov, S. M. Nosakova, R. M. Smirnova, S. B. Verbovetskaya, *Dokl. Akad. Nauk SSSR*, **234**, 1354 (1977).
188. M. Erdweg, Ph. D. Thesis, Aachen, 1979.
189. R. A. Fridman, S. M. Nosakova, L. G. Liberov and A. N. Bashkirov, *Izv. Akad. Nauk SSSR, Ser. Khim.*, **1977**, 678.
190. R. Streck, Internat. Symp. Metathesis, Belfast, 1981.
191. R. Streck and H. Weber, **Ger. Aus.** 2,157,405, 19.11.1971; *Chem. Abstr.*, **79**, 43066 (1973).
192. R. Streck and H. Weber, **Ger. Aus.** 2,166,867, 19.11.1971.
193. D. Zerpner and R. Streck, **Ger. Offen.** 2,922,335, 1.06.1979.
194. F. Pennella, R. L. Banks and G. C. Bailey, *J. Chem. Soc., Chem. Commun.*, **1968**, 1548.
195. J. A. Moulijn, H. J. Rietsma and C. Boelhouwer, *J. Catalysis*, **25**, 434 (1972).
196. A. V. Mushegyan, V. Kh. Ksipteridis, R. Kh. Dzhylakyan, N. A. Gevorkyan, G. A. Chukhadzhyan, *Arm. Khim. Zh.*, **28**, 672 (1975).
197. A. Mortreux, F. Petit and M. Blanchard, *Tetrahedron Letters*, **1978**, 4967.
198. A. V. Mushegyan, V. Kh. Ksipteridis, G. A. Chukhadzhyan, *Arm. Khim. Zh.*, **28**, 674 (1975).
199. A. Mortreux, F. Petit and M. Blanchard, *J. Mol. Catalysis*, **8**, 97 (1980).
200. A. Ben Cheick, M. Petit, A. Mortreux and F. Petit, *Internat. Symp. Metathesis*, Belfast, 1981.
201. J. H. Wengrovius, J. Sancho and R. R. Schrock, *J. Amer. Chem. Soc.*, **103**, 3932 (1981).
202. E. Wasserman, D. A. Ben-Efraim and R. Wolovsky, *J. Amer. Chem. Soc.*, **90**, 3286 (1968).
203. K. W. Scott, N. Calderon, E. A. Ofstead, W. A. Judy and J. P. Ward, *Adv. Chem. Ser.*, **91**, 399 (1969).
204. N. Calderon, **U.S.** 3,439,056, 15.04.1969; *Chem. Abstr.*, **71**, 38438c (1969).
205. N. Calderon, **U.S.** 3,439,057, 15.04.1969; *Chem. Abstr.*, **71**, 80807x (1969).
206. H. Höcker and R. Musch, *Makromol. Chem.*, **157**, 201 (1972).
207. H. Höcker, W. Reimann, K. Riebel and Zs. Szentivanyi, *Makromol. Chem.*, **177**, 1707 (1976).
208. H. Höcker, L. Reif, W. Reimann and K. Riebel, *Rec. Trav. Chim.*, **96**, M47 (1977).
209. H. Höcker and F. R. Jones, *Makromol. Chem.*, **161**, 251 (1972).
210. R. Wolovsky and Z. Nir, *Synthesis*, **1972**, 134.
211. D. A. Ben-Efraim, C. Batich and E. Wasserman, *J. Amer. Chem. Soc.*, **92**, 2133 (1970).
212. R. Wolovsky, *J. Amer. Chem. Soc.*, **92**, 2132 (1970).
213. R. Wolovsky, **Ger. Offen.** 2,108,965, 9.09.1971; *Chem. Abstr.*, **76**, 3466m (1972).
214. R. Wolovsky, **Israeli** 33,970, 31.05.1973; *Chem. Abstr.*, **79**, 52876 (1973).
215. Ullmanns Encyklopädie der technischen Chemie, Dritte Auflage, Urban und Schwarzenberg, Berlin, 1957, Vol. 9, pp. 321.
216. G. Natta, G. Dall'Asta, G. Mazzanti and G. Motroni, *Makromol. Chem.*, **69**, 163 (1963).

217. G. Natta, G. Dall'Asta and G. Motroni, *J. Polymer. Sci., B*, **2**, 349 (1964).
218. G. Natta, G. Dall'Asta and L. Porri, *Makromol. Chem.*, **81**, 253 (1965).
219. T. J. Katz, S. J. Lee and N. Acton, *Tetrahedron Letters*, **1976**, 4247.
220. N. Calderon, *J. Macromol. Sci.—Revs. Macromol. Chem.*, **C7**, 105 (1972).
221. W. Brenner, P. Heimbach, E. W. Muller and G. Wilke, *Ann.*, **727**, 161 (1969).
222. G. Wilke, *Angew. Chem.*, **69**, 397 (1957).
223. T. J. Katz, J. McGinnis and C. Altus, *J. Amer. Chem. Soc.*, **98**, 606 (1976).
224. P. Günther, F. Haas, G. Marwede, K. Nützel, W. Oberkirch, G. Pampus, N. Schön and J. Witte, *Angew. Makromol. Chem.*, **14**, 87 (1970) ; 16/17, 27 (1971).
225. F. Haas and D. Theisen, *Kautch. Gummi Kunstst.*, **23**, 502 (1970).
226. W. Graulich, W. Swodenk and D. Theisen, *Hydrocarbon Processing*, **51**, 74 (1972).
227. A. J. Amass, *Brit. Polymer J.*, **4**, 327 (1972).
228. G. Dall'Asta, *Rubber Chem. Technol.*, **47**, 511 (1974).
229. G. Natta, G. Dall'Asta and G. Mazzanti, *Angew. Chem.*, **76**, 765 (1964) ; *Internat. Ed.*, **3**, 723 (1964).
230. G. Natta, G. Dall'Asta and G. Mazzanti, **Fr.** 1,394,380, 2.04.1965 ; *Chem. Abstr.*, **68**, 3656h (1968).
231. P. Günther, W. Oberkirch and G. Pampus, **Fr.** 2,065,300, 9.10.1969 ; *Chem. Abstr.*, **75**, 21485 (1971).
232. P. Günther, W. Oberkirch and G. Pampus, **Ger. Offen.** 1,812,338, 18.06.1970 ; *Chem. Abstr.*, **73**, 56973 (1970).
233. F. Haas, K. Nützel, G. Pampus and D. Theisen, *Rubber Chem. Technol.*, **43**, 1126 (1970).
234. K. Nützel, K. Dinges and F. Haas, **Ger. Offen.** 1,720,797, 7.03.1968 ; *Chem. Abstr.*, **72**, 79656 (1970).
235. K. Nützel, F. Haas, K. Dinges and W. Graulich, **Ger. Offen.** 1,720,791, 24.02.1968 ; *Chem. Abstr.*, **72**, 91326 (1970).
236. K. Nützel, F. Haas and G. Marwede, **Ger. Offen.** 1,919,047, 15.04.1969 ; *Chem. Abstr.*, **74**, 3253s (1971).
237. K. Nützel, F. Haas and G. Marwede, **Ger. Offen.** 1,919,048, 15.04.1969 ; *Chem. Abstr.*, **74**, 64592e (1971).
238. K. Nützel and F. Haas, **Fr.** 2,008,180, 9.05.1968 ; *Chem. Abstr.*, **73**, 46423 (1970).
239. W. Oberkirch, P. Günther and G. Pampus, **Ger. Offen.** 1,811,653, 17.09.1970 ; *Chem. Abstr.*, **73**, 110694 (1970).
240. W. Oberkirch, P. Günther and G. Pampus, **Ger. Offen.** 1,957,025, 19.05.1971 ; *Chem. Abstr.*, **75**, 64648 (1971).
241. W. Oberkirch, P. Günther and G. Pampus, **U.S.** 3,787,381, 22.01.1974 ; *Chem. Abstr.*, **81**, 38655 (1974).
242. G. Pampus and J. Witte, **Ger. Offen.** 1,770,143, 6.04.1968 ; *Chem. Abstr.*, **72**, 133851 (1979).
243. G. Pampus, N. Schön, J. Witte and G. Marwede, **Ger. Offen.** 1,961,865, 10.12.1969 ; *Chem. Abstr.*, **75**, 77954 (1971).
244. G. Pampus, J. Witte and M. Hoffmann, *Rev. Gen. Caout. Plast.*, **47**, 1343 (1970).
245. M. F. Farona and C. Tsonis, *J. Chem. Soc., Chem. Commun.*, **1977**, 363.
246. G. Natta, A. Zambelli, I. Pasquon and F. Ciampelli, *Makromol. Chem.*, **79**, 161 (1964).
247. A. Kawasaki, I. Maruyama, M. Taniguchi, R. Hirai and J. Furukawa, *J. Polymer. Sci., B*, **7**, 613 (1969).
248. J. Furukawa, *Angew. Makromol. Chem.*, **23**, 189 (1972).
249. G. Natta, G. Dall'Asta and G. Mazzanti, **Ital.** 733,857, 17.01.1964.
250. G. Natta, G. Dall'Asta and G. Mazzanti, **Fr.** 1,425,601, 21.01.1966 ; *Chem. Abstr.*, **65**, 10692 (1966).
251. G. Natta, G. Dall'Asta, I. W. Bassi and G. Carella, *Makromol. Chem.*, **91**, 87 (1966).
252. N. Calderon, E. A. Ofstead and W. A. Judy, *J. Polymer Sci.*, *A-1*, **5**, 2209 (1967).
253. G. Dall'Asta and R. Manetti, **Ital.** 932,461, 1.07.1971.
254. A. Dräxler, *Kautsch. Gummi Kunstst.*, **81**, 185 (1980).
255. G. Dall'Asta and R. Manetti, *European Polymer J.*, **4**, 145 (1968).
256. G. Dall'Asta, *Makromol. Chem.*, **154**, 1 (1972).
257. G. Gianotti, G. Dall'Asta, A. Valvassori and V. Zamboni, *Makromol. Chem.*, **149**, 117 (1971).

258. G. Crespi, A. Valvassori, V. Zamboni and U. Flisi, *Chimica e Industria (Milan)*, **55**, 130 (1973).
259. P. Le Delliou, *Rev. Gen. Caout. Plast.*, **54**, 77 (1977).
260. E. A. Ofstead, cited after ref. 220.
261. G. Dall'Asta, G. Motroni and G. Carella, **Fr.** 1,545,643, 15.07.1966; *Chem. Abstr.*, **71**, 13733 (1969).
262. G. Motroni, G. Dall'Asta and I. W. Bassi, *European Polymer J.*, **9**, 257 (1973).
263. G. Pampus, N. Schön, J. Witte and G. Marwede, **Fr.** 2,070,802, 10.12.1969.
264. J. D. Brown and C. A. Uraneck, **Fr.** 2,103,750, 27.07.1970.
265. E. A. Ofstead, **U.S.** 3,597,403, 27.10.1969.
266. K. W. Scott and N. Calderon, **Ger. Offen.** 2,058,198, 4.12.1969.
267. K. W. Scott and N. Calderon, **U.S.** 4,010,224, 1.03.1977; *Chem. Abstr.*, **86**, 140943 (1977).
268. G. Pampus, J. Witte and M. Hoffmann, **Ger. Offen.** 2,016,840, 21.10.1970; *Chem. Abstr.*, **76**, 72998 (1972).
269. D. Medema, H. J. Alkema and R. van Helden, **U.S.** 3,649,709, 14.03.1972; *Chem. Abstr.*, **77**, 6963c (1972).
270. G. Pampus, J. Witte and M. Hoffmann, **Ger. Offen.** 1,954,092, 28.12.1969; *Chem. Abstr.*, **75**, 37611z (1971).
271. G. Pampus, J. Witte and M. Hoffmann, **Fr.** 2,066,701, 28.10.1969.
272. G. Pampus, N. Schön, W. Oberkirch and P. Günther, **Ger. Offen.** 2,009,740, 30.09.1971; *Chem. Abstr.*, **76**, 47144 (1972).
273. G. Pampus, J. Witte and M. Hoffmann, **Ger. Offen.** 2,016,471, 28.10.1961; *Chem. Abstr.*, **76**, 60711 (1972).
274. G. Dall'Asta and G. Motroni, **Fr.** 2,149,810, 2.08.1971.
275. K. Hummel and W. Ast, *Makromol. Chem.*, **166**, 39 (1973).
276. K. Hummel, D. Wewrka, F. Lorber and G. Zeplichal, *Makromol. Chem.*, **166**, 45 (1973).
277. G. Dall'Asta, **Belg.** 691,503, 20.06.1967.
278. G. Dall'Asta and R. Manetti, **Ital.** 789,585, 21.12.1965.
279. K. Nützel, F. Haas and G. Marwede, **Fr.** 2,039,195, 15.04.1969.
280. R. Streck and H. Weber, **Ger. Offen.** 1,195,358, 8.09.1969.
281. H. J. Alkema and R. van Helden, **U.S.** 3,649,709, 17.06.1968.
282. K. Hummel and G. Raithofer, *Angew. Makromol. Chem.*, **50**, 183 (1976).
283. K. Hummel, F. Stelzer, W. Kathan, H. Demel, O. A. Wedam and G. Karaoulis, *Rec. Trav. Chim.*, **96**, M75 (1977).
284. W. Ast, C. Zott, H. Bosch and R. Kerber, *Rec. Trav. Chim.*, **96**, M81 (1977).
285. W. Ast, H. Bosch, K. Grundmeaer and R. Kerber, *Internat. Symp. Metathesis*, Lyons, 1979, Abstracts, 27.
286. H. Abendroth and E. Canji, *Makromol. Chem.*, **176**, 775 (1975).
287. E. Canji and H. Perner, *Rec. Trav. Chim.*, **96**, M70 (1977).
288. a. K. Hummel, F. Stelzer, P. Heiling, O. A. Wedam and H. Griesser, *J. Mol. Catalysis*, **8**, 253 (1980); b. K. Hummel, O. A. Wedam, W. Kathan, H. Demel, *Makromol. Chem.*, **179**, 1159 (1978).
289. G. Dall'Asta, G. Scigliani, A. Greco and L. Motta, *Chimica e Industria (Milan)*, **55**, 142 (1973).
290. V. Drăguţan, *Studii Cercetări Chim.*, **22**, 239 (1974).
291. H. J. Jahn, cited after ref. 228.
292. M. Hoffmann, *Makromol. Chem.*, **144**, 309 (1971).
293. H. Schencko and R. Caspary, *Kautsch. Gummi Kunstst.*, **25**, 309 (1972).
294. G. Dall'Asta and P. Scaglione, *Rubber Chem. Technol.*, **42**, 1235 (1969).
295. N. Calderon and M. Morris, *J. Polymer Sci.*, A-2, **5**, 1283 (1967).
296. W. Holtrup, F. W. Küpper, W. Meckstroth, H. H. Meyer, K. H. Nordsiek, *Lichtbogen*, **174**, 16 (1974).
297. A. Dräxler, *Lichbogen*, **198**, 4 (1980).
298. Chemische Werke Hüls, *Data Sheet*, 2247e, 1980.
299. P. Le Delliou, *Hule Mex. Plast.*, **33**, 17, 20 (1977).
300. P. Le Delliou and H. D. Manager, *Gummi Asbest Kunstst.*, **30**, 518 (1977).
301. P. Le Delliou, *Internat. Rubber Conf.*, **1977**, 1.

302. T. J. Katz, S. J. Lee and M. A. Shippey, *J. Mol. Catalysis*, **8**, 219 (1980).
303. N. Calderon, E. A. Ofstead and W. A. Judy, *J. Polymer Sci.*, *A-1*, **5**, 2209 (1967).
304. E. A. Ofstead, **U.S.** 3,801,559, 25.09.1968.
305. A. J. Bell, **Ger. Offen.** 2,049,026, 1.10.1970.
306. W. J. Feast and B. Wilson, *Polymer*, **20**, 1182 (1979).
307. W. J. Feast and B. Wilson, *J. Mol. Catalysis*, **8**, 277 (1980).
308. A. B. Alimuniar, J. H. Edwards and W. J. Feast, *Internat. Symp. Metathesis*, Belfast, 1981.
309. F. Kanega, M. Shitara, F. Shyoji, H. Kogame and T. Fuji, **Japan. Kokai** 76 111,300, 1.10.1976; *Chem. Abstr.*, **86**, 44222 (1977).
310. Y. Kobayashi, T. Ueshima, Y. Tanaka, S. Kurosawa and S. Kobayashi, **Ger. Offen.** 2,540,553, 17.03.1977; *Chem. Abstr.*, **86**, 156245 (1977).
311. T. Nagaoka, N. Kuwabara and T. Ueshima, **Japan. Kokai** 76 28,146, 9.06.1976; *Chem. Abstr.*, **85**, 64026 (1976).
312. J. Nakamura, H. Konuma, S. Kokuryo, T. Ueshima and C. Tsuge, **Japan. Kokai** 75 142,661, 15.11.1975; *Chem. Abstr.*, **84**, 90959 (1976).
313. J. Nakamura, H. Konuma, S. Kokuryo, T. Ueshima, **Japan. Kokai** 76 112,866, 29.03. 1976; *Chem. Abstr.*, **86**, 56152 (1977).
314. H. Ikeda, S. Matsumoto, K. Makiuo, **Ger. Offen.** 2,447,687, 17.04.1975; *Chem. Abstr.*, **83**, 44046 (1975).
315. F. Imaizumi, H. Nobuyo, M. Kawakami, **Japan. Kokai** 77 87,449, 21.07.1977.
316. Y. Kobayashi, T. Ueshima, S. Kobayashi, **Japan. Kokai** 77 36,200, 19.03.1977; *Chem. Abstr.*, **87**, 53856 (1977).
317. Y. Kobayashi, T. Ueshima, S. Kobayashi, **Japan. Kokai** 77 26,599, 28.02.1977; *Chem. Abstr.*, **87**, 40076 (1977).
318. Y. Kobayashi, T. Ueshima, S. Kobayashi, **Japan. Kokai** 77 42,600, 2.04.1977; *Chem. Abstr.*, **87**, 53860 (1977).
319. Y. Kobayashi, T. Ueshima, S. Kobayashi, **Japan. Kokai** 77 45,698, 11.04.1977; *Chem. Abstr.*, **87**, 53875 (1977).
320. Y. Kobayashi, T. Ueshima, S. Kobayashi, **Japan. Kokai** 77 44,899, 8.04.1977; *Chem. Abstr.*, **87**, 53857 (1977).
321. Y. Kobayashi, T. Ueshima, S. Kobayashi, **Japan. Kokai** 77 37,998, 24.03.1977; *Chem. Abstr.*, **87**, 40075 (1977).
322. T. Kotani, S. Matsumoto, F. Imaizumi, K. Igarashi, K. Takahashi, **Japan. Kokai** 77 32,100, 10.03.1977; *Chem. Abstr.*, **87**, 53832 (1977).
323. S. Kurosawa, T. Ueshima, S. Kobayashi, **Japan. Kokai** 75 110,000, 29.08.1975; *Chem. Abstr.*, **84**, 31737 (1976).
324. S. Kurosawa, T. Ueshima, Y. Tanaka, S. Kobayashi, T. Saito, **Japan. Kokai** 75 61,500, 27.05.1975; *Chem. Abstr.*, **83**, 179939 (1975).
325. S. Kurosawa, T. Ueshima, Y. Tanaka, S. Kobayashi, **Ger. Offen.** 2,529,963, 15.01.1976; *Chem. Abstr.*, **84**, 122565 (1976).
326. M. Shitara, F. Shoji, H. Kougame, F. Kanega, **Japan. Kokai** 76 11,898, 30.01.1976; *Chem. Abstr.*, **86**, 55927 (1977).
327. Y. Tanaka, T. Ueshima, S. Kobayashi, **Japan. Kokai** 75 159,598, 24.12.1975; *Chem. Abstr.*, **84**, 165412 (1976).
328. Y. Tanaka, T. Ueshima, S. Kurosawa, **Japan. Kokai** 75 160,400, 25.12.1975; *Chem. Abstr.*, **84**, 151261 (1976).
329. J. Nakamura, S. Kurosawa and S. Kokuryo, **Japan. Kokai** 76 107,351, 19.03.1976; *Chem. Abstr.*, **86**, 17577 (1977).
330. J. Nakamura, S. Kurosawa and S. Kokuryo, **Japan. Kokai** 76 112,865, 5.10.1976; *Chem. Abstr.*, **86**, 90838 (1977).
331. F. Imaizumi, K. Suzuki and M. Kawakami, **Japan. Kokai** 76 136,800, 26.11.1976; *Chem. Abstr.*, **87**, 23950 (1977).
332. H. Lammens, G. Sartori, J. Siffert and N. Sprecher, *Polymer Letters*, **9**, 341 (1971).
333. W. Ast and G. Rheinewald, **Ger. Offen.** 2,535,496, 17.02.1977; *Chem. Abstr.*, **86**, 156201 (1977).
334. C. T. Thu, T. Bastelberger and H. Höcker, *Makromol. Chem., Rapid Commun.*, **2**, 383 (1981).

335. R. Streck, H. K. Nordsiek, H. Weber, K. Mayer, **Ger. Offen.** 2,131,355, 11.01.1973 ; *Chem. Abstr.*, **78**, 112522 (1973).
336. K. W. Scott, *Polymer Preprints*, **13**, 874 (1972).
337. Yu. V. Korshak, B. A. Dolgoplosk and M. A. Tlenkopachev, *Rec. Trav. Chim.*, **96**, M64 (1977).
338. Yu. V. Korshak, M. A. Tlenkopachev, B. A. Dolgoplosk, E. G. Avdeikina and D. F. Kutepov, *Internat. Symp. Metathesis*, Belfast, 1981.
339. K. Hummel, R. Streck and H. Weber, *Naturwissenschaften*, **57**, 194 (1970).
340. K. Hummel and W. Ast, *Naturwissenschaften*, **57**, 545 (1970).
341. L. Mikhailov and H. J. Harwood, *16 0th National ACS Meeting*, Chicago, 1970.
342. K. Hummel and W. Ast, *Makromol. Chem.*, **166**, 39 (1973).
343. K. Hummel, D. Wewerka, F. Lorber and G. Zeplichal, *Makromol. Chem.*, **166**, 45 (1973).
344. K. Hummel, H. Demel and D. Wewerka, *Makromol. Chem.*, **178**, 19 (1977).
345. E. Canji and H. Perner, *Makromol. Chem.*, **179**, 567 (1978).
346. E. Thorn-Csanyi, H. Abendroth and H. Perner, *Makromol. Chem.*, **181**, 2081 (1980).
347. E. Thorn-Csanyi, C. Hennemann-Perner, H. Perner, *Makromol. Chem.*, in press.
348. E. Thorn-Csanyi and H. Perner, *Makromol. Chem.*, **180**, 919 (1979).
349. D. P. Montgomery, R. N. Moore and R. W. Knox, **U.S.** 3,965,206, 22.06.1976.
350. T. J. Katz and S. J. Lee, *J. Amer. Chem. Soc.*, **102**, 422 (1980).
351. J. H. Edwards and W. J. Feast. *Polymer*, **21**, 595 (1980).
352. A. B. Alimuniar, J. H. Edwards, K. Harper and W. J. Feast, *Internat. Symp. Metathes is*, Belfast, 1981.
353. S. R. Wilson and D. E. Schalk, *J. Org. Chem.*, **41**, 3928 (1976).
354. R. Rossi, *Chimica e Industria (Milan)*, **57**, 242 (1975).
355. F. W. Küpper and R. Streck, *Chem. Ztg.*, **99**, 464 (1975).
356. F. W. Küpper and R. Streck, *Z. Naturforsch.*, **31B**, 1256 (1976).
357. F. W. Küpper and R. Streck, **Ger. Offen.** 2,531,959, 20.01.1977 ; *Chem. Abstr.*, **87**, 22390 (1977).
358. F. W. Küpper and R. Streck, **Ger. Offen.** 2,533,247, 10.02.1977 ; *Chem. Abstr.*, **86**, 155197 (1977).
359. J. Levisalles and D. Villemin, *Tetrahedron*, **36**, 3181 (1980).
360. D. Villemin, *Tetrahedron Letters*, **1980**, 1715.
361. J. Tsuji and Sh. Hashiguchi, *Tetrahedron Letters*, **1980**, 2955.
362. K. Maruyama, K. Terada and Y. Yamamoto, *J. Org. Chem.*, **45**, 437 (1980).

Chapter 9

TABULAR SURVEY OF OLEFIN METATHESIS AND RING-OPENING POLYMERIZATION

The present chapter provides a comprehensive tabular survey of the metathetical catalytic systems and unsaturated substrates (alkenes, cyclo-alkenes and alkynes) for which metathesis reactions or ring-opening polymerizations have been reported. Tables 1 and 2 present the catalytic systems used in metathesis and ring-opening polymerization, respectively, tables 3 and 4 survey the unsaturated substrates which undergo metathesis and ring-opening polymerization, respectively, while table 5 covers alkyne metathesis and polymerization reactions. In the final section, the cited literature published up to 1982 is arranged in the following categories: reviews and general papers, full papers, communications and Ph. D. theses, and patents. In each of these groups the references are ordered alphabetically according to the first-named author or company. More recent pertinent publications have been included under item D, page 523.

TABLE I. Catalytic systems for metathesis

Catalytic system	Alkene or cyclo-alkene	References
1	2	3

A. Heterogeneous catalytic systems

A.1. Unicomponent heterogeneous systems

Al_2O_3	propene	110a, 516, 1420
	butadiene	943
	ethene + alkenes $C_3 - C_{20}$	1505
SiO_2	propene	1414
MgO	propene	1211, 1422
Aluminosilicates	propene	739, 1699
Zeolite Y	1-butene	139
	isobutene	628
Zeolite Y + CO_2	isobutene	629

1	2	3

A.2. Binary heterogeneous systems

$MoO_3 \cdot Al_2O_3$	ethene	669
	ethene + ethene-d_4	784
	propene	131, 339, 372, 383, 384, 386, 414, 415, 552, 750, 758, 779, 780, 781, 782, 784, 818, 819, 820, 941, 942, 958, 990, 1064, 1182, 1519, 1750, 1916, 1918.
	1-butene	131, 582, 583, 784, 958.
	isobutene	784, 1182
	isobutene + propene	553
	isobutene + 2-butene	778, 784, 1182
	butenes + ethene	1998
	butadiene	943
	1-pentene	131
	2-pentene	131, 517
	1-hexene	131, 1856
	1-heptene	141, 142, 143, 144, 386.
	1-octene	131, 582, 583, 1856.
	4-methyl-1-octene + 1-dodecene	1626
	1-decene	1856
	methylenecyclobutane	871
	cyclopentene + ethene	1819
	cyclohexene + ethene	1819
	4-vinylcyclohexene	1321
	cyclo-octene + ethene	1819
	1,5-cyclo-octadiene + ethene	1819
	vinyltrimethylsilane	386
	stilbene + ethene	1713
$MoO_3 \cdot SiO_2$	propene	372, 375, 396, 450, 451, 539, 678, 750, 758, 780, 781, 782, 941, 942, 989, 1075, 1428, 1816.
	propene + isobutene	376
	1-butene	1428
	2-butene	1428
	pentene	1219
$Mo_2O_3 \cdot SiO_2$	propene	1206
$MoO_3 \cdot AlPO_4$	propene	451, 678, 1417
	pentenes	1219
	pentene + ethene	1219
$WO_3 \cdot Al_2O_3$	propene	542, 581, 678, 679, 748, 750, 930, 942, 1064, 1385, 1918, 1919.
	alkenes C_3-C_8	1918, 1919.
	propene + ethene	581

1	2	3
	pentenes	1219
	1-pentene + 2-pentene	1270
	4-methyl-2-pentene	1263
	methylenecyclobutane	871
	4-vinylcyclohexene	1321
$WO_3 \cdot SiO_2$	ethene	1830
	propene	159, 448, 449, 450, 451, 539, 678, 683, 684, 685, 750, 990, 1128, 1203, 1206, 1218, 1348, 1428, 1666.
	propene + isobutene	132, 992
	1-butene	995, 996
	1-butene + isobutene	993
	2-butene + ethene	552
	2-butene + 1-butene	993
	2-butene + isobutene	907, 908, 993
	n-butenes + propene	994
	butadiene	133, 452
	butadiene + propene	452
	butadiene + 2-butene	452
	butadiene + isobutene	452
	stilbene + ethene	1713
	cyclo-octene + ethene	1958, 1959
	1,5-cyclo-octadiene + ethene	1957
	1,5,9-cyclododecatriene + ethene	1959
$WO_3 \cdot ThO_2$	propene	451, 1429
$WO_3 \cdot ZrO_2$	propene	451, 1425
$WO_3 \cdot AlPO_4$	propene	451, 678, 1417, 1427,
	pentenes	1219
	pentenes + ethene	1219
$WO_3 \cdot Ca_3(PO_4)_2$	propene	451
$WO_3 \cdot Mg_3(PO_4)_2$	propene	451
$WO_3 \cdot Ti_3(PO_4)_4$	propene	451
$WO_3 \cdot Zr_3(PO_4)_4$	propene	451
$Re_2O_7 \cdot Al_2O_3$	propene	372, 386, 739, 749, 750, 787, 827, 869, 948, 958, 1064, 1065, 1066, 1182, 1278, 1286, 1456, 1916, 1918, 1919.
	1-butene	325, 827, 948, 958, 1065, 1068, 1273, 1277, 1920.
	1-butene + 2-butene	1065, 1068, 1671, 1673, 1675.
	1-butene + 2-butene + isobutene	1671
	2-butene	1274
	2-butene + ethene	1274
	n-butenes	742, 833
	4-methyl-2-butene	760, 975
	2-methyl-1-butene + 2-methyl-2-butene	1065

1	2	3
	1-pentene	975
	n-pentenes	948
	isopentene	1982
	1-pentene + 2-pentene	1065
	1-pentene + isopentene	1070
	2-methyl-2-penthene + ethene	373
	2-methyl-2-pentene + propene	373
	4-methyl-2-pentene + ethene	373
	4-methyl-2-pentene + propene	373
	heptene	1065
	7-methyl-1-octene + 1-dodecene	1626
	1-decene + 1-octene + 1-hexene + + 2-butene	1859
	stilbene + ethene	386
	oleil chloride	788
	vinyltrimethylsilane	386
	methylenecyclobutane	379, 975
	methylenecyclobutane + ethene	383
	methylenecyclobutane + propene	141, 144, 383, 386
	vinylcyclohexane	379
	cycloheptatriene + ethene	107
	1-methylcyclo-octene + 2-butene	1341
	1-methylcyclo-octene + isobutene	1341
$Re_2O_7 \cdot TiO_2$	propene	1276
$Re_2O_7 \cdot SnO_2$	propene	1264, 1276
	2-butene + ethene	1264
	alkenes $C_3 - C_8$	1264
$Re_2O_7 \cdot SiO_2$	ethene	674
	propene	451, 750, 1428
$ReO_4 \cdot Al_2O_3$	pentenes	1219
$ReO_4 \cdot AlPO_4$	pentenes	1219
$Re_2O_7 \cdot Al_2O_3$	methylenecyclobutane	871
$Re_2O_7 \cdot$ aluminosilicates	propene	1071, 1072, 1946
	1-butene	1071
$Re_2O_7 \cdot ZrO_2$	propene	1276
$Re_2O_7 \cdot ThO_2$	propene	1276
$Re_2O_7 \cdot Fe_3O_4$	propene	1915
$Re_2O_7 \cdot CoO$	propene	1915
$Re_2O_7 \cdot NiO$	propene	1915
$Re_2O_7 \cdot WO_3$	propene	1915
$Re_2O_7 \cdot AlPO_4$	propene	1915
$Re_2O_7 \cdot Ca_3(PO_4)_2$	propene	1915
$Re_2O_7 \cdot Mg_3(PO_4)_2$	propene	1915
$Re_2O_7 \cdot Ti_3(PO_4)_4$	propene	1915
$Re_2O_7 \cdot Zr_3(PO_4)_4$	propene	1915
$ReO_4 \cdot Al_2O_3$	pentene	1219
$ReO_4 \cdot AlPO_4$	pentene	1219
$Ta_2O_5 \cdot SiO_2$	ethene	1208
	propene	1208
$Ta_2O_5 \cdot ZrO_2$	propene	1208
$Ta_2O_5 \cdot AlPO_4$	propene	1208
$Ta_2O_5 \cdot Ca_3(PO_4)_2$	propene	1208
$Ta_2O_5 \cdot Mg_3(PO_4)_2$	propene	1208
$Ta_2O_5 \cdot Ti_3(PO_4)_4$	propene	1208
$Ta_2O_5 \cdot Zr_3(PO_4)_4$	propene	1208

1	2	3
$Nb_2O_5 \cdot SiO_2$	ethene	1208
	propene	451, 1208
$Nb_2O_5 \cdot ZrO_2$	propene	451, 1208
$Nb_2O_5 \cdot AlPO_4$	propene	451, 1208
$Nb_2O_5 \cdot Ca_3(PO_4)_2$	propene	451, 1208
$Nb_2O_5 \cdot Mg_3(PO_4)_2$	propene	451, 1208
$Nb_2O_5 \cdot Ti_3(PO_4)_4$	propene	451, 1208
$Nb_2O_5 \cdot Zr_3(PO_4)_4$	propene	451, 1208
$TeO_3 \cdot SiO_2$	propene	451
$La_2O_3 \cdot Al_2O_3$	propene	1914
$Ru_2O_3 \cdot SiO_2$	propene	1822
	1-butene	1822
$SnO_2 \cdot Al_2O_3$	propene	1914
$IrO_2 \cdot Al_2O_3$	propene	1914
$Rh_2O_3 \cdot Al_2O_3$	propene	1914
$OsO_2 \cdot Al_2O_3$	propene	1914
$V_2O_5 \cdot ZrO_2$	propene	1914
$V_2O_5 \cdot AlPO_4$	propene	1914
$V_2O_5 \cdot Ca_3(PO_4)_2$	propene	1914
$V_2O_5 \cdot Mg_3(PO_4)_2$	propene	1914
$V_2O_5 \cdot Ti_3(PO_4)_4$	propene	1914
$V_2O_5 \cdot Zr_3(PO_4)_4$	propene	1914
$Al_2O_3 \cdot SiO_2$	propene	810, 976a
$Mo(CO)_6 \cdot Al_2O_3$	ethene	830, 1204
	propene	131, 207, 283, 284, 285, 480, 678, 750, 954, 955, 1200, 1204, 1275, 1666.
	1-butene	131, 286, 1200, 1204, 1275
	2-butene	286
	1-pentene	131, 1200, 1204, 1275.
	2-pentene	131, 1200
	1-hexene	131, 1200, 1204
	1-octene	131, 1200, 1204
	2-octene + ethene	1322
	dodecene + ethene	1322
	cyclopentene + ethene	1819
	cyclohexene + ethene	1819
	cyclo-octene + ethene	1819
	1,5-cyclo-octadiene + ethene	1819
$Mo(CO)_6 \cdot SiO_2$	propene	480, 750, 954, 1204, 1666
$Mo(CO)_6 \cdot ZrO_2$	propene	1204
$Mo(CO)_6 \cdot ThO_2$	propene	1204
$Mo(CO)_6 \cdot MgO$	propene	480, 954
$Mo(CO)_6 \cdot AlPO_4$	propene	1204
$Mo(CO)_6 \cdot Ca_3(PO_4)_2$	propene	1204
$Mo(CO)_6 \cdot Mg_3(PO_4)_2$	propene	1204
$Mo(CO)_6 \cdot Ti_3(PO_4)_4$	propene	1204
$Mo(CO)_6 \cdot Zr_3(PO_4)_4$	propene	1204
$W(CO)_6 \cdot Al_2O_3$	propene	131, 678, 750, 769, 1274, 1275, 1666.
	1-butene	131, 1275

1	2	3
	1-pentene	131, 1275
	2-pentene	131, 1275
	1-hexene	131, 1275
	1-octene	131, 1275
$W(CO)_6 \cdot SiO_2$	ethene	1204
	propene	750, 1204, 1666
	1-butene	1204
	1-pentene	1204
	1-hexene	1204
$W(CO)_6 \cdot ZrO_2$	propene	1203
$W(CO)_6 \cdot ThO_2$	propene	1203
$W(CO)_6 \cdot AlPO_4$	propene	1203
$W(CO)_6 \cdot Ca_3(PO_4)_2$	propene	1203
$W(CO)_6 \cdot Mg_3(PO_4)_2$	propene	1203
$W(CO)_6 \cdot Ti_3(PO_4)_4$	propene	1203
$W(CO)_6 \cdot Zr_3(PO_4)_4$	propene	1203
$MoS_2 \cdot Al_2O_3$	propene	151, 824
	1-butene	824
	2-butene + 2-butene-d_8	824
$MoS_2 \cdot SiO_2$	propene	451
$MoS_3 \cdot Al_2O_3$	propene	1414, 1417
$MoS_3 \cdot SiO_2$	propene	1414, 1417
$MoS_3 \cdot ZrO_2$	propene	1414, 1417
$MoS_3 \cdot ThO_2$	propene	1414, 1417
$MoS_3 \cdot AlPO_4$	propene	1414, 1417
$MoS_3 \cdot Ca_3(PO_4)_2$	propene	1414, 1417
$MoS_3 \cdot Mg_3(PO_4)_4$	propene	1414, 1417
$MoS_3 \cdot Ti_3(PO_4)_4$	propene	1414, 1417
$MoS_3 \cdot Zr_3(PO_4)_4$	propene	1414, 1417
$WS_2 \cdot Al_2O_3$	propene	451, 1414, 1417
$WS_2 \cdot SiO_2$	propene	451, 1414, 1417
$WS_3 \cdot Al_2O_3$	propene	451
$WS_3 \cdot SiO_2$	propene	451
$WS_3 \cdot ZrO_2$	propene	451
$WS_3 \cdot ThO_2$	propene	451
$WS_3 \cdot AlPO_4$	propene	451
$WS_3 \cdot Ca_3(PO_4)_2$	propene	451
$WS_3 \cdot Mg_3(PO_4)_2$	propene	451
$WS_3 \cdot Ti_3(PO_4)_4$	propene	451
$WS_3 \cdot Zr_3(PO_4)_4$	propene	451
$MoO_3 \cdot Al_2O_3(+Ag, Cu)$	1-octene	1365
	1-hexadecene	1369
$MoO_3 \cdot Al_2O_3(+Na)$	1-butene	1428, 1913
	2-butene	1428
	isobutene	
$MoO_3 \cdot Al_2O_3(+Na, K, Rb, Cs, Tl, Ba)$	propene	1415, 1428
	2-butene + isobutene	1428
	isobutene + propene	1428
	4-vinylcyclohexene	1319
$MoO_3 \cdot Al_2O_3(+NaOH, LiOH, KOH, RbOH, CsOH)$	propene	723
	2-butene + isobutene	723
	isobutene + propene	263
	2,3-dimethyl-2-butene + ethene	242, 263

1	2	3
	3-heptene	262
	1-octene	262, 263
	2-octene	262, 263
	vinylcyclohexene	263, 1618
$MoO_3 \cdot Al_2O_3(+NaCl, NaF, KF, CdF_2, ZnF_2)$	propene	1180
	2-butene	1180
	2-methyl-2-butene	1180
	isobutene + 1-butene	1738
$MoO_3 \cdot Al_2O_3(+Ag, Bi, Sb, Pb, Co, Sn, Zn, Cd)$	1-butene + isobutene	1738
$MoO_3 \cdot Al_2O_3(+Na_2O, K_2O, Cs_2O, BaO)$	propene	1423
	1-butene	1423
	2-butene	1423
	isobutene	1423
$MoO_3 \cdot Al_2O_3(+O_2)$	propene	686
$WO_3 \cdot Al_2O_3(+H_2)$	propene	1213
$WO_3 \cdot Al_2O_3(+CO)$	propene	1213
	1-butene	1213
$WO_3 \cdot Al_2O_3(+K)$	2-butene + isobutene	1858
$WO_3 \cdot Al_2O_3(+K_2O)$	stilbene + ethene	1713
$WO_3 \cdot Al_2O_3(+CuO, Fe_2O_3, CoO, NiO)$	propene	1372
	alkenes C_3-C_{30}	1372
$WO_3 \cdot SiO_2(+Na, K, Cs)$	1-butene	995
$WO_3 \cdot SiO_2(+Na_2O)$	propene	1422, 1428
	1-butene	1428
	isobutene	1428
	2-butene	1428
	2-butene + isobutene	1428
	isobutene + propene	1428
	butadiene + 2-butene	1421
	butadiene + isobutene	1421
$WO_3 \cdot SiO_2(BaO, K_2O, CsO)$	propene	1428
$WO_3 \cdot SiO_2(+KOH)$	3-heptene	1826
$WO_3 \cdot SiO_2(+Nd_2O_3, Pr_2O_3, Sm_2O_3)$	propene	1221
$WO_3 \cdot SiO_2(+Bi)$	propene	1373
$WO_3 \cdot SiO_2$ (+ vinyl chloride, propyl chloride, chlorobenzene, hydrochloride)	propene	1807
$WO_3 \cdot SiO_2(+F^-)$	1-butene	1823, 1825
$Re_2O_7 \cdot Al_2O_3(+Na)$	1-butene	1672
	1-butene + 2-butene	1672
	1-butene + 2-butene + isobutene	1672
$Re_2O_7 \cdot Al_2O_3(+K)$	butenes	1265
	2-butene	1265
	2-pentene + 3-hexene	1265
$Re_2O_7 \cdot Al_2O_3(+F^-, SO_4^{2-}, PO_4^{3-})$	propene	1794
	2-butene + isobutene	1794
$Mo(CO)_6 \cdot SiO_2(+K, Na, Ca, Mg)$	n-butenes	1286
	n-butenes + isobutene	1286
$Mo(CO)_6 \cdot Al_2O_3(+K, Na, Ca, Mg)$	n-butenes	1286
	n-butenes + isobutene	1286

1	2	3
$W(CO)_6 \cdot Al_2O_3(+K, Na, Ca, Mg)$	n-butenes	1286
	n-butenes + isobutene	1286
$W(CO)_6 \cdot SiO_2(+K, Na, Ca, Mg)$	n-butenes	1286
	n-butenes + isobutene	1286
$Re_2(CO)_{10} \cdot Al_2O_3$	propene	1275, 1969, 1970
	1-pentene	1275
	1-hexene	1275
	1-heptene	1275
	1-octene	1275
$Re_2(CO)_{10} \cdot SiO_2(+K, NA, Ca, Mg)$	n-butenes	1286
	n-butenes + isobutene	1286
$Re_2(CO)_{10} \cdot Al_2O_3(+K, Na, Ca, Mg)$	n-butenes	1286
	n-butenes + isobutene	1286
$(\pi\text{-allyl})_4Mo \cdot SiO_2(Al_2O_3)$	propene	358, 361
	1-butene	1716
	3,3-dimethyl-1-butene	1716
	1-hexene	1716
	3-heptene	1716
$(\pi\text{-allyl})_4W \cdot SiO_2(Al_2O_3)$	propene	358, 1716
	1-butene	1716
	3,3-dimethyl-1-butene	1716
	1-hexene	1716
	3-heptene	1716
$Me_6W \cdot SiO_2(Al_2O_3)$	propene	769, 955
$Me_6Mo \cdot SiO_2(Al_2O_3)$	propene	955
$W(acetate)_n \cdot SiO_2(Al_2O_3)$	propene	955
$Mo(acetate)_n \cdot SiO_2(Al_2O_3)$	propene	955
$WCl_6 \cdot SiO_2(Al_2O_3)$	2-pentene	1169
$H_3(PW_{12}O_{40}) \cdot 5H_2O \cdot Al_2O_3$	n-butenes	1735
	alkenes C_2-C_4	1735

A.3. Ternary heterogeneous systems

$MoO_3 \cdot CoO \cdot Al_2O_3$	ethene	830, 1814
	propene	71, 110a, 131, 451, 678, 679, 747, 748, 779, 782, 830, 1139, 1163, 1199, 1200, 1347, 1666, 1667, 1860, 1870, 1879.
	1-butene	131, 197, 258, 1179, 1200, 1201, 1202, 1217, 1346.
	2-butene	71, 131, 1199, 1200, 1202, 1346
	2-butene + ethene	197
	isobutene	71
	1-pentene	78, 131, 1199, 1200, 1202.
	2-pentene	78, 131, 517, 1199, 1200.
	2-pentene + 1-pentene	78
	pentenes + propene + ethene	1579
	1-hexene	131, 1199, 1200, 1202.
	1-hexene + 1-pentene	78

1	2	3
	2-heptene	106, 1202
	3-heptene	1202
	1-octene	106, 131, 1199, 1202.
	2-nonene	1202
	1-dodecene	1200, 1202
	2-tetradecene	1202
	1-hexadecene	1202
	2-methyl-1-butene	1200, 1202
	2-methyl-2-butene	1202
	3-methyl-1-butene	1200, 1202
	1-phenyl-2-butene	1202
	cyclopentene + ethene	1662
	4-vinylcyclohexene	1620
	cyclo-octene + ethene	1662
	1,5-cyclo-octadiene + ethene	1662
	bicyclo[2.2.1]hept-2-ene + ethene	1662
	bicyclo[2.2.1]hept-2,5-diene + + ethene	1662
$MoO_3 \cdot Al_2O_3 \cdot SiO_2$	propene	1418, 1437, 1816.
	2-butene	1437
$MoO_3 \cdot Al_2O_3 \cdot TiO_2$	propene	451, 458
	1-butene	458
$MoO_3 \cdot MgO \cdot TiO_2$	propene	451, 1421
$MoO_3 \cdot AgO \cdot Al_2O_3$	propene	451
$MoO_3 \cdot SiO_2 \cdot P_2O_5$	propene	1661
$MoO_3 \cdot SiO_2 \cdot Me_3Al$	propene	941
	2-pentene	940
$MoO_3 \cdot SiO_2 \cdot Et_3Al$	propene	940
$MoO_3 \cdot SiO_2 \cdot Pr_3Al$	propene	940
$MoO_3 \cdot Al_2O_3 \cdot Me_3Al$	propene	940, 944
$MoO_3 \cdot Al_2O_3 \cdot Pr_3Al$	propene	372
$MoO_3 \cdot Al_2O_3 \cdot EtAlCl_2$	propene	940, 2001
$MoO_3 \cdot Al_2O_3 \cdot Et_2AlCl$	propene	940, 2001
$MoO_3 \cdot Al_2O_3 \cdot Me_3Al_2Cl_3$	propene	940, 2001
$MoO_3 \cdot Al_2O_3 \cdot Bu_4Sn$	propene	386
	1-heptene	386
	vinylmethylsilane	386
$MoO_3 \cdot Al_2O_3 \cdot Et_4Sn$	propene	386
	1-heptene	386
$Mo_2O_3 \cdot CoO \cdot Al_2O_3$	propene	1844
	isobutene	1844
	isobutene + propene	1844
$Mo_2O_3 \cdot CoO \cdot SiO_2$	propene	1844
	isobutene	1844
	isobutene + propene	1844
$Mo_2O_3 \cdot CoO \cdot MgO$	propene	1844
	isobutene	1844
	isobutene + propene	1844
$MoO_3 \cdot Al_2O_3 \cdot Et_2AlOSiMe$	1-hexene	1981
	1-dodecene	1981
	1-eicosene	1981
	6-chloro-1-hexene	1981
$MoO_3 \cdot Al_2O_3 \cdot Cr_2O_3$	propene	123, 868, 1185.
	1-hexene	868

1	2	3
$MoO_3 \cdot Al_2O_3 \cdot Re_2O_7$	propene	1181, 1182
	1-butene	1181, 1182
$WO_3 \cdot Al_2O_3 \cdot K_2O$	propene	1864
$WO_3 \cdot Al_2O_3 \cdot ThO_2$	propene	451
$WO_3 \cdot Al_2O_3 \cdot Me_3Al_2Cl_3$	propene	2001
$WO_3 \cdot Al_2O_3 \cdot MeAl(OMe)_2$	propene	940
$WO_3 \cdot Al_2O_3 \cdot EtAlCl_2$	propene	2001
	1-pentene	2001
	1-octene	2001
	cyclopentene	2001
$WO_3 \cdot Al_2O_3 \cdot Et_2AlCl$	propene	2001
	1-pentene	2001
	1-octene	2001
	cyclopentene	2001
$WO_3 \cdot Al_2O_3 \cdot Pr_3Al$	propene	940
$WO_3 \cdot SiO_2 \cdot Na_2O(K_2O)$	butadiene	943
$WO_3 \cdot SiO_2 \cdot ZnO$	1-butene	1824
$WO_3 \cdot SiO_2 \cdot MgO$	propene	1528
	1-butene	1824
	isobutene	1832
	2-pentene	1205, 1216
	neohexene	1833
	3-heptene	1835
	pentenes + ethene	1216
	hexenes + ethene	1216
$WO_3 \cdot SiO_2 \cdot CdO$	1-butene	1824
$WO_3 \cdot SiO_2 \cdot NiO$	ethene	1830
$WO_3 \cdot SiO_2 \cdot CoO$	propene	1830
$WO_3 \cdot SiO_2 \cdot P_2O_5$	propene	1661
$WO_3 \cdot SiO_2 \cdot Pr_3Al$	propene	372
$WO_3 \cdot SiO_2 \cdot NH_3$	1-butene	1894, 1897
$WO_3 \cdot SiO_2 \cdot EtNH_2$	1-butene	1896
$WO_3 \cdot SiO_2 \cdot Et_2NH$	1-butene	1896
$WO_3 \cdot SiO_2 \cdot Et_3N$	1-butene	1896
$WO_3 \cdot SiO_2 \cdot PrNH_2$	1-butene	1896
$WO_3 \cdot SiO_2 \cdot BuNH_2$	1-butene	1896
$WO_3 \cdot SiO_2 \cdot C_5H_{10}N$	1-butene	1896
$WO_3 \cdot SiO_2 \cdot C_5H_5N$	1-butene	1894
	2-butene	1894
	isobutene	1894
$WO_3 \cdot SiO_2 \cdot PhNH_2$	1-butene	1894
	2-butene	1894
	isobutene	1894
$WO_3 \cdot SiO_2 \cdot PhNR_2$	1-butene	1894
(R = alkyl)	2-butene	1894
	isobutene	1894
$WO_3 \cdot SiO_2 \cdot CHCl_3$	1-butene	1895
$WO_3 \cdot SiO_2 \cdot CCl_4$	1-butene	1895
$WO_3 \cdot SiO_2 \cdot Bu_3P$	2-pentene	1416, 1424
$WO_3 \cdot SiO_2 \cdot Bu_3As$	2-pentene	1424
$WO_3 \cdot SiO_2 \cdot Bu_3Sb$	2-pentene	1424
$WO_3 \cdot SiO_2 \cdot Bu_3Bi$	2-pentene	1424
$WO_3 \cdot TiO_2 \cdot CoO$	propene	1426
$WO_3 \cdot TiO_2 \cdot MgO$	propene	1426
$WO_3 \cdot MgO \cdot CoO$	propene	1426

1	2	3
$Re_2O_7 \cdot Al_2O_3 \cdot K_2CO_3$	trimethyltetradeca-1,13-diene + + ethene	1443
	trimethylcyclodecene + 2-butene	1443
$Re_2O_7 \cdot Al_2O_3 \cdot CoO$	propene	1067
$Re_2O_7 \cdot Al_2O_3 \cdot Co_2O_3$	2-butene + isobutene	1715
$Re_2O_7 \cdot Al_2O_3 \cdot NiO$	propene	1067
$Re_2O_7 \cdot Al_2O_3 \cdot SnO_2$	2-butene + isobutene	1795
$Re_2O_7 \cdot Al_2O_3 \cdot V_2O_5$	allyl chloride	788
$Re_2O_7 \cdot Al_2O_3 \cdot MoO_3$	allyl chloride	788
$Re_2O_7 \cdot Al_2O_3 \cdot WO_3$	allyl chloride	788
$Re_2O_7 \cdot Al_2O_3 \cdot GeO_2$	allyl chloride	788
$Re_2O_7 \cdot Al_2O_3 \cdot Me_4Sn$	methyl pentenoate	1097
$Re_2O_7 \cdot Al_2O_3 \cdot Et_4Sn$	propene	386
	1-heptene	386
	vinyltrimethylsilane	386
$Re_2O_7 \cdot Al_2O_3 \cdot Bu_4Sn$	propene	386
	1-heptene	386
	vinyl chloride	387
	vinyl bromide	387
	vinyltrimethylsilane	386
	allyl chloride	387
	allyl bromide	387
	5-bromo-1-hexene	387
	vinyl chloride + propene	387
	vinyl chloride + 2-butene	387
	vinyl chloride + cyclohexene	387
	vinyl chloride + ethene-d_4	387
	vinyl bromide + propene	387
$Re_2O_7 \cdot Al_2O_3 \cdot R_4Sn$	1-heptene	1225
(R = alkyl)	halo-alkenes	1226
$Re_2O_7 \cdot Al_2O_3 \cdot$ oxasilacyclobutane	1-heptene	1224
$Re_2O_7 \cdot Al_2O_3 \cdot$ dicyclopentadiene	propene	1217
$Re_2O_7 \cdot Al_2O_3 \cdot (NH_4)_2HPO_4$	2-butene + isobutene	1655
$RhO_2 \cdot Al_2O_3 \cdot R_4Sn$	ethene + piperilene	1227
(R = alkyl)		
$V_2O_5 \cdot WO_3 \cdot Me_3Al$	propene	1754
$Mo(CO)_6 \cdot Al_2O_3 \cdot SiO_2$	propene	1666
$Mo(CO)_6 \cdot SiO_2 \cdot Al_2O_3$	ethene	1204
	propene	954, 1204
	1-butene	1204
	1-pentene	1204
	1-hexene	1204
$W(CO)_6 \cdot Al_2O_3 \cdot SiO_2$	propene	1666
$WCl_6 \cdot Al_2O_3 \cdot EtAlCl_2$	2-pentene	1269
$WCl_6 \cdot Al_2O_3 \cdot Bu_3Al$	2-pentene	1269
$WCl_6 \cdot Al_2O_3 \cdot i\text{-}Bu_3Al$	2-pentene	1269
$WCL_6 \cdot Al_2O_3 \cdot Bu_4Sn$	2-pentene	1269
$WCl_6 \cdot AlO_2 \cdot EtAlCl_2$	2-pentene	1269
$WCl_6 \cdot SiO_2 \cdot Bu_3Al$	2-pentene	1269
$WCl_6 \cdot SiO_2 \cdot i\text{-}Bu_3Al$	2-penetne	1269
$WCl_6 \cdot SiO_2 \cdot Bu_4Sn$	2-pentene	1269
$WCl_6 \cdot SiO_2 \cdot RMgX$	2-pentene	1852
(R = alkyl $C_3 - C_{10}$)		
$WCl_6 \cdot SiO_2 \cdot i\text{-}Bu_2AlCl$	2-pentene	935

1	2	3
$WCl_6 \cdot SiO_2 \cdot$ n-BuLi	2-pentene	935
	2-pentene + 1-pentene	935
$WCl_6 \cdot SiO_2 \cdot C_5H_{11}MgBr$	2-pentene	935
$WCl_6 \cdot SiO_2 \cdot EtAlCl_2$	2-pentene	935
$WCl_6 \cdot SiO_2 \cdot MgBr_2$	2-pentene	935
$WCl_6 \cdot SiO_2 \cdot TiCl_4$	2-pentene	935
$MoCl_5 \cdot SiO_2 \cdot$ n-BuLi	2-pentene	935
	2-pentene + 1-pentene	935
$MoCl_5 \cdot SiO_2 \cdot C_5H_{11}MgBr$	2-pentene	935
$MoCl_5 \cdot SiO_2 \cdot EtAlCl_2$	2-pentene	935
$MoCl_5 \cdot SiO_2 \cdot MgBr_2$	2-pentene	935
$MoCl_5 \cdot SiO_2 \cdot TiCl_4$	2-pentene	935

A.4. Multicomponent heterogeneous systems

$MoO_3 \cdot CoO \cdot Al_2O_3$ + KOH	propene	1166, 1949
	1-butene	1166
$MoO_3 \cdot CoO \cdot Al_2O_3 + Na_2O, K_2O$	2-heptene	106
	1-octene	106
$MoO_3 \cdot Al_2O_3 \cdot MeAlCl_2 \cdot Me_2AlCl$	propene	1966
	1-pentene	1966
	2-pentene	1966
	1-octene	1966
$MoO_3 \cdot CoO \cdot Al_2O_3 \cdot Et_3Al +$	1-butene	1168
+ K, Rb, Cs	2-butene + 1-hexene + 1-octene +	
	1-decene	1382
$WO_3 \cdot SiO_2MgO + KOH$	n-butenes	1834, 1836
$WO_3 \cdot SiO_2 \cdot MgO + CO$	2,2,4-trimethyl-2-pentene + ethene +	1711
$WO_3 \cdot SiO_2 \cdot Al_2O_3 \cdot NiO$	ethene	1830
$WO_3 \cdot Al_2O_3 \cdot MeAlCl_2 \cdot Me_2AlCl$	propene	1996
	1-pentene	1996
	2-pentene	1996
	1-octene	1996
$WO_3 \cdot Al_2O_3 \cdot SiO_2 \cdot Et_2AlCl \cdot$		
$EtAlCl_2 \cdot AcOH$	2-pentene	1874
$Re_2O_7 \cdot Al_2O_3 \cdot La_2O_3 \cdot Pr_6O_{11} \cdot$		
$Sm_2O_3 \cdot Gd_2O_3 \cdot Ce_2O_3 \cdot Y_2O_3$	isobutene + 2-butene	1950

B. Homogeneous catalytic systems

B.1. Unicomponent homogeneous systems

$PhWCl_3$	2-pentene	832
$Ph_2CW(CO)_5$	isobutene	224
	2-butene	224
	2-pentene	560
	1,3-pentadiene	585
	2,3-diphenylbutadiene	585
	1-hexene + 1-octene-1,1-d_2	727
$Tol_2CW(CO)_5$	isobutene	225
	2-butene	225
	1-pentene	225
	styrene	225
	1-methyl-2-hexene	225

1	2	3
$(Ph_3P)_3RhCl$	$\begin{bmatrix} NR_1 \\ \diagdown \\ NR_1 \end{bmatrix} = \begin{matrix} NR_2 \\ \diagup \\ NR_2 \end{matrix}$ a. $R_1 = R_2 = Ph$ b. $R_1 = R_2 = p - MeC_6H_4$ c. $R_1 = Ph$; $R_2 = p - MeC_6H_4$	219
$(CO)\,(Ph_3P)_2RhCl$	idem	219
$W(CO)_6$	styrene	670
	1,7-octadiene	670
	2-pentene	79, 149, 606. 607
	3-heptene	79, 670
	4-nonene	79, 670
	5-methyl-2-pentene	79
$W(CO)_3Mes_3$	3-heptene	79
$W(CO)_6MeCN$	3-heptene	79
$W(CO)_5MeCN$	1-pentene	421
	2-pentene	421
	2-methyl-2-butene	421
$W(CO)_4Cl_2$	2-pentene	80
$\pi\text{-}C_3H_5)_6Re_2$	1-hexene	1300
$(\pi\text{-}C_3H_5)_4Mo_2$	1-hexene	1300
$(\pi\text{-}C_4H_8)_4Mo_2$	1-hexene	1300
$Ph(CH_3CO)C=W(CO)_5$	2-pentene	556
$PhMeCW(CO)_5$	2-pentene	226
	styrene	226
	p-methylphenol-ethene	226
$WOCl_4$	2-pentene	745
$WOCl_2$	ethene + 2-butene	1452
$(\pi\text{-allyl})MoCl$	1-butene	1476
$(\pi\text{-crotyl})MoCl$	1-hexene	1476
$(\pi\text{-allyl})WCl$	3-heptene	1476
$(\pi\text{-crotyl})WCl$	3,3-dimethyl-1-butene	1476
$(\pi\text{-allyl})ReCl$	cyclopentene	1476
$(\pi\text{-crotyl})ReCl$	norbornene	1476
$Mo(CO)_6$	2-pentene	670
	4-octene	670
	1,7-octadiene	670
	styrene	670
$(Tol)Mo(CO)_3$	2-pentene	670
	4-octene	670
	1,7-octadiene	670
	styrene	670
$(cyclooctene)_2IrCl_2$	norbornene + 2-butene	897
	norbornene + isobutene	897
	norbornene + 1-pentene	897
	norbornene + 2-pentene	897
	norbornene + 2-methyl-2-butene	897

1	2	3
B.2. Binary homogeneous systems		
$WCl_6 \cdot Me_3Al_2Cl_3$	methyl oleate	786
$WCl_6 \cdot EtAlCl_2$	1-pentene	1053
	1-butene	1053
	2-pentene	1252
	2-hexene	472
	4-octene	619
	1,6-cyclodecadiene	493, 497, 498
	1-hexene	1360, 1361, 473, 474, 476, 602, 745, 770, 771
	1-heptene	488
	1-octene	488
	1-nonene	488, 1053
	1-undecene	488
	polybutadiene + 4-octene	490
	polybutadiene + 6-dodecene	491
	erucic acid + 7-tetradecene + 4-octene	494
	polypentenamer + polybutadiene	495
	polypentenamer + polybutadiene + 4-octene	495
	1-pentene + 2-pentene	569
	1-pentene + 1-pentene-d_{10}	569
	1-pentene + cyclopentene	569
	2-methyl-2-butene	1053
	partially hydrogenated polybutadiene	1129
$WCl_6 \cdot Me_4Sn$	methyl oleate + 3-hexene	1079
	methyl oleate	1079, 1096
	methyl erucate	1079
	methyl undecenate	1079
	unsatd. triglicerides	1079
	2,8-decadiene	436
	methyl linoleate	113, 117
	polybutadiene + 4-octene	453
$WCl_6 \cdot Et_4Sn$	1-hexene	176
	methyl oleate	507
	methyl oleate + 1-decene	507
$WCl_6 \cdot Bu_4Sn$	1-hexene	176
	2-pentene	1266
	2-heptene	988, 509, 511, 986
	methyl oleate	507
	methyl oleate + 1-decene	507
	1-heptene	617
	1-octene	722
$WCl_6 \cdot Ph_4Sn$	2-pentene	1077
	1-hexene	176
	2-heptene	509, 510, 506
$WCl_6 \cdot (Me_3SiCH_2)_4Sn$	1-hexene	176
$WCl_6 \cdot Ph_3SnEt$	1-hexene + 1-octene-1,1-d_2	727
$WCl_6 \cdot Ph_4Pb$	2-heptene	510
	2-pentene	1077

1	2	3
$WCl_6 \cdot Et_2Zn$	2-pentene	745, 770, 771
	2-heptene	511
	1-hexene + 1-pentene-d_{10}	744
$WCl_6 \cdot Ph_3Sb$	2-pentene	1077
$WCl_6 \cdot Ph_3Bi$	2-heptene	510
	2-pentene	1077
$WBr_5 \cdot EtAlCl_2$	2-pentene	1102
$WCl_6 \cdot$ n-PrLi	2-pentene	1078
	3-hexene + 4-octene	1078
$WCl_6 \cdot$ n-BuLi	1,7-octadiene	432
	1,7-octadiene + 1,7-octa- diene-1,1,8,8-d_4	433, 434
$WCl_6 \cdot$ n-OctLi	ethene	435
$WCl_6 \cdot C_5H_5Na$	2-heptene	512
$WCl_6 \cdot PhC_2Na$	2-heptene	512
$WCl_6 \cdot LiAlH_4$	2-pentene	246
	3-heptene	723
	1,5-cyclononadiene	899, 901
$WCl_6 \cdot NaBH_4$	2-pentene	246
$WCl_6 \cdot Et_3SiH$	⎧ 1-pentene	796
	⎪ 2-pentene	796
	⎨ 1-hexene	796
	⎪ 2-hexene	796
	⎩ 1-heptene	796
$WCl_6 \cdot Ph_2SiH$	idem	796
$WCl_6 \cdot (MeSiH)_2O$	idem	796
$WCl_6 \cdot (MeSiH)_2NH$	idem	796
$WCl_6 \cdot EtAlCl_2$	7-methyl-1-octene + 1-dodecene	620, 621
	1-decene + 1-pentadecene	620, 621
	1-tetradecene + 1-pentadecene	620, 621
	1-tetradecene + 1-heptadecene	620, 621
	1-pentadecene + 1-hexadecene	620, 621
	polybutadiene + 2-butene + 2-pen- tene	1683
	polyisoprene + cyclo-octene	1683
	polybutadiene + 2-butene	1737
	methyl oleate	1733
	allyl acetate	1733
	vinyl acetate	1733
	p-chlorophenylethene + 2-pentene	1785
	5-bromo-1-pentene + 2-pentene	1785
	1,5-cyclo-octadiene + 4-methyl-4- octene	1785
	allyl acetate + 1-hexene	1733
	methyl-9-decenoate + 1-hexene	1733
	cyclododecene	1137, 1138, 1976, 1977
$WCl_6 \cdot Et_3Al$	2-hexene	1925
	2-heptene	509, 511, 506
	2-pentene	1052, 1925
	1-pentene	1053, 1925
	1-butene	1053
	1-octene	1053, 1054
	methyl oleate	507
	methyl oleate + 1-decene	507
	2-methyl-2-butene	1052

1	2		3
$WCl_6 \cdot Et_3Al_2Cl_3$	propene		732
	2-pentene		732
$WCl_6 \cdot Me_4Sn$	methyl methacrylate		1098
	methyl 2-methyl-2-butenoate		1098
	methyl 2-butenoate		1098
	methyl 3-pentenoate		1098
	methyl acrylate		1098
	methyl 1-butenoate		1098
	methyl 1-pentenoate		1098
	methyl 2-pentenoate		1098
	arylpolybutadiene + 4-octene		554
	1,2-disubstituted olefins		960
	methyl oleate		1076, 1098
	methyl oleate + 3-hexene		1076
	methyl linoleate		1095
	methyl linolenate		1095
	geranyl acetate + 1-methylcyclo-butene		1133
$WCl_6 \cdot Me_3SnCl$	methyl oleate		1079
$WCl_6 \cdot Ph_4Sn$	2-heptene		988
$WCl_6 \cdot MeLi$	1,2-disubstituted olefins		960
$WCl_6 \cdot n\text{-}BuLi$	2-pentene		970, 1078, 1106, 1108
	2-heptene		509, 511
	1-hexene + 1-octene-1,1-d_2		727
	3-hexene + 4-octene		1078
$WCl_6 \cdot R_3Al$	$Me(CH_2)_7CH = CH(CH_2)_7R$ $(R = CO_2Me, CO_2Bu, CN, COEt, CONET_2)$		1510
$(MoCl_5) \cdot R_2AlCl$	$CH_2 = CHCH_2OR$ (R = Ac, Ph, SiMe)		1510
$WOCl_4 \cdot EtAlCl_2$	2-pentene		745
$WOCl_4 \cdot Et_2Zn$	2-pentene		745
$WOCl_4 \cdot n\text{-}BuLi$	2-pentene		745
$WOCl_2 \cdot EtAlCl_2$	2-pentene		745
$WCl_6 \cdot Et_3Al_2Cl_3$	methyl oleate		1733
	allyl acetate		1733
	ethylvinylacetate		1733
	oleil acetate + 1-hexene		1733
	methyl 9-decenoate + 1-hexene		1733
$WCl_6 \cdot Et_2AlCl$	methyl oleate		1733
	allyl acetate		1733
	ethylvinylacetate		1733
$WCl_6 \cdot (C_8H_{17})_3Al$	methyl oleate		1733
	allyl acetate		1733
	ethylvinylacetate		1733
	oleil acetate + 1-hexene		1733
	methyl 9-decenoate + 1-hexene		1733
$ReCl_5 \cdot Et_3Al_2Cl_3$	methyl oleate		1733
	allyl acetate		1733
	ethylvinyl acetate		1733
	oleil acetate + 1-hexene		1733
	methyl 9-decenoate + 1-hexene		1733
$ReCl_5 \cdot (C_8H_{17})_3Al$	methyl oleate		1733
	allyl acetate		1733
	ethylvinyl acetate		1733

1	2	3
$Mo(OEt)_2Cl_2 \cdot Et_3Al_2Cl_3$	methyl oleate	1733
	allyl acetate	1733
	ethylvinyl acetate	1733
	oleil acetate + 1-hexene	1733
	methyl 9-decenoate + 1-hexene	1733
$Mo(OEt)_2Cl_2 \cdot (C_8H_{17})_3Al$	methyl oleate	1733
	allyl acetate	1733
	ethylvinyl acetate	1733
	oleil acetate + 1-hexene	1733
	methyl 9-decenoate + 1-hexene	1733
$WCl_6 \cdot Bu_4Sn$	2-pentene	903
	2-octene	903
	4-methyl-2-hexene	903
	6-methyl-2-octene	903
$WCl_6 \cdot Bu_3SnCl$	2-heptene	986
$WCl_6 \cdot Bu_2SnCl_2$	2-heptene	986
$WCl_6 \cdot LiAlH_4$	1-pentene	900
	2-pentene	900
	2-octene	900
$WCl_6 \cdot Ph_3SnCl$	2-pentene	1077
$WCl_6 \cdot Ph_2SnCl_2$	2-pentene	1077
$WCl_6 \cdot Ph_6Sn_2$	2-pentene	1077
$WCl_6 \cdot Ph_3PbCl$	2-pentene	1077
$WCl_6 \cdot Ph_6Pb_2$	2-pentene	1077
$WCl_6 \cdot Ph_2Hg$	2-pentene	1077
$WCl_6 \cdot PrMgCl$	2-pentene + 1-pentene	702
$WCl_6 \cdot PrMgBr$	2-pentene + 1-pentene	702
$WCl_6 \cdot BuMgCl$	2-pentene	577, 701, 703, 704
	2-pentene + 1-pentene	702
$WCl_6 \cdot BuMgBr$	2-pentene	577, 701, 703, 704
	2-pentene + 1-pentene	702
$WCl_6 \cdot BuMgI$	2-heptene	509, 985, 987, 988
$WCl_6 \cdot PrMgCl$	2-pentene	577
	2-pentene + 1-pentene	702
$WCl_6 \cdot PrMgBr$	2-pentene	577
	2-pentene + 1-pentene	702
$WCl_6 \cdot HexMgX$ (X=Cl, Br)	2-pentene + 1-pentene	702
$WCl_6 \cdot OctMgX$ (X=Cl, Br)	2-pentene + 1-pentene	702
$WCl_6 \cdot PhMgX$ (X=Cl, Br)	2-pentene + 1-pentene	702
$WCl_6 \cdot i\text{-}BuMgX$ (X=Cl, Br)	2-pentene + 1-pentene	702
$WCl_6 \cdot R_2Sn \overset{\displaystyle -SiMe_2-}{\underset{\displaystyle \_SiMe_2\_}{}} O$ (R=Me, Et, Pr, Bu)	1-hexene	124
$WCl_6 \cdot CH_2 \overset{\displaystyle -GeMe_2-}{\underset{\displaystyle \_GeMe_2\_}{}} CH_2$	1-hexene	125

1	2	3
WCl$_6$ · Et$_3$B	9-octadecenyl acetate	785
WCl$_6$ · AlCl$_3$	α-olefins	459
	β-olefins	459
WCl$_2$ · AlCl$_3$	α-olefins	459
	β-olefins	459
WBr$_5$ · AlCl$_3$	α-olefins	459
	β-olefins	459
	2-hexene	1252
WOCl$_4$ · AlCl$_3$	α-olefins	459
	β-olefins	459
WOCl$_3$ · AlCl$_3$	α-olefins	459
	β-olefins	459
WBz$_4$ · AlCl$_3$	2-pentene	831
WBz$_4$ · AlBr$_3$	2-pentene	167
WBz$_3$Cl · AlX$_3$ (X=Cl, Br)	2-pentene	167
WPhCl$_3$ · AlCl$_3$	2-butene	402
	2-butene + ethene	401
	1,7-octadiene +	
	+ 1,7-octadiene-1,1, 8,8-d$_4$	433, 434
	1,4-cyclohexadiene	402
	ethylcyclopropane	403
	butylcyclopropane	403
	isobutylcyclopropane	403
	bicyclo[2.1.0]pentane	403
	ethylcyclopropane + ethyl acetate	404
	ethylcyclopropane + ethyl crotonate	404
	ethylcyclopropane + acrylonitrile	404
	ethylcyclopropane + ethyl cinnamate	404
	ethylcyclopropane + methyl 3-methylcrotonate	404
	ethylcyclopropane + methyl 2-methylcrotonate	404
WCl$_6$ · Me$_2$Si ⌐CH$_2$⌐ SiMe$_2$ ⌊CH$_2$⌋	2-pentene	797
	1-hexene	797
WCl$_6$ · (CH$_2$)$_3$SiMe$_2$	2-pentene	797
	1-hexene	797
(Ph$_3$P)W(CO)$_5$ · EtAlCl$_2$	2-pentene	878
W(CO)$_4$(1,5-cyclo-octadiene) · · i-BuAlCl$_2$	3-heptene	1111
W(CO)$_4$(norbornadiene) · i-BuAlCl$_2$	3-heptene	1111
π-C$_5$H$_5$W(CO)$_3$GeMe$_3$ · i-PrAlCl$_2$ (+O$_2$)	2-hexene	1112
	3-heptene	1112
	4-octene	1112
π-C$_5$H$_5$W(CO)$_3$GePh$_3$ · i-PrAlCl$_2$ (+O$_2$)	2-hexene	1112
	3-heptene	1112
	4-octene	1112

1	2	3
π-$C_5H_5W(CO)_3SnMe_3 \cdot$ i-PrAlCl$_2$ ($+O_2$)	2-hexene	1112
	3-heptene	1112
	4-octene	1112
π-$C_5H_5W(CO)_3SnPh_3 \cdot$ i-PrAlCl$_2$ ($+O_2$)	2-hexene	1112
	3-heptene	1112
	4-octene	1112
$W(CO)_6 \cdot$ i-BuAlCl$_2$	2-hexene	1110
$W(CO)_6 \cdot$ EtAlCl$_2$	2-pentene	181
$W(CO)_5P(OPh)_3 \cdot$ EtAlCl$_2$	2-pentene	182
$W(CO)_5PBu_3 \cdot$ EtAlCl$_2$	2-pentene	149, 181
$W(CO)_5PPh_3 \cdot$ EtAlCl$_2$	2-pentene	147, 181
$W(CO)_5Py_2 \cdot$ EtAlCl$_2$	2-pentene	182
$W(CO)_5Ph_3 \cdot$ AlCl$_3$	α-and β-olefins	459
$W(CO)_6 \cdot R_mAlX_n$	olefins, $C > 3$	1142
$W(CO)_6 \cdot Et_3Al_2Cl_3$	1-octene	1142
$W(CO)_3Cl_2(AsPh_3)_2 \cdot$ AlCl$_3$	2-pentene	168, 1014
$W(CO)_3Cl_2(PPh_3)_2 \cdot$ AlCl$_3$	2-pentene	168
$W(CO)_3Cl_2(SbPh_3)_2 \cdot$ AlCl$_3$	2-pentene	168
$W(CO)_3Cl_2(PPh_3)_2 \cdot$ EtAlCl$_2$	2-pentene	1251
	2-hexene	1252
$W(NO)_2Cl_2(PPh_3)_2 \cdot Me_3Al_2Cl_3$	2-pentene	1995
$W(NO)_2Cl_2(PPh_3)_2 \cdot$ EtAlCl$_2$	1-pentene	1995
	ω-aryl-α-olefins	256
$W(CO)_5PPh_3 \cdot$ EtAlCl$_2 + O_2$	2-pentene	146, 181
$Mo(CO)_5PPh_2CH_2R \cdot$ EtAlCl$_2 + O_2$	2-pentene	148, 181
$EtCHWPhCl_3 \cdot$ AlCl$_3$	1-butene	405
	1-pentene	405
	1-hexene	405
$WO_3 \cdot$ AlCl$_3$	α- and β-olefins	1446
$H_2WO_4 \cdot$ AlCl$_3$	α- and β-olefins	1446
$K_2WO_4 \cdot Al_2O_3$	α- and β-olefins	1446
$(NH_4)_{10}W_{12}O_{41} \cdot Al_2O_3$	α- and β-olefins	1446
$H_3P(W_3O_{10})_4 \cdot$ AlCl$_3$	α- and β-olefins	1446
$(NH_4)_4(GeW_{12}O_{40}) \cdot H_2O \cdot$ AlCl$_3$	α- and β-olefins	1446
$H_8(GeW_{11}O_{39}) \cdot$ AlCl$_3$	α- and β-olefins	1446
$WPy_2Cl_4 \cdot$ EtAlCl$_2$	2-hexene	1252
$WPyCl_4 \cdot$ EtAlCl$_2 + CO$	2-pentene	164
$W(PPh_3)_2Cl_4 \cdot$ EtAlCl$_2 + CO$	2-pentene	164
$WC_2H_4(PPh_3)_2Cl_4 \cdot$ EtAlCl$_2 + CO$	2-pentene	164
$MoPy_2Cl_4 \cdot$ EtAlCl$_2 + CO$	2-pentene	164
$Mo(PPh_3)_2Cl_4 \cdot$ EtAlCl$_2 + CO$	2-pentene	164
$MoC_2H_4(PPh_3)_2Cl_4 \cdot$ EtAlCl$_2 + CO$	2-pentene	164
$W(OR)_2Cl_4 \cdot$ EtAlCl$_2$	polybutadiene + 1,5-cyclo-octadiene	595
	polybutadiene + cyclopentene	595
$W(CO)_3Mes \cdot$ EtAlCl$_2$	propenylamine	814
	propenylcyclohexylamine	814
	propenyldimethylamine	814
	propenylpropylamine	814
	butenylpropylamine	1754
	pentenylpropylamine	1755
$W(CO)_3Mes \cdot$ EtAlCl$_2$	hexenylpropylamine	814
	undecenylpropylamine	814

1	2	3
$MoCl_5 \cdot Me_3Al_2Cl_3$	1-pentene	1157
$MoCl_5 \cdot Et_2AlCl$	1-pentene	1924
$MoCl_5 \cdot Et_3Al$	1-pentene	1924
$MoOCl_3 \cdot Me_3Al_2Cl_3$	1-pentene	1157
$Mo(acac)_3 \cdot Et_3Al$	1-pentene	1924
$MoO_2(acac)_2 \cdot Me_3Al_2Cl_3$	1-pentene	1157
$MoCl_5 \cdot Ph_4Pb$	2-pentene	1077
$MoCl_5 \cdot Ph_4Sn$	2-pentene	1077
$Mo(OEt)_2Cl_3 \cdot Et_3B$	9-octadecenyl acetate	785
$Mo(OEt)_2Cl_3 \cdot Me_3Al_2Cl_3$	methyl oleate	786, 1733
$Mo(NO)Cl_3 \cdot EtAlCl_2$	2-pentene	1016, 1018
$Mo(NO)Cl_3(OPPh_3)_2 \cdot EtAlCl_2$	2-pentene	1016, 1017, 1018, 1019
$Mo(NO)_2Cl_2(OPPh_3)_2 \cdot EtAlCl_2$	2-pentene	1016
$Mo(NO)_2Cl_2(PPh_3)_2 \cdot Me_3Al_2Cl_3$	1-pentene	1158, 1157
	2-heptene	1158, 1157
	1-octene	1158, 1157
$Mo(NO)_2Cl_2(PPh_3)_2 \cdot EtAlCl_2$	1-pentene	1158, 1157
	2-heptene	1158, 1157
	1-octene	1158, 1157
	arylolefins	291
	1-decene + 1-tetradecene	296
$Mo(NO)_2Cl_2(PPh_3)_2 \cdot Et_3Al_2Cl_3$	1-pentene	1158, 1157
	2-heptene	1158, 1157
	1-octene	1158, 1157
$Mo(NO)_2Cl_2(PPh_3)_2 \cdot Et_2AlCl$	1-pentene	1158, 1157
	2-heptene	1158, 1157
	1-octene	1158, 1157
$Mo(NO)_2Cl_2Py_2 \cdot Me_3Al_2Cl_3$	1-octene	1157
	2-pentene	486
	4-nonene	486
$Mo(NO)_2Cl_2Py_2 \cdot EtAlCl_2$	1-octene	1157
	2-pentene	485, 1019
	4-nonene	486
$Mo(NO)_2Cl_2Py_2 \cdot Et_3Al_2Cl_3$	1-octene	1157
	2-pentene	486
	4-nonene	486
$MoCl_4(PPh_3)_2 \cdot Me_3Al_2Cl_3$	1-octene	1157
$MoCl_4Py_2 \cdot Me_3Al_2Cl_3$	1-octene	1157
$MoCl_4 \cdot C_3H_7CN \cdot Me_3Al_2Cl_3$	1-octene	1157
$Mo(CO)_5C_5H_5 \cdot Me_3Al_2Cl_3$	1-octene	1157
$MoCl_3(PhCO_2)_2 \cdot Me_3Al_2Cl_3$	1-pentene	1157
$Mo(NO)_2Cl_2(EtCN)_2 \cdot EtAlCl_2$	2-pentene	1017
$Mo(NO)_2(PPh_3)_2(PhMe) \cdot AlCl_3$	2-pentene	1926
$Mo(CO)(PPh_3)_2 \cdot AlCl_3$	2-pentene	1926
$Mo(NO)_2Cl_2(PPh_3)_2 \cdot Me_3Al_2Cl_3$	1,7-octadiene + 1,7-octadiene-1,1,8,8-d_4	434
	2,8-decadiene	436
	cyclo-octene + 2-butene + 4-octene	555, 559
	cyclo-octene + 2-hexene	555
	2-pentene	555
	1-hexene + 1-octene-1,1-d_2	727
$Mo(NO)_2Cl_2(POct_3)_2 \cdot Me_3Al_2Cl_3$	2,2'-divinylbiphenyl	558

1	2	3
$Mo(NO)_2Cl_2(PPh_3)_2 \cdot Me_3Al_2Cl_3$ $(EtAlCl_2, Et_2AlCl, Et_3Al_2Cl_3)$	1-pentene	1146, 1995, 1157
	2-heptene	1995, 1157
	1-octene	1995, 1157
	ethene + 2-heptene	1995, 1157
	ethene + 2-butene + 5-decene	1995, 1157
$Mo(NO)_2Cl_2(PPh_3)_2 \cdot EtAlCl_2$	7-methyl-1-octene + 1-dodecene	1626
$Mo(NO)_2Cl_2Py_2 \cdot Me_3Al_2Cl_3$ $(EtAlCl_2, Et_3Al_2Cl_3)$	1-octene	1157, 1147
$MoCl_4(PPh_3)_2 \cdot Me_3Al_2Cl_3$	1-octene	1157
$MoCl_4Py_2 \cdot Me_3Al_2Cl_3$ $(EtAlCl_2)$	1-octene	1157
$MoCl_4 \cdot C_3H_7CN \cdot Me_3Al_2Cl_3$	1-octene	1157
$Mo(CO)_5C_5H_5 \cdot Me_3Al_2Cl_3$	1-octene	1157
$MoCl_3(PhCO_2) \cdot Me_3Al_2Cl_3$	1-pentene	1157
$Bu_4N[(CO)_5Mo\text{-}Mo(CO)_5] \cdot MeAlCl_2$	1-pentene	614, 1612
(Na)	1,7-octadiene	614
	2-butene + 4-octene	614
$Bu_4N[(CO)_5W-W(CO)_5] \cdot MeAlCl_2$ (Na)	idem	614
$Bu_4N[(CO)_5Mo-Re(CO)_5] \cdot MeAlCl_2$ (Na)	idem	614
$Bu_4N[(CO)_5Mo-Mn(CO)_5] \cdot MeAlCl_2$ (Na)	idem	614
$Bu_4N[M(CO)_5X] \cdot MeAlCl_2$	1-pentene	1357, 320
(M = Mo, Cr, W)	3-methyl-2-pentene	320
(X = Br, Cl)	1,7-octadiene	320
	4-vinylcyclohexene	320
$Bu_4N[M(CO)_5COR] \cdot MeAlCl_2$	1-pentene	613, 1611
(R = Me, Ph)	2-pentene	613, 1611
(M = Mo, W)	4-methyl-1-pentene	613, 1611
	1,7-octadiene	613, 1611
	1-octene	1611
$EtW(CO)_5(COMe) \cdot MeAlCl_2$	1-pentene	613
$MeW(CO)_5(CNMe_2) \cdot MeAlCl_2$	1-pentene	613
$Bu_4NMo(CO)_5Cl \cdot EtAlCl_2$	7-methyl-1-octene + 1-dodecene	1626
$MoCl_5 \cdot EtAlCl_2$	polybutadiene + 2-butene +	
	2-pentene	1683
$MoCl_2(NO)_2(PPh_3)_2 \cdot Me_3Al_2Cl_3$	propene	2002
	1-pentene	2002
	1,7-octadiene	2002
$Mo(NO)_2Cl_2(PPh_3)_2 \cdot EtAlCl_2$	1-undecene + 1-tetradecene	898
	4-methyl-1-pentene + 1-tetra-undecene	898
$MoCl_4(PPh_3)_2 \cdot EtAlCl_2$	2-pentene	1091
$MoCl_3(PPh_3)_2 \cdot EtAlCl_2$	2-pentene	1091
$MoCl_4(PPh_3)_2 \cdot EtAlCl_2 + CO$	2-pentene	1091
$MoCl_3(PPh_3)_2 \cdot EtAlCl_2 + CO$	2-pentene	1091
$MoCl_3(NO) \cdot EtAlCl_2$	2-pentene	1018
$MoCl_2(NO)_2 \cdot EtAlCl_2$	2-pentene	1016, 1017
$MoCl_2(NO)_2 \cdot n\text{-}BuLi$	2-pentene	1017
$MoCl_2(NO)_2 \cdot AlBr_3$	2-pentene	1017
$Mo(NO)_2Cl_2(OPPh_3)_2 \cdot n\text{-}BuLi$	2-pentene	1017
$Mo(NO)_2Cl_2(OPPh_3)_2 \cdot AlBr_3$	2-pentene	1017
$MoCl_2(NO)_2(EtCN)_2 \cdot n\text{-}BuLi$	2-pentene	1017

1	2	3
$MoCl_2(NO)_2(EtCN)_2 \cdot AlBr_3$	2-pentene	1017
$MoCl_3(NO)(OPPh_3)_2 \cdot AlCl_3$	2-pentene	1019
$MoCl_3(NO)(OPPh_3)_2 \cdot AlEt_3$	2-pentene	1019
$MoCl_2(NO)_2Py_2 \cdot AlCl_3$	2-pentene	1019
$MoCl_2(NO)_2Py_2 \cdot AlEt_3$	2-pentene	1019
$MoCl_3(NO)Py_2 \cdot EtAlCl_2$	2-pentene	1018
$MoCl_3(NO)(CH_3CN) \cdot EtAlCl_2$	2-pentene	1018
$MoCl_3(NO)(DMF) \cdot EtAlCl_2$	2-pentene	1018
$MoCl_3(NO)(DMSO) \cdot EtAlCl_2$	2-pentene	1018
$MoCl_3(NO)(HMPT) \cdot EtAlCl_2$	2-pentene	1018
$MoCl_3(NO)Py_2 \cdot EtAlCl_2$	2-pentene	1018
$MoCl(salen)(NO) \cdot EtAlCl_2$	2-pentene	1018
$MoCl_5Py_x \cdot Me_3Al_2Cl_3$	1-pentene	1157
$MoCl_5(Ph_3PO)_x \cdot Me_3Al_2Cl_3$	1-pentene	1157
$MoCl_5(n\text{-}Oct_3PO)_x \cdot Me_3Al_2Cl_3$	1-pentene	1157
$MoCl_5(n\text{-}Bu_3P)_x \cdot Me_3Al_2Cl_3$	1-pentene	1157
$Et_4N(MoCl_6) \cdot MgBr_2$	2-hexene	1252
$Bu_4N[Mo(CO)_5Py] \cdot EtAlCl_2$	1,7-octadiene	364
$Mo(CO)_6 \cdot PhOH$	acetylenes	759
$Mo(CO)_3Mes \cdot PhOH$	acetylenes	759
$WO_3 \cdot EtAlCl_2$	2-pentene	745
$ReCl_5 \cdot Et_2AlCl + O_2$	1-pentene	1925
	2-pentene	1925
	2-hexene	1925
$ReCl_5 \cdot Et_3Al + O_2$	1-pentene	1060, 1925
	2-pentene	1060, 1925
	2-hexene	1060, 1925
$WCl_6 \cdot Et_3Al + O_2$	idem	1060, 1925
$MoCl_5 \cdot Et_3Al + O_2$	idem	1060
$ReCl_5 \cdot Et_3Al$	1-pentene	1925
	2-pentene	1925
	2-hexene	1925
$ReCl_5 \cdot Et_2AlCl$	idem	1925
$ReCl_5 \cdot Bu_4Sn$	2-pentene	767
$Re(CO)_6 \cdot R_mAlX_n$	α-olefins	1980
$Re(CO)_{10} \cdot Et_3AlCl_2$	1-butene	1980
$Re(CO)_5Cl \cdot EtAlCl_2$	propene	363
	2-pentene	363
	1-hexene	363
	1,7-octadiene	363, 429
	4-nonene	363
	4-octene	429
$Re(CO)_5Cl \cdot MeAlCl_2$	1,7-octadiene	429
	4-octene	429
$SmCl_3 \cdot EtAlCl_2$	2-pentene	1541
	2-heptene	1541
$SmCl_3 \cdot Me_3Al_2Cl_3$	2-pentene	1541
	2-heptene	1541, 1539
$(Ph_3P)_3CuCl \cdot Me_3Al_2Cl_3$	2-heptene	1613
$(Ph_3P)_3Cu_2Cl_2 \cdot EtAlCl_2$	2-pentene	1617
$(acac)_4Zr \cdot Me_3Al_2Cl_3$	2-pentene	1615
$Py_2TiCl_4 \cdot EtAlCl_2$	2-heptene	1615
$(Ph_3P)_2CoCl_2 \cdot EtMgBr$	2-hexene	1252

1	2	3
$(NO)_2RhCl \cdot Me_3Al_2Cl_3$	1-pentene	1469
	2-pentene	1469
	1-hexene	1469
$TiCl_4 \cdot EtAlCl_2$	1-octene	722
$(\pi\text{-allyl})_4W \cdot Cl_3CCOOH$	1-hexene	833

B.3. Ternary homogeneous systems

1	2	3
$WCl_6 \cdot EtOH \cdot EtAlCl_2$	2-pentene	210, 213, 744, 770, 771, 1291
	2-pentene + 6-dodecene	213
	2-butene + 2-butene-d_8	213, 1291
	2-butene-d_8 + 3-hexene	213
	2-methyl-2-pentene	1291
	stylbene	1291
	2-pentene + styrene	1291
	4-methyl-2-pentene + 2-hexene	1291
	2-hexene	1291
	2-heptene	1291
	2-octene	1291
	2-decene	1291
	2-dodecene	1291
	cyclododecene	213
$WCl_6 \cdot EtOH \cdot Et_3Al_2Cl_3$	2-pentene	1156
$WCl_6 \cdot n\text{-BuLi} \cdot AlCl_3$	2-pentene	1107
$WCl_6 \cdot AcOH \cdot EtAlCl_2$	1-hexene	1856
	1-octene	1856
	1-decene	1856
$WCl_6 \cdot Et_2O \cdot R_2AlO(AlR)_nOAlR_n$	1-hexene	1595
$WCl_6 \cdot Et_3N \cdot R_2AlO(AlR)_nOAlR_n$	1-hexene	1595
$WOCl_4 \cdot C_6H_4(CN)_2 \cdot EtAlCl_2$	2-pentene	1399
$WCl_6 \cdot BuMgCl \cdot AlCl_3$	2-pentene	1530
$WCl_6 \cdot 1,3\text{-}Cl_2\text{-i-PrOH} \cdot Et_3Al_2Cl_3$	1,4-pentadiene	280
	1,5-hexadiene	280
	1,7-octadiene	280
$Mo(CO)_5Py \cdot Bu_4NCl \cdot EtAlCl_2$	1,7-octadiene	765
$Mo(CO)_5Py \cdot Bu_4NCl \cdot MeAlCl_2$	1,7-octadiene	765
$WCl_6 \cdot CH_3COOR \cdot Bu_4Sn$	1-pentene	508
	1-hexene	508
	1-heptene	508
	1-octene	508
	1-decene	508
$WCl_6 \cdot CH_3CN \cdot Bu_4Sn$	1-pentene	508
	1-hexene	508
	1-heptene	508
	1-octene	508
	1-decene	508
$NbCl_5 \cdot BzOH \cdot Me_3Al_2Cl_3$	2-pentene	1471
$NbCl_5 \cdot BzOH \cdot Me_3Al_2Cl_3 + NO$	2-pentene	1470
$VCl_5 \cdot BzOH \cdot Me_3AlCl_3 + NO$	2-pentene	1470
$WCl_6 \cdot EtOH \cdot i\text{-Bu}_2AlCl$	2-pentene	1291
$(H_2O) \cdot (i\text{-Bu}_3Al)$	2-hexene	1291
	2-heptene	1291
	2-octene	1291

1	2	3
	2-methyl-2-pentene	1291
	2-pentene + styrene	1291
	stylbene	1291
	4-methyl-2-pentene + 2-hexene	1291
	2-butene-d_8 + 2-butene	1291
$WCl_6 \cdot EtOH \cdot EtAlCl_2$	1,5-cyclo-octadiene + 4-octene	860
	1,6-cyclodecadiene	1623
	1-hexene + cyclo-octene	626
	1-hexene + 1-octene-1,1-d_2	727
$WCl_6 \cdot$ epichlorohydrin $\cdot Et_2AlCl$	polypentenamer + 2-butene	1800
	polypentenamer	1801
	polybutadiene	1801
	polypentenamer + polybutadiene	1803
	polybutadiene + bromocrotyl	1804
$WCl_6 \cdot ClCH_2CH_2OH \cdot Et_2AlCl$	polybutadiene	1798
	polybutadiene + bromocrotyl	1804
$WCl_6 \cdot AcOH \cdot EtAlCl_2$	2-pentene	1680
$WCl_6 \cdot 2Et_2O \cdot Bu_4Sn$	1-hexene + cyclo-octene	626
$Bu_4NCl \cdot W(CO)_5C(OMe)Et \cdot MeAlCl_2$	1-pentene	613, 1611
	2-pentene	1611
	4-methyl-1-pentene	1611
	1-octene	1611
	1,7-octadiene	1611
$Bu_4NCl \cdot W(CO)_5C(NMe_2)Me \cdot$ $MeAlCl_2$	1-pentene	613
$WCl_6 \cdot PhNH_2 \cdot Et_3Al_2Cl_3$	2-pentene	731
$W(CO)_5PPh_3 \cdot Et_2O \cdot EtAlCl_2 + O_2$	2-pentene	1719
$Mo(CO)_6 \cdot Pr_4NCl \cdot Me_3Al_2Cl_3$ (Bu_4NCl)	1-pentene	1839
	2-hexene	1839
	4-phenol-1-butene	1839
	3-methyl-1-butene	1839
	1,7-octadiene	1839
$Mo(CO)_4Py_2 \cdot Pr_4NCl \cdot Me_3Al_2Cl_3$ (Bu_4NCl)	idem	1839
$Mo(CO)_6 \cdot Pr_4NCl \cdot MeAlCl_2$ (Bu_4NCl)	idem	1839
$Mo(CO)_4Py_2 \cdot Pr_4NCl \cdot MeAlCl_2$ (Bu_4NCl)	idem	1839
$ReCl_4 \cdot PPh_3 \cdot EtAlCl_2$	2-pentene	1538
$ReOCl_3 \cdot PPh_3 \cdot EtAlCl_2$	2-pentene	1538
$ReOBr_3 \cdot PPh_3 \cdot EtAlCl_2$	2-pentene	1538
$MnCl_4 \cdot PPh_3 \cdot EtAlCl_2$	2-pentene	1538
$TcCl_4 \cdot PPh_3 \cdot EtAlCl_2$	2-pentene	1538
$PdBr_2 \cdot PPh_3 \cdot EtAlCl_2$	1-butene	1817
	4-methyl-2-butene	1817
	$C_3 - C_{30}$ olefins	1817
$MoCl_2(NO)_2(PPh_3)_2 \cdot NiCl_2(PPh_3)_2 \cdot$ $EtAlCl_2$	propene	2000
$MoCl_5 \cdot Py \cdot Me_3Al_2Cl_3 + NO$	1-pentene	1157
$MoCl_5 \cdot Ph_3PO \cdot Me_3Al_2Cl_3 + NO$	1-pentene	1157
$MoCl_5 \cdot n\text{-}Oct_3PO \cdot Me_3Al_2Cl_3 + NO$	1-pentene	1157
$MoCl_5 \cdot n\text{-}Bu_3P \cdot Me_3Al_2Cl_3 + NO$	1-pentene	1157
$WCl_6 \cdot Py \cdot Me_3Al_2Cl_3 + NO$	1-octene	1157
$WCl_6 \cdot Py \cdot EtAlCl_2 + NO$	2-heptene + ethylene	1157

1	2	3
Polymer$-C_6H_4CH_2(PPh_3)W(CO)_5 \cdot$ $\cdot$i-BuAlCl$_2$ + O$_2$	3-heptene	1113
Polymer$-C_6H_4CH_2(\eta C_5H_5)W(CO)_5 \cdot$ $\cdot$i-BuAlCl$_2$ + O$_2$	3-heptene	1113

B.4. Multicomponent homogeneous systems

WCl$_6 \cdot$ [O(MeO$_2$C)C$_6$H$_4$O] WCl$_5 \cdot$ EtAlCl$_2 \cdot$ PhNH$_2$	propene	1965
	1-pentene	1965
WCl$_6 \cdot$ [O(MeO$_2$C)C$_6$H$_4$O] WCl$_5 \cdot$ Et$_3$Al$_2$Cl$_3 \cdot$ PhNH$_2$	propene	1965
	1-pentene	1965
WCl$_6 \cdot$ polyethylene $\cdot$ BuMgCl $\cdot$ AlCl$_3$	2-pentene	1530
WO(OPh)$_4 \cdot$ AlCl$_3 \cdot$ Et$_3$Al	n-butene	1543
	2-pentene	1543
	3-methyl-1-butene	1543
	2-hexene	1543

TABLE II. Catalytic systems for ring-opening polymerization

Catalytic system	Cyclo-alkene	References
1	2	3

A. Heterogeneous catalytic systems

A.1. Unicomponent heterogeneous systems

MoO$_3$	cyclopentene	1764
MoO$_3$	bicyclo[2.2.1]heptene	1764
WO$_3$	cyclopentene	1764
WO$_3$	bicyclo[2.2.1]heptene	1764
Al$_2$O$_3$	cyclopentene	1764
SiO$_2$	cyclopentene	1764

A.2. Binary heterogeneous systems

SiO$_2 \cdot$ Al$_2$O$_3$	1-methylcyclopentene	129a
	1-ethylcyclopentene	129a
	1,2-dimethylcyclopentene	129a
	2,3-dimethylcyclopentene	129a
MoO$_3 \cdot$ Al$_2$O$_3$(TiO$_2$, ZrO$_2$)	cyclobutene	1364
	cyclopentene	1364
	cyclohexene	188, 1364
	cyclo-octene	1364
	bicyclo[2.2.1]heptene	1364
	bicyclopentadiene	1364
WO$_3 \cdot$ Al$_2$O$_3$(TiO$_2$, ZrO$_2$)	cyclobutene	1364
	cyclopentene	1364
	cyclohexene	1364
	cyclo-octene	1364
	bicyclo[2.2.1]heptene	1364
	bicyclopentadiene	1364
	2-carbomethoxy-5-bicyclo-[2.2.1] heptene	1492

1	2	3
$Re_2O_7 \cdot Al_2O_3$	cyclopentene	1911
	cyclo-octene	1917
	1,5,7-cyclododecatriene	1917
$Cr_2O_3 \cdot Al_2O_3(TiO_2, ZrO_2)$	cyclobutene	1364
	cyclopentene	1364
	cyclohexene	1364
	cyclo-octene	1364
	bicyclo[2.2.1]heptene	1364
	bicyclopentadiene	1364
$UO_2 \cdot Al_2O_3(TiO_2, ZrO_2)$	cyclobutene	1364
	cyclopentene	1364
	cyclohexene	1364
	cyclo-octene	1364
	bicyclo[2.2.1]heptene	1364
	bicyclopentadiene	1364
$WO_3 \cdot AlCl_3$	cyclopentene	1246
	cyclo-octene	1246
$WO_3EtAlCl_2$	cyclopentene	1246
	cyclo-octene	1246
$(\pi\text{-allyl})_4Mo \cdot Al_2O_3$	cyclopentene	1716
	bicyclo[2.2.1]heptene	1716
$(\pi\text{-allyl})_4W \cdot SiO_2$	cyclopentene	1716
	bicyclo[2.2.1]heptene	1716
$(\pi\text{-allyl})_4Mo \cdot Al_2O_3$	cyclopentene	1716
	bicyclo[2.2.1]heptene	1716
$(\pi\text{-allyl})_4W \cdot SiO_2$	cyclopentene	1716
	bicyclo[2.2.1]heptene	1716
$(\pi\text{-allyl})_4W \cdot$ alumino-silicates	1,5,9-caclododecatriene	833

A.3. Ternary heterogeneous systems

1	2	3
$CoO \cdot MoO_3 \cdot Al_2O_3$	cyclopentene	1323
	cyclohexene	1323
	cyclo-octene	1323
	1,5-cyclo-octadiene	1323
$CoO \cdot MoO_3 \cdot Al_2O_3$	cyclopentene	1323
	cyclohexene	1323
	cyclo-octene	1323
	1,5-cyclo-octadiene	1323
$CoO \cdot MoO_3 \cdot ThO_2$	cyclopentene	1323
	cyclohexene	1323
	cyclo-octene	1323
	1,5-cyclo-octadiene	1323
$CoO \cdot MoO_3 \cdot AlPO_4$	cyclopentene	1323
$Zr_3(PO_4)_4$, $Ti_3(PO_4)_4$,	cyclohexene	1323
$Ca_3(PO_4)_4$, $Mg_3(PO_4)_4$)	cyclo-octene	1323
	1,5-cyclo-octadiene	1323
$MoO_3 \cdot Al_2O_3 \cdot Me_3Al_2Cl_3$	cyclopentene	2001
$WO_3 \cdot Al_2O_3 \cdot Me_3Al_2Cl_3$	cyclopentene	2001
$MoO_3 \cdot Al_2O_3 \cdot EtAlCl_2$	cyclopentene	2001
$MoO_3 \cdot Al_2O_3 \cdot Et_2AlCl$	cyclopentene	2001
$WO_3 \cdot Al_2O_3 \cdot EtAlCl_2$	cyclopentene	2001
$WO_3 \cdot Al_2O_3 \cdot Et_2AlCl$	cyclopentene	2001
$WO_3 \cdot AlCl_3 \cdot Et_2AlCl$	cyclopentene	2001
$MoO_3 \cdot PCl_5 \cdot$ Et Acetate	cyclopentene	2001

1	2	3
$MoO_3 \cdot POCl_3 \cdot$ Et Acetate	cyclopentene	2001
$MoO_3 \cdot Et_2AlCl \cdot$ Et Acetate	cyclopentene	2001
$WO_3 \cdot PCl_5 \cdot$ Et Acetate	cyclopentene	2001
$WO_3 \cdot POCl_3 \cdot$ Et Acetate	cyclopentene	2001
$WO_3 \cdot Et_2AlCl \cdot$ Et Acetate	cyclopentene	2001
$MoO_3 \cdot Al_2O_3 \cdot LiAlH_4$	bicyclo[2.2.1]heptene	1363
$WO_3 \cdot Al_2O_3 \cdot LiAlH_4$	2-carbomethoxy-5-bicyclo[2.2.1] heptene	1492

A.4. Multicomponent heterogeneous systems

$MoO_3 \cdot Al_2O_3 \cdot MeAlCl_2 \cdot Me_2AlCl$	cyclopentene	2001
	cyclo-octene	2001
$WO_3 \cdot Al_2O_3 \cdot MeAlCl_2 \cdot Me_2AlCl$	cyclopentene	2001
	cyclo-octene	2001

B. Homogeneous catalytic systems

B.1. Unicomponent homogeneous systems

WCl_6	cyclopentene	91, 1327
	cyclohexene	91
	1,3-cyclohexadiene	632
	cycloheptene	91
	1,3-cycloheptadiene	632
	1,3-cyclo-octadiene	91, 632
	bicyclo[2.2.1]heptadiene	91, 839, 1586
	bicyclo[2.2.2]octene	632
$WOCl_4$	cyclopentene	1327
$WOCl_2$	cyclopentene	1327
WCl_5	cyclopentene	1327
$MoCl_5$	cyclopentene	1327
$ReCl_5$	cyclopentene	1327
$PhWCl_3$	cyclopentene	422
$PhWCl_5$	cyclopentene	1907
$PhMoCl_5$	cyclopentene	1907
$Me_3SiCH_2 \cdot WCl_5$	cyclopentene	306
	1,5-cyclo-octadiene	306
$Ph_2C=WCl_5$	1-methylcyclo-octene	650
$Ph_2C=W(CO)_5$	cyclobutene	561
	1-methylcyclobutene	557
	cyclopentene	561
	cycloheptene	561
	cyclo-octene	561
	bicyclo[2.2.1]heptene	561
$RuCl_3$	cyclo-octene	875
$IrCl_3$	cyclo-octene	875
$RhCl_3 \cdot H_2O$	cyclobutene	801
	bicyclo[4.2.0]-7-octene	270
	bicyclo[3.2.0]-2,6-heptadiene	270
	dicyclopentadiene	274
$RuCl_3 \cdot H_2O$	cyclobutene	802
	3-methylcyclobutene	802
	bicyclo[4.2.0]-7-octene	270

1	2	3
	bicyclo[3.2.0]-2,6-heptadiene	270
	dicyclopentadiene	274
	bicyclo[2.2.1]heptene	738, 895
IrCl₃	idem	idem
PdCl₂	idem	idem
π-allyl-NiBr₂	idem	270, 274
OsCl₃	bicyclo[2.2.1]heptene	738
IrCl₃(cyclo-octadiene)	bicyclo[2.2.1]heptene	632, 1942
(cyclo-octene)	bicyclo[2.2.1]heptadiene	1942
	5-carbomethoxy-bicyclo[2.2.1]-2-heptene	1942
	1,3-cyclohexadiene	1942
IrBr₃ (cyclo-octadiene) (cyclo-octene)	idem	1942
(π-allyl)₄Zr	cyclobutene	591
	cyclopentene	591
	cyclo-octene	591
	cyclododecene	591
	bicyclo[2.2.1]heptene	591
(π-allyl)₃Cr	idem	591
(π-allyl)₄Mo	idem	591
(π-allyl)₄W	idem	591
(π-allyl)NiBr₂	cyclopentene	872
ReCl₅	bicyclo[2.2.1]heptene	632, 839, 841
MoCl₅	idem	idem
chloroiridium(IV)acid	bicyclo[2.2.1]heptene	401
	bicyclo[2.2.1]heptene + cyclopentene	401
	bicyclo[2.2.1]heptene + cyclo-octene	401
(cyclo-octene)₂IrCl₂	cyclopentene	874
	cyclo-octene	874
	bicyclo[2.2.1]heptene	874
	bicyclo[2.2.1]heptene + cyclopentene	874
	bicyclo[2.2.1]heptene + cyclo-octene	874
(cyclo-octene)₂Ir(CO)Cl	idem	874
(cyclo-octene)₂Ir(CF₃CO₂)	cyclopentene	873, 874
	cyclo-octene	873, 874
	bicyclo[2.2.1]heptene	873, 874
	bicyclo[2.2.1]heptene + cyclopentene	873, 874
(2,7-Me₂octa-2,6-dien-1,8-yl)RuCl₂	bicyclo[2.2.1]heptene + cyclo-octene	873, 874
(2,7-Me₂octa-2,6-dien-1,8-yl)Ru (CF₃CO₂)₂	bicyclo[2.2.1]heptene	873, 874
	bicyclo[2.2.1]heptene + cyclopentene	873, 874
	bicyclo[2.2.1]heptene + cyclo-octene	873, 874
(2,7-Me₂octa-2,6-dien-1,8-yl)Ru (CF₃CO₂)₂ + H₂	cyclopentene	873

1	2	3

B.2. Binary homogeneous systems

$WCl_6 \cdot EtAlCl_2$	1,5-cyclodecadiene	475, 569
$WCl_6 \cdot Et_2AlCl$	cyclobutene	269
	1-methylcyclobutene	267
	3-methylcyclobutene	269
	cyclopentene	1743, 1746, 803
	cyclohexene	803
	cycloheptene	803, 1747
	cyclododecene	1743, 1746, 803
	1,5-cyclo-octadiene	569
	5,6-bis(chloromethyl)-bicyclo[2.2.1]	
	heptene	1586
	benzobicyclo[2.2.1]heptadiene	1632
	bicyclo[6.1.0]-4-nonene	663
	9,9-dichlorobicyclo[6.1.0]-4-nonene	663, 664
	9,9-dibromobicyclo[6.1.0]-4-nonene	664
	9-bromo-9-chlorobicyclo[6.1.0]-4-	
	nonene	664
	5-phenyl-bicyclo[2.2.1]-2-heptene	781
	5-phenyl-bicyclo[2.2.1]-2-heptene +	
	+ 5-cyano-bicyclo[2.2.1]-2-heptene	782
	5-cyanobicyclo[2.2.1]-2-heptene	820
$WCl_6 \cdot Et_3Al$	5-tribromophenylcarbomethoxy-	
	bicyclo[2.2.1]heptene	1485
	2-phenylbicyclo[2.2.1]heptene	1590
	cyclobutene	799, 803
	cyclopentene	1687, 1743, 800
	cyclohexene	803, 1747
	cycloheptene	803, 1747
	cyclo-octene	1743, 803
	cyclododecene	1743, 803
$WCl_6 \cdot Et_2AlI$	cyclopentene + dicyclopentadiene	1692
	tetrahydroindene	271
	2-methylcyclobutene +	
	+ 1-C^{14}-cyclobutene	278
	5-cyano-bicyclo[2.2.1]heptene	1524
	polypentenamer + butadiene	1328
	polyoctenamer + butadiene	1328
$WCl_6 \cdot i\text{-}Pr_3Al$	cyclopentene	447
$WCl_6 \cdot i\text{-}Bu_2AlCl$	cyclo-octene	300
	1,5-cyclo-octadiene	300, 1088
	1,5,9-cyclododecatriene	300, 1087
	bicyclo[2.2.1]heptene	300
	cyclo-octene + 1,5-cyclo-octadiene	1090
	cyclo-octadiene + 1,5,7-cyclododeca-	
	catriene	1090
$WCl_6 \cdot i\text{-}Bu_3Al$	1,5-cyclo-octadiene	300, 1298
	bicyclo[2.2.1]heptene	1298
	cyclopentene	34
$WCl_6 \cdot Me_4Sn$	lactones	1183

1	2	3
$WCl_6 \cdot Ph_4Sn$	cyclopentene	455
	bicyclo[2.2.1]heptene	455
	cyclopentene + ethylene	455
$WOCl_4 \cdot Et_2AlCl$	cyclopentene + 1-pentene	460
	cyclopentene + 2-pentene	460
	cyclopentene + 2-butene	460
	cyclopentene + propene	460
	cyclo-octene + 1-pentene	460
	cyclo-octene + 2-pentene	460
	cyclo-octene + 2-butene	460
	cyclo-octene + propene	460
	1,5-cyclo-octadiene + 2-pentene	460
	1,5,9-cyclododecatriene + 2-pentene	460
$WOCl_4 \cdot Bu_4Sn$	cyclopentene	460
	bicyclo[2.2.1]heptene	460
	cyclopentene + ethylene	460
$WCl_6 \cdot Et_3SnH$	cyclopentene	1758
$WOCl_4 \cdot Et_3Al$	polypentenamer + butadiene	1328
	polyoctenamer + butadiene	1328
$WCl_6 \cdot AlX_3$ (X=Cl, Br, I)	cyclopentene	1327
$WCl_5 \cdot AlX_3$	cyclopentene	1327
$WBr_5 \cdot AlX_3$	cyclopentene	1327
$WOCl_4 \cdot AlX_3$	cyclopentene	1327
$WF_6 \cdot EtAlCl_2$	cyclo-octene	1333
$KWCl_6 \cdot AlCl_3$	cyclo-octene	1515
	1,5,9-cyclododecatriene	1515
$KWCl_6 \cdot AlBr_3$	cyclo-octene	1515
	1,5,9-cyclododecatriene	1515
$KWCl_6 \cdot EtAlCl_2$	cyclo-octene	1515
	1,5,9-cyclododecatriene	1515
$Na_3WPh_6 \cdot SnCl_4$	cyclopentene	1755
$WCl_6 \cdot R_nSiCH_2Y$ (R = alkyl; Y = aryl)	cyclopentene	1354, 1355
	cycloheptene	1354
$WCl_6 \cdot C_6H_5CHN_2$	cyclopentene	302, 306
	cyclo-octadiene	302, 306
$WCl_6 \cdot EtOOCCN_2$	cyclopentene	419
	1,5-cyclo-octadiene	419
$NMe_4(CO)_5WCOPh \cdot EtAlCl_2$	cyclopentene	109
	cyclo-octene	109
$WF_6 \cdot Et_3Al$	cyclopentene	1743
	cyclo-octene	1743
	cyclododecene	1743
$WF_6 \cdot Et_2AlCl$	cyclopentene	1743, 1765
	cyclo-octene	1743
	cyclododecene	1743
$WF_6 \cdot Et_3Al_2Cl_3$	cyclopentene	1765
$MoCl_5 \cdot Et_3Al$	cyclopentene	1743, 1744
$(MoCl_4)$	cyclo-octene	1743
	cyclododecene	1743
$MoCl_5 \cdot Et_2AlCl$	cyclopentene	1743
	cyclo-octene	1743
	cyclododecene	1743
$WOCl_4 \cdot Et_3Al$	cyclopentene	1743
(Et_2AlCl)	cyclo-octene	1743
	cyclododecene	1743

1	2	3
W(OPh)$_6$ · Et$_3$Al	cyclopentene	1743
(Et$_2$AlCl)	cyclo-octene	1743
	cyclododecene	1743
W(OPh)$_4$Cl$_2$ · MeAlCl$_2$	cyclopentene	1857
W(OPh)$_6$ · MeAlCl$_2$	cyclopentene	1857
Na$_3$WPh$_5$ · WCl$_6$	cyclopentene	34
AlWPh$_5$ · TiCl$_4$	cyclopentene	34
(HSiOMe)$_4$ · WCl$_6$	cyclopentene	34
(HSiOMe)$_8$ · WCl$_6$	cyclopentene	34
Li$_3$WPh$_6$ · TiCl$_4$	cyclopentene	34
Li$_3$WEt$_6$ · WCl$_6$	cyclopentene	34
Li$_3$WPh$_6$ · SnCl$_4$	cyclopentene	34
Li$_3$WPh$_6$ · BCl$_3$	cyclopentene	34, 1755
Na$_3$WPh$_6$ · SnCl$_4$	cyclopentene	34, 1755
Ca · WCl$_6$	cyclopentene	34
Li · WCl$_6$	cyclopentene	34
Ca · WCl$_5$OX	cyclopentene	34
Al · WCl$_6$	cyclopentene	34
AlCl$_3$ · WCl$_6$	cyclopentene	34
Ca · WBr$_5$	cyclopentene	34
Li$_3$WPh$_6$ · MoCl$_5$	cyclopentene	1755
Li$_3$MoPh$_6$ · BF$_3$	cyclopentene	1755
Ph$_4$W · SnCl$_4$	cyclopentene	1755
Ph$_2$Ti · WOCl$_4$	cyclopentene	1755
Ph$_3$Cr · WCl$_6$	cyclopentene	1755
1,3,5,7-Me$_4$-cyclotetrasiloxane · WCl$_6$	cyclopentene	1755
WCl$_6$ · Et$_2$Be	cyclopentene	1744
WCl$_6$ · LiCH$_2$SiMe$_3$	cyclopentene	836, 949
	1,5-cyclo-octadiene	836
WCl$_6$ · LiCH$_2$GeMe$_3$	cyclopentene	836
	1,5-cyclo-octadiene	836
WCl$_6$ · N$_2$CHSiMe$_3$	cyclopentene	949
WCl$_6$ · Ph$_3$SnEt	1-methylcyclobutene	557
WCl$_6$ · n-BuLi	1-methylcyclobutene	557
WCl$_4$(OR) · EtAlCl$_2$	cyclopentene	949
W(CO)$_6$ · GaBr$_3$	1,5-cyclo-octadiene	1983
WCl$_6$ · AlCl$_3$	cyclopentene	1513
(WOCl$_4$)	cyclo-octene	1513
	1,5,9-cyclododecatriene	1513, 1665
	1,5-cyclo-octadiene	1665
	cyclododecene	1665
WCl$_6$ · AlBr$_3$	cyclo-octene	708
	1,5-cyclo-octadiene	1655
	cyclododecene	708
	1,5,7-cyclododecatriene	708
MoCl$_5$ · AlCl$_3$ (AlBr$_3$, AlI$_3$)	1,5-cyclo-octadiene	1665
(MoOCl$_2$)	cyclododecene	1665
	1,5,7-cyclododecatriene	1665
WCl$_5$ · AlCl$_3$(AlBr$_3$, AlI$_3$)	1,5-cyclo-octadiene	1665
	cyclododecene	1665
	1,5,7-cyclododecatriene	1665
WCl$_6$ · LiAlH$_4$	cyclo-octene	900
	cyclononene	900
TiCl$_3$ · Et$_3$Al	cyclobutene	799

1	2	3
$TiCl_4 \cdot Et_3Al$	cyclobutene	799
	cyclopentene	800
	bicyclo[2.2.1]heptene	910, 916, 1050
	5-methylbicyclo[2.2.1]-heptene	984
$TiCl_4 \cdot Et_3Al$	cyclopentene	800
$VCl_4 \cdot Et_3Al$	cyclobutene	799
	cyclopentene	800
$ZrCl_4 \cdot Et_3Al$	cyclopentene	800
$V(acac)_3 \cdot Et_2AlCl$	cyclobutene	799
	cyclopentene	800
$Cr(acac)_3 \cdot Et_2AlCl$	cyclobutene	799
	cyclopentene	800
$CrBr_3 \cdot Et_3Al$	cyclopentene	800
$CrO_2Cl_2 \cdot Et_3Al$	cyclopentene	800
$MoCl_3 \cdot Et_3Al$	cyclobutene	799
$MoO_2(acac)_2 \cdot Et_2AlCl$	cyclobutene	799
$VF_4 \cdot Et_3Al$	cyclopentene	800
$VO_2Cl_2 \cdot Et_2AlCl$	cyclopentene	800
$MnCl_2 \cdot Et_3Al$	cyclopentene	800
$FeCl_3 \cdot Et_3Al$	cyclopentene	800
$CoCl_2Py_2 \cdot Et_2AlCl$	cyclopentene	800
$BaFeO_4 \cdot Et_2AlCl$	cyclopentene	800
$MoCl_5 \cdot Et_3Al$	cyclobutene	99
	cyclopentene	800, 803
	tetrahydroindene	274
	tetrahydroindane	274
	cyclohexene	803
	cycloheptene	803
	bicyclo[2.2.1]heptene	916
	bicyclo[2.2.1]heptadiene	916
	cyclo-octene	803
	cyclododecene	803
$MoCl_5 \cdot Et_2AlCl$	bicyclo[2.2.1]heptene	916
$MoCl_5 \cdot i\text{-}Bu_3Al$	cyclopentene	1944
$MoCl_6 \cdot Bu_4Sn$	cyclo-octene	1268
$MoCl_3distearate \cdot Et_2AlCl$	cyclopentene	846
$MoCl_3distearate \cdot i\text{-}Bu_3Al$	cyclopentene	846
$MoCl_3(laurate)_2 \cdot Et_2AlCl$	cyclopentene + cyclopentadiene	1356
$(EtO)MeC=M(CO)_5 \cdot EtAlCl_2$ $(AlCl_3)(M = Mo, W)$	cyclopentene	1314
$(MeO)MeC=M(CO)_5 \cdot EtAlCl_2$ $(AlCl_3)$	cyclopentene	1314
$(EtO)PhC=M(CO)_5 \cdot EtAlCl_2$ $(AlCl_3)$	cyclopentene	1314
$(MeO)PhC=M(CO)_5 \cdot EtAlCl_2$ $(AlCl_3)$	cyclopentene	1314
$(EtO)RC=W(CO)_5 \cdot TiCl_4$	cyclopentene	260
$ReCl_5 \cdot i\text{-}Bu_3Al$	cyclopentene	1404
$ReCl_6 \cdot Et_3Al$	cyclopentene	1332
$ReCl_6 \cdot i\text{-}Bu_3Al$	cyclopentene	1332
$ReCl_6 \cdot Et_2AlCl$	cyclopentene	1332
$Re(CO)_5Cl \cdot EtAlCl_2$	cyclohexene	365
$TaCl_5 \cdot i\text{-}Bu_3Al$	cyclopentene	1766, 1944
$NbCl_5 \cdot i\text{-}Bu_3Al$	cyclopentene	1944
$TiCl_4 \cdot EtMgBr$	bicyclo[2.2.1]heptene	1176
$VCl_4 \cdot Hex_3Al$	cyclobutene	265
	cyclopentene + ethylene	266
	cyclohexene + ethylene	266

1	2	3
	cycloheptene + ethylene	266
	cyclo-octene + ethylene	266
	cyclopentene	246
$(\pi\text{-allyl})_2 M \cdot AlBr_3$ (M=Ni, Co, Zr)	cyclopentene	1591
	cyclo-octene	1591
	bicyclo[2.2.1]heptene	1591
$(\pi\text{-allyl})_3 M \cdot AlBr_3$ (M=W, Cr)	idem	1591
$(\pi\text{-allyl})_2 M \cdot WF_6$ (M=Ni, Co, Zr)	idem	1591
$(\pi\text{-allyl})_3 \cdot WF_6$ (M=W, Cr)	idem	1591
$(\pi\text{-allyl})_2 M \cdot TiCl_4$ (M= Ni, Co, Zr)	idem	1591
$(\pi\text{-allyl})_3 M \cdot TiCl_4$ (M= W, Cr)	idem	1591
$(\pi\text{-allyl})_2 M \cdot MoCl_5$ (M= Ni, Co, Zr)	idem	1591
$(\pi\text{-allyl})_3 M \cdot MoCl_5$ (M= W, Cr)	idem	1591
$(\pi\text{-allyl})_3 Cr \cdot WCl_6$	cyclopentene	34
$(\pi\text{-allyl})_4 W_2 \cdot WCl_6$	cyclopentene	34
$(\pi\text{-allyl})_4 W \cdot BCl_3$	cyclopentene	1758
$MoCl_3(\text{octanoate})_2 \cdot Et_2AlCl$	cyclopentene + 1-hexene +5-vinyl-bicyclo[2.2.1]-2-heptene	1945
	cyclopentene + 1-hexene + dicyclopentadiene	1945
	cyclopentene + 1-hexene	1945
$MoCl_2(NO)_2(PPh_3)_2 \cdot Me_3Al_2Cl_3$	1-methylcyclobutene	557
	1-methylcyclopentene	557
	cyclo-octene	2002
$Bu_4NMo(CO)_5(COPh) \cdot MeAlCl_2$	cyclopentene	1611
$TiCl_4 \cdot Hex_3Al$	cyclobutene + ethylene	798
	cyclobutene + propene	798
	cyclopentene + ethylene	798, 1745
	cyclopentene + propene	1745, 798
$V(\text{acac})_2 \cdot Et_2AlCl$	cyclobutene + ethylene	798
	cyclobutene + ethylene-$^{14}C_2$	1741
	cyclobutene + propene	798
	cyclopentene + ethylene	798
	cyclopentene + propene	798
$(\pi\text{-allyl})_4 W \cdot Cl_3CCOOH$	1,5,9-cyclododecatriene	833
$(\pi\text{-crotyl})_4 W \cdot AlBr_3$	1,5,9-cyclododecatriene	1984, 1985
$TiCl_4 \cdot Et_2AlCl$	bicyclo[4.2.0]-7-octene	270
	bicyclo[3.2.0]-2,6-heptadiene	270
	dicyclopentadiene	274
$TiCl_3 \cdot Et_2AlCl$	idem	270, 274
$VCl_4 \cdot Et_2AlCl$	idem	270, 274
$Cr(\text{acac})_3 \cdot Et_2AlCl$	idem	270, 274
$WCl_6 \cdot Et_2AlCl$	idem	270, 274
$TiCl_3 \cdot Et_3Al$	cyclobutene	265
	tetrahydroindene	271
$TiCl_4 \cdot Et_3Al$	cyclobutene	265
	tetrahydroindene	271
	tetrahydroindane	271
$VCl_4 \cdot Et_3Al$	tetrahydroindene	271
$VOCl_3 \cdot Et_3Al$	tetrahydroindene	271
$V(\text{acac})_3 \cdot Et_2AlCl$	cyclobutene	265
	tetrahydroindene	271
$Cr(\text{acac})_3 \cdot Et_2AlCl$	tetrahydroindene	271
$CrCl_3Py_3 \cdot Et_3Al$	tetrahydroindene	271
$MoO_2(\text{acac})_2 \cdot Et_3Al$	tetrahydroindene	271

1	2	3
$CrO_2Cl_2 \cdot Et_2AlCl$	tetrahydroindene	271
$TiCl_4 \cdot LiAl(C_{10}H_{22})_4$	5-butenyl-bicyclo[2.2.1]heptene	288
$TiCl_4 \cdot LiAl(C_7H_{15})_4$	bicyclo[2.2.1]heptene	1043
$RuCl_3 \cdot PPh_3$	bicyclo[2.2.1]heptene	463
o-Phenanthioline $W(CO)_4 \cdot AlBr_3$	cyclo-octene	1517

B.3. Ternary homogeneous systems

1	2	3
$WCl_6 \cdot EtOH \cdot EtAlCl_2$	cyclopentene	214, 815, 928, 1883, 1885
	cyclo-octene	211, 212, 464, 465, 1292, 1293, 469, 815, 927, 1850
	cyclododecene	211
	1,5-cyclo-octadiene	126, 927, 1849, 211 1247, 1289
	1,5,9-cyclododecatriene	211, 920, 927, 1247, 1289, 1849, 1850
	3-methylcyclo-octene	211
	5-methylcyclo-octene	211
	3-phenylcyclo-octene	211
	5-phenylcyclo-octene	211
	1-methyl-1,5-cyclo-octadiene	1297
	bicyclo[2.2.1]heptene	1150
	cyclotetraeicosene	815
	3,3′-bicyclopentene	815
	bicyclo[3.3.0]-octene	815
	bicyclo[2.2.2]octene	815
	bicyclo[4.3.0]-3,7-nonadiene	815
	tricyclo[5.2.1.0^{2,6}]-8-decene	815
$WCl_6 \cdot H_2O \cdot EtAlCl_2$	cyclo-octene	1290
$WCl_6 \cdot MeOH \cdot Et_2AlCl$	cyclo-octene + 1,5-cyclo-octadiene	1290
$WCl_6 \cdot EtOH \cdot Et_3Al_2Cl_3$	cyclo-octene + 1,5,9-cyclododeca-triene	1290, 1849
$WCl_6 \cdot PrOH \cdot i\text{-}Bu_2AlCl$	poly(diene · ethylene · propene)+ cyclo-octadiene	1849
$WCl_6 \cdot i\text{-}BuOH \cdot i\text{-}Bu_3Al$	polybutadiene + cyclo-octene	1849
$WCl_6 \cdot t\text{-}BuOH \cdot Et_2Zn$	polybenzobicyclo[4.2.1]-2,6-nonadiene	1849
$WCl_6 \cdot PhOH \cdot EtAlCl_2$	cyclo-octene	1849
$WCl_6 \cdot BzOH \cdot EtAlCl_2$	5-chloro-1,5-cyclo-octadiene	927
$WCl_6 \cdot PhCMe_2OH \cdot EtAlCl_2$	polycyclo-octene+polycyclododeca-triene	1850
$WCl_6 \cdot O_2 \cdot EtAlCl_2$		
$WCl_6 \cdot Cum_2O_2 \cdot EtAlCl_2$		
$WCl_6 \cdot PhSH \cdot EtAlCl_2$		
$WCl_6 \cdot t\text{-}C_9H_{19}SH \cdot EtAlCl_2$		
$WCl_6 \cdot Bz_2O_2 \cdot EtAlCl_2$	3-methylcyclo-octene	408, 1335
$WCl_6 \cdot Cum_2O_2 \cdot EtAlCl_2$	3-methylcyclo-octene	1335
$WCl_6 \cdot Bz_2O_2 \cdot Et_2AlCl$	cyclopentene	1329
	cyclopentene + cycloheptene	1331
	cyclopentene + cyclo-octene	1331, 764
	cyclo-octene + cycloheptene	1331
	cyclopentadiene	1331

1	2	3
	5-carbomethoxy-bicyclo[2.2.1]-heptene	1862
$WCl_6 \cdot Cum_2O_2 \cdot Et_2AlCl$	cyclopentene	1329
	cyclopentene + cycloheptene	1331
	cyclopentene + cyclo-octene	764, 1331
	cyclo-octene + cycloheptene	1331
	cyclopentadiene	1331
$WOCl_4 \cdot Cum_2O_2 \cdot Et_2AlCl$	cyclopentene	1329
	cyclopentene + cycloheptene	1331
	cyclopentene + cyclo-octene	1331
	cyclo-octene + cycloheptene	1331
	cyclopentadiene	1331
	cyclodecene	268
$WCl_6 \cdot Bz_2O_2 \cdot Et_3Al$	cyclopentene	1329
	cyclopentene + cyclo-octene	764
$WCl_6 Cum_2O_2 \cdot Et_3Al$	cyclopentene + cyclo-octene	764
$WCl_6 \cdot Bz_2O_2 \cdot (C_6H_{13})_3Al$	cyclopentene	1329
$WCl_6 \cdot Bz_2O_2 \cdot i\text{-}Bu_2AlH$	cyclopentene	1329
$WCl_6 \cdot Bz_2O_2 \cdot Et_2Be$	cyclopentene	1329
$WOCl_4 \cdot Bz_2O_2 \cdot Et_3Al$	cyclopentene	1329
$WOCl_4 \cdot Bz_2O_2 \cdot Et_2AlCl$	cyclopentene	1329
	cyclopentene + cycloheptene	1331
	cyclopentene + cyclo-octene	276, 1331
	cyclo-octene + cycloheptene	1331
	cyclopentadiene	1331
$WOCl_4 \cdot Bz_2O_2 \cdot (C_6H_{13})_3Al$	cyclopentene	1329
$WOCl_4 \cdot Bz_2O_2 \cdot i\text{-}Bu_2AlH$	cyclopentene	1329
$WOCl_4 \cdot Bz_2O_2 \cdot Et_2Be$	cyclopentene	1329
$WCl_6 \cdot AlCl_3 \cdot CO_2$	cyclopentene	1337
$MoCl_5 \cdot AlCl_3 \cdot CO_2$	cyclopentene	1337
$WCl_6 \cdot CO_2 Et_2AlCl$	cyclopentene	1336
$MoCl_5 \cdot CO_2 \cdot Et_2AlCl$	cyclopentene	1337
$WCl_6 \cdot PPh_3 \cdot Et_2AlCl$	5-cyanobicyclo[2.2.1]heptene	1524
$WF_6 \cdot EtAlCl_2 \cdot CO_2 \ (O_2, H_2O)$	cyclo-octene	1333
$WCl_6 \cdot Na_2O_2 \cdot i\text{-}Bu_3Al$	cyclopentene	1756
$WCl_6 \cdot BaO_2 \cdot i\text{-}Bu_3Al$	cyclopentene	1756
$WCl_6 \cdot t\text{-}BuClO \cdot EtAlCl_2$	cyclopentene	815
$WCl_6 \cdot$ epichlorhydrin $\cdot Et_2AlCl$	cyclopentene	852, 1803
	cyclopentene + bicyclo[2.2.1]-heptadiene	1799
$WCl_6 \cdot$ epichlorhydrin $\cdot i\text{-}Bu_3Al$	cyclopentene	1796, 1847
$WCl_6 \cdot$ epichlorhydrin $\cdot$ sodium	cyclopentene	1405
$WCl_6 \cdot H_2O \cdot$ sodium	cyclopentene	1405
$WCl_6 \cdot ClCH_2CH_2OH \cdot Et_2AlCl$	cyclopentene	852, 1972, 1973, 1974
	polybutadiene + cyclopentene	1798, 1802
	polybutadiene	1798
	cyclopentene + bicyclo[2.2.1]-heptene	1972
$WCl_6 \cdot ClCH_2CH_2OH \cdot i\text{-}Bu_3Al$	cyclopentene	1972
	cyclopentene + bicyclo[2.2.1]-heptadiene	1972
$WCl_6 \cdot$ cyclopentyl $\cdot i\text{-}Bu_3Al$ hydroperoxide	cyclopentene	1758
	1,5-cyclo-octadiene	1758

1	2	3
	bicyclo[2.2.1]heptadiene	1758
	bicyclo-octene	1758
	cyclopentene + bicyclo[2.2.1]heptene	1758
WCl$_6$ · ethylene oxide · i-Bu$_3$Al	cyclopentene	1847
WCl$_6$ · propylene oxide · i-Bu$_3$Al	cyclopentene	1796, 1797
WCl$_6$ · 1-butylene oxide · i-Bu$_3$Al	cyclopentene	1847
WCl$_6$ · vinyl chloride · i-Bu$_3$Al	cyclopentene	1766
WCl$_6$ · vinyl fluoride · EtAlCl$_2$	cyclopentene	1884
WCl$_6$ · HOCH$_2$CH$_2$CN · EtAlCl$_2$	cyclopentene	1771, 1772, 1775, 1776
WCl$_6$ · ClCH$_2$CH$_2$OSiMe$_3$ · EtAlCl$_2$	cyclopentene	1777
	1,5-cyclo-octadiene	1777
	dicyclopentadiene	1777
WCl$_6$ · C$_6$H$_4$ClNO$_2$-o · i-Bu$_3$Al	cyclopentene	1759
WCl$_6$ · C$_6$H$_2$Cl$_2$(NO$_2$)$_2$-1,2 · i-Bu$_3$Al	cyclopentene	1759
WCl$_6$ · C$_6$H$_2$Cl(NO$_2$)$_3$-2,4,6 · i-Bu$_3$Al	cyclopentene	1759
WCl$_6$ · p-benzoquinone · EtAlCl$_2$	cyclopentene	1188
WCl$_6$ · AcOH · EtAlCl$_2$	cyclo-octene	1621
WCl$_6$ · (MeO)$_2$CH$_2$ · Et$_2$AlCl	cyclopentene	1848
WCl$_6$ · (EtO)$_2$CH$_2$ · Et$_2$AlCl	cyclopentene	1848
	5-cyano-bicyclo[2.2.1]-2-heptene	1937
WCl$_6$ · (βClEtO)$_2$CH$_2$ · Et$_2$AlCl	cyclopentene	1848
WCl$_6$ · (MeO)$_2$CHCCl$_3$ · Et$_2$AlCl	cyclopentene	1848
WCl$_6$ · (EtO)$_2$CHC$_6$H$_5$ · Et$_2$AlCl	cyclopentene	1848
WCl$_6$ · (EtO)$_2$CHCH$_3$ · Et$_2$AlCl	5-cyano-bicyclo[2.2.1]-2-heptene	1177, 1723, 1727
WCl$_6$ · thipohene · i-Bu$_2$AlCl	bicyclo[2.2.1]heptene	300
	1,5-cyclo-octadiene	300
WCl$_6$ · Bz$_2$O$_2$ · i-Bu$_3$Al	cyclopentene	1235
	1,5-cyclo-octadiene	1235
	bicyclo[2.2.1]heptadiene	1235
	bicyclo-octene	1235
	cyclopentene + bicyclo[2.2.1]heptene	1235
WCl$_6$ · t-Bu$_2$O$_2$ · i-Bu$_3$Al	idem	1235
WCl$_6$ · t-BuOOH · i-Bu$_3$Al	idem	1235
WCl$_6$ · CumOOH · i-Bu$_3$Al	idem	1235
WCl$_6$ · phenoxypropylene oxide	cyclopentene	1796, 1797
EtWCl$_4$ · Et$_2$O · Bu$_4$Sn	cyclopentene	853
WCl$_6$ · Et$_2$O · Bu$_4$Sn	cyclopentene	854
WCl$_6$ · 3,5-(NO$_2$)$_2$C$_6$H$_3$COCl · i-Bu$_3$Al	cyclopentene	1759
WCl$_6$ · 2,4-(NO$_2$)$_2$C$_6$H$_3$NHNH$_2$ · i-Bu$_3$Al	cyclopentene	1759
WCl$_6$ · Cr(acac)$_3$ · Et$_3$Al	5-carbomethoxy-bicyclo-[2.2.1]heptene	1478
WCl$_6$ · Al(OEt)$_3$ · Et$_2$AlCl	5-cyano-bicyclo[2.2.1]-2-heptene	1565
WCl$_6$ · Al(OH)$_3$ · Et$_2$AlCl	5-cyano-bicyclo[2.2.1]-2-heptene	1568
WCl$_6$ · APO$_4$ · Et$_2$AlCl	cyclopentene	1762
	5-cyano-bicyclo[2.2.1]-2-heptene	1569
WCl$_6$ · CPDCl$_6$ · EtAlCl$_2$	cyclopentene	1762, 1764
WCl$_6$ · 2-chlorobutadiene · EtAlCl$_2$	cyclopentene	1762
WCl$_6$ · Al · AlCl$_3$	cyclo-octene	1514
	1,5-cyclo-octadiene	1514
	1,5,9-cyclododecatriene	1514
WO(OPh)$_4$ · AlCl$_3$ · EtAlCl$_2$	cyclopentene	1543
WO(OPh)$_4$ · AlCl$_3$ · Et$_3$Al	cyclopentene	1543

1	2	3
$WO_3 \cdot TlCl_5 \cdot Et_2AlCl$	cyclopentene	1563
	1,5-cyclo-octadiene	1563
	5-cyano-bicyclo[2.2.1]-2-heptene	1563
	5-acetoxy-bicyclo[2.2.1]-2-heptene	1563
$MoCl_3$distearate $\cdot Et_2AlCl \cdot Et_3N$	cyclopentene	1287
$TaCl_5 \cdot$ ethanol $\cdot EtAlCl_2$	cyclopentene	1402
	1,5-cyclo-octadiene	1402
	1,5,9-cyclododecatriene	1402
	bicyclo[2.2.1]heptene	1402
	dicyclopentadiene	1402
$TaCl_5 \cdot$ ethanol $\cdot$ i-Bu_3Al	idem	1402
$TaCl_5 \cdot$ epichlorhydrin $\cdot EtAlCl_2$	cyclopentene	1238
$TaCl_5 \cdot$ 3-chlorocyclopentene $\cdot$ $EtAlCl_2$	cyclopentene	1238
$TaCl_5 \cdot$ 3-chlorocyclopentene $\cdot$ $Et_3Al_2Cl_3$	cyclopentene	1238
$TaCl_5 \cdot ClCH_2COMe \cdot EtAlCl_2$	cyclopentene	1238
$TaCl_5 \cdot ClCH_2COMe \cdot Et_3Al_2Cl_3$	cyclopentene	1238
$TaCl_5 \cdot$ epichlorhydrin $\cdot Et_3Al_2Cl_3$	cyclopentene	1238
$TaCl_5 \cdot$ epichlorhydrin $\cdot$ i-Bu_3Al	cyclopentene	1238
$TaCl_5 \cdot$ ethylene oxide $\cdot$ i-Bu_3Al	cyclopentene	1238
$TaCl_5 \cdot$ butylene oxide $\cdot$ i-Bu_3Al	cyclopentene	1238
$TaCl_5 \cdot$ vinyl chloride $\cdot$ i-Bu_3Al	cyclopentene	1238
$TaCl_5 \cdot CPOCl_6 \cdot EtAlCl_2$	cyclopentene	1764
$TaCl_5 \cdot CPOCl_6 \cdot$ i-Bu_3Al	cyclopentene	1764
$TaCl_5 \cdot CuCl_2 \cdot EtAlCl_2, Et_3Al_2Cl_3$	cyclopentene	1764
$TaCl_5 \cdot FeCl_3 \cdot EtAlCl_2, Et_3Al_2Cl_3$	cyclopentene	1764
$ZrCl_4 \cdot$ t-$BuOCl \cdot Et_3AlCl_3$	cyclopentene	1403
$TaCl_5 \cdot Ph_3PO$		
$CoCl_2Ph_3P$		
$Co(acac)_3Ph_2(CH_2)_2PPh_2$		
$Ni(acac)_2 \ (Me_2N)_3PO$		
$NiBr_2$		
$WCl_6 \cdot PCl_5 \cdot Et_2AlCl$	5-cyanobicyclo[2.2.1]-2-heptene	1564
$WCl_6 \cdot POCl_3 \cdot Et_2AlCl$	idem	1564
$WCl_6 \cdot$ ethylenediaminetetra-acetate Mg-$Na_2 \cdot Et_2AlCl$	idem	1566
$WCl_6 \cdot AlPO_4 \cdot Et_2AlCl$	5-cyano-bicyclo[2.2.1]-2-heptene	1569
$WCl_6 \cdot$ polyvinyl alcohol $\cdot Et_2AlCl$	idem	1570
Co octanoate $\cdot$ phosphoric acid, hexamethyl triamid $\cdot EtAlCl_2$	1,3-pentadiene	1406
$WCl_6 \cdot Et_3N \cdot$ dichloroalane	cyclopentene	1594
$WCl_6 \cdot Et_3N \cdot Et_2AlCl$	5-(2-pyridyl)-bicyclo[2.2.1]-2-heptene	1634
$W(CO)_5(COMe)Et \cdot Bu_4NCl \cdot MeAlCl_2$	cyclopentene	1611
$WCl_6 \cdot$ Li palmitate $\cdot EtAlCl_2$	cyclo-octene	1622
$WCl_6 \cdot Al(s$-$BuO)_3 \cdot EtAlCl_2$	cyclo-octene	1622
$WCl_6 \cdot Et_3Al \cdot AlCl_3$	cyclopentene	1659
$MoCl_5 \cdot Et_3Al \cdot BF_3$-$Et_2O$	cyclopentene	1659
$WCl_6 \cdot EtAlCl_2 \cdot HCl$	cyclopentene	1658
$TiCl_4 \cdot LiAlEt_3Bu \cdot$ amines	bicyclo[2.2.1]heptene	1051
$TiCl_4 \cdot LiAlEt_3Bu \cdot$ dioxane	idem	1051
$TiCl_4 \cdot LiAlEt_3Bu \cdot$ pyridine	idem	1051
$TiCl_4 \cdot LiAlEt_3Bu \cdot Bu_2S$	idem	1051

1	2	3
TiCl$_4$ · Et$_3$Al · amines	bicyclo[2.2.1]heptene	910, 1050, 1051
TiCl$_4$ · Et$_3$Al · dioxane	idem	1051
TiCl$_4$ · Et$_3$Al · Bu$_2$S	idem	1051
TiCl$_4$ · Et$_3$Al · pyridine	idem	1051
WO$_3$ · AlCl$_3$ · i-Bu$_3$Al + O$_2$	cyclopentene	1892
W$_2$O$_3$ · AlCl$_3$ · i-Bu$_3$Al	cyclopentene + dicyclopentadiene	1960

B.4. Multicomponent homogeneous systems

W(CO)$_4$(1,5-COD) · Y · EtAlCl$_2$ · AlCl$_3$ (Y = O$_2$, Br$_2$, I$_2$, BrCN)	cyclo-octene	1768
W(CO)$_4$(1,5-COD) · Y · EtAlCl$_2$ · AlBr$_3$	cyclo-octene	1768
Mo(CO)$_4$(NBD) · Y · EtAlCl$_2$ · AlCl$_3$	cyclo-octene	1768
Mo(CO)$_4$(NBD) · Y · EtAlCl$_2$ · AlBr$_3$	cyclo-octene	1768
WCl$_6$ · ClCH$_2$CH$_2$OH · I$_2$ · Et$_2$AlCl	cyclopentene	1245
WCl$_6$ · (EtO)$_2$CH$_2$ · Et$_3$Al · Me$_3$SiOH	5-cyano-bicyclo[2.2.1]-2-heptene	1939

TABLE III. Alkene metathesis reactions

Alkene	Catalytic system	References
1	2	3

A. Acyclic alkenes

Alkene	Catalytic system	References
Ethylene	WCl$_6$ · n-OctLi	435
	MoO$_3$ · Al$_2$O$_3$	784
	WO$_3$ · SiO$_2$	1830
	Re$_2$O$_7$ · SiO$_2$	674
	Ta$_2$O$_5$ · SiO$_2$	1208
	Nb$_2$O$_5$ · SiO$_2$	1208
	Mo(CO)$_6$ · Al$_2$O$_3$	830, 1204
	W(CO)$_6$ · SiO$_2$	1204
	MoO$_3$ · CoO · Al$_2$O$_3$	830, 1814
	Mo(CO)$_6$ · SiO$_2$ · Al$_2$O$_3$	1204
	WO$_3$ · SiO$_2$ · Al$_2$O$_2$ · NiO	1830
Ethylene + ethylene-d$_4$	MoO$_3$ · Al$_2$O$_3$	669
Ethylene + propene	WO$_3$ · Al$_2$O$_3$	581
Propene	Al$_2$O$_3$	1420, 516, 1414
	SiO$_2$	1211, 1422
	MgO	739, 1699, 776
	MoO$_3$ · Al$_2$O$_3$	131, 1182, 339, 372, 383, 384, 386, 414, 415, 1519, 552, 678, 679, 750, 758, 779, 782, 784, 1750, 818, 819, 820, 941, 942, 958, 990, 1918, 1919, 1916, 1064, 188 a
	MoO$_3$ · SiO$_2$	372, 375, 396, 450, 451, 539, 678, 750, 758, 780, 781, 782, 941, 942, 989, 1075, 1816

1	2	3
	$Mo_2O_3 \cdot SiO_2$	1206
	$MoO_3 \cdot AlPO_4$	451, 678, 1417
	$MoO_3 \cdot ZrO_2$	1425
	$WO_3 \cdot Al_2O_3$	1385, 542, 581, 678, 679, 748 750, 942, 990, 1918, 1919
	$WO_3 \cdot SiO_2$	159, 1203, 1206, 1218, 1348, 448, 449, 450, 539, 1422, 451, 1428, 1528, 678, 683, 684, 685, 1666, 750, 857, 1816, 990, 1128
	$WO_3 \cdot AlPO_4$	451, 678, 1417, 1427
	$WO_3 \cdot ZrO_2$	451, 1425
	$WO_3 \cdot Ca_3(PO_4)_2$	451
	$WO_3 \cdot Mg_3(PO_4)_2$	451
	$WO_3 \cdot Ti_3(PO_4)_4$	451
	$WO_3 \cdot Zr_3(PO_4)_4$	451
	$Re_2O_7 \cdot Al_2O_3$	372, 386, 739, 749, 750, 787, 827, 869, 948, 958, 1064, 1065, 1066, 1182, 1278, 1286, 1435, 1437, 1454, 1456
	$Re_2O_7 \cdot SiO_2$	451, 750, 1428
	$Re_2O_7 \cdot SnO_2$	1264, 1276
	$Re_2O_7 \cdot TiO_2$	1276
	$Re_2O_7 \cdot$ zeolite	1071, 1072, 1946
	$Re_2O_7 \cdot ZrO_2$	1276
	$Re_2O_7 \cdot ThO_2$	1276
	$Re_2O_7 \cdot Fe_3O_4$	1915
	$Re_2O_7 \cdot CoO$	1915
	$Re_2O_7 \cdot NiO$	1915
	$Re_2O_7 \cdot WO_3$	1915
	$TeO_3 \cdot SiO_2$	451
	$La_2O_3 \cdot Al_2O_3$	1914, 1915
	$Ta_2O_5 \cdot SiO_2$	451, 1208
	$Na_2O_5 \cdot SiO_2$	451, 1208
	$Ru \cdot SiO_2$	1822
	$Ru_2O_3 \cdot SiO_2$	1822
	$SnO_2 \cdot Al_2O_3$	1914, 1915
	$IrO_2 \cdot Al_2O_3$	1914, 1915
	$Rh_2O_3 \cdot Al_2O_3$	1914, 1915
	$OsO_2 \cdot Al_2O_3$	1914, 1915
	$Al_2O_3 \cdot SiO_2$	810, 976a
	$Mo(CO)_6 \cdot Al_2O_3$	131, 1200, 1204, 1275, 206, 207, 283, 284, 285, 480, 678, 1666, 750, 830, 954
	$Mo(CO)_6 \cdot SiO_2$	480, 954, 750, 1666
	$W(CO)_6 \cdot Al_2O_3$	131, 1274, 1275, 678, 1666, 750, 769
	$W(CO)_6 \cdot SiO_2$	750, 1204, 1666
	$Mo(CO)_6 \cdot MgO$	480, 954

1	2	3
	$Re_2(CO)_{10} \cdot Al_2O_3$	1275, 1969, 1970
	$MoS_2 \cdot Al_2O_3$	451, 824
	$MoS_2 \cdot SiO_2$	451
	$MoS_3 \cdot Al_2O_3$	1414, 1417
	$MoS_3 \cdot SiO_2$	1414, 1417
	$WS_2 \cdot Al_2O_3$	451, 1414, 1417, 1485
	$WS_2 \cdot SiO_2$	451, 1414, 1417
	$WS_3 \cdot SiO_2$	451
	$WO_3 \cdot SiO_2$ + vinyl chloride	1807
	$WO_3 \cdot SiO_2$ + propyl chloride	1807
	$WO_3 \cdot SiO_2$ + chlorobenzene	1807
	$WO_3 \cdot SiO_2$ + hydrogen chloride	1807
	$MoO_3 \cdot Al_2O_3$ + Na, K, Re, Cs, Tl, Ba	1415, 1428
	$MoO_3 \cdot Al_2O_3$ + LiOH, NaOH, KOH, RbOH, CsOH	783
	$MoO_3 \cdot Al_2O_3$ + NaCl, NaF, KF, CdF_2, ZnF_2	1180
	$MoO_3 \cdot Al_2O_3$ + O_2	686
	$WO_3 \cdot SiO_2$ + Na_2O	1421, 1422, 1428
	$WO_3 \cdot SiO_2$ + K_2O, Cs_2O, BaO	1428
	$WO_3 \cdot SiO_2$ + Bi	1373
	$WO_3 \cdot SiO_2$ + Nd_2O_3, Pr_2O_3, Sm_2O_3	1221
	$WO_3 \cdot Al_2O_3$ + Cs_2O, Tl_2O	956, 957
	$WO_3 \cdot Al_2O_3$ + Na_2O, K_2O, Rb_2O PbO, BiO, NiO, CoO	956
	$WO_3 \cdot Al_2O_3$ + CO	1213
	$WO_3 \cdot Al_2O_3$ + H_2	1213
	$(\pi\text{-allyl})_4Mo \cdot SiO_2$	358, 361
	$(\pi\text{-allyl})_4W \cdot SiO_2$	358
	$Me_6W \cdot Al_2O_3$	769, 955
	$Me_6W \cdot SiO_2$	769, 955
	$Me_5W \cdot SiO_2$	955
	$(acetate)_6W \cdot SiO_2$, Al_2O_3	955
	$(acetate)_5Mo \cdot SiO_2$, Al_2O_3	955
	$MoO_3 \cdot CoO \cdot Al_2O_3$	71, 110a, 131, 258, 451, 507, 672, 673, 678, 679, 747, 748, 779, 782, 830, 1139, 1163, 1199, 1200, 1201, 1202, 1217, 1346, 1347 1666, 1667, 1814, 1860, 1879, 1970.
	$Mo_2O_3 \cdot CoO \cdot Al_2O_3$	1844
	$Mo_2O_3 \cdot CoO \cdot SiO_2$	1844
	$Mo_2O_3 \cdot CoO \cdot MgO$	1844
	$MoO_3 \cdot Cr_2O_3 \cdot Al_2O_3$	123, 868, 1185
	$MoO_3 \cdot Re_2O_7 \cdot Al_2O_3$	1181, 1182
	$MoO_3 \cdot SiO_2 \cdot P_2O_5$	1661
	$WO_3 \cdot SiO_2 \cdot P_2O_5$	1661
	$WO_3 \cdot SiO_2 \cdot MgO$	1528
	$WO_3 \cdot Al_2O_3 \cdot ThO_2$	451
	$WO_3 \cdot Al_2O_3 \cdot K_2O$	1864
	$WO_3 \cdot TiO_2 \cdot MgO$	1426

1	2	3
	$WO_3 \cdot Al_2O_3 \cdot MeAl(OMe)_2$	940
	$WO_3 \cdot SiO_2 \cdot Pr_3Al$	372
	$WO_3 \cdot Al_2O_3 \cdot Me_3Al$	944
	$MoO_3 \cdot Al_2O_3 \cdot Me_3Al_2Cl_3$	2001
	$MoO_3 \cdot Al_2O_3 \cdot Et_2AlCl$	2001
	$MoO_3 \cdot Al_2O_3 \cdot EtAlCl_2$	2001
	$WO_3 \cdot Al_2O_3 \cdot Et_2AlCl$	2001
	$WO_3 \cdot Al_2O_3 \cdot EtAlCl_2$	2001
	$Re_2O_7 \cdot Al_2O_3 \cdot Et_4Sn$	386
	$Re_2O_7 \cdot Al_2O_3 \cdot Bu_4Sn$	386
	$Re_2O_7 \cdot Al_2O_3 \cdot DCPD$	386
	$V_2O_5 \cdot WO_3 \cdot Me_3Al$	1754
	$Mo(CO)_6 \cdot SiO_2 \cdot Al_2O_3$	954, 1204
	$W(CO)_6 \cdot SiO_2 \cdot Al_2O_3$	1666
	$Re_2O_7 \cdot CoO \cdot Al_2O_3$	1067
	$Re_2O_7 \cdot NiO \cdot Al_2O_3$	1067
	$MoO_3 \cdot CoO \cdot Al_2O_3 + KOH$	1166, 1949
	$WO_3 \cdot Al_2O_3 \cdot MeAlCl_2 \cdot Me_2AlCl$	1996
	$MoO_3 \cdot Al_2O_3 \cdot MeAlCl_2 \cdot Me_2AlCl$	1996
	$WCl_6 \cdot [O(MeO_2C)C_6H_4O]WCl_5 \cdot$ $EtAlCl_2 \cdot PhNH_2$	1965
	$WCl_6 \cdot [O(MeO_2C)C_6H_4O]WCl_5 \cdot$ $Et_3Al_2Cl_3 \cdot PhNH_2$	1965
propene-2-d_1	$Re_2O_7 \cdot Al_2O_3$	751
1-butene	zeolite	139
	$MoO_3 \cdot Al_2O_3$	131, 582, 583, 784, 958
	$MoO_3 \cdot SiO_2$	1428
	$WO_3 \cdot SiO_2$	995, 996
	$Re_2O_7 \cdot Al_2O_3$	325, 827, 948, 958, 1065, 1068, 1273, 1277, 1920
	$Re_2O_7 \cdot$ aluminosilicates	1071
	$Ru \cdot SiO_2$	1822
	$Ru_2O_3 \cdot SiO_2$	1822
	$Mo(CO)_6 \cdot Al_2O_3$	131, 286, 1200, 1204, 1275
	$W(CO)_6 \cdot Al_2O_3$	131, 1275
	$W(CO)_6 \cdot SiO_2$	1204
	$MoS_2 \cdot Al_2O_3$	824
	$MoO_3 \cdot Al_2O_3 + Na$	1913
	$MoO_3 \cdot Al_2O_3 + Na, K, Rb, Cs,$ Tl, Ba	1428
	$MoO_3 \cdot Al_2O_3 + Na_2O, K_2O, Cs_2O,$ BaO	1423
	$MoO_3 \cdot Al_2O_3 + Cs_2O, Tl_2O$	956, 957
	$WO_3 \cdot Al_2O_3 + Na_2O, K_2O, Rb_2O,$ PbO, BiO, NiO, CoO	956
	$WO_3 \cdot SiO_2 + Na_2O$	1428
	$WO_3 \cdot Al_2O_3 + CO$	1213
	$WO_3 \cdot Al_2O_3 + H_2$	1213
	$WO_3 \cdot SiO_2 + Na, K, Cs$	995
	$WO_3 \cdot SiO_2 + F^-$	1823, 1825
	$(\pi\text{-allyl})_4Mo \cdot SiO_2$	1716
	$(\pi\text{-allyl})_4W \cdot SiO_2$	1716

1	2	3
	$MoO_3 \cdot CoO \cdot Al_2O_3$	71, 131, 197, 1199, 1200
	$MoO_3 \cdot Re_2O_7 \cdot Al_2O_3$	1181, 1182
	$WO_3 \cdot SiO_2 \cdot ZnO$	1824
	$WO_3 \cdot SiO_2 \cdot MgO$	1824
	$WO_3 \cdot SiO_2 \cdot NH_3$	1897
	$WO_3 \cdot SiO_2 \cdot EtNH_2$	1896
	$WO_3 \cdot SiO_2 \cdot Et_2NH$	1896
	$WO_3 \cdot SiO_2 \cdot Et_3N$	1896
	$WO_3 \cdot SiO_2 \cdot PrNH_2$	1896
	$WO_3 \cdot SiO_2 \cdot n\text{-}BuNH_2$	1896
	$WO_3 \cdot SiO_2 \cdot i\text{-}BuNH_2$	1896
	$WO_3 \cdot SiO_2 \cdot C_5H_{10}N$	1896
	$WO_3 \cdot SiO_2 \cdot C_5H_5N$	1894
	$WO_3 \cdot SiO_2 \cdot PhNH_2$	1894
	$WO_3 \cdot SiO_2 \cdot CHCl_3$	1894
	$WO_3 \cdot SiO_2 \cdot CCl_4$	1894
	$WO_3 \cdot SiO_2 \cdot PhNR_2$ (R = alkyl)	1894
	$Mo(CO)_6 \cdot SiO_2 \cdot Al_2O_3$	1204
	$MoO_3 \cdot CoO \cdot Al_2O_3 + KOH$	1166
	$MoO_3 \cdot CoO \cdot Al_2O_3 \cdot Et_3Al + K,$ Rb, Cs	1168
	$(\pi\text{-allyl})_4Mo$	1476
	$(\pi\text{-crotyl})_4Mo$	1476
	$(\pi\text{-allyl})_4W$	1476
	$(\pi\text{-crotyl})_4W$	1476
	$WCl_6 \cdot EtAlCl_2$	1053
	$WCl_6 \cdot Et_3Al$	1053
	$EtCHWPhCl_3 \cdot AlCl_3$	405
	$PdBr_2 \cdot PPh_3 \cdot EtAlCl_2$	1817
2-butene	$MoO_3 \cdot SiO_2$	774, 1041
	$Re_2O_7 \cdot Al_2O_3$	1274
	$Mo(CO)_6 \cdot Al_2O_3$	286
	$MoO_3 \cdot Al_2O_3 + Na$	1428
	$Re_2O_7 \cdot Al_2O_3 + K$	1265
	$MoO_3 \cdot CaO \cdot Al_2O_3$	71, 131, 1199, 1200, 1202, 1346
	$MoO_3 \cdot Al_2O_3 \cdot SiO_2$	1437
	$WO_3 \cdot SiO_2 \cdot C_5H_5N$	1894
	$WO_3 \cdot SiO_2 \cdot PhNH_2$	1894
	$Ph_2C = W(CO)_5$	224
	$Tol_2C = W(CO)_5$	225
	$PhWCl_3 \cdot AlCl_3$	402
isobutene	zeolite	628
	zeolite $+ CO_2$	629
	$MoO_3 \cdot Al_2O_3$	784, 1182
	$MoO_3 \cdot Al_2O_3 + F^-$	1182
	$MoO_3 \cdot Al_2O_3 + K$	1428
	$MoO_3 \cdot SiO_2 + Na_2O, K_2O, Cs_2O,$ BaO	1423
	$WO_3 \cdot SiO_2 + Na_2O$	1428
	$MoO_3 \cdot CoO \cdot Al_2O_3$	71
	$Mo_2O_3 \cdot CoO \cdot Al_2O_3$	1844
	$Mo_2O_3 \cdot CoO \cdot SiO_2$	1844
	$Mo_2O_3 \cdot CoO \cdot MgO$	1844

1	2	3
	$WO_3 \cdot SiO_2 \cdot MgO$	1832
	$WO_3 \cdot SiO_2 \cdot C_5H_5N$	1894
	$WO_3 \cdot SiO_2 \cdot PhNH_2$	1894
	$WO_3 \cdot SiO_2 \cdot PhNR_2$ (R = alkyl)	1894
	$Ph_2C=W(CO)_5$	224
	$Tol_2C=W(CO)_5$	225
n-butenes	$WO_3 \cdot SiO_2 \cdot MgO$	1831, 1836
	$H_3(PW_{12}O_{40}) \cdot 5H_2O \cdot Al_2O_3$	1835
	$WO_3 \cdot SiO_2 \cdot MgO + KOH$	1833, 1836
	$Re_2O_7 \cdot Al_2O_3$	948, 1070
	$WO(OPh)_4 \cdot AlCl_3 \cdot Et_3Al$	1543
n-butenes + ethylene	$MoO_3 \cdot Al_2O_3$	2000
	$WO_3 \cdot SiO_2$	2000
n-butenes + propene	$WO_3 \cdot SiO_2$	994
n-butenes + isobutene	$Mo(CO)_6 \cdot SiO_2 + K, Na, Ca, Mg$	1286
	$Mo(CO)_6 \cdot Al_2O_3 + K, Na, Ca, Mg$	1286
	$W(CO)_6 \cdot SiO_2 + K, Na, Ca, Mg$	1286
	$W(CO)_6 \cdot Al_2O_3 + K, Na, Ca, Mg$	1286
	$Re_2(CO)_{10} \cdot SiO_2 + K, Na, Ca, Mg$	1286
	$Re_2(CO)_{10} \cdot Al_2O_3 + K, Na, Ca, Mg$	1286
isobutene + propene	$MoO_3 \cdot Al_2O_3$	553
	$MoO_3 \cdot SiO_2$	773
	$WO_3 \cdot SiO_2$	132, 992
	$MoO_3 \cdot Al_2O_3 + Na, K, Rb, Cs,$ Tl, Ba	1428
	$MoO_3 \cdot Al_2O_3 + LiOH, NaOH,$ KOH, RbOH, CsOH	263
	$WO_3 \cdot SiO_2 + Na_2O$	1428
	$Mo_2O_3 \cdot CoO \cdot Al_2O_3$	1844
	$Mo_2O_3 \cdot CoO \cdot SiO_2$	1844
	$Mo_2O_3 \cdot CoO \cdot MgO$	1844
1-butene + ethylene	$MoO_3 \cdot Al_2O_3$	1880
	$WO_3 \cdot Al_2O_3$	1880
2-butene + ethylene	$WO_3 \cdot SiO_2$	552
	$Re_2O_7 \cdot Al_2O_3$	1274
	$Re_2O_7 \cdot SnO_2$	1264
	$MoO_3 \cdot CoO \cdot Al_2O_3$	197
2-butene + 1-butene	$WO_3 \cdot SiO_2$	1348, 1428, 993
	$Re_2O_7 \cdot Al_2O_3$	758, 759, 1056, 1068, 1673
2-butene + 2-butene-d_8	$MoS_2 \cdot Al_2O_3$	824
butadiene	$MoO_3 \cdot Al_2O_3$	943
	$WO_3 \cdot SiO_2$	133, 452
	Al_2O_3	943
butadiene + propene	$WO_3 \cdot SiO_2$	452
butadiene + isobutene	$WO_3 \cdot SiO_2$	452
	$WO_3 \cdot SiO_2 + Na_2O$	1421
butadiene + 2-butene	$WO_3 \cdot SiO_2$	452
	$WO_3 \cdot SiO_2 + Na_2O$	1421
1-butene + 2-butene + isobutene	$Re_2O_7 \cdot Al_2O_3$	1671
	$Re_2O_7 \cdot Al_2O_3 + Na$	1672
1-butene + isobutene	$WO_3 \cdot SiO_2$	993
	$BaO \cdot AgO \cdot Al_2O_3 + SiO_2 + Na_2O$	1736
	$SrO_2 \cdot AgO \cdot Al_2O_3 + SiO_2 + Na_2O$	1736

1	2	3
2-butene + isobutene	$MoO_3 \cdot Al_2O_3$	778, 907, 908, 993, 1182
	$WO_3 \cdot SiO_2$	1172, 1173, 778, 787, 1182, 1714
	$Re_2O_7 \cdot Al_2O_3$	1794
	$Re_2O_7 \cdot Al_2O_3 + F^-$	1794
	$Re_2O_7 \cdot Al_2O_3 + SO_4^{2-}$	1794
	$Re_2O_7 \cdot Al_2O_3 + PO_4^{2-}$	1794
	$MoO_3 \cdot Al_2O_3 + Na, K, Rb, Cs, Tl, Ba$	1428
	$MoO_3 \cdot Al_2O_3 + LiOH, NaOH, KOH, RbOH, CsOH$	783
	$MoO_3 \cdot SiO_2 + Na_2O$	1428
	$WO_3 \cdot Al_2O_3 + K$	1858
	$Re_2O_7 \cdot Al_2O_3 \cdot (NH_4)_2HPO_4$	1655
	$Re_2O_7 \cdot Co_2O_3 \cdot Al_2O_3$	1715
	$Re_2O_7 \cdot Al_2O_3 \cdot SnO_2$	1795
	$Re_2O_7 \cdot Co_2O_3 \cdot Al_2O_3 \cdot TiO_2$	1715
	$Re_2O_7 \cdot Co_2O_3 \cdot Al_2O_3 \cdot SiO_2$	1715
	$Re_2O_7 \cdot Al_2O_3 \cdot La_2O_3 \cdot Pr_6O_{11} \cdot Sm_2O_3 \cdot Gd_2O_3 \cdot Ge_2O_3 \cdot Y_2O_3$	1950
2-methyl-1-butene	$MoO_3 \cdot CoO \cdot Al_2O_3$	1199, 1202
2-methyl-2-butene	$MoO_3 \cdot CoO \cdot Al_2O_3$	1202
	$W(CO)_5MeCN$	421
3-methyl-1-butene	$MoO_3 \cdot CoO \cdot Al_2O_3$	1199, 1202
	$Mo(CO)_6 \cdot Pr_4NCl \cdot Me_3Al_2Cl_3$	1839
	$Mo(CO)_6 \cdot Bu_4NCl \cdot Me_3Al_2Cl_3$	1839
	$Mo(CO)_4Py_2 \cdot Pr_4NCl \cdot Me_3Al_2Cl_3$	1839
	$Mo(CO)_4Py_2 \cdot Pr_4NCl \cdot MeAlCl_2$	1839
	$WO(OPh)_4 \cdot AlCl_3 \cdot Et_3Al$	1543
1-pentene	$MoO_3 \cdot Al_2O_3$	131
	$Re_2O_7 \cdot Al_2O_3$	975
	$Mo(CO)_6 \cdot Al_2O_3$	131, 1200, 1204, 1275
	$W(CO)_6 \cdot Al_2O_3$	131, 1275
	$W(CO)_6 \cdot SiO_2$	1204
	$Re_2(CO)_{10} \cdot Al_2O_3$	1275
	$MoO_3 \cdot CoO \cdot Al_2O_3$	78, 131, 1199, 1200, 1202
	$MoO_3 \cdot Al_2O_3 \cdot Me_3Al_2Cl_3$	2001
	$MoO_3 \cdot Al_2O_3 \cdot EtAlCl_2$	2001
	$MoO_3 \cdot Al_2O_3 \cdot Et_2AlCl$	2001
	$WO_3 \cdot Al_2O_3 \cdot EtAlCl_2$	2001
	$WO_3 \cdot Al_2O_3 \cdot Et_2AlCl$	2001
	$Mo(CO)_6 \cdot SiO_2 \cdot Al_2O_3$	1204
	$MoO_3 \cdot Al_2O_3 \cdot MeAlCl_2 \cdot Me_2AlCl$	1996
	$WO_3 \cdot Al_2O_3 \cdot MeAlCl_2 \cdot Me_2AlCl$	1996
	$Mo(CO)_5MeCN$	421
	$Tol_2C = W(CO)_5$	225
	$WCl_6 \cdot EtAlCl_2$	1053
	$WCl_6 \cdot Et_3SiH$	796
	$WCl_6 \cdot (MeSiH)_2O$	796
	$WCl_6 \cdot (MeSiH)_2NH$	796
	$WCl_6 \cdot Ph_3SiH$	796
	$WCl_6 \cdot Et_3Al$	1053, 1054

1	2	3
	$WCl_6 \cdot LiAlH_4$	900
	$WCl_2(NO)_2(PPh_3)_2 \cdot Me_3Al_2Cl_3$	1157, 1158
	$WCl_2(NO)_2(PPh_3)_2 \cdot EtAlCl_2$	1157, 1158
	$EtCH{=}WPhCl_3 \cdot AlCl_3$	405
	$MoCl_5 \cdot Me_3Al_2Cl_3$	1158
	$MoCl_5 \cdot Et_2AlCl$	1924
	$MoCl_5 \cdot Et_3Al$	1924
	$MoOCl_3 \cdot Me_3Al_2Cl_3$	1158
	$Mo(acac)_3 \cdot Et_3Al$	1924
	$Mo(acac)_3 \cdot Me_3Al_2Cl_3$	1158
	$MoCl_2(NO)_2(PPh_3)_2 \cdot Me_3Al_2Cl_3$	1157, 1158
	$MoCl_2(NO)_2(PPh_3)_2 \cdot EtAlCl_2$	1157, 1158
	$MoCl_2(NO)_2(PPh_3)_2 \cdot Et_2AlCl$	1157, 1158
	$MoCl_2(NO)_2(PPh_3)_2 \cdot Et_3Al_2Cl_3$	1157, 1158
	$MoCl_3(PhCO_2) \cdot Me_3Al_2Cl_3$	1158
	$MoCl_5Py_x \cdot Me_3Al_2Cl_3$	1158
	$MoCl_5(OPPh_3)_x \cdot Me_3Al_2Cl_3$	1158
	$MoCl_5(n\text{-}Oct_3P)_x \cdot Me_3Al_2Cl_3$	1158
	$MoCl_5(n\text{-}Bu_3P)_x \cdot Me_3Al_2Cl_3$	1158
	$Bu_4N[(CO)_5Mo{-}Mo(CO)_5] \cdot MeAlCl_2$	614, 1612
	$Na[(CO)_5Mo{-}Mo(CO)_5] \cdot MeAlCl_2$	614, 1612
	$Bu_4N[(CO)_5W{-}W(CO)_5] \cdot MeAlCl_2$	614
	$Bu_4N[(CO)_5Mo{-}Re(CO)_5] \cdot MeAlCl_2$	614
	$Bu_4N[(CO)_5Mo{-}Mn(CO)_5] \cdot MeAlCl_2$	614
	$Bu_4N[Mo(CO)_5Cl] \cdot MeAlCl_2$	320, 1357
	$Bu_4N[W(CO)_5Cl] \cdot MeAlCl_2$	320, 1357
	$Bu_4N[Cr(CO)_5Cl] \cdot MeAlCl_2$	320, 1357
	$Bu_4N[Mo(CO)_5COR] \cdot MeAlCl_2$ $(R = Me, Ph)$	613, 1611
	$Bu_4N[W(CO)_5COR] \cdot MeAlCl_2$ $(R = Me, Ph)$	613, 1611
	$EtW(CO)_5(COMe) \cdot MeAlCl_2$	613
	$MeW(CO)_5(CNMe_2) \cdot MeAlCl_2$	613
	$ReCl_5 \cdot Et_2AlCl + O_2$	1925
	$ReCl_5 \cdot Et_3Al + O_2$	1060, 1925
	$WCl_6 \cdot Et_3Al + O_2$	1060, 1925
	$MoCl_5 \cdot Et_3Al + O_2$	1060
	$ReCl_5 \cdot Et_3Al$	1925
	$ReCl_5 \cdot Et_2AlCl$	1925
	$RhCl(NO)_2 \cdot Me_3Al_2Cl_3$	1469
	$WCl_6 \cdot CH_3COOR \cdot Bu_4Sn$	508
	$WCl_6 \cdot CH_3CN \cdot Bu_4Sn$	508
	$Bu_4NCl \cdot W(CO)_5(COMe)Et \cdot MeAlCl_2$	613, 1611
	$Bu_4NCl \cdot W(CO)_5(CNMe_2)Me \cdot$ $MeAlCl_2$	613
	$Mo(CO)_6 \cdot Pr_4NCl \cdot Me_3Al_2Cl_3$	1839
	$Mo(CO)_6 \cdot Bu_4NCl \cdot Me_3Al_2Cl_3$	1839
	$Mo(CO)_4Py_2 \cdot Pr_4NCl \cdot Me_3Al_2Cl_3$	1839
	$Mo(CO)_6 \cdot Pr_4NCl \cdot MeAlCl_2$	1839
	$Mo(CO)_4Py_2 \cdot Pr_4NCl \cdot MeAlCl_2$	1839
	$MoCl_5 \cdot Py \cdot Me_3Al_2Cl_3 + NO$	1158
	$MoCl_5 \cdot Ph_3PO \cdot Me_3Al_2Cl_3 + NO$	1158
	$MoCl_5 \cdot n\text{-}Oct_3PO \cdot Me_3Al_2Cl_3 + NO$	1158
	$MoCl_5 \cdot n\text{-}Bu_3P \cdot Me_3Al_2Cl_3 + NO$	1158

1	2	3
	$WCl_6 \cdot [O(MeO_2C)C_6H_4O]WCl_5 \cdot$	
	$EtAlCl_2 \cdot PHNH_2$	1965
	$WCl_6 \cdot [O(MeO_2C)C_6H_4O]WCl_6 \cdot$	
	$Et_3Al_2Cl_3 \cdot PhNH_2$	1965
2-pentene	$MoO_3 \cdot Al_2O_3$	131, 517
	$WO_3 \cdot SiO_2$	1205, 1219, 1424
	$Mo(CO)_6 \cdot Al_2O_3$	131, 1200
	$W(CO)_6 \cdot Al_2O_3$	131, 1275
	$WCl_6 \cdot Al_2O_3$	1169
	$WCl_6 \cdot SiO_2$	1169
	$MoO_3 \cdot CoO \cdot Al_2O_3$	78, 131, 517, 1199, 1202
	$MoO_3 \cdot SiO_2 \cdot Me_3Al$	940
	$WO_3 \cdot SiO_2 \cdot MgO$	1205, 1216
	$WO_3 \cdot SiO_2 \cdot Bu_3P$	1416, 1424
	$WO_3 \cdot SiO_2 \cdot Bu_3As$	1424
	$WO_3 \cdot SiO_2 \cdot Bu_3Sb$	1424
	$WO_3 \cdot SiO_2 \cdot Bu_3Bi$	1424
	$WCl_6 \cdot Al_2O_3 \cdot EtAlCl_2$	1269
	$WCl_6 \cdot Al_2O_3 \cdot n\text{-}Bu_3Al$	1269
	$WCl_6 \cdot Al_2O_3 \cdot i\text{-}Bu_3Al$	1269
	$WCl_6 \cdot Al_2O_3 \cdot Bu_4Sn$	1269
	$WCl_6 \cdot SiO_2 \cdot EtAlCl_2$	1269
	$WCl_6 \cdot SiO_2 \cdot n\text{-}Bu_3Al$	1269
	$WCl_6 \cdot SiO_2 \cdot i\text{-}Bu_3Al$	1269
	$WCl_6 \cdot SiO_2 \cdot Bu_4Sn$	1269
	$WCl_6 \cdot SiO_2 \cdot RMgX \ (R = C_3-C_{10})$	1852
	$WCl_6 \cdot SiO_2 \cdot i\text{-}Bu_2AlCl$	1269
	$WCl_6 \cdot SiO_2 \cdot n\text{-}BuLi$	935
	$MoCl_5 \cdot SiO_2 \cdot n\text{-}BuLi$	935
	$MoCl_5 \cdot SiO_2 \cdot C_5H_{11}MgBr$	935
	$MoCl_5 \cdot SiO_2 \cdot EtAlCl_2$	935
	$MoCl_5 \cdot SiO_2 \cdot MgBr_2$	935
	$MoCl_5 \cdot SiO_2 \cdot TiCl_4$	935
	$WCl_6 \cdot n\text{-}BuLi \cdot AlCl_3 \cdot$ polyethylene	934
	$WCl_6 \cdot BuMgCl \cdot AlCl_3 \cdot$ polyethylene	934
	$WCl_6 \cdot C_5H_{11}MgCl \cdot AlCl_3 \cdot$ poly-ethylene	934
	$WCl_6 \cdot n\text{-}BuLi \cdot SbCl_3 \cdot$ polyethylene	934
	$WCl_6 \cdot n\text{-}BuLi \cdot SnCl_2 \cdot$ polyethylene	934
	$WCl_6 \cdot n\text{-}BuLi \cdot TiCl_4 \cdot$ polyethylene	934
	$WCl_6 \cdot n\text{-}BuLi \cdot ZnCl_2 \cdot$ polyethylene	934
	$WCl_6 \cdot n\text{-}BuLi \cdot AlCl_3 \cdot$ polyisoprene	934
	$WCl_6 \cdot n\text{-}BuLi \cdot AlCl_3 \cdot$ polycyclo-pentadiene	934
	$MoCl_5 \cdot n\text{-}BuLi \cdot AlCl_3 \cdot$ polyethylene	934
	$WO_3 \cdot Al_2O_3 \cdot MeAlCl_2 \cdot Me_2AlCl$	1996
	$MoO_3 \cdot Al_2O_3 \cdot MeAlCl_2 \cdot Me_2AlCl$	1996
	$WO_3 \cdot Al_2O_3 \cdot SiO_2 \cdot Et_2AlCl_2 \cdot$ $Et_2AlCl \cdot AcOH$	1874
	$PhWCl_3$	832
	$Ph_2C = W(CO)_5$	224
	$W(CO)_6$	79, 149, 606, 607, 670
	$W(CO)_4Cl_2$	80

1	2	3
	$W(CO)_5MeCN$	421
	$Ph(MeCO)C=W(CO)_5$	556
	$PhMeC=W(CO)_5$	226
	$WOCl_4$	745
	$Mo(CO)_6$	670
	$TolMo(CO)_2$	670
	$WCl_6 \cdot EtAlCl_2$	1360, 1361, 473, 474, 476, 602, 745, 770, 771, 1052
	$WCl_6 \cdot Et_3Al$	1052, 1925
	$WCl_6 \cdot Et_3Al_2Cl_3$	732
	$WCl_6 \cdot n\text{-}BuLi$	770, 1078, 1106, 1108
	$WCl_6 \cdot Bu_4Sn$	903, 1266
	$WCl_6 \cdot Ph_4Sn$	1077
	$WCl_6 \cdot Ph_4Pb$	1077
	$WCl_6 \cdot Et_2Zn$	745, 770, 771
	$WCl_6 \cdot Ph_3Sb$	1077
	$WCl_6 \cdot Ph_3Bi$	1077
	$WBr_5 \cdot EtAlCl_2$	1102
	$WCl_6 \cdot n\text{-}PrLi$	1078
	$WCl_6 \cdot LiAlH_4$	246, 900
	$WCl_6 \cdot Et_3SiH$	796
	$WCl_6 \cdot Ph_3SiH$	796
	$WCl_6 \cdot (MeSiH)_2O$	796
	$WCl_6 \cdot (MeSiH)_2NH$	796
	$WOCl_4 \cdot EtAlCl_2$	745
	$WOCl_4 \cdot Et_2Zn$	745
	$WOCl_4 \cdot n\text{-}BuLi$	745
	$WOCl_2 \cdot EtAlCl_2$	745
	$WCl_6 \cdot Ph_3SnCl$	1077
	$WCl_6 \cdot Ph_2SnCl_2$	1077
	$WCl_6 \cdot Ph_6Sn_2$	1077
	$WCl_6 \cdot Ph_3PbCl$	1077
	$WCl_6 \cdot Ph_6Pb_2$	1077
	$WCl_6 \cdot Ph_2Hg$	1077
	$WCl_6 \cdot BuMgCl$	577, 701, 703, 704
	$WCl_6 \cdot BuMgBr$	577, 701, 703, 704
	$WCl_6 \cdot PeMgCl$	577
	$WCl_6 \cdot PeMgBr$	577
	$WBz_4 \cdot AlCl_3$	831
	$WBz_4 \cdot AlX_3 \ (X= Cl, Br)$	167
	$WBz_3Cl \cdot AlX_3$	167
	$WPhCl_3 \cdot AlCl_3$	167
	$WPhCl_3 \cdot AlBr_3$	167, 846
	$WBz_4 \cdot EtAlCl_2$	167
	$WBz_3Cl \cdot EtAlCl_2$	167
	$WPhCl_3 \cdot EtAlCl_2$	167
	$WCl_6 \cdot Me_2Si\underset{CH_2}{\overset{CH_2}{<\,>}}SiMe_2$	797
	$WCl_6 \cdot (CH_2)_3SiMe_2$	797
	$WCl_6 \cdot (CH_2)_3SiNH$	797
	$WCl_6 \cdot (CH_2)_3SiMeO$	797
	$WCl_4 \cdot AlCl_3$	1873

1	2	3
	$MoCl_4 \cdot AlBr_3$	1873
	$W(CO)_5(PPh_3) \cdot EtAlCl_2$	878
	$W(CO)_6 \cdot EtAlCl_2$	181
	$W(CO)_5P(OPh)_3 \cdot EtAlCl_2$	182
	$W(CO)_5PBu_3 \cdot EtAlCl_2$	149, 181
	$W(CO)_5PPh_3 \cdot EtAlCl_2$	147, 181
	$W(CO)_5Py_2 \cdot EtAlCl_2$	182
	$WCl_2(CO)_3(AsPh_3)_2 \cdot AlCl_3$	168, 1014
	$WCl_2(CO)_3(PPh_3)_2 \cdot AlCl_3$	168
	$WCl_2(CO)_3(SbPh_3)_2 \cdot AlCl_3$	168
	$WCl_2(CO)_3(PPh_3)_2 \cdot EtAlCl_2$	1251
	$W(CO)_5PPh_3 \cdot EtAlCl_2 + O_2$	146, 181
	$Mo(CO)_5PPh_2CH_2R \cdot EtAlCl_2 + O_2$	148, 181
	$WCl_4Py_2 \cdot EtAlCl_2 + CO$	164
	$WCl_4(PPh)_2 \cdot EtAlCl_2 + CO$	164
	$WCl_2 \cdot (C_2H_4)(PPh_3)_2 \cdot EtAlCl_2 + CO$	164
	$MoCl_4Py_2 \cdot EtAlCl_2 + CO$	164
	$MoCl_4(PPh_3)_2 \cdot EtAlCl_2 + CO$	164
	$MoCl_4(C_2H_4)(PPh_3)_3 \cdot EtAlCl_2 + CO$	164
	$MoCl_5 \cdot Ph_4Pb$	1077
	$MoCl_5 \cdot Ph_4Sn$	1077
	$MoCl_3(NO) \cdot EtAlCl_2$	1016, 1018
	$MoCl_3(NO)(OPPh_3)_2 \cdot EtAlCl_2$	1016, 1017, 1018, 1019
	$MoCl_2(NO)_2(OPPh_3)_2 \cdot EtAlCl_2$	1016
	$MoCl_2(NO)_2 \cdot EtAlCl_2$	1016, 1017
	$MoCl_2(NO)_2 \cdot n\text{-}BuLi$	1017
	$MoCl_2(NO)_2 \cdot AlBr_3$	1017
	$MoCl_2(NO)_2(OPPh_3)_2 \cdot n\text{-}BuLi$	1017
	$MoCl_2(NO)_2(OPPh_3)_2 \cdot AlBr_3$	1017
	$MoCl_2(NO)_2(EtCN)_2 \cdot n\text{-}BuLi$	1017
	$MoCl_2(NO)_2(EtCN)_2 \cdot AlBr_3$	1017
	$MoCl_3(NO)(OPPh_3)_2 \cdot AlCl_3$	1019
	$MoCl_3(NO)(OPPh_3)_2 \cdot AlEt_3$	1019
	$MoCl_2(NO)_2Py_2 \cdot AlCl_3$	1019
	$MoCl_2(NO)_2Py_2 \cdot Me_3Al_2Cl_3$	486
	$MoCl_2(NO)_2Py_2 \cdot EtAlCl_2$	485, 1019
	$MoCl_2(NO)_2Py_2 \cdot Et_3Al_2Cl_3$	486
	$MoCl_2(NO)_2(EtCN)_2 \cdot EtAlCl_2$	1017
	$Mo(N_2)(PPh_3)_2(PhMe) \cdot AlCl_3$	1926
	$Mo(CP)(PPh_3)_2 \cdot AlCl_3$	1926
	$MoCl_4(PPh_3)_2 \cdot EtAlCl_2$	1091
	$MoCl_3(PPh_3)_2 \cdot EtAlCl_2$	1091
	$MoCl_4(PPh_3)_2 \cdot EtAlCl_2 + CO$	1091
	$Bu_4N[Mo(CO)_5COR] \cdot MeAlCl_2$ (R = Me, Ph)	613, 1611
	$Bu_4N[W(CO)_5COR] \cdot MeAlCl_2$ (R = Me, Ph)	613, 1611
	$MoCl_2(NO)_2Py_2 \cdot AlEt_3$	1019
	$MoCl_3(NO)Py_2 \cdot EtAlCl_2$	1018
	$MoCl_3(NO)(MeCN) \cdot EtAlCl_2$	1018
	$MoCl_3(NO)(DMF) \cdot EtAlCl_2$	1018
	$MoCl_3(NO)(DMSO) \cdot EtAlCl_2$	1018
	$MoCl_3(NO)(HMPT) \cdot EtAlCl_2$	1018
	$MoCl_5(NO)Py_2 \cdot EtAlCl_2$	1018

1	2	3
	MoCl (salen) (NO) · EtAlCl$_2$	1018
	WO$_3$ · EtAlCl$_2$	745
	ReCl$_5$ · Et$_2$AlCl + O$_2$	1925
	ReCl$_5$ · Et$_3$Al + O$_2$	1060, 1925
	WCl$_6$ · Et$_3$Al + O$_2$	1060, 1925
	MoCl$_5$ · Et$_3$Al + O$_2$	1060
	ReCl$_5$ · Et$_3$Al	1925
	ReCl$_3$ · Et$_2$AlCl	1925
	ReCl$_5$ · Bu$_4$Sn	767
	Re(CO)$_5$Cl · EtAlCl$_2$	363
	Re(CO)$_5$Br · EtAlCl$_2$	363
	SmCl$_3$ · EtAlCl$_2$	1541
	SmCl$_3$ · Me$_3$Al$_2$Cl$_3$	1541
	Cu$_2$Cl$_2$(PPh$_3$) · EtAlCl$_2$	1616
	Zr(acac)$_4$ · Me$_3$Al$_2$Cl$_3$	1615
	RhCl(NO)$_2$ · Me$_3$Al$_2$Cl$_3$	1469
	WCl$_6$ · EtOH · EtAlCl$_2$	210, 213, 744, 770, 1291
	WOCl$_4$ · C$_6$H$_4$(CN)$_2$ · EtAlCl$_2$	1399
	WCl$_6$ · BuMgCl · AlCl$_3$	1530
	NbCl$_5$ · BzOH · Me$_3$Al$_2$Cl$_3$	1471
	NbCl$_5$ · BzOH · Me$_3$Al$_2$Cl$_3$ + NO	1470
	VCl$_5$ · BzOH · Me$_3$Al$_2$Cl$_3$ + NO	1470
	WCl$_6$ · EtOH · i-Bu$_2$AlCl	1291
	WCl$_6$ · EtOH · i-Bu$_3$Al	1291
	WCl$_6$ · H$_2$O · i-Bu$_2$AlCl	1291
	WCl$_6$ · AcOH · EtAlCl$_2$	1680
	Bu$_4$NCl · W(CO)$_5$ · (COMe) Et · MeAlCl$_2$	1611
	WCl$_6$ · PhNH$_2$Et$_3$Al$_2$Cl$_3$	731
	W(CO)$_5$ · PPh$_3$ · Et$_2$O · EtAlCl$_2$ + O$_2$	1719
	W(OPh)$_4$ · AlCl$_3$ · Et$_3$Al	1543
	ReCl$_4$ · PPh$_3$ · EtAlCl$_2$	1538
	ReOCl$_3$ · PPh$_3$ · EtAlCl$_2$	1538
	ReOBr$_3$ · PPh$_3$ · EtAlCl$_2$	1538
	MnCl$_4$ · PPh$_3$ · EtAlCl$_2$	1538
	TeCl$_4$ · PPh$_3$ · EtAlCl$_2$	1538
	WCl$_6$ · polyethylene · BuMgCl · AlCl$_3$	1530
	PdBr$_2$ · PPh$_3$ · EtAlCl$_2$	1817
n-pentenes	MoO$_3$ · Al$_2$O$_3$	1219
	MoO$_3$ · AlPO$_4$	1219
	WO$_3$ · Al$_2$O$_3$	1219
	WO$_3$ · AlPO$_4$	1219
	Re$_2$O$_7$ · Al$_2$O$_3$	948
	ReO$_4$ · Al$_2$O$_3$	1219
	ReO$_4$ · AlPO$_4$	1219
n-pentenes + ethylene	WO$_3$ · SiO$_2$	1218, 1219, 1220
	MoO$_3$ · AlPO$_4$	1219
	WO$_3$ · SiO$_2$ · MgO	1219
2-methyl-2-pentene	WCl$_6$ · EtOH · EtAlCl$_2$	1291
	WCl$_6$ · EtOH · i-Bu$_3$Al	1291
	WCl$_6$ · EtOH · i-Bu$_2$Al	1291
1-pentene + 2-butene	WCl$_4$[HOCH(CH$_2$Cl)$_2$]$_2$ · i-Bu$_2$AlCl	309
1-pentene + 2-pentene	WCl$_6$ · PrMgX (X = Cl, Br)	702
	WCl$_6$ · BuMgX (X = Cl, Br)	702

1	2	3
	$WCl_6 \cdot PeMgX$ (X= Cl, Br)	702
	$WCl_6 \cdot HexMgX$ (X = Cl, Br)	702
	$WCl_6 \cdot OctMgX$ (X = Cl, Br)	702
	$WCl_6 \cdot Dec\ MgX$ (X = Cl, Br)	702
	$WCl_6 \cdot PhMgX$ (X = Cl, Br)	702
	$WCl_6 \cdot i\text{-}BuMgX$ (X=Cl, Br)	702
3-methyl-1-pentene	$Bu_4N[Mo(CO)_5X] \cdot MeAlCl_2$	
	(X = Cl, Br)	320
	$Bu_4N[Cr(CO)_5X] \cdot MeAlCl_2$	
	(X = Cl, Br)	320
	$Bu_4N[W(CO)_5X] \cdot MeAlCl_2$	
	(X = Cl, Br)	320
4-methyl-1-pentene	$Bu_4N[Mo(CO)_5COR] \cdot MeAlCl_2$	
	(R = Me, Ph)	613, 1611
	$Bu_4N[W(CO)_5COR] \cdot MeAlCl_2$	
	(R = Me, Ph)	613, 1611
2,2,4-trimethyl-2-pentene + ethylene	$WO_3 SiO_2 \cdot MgO + CO$	1711
1,3-pentadiene	$Ph_2C{=}W(CO)_5$	585
1,3-pentadiene + ethylene	$Ph_2O \cdot Al_2O_3 \cdot R_4Sn$	1227
1-hexene	$MoO_3 \cdot Al_2O_3$	131, 1856
	$Mo(CO)_6 \cdot Al_2O_3$	131, 1200, 1275
	$W(CO)_6 \cdot Al_2O_3$	131, 1275
	$W(CO)_6 \cdot SiO_2$	1204
	$Re_2(CO)_{10} \cdot Al_2O_3$	1275
	$(\pi\text{-allyl})_4Mo \cdot SiO_2$	1716
	$(\pi\text{-allyl})_4W \cdot SiO_2$	1716
	$(\pi\text{-allyl})_4W \cdot$ aluminosilicates	833
	$MoO_3 \cdot CoO \cdot Al_2O_3$	131, 1199, 1200, 1202
	$MoO_3 \cdot Cr_2O_3 \cdot Al_2O_3$	868
	$MoO_3 \cdot Al_2O_3 \cdot Et_2AlOSiMe$	1981
	$Mo(CO)_6 \cdot SiO_2 \cdot Al_2O_3$	1204
	$(\pi\text{-allyl})_6Re_2$	1300
	$(\pi\text{-allyl})_4Mo_2$	1300
	$(\pi\text{-crotyl})_4Mo_2$	1300
	$(\pi\text{-allyl})_4Mo$	1476
	$(\pi\text{-allyl})_4W$	1476
	$(\pi\ crotyl)_4Mo$	1476
	$(\pi\text{-crotyl})_4W$	1476
	$WCl_6 \cdot EtAlCl_2$	488
	$WCl_6 \cdot Bu_4Sn$	176
	$WCl_6 \cdot (Me_3SiCH_2)_4Sn$	176
	$WCl_6 \cdot Et_3SiH$	796
	$WCl_6 \cdot (MeSiH)_2O$	796
	$WCl_6 \cdot (Me_3SiH)_2NH$	796
	$WCl_6 \cdot R_2Sn\overset{SiMe_2}{\underset{SiMe_2}{<\quad>}}O$(R=Me, Et, Pr, Bu)	124
	$WCl_6 \cdot H_2C\overset{GeMe_2}{\underset{GeMe_2}{<\quad>}}CH_2$	125
	$WCl_6 \cdot Me_2Ge\overset{CH_2}{\underset{CH_2}{<\quad>}}SiMe_2$	125

1	2	3
	$WCl_6 \cdot$ i-$Bu_2Al(OAli-Bu)_{12}OAli-Bu_2$	250, 251
	$WCl_6 \cdot Me_2Si\langle\square\rangle SiMe_2$	797
	$WCl_6 \cdot \langle\square\rangle SiMe_2$	797
	$WCl_6 \cdot \langle\square\rangle SiNHMe$	797
	$WCl_6 \cdot \langle\square\rangle SiOMe$	797
	$EtWPhCl_3 \cdot AlCl_3$	405
	$RhCl(NO)_2 \cdot Me_3Al_2Cl_3$	1469
	$WCl_6 \cdot Et_2O \cdot R_2AlO(AlR)_nOAlR_n$	1595
	$WCl_6 \cdot Et_3N \cdot R_2AlO(AlR)_nOAlR_n$	1595
	$WCl_6 \cdot CH_3COOR \cdot Bu_4Sn$	508
	$WCl_6 \cdot CH_3CN \cdot Bu_4Sn$	508
2-hexene	$WCl_6 \cdot Et_3SiH$	796
	$WCl_6 \cdot Ph_2SiH$	796
	$WCl_6 \cdot (MeSiH)_2O$	796
	$WCl_6 \cdot (MeSiH)_2NH$	796
	$WCl_6 \cdot Et_3Al$	1925
	$W(CO)_6 \cdot$ i-$BuAlCl_2$	1110
	$W(CO)_3Cl_2 \cdot (PPh_3)_2 \cdot EtAlCl_2$	1252
	$ReCl_5 \cdot Et_3Al$	1925
	$ReCl_5 \cdot Et_2AlCl$	1925
	$ReCl_5 \cdot Et_3Al + O_2$	1925
	$ReCl_5 \cdot Et_2AlCl + O_2$	1060, 1925
	$WCl_6 \cdot Et_3Al + O_2$	1060, 1925
	$MoCl_5 \cdot Et_3Al + O_2$	1060, 1925
	$MoCl_6 \cdot Et_4N \cdot MgBr_2$	1252
	$CoCl_2(PPh_3)_2 \cdot EtMgBr$	1252
	$WCl_6 \cdot EtOH \cdot EtAlCl_2$	1291
	$WCl_6 \cdot EtOH \cdot$ i-Bu_2AlCl	1291
	$WCl_6 \cdot EtOH \cdot$ i-Bu_3Al	1291
	$Mo(CO)_6 \cdot Pr_4NCl \cdot Me_3Al_2Cl_3$	1839
	$Mo(CO)_6 \cdot Bu_4NCl \cdot Me_3Al_2Cl_3$	1839
	$Mo(CO)_6 \cdot Pr_4NCl \cdot MeAlCl_2$ (Bu_4NCl)	1839
	$Mo(CO)_4Py_2 \cdot Pr_4NCl \cdot Me_3Al_2Cl_3$ (Bu_4NCl)	1839
	$Mo(CO)_4Py_2 \cdot Pr_4NCl \cdot MeAlCl_2$ (Bu_4NCl)	1839
	$W(OPh)_4 \cdot AlCl_3 \cdot Et_3Al$	1543
	π-C_5H_5 $W(CO)_3GeMe_3 \cdot$ i$PrAlCl_2 +$ $+ O_2$	1112
1,5-hexadiene	$WCl_6 \cdot [HOCH(CH_2Cl)_2] \cdot Et_3Al_2Cl_3$	280
neohexene	$WO_3 \cdot SiO_2 \cdot MgO$	1146
	$(\pi$-allyl$)_4$ Mo	1476
	$(\pi$-allyl$)_4W$	1476
	$(\pi$-allyl$)_4Re$	1476
	$(\pi$-allyl$)_4Mo \cdot SiO_2(Al_2O_3)$	1716
	$(\pi$-allyl$)_4W \cdot SiO_2(Al_2O_3)$	1716
n-hexene + ethylene	$WO_3 \cdot SiO_2 \cdot MgO$	1216
1-hexene + 1-pentene-d_{10}	$WCl_6 \cdot Et_2Zn$	744

1	2	3
1-heptene	$MoO_3 \cdot Al_2O_3$	141, 142, 143, 144, 386
	$Re_2O_7 \cdot Al_2O_3$	141, 144, 386
	$Re_2(CO)_{10} \cdot Al_2O_3$	1275
	$Re_2O_7 \cdot Al_2O_3 \cdot Et_4Sn$	386
	$Re_2O_7 \cdot Al_2O_3 \cdot Bu_4Sn$	386
	$Re_2O_7 \cdot Al_2O_3 \cdot R_4Sn$	386, 1225
	$Re_2O_7 \cdot Al_2O_3 \cdot$ oxasilacyclobutane	1224
	$WCl_6 \cdot EtAlCl_2$	488
	$WCl_6 \cdot Et_3SiH$	796
	$WCl_6 \cdot Ph_2SiH$	796
	$WCl_6 \cdot (MeSiH)_2O$	796
	$WCl_6 \cdot (MeSiH)NH$	796
	$WCl_6 \cdot CH_3COOR \cdot Bu_4Sn$	508
	$WCl_6 \cdot CH_3CN \cdot Bu_4Sn$	508
	$MoO_3 \cdot Al_2O_3 \cdot Et_4Sn$	386
	$MoO_3 \cdot Al_2O_3 \cdot Bu_4Sn$	386
2-heptene	$MoO_3 \cdot CoO \cdot Al_2O_3$	106, 1202
	$MoO_3 \cdot CoO \cdot Al_2O_3 + Na_2O, K_2O$	106
	$WCl_6 \cdot Et_3Al$	1052, 1925
	$WCl_6 \cdot Bu_4Sn$	506, 509, 986, 988
	$WCl_6 \cdot Ph_4Sn$	506, 509, 510
	$WCl_6 \cdot Ph_4Pb$	510
	$WCl_6 \cdot Et_2Zn$	511
	$WCl_6 \cdot Ph_3Bi$	510
	$WCl_6 \cdot C_5H_5Na$	512
	$WCl_6 \cdot PhC_2Na$	512
	$WCl_6 \cdot n\text{-}BuLi$	509, 511, 988
	$WCl_6 \cdot Bu_3SnCl$	986
	$WCl_6 \cdot Bu_2SnCl_2$	986
	$WCl_6 \cdot BuMgI$	509, 985, 987, 988
	$MoCl_2(NO)_2(PPh_3)_2 \cdot Me_3Al_2Cl_3$	1157, 1158
	$MoCl_2(NO)_2(PPh_3)_2 \cdot EtAlCl_2$	1157, 1158
	$MoCl_2(NO)_2(PPh_3)_2 \cdot Et_3Al_2Cl_3$	1157, 1158
	$MoCl_2(NO)_2(PPh_3)_2 \cdot Et_2AlCl$	1157, 1158
	$SmCl_3 \cdot EtAlCl_2$	1541
	$SmCl_3 \cdot Me_3Al_2Cl_3$	1539, 1541
	$CuCl(PPh_3)_3 \cdot M_3Al_2Cl_3$	1613
	$TiCl_4Py_2 \cdot EtAlCl_2$	1615
	$WCl_6 \cdot EtOH \cdot EtAlCl_2$	1291
	$WCl_6 \cdot EtOH \cdot i\text{-}Bu_2AlCl$	1291
	$WCl_6 \cdot EtOH \cdot i\text{-}Bu_3Al$	1291
2-heptene + ethylene	$WCl_6 \cdot Py \cdot EtAlCl_2 + NO$	1158
3-heptene	$MoO_3 \cdot Al_2O_3 + LiOH, NaOH, KOH, RbOH, CsOH$	262
	$WO_3 \cdot SiO_2 + KOH$	1826
	$(\pi\text{-allyl})_4Mo \cdot SiO_2$	1716
	$(\pi\text{-allyl})_4W \cdot SiO_2$	1716
	$MoO_3 \cdot CoO \cdot Al_2O_3$	1202
	$WO_3 \cdot SiO_2 \cdot MgO$	1835
	$W(CO)_6$	79
	$W(CO)_3Mes_3$	79
	$W(CO)_6 \cdot MeCN$	79
	$WCl_6 \cdot LiAlH_4$	723
	$W(CO)_4(1,5\text{-COD}) \cdot i\text{-}BuAlCl_2$	1111

1	2	3
	$W(CO)_4(NBD) \cdot i\text{-}BuAlCl_2$	1111
	$\pi\text{-}C_5H_5W(CO)_3 \cdot GeMe_3 \cdot i\text{-}PrAlCl_2 +$	
	O_2	1112
	$\pi\text{-}C_5H_5 \; W(CO)_3 \; GePh_3 \cdot i\text{-}PrAlCl_2 +$	
	$+ O_2$	1112
	$\pi\text{-}C_5H_5W(CO)_3 \; SnMe_3 \cdot i\text{-}PrAlCl_2 +$	
	$+ O_2$	1112
	$\pi\text{-}C_5H_5W(CO)_3SnPh_3 \cdot i\text{-}PrAlCl_2 + O_2$	1112
	$Polymer\text{-}C_6H_4 - CH_2(PPh_3) \; W(CO)_5 \cdot$	
	$\cdot i\text{-}BuAlCl_2 + O_2$	1113
	$Polymer\text{-}C_6H_4\text{-}CH_2(\pi\text{-}C_5H_5) \; W(CO)_5 \cdot$	
	$\cdot i\text{-}BuAlCl_2 + O_2$	1113
1-octene	$MoO_3 \cdot A_2O_3$	131, 582, 583, 1856, 1365
	$Re_2(CO)_{10} \cdot Al_2O_3$	1275
	$MoO_3 \cdot Al_2O_3 + LiOH, \; NaOH,$	
	$KOH, \; RbOH, \; CsOH$	262, 263
	$WO_3 \cdot SiO_2 + Nd_2O_3, \; Pr_2O_3, \; Sm_2O_3$	1221
	$MoO_3 \cdot CoO \cdot Al_2O_3$	106, 131, 1199, 1202, 1585
	$MoO_3 \cdot Al_2O_3 \cdot Me_3Al_2Cl_3$	2001
	$MoO_3 \cdot Al_2O_3 \cdot EtAlCl_2$	2001
	$MoO_3 \cdot Al_2O_3 \cdot Et_2AlCl$	2001
	$WO_3 \cdot Al_2O_3 \cdot EtAlCl_2$	2001
	$WO_3 \cdot Al_2O_3 \cdot Et_2AlCl$	2001
	$MoO_3 \cdot CoO \cdot Al_2O_3 + Na_2O, \; K_2O$	106
	$MoO_3 \cdot Al_2O_3 \cdot MeAlCl_2 \cdot Me_2AlCl$	1996
	$WO_3 \cdot Al_2O_3 \cdot MeAlCl_2 \cdot Me_2AlCl$	1996
	$WCl_6 \cdot EtAlCl_2$	488, 1053
	$WCl_6 \cdot Et_3Al$	1053, 1054
	$WCl_6 \cdot Bu_4Sn$	722
	$MoCl_2(NO)_2(PPh_3)_2 \cdot Me_3Al_2Cl_3$	1157, 1158
	$MoCl_2(NO)_2(PPh_3)_2 \cdot EtAlCl_2$	1157, 1158
	$MoCl_2(NO)_2(PPh_3)_2 \cdot Et_3Al_2Cl_3$	1157, 1158
	$MoCl_2(NO)_2(PPh_3)_2 \cdot Et_2AlCl$	1157, 1158
	$MoCl_2(NO)_2Py_2 \cdot Me_3Al_2Cl_3$	1158
	$MoCl_2(NO)_2Py_2 \cdot EtAlCl_2$	1158
	$MoCl_2(NO)_2Py_2 \cdot Et_3Al_2Cl_3$	1158
	$MoCl_4(PPh_3)_2 \cdot Me_3Al_2Cl_3$	1158
	$MoCl_4Py_2 \cdot Me_3Al_2Cl_3$	1158
	$MoCl_4Py_2 \cdot EtAlCl_2$	1158
	$MoCl_4PrCN \cdot Me_3Al_2Cl_3$	1158
	$Mo(CO)_5(\pi\text{-}C_5H_5) \cdot Me_3Al_2Cl_3$	1158
	$Bu_4N[Mo(CO)_5COR] \cdot MeAlCl_2$	
	$(R = Me, \; Ph)$	1611
	$Bu_4N[W(CO)_5COR] \cdot MeAlCl_2$	
	$(R = Me, \; Ph)$	1611
	$TiCl_4 \cdot EtAlCl_2$	722
	$WCl_6 \cdot CH_3COOR \cdot Bu_4Sn$	508
	$WCl_6 \cdot CH_3CN \cdot Bu_4Sn$	508
	$WCl_6 \cdot Py \cdot Me_3Al_2Cl_3 + NO$	1158
1-octene + 1-hexene	$MoO_3 \cdot Al_2O_3$	1880
	$WO_3 \cdot Al_2O_3$	1880
1-octene + 1-hexene + ethylene	$MoO_3 \cdot Al_2O_3$	1880
	$WO_3 \cdot Al_2O_3$	1880

1	2	3
1-octene-1,1-d_2 + 1-hexene	$WCl_6 \cdot$ n-BuLi	727
	$Ph_2C = W(CO)_5$	727
2-octene	$MoO_3 \cdot Al_2O_3$ + LiOH, NaOH, KOH, RbOH, CsOH	262, 263
	$WCl_6 \cdot Bu_4Sn$	901, 903
	$WCl_6 \cdot LiAlH_4$	900
	$WCl_6 \cdot EtOH \cdot EtAlCl_2$	1398
6-methyl-2-octene	$WCl_6 \cdot Bu_4Sn$	901, 903
2-octene + ethylene	$Mo(CO)_6 \cdot Al_2O_3$	262, 263, 1322
	$MoO_3 \cdot Al_2O_3$ + LiOH, NaOH, KOH, RbOH, CsOH	262, 263
4-octene	$Mo(CO)_6$	670
	$(Tol)Mo(CO)_2$	670
	$WCl_6 \cdot EtAlCl_2$	472
	$Re(CO)_5Cl \cdot EtAlCl_2$	429
4-octene + 2-butene	$Bu_4N[(CO)_5Mo-Mo(CO)_5] \cdot$ $\cdot MeAlCl_2$	614
	$Na[(CO)_5Mo-Mo(CO)_5] \cdot MeAlCl_2$	614
	$Bu_4N[(CO)_5W-W(CO)_5] \cdot MeAlCl_2$	614
	$Bu_4N[(CO)_5Mo-Re(CO)_5] \cdot MeAlCl_2$	614
	$Bu_4N[(CO)_5Mo-Mn(CO)_5] \cdot MeAlCl_2$	614
4-octene + 3-hexene	WCl_6-n-BuLi	1078
1,7-octadiene	WCl_6-n-BuLi	432
	$Bu_4N[(Co)_5Mo-Mo(CO)_5] \cdot MeAlCl_2$	614
	$Na[(Co)_5Mo-Mo(CO)_5] \cdot MeAlCl_2$	614
	$Bu_4N[(CO)_5W-W(Co)_5] \cdot MeAlCl_2$	614
	$Bu_4N[CO)_5Mo \cdot Re(CO)_5] \cdot MeAlCl_2$	614
	$Bu_4N[(CO)_5Mo \cdot Mn(CO)_5] \cdot MeAlCl_2$	614
	$Bu_4N[(Mo(CO)_5X] \cdot MeAlCl_2$ (X = Cl, Br)	320
	$Bu_4N[Cr(CO)_5X] \cdot MeClAl_2$ (X = Cl, Br)	320
	$Bu_4N[W(CO)_5X] \cdot MeAlCl_2$ (X = Cl, Br)	320
	$Bu_4N[Mo(CO)_5 COR] \cdot MeAlCl_2$ (R = Oct, Ph)	613, 1611
	$Bu_4N[W(CO)_5COR] \cdot MeAl \cdot Cl_2$ (R = Oct, Ph)	613, 1611
	$MoCl_2(NO)_2(PPh_3)_2 \cdot Me_3Al_2Cl_3$	2002
	$Bu_4N[Mo(CO)_5Py] \cdot EtAlCl_2$	364
	$Re(CO)_5Cl \cdot EtAlCl_2$	363, 429
	$Re(CO)_5Cl \cdot MeAlCl_2$	429
	$Mo(CO)_5Py \cdot Bu_4NCl \cdot EtAlCl_2$	765
	$Mo(CO)_5Py \cdot Bu_4NCl \cdot MeAlCl_2$	765
	$Bu_4NCl \cdot W(CO)_5(COMe)Et \cdot MeAlCl_2$	1611
	$Bu_4NCl \cdot Mo(CO)_6 \cdot Me_3Al_2Cl_3$	1839
	$Pr_4NCl \cdot Mo(CO)_6 \cdot Me_3Al_2Cl_3$	1839
	$Bu_4NCl \cdot Mo(CO)_4Py_2 \cdot Me_3Al_2Cl_3$	1839
	$Bu_4NCl \cdot Mo(CO)_6 \cdot MeAlCl_2$	1839
	$Bu_4NCl \cdot Mo(CO)_4Py_2 \cdot MeAlCl_2$	1839
	$WCl_6 \cdot [HOCH(CH_2Cl)_2] \cdot Et_3Al_2Cl_3$	280
1,7-octadiene + 1,7-octadiene-1,1,8,8-d_4	$WCl_6 \cdot$ n-BuLi	433, 434
	$PhWCl_3 \cdot AlCl_3$	433, 434
	$MoCl_2(NO)_2(PPh_3)_2 \cdot Me_3Al_2Cl_3$	434

1	2	3
1-nonene	$WCl_6 \cdot EtAlCl_2$	744
2-nonene	$MoO_3 \cdot CoO \cdot Al_2O_3$	1202
4-nonene	$MoCl_3(NO)Py_2 \cdot Me_3Al_2Cl_3$	486
	$MoCl_2(NO)_2Py_2 \cdot EtAlCl_2$	486
	$MoCl_2(NO)_2Py_2 \cdot Et_3Al_2Cl_3$	486
	$Re(CO)_5Cl \cdot EtAlCl_2$	363
1-decene	$MoO_3 \cdot Al_2O_3$	1856
	$WCl_6 \cdot CH_3COOR \cdot Bu_4Sn$	508
	$WCl_6 \cdot CH_3CN \cdot Bu_4Sn$	508
1-decene + 1-octene + 1-hexene + + 2-butene	$Re_2O_7 \cdot Al_2O_3$	1859
2-decene	$WCl_6 \cdot EtOH \cdot EtAlCl_2$	1398
5-decene + 2-butene + ethylene	$MoCl_2(NO)_2(PPh_3)_2 \cdot Me_3Al_2Cl_3$	1157, 1158
2,8-decadiene	$MoCl_2(NO)_2(PPh_3)_2 \cdot Me_3Al_2Cl_3$	436
1-undecene	$WCl_6 \cdot EtAlCl_2$	488
1-dodecene	$MoO_3 \cdot CoO \cdot AlO_3$	1199, 1202
	$MoO_3 \cdot Al_2O_3 \cdot Et_2AlOSiMe$	1981
2-dodecene	$WCl_6 \cdot EtOH \cdot EtAlCl_2$	1398
1-dodecene + ethylene	$Mo(CO)_6 \cdot Al_2O_3$	1322
1-dodecene + 7-methyl-1-octene	$Re_2O_7 \cdot Al_2O_3$	1626
	$WCl_6 \cdot EtAlCl_2$	620, 621
	$MoCl_2(NO)_2(PPh_3)_2 \cdot EtAlCl_2$	1626
trimethyl-1,13-tetradecadiene + + ethylene	$Re_2O_7 \cdot Al_2O_3 \cdot K_2CO_3$	1443
2-tetradecene	$MoO_3 \cdot CoO \cdot Al_2O_3$	1202
1-tetradecene + 1-decene	$MoCl_2(NO)_2(PPh_3)_2 \cdot EtAlCl_2$	296
1-tetradecene + 1-undecene	$MoCl_2(NO)_2(PPh_3)_2 \cdot EtAlCl_2$	898
1-tetradecene + 4-methyl-1- -pentene	$MoCl_2(NO)_2(PPh_3)_2 \cdot EtAlCl_2$	898
1-pentadecene + 1-decene	$WCl_6 \cdot EtAlCl_2$	620, 621
1-pentadecene + 1-tetradecene	$WCl_6 \cdot EtAlCl_2$	620, 621
1-hexadecene	$MoO_3 \cdot Al_2O_3 + Ag, Cu$	1369
	$MoO_3 \cdot CoO \cdot Al_2O_3$	1202
1-hexadecene + 1-pentadecene	$WCl_6 \cdot EtAlCl_2$	620, 621
2-hexadecene + 1-tetradecene	$MoO_3 \cdot CoO \cdot Na_2O \cdot Fe_2O_3$	1627
1-heptadecene + 1-tetradecene	$WCl_6 \cdot EtAlCl_2$	620, 621
1-eicosene	$MoO_3 \cdot Al_2O_3 \cdot Et_2AlOS_2Me$	1981
vinylcyclohexane	$Re_2O_7 \cdot Al_2O_3$	379
	$MoO_3 \cdot Al_2O_3$	1321
	$WO_3 \cdot Al_2O_3$	1321
	$MoO_3 \cdot Al_2O_3 + Na, K, Rb, Co, Te, Ba$	1319
	$MoO_3 \cdot Al_2O_3 + LiOH, NaOH, KOH, RbOH, CsOH$	263, 1618
	$MoO_3 \cdot CoO \cdot Al_2O_3$	1620
styrene	$PhMeC = W(CO)_5$	226
	$Tol_2C = W(CO)_5$	225
	$Mo(CO)_6$	670
	$Tol_2Mo(CO)_2$	670
styrene + 2-pentene	$WCl_6 \cdot EtOH \cdot i\text{-}Bu_2AlCl$	1291
	$WCl_6 \cdot EtOH \cdot i\text{-}Bu_3Al$	1291
p-methylphenylethylene	$PhMeC = W(CO)_5$	226
stilbene	$WCl_6 \cdot EtOH \cdot i\text{-}Bu_2AlCl$	1291
	$WCl_6 \cdot EtOH \cdot i\text{-}Bu_3Al$	1291

1	2	3
stilbene + ethylene	$MoO_3 \cdot Al_2O_3$	1713
	$WO_3 \cdot SiO_2$	103, 1713
2,2'-divinylbiphenyl	$MoCl_2(NO)_2(Oct_3P)_2 \cdot Me_3Al_2Cl_3$	558
2,3-diphenylbutadiene	$Ph_2C=W(CO)_5$	585
4-phenyl-1-butene	$Mo(CO)_6 \cdot Pr_4NCl \cdot Me_3Al_2Cl_3$	1839
	$Mo(CO)_6 \cdot Bu_4NCl \cdot Me_3Al_2Cl_3$	1839
	$Mo(CO)_4Py_2 \cdot Bu_4NCl \cdot Me_3Al_2Cl_3$	1839
	$Mo(CO)_4Py_2 \cdot Bu_4NCl \cdot MeAlCl_2$	1839
1-phenyl-2-butene	$MoO_3 \cdot Al_2O_3 \cdot CoO$	1202
ω-arylolefins	$WCl_2(NO)_2(PPh_3)_2 \cdot EtAlCl_2$	256
	$MoCl_2(NO)_2(PPh_3)_2 \cdot EtAlCl_2$	291
ethylcyclopropane	$WPhCl_3 \cdot AlCl_3$	403
butylcyclopropane	$WPhCl_3 \cdot AlCl_3$	403
methylenecyclobutane	$MoO_3 \cdot Al_2O_3$	871
	$WO_3 \cdot Al_2O_3$	871
	$Re_2O_7 \cdot Al_2O_3$	379, 841, 975
methylenecyclobutane + ethylene	$Re_2O_7 \cdot Al_2O_3$	383
methylenecyclobutane + propene	$Re_2O_7 \cdot Al_2O_3$	141, 144, 386
3-vinyl-1-butene	$MoO_3 \cdot SiO_2$	773
vinyltrimethylsilane	$MoO_3 \cdot Al_2O_3$	386
	$Re_2O_7 \cdot Al_2O_3$	386
	$MoO_3 \cdot Al_2O_3 \cdot Bu_4Sn$	386
	$Re_2O_7 \cdot Al_2O_3 \cdot Et_4Sn$	386
	$MoO_3 \cdot Al_2O_3 \cdot Et_4Sn$	386
unsaturated triglycerides	$WCl_6 \cdot Me_4Sn$	1079
halo-olefins	$Re_2O_7 \cdot Al_2O_3 \cdot$ silanes	1226
vinyl chloride	$Re_2O_7 \cdot Al_2O_3 \cdot Bu_4Sn$	387
vinyl chloride + ethylene-d_4	$Re_2O_7 \cdot Al_2O_3 \cdot Bu_4Sn$	387
vinyl chloride + propene	$Re_2O_7 \cdot Al_2O_3 \cdot Bu_4Sn$	387
vinyl chloride + 2-butene	$Re_2O_7 \cdot Al_2O_3 \cdot Bu_4Sn$	387
vinyl chloride + cyclohexene	$Re_2O_7 \cdot Al_2O_3 \cdot Bu_4Sn$	387
vinyl bromide	$Re_2O_7 \cdot Al_2O_3 \cdot Bu_4Sn$	387
vinyl bromide + propene	$Re_2O_7 \cdot Al_2O_3 \cdot Bu_4Sn$	387
allyl chloride	$Re_2O_7 \cdot Al_2O_3 \cdot Bu_4Sn$	387
	$Re_2O_7 \cdot Al_2O_3 \cdot V_2O_5$	788
	$Re_2O_7 \cdot Al_2O_3 \cdot MoO_3$	788
	$Re_2O_7 \cdot Al_2O_3 \cdot WO_3$	788
	$Re_2O_7 \cdot Al_2O_3 \cdot GeO_2$	788
allyl bromide	$Re_2O_7 \cdot Al_2O_3 \cdot Bu_4Sn$	387
5-bromo-1-pentene + 2-pentene	$WCl_6 \cdot EtAlCl_2$	1785
5-bromo-1-hexene	$Re_2O_7 \cdot Al_2O_3 \cdot Bu_4Sn$	379
6-chloro-1-hexene	$WO_3 \cdot Al_2O_3 \cdot Et_2AlOSiMe$	1981
p-chlorophenylethylene + 2-pentene	$WCl_6 \cdot EtAlCl_2$	1785
allyl phenyl ether	$WCl_6 \cdot Et_3Al$	1510
allylmethylsilylether	$WCl_6 \cdot Et_3Al$	1510
propenylamine	$W(CO)_3Mes \cdot EtAlCl_2$	814
propenylcyclohexylamine	$W(CO)_3Mes \cdot EtAlCl_2$	814
propenyldimethylamine	$W(CO)_3Mes \cdot EtAlCl_2$	814
propenylpropylamine	$W(CO)_3Mes \cdot EtAlCl_2$	814
butenylpropylamine	$W(CO)_3Mes \cdot EtAlCl_2$	814
pentenylpropylamine	$W(CO)_3Mes \cdot EtAlCl_2$	814
hexenylpropylamine	$W(CO)_3Mes \cdot EtAlCl_2$	814
undecenylpropylamine	$W(CO)_3Mes \cdot EtAlCl_2$	814
allyl acetate	$WCl_6 \cdot EtAlCl_2$	1733
	$WCl_6 \cdot Et_3Al$	1510

1	2	3
	$WCl_6 \cdot Oct_3Al$	1733
	$ReCl_5 \cdot Et_3Al_2Cl_3$	1733
	$ReCl_5 \cdot Oct_3Al$	1733
	$Mo(OEt)_2Cl_2 \cdot Et_2Al_2Cl_3$	1733
	$Mo(OEt)_2Cl_2 \cdot Oct_3Al$	1733
allyl acetate + 1-hexene	$WCl_6 \cdot EtAlCl_2$	1733
methyl acrylate	$WCl_6 \cdot Me_4Sn$	1098
methyl methacrylate	$WCl_6 \cdot Me_4Sn$	1098
methyl 1-butenoate	$WCl_6 \cdot Me_4Sn$	1098
methyl 2-butenoate	$WCl_6 \cdot Me_4Sn$	1098
methyl 2-methyl-2-butenoate	$WCl_6 \cdot Me_4Sn$	1098
methyl 1-pentenoate	$Re_2O_7 \cdot Al_2O_3 \cdot Me_4Sn$	1097
methyl 2-pentenoate	$WCl_6 \cdot Me_4Sn$	1098
	$WCl_6 \cdot Me_4Sn$	1089
methyl 9-decenoate + 1-hexene	$WCl_6 \cdot EtAlCl_2$	1733
methyl oleate	$WCl_6 \cdot Et_3Al$	1510
	$WCl_6 \cdot Me_3Al_2Cl_3$	786
	$WCl_6 \cdot Me_4Sn$	1079, 1733
	$WCl_6 \cdot Et_4Sn$	507
	$WCl_6 \cdot Bu_4Sn$	507
	$WCl_6 \cdot Et_3Al_2Cl_3$	1733
	$WCl_6 \cdot Et_2AlCl$	1733
	$WCl_6 \cdot Oct_3Al$	1733
	$ReCl_5 \cdot Et_3Al_2Cl_3$	1733
	$ReCl_5 \cdot Oct_3Al$	1733
	$Mo(OEt)_2Cl_2 \cdot Et_3Al_2Cl_3$	1733
	$Mo(OEt)_2Cl_2 \cdot Oct_3Al$	1733
oleyl chloride	$Re_2O_7 \cdot Al_2O_3$	1098
methyl oleate + 3-hexene	$WCl_6 \cdot Me_4Sn$	1076, 1079
methyl oleate + 1-decene	$WCl_6 \cdot Et_4Sn$	507
	$WCl_6 \cdot Bu_4Sn$	507
ethyl oleate	$WCl_6 \cdot Et_3Al$	1510
butyl oleate	$WCl_6 \cdot Et_3Al$	1510
methyl linoleate	$WCl_6 \cdot Me_4Sn$	113, 117, 1095
methyl linolenate	$WCl_6 \cdot Me_4Sn$	1095
methyl brucate	$WCl_6 \cdot Me_4Sn$	1079
erucic acid + 7-tetradecene + 4-octene	$WCl_6 \cdot EtAlCl_2$	494
9-octadecenyl acetate	$Mo(OEt)_2Cl_3 \cdot Et_3B$	785
	$WCl_6 \cdot Et_3B$	785
geranyl acetate + 1-methylcyclo-butene	$WCl_6 \cdot Me_4Sn$	1133
$CH_3(CH_2)_7CH = CH(CH_2)_7-CN$	$WCl_6 \cdot Et_3Al$	1510
$CH_3(CH_2)_7CH = CH(CH_2)_7-COEt$	$WCl_6 \cdot Et_3Al$	1510
$CH_3(CH_2)_7CH=CH(CH_2)_7-CONEt_2$	$WCl_6 \cdot Et_3Al$	1510
cyclopentene + 1-pentene	$WCl_6 \cdot EtAlCl_2$	569
cyclopentene + ethylene	$MoO_3 \cdot Al_2O_3$	1819
	$Mo(CO)_6 \cdot Al_2O_3$	1819
	$MoO_3 \cdot CoO \cdot Al_2O_3$	1662
	$MoO_3 \cdot CoO \cdot Al_2O_3 + K$	1662
cyclohexene + ethylene	$MoO_3 \cdot Al_2O_3$	1819
	$Mo(CO)_6 \cdot Al_2O_3$	1819
1,4-cyclohexadiene	$WCl_3Ph \cdot AlCl_3$	401
cycloheptatriene + ethylene	$Re_2O_7 \cdot Al_2O_3$	107
cyclo-octene	$WCl_6 \cdot EtOH \cdot EtAlCl_2$	1879

1	2	3
cyclo-octene + 2-butene	$WCl_4 \cdot AlCl_3$	1879
	$MoCl_4 \cdot AlCl_3$	1879
cyclo-octene + ethylene	$MoO_3 \cdot Al_2O_3$	1819
	$WO_3 \cdot SiO_2$	1958, 1959
	$Mo(CO)_6 \cdot Al_2O_3$	1819
	$MoO_3 \cdot CoO \cdot Al_2O_3$	1662
	$MoO_3 \cdot CoO \cdot Al_2O_3 + K$	1662
cyclo-octene + 1-hexene	$WCl_6 \cdot EtOH \cdot EtAlCl_2$	626
	$WCl_6 \cdot Et_2O \cdot Bu_4Sn$	626
cyclo-octene + 2-hexene	$MoCl_2(NO)_2(PPh_3)_2Me_3Al_2Cl_3$	555
cyclo-octene + 4-octene + 2-butene	$MoCl_2(NO)_2(PPh_3)_2 \cdot Me_3Al_2Cl_3$	555, 559
1-methylcyclo-octene + isobutene	$Re_2O_7 \cdot Al_2O_3$	1341
1-methylcyclo-octene + 2-butene	$Re_2O_7 \cdot Al_2O_3$	1341
1,5-cyclo-octadiene + ethylene	$MoO_3 \cdot Al_2O_3$	1819
	$WO_3 \cdot SiO_2$	1957
	$Mo(CO)_6 \cdot Al_2O_3$	1819
	$MoO_3 \cdot CoO \cdot Al_2O_3$	1662
	$MoO_3 \cdot CoO \cdot Al_2O_3 + K$	1662
1,5-cyclo-octadiene + 4-octene	$WCl_6 \cdot EtOH \cdot EtAlCl_2$	860
1,5-cyclo-octadiene + 4-methyl-4-octene	$WCl_6 \cdot EtAlCl_2$	864
1,5-cyclononadiene	$WCl_6 \cdot LiAlH_4$	899, 902
trimethylcyclodecene + 2-butene	$Re_2O_7 \cdot Al_2O_3 \cdot K_2CO_3$	1443
1,6-cyclodecadiene	$WCl_6 \cdot EtAlCl_2$	619
cyclododecene	$WCl_6 \cdot EtAlCl_2$	1137, 1138, 1976, 1977
	$WCl_6 \cdot EtOH \cdot EtAlCl_2$	1135
1,5,9-cyclododecatriene + ethylene	$WO_3 \cdot SiO_2$	1959
bicyclo[2.2.1]heptene + ethylene	$MoO_3 \cdot CoO \cdot Al_2O_3$	1662
	$MoO_3 \cdot CoO \cdot Al_2O_3 + K$	1662
bicyclo[2.2.1]heptene + 2-butene	$WCl_6 \cdot AlCl_3$	1873
	$MoCl_4 \cdot AlCl_3$	1873
bicyclo[2.2.1]heptene + isobutene	$IrCl_2(cyclo-octene)_2$	897
bicyclo[2.2.1]heptene + 1-pentene	$IrCl_2(cyclo-octene)_2$	897
bicyclo[2.2.1]heptene + 2-pentene	$IrCl_2(cyclo-octene)_2$	897
bicyclo[2.2.1]heptene + 2-methyl-2-butene		
	$IrCl_2 (cyclo-octene)_2$	897
	$IrCl_2 (cyclo-octene)_2$	897
bicyclo[2.2.1]heptadiene + ethylene	$MoO_3 \cdot CoO \cdot Al_2O_3$	1662
	$MoO_3 \cdot CoO \cdot Al_2O_3 + K$	1662
bicyclo[2.1.0]-pentane	$WPhCl_3 \cdot AlCl_3$	510
ethylcyclopropane + ethyl acrylate	$WPhCl_3 \cdot AlCl_3$	510
ethylcyclopropane + ethyl crotonate	$WPhCl_3 \cdot AlCl_3$	510
ethylcyclopropane + acrylonitrile	$WPhCl_3 \cdot AlCl_3$	510
ethylcyclopropane + ethyl cinnamate	$WPhCl_3 \cdot AlCl_3$	510
ethylcyclopropane + ethyl 2-methylcrotonate	$WPhCl_3 \cdot AlCl_3$	510
ethylcyclopropane + ethyl 3-methylcrotonate	$WPhCl_3 \cdot AlCl_3$	510
polybutadiene	$WCl_6 \cdot epichlorhydrin \cdot Et_2AlCl$	1801
	$WCl_6 \cdot ClCH_2CH_2OH \cdot Et_2AlCl$	1798
polybutadiene + 2-butene	$WCl_6 \cdot EtAlCl_2$	737

1	2	3
polybutadiene + 4-octene	$WCl_6 \cdot Me_4Sn$	453
	$WCl_6 \cdot EtAlCl_2$	490, 492, 493, 497, 498, 964, 1039
polybutadiene + 6-dodecene	$WCl_6 \cdot EtAlCl_2$	491
polybutadiene + 2-butene + 2-pentene	$WCl_6 \cdot EtAlCl_2$	1683
	$MoCl_5 \cdot EtAlCl_2$	1683
polybutadiene + crotyl bromide	$WCl_6 \cdot ClCH_2CH_2OH \cdot Et_2AlCl$	1804
partially hydrogenated polybutadiene + 4-octene		
	$WCl_6 \cdot EtAlCl_2$	1129
polypentenamer + polybutadiene	$WCl_6 \cdot EtAlCl_2$	495
polypentenamer + polybutadiene + + 4-octene	$WCl_6 \cdot EtAlCl_2$	495

TABLE IV. Metathesis reactions and ring-opening polymerization of cyclo-olefins

Cyclo-olefin	Catalytic system	References
1	2	3
Cyclobutene	$Ph_2C = W(CO)_5$	557
	$RhCl_3 \cdot x \ H_2O$	801
	$RuCl_3 \cdot x \ H_2O$	802
	$(\pi\text{-allyl})_4Zr$	591
	$(\pi\text{-allyl})_3Cr$	591
	$(\pi\text{-allyl})_4Mo$	591
	$(\pi\text{-allyl})_4W$	591
	$WCl_6 \cdot Et_2AlCl$	269
	$WCl_6 \cdot Et_3Al$	799
	$TiCl_3 \cdot Et_3Al$	799, 265
	$TiCl_4 \cdot Et_3Al$	799, 265
	$VCl_4 \cdot Et_3Al$	799
	$V(acac)_3 \cdot Et_2AlCl$	799, 265
	$Cr(acac)_3 \cdot Et_2AlCl$	799
	$MoCl_3 \cdot Et_3Al$	799
	$MoO_2(acac)_2 \cdot Et_2AlCl$	799
	$MoCl_5 \cdot Et_3Al$	799
	$VCl_4 \cdot Hex_3Al$	265
	$Re_2O_7 \cdot Al_2O_3$	1911
	$MoO_3 \cdot Al_2O_3 \ (TiO_2, ZrO_2)$	1364
	$WO_3 \cdot Al_2O_3 \ (TiO_2, ZrO_2)$	1364
	$Cr_2O_3 \cdot Al_2O_3 \ (TiO_2, ZrO_2)$	1364
	$VO_2 \cdot Al_2O_3 \ (TiO_2, ZrO_2)$	1364
cyclobutene + ethylene	$TiCl_4 \cdot Hex_3Al$	798
	$V(acac)_2 \cdot Et_2AlCl$	798
cyclobutene + ethylene-C_2^{14}	$V(acac)_2 \cdot Et_2AlCl$	1741
cyclobutene + propene	$TiCl_4 \cdot Hex_3Al$	798
	$V(acac)_2 \cdot Et_2AlCl$	1741
1-methylcyclobutene	$Ph_2C = WC(O)_5$	557
	$WCl_6 \cdot Et_2AlCl$	267
	$WCl_6 \cdot n\text{-BuLi}$	557
	$WCl_6 \cdot Ph_3SnEt$	557
	$MoCl_2(NO)_2(PPh_3)_2 \cdot Me_3Al_2Cl_3$	557

1	2	3
3-methylcyclobutene	$RuCl_3 \cdot xH_2O$	802
	$WCl_6 \cdot Et_2AlCl$	269
3-methylcyclobutene+ cyclobutene-1-C^{14}	$WCl_6 \cdot Et_2AlCl$	278
cyclopentene	WCl_6	91, 1327
	$WOCl_4$	1327
	WCl_5	1327
	$PhWCl_3$	422
	$PhWCl_5$	1907
	$Me_3SiCH_2WCl_5$	306
	$Ph_2C=W(CO)_5$	561
	$(\pi\text{-allyl})_4Zr$	591
	$(\pi\text{-allyl})_3Cr$	591
	$(\pi\text{-allyl})_4Mo$	591
	$(\pi\text{-allyl})_4W$	591
	$(\pi\text{-allyl})NiBr_2$	872
	$IrCl_2(\text{cyclo-octene})_2$	874
	$IrCl(CO)\ (\text{cyclo-octene})_2$	874
	$Ir(CF_3CO_2)\ (\text{cyclo-octene})_2$	873, 874
	$RuCl_2(2,7\text{-Me}_2\text{-}2,6\text{-octadien-1,8-yl})$	874
	$Ru(CF_3CO_2)\ (2,7\text{-Me}_2\text{-}2,6\text{-octadien-1,8-yl})$ $+ H_2$	874
	$WCl_6 \cdot Et_2AlCl$	803, 1743, 1746
	$WCl_6 \cdot Et_3Al$	800, 803, 1687, 1743
	$WCl_6 \cdot i\text{-}Pr_3Al$	447
	$WCl_6 \cdot i\text{-}Bu_3Al$	34
	$WCl_6 \cdot Ph_4Sn$	455
	$WOCl_4 \cdot Et_2AlCl$	455
	$WOCl_4 \cdot Et_3Al$	1743
	$WOCl_4 \cdot Bu_4Sn$	455
	$WCl_6 \cdot Et_3SnH$	34, 1758
	$WCl_6 \cdot R_3SiCH_2Y\ (R=\text{alkyl};\ Y=\text{aryl})$	1354, 1355
	$WCl_6 \cdot C_6H_5CHN_2$	302, 306
	$WCl_6 \cdot EtOOCCN_2$	419
	$Me_4N\ W(CO)_5COPh \cdot EtAlCl_2$	109
	$WCl_6 \cdot AlX_3\ (X=Cl,\ Br,\ I)$	1327
	$WCl_5 \cdot AlX_3$	1327
	$WBr_5 \cdot AlX_3$	1327
	$WOCl_4 \cdot AlX_3$	1327
	$WCl_6 \cdot AlCl_3$	1513
	$WF_6 \cdot Et_3Al$	1743
	$WF_6 \cdot Et_2AlCl$	1743, 1765
	$WF_6 \cdot Et_3Al_2Cl_3$	1743
	$MoCl_5 \cdot Et_3Al$	800, 803, 1743, 1744
	$MoCl_5 \cdot Et_2AlCl$	1743
	$W(OPh)_6 \cdot Et_3Al$	1743
	$W(OPh)_6 \cdot Et_2AlCl$	1743
	$W(OPh)_4Cl_2 \cdot MeAlCl_2$	1857
	$W(OPh)_6 \cdot MeAlCl_2$	1857
	$WCl_6 \cdot Na_3WPh_5$	34
	$TiCl_4 \cdot AlWPh_5$	34
	$WCl_6 \cdot (HSiOMe)_4$	34
	$TiCl_4 \cdot Li_3WPh_6$	34

1	2	3
	$WCl_6 \cdot Li_3WEt_6$	34
	$SnCl_4 \cdot Li_3WPh_6$	34
	$BCl_3 \cdot Li_3WPh_6$	34, 1755
	$SnCl_4 \cdot Na_3WPh_6$	34, 1755
	$WCl_6 \cdot Ca$	34
	$WCl_6 \cdot Li$	34
	$WCl_5OX \cdot Ca$	34
	$WCl_6 \cdot Al$	34
	$WCl_6 \cdot AlCl_3$	34
	$WBr_5 \cdot Ca$	34
	$MoCl_5 \cdot Li_3WPh_6$	1755
	$BF_3 \cdot LiMoPh_6$	1755
	$SnCl_4 \cdot Ph_4W$	1755
	$WOCl_4 \cdot Ph_2Ti$	1755
	$WCl_6 \cdot Ph_3Cr$	1755
	$WCl_6 \cdot 1,3,5,7\text{-}Me_4\text{-cyclotetrasiloxane}$	1755
	$WCl_6 \cdot Et_2Be$	1744
	$WCl_6 \cdot LiCH_2SiMe_3$	835, 949
	$WCl_6 \cdot LiCH_2GeMe_3$	836
	$WCl_6 \cdot N_2CHSiMe_3$	949
	$WCl_4(OR) \cdot EtAlCl_2$	949
	$TiCl_4 \cdot Et_3Al$	800
	$TiBr_4 \cdot Et_3Al$	800
	$VCl_4 \cdot Et_3Al$	800
	$ZrCl_4 \cdot Et_3Al$	800
	$V(acac)_3 \cdot Et_2AlCl$	800
	$Cr(acac)_3 \cdot Et_2AlCl$	800
	$CrBr_3 \cdot Et_3Al$	800
	$CrO_2Cl_2 \cdot Et_3Al$	800
	$UF_4 \cdot Et_3Al$	800
	$UO_2Cl_2 \cdot Et_2AlCl$	800
	$MnCl_2 \cdot Et_3Al$	800
	$FeCl_3 \cdot Et_3Al$	800
	$CoCl_2Py_2 \cdot Et_2AlCl$	800
	$BaFeO_4 \cdot Et_2AlCl$	800
	$MoCl_5 \cdot i\text{-}Bu_3Al$	1944
	$MoCl_3$ distearate $\cdot Et_2AlCl$	846
	$MoCl_3$ distearate $\cdot i\text{-}Bu_3Al$	846
	$ReCl_5 \cdot i\text{-}Bu_3Al$	1404
	$ReCl_6 \cdot Et_3Al$	1332
	$TaCl_5 \cdot i\text{-}Bu_3Al$	1766, 1944
	$NbCl_5 \cdot i\text{-}Bu_3Al$	1944
	$Mo(CO)_5 = C(OEt)Me \cdot EtAlCl_2(M = Mo,$	
	W)	1314
	$Mo(CO)_5 = C(OMe)$ Me $\cdot EtAlCl_2$	1314
	$Mo(CO)_5 = C(OEt)Ph \cdot EtAlCl_2$	1314
	$W(CO)_5 = C(OEt)R \cdot TiCl_4$	260
	$PdCl_2(PPh_3) = C(OMe)NHPh \cdot AlCl_3$	246
	$(\pi\text{-allyl})_3Cr \cdot WCl_6$	34
	$(\pi\text{-allyl})_4W_2 \cdot WCl_6$	34
	$(\pi\text{-allyl})_4W \cdot BCl_3$	1758
	$(\pi\text{-allyl})_2M \cdot AlBr_3$ $(WF_6, TiCl_4, MoCl_5)$	
	(M = Ni, Co, Zr)	1591
	$(\pi\text{-allyl})_3M \cdot AlBr_3$ $(WF_6, TiCl_4, MoCl_5)$	
	(M = W, Cr)	1591

1	2	3
	Bu₄N Mo(CO)₅COPh ·MeAlCl₂	1611
	WCl₆ ·EtOH ·EtAlCl₂	214, 815, 928, 1883, 1885
	WCl₆ ·t-BuClO ·EtAlCl₂	1761
	WCl₆ ·Bz₂O₂ ·Et₂AlCl	1329
	WCl₆ ·Cum₂O₂ ·Et₂AlCl	1329
	WCl₆ ·Bz₂O₂ ·Hex₃Al	1329
	WCl₆ ·Bz₂O₂ ·Et₂Be	1329
	WOCl₄ ·Bz₂O₂ ·Et₃Al	1329
	WOCl₄ ·Bz₂O₂ ·Et₂Be	1329
	WCl₆ ·CO₂ ·Et₂AlCl	1336
	MoCl₅ ·Co₂ ·Et₂AlCl	1337
	WCl₆ ·Na₂O₂ ·i-Bu₃Al	1756
	WCl₆ ·BaO₂ ·i-Bu₃Al	1756
	WCl₆ ·CO₂ ·AlCl₃	1337
	MoCl₅ ·Co₂ ·AlCl₃	1337
	WCl₆ ·epichlorhydrin ·Et₂AlCl	852, 1803
	WCl₆ ·epichlorhydrin ·i-Bu₃Al	1796, 1797, 1847,
	WCl₆ ·ClCH₂CH₂OH ·Et₂AlCl	852, 1972, 1973, 1974
	WCl₆ ·cylopentyl hydroperoxide ·i-Bu₃Al	1235, 1758
	WCl₆ ·ethylene oxide ·i-Bu₃Al	1847
	WCl₆ ·propylene oxide ·i-Bu₃Al	1796, 1797
	WCl₆ ·1-butylene oxide ·i-Bu₃Al	1847
	WCl₆ ·HOCH₂CN ·EtAlCl₂	1771
	WCl₆ ·HOCH₂CH₂CN ·EtAlCl₂	1771, 1772, 1775, 1776
	WCl₆ ·ClCH₂CH₂OSiMe₃ ·EtAlCl₂	1777
	WCl₆ ·epichlorhydrin ·Na	1405
	WCl₆ ·H₂O₂ ·Na	1405
	WCl₆ ·(MeO)₂CH₂ ·Et₂AlCl	1848
	WCl₆ ·(β-ClEtO)₂CH₂ ·Et₂AlCl	1848
	WCl₆ ·(MeO)₂CHCCl₃ ·Et₂AlCl	1848
	WCl₆ ·(EtO)₂CHC₆H₅ ·Et₂AlCl	1848
	WCl₆ ·2-chlorobutadiene ·EtAlCl₂	1762
	WCl₆ ·Cl₆CPD ·EtAlCl₂	1762, 1764
	WCl₆ ·vinyl chloride ·EtAlCl₂	1762
	WCl₆ ·vinyl chloride ·i-Bu₃Al	1766
	WCl₆ ·vinyl fluoride ·EtAlCl₂	1884
	WCl₆ ·Bz₂O₂ ·i-Bu₃Al	1235
	WCl₆ ·t-BuOOH ·i-Bu₃Al	1235
	WCl₆ ·phenoxypropylene oxide ·i-Bu₃Al	1796, 1797
	WCl₆ ·Et₂O ·Bu₄Sn	854
	EtWCl₆ ·Et₂O ·Bu₄Sn	853
	WCl₆ ·o-NO₂C₆H₄Cl ·i-Bu₃Al	1759
	WCl₆ ·3,5-(NO₂)₂C₆H₃COCl ·i-Bu₃Al	1759
	WCl₆ ·2,4-(NO₂)₂C₆H₃NHNH₂ ·i-Bu₃Al	1759
	WO(OPh)₄ ·AlCl₃ ·EtAlCl₂	1543
	WO₃ ·TlCl₅ ·Et₂AlCl	1563
	MoCl₃(stearate)₂ ·Et₃N ·Et₂AlCl	1287
	TaCl₅ ·EtOH ·EtAlCl₂	1402
	TaCl₅ ·epichlorhydrin ·EtAlCl₂	1238
	TaCl₅ ·Cl₆CPD ·EtAlCl₂	1764
	WCl₆ ·Et₃N ·dichloroalane	1594
	W(CO)₅(COMe)Et ·Bu₄NCl ·MeAlCl₂	1611

1	2	3
	$WCl_6 \cdot AlCl_3 \cdot Et_3Al$	1659
	$WCl_6 \cdot HCl \cdot EtAlCl_2$	1658
	$MoCl_5 \cdot BF_3\text{-}Et_2O \cdot Et_3Al$	1659
	$WO_3 \cdot AlCl_3 \cdot i\text{-}Bu_3Al + O_2$	1892
	$WCl_6 \cdot ClCH_2CH_2OH \cdot I_2 \cdot Et_2AlCl$	1245
	$ZrCl_4 \cdot t\text{-}BuOCl \cdot Et_3Al_2Cl_3$	1403
	$(OPPh_3, PPh_3, Ph_2P(CH_2)_2PPh_2,$	
	$OP(Me_2N)_3)$	
	$TaCl_3 \cdot t\text{-}BuOCl \cdot Et_3Al_2Cl_3$	1403
	$CoCl_2 \cdot t\text{-}BuOCl \cdot Et_3Al_2Cl_3$	1403
	$Co(acac)_3 \cdot t\text{-}BuOCl \cdot Et_3Al_2Cl_3$	1403
	$NiBr_2 \cdot t\text{-}BuOCl \cdot Et_3Al_2Cl_3$	1403
	$CuCl_2 \cdot t\text{-}BuOCl \cdot Et_3Al_2Cl_3$	1403
	$WO_3 \cdot Al_2O_3$	1364
	$WO_3 \cdot TiO_2$	1364
	$WO_3 \cdot ZrO_2$	1364
	$Re_2O_7 \cdot Al_2O_3$	1911
	$Cr_2O_3 \cdot Al_2O_3$	1364
	$Cr_2O_3 \cdot ZrO_2$	1364
	$VO_2 \cdot Al_2O_3$	1364
	$VO_2 \cdot ZrO_2$	1364
	$VO_2 \cdot TiO_2$	1364
	$CoO \cdot MoO_3 \cdot Al_2O_3$	1323
	$CoO \cdot MoO_3 \cdot SiO_2$	1323
	$WO_3 \cdot AlCl_3$	1246
	$MoO_3 \cdot Al_2O_3$	1364
	$MoO_3 \cdot TiO_2$	1364
	$MoO_3 \cdot ZrO_2$	1364
	$(\pi\text{-allyl})_4Mo \cdot SiO_2(Al_2O_3)$	1716
	$(\pi\text{-allyl})_4W \cdot SiO_2 (Al_2O_3)$	1716
	$MoO_3 \cdot Al_2O_3 \cdot Me_3Al_2Cl_3$	2001
	$MoO_3 \cdot Al_2O_3 \cdot EtAlCl_2 (Et_2AlCl)$	2001
	$WO_3 \cdot Al_2O_3 \cdot Me_3Al_2Cl_3$	2001
	$MoO_3 \cdot Al_2O_3 \cdot MeAlCl_2 \cdot Me_2AlCl$	1996
	$WO_3 \cdot Al_2O_3 \cdot MeAlCl_2 \cdot Me_2AlCl$	1996
1-methylcyclopentene	$WO_3 \cdot SiO_2$	129a
	$MoCl_2(NO)_2(PPh_3)_2 \cdot Me_3Al_2Cl_3$	557
1-ethylcyclopentene	$WO_3 \cdot SiO_2$	129a
3,3'-bicyclopentene	$WCl_6 \cdot EtOH \cdot EtAlCl_2$	815
1,2-dimethylcyclo-pentene	$WO_3 \cdot SiO_2$	129a
2,3-dimethylcyclo-pentene	$WO_3 \cdot SiO_2$	129a
cyclopentene+ethylene	$WOCl_4 \cdot Bu_4Sn$	455
	$VCl_4 \cdot Hex_3Al$	266
	$TiCl_4 \cdot Hex_3Al$	798, 1745
	$V(acac)_3 \cdot Et_2AlCl$	798
cyclopentene+propene	$WOCl_4 \cdot Et_2AlCl$	460
	$TiCl_4 \cdot Hex_3Al$	798, 1745
	$V(acac)_2 \cdot Et_2AlCl$	798
cyclopentene+2-butene	$WOCl_4 \cdot Et_2AlCl$	460
cyclopentene+1-pentene	$WOCl_4 \cdot Et_2AlCl$	460
cyclopentene+1-hexene	$MoCl_3(octanoate)_2 \cdot Et_2AlCl$	1945
cyclopentene+cyclopentadiene	$MoCl_3(laurate)_2 \cdot Et_2AlCl$	846
cyclopentene+cycloheptene	$WCl_6 \cdot Bz_2O_2 \cdot Et_2AlCl$	1331
	$WCl_6 \cdot Cum_2O_2 \cdot Et_2AlCl$	1331
	$WOCl_4 \cdot Cum_2O_2 \cdot Et_2AlCl$	1331

1	2	3
cyclopentene + cyclo-octene	$WCl_6 \cdot Bz_2O_2 \cdot Et_2AlCl$	738, 1331
	$WOCl_4 \cdot Cum_2O_2 \cdot Et_2AlCl$	1331
	$WCl_6 \cdot Bz_2O_2 \cdot Et_3Al$	738
	$WOCl_4 \cdot Bz_2O_2 \cdot Et_2AlCl$	1331
	$WCl_6 \cdot EtOH \cdot Et_2AlCl$	195
1-C^{14}-cyclopentene + cyclo-octene	$WOCl_4 \cdot Bz_2O_2 \cdot Et_2AlCl$	276
cyclopentene + bicyclo [2.2.1] heptene	$WCl_6 \cdot$ cyclopentyl hydroperoxide. $\cdot$ i-Bu_3Al	2135, 1758
cyclopentene + bicyclo [2.2.1] heptadiene	$WCl_6 \cdot ClCH_2CH_2OH \cdot Et_2AlCl$	1972
	$WCl_6 \cdot ClCH_2CH_2OH \cdot$ i-Bu_3Al	1972
cyclohexene + ethylene	$VCl_4 \cdot Hex_3Al$	266
cyclohexene	$WO_3 \cdot Al_2O_3 (TiO_2, ZrO_2)$	1364
	$MoO_3 \cdot Al_2O_3 (TiO_2, ZrO_2)$	1364
	$Cr_2O_3 \cdot Al_2O_3 (TiO_2, ZrO_2)$	1364
	$VO_2 \cdot Al_2O_3 (TiO_2, ZrO_2)$	1364
	WCl_6	91
	$WCl_6 \cdot Et_2AlCl$	803
	$WCl_6 \cdot Et_3Al$	803
	$MoCl_5 \cdot Et_3Al$	803
	$Re(CO)_5Cl \cdot EtAlCl_2$	365
1,3-cyclohexadiene	$IrCl_3$(cyclo-octene)	1942
	$IrCl_3$(COD)	1942
cycloheptene	WCl_6	91
	$Ph_2C = W(CO)_5$	561
	$NiBr_2(\pi$-allyl)	872
	$WCl_6 \cdot RnSiCH_2Y(R = alkyl; Y = aryl)$	1354
	$WCl_6 \cdot Et_2AlCl$	803, 1747
	$WCl_6 \cdot Et_3Al$	803
	$MoCl_5 \cdot Et_3Al$	803
cycloheptene + ethylene	$VCl_4 \cdot Hex_3Al$	266
cycloheptadiene	WCl_6	632
	$WOCl_4 \cdot Bz_2O_2 \cdot Et_2AlCl$	1331
cyclo-octene	$WO_3 \cdot Al_2O_3 (TiO_2, ZrO_2)$	1364
	$Cr_2O_3 \cdot Al_2O_3 (TiO_2, ZrO_2)$	1364
	$VO_2 \cdot Al_2O_3 (TiO_2, ZrO_2)$	1364
	$MoO_3 \cdot Al_2O_3 (TiO_2, ZrO_2)$	1364
	$Re_2O_7 \cdot Al_2O_3$	1917
	$CoO \cdot MoO_3 \cdot Al_2O_3$	1322
	$WO_3 \cdot AlCl_3$	1246
	$MoO_3 \cdot Al_2O_3 \cdot MeAlCl_2 \cdot Me_2AlCl$	1996
	$WO_3 \cdot Al_2O_3 \cdot MeAlCl_2 \cdot Me_2AlCl$	1996
	$Ph_2C = W(CO)_5$	561
	$RuCl_3 \cdot xH_2O$	875
	$IrCl_3 \cdot xH_2O$	875
	$(\pi$-allyl)$_4Zr$	591
	$(\pi$-allyl)$_4Cr$	591
	$(\pi$-allyl)$_4W$	591
	$(\pi$-allyl)$_4Mo$	591
	$IrCl_2$ (cyclo-octene)$_2$	874
	$IrCl(CO)$ (cyclo-octene)$_2$	874
	$Ir(CF_3CO_2)$ (cyclo-octene)$_2$	873, 874
	$CuCl_2(2,7$-Me_2-2,6-octadien-1,8-yl)	874
	$WCl_6 \cdot$ i-Bu_2AlCl	300
	$WF_6 \cdot EtAlCl_2$	1333

1	2	3
	KWCl₆ ·AlCl₃	1515
	KWCl₆ ·EtAlCl₂	1515
	Me₄N(CO)₅W(COPh) ·EtAlCl₂	109
	WF₆ ·Et₃Al	1743
	MoCl₅ ·Et₃Al	1743
	MoCl₃ ·Et₃Al	803, 1743
	MoCl₅ ·Et₂AlCl	1743
	WOCl₄ ·Et₃Al	1743
	W(OPh)₆ ·Et₃Al	1743
	WCl₆ ·Et₂AlCl	803, 1743, 1746
	WCl₆ ·Et₃Al	1743
	WCl₆ ·AlCl₃	1513
	WCl₆ ·AlBr₃	708
	WCl₆ ·LiAlH₄	900
	MoCl₆ ·Bu₄Sn	1268
	(π-allyl)₂M ·AlBr₃ (WF₆, TiCl₄, MoCl₅) (M= Ni, Co, Zr)	1591
	(π-allyl)₃M ·AlBr₃ (WF₆, TiCl₄, MoCl₅) (M= W, Ca)	1591
	MoCl₂(NO)₂(PPh₃)₂ ·Me₃Al₂Cl₃	2002
	W(CO)₄(o-phenantroline) ·AlBr₃	1517
	WCl₆ ·EtOH ·EtAlCl₂	11, 211, 212, 465, 469, 815, 927, 1292, 1293, 1850
	WF₆ ·EtAlCl₂ ·CO₂(O₂, H₂O)	1333
	WCl₆ ·H₂O ·EtAlCl₂(Et₃AlCl, Et₃Al₂Cl₃, i-Bu₂AlCl, i-Bu₃Al, Et₂Zn)	1290
	WCl₆ ·MeOH ·EtAlCl₂	1290
	WCl₆ ·PrOH ·EtAlCl₂	1290
	WCl₆ ·i-BuOH ·EtAlCl₂	1290
	WCl₆ ·PhOH ·EtAlCl₂	1290
	WCl₆ ·PhCMe₂OH ·EtAlCl₂	1290
	WCl₆ ·O₂ ·EtAlCl₂	1290
	WCl₆ ·Cum₂O₂ ·EtAlCl₂	1290
	WCl₆ ·PhSH ·EtAlCl₂	1290
	WCl₆ ·AcOH ·EtAlCl₂	1621
	WCl₆ ·Al ·AlCl₃	1514
	WCl₆ ·lithium palmitate ·EtAlCl₂	1622
	WCl₆ ·Al(s-BuO)₃ ·EtAlCl₂	1622
	W(CO)₄(1,5-COD) ·Y ·EtAlCl₂ ·AlCl₃ (Y= O₂, Br₂, I₂, BrCN)	1768
	W(CO)₄(1,8-COD) ·Y ·EtAlCl₂ ·AlBr₃	1768
	Mo(CO)₄(NBD) ·Y ·EtAlCl₂ ·AlCl₃	1768
	Mo(CO)₄(NBD) ·Y ·EtAlCl₂ ·AlBr₃	1768
1-methylcyclo-octene	Ph₂C=WCl₅	680
3-methylcyclo-octene	WCl₆ ·EtOH ·EtAlCl₂	211
3-phenylcyclo-octene	WCl₆ ·EtOH ·EtAlCl₂	211
5-methylcyclo-octene	WCl₆ ·EtOH ·EtAlCl₂	211
5-phenylcyclo-octene	WCl₆ ·EtOH ·EtAlCl₂	211
cyclo-octene+propene + 2-butene, + 1-pentene + 2-pentene	WOCl₄ ·Et₂AlCl	460
1,3-cyclo-octadiene	WCl₆	91, 632

1	2	3
1,5-cyclo-octadiene	$CoO \cdot MoO_3 \cdot Al_2O_3$	1323
	$Me_3SiCH_2WCl_5$	306
	$WCl_6 \cdot EtAlCl_2$	475, 569
	$WCl_6 \cdot iBu_2AlCl$	300, 1088
	$WCl_6 \cdot iBu_3Al$	1298. 300
	$WCl_6 \cdot EtOOCCN_2$	419
	$WCl_6 \cdot Et_2AlCl$	569
	$WCl_6 \cdot LiCH_2SiMe_3$	836
	$WCl_6 \cdot LiCH_2GeMe_3$	836
	$W(CO)_6 \cdot GaBr_3$	1983
	$WCl_6 \cdot AlCl_3$	1665
	$WCl_6 \cdot AlBr_3$	1665
	$MoCl_5 \cdot AlCl_3 \ (AlBr_3, AlI_3)$	1665
	$WCl_5 \cdot AlCl_3 \ (AlBr_3, AlI_3)$	1665
	$WCl_6 \cdot EtOH \cdot EtAlCl_2$	211, 1289, 1297, 1347, 1348
	$WCl_6 \cdot CP \ hydroperoxide \cdot i\text{-}Bu_3Al$	1747, 1758
	$WCl_6 \cdot ClCH_2CH_2OSiMe_3 \cdot EtAlCl_2$	1777
	$WCl_6 \cdot thiophene \cdot i\text{-}Bu_2AlCl$	300
1-methyl-1,5-cyclo-octadiene	$WCl_6 \cdot EtOH \cdot EtAlCl_2$	1297
1,5-cyclo-octadiene + 2-pentene	$WOCl_4 \cdot Et_2AlCl$	460
1,5-cyclo-octadiene + cyclo-octene	$WCl_6 \cdot i\text{-}Bu_2AlCl$	1090
	$WCl_6 \cdot EtOH \cdot EtAlCl_2$	1290
1,5-cyclo-octadiene	$WCl_6 \cdot Bz_2O_2 \cdot i\text{-}Bu_3Al$	1235
	$WCl_6 \cdot t\text{-}Bu_2O_2 \cdot i\text{-}Bu_3Al$	1235
	$WCl_6 \cdot t\text{-}BuOOH \cdot i\text{-}Bu_3Al$	1235
	$WCl_6 \cdot CumOOH \cdot i\text{-}Bu_3Al$	1235
	$WCl_6 \cdot Al \cdot AlCl_3$	1514
	$WO_3 \cdot TlCl_5 \cdot Et_2AlCl$	1563
	$TaCl_5 \cdot EtOH \cdot EtAlCl_2$	1402
	$TaCl_5 \cdot EtOH \cdot iBu_3Al$	1402
cyclononene	$WCl_6 \cdot LiAlH_4$	900
1,5-cyclononadiene	$WCl_6 \cdot LiAlH_4$	899
cyclodecene	$WOCl_4 \cdot Cum_2O_2 \cdot Et_2AlCl$	268
1,5-cyclodecadiene	$WCl_2 \cdot EtAlCl_2$	569
cyclododecene	$(\pi\text{-allyl})_4Zr$	591
	$(\pi\text{-allyl})_4Mo$	591
	$(\pi\text{-allyl})_4W$	591
	$(\pi\text{-allyl})_3Cr$	591
	$WCl_6 \cdot Et_2AlCl$	1743, 1746, 803
	$WCl_6 \cdot Et_3Al$	1743, 803
	$WCl_6 \cdot AlCl_3$	1665
	$WOCl_4 \cdot AlCl_3$	1665
	$WCl_6 \cdot AlBr_3$	708, 1665
	$MoCl_5 \cdot AlCl_3$	1665
	$MoOCl_2 \cdot AlCl_3$	1665
	$WCl_5 \cdot AlCl_3$	1665
	$MoCl_5 \cdot Et_3Al$	803
	$WCl_6 \cdot EtOH \cdot EtAlCl_2$	211
1,5,9-cyclododecatriene	$Re_2O_7 \cdot Al_2O_3$	1917
	$(\pi\text{-allyl})_4W.aluminosilicates$	833
	$WCl_6 \cdot i\text{-}Bu_2AlCl$	1087
	$KWCl_6 \cdot AlCl_3$	1515
	$KWCl_6 \cdot AlBr_3$	1515
	$KWCl_6 \cdot EtAlCl_2$	1515
	$WCl_6 \cdot AlCl_3$	1513, 1665

1	2	3
	$WOCl_4 \cdot AlCl_3$	1513, 1665
	$WCl_6 \cdot AlBr_3$	708
	$MoCl_5 \cdot AlCl_3$	1665
	$WCl_5 \cdot AlCl_3$	1665
	$MoOCl_2 \cdot AlCl_3$	1665
	$(\pi\text{-allyl})_4 W \cdot Cl_3CCOOH$	833
	$(\pi\text{-crotyl})_4 W.AlBr_3$	1895
	$WCl_6 \cdot EtOH \cdot EtAlCl_2$	211, 1289, 1347, 1348
	$WCl_6 \cdot Al \cdot AlCl_3$	1514
	$TaCl_5 \cdot EtOH \cdot EtAlCl_2$	1402
	$TaCl_5 \cdot EtOH \cdot i\text{-}Bu_3Al$	1402
1,5,9-cyclododecatriene+2-pentene	$WOCl_4 \cdot Et_2AlCl$	460
1,5,9-cyclododecatriene+ 1,5-cyclo-octadiene	$WCl_6 \cdot i\text{-}Bu_2AlCl$	1090
1,9,17-cyclotetraeicosatriene	$WCl_6 \cdot EtOH \cdot EtAlCl_2$	815
bicyclo [2.2.1]heptene	$WO_3 \cdot Al_2O_3 \ (TiO_2, ZrO_2)$	1364
	$Cr_2O_3 \cdot Al_2O_3 \ (TiO_2, ZrO_2)$	1364
	$MoO_3 \cdot Al_2O_3 \ (TiO_2, ZrO_2)$	1364
	$UO_2 \cdot Al_2O_3 \ (TiO_2, ZrO_2)$	1364
	$MoO_3 \cdot Al_2O_3 \cdot LiAlH_4$	1363
	$(\pi\text{-allyl})_4 Mo \cdot SiO_2(Al_2O_3)$	1716
	$(\pi\text{-allyl})_4 W \cdot SiO_2 \ (Al_2O_3)$	1716
	WCl_6	91, 632, 841
	$Ph_2C=W(CO)_5$	561
	$RuCl_3 \cdot xH_2O$	738, 895
	$IrCl_3$	738, 895
	$OsCl_3$	738
	$IrCl_3 \ (1,5\text{-COD})$	632, 1942
	$IrCl_3 \ (\text{cyclo-octene})$	632, 1942
	$IrBr_3(1,5\text{-COD})$	1942
	$IrBr_3 \ (\text{cyclo-octene})$	1942
	$(\pi\text{-allyl})_4 Zr$	591
	$(\pi\text{-allyl})_4 Mo$	591
	$(\pi\text{-allyl})_4 W$	591
	$(\pi\text{-allyl})_3 Cr$	591
	$ReCl_5$	632, 841
	$MoCl_5$	632, 839, 841
	chloroiridic acid (IV)	401
	$IrCl_2 \ (\text{cyclo-octene})$	874
	$IrCl(CO)(\text{cyclo-octene})$	874
	$Ir(CF_3COO)_2\text{cyclo-octene}$	873, 874
	$RuCl_2(2,7\text{-}Me_2\text{-}2,6\text{-octadien-1,8-yl})$	873, 874
	$Ru(CF_3CO_2)_2 \ (2,7\text{-}Me_2\text{-}2,6\text{-octadien-1,8-yl})$	873, 874
	$WCl_6 \cdot i\text{-}Bu_2AlCl$	300
	$WCl_6 \cdot i\text{-}Bu_3Al$	300
	$WCl_6 \cdot Ph_4Sn$	455
	$WOCl_4 \cdot Bu_4Sn$	455
	$TiCl_4 \cdot Et_3Al$	910, 916, 1050
	$MoCl_5 \cdot Et_3Al$	916, 917
	$MoCl_5 \cdot Et_2AlCl$	916
	$TiCl_4 \cdot EtMgBr$	1176
	$(\pi\text{-allyl})_2 M \cdot AlBr_3 \ (WF_6, TiCl_4, MoCl_5)$ $(M=Ni, Co, Zr)$	1591
	$(\pi\text{-allyl})_3 M \cdot AlBr_3 \ (WF_6, TiCl_4, MoCl_5) \ (M=W, Cr)$	1591

1	2	3
	$RuCl_3 \cdot PPh_3$	463
	$TiCl_4 \cdot LiAlHep_4$	1043
	$WCl_6 \cdot EtOH \cdot EtAlCl_2$	1150
	$WCl_6 \cdot$ thiophene $\cdot$ i-Bu_2AlCl	300
	$TaCl_5 \cdot EtOH \cdot EtAlCl_2$	1402
	$TaCl_5 \cdot EtOH \cdot$ i-Bu_3Al	1403
	$TiCl_4 \cdot LiAlEt_3Bu \cdot$ amines	1051
	$TiCl_4 \cdot LiAlEt_3Bu \cdot$ dioxane	1051
	$TiCl_4 \cdot LiAlEt_3Bu \cdot$ pyridine	1051
	$TiCl_4 \cdot LiAlEt_3Bu \cdot Bu_2S$	1051
	$TiCl_4 \cdot Et_3Al \cdot$ amines	910, 1050, 1051
	$TiCl_4 \cdot Et_3Al \cdot$ dioxane	1051
	$TiCl_4 \cdot Et_3Al \cdot$ pyridine	1051
	$TiCl_4 \cdot Et_3Al \cdot Bu_2S$	1051
5-methylbicyclo [2.2.1]-heptene	$TiCl_4 \cdot Et_3Al$	984
5-butenylbicyclo [2.2.1]-heptene	$TiCl_4 \cdot LiAl(C_{10}H_{22})_4$	288
2-phenylbicyclo [2.2.1]-heptene	$WCl_6 \cdot Et_3Al$	1590
5-phenyl-bicyclo [2.2.1]-heptene	$WCl_6 \cdot Et_2AlCl$	1900
5-carbomethoxy-bicyclo-[2.2.1]-2-heptene	$WO_3 \cdot Al_2O_3$	1492
	$WO_3 \cdot Al_2O_3 \cdot LiAlH_4$	1492
	$IrCl_3$ (1,5-cyclo-octadiene)	1942
	$IrCl_3$ (cyclo-octene)	1942
	$WCl_6 \cdot Bz_2O_2 \cdot Et_2AlCl$	1862
	$WCl_6 \cdot Cr(acac)_3 \cdot Et_3Al$	1478
5-acetoxy-bicyclo [2.2.1]-heptene	$WO_3 \cdot TiCl_5 \cdot Et_3AlCl$	1563
bicyclo [2.2.1] heptenyl acetate	$WCl_6 \cdot EtOH \cdot Et_3Al_2Cl_3$	1444
5-cyano-bicyclo[2.2.1]-heptene	$WCl_6 \cdot EtAlCl_2$	1524
	$WCl_6 \cdot Et_2AlCl$	1928
	$WCl_6 \cdot PPh_3 \cdot Et_2AlCl$	1524
	$WCl_6 \cdot$ paraldehyde $\cdot$ i-Bu_3Al	1598, 1600
	$WCl_6 \cdot (EtO)_2CH_2 \cdot Et_2AlCl$	1937
	$WCl_6 \cdot (EtO)_2CHCH_3 \cdot Et_2AlCl$	1723, 1727
	$WCl_6 \cdot Al(OEt)_3 \cdot Et_2AlCl$	1565
	$WCl_6 \cdot Al(OH)_3 \cdot Et_2AlCl$	1568
	$WO_3 \cdot TiCl_5 \cdot Et_2AlCl$	1563
	$WCl_6 \cdot$ Na-ethylenediaminetetra-acetate-Mg $\cdot Et_2AlCl$	1566
	$WCl_6 \cdot$ polyvinyl alcohol $\cdot Et_2AlCl$	1570
	$WCl_6 \cdot AlPO_4 \cdot Et_2AlCl$	1569
	$WCl_6 \cdot PCl_5 \cdot Et_2AlCl$	1564
	$WCl_6 \cdot POCl_3 \cdot Et_2AlCl$	1564
	$WCl_6 \cdot (EtO)_2CH_2 \cdot Et_3Al \cdot Me_3SiOH$	1675
2-cyano-bicyclo [2.2.1]-heptene	$WCl_6 \cdot EtOH \cdot Et_3Al_2Cl_3$	1445
5-(2-pyridyl)-bicyclo-[2.2.1] heptene	$WCl_6 \cdot Et_3N \cdot Et_2AlCl$	1634
2,3-dicarboxyanhydride-bicyclo [2.2.1]-5-heptene	$WCl_6 \cdot (EtO)_2CHCH_3 \cdot Et_2AlCl$	1635
5,6-bis-chloromethyl-bicyclo [2.2.1] heptene	$WCl_6 \cdot Et_2AlCl$	1586
	$WCl_6 \cdot BuOH \cdot Et_2AlCl$	1724, 1726
2,5-dimethyl-5-carbomethoxy-bicyclo [2.2.1] heptene	$WCl_6 \cdot EtOH \cdot Et_3Al_2Cl_3$	1444
bicyclo [2.2.1] heptadiene	$MoCl_5 \cdot Et_3Al$	917
benzobicyclo [2.2.1] heptene	$WCl_6 \cdot Et_2AlCl$	1632

1	2	3
bicyclo [2.2.1] heptene + cyclopentene	$WCl_6 \cdot t\text{-BuOOH} \cdot i\text{-Bu}_3Al$	1235
	$WCl_6 \cdot Cum_2O_2 \cdot i\text{-Bu}_3Al$	1235
	chloroiridic (IV) acid	401
	$IrCl_2(\text{cyclo-octene})_2$	874
	$IrCl(CO)(\text{cyclo-octene})_2$	874
	$Ir(CF_3CO_2)$ (cyclo-octene)$_2$	873, 874
	$RuCl_3(2,7\text{-Me}_2\text{-2,6-octadien-1,8-yl})$	874
	$Ru(CF_3CO_2)_2(2,7\text{-Me}_2\text{-2,6-} \\ \text{octadien-1,8-yl})$	873, 874
	$WCl_6 \cdot \text{cyclopentyl hydroperoxide} \cdot i\text{-Bu}_3Al$	1235, 1758
	$WCl_6Bz_2O_2 \cdot i\text{-Bu}_3Al$	1235
	$WCl_6 \cdot t\text{-Bu}_2O_2 \cdot i\text{-Bu}_3Al$	1235
bicyclo [2.2.1] heptene + chloro cyclo-octene	chloroiridic (IV)acid	401
	$IrCl_2(\text{cyclo-octene})_2$	874
	$Ir(CF_3CO_2)$ (cyclo-octene)$_2$	873, 874
	$RuCl_2(2,7\text{-Me}_2\text{-2,6-octadien-1,8-yl})$	874
5-phenyl-bicyclo [2.2.1]- 2-heptene+5-cyanobicyclo- [2.2.1]-2-heptene	$WCl_6 \cdot Et_2AlCl$	1901
bicyclo [2.2.1]heptadiene + cyclopentene	$WCl_6 \cdot \text{epichlorhydrin} \cdot Et_2AlCl$	1799
	$WCl_6 \cdot ClCH_2CH_2OH \cdot Et_2AlCl$	1972
	$WCl_6 \cdot ClCH_2CH_2OH \cdot i\text{-Bu}_3Al$	1972
	$WO_3 \cdot Al_2O_3(TiO_2, ZrO_2)$	1764
	$MoO_3 \cdot Al_2O_3$	1764
	$Cr_2O_3 \cdot Al_2O_3$	1764
	$UO_2 \cdot Al_2O_3$	1764
dicyclopentadiene	$RhCl_3 \cdot xH_2O$	274
	$IrCl_3$	274
	$PdCl_2$	274
	$NiBr(\pi\text{-allyl})$	274
	$TiCl_4 \cdot Et_2AlCl$	270
	$TiCl_3 \cdot Et_2AlCl$	270
	$VCl_4 \cdot Et_2AlCl$	270, 274
	$Cl(acac)_3 \cdot Et_2AlCl$	270, 274
	$WCl_6 \cdot Et_2AlCl$	270, 274
bicyclo[3.2.0]-2,6-heptadiene	$RhCl_3 \cdot xH_2O$	270
	$RuCl_3 \cdot xH_2O$	270
	$IrCl_3$	270
	$PdCl_2$	270
	$NiBr(\pi\text{-allyl})$	270
	$TiCl_4 \cdot Et_2AlCl$	270
	$TiCl_3 \cdot Et_2AlCl$	270
	$VCl_4 \cdot Et_2AlCl$	270, 274
	$Cr(acac)_3 \cdot Et_2AlCl$	270, 274
	$WCl_6 \cdot Et_2AlCl$	270, 274
bicyclo[2.2.2.]octene	WCl_6	632
	$WCl_6 \cdot EtOH \cdot EtAlCl_2$	815
bicyclo [4.2.0]-7-octene	$RhCl_3 \cdot xH_2O$	270
	$RuCl_3 \cdot xH_2O$	270
	$IrCl_3$	270
	$PdCl_2$	270
	$NiBr(\pi\text{-allyl})$	270
	$TiCl_4 \cdot Et_2AlCl$	270
	$TiCl_3 \cdot Et_2AlCl$	270

1	2	3
	VCl$_4$·Et$_2$AlCl	270, 274
	Cr(acac)$_3$·Et$_2$AlCl	270, 274
	WCl$_6$·Et$_2$AlCl	270, 274
bicyclo [3.3.0] octene	WCl$_6$·EtOH·EtAlCl$_2$	815
bicyclo [6.1.0]-4-nonene	WCl$_6$·Et$_2$AlCl	861
9,9-dichlorobicyclo [6.1.0]-4-nonene	WCl$_6$·Et$_2$AlCl	861, 863
9,9-dibromobicyclo[6.1.0]-4-nonene	WC$_6$·Et$_2$AlCl	863
9-bromo-9-chlorobicyclo-[6.1.0.]--4-nonene	WCl$_6$·Et$_2$AlCl	863
bicyclo [4.3.0]-3,7-nonadiene	WCl$_6$·EtOH·EtAlCl$_2$	815
tricyclo [5.2.1.0.^{2,6}]8-decene	WCl$_6$·EtOH·EtAlCl$_2$	815
tetrahydroindene	WCl$_6$·Et$_2$AlCl	271
	MoCl$_5$·Et$_3$Al	803
	TiCl$_4$·Et$_3$Al	271
	VCl$_4$·Et$_3$Al	271
	VOCl$_3$·Et$_3$Al	271
	V(acac)$_3$·Et$_2$ACl	271
	Cr(acac)$_3$·Et$_2$AlCl	271
	CrCl$_3$Py$_3$·Et$_3$Al	271
	MoO$_2$(acac)$_2$·Et$_3$Al	271
	CrO$_2$Cl·Et$_2$AlCl	271
tetrahydroindane	MoCl$_5$·Et$_3$Al	803
	TiCl$_3$·Et$_3$Al	271
	TiCl$_4$·Et$_3$Al	271
polypentenamer+butadiene	WCl$_6$·Et$_2$AlCl	1382
	WCl$_6$·Et$_2$Be	1382
	WOCl$_4$·Et$_3$Al	1382
polyoctenamer+butadiene	WCl$_6$·Et$_2$AlCl	1382
	WCl$_6$·Et$_2$Be	1382
	WOCl$_4$·Et$_3$Al	1382
poly(diene+ethylene+propene)+1-,5-cyclo-octadiene	WCl$_6$·EtOH·EtAlCl$_2$	1849
polybutadiene+cyclopentene	WCl$_6$·ClCH$_2$CH$_2$OH·Et$_2$AlCl	1798, 1802
polybutadiene+ cyclo-octene	WCl$_6$·EtOH·EtAlCl$_2$	1849
polycyclododecatriene+poly-cyclo-octene	WCl$_6$·EtOH·EtAlCl$_2$	1850
polybenzobicyclo[4.2.1]-2,6-nonadiene+cyclo-octene	WCl$_6$·EtOH·EtAlCl$_2$	1849

TABLE V — Alkyne metathesis reactions

Alkyne	Catalytic system	References
1	2	3
Acetylene	MoO$_3$·Al$_2$O$_3$	759
phenylacetylene	MoO$_3$·SiO$_2$	773
propyne	WO$_3$·Al$_2$O$_3$	768
propargyl alcohol	MoO$_3$·SiO$_2$	774, 1041
1-butyne	WO$_3$·Al$_2$O$_3$	768
3-vinyl-1-butyne	MoO$_3$·SiO$_2$	773
1-pentyne	WO$_3$·Al$_2$O$_3$	768
2-pentyne	WO$_3$·SiO$_2$	856
1-hexyne	MoO$_3$·SiO$_2$	773

References

A. Reviews and General Papers

1. A. J. Amass, *Brit. Polymer J.*, **4**, 327 (1972).
2. T. Asahara, M. Seno and M. Tobayashi, *Yukei Gosei Kagaku Kyokaishi*, **30**, 424 (1972).
3. B. D. Babitsky and V. A. Kormer, *Sint. Kauch.* **1976**, 317.
4. G. C. Bailey, *Catalysis Revs.*, **3**, 37 (1969).
5. R. L. Banks, *Top. Curr. Chem.*, **25**, 39 (1972).
6. R. L. Banks, *Chemtech.*, **9**, 494 (1979).
7. R. L. Banks, *Catalysis Chem. Soc. Specialist Report*, **4**, 100 (1981).
8. J. M. Basset and M. Leconte, *Chemtech.*, **10**, 762 (1980).
9. L. Bencze and L. Markó, *Magy. Kem. Lapja*, **27**, 213 (1972).
10. C. Boelhouwer and E. Verkujlen, *Riv. Ital. Sostanze Grasse*, **53**, 237 (1976).
11. N. Calderon, *Accounts Chem. Res.*, **5**, 127 (1972).
12. N. Calderon, *J. Macromol. Rev.-C*, **7**, 105 (1972).
13. N. Calderon in "The Chemistry of Double-Bonded Functional Groups", Part 2, S. Patai (Ed.), Wiley, London, 1977, pp 913.
14. N. Calderon, J. P. Lawrence and E. A. Ofstead, *Adv. Organomet. Chem.*, **17**, 449 (1979).
15. C. P. Casey in "Transition Metal Organometallics in Organic Synthesis", H. Alper (Ed.), vol. 1, ch. 3, Academic Press, New York, 1976.
16. Y. Chauvin and M. Commereuc, *Inst. Fr. Pétrole, Ann. Combust. Liquid.*, **29**, 73 (1974).
16a. *Chem. Eng. News*, **45**, 51 (25 Sep 1967).
16b. *Chem. Eng. News*, **46**, 54 (15 Apr 1968).
17. G. Dall'Asta, *Rubber Chem. Technol.*, **47**, 515 (1974).
18. G. Dall'Asta, *Stereo Rubber*, **1977**, 285.
19. B. A. Dolgoplosk, *Vysokomol. Soedin.*, *A*, **16**, 1171 (1974).
20. B. A. Dolgoplosk, *Usp. Khim.*, **46**, 2027 (1977).
21. B. A. Dolgoplosk, *Polymer Symposia*, **67**, 99 (1980).
22. V. Drăguțan, *Studii Cerc. Chim.*, **20**, 1171 (1972).
23. V. Drăguțan, *Studii Cerc. Chim.*, **22**, 293 (1974).
24. V. Drăguțan, *Studii Cerc. Chim.*, **22**, 455 (1974).
25. M. F. Farona, *Chemtech.*, **8**, 40 (1978).
26. V. S. Fel'dblyum, *Zh. Vses. Khim. O-va*, **22**, 23 (1977).
27. V. S. Fel'dblyum, *Neftekhimyia*, **17**, 375 (1977).
28. V. S. Fel'dblyum, "Dimerizatsiya i Disproportsionirovanie Olefinov", Khimiya, Moscow. 1978.
29. A. Frantisek and J. Novansky, *Ropa Uhlie*, **20**, 439 (1978).
30. W. Graulich, *Chimia*, **28**, 534 (1974).
31. R. H. Grubbs, *J. Organomet. Chem. Libr.*, **1**, 423 (1976).
32. R. H. Grubbs, *Progress Inorg. Chem.*, **24**, 1 (1978).
33. R. H. Grubbs in "Comprehensive Organometallic Chemistry", G. Wilkinson (Ed.), vol. 5, ch. 11, Pergamon Press, Oxford, 1981.
34. P. Günther. F. Haas, G. Marwede, K. Nützel, W. Oberkirch, G. Pampus, N. Schön and J. Witte, *Angew. Makromol. Chem.*, **14**, 87 (1970).
35. R. J. Haines and G. J. Leigh, *Chem. Soc. Revs.*, **4**, 115 (1975).
36. L. Hocks, *Bull. Soc. Chim. France*, **1975**, 1893.
37. W. B. Hughes, *Organomet. Chem. Synth.*, **1**, 341 (1972).
38. W. B. Hughes, *Adv. Chem. Ser.*, **132**, 192 (1974).
39. W. B. Hughes, *Ann. New York Acad. Sci.*, **295**, 271 (1977).
40. K. Hummel, *Pure Appl. Chem.*, **54**, 351 (1982).
40a. K. Hummel, 5th *Internat. Symp. Olefin Metathesis*, Graz, August 22—26, 1983.
41. T. Ikariya and S. Yoshikawa, *Kagaku no Ryoichi*, **36**, 456 (1982).
42. C. Inoue and K. Hirota, *Yuki Gosei Kagaku Kyokai Shi*, **28**, 744 (1970).
43. K. J. Ivin, J. J. Rooney, L. Bencze, J. C. Hamilton, L.-M. Lam, G. Lapienis, B. S. R. Reddy and H. H. Thoi, *Pure Appl. Chem.*, **54**, 447 (1982).

44. Y. Kamiya, *Sekiyu Gakkai Shi*, **16**, 540 (1973).
45. T. J. Katz, *Adv. Organomet. Chem.*, **16**, 283 (1977).
46. M. L. Khidekel', A. D. Shebaldova and I. V. Kalechits, "Kataliticheskoe Disproportsio-nirovanie Uglevodorodov", Nauka, Moscow, 1970.
47. M. L. Khidekel', A. D. Shebaldova and I. V. Kalechits, *Russ. Chem. Rev.*, **40**, 669 (1971).
48. A. A. Makarenya, *Nauka Tekhn. Vspr. Ist. Teor.*, **1973**, 104.
49. A. A. Makarenya, *Sh. Nauch. Met. Statei po Khim.*, **1977**, 33.
50. K. L. Makovetskii, *Itogi Nauk Tekh., Khim. Technol. Vysokomol. Soedin.* **9**, 129 (1977).
51. K. Miyahara, *J. Res. Inst. Catal. (Hokkaido University)*, **28**, 279 (1980).
52. J. C. Mol and J. A. Moulijn, *Adv. Catalysis*, **24**, 131 (1971).
53. J. C. Mol, *Chem. Weekbl. Mag.*, **1978**, 467.
54. R. Nakamura and E. Echigoya, *Sekiyu Gakkai Shi*, **19**, 707 (1976).
55. R. Nakamura and E. Echigoya, *Sekiyu Gakkai Shi*, **19**, 805 (1976).
56. R. Nakamura, *Petrotechn.*, **4**, 623 (1981).
57. E. A. Ofstead in "Encyclopedia of Chemical Technology" (Kirk-Othmer Eds.), 3rd Ed., vol. 8, p. 592, Elsevier, Amsterdam, 1977.
58. R. Ohm and C. Stein, in "Encyclopedia of Chemical Technology" (Kirk-Othmer Eds.), 3rd Ed., vol. 18, p. 436, Elsevier, Amsterdam, 1982.
59. M. Okada, *Kobunshi*, **25**, 477 (1976).
60. J. Otton, *Inf. Chem.*, **201**, 161 (1980).
61. R. Pearce, *Catal. Chem. Processes*, **1981**, 194.
62. J. J. Rooney and A. Stewart, *Catalysis, (Chem. Soc. Specialist Report)*, **1**, 277 (1977).
63. T. Saegusa, *Macromol., Main Lect. Int. Symp., 27th*, **1981**, 27.
64. R. Streck, *Chem.-Ztg.*, **99**, 397 (1975).
65. S. Takahashi and N. Haghihara, *Kagaku*, **31**, 561 (1976).
66. T. Tatsumi, *Sekiyu Gakkai Shi*, **19**, 996 (1976).
67. R. Taube and K. Seyferth, *Wiss. Z. Tech. Hochsch. Chem. Leuna-Merseburg*, **17**, 350 (1975).
68. C. P. Tsonis, *J. Appl. Polymer Sci.*, **26**, 3525 (1981).
69. J. Tsuji, *Kagaku No Ryoiki, Zokan*, **89**, 169 (1970).
70. H. Weber, *Chem. Unserer Zeit*, **11**, 22 (1977).
71. S. Yoshitomi, *Sekiyu Gakkai Shi*, **13**, 92 (1970).

B. Full Papers, Communications and Ph. D. Theses

72. H. Abendroth and E. Canji, *Makromol. Chem.*, **176**, 775 (1975).
73. E. L. Abramenko, B. D. Babitskii, T. T. Denisova, A. S. Khachaturov and A. I. Syatkovskii, *Vysokomol. Soedin., B*, **23**, 515 (1981).
74. G. J. A. Adam, S. G. Davis, K. A. Ford, M. Ephritikhine, P. F. Todd and M. L. H. Green, *J. Mol. Catalysis*, **8**, 15 (1980).
75. F. Adamek and J. Novansky, *Ropa Uhlie*, **21**, 440 (1979).
76. F. Adamek and J. Novansky, *Zh. pr. Chemickotechnol. Fac. SVST 1977/1978, Bratislava*, **1981**, 81.
77. F. Adamek and J. Novansky, *Zh. pr. Chemickotechnol. Fac. SVST 1977/1978, Bratislava*, **1981**, 89.
78. C. T. Adams and S. G. Brandenberger, *J. Catalysis*, **13**, 360 (1969).
79. A. Agapiou and E. McNelis, *J. Chem. Soc., Chem. Commun.*, **1975**, 187.
80. A. Agapiou and E. McNelis, *J. Organomet. Chem.*, **99**, C47 (1975).
81. A. Agapiou, *Ph. D. Thesis, Diss. Abstr. Int. B*, **38**, 682 (1979).
82. A. Aivasidis, *Ph. D. Thesis*, TH Aachen, 1979.
83. A. W. Aldag, C. J. Lin and A. Clark, *Rec. Trav. Chim.*, **96**, M27 (1977).
84. A. W. Aldag, C. J. Lin and A. Clark, *J. Catalysis*, **51**, 278 (1978).
85. R. J. Al-Essa, R. J. Puddephatt, M. A. Quyser and C. F. H. Tipper, *J. Amer. Chem. Soc.*, **101**, 364 (1979).
86. R. K. Aliev, A. A. Kadushin, I. L. Tsikovskaya and O. V. Krylov, *Izvest. Akad. Nauk S.S.S.R., Ser. Khim.*, **1977**, 1004.
87. R. K. Aliev, I. L. Tsikovskaya, A. A. Kadushin and O. V. Krylov, *React. Kin. Catal. Lett.*, **8**, 257 (1978).

88. R. K. Aliev, I. L. Tsikovskaya, A. A. Kadushin, K. N. Spiridonov and O. V. Krylov, *React. Kin. Catal. Lett.*, **8**, 347 (1978).
89. A. B. Alimuniar, J. H. Edwards, K. Harper and W. J. Feast, *4th Internat. Symp. Olefin Metathesis*, Belfast, September 1—4, **1981**.
90. A. B. Alimuniar, J. H. Edwards and W. J. Feast, *4th Internat. Symp. Olefin Metathesis*, Belfast, September 1—4, 1981.
90a. A. B. Alimuniar, J. H. Edwards, W. J. Feast and L. A. H. Shahaba, *5th Internat. Symp. Olefin Metathesis*, Graz, August 22—26, 1983.
91. A. J. Amass, T. A. McGourtey and C. N. Tuck, *European Polymer J.*, **12**, 93 (1976).
92. A. J. Amass and C. N. Tuck, *European Polymer J.*, **14**, 817 (1978).
93. A. J. Amass and J. A. Zurimendi, *J. Mol. Catalysis*, **8**, 243 (1980).
94. A. J. Amass and T. A. McGourtey, *European Polymer J.*, **16**, 235 (1980).
95. A. J. Amass and J. A. Zurimendi, *European Polymer J.*, **17**, 1 (1981).
96. A. J. Amass, M. Lotfipour, J. A. Zurimendi and D. Gregory, *4th Internat. Symp. Olefin Metathesis*, Belfast, September 1—4, 1981.
97. A. J. Amass and J. A. Zurimendi, *Polymer*, **23**, 211 (1982).
98. A. J. Amass, *Can. J. Chem.*, **60**, 6 (1982).
98a. A. J. Amass, D. Gregory and M. Lotfipour, *5th Internat. Symp. Olefin Metathesis*, Graz, August 22—26, 1983.
99. A. B. Amerik and V. A. Poletaev, *Nauchn. Osn. Neftekhim. Sint.*, **1977**, 148.
100. K. L. Anderson and T. D. Brown, *Hydrocarbon Process*, **55**, 119 (1976).
101. A. Andreev, R. Edreva-Kardjieva and N. Neshev, *Rec. Trav Chim.*, **96**, M23 (1977).
102. A. Andreini and J. C. Mol, *J. Colloid. Interface Sci.*, **84**, 57 (1981).
103. Anon., *Res. Discl.*, **143**, 8 (1976).
104. M. Anpo, I. Tanahashi and Y. Kubokawa, *J. Chem. Soc., Faraday Trans. 1*, **78**, 2121 (1982).
105. J. O. Apatu, D. C. Chapman and H. Heaney, *J. Chem. Soc., Chem. Commun.* **1981**, 1079.
106. Y. Arbab-Zadeh and B. Blouri, *Bull. Soc. Chim. France*, **1976**, 1575.
107. J. I. C. Archibald, J. J. Rooney and A. Stewart, *J. Chem. Soc., Chem. Commun.*, **1975**, 547.
108. J. I. C. Archibald, J. J. Rooney and A. Stewart, *4th Internat. Symp. Olefin Metathesis*, Belfast, September 1—4, 1981.
109. J. P. Arlie, Y. Chauvin, D. Commerauc and J. P. Souflet, *Makromol. Chem.*, **175**, 861 (1974).
110. D. R. Armstrong, *Rec. Trav. Chim.*, **96**, M17 (1977).
111. W. Ast and K. Hummel, *Naturwiss.*, **57**, 54 (1970).
112. W. Ast and K. Hummel, *Kautch. Gummi Kunstst.*, **24**, 220 (1971).
113. W. Ast, G. Rheinwald and R. Kerber, *Makromol. Chem.*, **177**, 39 (1976).
114. W. Ast, G. Rheinwald and R. Kerber, *Makromol. Chem.*, **177**, 1341 (1976).
115. W. Ast, G. Rheinwald and R. Kerber, *Makromol. Chem.*, **177**, 1349 (1976).
116. W. Ast, C. Zott, H. Bosch and R. Kerber, *Rec. Trav. Chim.*, **96**, M81 (1977).
117. W. Ast, G. Rheinwald and R. Kerber, *Rec. Trav. Chim.*, **96**, M127 (1977).
118. W. Ast, H. Bosch, K. Grundmeyer and R. Kerber, *3rd Internat. Symp. Olefin Metathesis, Abstracts*, p. 27, Lyons, September 10—12, 1979.
119. W. Ast, C. Zott and R. Kerber, *Makromol. Chem.*, **180**, 315 (1979).
120. W. Ast, H. Bosch and R. Kerber, *Angew. Makromol. Chem.*, **76/77**, 67 (1979).
121. V. V. Atlas, I. I. Pis'man and A. M. Bakhshi-Zade, *Neftepererab. Neftekhim.* **5**, 30 (1969).
122. V. V. Atlas, I. I. Pis'man and A. M. Bakhshi-Zade, *Khim. Prom.*, **10**, 17 (1969).
123. V. V. Atlas, I. I. Pis'man and A. M. Bakhshi-Zade, *Khim. Prom.*, **45**, 734 (1969).
124. E. D. Babich, B. N. Bespalova, V. M. Vdovin, V. F. Mironov, V. I. Shiryaev and V. P. Kochergin, *Izvest. Akad. Nauk SSSR, Ser. Khim.*, **1976**, 2609.
125. E. D. Babich, N. B. Bespalova, V. M. Vdovin, T. K. Gar and V. F. Mironov, *Izvest. Akad. Nauk SSSR, Ser. Khim.*, **1977**, 960.
126. B. D. Babitskii, L. V. Gavrilova, T. T. Denisova and A. I. Syatkovskii, *Dokl. Akad. Nauk SSSR.*, **257**, 378 (1981).
127. E. R. Badamshina, M. A. Tlenkopachev, Yu. V. Korshak and B. A. Dolgoplosk, *Vysokomol. Soedin. Ser. B*, **23**, 828 (1981).

128. E. R. Badamshina, G. I. Timofeeva, Yu. V. Korshak, A. A. Berlin, V. M. Vdovin, D. F. Kutepov and S. A. Pavlova, *Vysokomol. Soedin. Ser. A*, **24**, 143 (1982).

129. R. Baker and M. J. Grimmin, *Tetrahedron Letters*, **1977**, 441.

130. H. Balcar and B. Matyska, *React. Kinet. Catal. Lett.*, **20**, 187 (1982).

130a. D. S. Banascak, *5th Internat. Symp. Olefin Metathesis*, Graz, August 22−26, 198.3.

131. R. L. Banks and G. C. Bailey, *Ind. Eng. Chem. Prod. Res. Develop.*, **3**, 170 (1964)

132. R. L. Banks and R. B. Regier, *Ind. Eng. Chem. Prod. Res. Develop.*, **10**, 46 (1971).

133. R. L. Banks and L. F. Heckelsberg, *Ind. Eng. Chem. Prod. Res. Develop.* **14**, 33 (1975).

134. R. L. Banks and S. F. Cook, *Sovrem. Probl. Nauki o Katalize, Sib. Chteniya po Katalizii*, **1977**, Novosibirsk (Pub. **1978**), 149.

135. R. L. Banks, *Chemtech.*, **1979**, 494.

136. R. L. Banks, *Catal. Org. Synth.*, **1978** (Pub. 1980), 233.

137. R. L. Banks, *J. Mol. Catalysis*, **8**, 269 (1980).

138. R. L. Banks, D. S. Banasiak, P. S. Hudson and J. R. Norell, *J. Mol. Catalysis*, **15**, 21 (1982).

138a. R. L. Banks and S. G. Kukes, *5th Internat. Symp. Olefin Metathesis*, Graz, August 22−26, 1983.

139. A. Baranski, S. Ceckiewicz and J. Galuszka, *Bull. Acad. Pol. Sci., Ser. Sci. Chim.*, **24**, 645 (1976).

140. S. J. Barer, *Ph. D. Thesis, Diss. Absr. Int. B*, **35**, 2647 (1974).

141. A. N. Bashkirov, R. A. Fridman, S. M. Nosakova and L. G. Liberov, *Deposited Doc. VINTI*, **1974**, 1125.

142. A. N. Bashkirov, R. A. Fridman, S. M. Nosakova, G. L. Liberov and S. I. Beilin, *Izvest. Akad. Nauk SSSR, Ser. Khim.*, **1975**, 2132.

143. A. N. Bashkirov, R. A. Fridman, S. M. Nosakova, L. G. Liberov, E. D. Babich and V.M. Vdovin, *Kinet. Katal.*, **16**, 1353 (1975).

144. A. N. Bashkirov, R. A. Fridman, S. M. Nosakova, L. G. Liberov, E. D. Babich, V. M. Vdovin, G. M. Larin and V. A. Kolosov, *Neftekhim.*, **16**, 823 (1976).

145. P. Basilio, *Chim. Ind.*, **48**, 1056 (1966).

146. J. M. Basset, G. Coudurier, R. Mutin and M. Praliaud, *J. Catalysis*, **34**, 152 (1974).

147. J. M. Basset, G. Coudurier, R. Mutin, H. Praliaud and Y. Trambouze, *J. Catalysis*, **34**, 196 (1974).

148. J. M. Basset, R. Mutin, G. Descotes and D. Sinou, *Compt. Rend. Ser. C* **280**, 1181 (1975).

149. J. M. Basset, J. L. Bilhou, R. Mutin and A. Theolier, *J. Amer. Chem. Soc.*, **97**, 7376 (1975).

150. J. M. Basset, J. L. Bilhou, Y. Ben Taarit and H. Praliaud, *Rol Koordinatsii v Katalize*, **1976**, 171.

151. J. M. Basset, Y. Ben Taarit, J. L. Bilhou, J. Bousquet, R. Mutin and A. Theolier, *Proc. Internat. Congr. Catal., 6th, 1976* **(Pub. 1977)**, **1**, 570.

152. J. M. Basset and M. Leconte, *Fundam. Res. Homogeneous Catal.*, **3**, 285 (1979).

153. J. M. Basset and M. Leconte, *Chemtech.*, **10**, 762 (1980).

153a. J. M. Basset, C. Larroche, J. P. Laval and A. Lattes, *5th Internat. Symp. Olefin Metathesis*, Graz, August 22−26, 1983.

154. C. D. Batich, *Ph. D. Thesis. Diss. Abstr. Int. B*, **35**, 2647 (1974).

155. S. H. Bauer and P. Jeffers, *J. Amer. Chem. Soc.*, **87**, 3278 (1965).

156. S. H. Bauer, *Ann. Rev. Phys. Chem.*, **30**, 271 (1979).

157. J. Beger, R. Sass and G. Zimmermann, *J. Prakt. Chem.*, **319**, 790 (1977).

158. J. Beger, R. Sass and G. Zimmermann, *J. Prakt. Chem.*, **320**, 283 (1978).

159. J. W. Begley and R. T. Wilson, *J. Catalysis*, **9**, 375 (1967).

160. S. I. Beilin, B. A. Dolgoplosk, I. N. Markewich, A. N. Bashkirov, Yu. B. Kryukov, S. M. Nosakova, R. A. Fridman and L. G. Liberov, *Izvest. Akad. Nauk SSSR, Ser. Khim.*, **1974**, 947.

161. A. BenCheick, M. Petit, A. Mortreux and F. Petit, *J. Mol. Catalysis*, **15**, 93 (1982).

162. L. Bencze and L. Markó, *J. Organomet. Chem.*, **28**, 271 (1971).

163. L. Bencze, *J. Organomet. Chem.*, **37**, C37 (1972).

164. L. Bencze, A. Redey and L. Markó, *Hung. J. Ind. Chem.*, **4**, 453 (1973).

165. L. Bencze and L. Markó, *J. Organomet. Chem.*, **69**, C19 (1974).

166. L. Bencze and L. Markó, *Proc. Internat. Conf. Coord. Chem., 16th*, **4**, 32 (1974).

167. L. Bencze L. Markó, R. Opitz and K.-H. Thiele, *Hung. J. Ind. Chem.*, **4**, 15 (1976).
168. L. Bencze, J. Pallos and L. Markó, *J. Mol. Catalysis*, **2**, 139 (1977).
169. L. Bencze, K.-H. Thiele and V. Marquardt, *Rec. Trav. Chim.*, **96**, 8 (1977).
170. L. Bencze, K. J. Ivin and J. J. Rooney, *3rd Internat. Symp. Olefin Metathesis, Abstracts*, p. 30, Lyons, September, 10—12, 1979.
171. L. Bencze, K. J. Ivin and J. J. Rooney, *3rd Internat. Symp. Olefin Metathesis, Abstracts*, p. 59, Lyons, September, 10—12, 1979.
172. L. Bencze, K. J. Ivin and J. J. Rooney, *J. Chem. Soc., Chem. Commun.*, **1980**, 834.
173. L. Bencze and J. Engelhardt, *J. Mol. Catalysis*, **15**, 123 (1982).
174. L. Bencze, cited after reference 173.
174a. L. Bencze and A. Vass, *5th Internat. Symp. Olefin Metathesis*, Graz, August 22—26, 1983.
175. D. A. Ben-Efraim, C. Batich and E. Wasserman, *J. Amer. Chem. Soc.*, **92**, 2133 (1970).
176. N. B. Bespalova, E. D. Babich, V. M. Vdovin and N. S. Nametkin, *Dokl. Akad. Nauk SSSR*, **225**, 1071 (1977).
177. N. B. Bespalova and L. V. Efstifeev, *Nov. Aspekty Neftekhim. Sint.*, **1978**, 19.
178. N. B. Bespalova, E. D. Babich and V. M. Vdovin, *Izvest. Akad. Nauk SSSR*, **1979**, 1403.
179. C. G. Bielfeld, H. A. Eick and R. H. Grubbs, *Inorg. Chem.*, **12**, 2166 (1973).
180. J. L. Bilhou, J. M. Basset, R. Mutin and W. F. Graydon, *J. Chem. Soc., Chem. Commun.*, **1976**, 970.
181. J. L. Bilhou, J. M. Basset and R. Mutin, *J. Organomet. Chem.*, **87**, C4 (1975).
182. J. L. Bilhou, J. M. Basset, R. Mutin and W. F. Graydon, *J. Amer. Chem. Soc.*, **99**, 4083 (1977).
183. J. L. Bilhou, R. Mutin, M. Leconte and J. M. Basset, *Rec. Trav. Chim.* **96**, M5 (1977).
184. J. L. Bilhou and J. M. Basset, *J. Organomet. Chem.*, **132**, 395 (1977).
185. J. L. Bilhou, A. K. Smith and J. M. Basset, *J. Organomet. Chem.*, **148**, 53 (1978).
186. J. L. Bilhou, A. Theolier, A. K. Smith and J. M. Basset, *J. Mol. Catalysis*, **3**, 245 (1978).
187. P. Biloen and G. T. Pott, *J. Catalysis*, **30**, 169 (1973).
188. K. Biniecki, A. Gieysztor and S. Malinovski, *Rocz. Chem.*, **49**, 1633 (1975).
188a. A. Blaga, R. Vlădea and L. Rusnac, *5th National Conference on General and Applied Physical Chemistry*, Bucharest, September, 1—4, 1976.
189. H. A. Bockmuelen and A. W. Parkins, *J. Chem. Soc., Dalton Trans.*, **1981**, 262.
190. E. I. Bogolepova, R. A. Fridman and A. N. Bashkirov, *Izvest. Akad. Nauk SSSR, Ser. Khim.*, **1978**, 2429.
191. E. I. Bogolepova, R. A. Fridman and A. N. Bashkirov, *Izvest. Akad. Nauk SSSR, Ser. Khim.*, **1979**, 1623.
192. E. I. Bogolepova, S. B. Verbovetskaya and A. N. Bashkirov, *Neftekhim.*, **22**, 207 (1982).
192a. D. Borowczak, T. Szymanska-Buzar and J. J. Ziolkowski, *5th Internat. Symp. Olefin Metathesis*, Graz, August 22—26, 1983.
193. R. H. A. Bosma, A. P. Kouwenhoven and J. C. Mol, *J. Chem. Soc., Chem. Commun.*, **1981**, 1085.
194. R. H. A. Bosma, F. van der Aardweg and J. C. Mol, *J. Chem. Soc., Chem. Commun.*, **1981**, 1132.
195. R. H. A. Bosma, X. D. Xu and J. C. Mol, *J. Mol. Catalysis*, **15**, 187 (1982).
196. R. G. Bowman, R. Nakamura and R. L. Burwell, Jr., *Prepr., Amer. Chem. Soc. Div. Pet. Chem.*, **26**, 17 (1981).
197. C. P. C. Bradshaw, E. J. Howman and L. Turner, *J. Catalysis*, **7**, 269 (1967).
198. A. Brenner and R. L. Burwell, Jr., *J. Amer. Chem. Soc.*, **97**, 2565 (1975).
199. A. Brenner, *Ph. D. Thesis, Diss. Abstr. Int. B*, **36**, 3403 (1976).
200. A. Brenner and R. L. Burwell, Jr., *J. Catalysis*, **52**, 364 (1978).
201. A. Brenner and D. A. Hucul, *Chem. Uses Molybdenum, Proc. Internat. Conf., 3rd*, **1979**, 194.
202. A. Brenner, D. A. Hucul and S. J. Hardwick, *Inorg. Chem.*, **18**, 1478 (1979).
203. F. J. Brown, *Progr. Inorg. Chem.*, **27**, 1 (1980).
203a. J. Brunthaler, G. Leising and F. Stelzer, *5th Internat. Symp. Olefin Metathesis*, Graz, August 22—26, 1983.
204. P. L. Burk, *Ph. D. Thesis, Diss. Abstr. Int. B*, **36**, 4488 (1976).
205. R. L. Burnett and T. R. Hughes, *J. Catalysis*, **31**, 55 (1973).

206. R. L. Burwell, Jr. and A. Brenner, *Internat. Symp. Catalysis, 1974 (Publ. 1975),* P. Delmon and G. James (Eds.), Elsevier, Amsterdam, pp 157.
207. R. L. Burwell, Jr. and A. Brenner, *J. Mol. Catalysis,* **1,** 77 (1976).
208. P. Buschmeyer, *Ph. D. Thesis,* TH Aachen, 1978.
209. P. Buschmeyer, *Diplomarbeit,* TH Aachen, 1976.
210. N. Calderon, H. Y. Chen and K. W. Scott, *Tetrahedron Letters,* **1967,** 3327.
211. N. Calderon, E. A. Ofstead and W. A. Judy, *J. Polymer Sci., A-1,* **5,** 2209 (1967).
212. N. Calderon and M. C. Morris, *J. Polymer Sci., A-2,* **5,** 1283 (1967).
213. N. Calderon, E. A. Ofstead, J. P. Ward, W. A. Judy and K. W. Scott, *J. Amer. Chem. Soc.,* **90,** 4133 (1968).
214. N. Calderon and N. L. Hinrichs, *Chem. Technol.,* **4,** 627 (1974).
215. N. Calderon, E. A. Ofstead and W. A. Judy, *Angew. Chem.,* **88,** 433 (1976).
216. G. L. Caldow and R. A. McGregor, *J. Chem. Soc., A,* **1971,** 1654.
217. E. Canji and H. Perner, *Rec. Trav. Chim.,* **96,** M70 (1977).
218. E. Canji and H. Perner, *Makromol. Chem.,* **179,** 567 (1978).
219. D. J. Cardin, M. J. Doyle and M. F. Lappert, *J. Chem. Soc., Chem. Commun.,* **1972,** 927.
220. D. J. Cardin, B. Cetinkaya, M. J. Doyle and M. F. Lappert, *Chem. Soc. Revs.,* **2,** 94 (1972).
220a. C. J. Carman and C. E. Wilkes, *Macromolecules,* **7,** 40 (1974).
221. D. D. Carr, *Ph. D. Thesis, Diss. Abstr. Int. B,* **38,** 1203 (1977).
222. C. P. Casey and T. J. Burkhardt, *J. Amer. Chem. Soc.,* **94,** 6543 (1972).
223. C. P. Casey and T. J. Burkhardt, *J. Amer. Chem. Soc.,* **95,** 5833 (1973).
224. C. P. Casey and T. J. Burkhardt, *J. Amer. Chem. Soc.,* **96,** 7808 (1974).
225. C. P. Casey, H. E. Tuinstra and M. C. Saeman, *J. Amer. Chem. Soc.,* **98,** 608 (1976).
226. C. P. Casey, L. D. Albin and T. J. Burkhardt, *J. Amer. Chem. Soc.,* **99,** 2533 (1977).
227. C. P. Casey, D. M. Scheck and A. J. Shusterman, *J. Amer. Chem. Soc.,* **101,** 4233 (1979).
228. C. P. Casey, D. M. Scheck and A. J. Shusterman, *Fundam. Res. Homogeneous Catal.,* **3,** 141 (1979).
229. C. P. Casey, *Adv. Pestic. Sci.,* **2,** 148 (1979).
230. C. P. Casey and A. J. Shusterman, *J. Mol. Catalysis,* **8,** 1 (1980).
231. C. P. Casey and P. J. Brondsema, *4th Internat. Symp. Olefin Metathesis,* Belfast, September 1—4, 1981.
232. K. F. Castner and N. Calderon, *J. Mol. Catalysis,* **15,** 47 (1982).
233. E. Ceauşescu, S. Bittman, M. Dimonie, A. Cornilescu, C. Cincu, E. Nicolescu, M. Popescu, S. Coca, C. Boghină and G. Hubcă, Paper presented at the 1st National Congress of Chemistry, Bucharest, September 11—14, 1978.
234. E. Ceauşescu, S. Bittman, M. Dimonie, A. Cornilescu, C. Boghină, E. Nicolescu, M. Popescu, S. Coca, C. Cincu and G. Hubcă in "Participation at Scientific Meetings", E. Ceauşescu (Ed.), Institute of Chemical Research, Ed. Academiei, Bucharest, 1979, p. 137.
235. E. Ceauşescu, S. Bittman, M. Dimonie, A. Cornilescu, C. Boghină, E. Nicolescu, M. Popescu, S. Coca, C. Cincu and G. Hubcă in "New Research in the Field of Macromolecular Compounds", E. Ceauşescu (Ed.), Ed. Academiei, Bucharest, 1981, p. 110.
236. E. Ceauşescu, S. Bittman, A. Cornilescu, E. Nicolescu, M. Popescu, C. Radovici, M. Dimonie, S. Coca and G. Stănescu, Paper presented at the 2nd National Congress of Chemistry, Bucharest, September 7—10, 1981.
237. E. Ceauşescu, A. Cornilescu, M. Dimonie, E. Nicolescu, M. Popescu and S. Coca, Paper presented at the 5th International Microsymposium on Ionic Polymerization, Praha, July 1982; *Acta Polymerica,* **11** (1983).
238. E. Ceauşescu, M. Dimonie, A. Cornilescu, M. Gheorghiu, M. Chipară, E. Nicolescu, M. Popescu, S. Coca, C. Belloiu and V. Drăguţan, Paper presented at the Microsymposium ICECHIM, Bucharest, January 28—29, 1983.
239. E. Ceauşescu, C. Radovici, A. Cornilescu, E. Nicolescu, M. Popescu and S. Coca, Paper presented at the Microsymposium ICECHIM, Bucharest, January 28—29, 1983.
240. E. Ceauşescu, M. Dimonie, V. Drăguţan, M. Chipară, A. Cornilescu, E. Nicolescu, M. Popescu, S. Coca, C. Belloiu, C. Oprescu and G. Hubcă, *5th International Symp. Olefin Metathesis,* Graz, August 22—26, 1983.

241. E. Ceaușescu, M. Dimonie, V. Drăguțan, M. Gheorghiu, M. Chipară, A. Cornilescu, E. Nicolescu, M. Popescu, S. Coca and C. Belloiu, *5th Internat. Symp. Olefin Metathesis*, Graz, August 22—26, 1983.
242. E. Ceaușescu, M. Dimonie, A. Cornilescu, E. Nicolescu, M. Popescu, C. Coca and C. Belloiu, Paper presented at the 29th Internat. Symp. Macromolecules, Bucharest, September 5—9, 1983.
243. E. Ceaușescu, M. Dimonie, A. Cornilescu, E. Burzo, M. Chipară, E. Nicolescu, M. Popescu, S. Coca and C. Belloiu, Paper presented at the 29th Internat. Symp. Macromolecules, Bucharest, September 5—9, 1983.
244. E. Ceaușescu, M. Dimonie, M. Gheorghiu, V. Drăguțan, A. Cornilescu, E. Nicolescu, M. Popescu, S. Coca and C. Belloiu, Paper presented at the 29th Internat. Symp. Macromolecules, Bucharest, September, 5—9, 1983.
245. M. Chalganov, N. A. Filikova, E. A. Demin, B. N. Kuznetsov and Yu. I. Ermakov, *Katal. Soderzh. Nanesen. Kompleksy*, **1977**, 231.
246. J. Chatt, R. J. Haines and G. J. Leigh, *J. Chem. Soc., Chem. Commun.*, **1972**, 1202.
247. Y. Chauvin, D. Commereuc and D. Cruypelnick, *Makromol. Chem.*, **177**, 2637 (1976).
248. Y. Chauvin, D. Commereuc and G. Zaborowski, *Rec. Trav. Chim.*, **96**, M131 (1977).
249. Y. Chauvin, D. Commereuc and G. Zaborowski, *Makromol. Chem.*, **179**, 1285 (1978).
250. E. V. Chechegoeva, Y. L. Lelyukhina, I. V. Kalechits, N. N. Korneev and N. M. Pugashova, *Kinet. Katal.*, **17**, 1614 (1976).
251. E. V. Chechegoeva, L. A. Zharkova, I. V. Kalechits, N. N. Korneev and Yu. L. Lelyukhina, *Kinet. Katal.*, **18**, 1339 (1977).
251a. R. Chemelli, K. Hummel, J. Kendlbacher and H. Zekoll, *5th Internat. Symp. Olefin Metathesis*, Graz, August 22—26, 1983.
252. H. Y. Chen, *J. Polymer Sci., Part B*, **12**, 85 (1974).
253. H. Y. Chen, *Polymer Prepr., Amer. Chem. Soc., Div. Polymer Chem.*, **17**, 688 (1976).
254. L. S. Cherezova, V. A. Iakovlev and B. A. Dolgoplosk, *Dokl. Akad. Nauk SSSR*, **254**, 1410 (1980).
255. L. S. Cherezova, V. A. Iakovlev and B. A. Dolgoplosk, *Dokl. Akad. Nauk SSSR*, **258**, 1380 (1981).
256. P. Chevalier, D. Sinou and G. Descotes, *Bull. Soc. Chim. France*, **1975**, 2254.
257. P. Chevalier, D. Sinou, G. Descotes, R. Mutin and J. M. Basset, *J. Organomet. Chem.*, **113**, 1 (1976).
258. A. Clark and C. Cook, *J. Catalysis*, **15**, 420 (1969).
259. M. Clark and M. Farona, *Polymer Bull.*, **7**, 445 (1982).
260. D. Commereuc, Y. Chauvin and Cruypelnik, *Makromol. Chem.*, **177**, 263 (1976).
261. F. A. Cotton, W. T. Hall, K. J. Cann and F. J. Karol, *Macromolecules*, **14**, 233 (1981).
262. D. L. Crain, *J. Catalysis*, **13**, 110 (1969).
263. D. L. Crain and R. E. Reusser, *Prepr., Amer. Chem. Soc., Div. Petrol. Chem.*, **17**, E80 (1972).
264. L. Cvarkov, J. Velickovic and R. Vesely, *Polymeri (Zagreb)*, **2**, 127 (1981).
265. G. Dall'Asta, G. Mazzanti, G. Natta and L. Porri, *Makromol. Chem.*, **56**, 224 (1962).
266. G. Dall'Asta and G. Mazzanti, *Makromol. Chem.*, **61**, 178 (1963).
267. G. Dall'Asta and R. Manetti, *Atti Accad. Naz. Lincei*, **41**, 351 (1966).
268. G. Dall'Asta and R. Manetti, *European Polymer J.*, **4**, 145 (1968).
269. G. Dall'Asta, *J. Polymer Sci., A-1*, **6**, 2397 (1968).
270. G. Dall'Asta and G. Motroni, *J. Polymer Sci., A-1*, **6**, 2405 (1968).
271. G. Dall'Asta and G. Motroni, *Chimica e Industria*, **50**, 172 (1968).
272. G. Dall'Asta, *Corsi Semin. Chim.*, **1968**, 89.
273. G. Dall'Asta and P. Scaglione, *Rubb. Chem. Technol.*, **42**, 1235 (1969).
274. G. Dall'Asta, G. Motroni, R. Manetti and C. Tossi, *Makromol. Chem.*, **130**, 153 (1969).
275. G. Dall'Asta and G. Motroni, *Angew. Makromol. Chem.*, **16/17**, 51 (1971).
276. G. Dall'Asta and G. Motroni, *European Polymer J.*, **7**, 707 (1971).
277. G. Dall'Asta, *Makromol. Chem.*, **154**, 1 (1972).
278. G. Dall'Asta, G. Motroni and L. Motta, *J. Polymer Sci., A-1*, **10**, 16, 1601 (1972).
279. G. Dall'Asta, G. Stigliani, A. Greco and L. Motta, *Prepr., Amer. Chem. Soc., Div. Polymer Chem.*, **13**, 910 (1972).
280. G. Dall'Asta, G. Stigliani, A. Greco and L. Motta, *Chimica e Industria*, **55**, 142 (1973).

281. A. C. Daly and M. A. McKervey, *4th Internat. Symp. Olefin Metathesis*, Belfast, 1981.
282. A. C. Daly and M. A. McKervey, *Tetrahedron Letters*, **1982**, 2997.
283. E. S. Davie, D. A. Whan and C. Kemball, *J. Chem. Soc., Chem. Commun.*, **1969**, 1430.
284. E. S. Davie, D. A. Whan and C. Kemball, *J. Chem. Soc., Chem. Commun.*, **1971**, 1202.
285. E. S. Davie, D. A. Whan and C. Kemball, *J. Catalysis*, **24**, 272 (1972).
286. E. S. Davie, D. A. Whan and C. Kemball, *Proc. Internat. Congr. Catal.*, *5th*, **2**, 1205, 1972 (Publ. 1973).
286a. P. F. Dearing, A. K. Coverdale and A. Ellison, *5th Internat. Symp. Olefin Metathesis*, Graz, August 22—26, 1983.
287. A. Dedieu and E. O. Eisenstein, *Nouveau J. Chim.*, **6**, 337 (1982).
288. H. G. G. Dekking, *J. Polymer Sci.*, **55**, 525 (1961).
289. H. Demel and K. Hummel, *Makromol. Chem.*, **178**, 1699 (1977).
290. H. Demel, H.-J. Griesser and K. Hummel, *Colloid. Polymer Sci.*, **255**, 1131 (1977).
291. G. Descotes, P. Chevalier and D. Sinou, *Synthesis*, **1974**, 564.
292. S. A. Devarayan, D. R. M. Walton and G. L. Leigh, *J. Organomet. Chem.*, **181**, 99 (1979).
293. J. L. K. F. DeVries and G. T. Pott, *Rec. Trav. Chim.*, **96**, M115 (1977).
294. M. Dimonie, G. Hubcă and C. Oprescu, *Mat. Plast.*, **10**, 239 (1973).
295. M. Dimonie, C. Oprescu and G. Hubcă, *Mat. Plast.*, **10**, 417 (1973).
296. L. DiNunno and S. Florio, *Chimica e Industria*, **57**, 242 (1973).
297. R. E. Dixon, J. F. Hutto, R. T. Wilson and R. L. Banks, *Chem. Age India*, **2**, 49 (1967).
298. H. T. Dodd and K. J. Rutt, *J. Mol. Catalysis*, **15**, 103 (1982).
299. B. A. Dolgoplosk, K. L. Makovetskii and E. I. Tinyakova, *Dokl. Akad. Nauk SSSR*, **202**, 871 (1972).
300. B. A. Dolgoplosk, *Vestn. Akad. Nauk SSSR*, **1974**, 40.
301. B. A. Dolgoplosk, K. L. Makovetskii, T. G. Golenko, Yu. V. Korshak and E. I. Tinyakova, *European Polymer J.*, **10**, 901 (1974).
302. B. A. Dolgoplosk, T. G. Golenko, K. L. Makovetskii, I. A. Oreshkin and E. I. Tinyakova, *Dokl. Akad. Nauk SSSR*, **216**, 807 (1974).
303. B. A. Dolgoplosk, *5th Sb.*, *11th Mendeleevsk S'ezd Obsch. Prikl. Khim.*, *Ref. Dokl. Soobsch.*, **1975**, 172.
304. B. A. Dolgoplosk, K. L. Makovetskii, I. A. Oreshkin, E. I. Tinyakova and V.I. Svergun, *Dokl. Akad. Nauk SSSR*, **223**, 1369 (1975).
305. B. A. Dolgoplosk, K. L. Makovetskii, I. A. Oreshkin, E. I. Tinyakova and V. I. Svergun, *Dokl. Akad. Nauk SSSR*, **225**, 1360 (1975).
306. B. A. Dolgoplosk, K. L. Makovetskii, E. I. Tinyakova, T. G. Golenko and I. A. Oreshkin, *Izvest. Akad. Nauk SSSR, Ser. Khim.*, **1976**, 1084.
307. B. A. Dolgoplosk, K. L. Makovetskii, E. I. Tinyakova and I. A. Oreshkin, *Dokl. Akad. Nauk SSSR*, **228**, 1109 (1976).
308. B. A. Dolgoplosk, E. I. Tinyakova and V. A. Yakovlev, *Dokl. Akad. Nauk SSSR*, **232**, 1075 (1977).
309. B. A. Dolgoplosk, K. L. Makovetskii, Yu. V. Korshak, I. A. Oreshkin, E. I. Tinyakova and V. A. Yakovlev, *Vysokomol. Soedin.*, *A*, **19**, 2464 (1977).
310. B. A. Dolgoplosk, K. L. Makovetskii, Yu. V. Korshak, I. A. Oreshkin, E. I. Tinyakova and V. A. Yakovlev, *Rec. Trav. Chim.*, **96**, M35 (1977).
311. B. A. Dolgoplosk, I. A. Oreshkin, K. L. Makovetskii, E. I. Tinyakova, I. Ya. Ostrovskaya, I. L. Kershenbaum and G. M. Chernenko, *J. Organomet. Chem.*, **128**, 339 (1977).
312. B. A. Dolgoplosk, I. A. Kopieva, I. A. Oreshkin and E. I. Tinyakova, *3rd Internat. Symp. Olefin Metathesis*, Lyons, September 10—12, 1979.
313. B. A. Dolgoplosk, S. A. Smirnov, I. L. Kershenbaum, I. A. Oreshkin and E. I. Tinyakova, *3rd Internat. Symp. Olefin Metathesis*, Lyons, September 10—12, 1979.
314. B. A. Dolgoplosk, I. A. Oreshkin, S. A. Smirnov, I. A. Kopieva and E. I. Tinyakova, *European Polymer J.*, **15**, 237 (1979).
315. B. A. Dolgoplosk, I. A. Kopieva, I. A. Oreshkin and E. I. Tinyakova, *European Polymer J.*, **16**, 547 (1980).
316. B. A. Dolgoplosk, *Kinetika i Kataliz*, **22**, 807 (1981).
317. B. A. Dolgoplosk, *Internat. Symp. Chem. Phys.*, Moscow—Erevan (1981), **M**, 111.
318. B. A. Dolgoplosk, *J. Mol. Catalysis*, **15**, 193 (1982).
319. D. A. Dowden, *Analés real. Soc. Españ. Fis. Quim.*, **61**, 326 (1965).

320. G. Doyle, *J. Catalysis*, **30**, 118 (1973).
321. A. Dräxler, *Lichtbogen*, **198**, 4 (1980).
322. A. Dräxler, *Kautsch. Gummi. Kunstst.*, **1981**, 185.
323. J. A. K. Duplessis, P. J. Heenop and J. J. Pienaar, *S. African J. Chem.*, **35**, 42 (1982).
324. A. F. Dyke, S. A. R. Knox, P. J. Naish and G. E. Taylor, *J. Chem. Soc., Chem. Commun.*, **1980**, 803.
325. R. M. Edreva-Kardjieva and A. A. Andreev, *React. Kinet. Catal. Lett.*, **5**, 465 (1976).
326. R. M. Edreva-Kardjieva and A. A. Andreev, *Rec. Trav. Chim.*, **96**, M141 (1977).
327. R. M. Edreva-Kardjieva and A. A. Andreev, *J. Neorg. Khim.*, **22**, 2007 (1977).
328. R. M. Edreva-Kardjieva, A. A. Andreev and N. M. Neshev, *3rd Internat. Symp. Olefin Metathesis*, Lyons, September 10—12, 1979.
329. R. M. Edreva-Kardjieva and A. A. Andreev, *4th Internat. Symp. Olefin Metathesis*, Lyons, September 10—12, 1979.
330. J. H. Edwards and W. J. Feast, *Polymer*, **21**, 595 (1980).
331. J. H. Edwards, K. Harper, W. J. Feast and F. Stelzer, *4th Internat. Symp. Olefin Metathesis*, Lyons, September 10—12, 1981.
332. C. Edwige, A. Lattes, J. P. Laval, R. Mutin, J. M. Basset and R. Nouguier, *J. Mol. Catalysis*, **8**, 297 (1980).
333. O. Eisenstein and R. Hoffmann, *4th Internat. Symp. Olefin Metathesis*, Belfast, September 1—4, 1981.
334. O. Eisenstein, R. Hoffmann and A. R. Rossi, *J. Amer. Chem. Soc.*, **103**, 5582 (1981).
335. K. F. Elgert, B. Stützel, P. Frenzel, H.-J. Cantow and R. Streck, *Makromol. Chem.*, **170**, 257 (1973).
336. I. F. A. F. El-Saafin and W. J. Feast, *J. Mol. Catalysis*, **15**, 61 (1982).
337. M. El-Sawi, A. Jannibelo, F. Morelli, G. Catalano, F. Intrieri and G. Giordano, *J. Chem. Technol. Biochem.*, **31**, 388 (1981).
338. J. Engelhardt, *Rec. Trav. Chim.*, **96**, M101 (1977).
339. J. Engelhardt, *Magy. Kem. Foly*, **83**, 311 (1977).
340. J. Engelhardt, *Magy. Kem. Foly*, **83**, 452 (1977).
341. J. Engelhardt, *React. Kinet. Catal. Lett.*, **9**, 35 (1978).
342. J. Engelhardt, *React. Kinet. Catal. Lett.*, **9**, 41 (1978).
343. J. Engelhardt, *React. Kinet. Catal. Lett.*, **11**, 229 (1979).
344. J. Engelhardt, *J. Catalysis*, **62**, 243 (1980).
345. J. Engelhardt, *J. Mol. Catalysis*, **8**, 119 (1980).
346. J. Engelhardt, *J. Catalysis*, **62**, 243 (1980).
347. J. Engelhardt, J. Goldwasser and W. K. Hall, *J. Catalysis*, **70**, 364 (1981).
348. J. Engelhardt and D. Kallo, *J. Catalysis*, **71**, 209 (1981).
349. J. Engelhardt, *J. Catalysis*, **72**, 179 (1981).
350. J. Engelhardt, J. Goldwasser and W. K. Hall, *J. Catalysis*, **76**, 48 (1982).
351. J. Engelhardt, J. Goldwasser and W. K. Hall, *J. Mol. Catalysis*, **15**, 173 (1982).
352. J. Engelhardt, *React. Kinet. Catal. Lett.*, **21**, 7 (1982).
352a. J. Engelhardt and I. Zsinka, *5th Internat. Symp. Olefin Metathesis*, Graz, August 22—26, 1983.
353. M. Ephritikhine, M. L. H. Green and R. E. McKenzie, *J. Chem. Soc., Chem. Commun.*, **1976**, 619.
354. M. Ephritikhine and M. L. H. Green, *J. Chem. Soc., Chem. Commun.*, **1976**, 926.
355. M. Ephritikhine, B. R. Francis, M. L. H. Green, R. E. McKenzie and M. J. Smith, *J. Chem. Soc., Dalton Trans.*, **1977**, 1131.
356. M. Ephritikhine, J. Levisalles, H. Rudler and D. Villemin, *J. Organomet. Chem.*, **124**, C1 (1977).
357. M. Erdweg, *Ph. D. Thesis*, TH Aachen, 1979.
358. Yu. I. Ermakov, B. N. Kuznetsov and A. N. Startsev, *Kin. Katal.*, **15**, 539 (1974).
359. Yu. I. Ermakov, B. N. Kuznetsov and Yu. A. Ryndin, *J. Catalysis*, **42**, 73 (1976).
360. Yu. I. Ermakov, *Catalysis Revs.*, **13**, 77 (1976).
361. Yu. I. Ermakov, B. N. Kuznetsov, Yu. P. Grabowski, A. N. Startsev, A. M. Lazutkin and V. A. Lazutkina, *J. Mol. Catalysis*, **1**, 93 (1976).
362. A. Farkas, *Hydrocarbon Process.*, **50**, 137 (1971).
363. M. F. Farona and W. S. Greenlee, *J. Chem. Soc., Chem. Commun.*, **1975**, 759.

364. M. F. Farona and V. W. Motz, *J. Chem. Soc., Chem. Commun.*, **1976**, 930.
365. M. F. Farona and C. Tsonis, *J. Chem. Soc., Chem. Commun.*, **1977**, 363.
366. M. F. Farona and C. Tsonis, *Fundam. Res. Homogeneous Catal.*, **3**, 409 (1979).
367. M. F. Farona and R. L. Tucker, *J. Mol. Catalysis*, **8**, 85 (1980).
368. J. Fathikalajahi, *Ph. D. Thesis, Diss. Abstr. Int. B*, **39**, 2409 (1978).
369. J. Fathikalajahi and G. B. Wills, *J. Mol. Catalysis*, **8**, 127 (1980).
370. W. J. Feast and B. Wilson, *Polymer*, **20**, 1182 (1979).
371. W. J. Feast and B. Wilson, *J. Mol. Catalysis*, **8**, 277 (1980).
371a. W. J. Feast, G. Pilcher and A. J. Tate, *5th Internat. Symp. Olefin Metathesis*, Graz, August 22−26, 1983.
371b. W. J. Feast and K. Harper, *5th Internat. Symp. Olefin Metathesis*, Graz, August 22−26, 1983.
372. V. S. Fel'dblyum, T. I. Baranova and T. A. Tsailingol'd, *Zh. Org. Khim.*, **9**, 870 (1973).
373. V. S. Fel'dblyum, T. I. Baranova, T. A. Tsailingol'd, N. V. Petrushanskaya, E. D. Tkachenko and G. A. Maksimenko, *Zh. Org. Khim.*, **9**, 874 (1973).
374. V. S. Fel'dblyum, A. M. Kut'in, T. I. Baranova, O. P. Karpov, D. A. Bol'shakov, V.F. Grigr'ev and A. I. Yablonskaya, *V Sb., XI Mendeleevsk. S'ezd Obsch. Prikl. Khim., Ref. Dokl. Soobsch.*, **1974**, 307.
375. V. S. Fel'dblyum, T. I. Baranova, O. P. Karpov, M. E. Basner, I. V. Schuk, D. A. Bol'-shakov, V. F. Grigor'ev and A. I. Yablonskaya, *Neftekhim.*, **16**, 194 (1976).
376. V. S. Fel'dblyum, T. I. Baranova and O. P. Karpov, *Neftekhim.*, **16**, 435 (1976).
377. A. Felzenstein, S. Sarel and Y. Yovell, *J. Chem. Soc., Chem. Commun.*, **1975**, 918.
378. M. C. C. Figueiredo and G. A. Souza, *J. Polymer. Sci., Polymer Chem. Ed.*, **17**, 2845 (1979).
379. E. S. Finkel'stein, B. S. Strel'chik, V. G. Zaikin and V. M. Vdovin, *Neftekhim.*, **15**, 667 (1975).
380. E. S. Finkel'stein, B. S. Strel'chik, E. B. Portnykh, V. M. Vdovin and N. S. Nametkin, *Dokl. Akad. Nauk SSSR*, **232**, 1322 (1977).
381. E. S. Finkel'stein, E. B. Portnykh, V. N. Ushakov and V. M. Vdovin, *Izvest. Akad. Nauk SSSR, Ser. Khim.*, **1979**, 474.
382. E. S. Finkel'stein, E. B. Portnykh, N. V. Ushakov and V. M. Vdovin, *Izvest. Akad. Nauk SSSR, Ser. Khim.*, **1981**, 641.
383. L. Fortina, R. Maggiore and G. Toscano, *Boll. Accad. Gioenia Sci. Nat. Catanio*, **11**, 13 (1973).
384. L. Fortina, R. Maggiore, G. Schembari, L. Solamio and G. Toscano, *Chimica e Industria*, **56**, 610 (1974).
385. R. A. Fridman, S. M. Nosakova, Yu. B. Kryukov, A. N. Bashkirov, N. S. Nametkin and V. M. Vdovin, *Izvest. Akad. Nauk SSSR, Ser. Khim.*, **1971**, 2100.
386. R. A. Fridman, S. M. Nosakova, L. G. Liberov and A. N. Bashkirov, *Izvest. Akad. Nauk SSSR, Ser. Khim.*, **1977**, 678.
387. R. A. Fridman, A. N. Bashkirov, L. G. Liberov, S. M. Nosakova, R. M. Smirnova and S. B. Verbovetskaya, *Dokl. Akad. Nauk SSSR*, **234**, 1354 (1977).
388. R. A. Fridman, L. G. Liberov, S. M. Nosakova and A. N. Bashkirov, *Izvest. Akad. Nauk SSSR, Ser. Khim.*, **1978**, 1225.
389. R. A. Fridman, L. G. Liberov, S. M. Nosakova, R. M. Smirnova and N. A. Bashkirov, *Izvest. Akad. Nauk SSSR, Ser. Khim.*, **1979**, 2816.
390. J. R. Fritch and K.P.C. Vollhardt, *Angew. Chem.*, **91**, 439 (1979).
391. J. R. Fritch, K. P. C. Vollhardt and M. R. Thompson, *J. Amer. Chem. Soc.*, **101**, 2768 (1979).
392. J. R. Fritch and K. P. C. Vollhardt, *Angew. Chem.*, **92**, 570 (1980).
393. K. Fukui and S. Inagaki, *J. Amer. Chem. Soc.*, **97**, 4445 (1975).
394. H. Funk and H. Naumann, *Z. Anorg. Algem. Chem.*, **343**, 294 (1966).
395. J. Furukawa and Y. Mizoe, *J. Polymer Sci., Polymer Lett. Ed.*, **11**, 263 (1973).
396. S. K. Gangwall, J. Fathikalajahi and G. B. Wills, *Ind. Eng. Chem., Prod. Res. Develop.*, **16**, 233 (1977).
397. S. K. Gangwall and G. B. Wills, *J. Catalysis*, **52**, 539 (1978).
398. F. Garnier, P. Krausz and J. E. Dubois, *J. Organomet. Chem.*, **170**, 195 (1979).
399. F. Garnier and P. Krausz, *J. Mol. Catalysis*, **8**, 91 (1980).

400. F. Garnier, P. Krausz and H. Rudler, *J. Organomet. Chem.*, **186**, 77 (1980).
401. P. G. Gassman and T. H. Johnson, *J. Amer. Chem. Soc.*, **98**, 861 (1976).
402. P. G. Gassman and T. H. Johnson, *J. Amer. Chem. Soc.*, **98**, 6055 (1976).
403. P. G. Gassman and T. H. Johnson, *J. Amer. Chem. Soc.*, **98**, 6057 (1976).
404. P. G. Gassman and T. H. Johnson, *J. Amer. Chem. Soc.*, **98**, 6058 (1976).
405. P. G. Gassman and T. H. Johnson, *J. Amer. Chem. Soc.*, **99**, 622 (1977).
406. G. Gatti and A. Carborano, *Makromol. Chem.*, **175**, 1627 (1974).
407. G. Gianotti and A. Capizzi, *European Polymer J.*, **6**, 743 (1970).
408. G. Gianotti, G. Dall'Asta, V. Valvassori and V. Zamboni, *Makromol. Chem.*, **149**, 117 (1971).
409. T. W. Gibson and L. Tulich, *J. Org. Chem.*, **46**, 1821 (1981).
410. R. Giezynski and A. Korda, *J. Mol. Catalysis*, **7**, 349 (1980).
411. M. Gilet, A. Mortreux, J. Nicole and F. Petit, *J. Chem. Soc., Chem. Commun.*, **1979**, 521.
412. M. Gilet, A. Mortreux, J. Nicole and F. Petit, *3rd Internat. Symp. Olefin Metathesis*, Lyons, September 10—12, 1979.
413. N. Giordano, A. Castellan, J. C. J. Bart, A. Vaghi and F. Campadelli, *J. Catalysis*, **37**, 204 (1971).
414. N. Giordano, A. Vaghi, J. C. J. Bart and A. Castellan, *Mech. Hydrocarb. Reacn., Symp.* 1973, **(Publ. 1975)**, 245, Elsevier, Amsterdam.
415. N. Giordano, M. Padovan, A. Vaghi, J. C. J. Bart and A. Castellan, *J. Catalysis*, **38**, 1 (1975).
416. N. Giordano, J. C. J. Bart, A. Vaghi, A. Castellan and G. Martinolli, *J. Catalysis*, **36**, 81 (1975).
417. N. Giordano, J. C. J. Bart and V. Ragaini, *Rev. Port. Quim.*, **19**, 31 (1977).
418. J. Goldwasser, J. Engelhardt and W. K. Hall, *J. Catalysis*, **70**, 275 (1981).
419. T. G. Golenko, B. A. Dolgoplosk, K. L. Makovetskii and I. Ya. Ostrovskaya, *Dokl. Akad. Nauk SSSR*, **220**, 863 (1975).
420. G. M. Graff, *Ph. D. Thesis, Diss. Abstr. Int. B*, **34**, 5900 (1974).
421. G. M. Graff and E. McNelis, *J. Catalysis*, **38**, 482 (1975).
422. W. Grahlert, K. Milowski, V. Langlein and E. Taeger, *Plaste Kautschuk*, **22**, 229 (1975).
422a. Ch. Graimann, H. Hönig, K. Hummel and F. Stelter, *5th Internat. Symp. Olefin Metathesis*, Graz, August 22—26, 1983.
423. W. Graulich, W. Swodenk and D. Theisen, *Hydrocarbon Process.*, **51**, 74 (1972).
424. A. Greco, F. Pirinoli and G. Dall'Asta, *J. Organomet. Chem.*, **60**, 115 (1973).
425. M. L. H. Green, *Pure Appl. Chem.*, **50**, 27 (1978).
426. M. L. H. Green, *Ann. New York Acad. Sci.*, **333**, 229 (1980).
427. J. Greene and M. D. Curtis, *J. Amer. Chem. Soc.*, **99**, 5176 (1977).
428. W. S. Greenlee, *Ph. D. Thesis, Diss. Abstr. Int. B*, **36**, 5575 (1976).
429. W. S. Greenlee and M. F. Farona, *Inorg. Chem.*, **15**, 2129 (1976).
430. H. Griesser and K. Hummel, *Colloid. Polymer Sci.*, **218**, 467 (1980).
431. H. Griesser, K. Hummel and H. Demel, *Monatsh. Chem.*, **112**, 1203 (1981).
432. R. H. Grubbs and T. K. Brunck, *J. Amer. Chem. Soc.*, **94**, 2538 (1972).
433. R. H. Grubbs, P. L. Burk and D. D. Carr, *J. Amer. Chem. Soc.*, **97**, 3265 (1975).
434. R. H. Grubbs, D. D. Carr, C. Hoppin and P. L. Burk, *J. Amer. Chem. Soc.*, **98**, 3478 (1976).
435. R. H. Grubbs, D. D. Carr and P. L. Burk, *Organotrans. Metal. Chem., Proc. Jpn.-Amer. Sem., 1st*, 1974 **(Publ. 1975)**, 135.
436. R. H. Grubbs and C. R. Hoppin, *J. Chem. Soc., Chem. Commun.*, **1977**, 634.
437. R. H. Grubbs and A. Miyashita, *J. Chem. Soc., Chem. Commun.*, **1977**, 864.
438. R. H. Grubbs, S. Swetnick and S.C.H. Su, *J. Mol. Catalysis*, **3**, 11 (1977).
439. R. H. Grubbs, *Inorg. Chem.*, **18**, 2623 (1979).
440. R. H. Grubbs, *Prepr., Div. Pet. Chem., Amer. Chem. Soc.*, **24**, 388 (1979).
441. R. H. Grubbs and C. R. Hoppin, *J. Amer. Chem. Soc.*, **101**, 1499 (1979).
442. R. H. Grubbs and S. J. Swetnick, *J. Mol. Catalysis*, **8**, 25 (1980).
443. R. H. Grubbs, *4th Internat. Symp. Olefin Metathesis*, Belfast, September 1—4, 1981.
443a. H. R. Grubbs, L. Gilliom and D. Meinhardt, *5th Internat. Symp. Olefin Metathesis*, Graz, August 22 — 26, 1983.

444. A. K. Gupta and R.S. Mann, *Indian J. Technol.*, **17**, 397 (1979).
445. A. K. Gupta and R. S. Mann, *Proc. Natl. Symp. Catal., 4th* 1978 **(Publ. 1980)**, 431.
446. F. Haas and D. Theisen, *Kautch. Gummi Kunstst.*, **23**, 502 (1970).
447. F. Haas, K. Nützel, G. Pampus and D. Theisen, *Rubber Chem. Technol.*, **43**, 1116 (1970).
447a. J. G. Hamilton, K. J. Ivin and J. J. Rooney, *5th Internat. Symp. Olefin Metathesis*, Graz, August 22 — 26, 1983.
447b. J. G. Hamilton, K. J. Ivin, G. M. McCann and J. J. Rooney, *5th Internat. Symp. Olefin Metathesis*, Graz, August 22 — 26, 1983.
448. U. R. Hattikudur and G. Thodos, *Adv. Chem. Ser.*, **133**, 80 (1974).
449. U. R. Hattikudur, *Ph. D. Thesis, Diss. Abstr. Int. B*, **35**, 2726 (1974).
450. L. F. Heckelsberg, R. L. Banks and G. C. Bailey, *Ind. Eng. Chem., Prod. Res. Develop.*, **7**, 29 (1968).
451. L. F. Heckelsberg, R. L. Banks and G. C. Bailey, *Ind. Eng. Chem., Prod. Res. Develop.*, **8**, 259 (1969).
452. L. F. Heckelsberg, R. L. Banks and G. C. Bailey, *J. Catalysis*, **13**, 99 (1969).
453. P. Heiling, D. Wewerka and K. Hummel, *Kautch. Gummi Kunstst.*, **29**, 238 (1976).
454. P. Heiling, H. Hönig and K. Hummel, *Monatsh. Chem.*, **111**, 575 (1980).
455. P. R. Hein, J. Polymer Sci., *Polymer Chem. Ed.*, **11**, 163 (1973).
456. H. Heinemann in "Catalysis", J. R. Anderson and M. Boudart (Eds.), Springer-Verlag, Vol. 1, Ch. 1, Berlin, 1981.

457. J. Heinzl and H. Heusinger, *Angew. Makromol. Chem.*, **107**, 191 (1982).
458. G. Henrici-Olivé and S. Olivé, *Angew. Chem.*, **85**, 149 (1973).
459. J. L. Hérisson, and Y. Chauvin and G. Lefebvre, *Compt. Rend.*, **269**, 661 (1969).
460. J. L. Hérisson and Y. Chauvin, *Makromol. Chem.*, **141**, 161 (1970).
461. J. Heveling, *Diplomarbeit*, TH Aachen, 1978.
462. J. Heveling, *Ph. D. Thesis*, TH Aachen, 1981.
463. K. Hiraki, A. Kuroiwa and H. Hirai, *J. Polymer Sci., A-1*, **9**, 2323 (1971).
464. H. Höcker and R. Musch, *Makromol. Chem.*, **157**, 201 (1972).
465. H. Höcker and F. R. Jones, *Makromol. Chem.*, **161**, 251 (1972).
466. H. Höcker, R. Musch, F. R. Jones and I. Lüderwald, *Angew. Chem. Internat. Ed.*, **12**, 430 (1973).
467. H. Höcker and R. Musch, *Makromol. Chem.*, **175**, 1395 (1974).
468. H. Höcker and R. Musch, *Makromol. Chem.*, **176**, 3117 (1975).
469. H. Höcker, W. Reimann, K. Riebel and Zs. Szentivanyi, *Makromol. Chem.*, **177**, 1707 (1976).
470. H. Höcker, L. Reif, W. Reimann and K. Riebel, *Rec. Trav. Chim.*, **96**, M47 (1977).
471. H. Höcker, W. Reimann, L. Reif and K. Riebel, *J. Mol. Catalysis*, **8**, 191 (1980).
471 a. H. Höcker, L. Reif and C. Tran-Thu, *5th Internat. Symp. Olefin Metathesis*, Graz, August 22 — 26, 1983.
472. L. Hocks, A. J. Hubert and P. Teyssié, *Tetrahedron Letters*, **1972**, 3687.
473. L. Hocks, A. J. Hubert and P. Teyssié, *Tetrahedron Letters*, **1973**, 2718.
474. L. Hocks, A. J. Hubert and P. Teyssié, *Tetrahedron Letters*, **1974**, 877.
475. L. Hocks, A. J. Hubert and P. Teyssié, *J. Polymer Sci., Polymer Lett.*, **13**, 391 (1975).
476. L. Hocks, A. Noels, A. J. Hubert and P. Teyssié, *J. Org. Chem.*, **41**, 1631 (1976).
477. V. A. Hodjemirov, V. A. Evdochimova and V. M. Cherebnichenko, *Vysokomol. Soedin.*, **13/14**, 727 (1972).
478. W. Holtrup, F. W. Küpper, W. Meckstroth, H.-H. Meyer and K. H. Nordsiek, *Lichtbogen*, **174**, 16 (1974).
479. T. R. Howard, J. B. Lee and R. H. Grubbs, *J. Amer. Che. Soc.*, **102**, 6876 (1980).
480. R. F. Howe, D. E. Davidson and D. A. Whan, *J. Chem. Soc., Faraday Trans.*, **60**, 2266 (1972).
481. C.-C. Hsu, *Ph. D. Thesis, Diss. Abstr. Int. B*, **41**, 2259 (1980).
482. J.-C. Hsu, *Ph. D. Thesis, Diss. Abstr. Int. B*, **40**, 1810 (1979).
483. J. C. Hubbs, *Ph. D. Thesis, Diss. Abstr. Int. B*, **42**, 4425 (1982).
484. R. T. Hughes, R. L. Burtnett and R. G. Wall, *Proc. Internat. Congr. Catal., 5th*, 1972 **(Publ. 1973)**, **2**, 1217.
485. W. B. Hughes, *J. Chem. Soc., Chem. Commun.*, **1969**, 431.
486. W. B. Hughes, *J. Amer. Chem. Soc.*, **92**, 532 (1970).

487. K. Hummel, R. Streck and H. Weber, *Naturwiss.*, **57**, 194 (1970).

488. K. Hummel and W. Ast, *Naturwiss.*, **57**, 245 (1970).

489. K. Hummel and A. Dürr, *Colloid.-Z. u Z. Polymere*, **244**, 337 (1971).

490. K. Hummel and W. Ast, *Makromol. Chem.*, **166**, 39 (1973).

491. K. Hummel, D. Wewerka, F. Lorber and G. Zeplichal, *Makromol. Chem.*, **166**, 45 (1973).

492. K. Hummel and F. Lorber, *Makromol. Chem.*, **171**, 257 (1973).

493. K. Hummel and A. Dürr, *Colloid. Polymer Sci.*, **252**, 928 (1974).

494. K. Hummel and Y. Imamoglu, *Colloid. Polymer Sci.*, **253**, 225 (1975).

495. K. Hummel and W. Ast, *Colloid. Polymer Sci.*, **253**, 474 (1975).

496. K. Hummel, *Colloid. Polymer Sci.*, **253**, 504 (1975).

497. K. Hummel and G. Reithofer, *Angew. Makromol. Chem.*, **50**, 183 (1976).

498. K. Hummel, H. Demel and D. Wewerka, *Makromol. Chem.*, **178**, 19 (1977).

499. K. Hummel, W. Kathan, I. Kovar and O. A. Wedam, *Kautch. Gummi Kunstst.*, **30**, 7 (1977).

500. K. Hummel, F. Stelzer, W. Kathan, H. Demel, O. A. Wedam and G. Karaoulis, *Rec. Trav. Chim.*, **96**, M75 (1977).

501. K. Hummel, O. A. Wedam, W. Kathan and H. Demel, *Makromol. Chem.*, **179**, 1159 (1978).

502. K. Hummel, F. Stelzer, P. Heiling, O. A. Wedam and H. Griesser, *J. Mol. Catalysis*, **8**, 253 (1980).

503. K. Kummel, P. Heiling, C. Karaoulis, W. Kathan and F. Stelzer, *Makromol. Chem.*, **181**, 1847 (1980).

504. K. Hummel and O. A. Wedam, *Makromol. Chem.*, **182**, 3041 (1981).

505. K. Hummel, S. Grayer and H. Lechner, *Kautsch. Gummi Kunstst.*, **35**, 731 (1982).

506. K. Ichikawa, T. Takagi and K. Fukuzumi, *Yukagaku*, **25**, 136 (1976).

507. K. Ichikawa and K. Fukuzumi, *Yukagaku*, **25**, 779 (1976).

508. K. Ichikawa and K. Fukuzumi, *J. Org. Chem.*, **41**, 2633 (1976).

509. K. Ichikawa, T. Takagi and K. Fukuzumi, *Trans. Met. Chem.*, **1**, 54 (1976).

510. K. Ichikawa, O. Watanabe and K. Fukuzumi, *Trans. Met. Chem.*, **1**, 183 (1976).

511. K. Ichikawa, O. Watanabe, T. Takagi and K. Fukuzumi, *J. Catalysis*, **44**, 416 (1976).

512. K. Ichikawa, T. Takagi and K. Fukuzumi, *Bull. Chem. Soc. Japan*, **49**, 750 (1976).

513. T. Ido, T. Tsuzuki and Sh. Goto, *Kagaku Kegaku Ronbunshi*, **8**, 226 (1982).

513a. I. N. Ikonitskii, N. A. Buzina, B. D. Babitskii and V. A. Kormer, *Vysokomol. Soedin.*, A, **21**, 2360 (1979).

514. Y. Imamoglu, *4th Internat. Symp. Olefin Metathesis*, Belfast, September 1—4, 1981.

514a. Y. Imamoglu and C. Kaya, *5th Internat. Symp. Olefin Metathesis*, Graz, August 22—26, 1983.

515. M. Imoto, *Setchaku*, **20**, 232 (1976).

516. G. V. Isagulyants and L. F. Rar, *Izvest. Akad. Nauk SSSR, Ser. Khim.*, **1969**, 1362.

517. A. Ismayel-Milanovic, J. M. Basset, H. Praliaud, M. Dufaux and I. DeMourgues, *J. Catalysis*, **31**, 408 (1973).

518. K. J. Ivin, D. T. Laverty, J. J. Rooney and P. Watt, *Rec. Trav. Chim.*, **96**, M54 (1977).

519. K. J. Ivin, D. T. Laverty and J. J. Rooney, *Makromol. Chem.*, **178**, 1545 (1977).

520. K. J. Ivin, D. T. Laverty and J. J. Rooney, *Makromol. Chem.*, **179**, 3 253 (1978).

521. K. J. Ivin, J. J. Rooney and C. D. Stewart, *J. Chem. Soc., Chem. Commun.*, **1978**, 603.

522. K. J. Ivin, J. J. Rooney, C. D. Stewart, M. L. H. Green and R. Mahtab, *J. Chem. Soc., Chem. Commun.*, **1978**, 604.

523. K. J. Ivin, S. Lillie and J. J. Rooney, *Makromol. Chem.*, **179**, 2787 (1978).

524. K. J. Ivin, J. H. O'Donnell, J. J. Rooney and C. D. Stewart, *Makromol. Chem.*, **180**, 1975 (1979).

525. K. J. Ivin, D. T. Laverty, J. H. O'Donnell, J. J. Rooney and C. D. Stewart, *Makromol. Chem.*, **180**, 1989 (1979).

526. K. J. Ivin, G. Lapienis, J. J. Rooney and C. D. Stewart, *Polymer*, **20**, 1308 (1979).

527. K. J. Ivin, G. Lapienis and J. J. Rooney, *Polymer*, **21**, 367 (1980).

528. K. J. Ivin, G. Lapienis and J. J. Rooney, *Polymer*, **21**, 436 (1980).

529. K. J. Ivin, G. Lapienis and J. J. Rooney, *J. Chem. Soc., Chem. Commun.*, **1980**, 1068.

530. K. J. Ivin, *Pure Appl. Chem.*, **52**, 1907 (1980).

531. K. J. Ivin, G. Lapienis, J. J. Rooney and C. D. Stewart, *J. Mol. Catalysis*, **8**, 203 (1980).

532. K. J. Ivin, B. S. R. Reddy and J. J. Rooney, *J. Chem. Soc., Chem. Commun.*, **1981**, 1062.
533. K. J. Ivin, L.-M. Lam and J. J. Rooney, *Makromol. Chem.*, **182**, 1847 (1981).
534. K. J. Ivin, J. J. Rooney and H. H. Thoi, *4th Internat. Symp. Olefin Metathesis*, Belfast, September 1—4, 1981.
535. K. J. Ivin, G. Lapienis and J. J. Rooney, *Makromol. Chem.*, **183**, 9 (1982).
536. Y. Iwasawa, S. Ogasawara and M. Soma, *Chem. Lett.*, **1978**, 1039.
537. Y. Iwasawa, H. Kubo, M. Yamagishi and S. Ogasawara, *Chem. Lett.*, **1980**, 1165.
538. Y. Iwasawa, H. Ichinose, S. Ogasawara and M. Soma, *J. Chem. Soc., Faraday Trans.*, **77**, 1763 (1981).
538a. Y. Iwasawa, H. Kubo and H. Hamamura, *5th Internat. Symp. Olefin Metathesis*, Graz, August 22 — 26, 1983.
539. F. Janowski, F. Wolf, A. Sophianos and G. Foester, *Chem. Techn.* **29**, 313 (1977).
540. *Japan Chemical Week*, **23(1149)**, 1 (1982).
541. P. H. Johnson, *Hydrocarbon Process.*, **46**, 149 (1967).
542. P. H. Johnson, *Proc. 7th World Petroleum Congr. 1967*, **5**, 247 (1968).
543. B. V. Kacharmin and N. B. Bespalova, *Issled. v Obl. Neftekhim.*, **1976**, 13.
544. A. A. Kadushin, R. K. Aliev, O. V. Krylov, A. A. Andreev, R. M. Edreva-Kardjieva and D. M. Shopov, *Kinet. Katal.*, **23**, 276 (1982).
545. A. A. Kadushin, V. A. Matyshak, N. A. Kutyreva, A. V. Sklyarov and A. G. Vlasenko, *Kinet. Katal.*, **23**, 1167 (1982).
546. Y. Kamiya, *Internat. Chem. Eng.*, **14**, 358 (1974).
547. F. Kapteijn, H. L. G. Bredt and J. C. Mol, *Rec. Trav. Chim.*, **96**, M139 (1977).
548. F. Kapteijn, *Ph. D. Thesis*, University of Amsterdam, 1980.
549. F. Kapteijn, H. L. G. Bredt, E. Homburg and J. C. Mol, *Ind. Eng. Chem., Prod. Res. Develop.*, **20**, 457 (1981).
550. F. Kapteijn and J. C. Mol, *J. Chem. Soc., Faraday Trans.*, **78**, 2583 (1982).
551. V. G. Karakashev, A. Z. Dorogochinskii and M. N. Zhavoronkov, *Neftekhim.* **19**, 51 (1979).
552. O. P. Karpov, N. P. Glazkova, V. S. Fel'dblyum, M. E. Basner, T. I. Baranova and E. D. Tkachenko, *Osnovn. Org. Synth. Neftekhim.*, **2**, 39 (1975).
553. O. P. Karpov, V. S. Fel'dblyum, T. L. Baranova and M. E. Basner, *Prom. Synth. Kautch.*, **1976**, 8.
554. W. Kathan, O. A. Wedam and K. Hummel, *Makromol. Chem.*, **178**, 1693 (1977).
555. T. J. Katz and J. McGinnis, *J. Amer. Chem. Soc.*, **97**, 1592 (1975).
556. T. J. Katz and N. Acton, *Tetrahedron Letters*, **1976**, 4251.
557. T. J. Katz, J. McGinnis and C. Altus, *J. Amer. Chem. Soc.*, **98**, 606 (1976).
558. T. J. Katz and R. Rotchild, *J. Amer. Chem. Soc.*, **98**, 2519 (1976).
559. T. J. Katz and J. McGinnis, *J. Amer. Chem. Soc.*, **99**, 1903 (1977).
560. T. J. Katz and W. H. Hersh, *Tetrahedron Letters*, **1977**, 585.
561. T. J. Katz, S. J. Lee and N. Acton, *Tetrahedron Letters*, **1977**, 4247.
562. T. J. Katz and S. J. Lee, *J. Amer. Chem. Soc.*, **102**, 422 (1980).
563. T. J. Katz, S. J. Lee and M. A. Shippey, *J. Mol. Catalysis*, **8**, 219 (1980).
564. T. J. Katz, S. J. Lee, M. Nair and E. B. Savage, *J. Amer. Chem. Soc.*, **102**, 7940 (1980).
565. T. J. Katz, E. B. Savage, S. J. Lee and M. Nair, *J. Amer. Chem. Soc.*, **102**, 7942 (1980).
566. T. J. Katz, E. B. Savage, S. J. Lee and M. Nair, *Report 1980 Order NoAD-A092* 73618.
567. T. J. Katz and Ch. Ch. Han, *Organometal.*, **1**, 1093 (1982).
567a. T. J. Katz, T. H. Ho, N. Y. Shih and C. C. Han, *5th Internat. Symp. Olefin Metathesis*, Graz, August 22 — 26, 1983.
568. W. J. Kelley and N. Calderon, *J. Macromol. Sci., Chem.*, A-9, 911 (1975).
569. W. J. Kelley and N. Calderon, *Ionic Polym: Unsolv. Probl., Jpn.-US Semin. Polym. Synth.*, 1974 **(Publ. 1976)**, 271.
570. F. P. J. M. Kerkhof, R. Thomas and J. A. Moulijn, *Rec. Trav. Chim.*, **96**, M121 (1977).
571. F. P. J. M. Kerkhof, J. A. Moulijn and R. Thomas, *J. Catalysis*, **56**, 279 (1979).
572. F. P. J. M. Kerkhof, J. A. Moulijn, R. Thomas and J. C. Qudejans, *Stud. Surf. Sci. Catal.*, **3**, 77 (1979).
573. F. P. J. M. Kerkhof, *Ph. D. Thesis*, University of Amsterdam, 1979.

574. I. L. Kershenbaum, L. N. Grebeniak, I. A. Oreshkin, B. A. Dolgoplosk and E. I. Tinyakova, *Dokl. Akad. Nauk SSSR*, **238**, 359 (1978).
575. I. L. Kershenbaum, I. A. Oreshkin, B. A. Dolgoplosk, E. I. Tinyakova and L. N. Grebeniak, *Dokl. Akad. Nauk SSSR*, **256**, 1400 (1981).
576. I. L. Kershenbaum, I. A. Oreshkin and B. A. Dolgoplosk, *Dokl. Akad. Nauk SSSR*, **259**, 119 (1981).
577. M. L. Khidekel', V. I. Mar'in, A. D. Shebaldova, T. A. Bolshinskaya and I. V. Kalechits, *Izvest. Akad. Nauk SSSR, Ser. Khim.*, **1971**, 663.
578. L. Kim, J. H. Raley and G. S. Bell, *Rec. Trav. Chim.*, **96**, M136 (1977).
579. A. Kinnemann, *Can. J. Chem.*, **60**, 7 (1982).
580. U. Klabunde, F. N. Tebbe, G. W. Parshall and R. L. Harlow, *J. Mol. Catalysis*, **8**, 37 (1980).
581. A. Kobayashi and E. Echigoya, *Nippon Kagaku Kaishi*, **1973**, 908.
582. T. P. Kobylinski and H. E. Swift, *J. Catalysis*, **26**, 416 (1972).
583. T. P. Kobylinski and H. E. Swift, *J. Catalysis*, **33**, 83 (1974).
584. S. Kokuryo, *Japan Plast. Age*, **15**, 26 (1977).
585. S. P. Kolesnikov, A. I. Joffe, N. I. Okhrinienko and O. M. Nefedov, *Izvest. Akad. Nauk SSSR, Ser. Khim.*, **1977**, 1206.
586. S. P. Kolesnikov, N. I. Povarova, A. Ya. Shteinshneider and O. M. Nefedov, *Izvest. Akad. Nauk SSSR, Ser. Khim.*, **1979**, 2398.
587. S. P. Kolesnikov, T. L. Mitenina, S. A. Radzinoki, A. P. Sheinker, A. D. Abrin and O. M. Nefedov, *Izvest. Akad. Nauk SSSR, Ser. Khim.*, **1981**, 2303.
588. S. P. Kolesnikov, T. L. Mitenina, A. Ya. Sheinschneider and O. M. Nefedov, *Izvest. Akad. Nauk SSSR, Ser. Khim.*, **1982**, 942.
589. I. A. Kop'eva, I. A. Oreshkin, E. I. Tinyakova and B. A. Dolgoplosk, *Dokl. Akad. Nauk SSSR*, **249**, 374 (1979).
590. A. Korda, R. Giezynski and S. Kricinski, *J. Mol. Catalysis*, **9**, 51 (1980).
591. V. A. Kormer, I. A. Poleyatova and T. L. Yufa, *J. Polymer Sci.*, A-1, **10**, 251 (1972).
592. Yu. V. Kroshak, L. M. Vardanyan and B. A. Dolgoplosk, *Dokl. Akad. Nauk SSSR*, **208**, 1138 (1973).
593. Yu. V. Korshak, L. M. Vardanyan and B. A. Dolgoplosk, *Mech. Hydrocarb. React.*, *Symp.* 1973 **(Publ. 1975)**, 527.
594. Yu. V. Korshak, M. A. Tlenkopachev, G. I. Timofeeva, S. A. Pavlova and B. A. Dolgoplosk, *Dokl. Akad. Nauk SSSR*, **226**, 1344 (1976).
595. Yu. V. Korshak, G. N. Bondarenko, V. M. Kutemikov and B. A. Dolgoplosk, *Dokl. Akad. Nauk SSSR*, **227**, 1383 (1976).
596. Yu. V. Korshak, B. A. Dolgoplosk and M. A. Tlenkopatchev, *Rec. Trav. Chim.*, **96**, M64 (1977).
597. Yu. V. Korshak, A. M. Shapiro, M. A. Tlenkopatchev and B. A. Dolgoplosk, *3rd International Symp. Olefin Metathesis*, Lyons, September 10—12, 1979.
598. Yu. V. Korshak, A. A. Berlin, E. R. Badamshina, G. I. Timofeeva and S. A. Pavlova, *Dokl. Akad. Nauk SSSR*, **248**, 372 (1979).
599. Yu. V. Korshak, B. S. Turov, L. M. Vardanyan, V. A. Efimov, M. A. Tlenkopatchev, A. Yu. Korshevnik and B. A. Dolgoplosk, *Vysokomol. Soedin.*, Ser. A, **22**, 781 (1980).
600. Yu. V. Korshak, M. A. Tlenkopatchev, B. A. Dolgoplosk, E. G. Avdeikina and D. F. Kutepov, *J. Mol. Catalysis*, **15**, 207 (1982).
601. O. L. Kosinskii, A. A. Kadushin, O. V. Krylov, F. K. Shmidt, N. Kolobova and Yu. V. Maksimov, *Geterog. Katal. 4th*, **1**, 403 (1979).
602. V. M. Kothari and J. J. Tazuma, *J. Org. Chim.*, **36**, 2951 (1971).
603. D. Kranz and M. Beck, *Angew. Makromol. Chem.*, **27**, 29 (1972).
604. V. S. Krasnobaeva, M. M. Mogilevich, V. A. Efimov and B. S. Turov, *Lakokros. Mater. Ikh. Primen.*, **3**, 10 (1982).
605. H.-L. Krauss and E. Humms, *Z. Naturforsch.*, **34B**, 1628 (1979).
605a. H. L. Krauss, E. Humms and K. Hagen, *5th Internat. Symp. Olefin Metathesis*, Graz, August 22—26, 1983.
606. P. Krausz, F. Garnier and J. E. Dubois, *J. Amer. Chem. Soc.*, **97**, 437 (1975).
607. P. Krausz, F. Garnier and J. E. Dubois, *J. Organomet. Chem.*, **108**, 197 (1976).
608. P. Krausz, F. Garnier and J. E. Dubois, *J. Organomet. Chem.*, **146**, 125 (1978).
609. P. Krausz, J. Guillerez and F. Garnier, *J. Organomet. Chem.*, **161**, 97 (1978).

610. J. Kress, M. J. M. Russel, M. G. Wesolek and J. A. Osborn, *J. Chem. Soc., Chem. Commun.*, **1980**, 431.
611. J. Kress, M. G. Wesolek, J. P. LeNy and J. A. Osborn, *J. Chem. Soc., Chem. Commun.*, **1981**, 1039.
612. J. Kress, M. G. Wesolek and J. A. Osborn, *J. Chem. Soc., Chem. Commun.*, **1982**, 514.
612a. J. Kress, M. Wesolek and J. A. Osborn, *5th Internat. Symp. Olefin Metathesis*, Graz, August 22—26, 1983.
613. W. R. Kroll and G. Doyle, *J. Chem. Soc., Chem. Commun.*, **1971**, 839.
614. W. R. Kroll and G. Doyle, *J. Catalysis*, **24**, 356 (1972).
615. E. N. Kropacheva, B. A. Dolgoplosk, D. E. Sterenzat and Yu. A. Patrushin, *Dokl. Akad. Nauk SSSR*, **175**, 1388 (1970).
616. E. N. Kropacheva, D. E. Sterenzat, Yu. A. Patrushin and B. A. Dolgoplosk, *Dokl. Akad. Nauk SSSR*, **206**, 878 (1972).
617. Y. B. Kryukov, V. M. Ignatov and E. V. Glebova, *Izvest. Vyssh. Uchebn. Zaved. Khim. Khim. Technol.*, **19**, 327 (1976).
618. V. N. G. Kumar and K. Hummel, *Angew. Makromol. Chem.*, **102**, 167 (1982).
619. F. W. Küpper and R. Streck, *Makromol. Chem.*, **175**, 2055 (1974).
620. F. W. Küpper and R. Streck, *Chem. Ztg.*, **99**, 464 (1975).
621. F. W. Küpper and R. Streck, *Z. Naturforsch B: Anorg. Chem., Org. Chem.*, **31B**, 1256 (1976).
622. F. W. Küpper, *Angew. Makromol. Chem.*, **80**, 207 (1979).
622a. A. W. Korobko, V. S. Turov and V. M. Cherebnichenko, *Vysokomol. Soedin.*, A, **17**, 195 (1975).
623. B. N. Kuznetsov, A. N. Startsev and Yu. I. Ermakov, *J. Mol. Catalysis*, **8**, 135 (1980).
624. S. M. Kuznetsova, L. S. Priss and V. A. Vorob'eva, *Kauch. Rezina*, **2**, 9 (1982).
625. J. Lal, R. R. Smith and J. M. O'Connor, *Polymer Prepr., Amer. Chem. Soc., Div. Polymer Chem.*, **13**, 714 (1972).
626. J. Lal and R. R. Smith, *J. Org. Chem.*, **40**, 775 (1975).
627. H. Lammens, G. Sartori, J. Siffert and N. Sprecher, *Polymer Letters*, **9**, 341 (1971).
628. A. L. Lapidus, L. N. Rudakova, V. P. Chemagina, Y. I. Isakov, Kh. M. Minachev and Y. T. Eidus, *Izvest. Akad. Nauk SSSR, Ser. Khim.*, **1973**, 1261.
629. A. L. Lapidus, L. N. Rudakova, Y. I. Isakov, Kh. M. Minachev and Y. T. Eidus, *Izvest. Akad. Nauk SSSR, Ser. Khim.*, **1973**, 1266.
630. C. Larroche, J. P. Laval, A. Lattes, M. Leconte, F. Quignard and J. M. Basset, *J. Org. Chem.*, **47**, 2019 (1982).
631. J. P. Laval, A. Lattes, R. Mutin and J. M. Basset, *J. Chem. Soc., Chem. Commun.*, **1977**, 502.
632. D. T. Laverty, A. M. McKervey, J. J. Rooney and A. Stewart, *J. Chem. Soc., Chem. Commun.*, **1976**, 193.
633. D. T. Laverty, J. J. Rooney and A. Stewart, *J. Catalysis*, **45**, 110 (1976).
634. B. V. Lebedev, I. B. Rabinovich, V. Ya. Lityagov, Yu. V. Korshak and V. M. Kuteinikov, *Vysokomol. Soedin. Ser. A*, **18**, 2444 (1976).
635. B. V. Lebedev, N. N. Mukhina, E. G. Kiparisova, V. E. Slievnina, Yu. V. Korshak and E. R. Badamshina, *Fiz.-Khim. Osnovy Sint. Pererab. Polim.*, **1979**, 36.
636. M. Leconte, J. L. Bilhou, W. Reimann and J. M. Basset, *J. Chem. Soc., Chem. Commun.*, **1978**, 3.
637. M. Leconte and J. M. Basset, *Nouveau J. Chim.*, **3**, 429 (1979).
638. M. Leconte and J. M. Basset, *J. Amer. Chem. Soc.*, **101**, 7296 (1979).
639. M. Leconte, Y. BenTaarit, J. L. Bilhou and J. M. Basset, *J. Mol. Catalysis*, **8**, 263 (1980).
640. M. Leconte and J. M. Basset, *Ann. New York Acad. Sci.*, **333**, 165 (1980).
640a. M. Leconte, A. Theolier and J. M. Basset, *5th Internat. Symp. Olefin Metathesis*, Graz, August 22—26, 1983.
641. P. LeDelliou, *Ind. Gomma*, **21**, 55 (1977).
642. P. LeDelliou, *Internat. Rubber Conf.*, **1977**, 1.
643. P. LeDelliou, *Hule Mex. Plast.*, **33**, 17 (1977).
644. P. LeDelliou, *Hule Mex. Plast.*, **33**, 20 (1977).
645. P. LeDelliou and H. D. Manager, *Gummi Asbest Kunstst.*, **30**, 518 (1977).

646. P. LeDelliou, *Rev. Gen. Caoutch. Plast.*, **54**, 77 (1977).
647. P. LeDelliou and V. Massara, *Internat. Rubber Conf.*, **1979**, 349.
648. P. LeDelliou, *Synth. Rubbers High Temp. Eng. Appl. (Jubilee Symp.)*, **1981**, 69.
649. J. B. Lee, G. J. Gajda, W. P. Schaefer, T. R. Howard, T. I. Karnja, D. A. Straus and R. H. Grubbs, *J. Amer. Chem. Soc.*, **103**, 7358 (1981).
650. S. J. Lee, J. McGinnis and T. J. Katz, *J. Amer. Chem. Soc.*, **98**, 7818 (1976).
651. G. Lehnert, D. Maertens, G. Pampus and M. Zimmermann, *Makromol. Chem.*, **175**, 2619 (1974).
652. G. J. Leigh, M .T. Rahman and D.R.M. Walton, *J. Chem. Soc., Chem. Commun.*, **1982**, 541.
653. J. Levisalles, H. Rudler and D. Willemin, *J. Organomet. Chem.*, **87**, C7 (1975).
654. J. Levisalles, H. Rudler and D. Willemin, *Rec. Trav. Chim.*, **96**, M138 (1977).
655. J. Levisalles, H. Rudler, F. Dahan and Y. Jeannin, *3rd Internat. Symp. Olefin Metathesis*, Lyons, September 10—12, 1979.
656. J. Levisalles, H. Rudler and D. Villemin, *J. Organomet. Chem.*, **146**, 259 (1978).
657. J. Levisalles, H. Rudler, D. Villemin, J. Daran, Y. Jeannin and L. Martin, *J. Organomet. Chem.*, **155**, C1 (1978).
658. J. Levisalles, H. Rudler and D. Villemin, *J. Organomet. Chem.*, **164**, 251 (1979).
659. J. Levisalles, H. Rudler, Y. Jeannin and F. Dahan, *J. Organomet. Chem.*, **178**, C8 (1979).
660. J. Levisalles and D. Villemin, *Tetrahedron*, **36**, 3181 (1980).
661. J. Levisalles, H. Rudler, F. Dahan and Y. Jeannin, *J. Organomet. Chem.*, **187**, 233 (1980).
662. J. Levisalles, H. Rudler, F. Dahan and Y. Jeannin, *J. Organomet. Chem.*, **188**, 193 (1980).
663. J. Levisalles, H. Rudler and D. Villemin, *J. Organomet. Chem.*, **192**, 195 (1980).
664. J. Levisalles, H. Rudler and D. Villemin, *J. Organomet. Chem.*, **192**, 375 (1980).
665. J. Levisalles, H. Rudler and D. Villemin, *J. Organomet. Chem.*, **193**, 69 (1980).
666. J. Levisalles, H. Rudler and D. Villemin, *J. Organomet. Chem.*, **193**, 235 (1980).
667. J. Levisalles, F. Rose-Munch, H. Rudler, J.-C. Daran, Y. Dromzee and Y. Jeannin, *J. Chem. Soc., Chem. Commun.*, **1981**, 152.
668. J. Levisalles, F. Rose-Munch, H. Rudler, J.-C. Daran, Y. Dromzee, Y. Jeannin, D. Ades and M. Fontanille, *J. Chem. Soc., Chem. Commun.*, **1981**, 1055.
668a. J. Levisalles, H. Rudler and T. Rull, *5th Internat. Symp. Olefin Metathesis*, Graz, August 22—26, 1983.
669. G. S. Lewandos and R. Pettit, *Tetrahedron Letters*, **1971**, 789.
670. G. S. Lewandos and R. Pettit, *J. Amer. Chem. Soc.*, **93**, 7087 (1971).
671. G. S. Lewandos, *Ph. D. Thesis, Diss. Abstr. Int. B.*, **29**, 3726 (1969).
672. M. J. Lewis and G. B. Wills, *J. Catalysis*, **15**, 140 (1969).
673. M. J. Lewis, *Ph. D. Thesis, Diss. Abstr. Int. B.*, **29**, 3726 (1969).
674. C. J. Lin, A. W. Aldag and A. Clark, *J. Catalysis*, **45**, 287 (1976).
675. C. J. Lin, *Ph. D. Thesis, Diss. Abstr. Int. B.*, **33**, 1527 (1972).
676. H.-C. Lin and S. W. Weller, *J. Catalysis*, **66**, 65 (1980).
677. H. Liu and H. Guo, *Cuihua Xuebao*, **3**, 151 (1982).
678. R. S. Logan and R. L. Banks, *Hydrocarbon Process.*, **47**, 135 (1968).
679. R. S. Logan and R. L. Banks, *Oil Gas J.*, **66**, 131 (1968).
680. E. A. Lombardo, M. LoJacono and W. K. Hall, *J. Catalysis*, **64**, 150 (1980).
681. F. Lorber and K. Hummel, *Makromol. Chem.*, **171**, 257 (1973).
682. B. Lorenz, H. Rommel and M. Wahren, *Z. Chem.*, **22**, 224 (1982).
683. R. C. Luckner, *Ph. D. Thesis, Diss. Abstr. Int. B.*, **33**, 1527 (1972).
684. R. C. Luckner, G. E. McConcie and G. B. Wills, *J. Catalysis*, **28**, 63 (1973).
685. R. C. Luckner and G. B. Wills, *J. Catalysis*, **28**, 83 (1973).
686. R. Maggiore, L. Fortina, G. Toscano, G. Schembari and L. Solamio, *Chimica e Industria*, **58**, 252 (1976).
687. W. Mahler and T. Fukunaga, *J. Chem. Soc., Chem. Commun.*, **1977**, 307.
688. K. L. Makovetskii, Yu. V. Korshak and T. G. Golenko and L. M. Vardanyan, *5th Sb., 11th Mendeleevsk. S'ezd Obsch. Prikl. Khim., Ref. Dokl. Soobsch.*, **1974**, 138.
689. K. L. Makovetskii and L. I. Red'nina, *Dokl. Akad. Nauk SSSR*, **231**, 143 (1976).
690. K. L. Makovetskii, B. A. Dolgoplosk, Yu. V. Korshak, I. A. Oreshkin and E. I. Tinyakova, *Zh. Vses. Khim. Obshch.*, **24**, 435 (1979).
691. K. L. Makovetskii, B. A. Dolgoplosk, Yu. V. Korshak, I. A. Oreshkin and E. I. Tinyakova, *Zh. Vses. Khim. Obshch.*, **24**, 513 (1979).

692. K. L. Makovetskii, L. L. Redkina and I. A. Oreshkin, *Izvest. Akad. Nauk SSSR, Ser. Khim.*, **1981**, 1928.
693. T. Mallat and J. Petro, *Magy. Kem. Foly*, **88**, 80 (1982).
694. F. D. Mango and J. H. Schachtschneider, *J. Amer. Chem. Soc.*, **89**, 2484 (1967).
695. F. D. Mango and J. H. Schachtschneider, *J. Amer. Chem. Soc.*, **91**, 1030 (1969).
696. F. D. Mango, *Adv. Catalysis*, **20**, 291 (1969).
697. F. D. Mango and J. H. Schachtschneider, *J. Amer. Chem. Soc.*, **93**, 1123 (1971).
698. F. D. Mango, *Polymer Prepr.*, *Amer. Chem. Soc.*, *Div. Polymer Chem.*, **13**, 903 (1972).
699. F. D. Mango, *J. Amer. Chem. Soc.*, **99**, 6117 (1977).
700. J. Margifalvi, *Magy. Chem. Foly*, **84**, 485 (1978).
701. V. I. Mar'in, A. D. Shebaldova and M. L. Khidekel', *Khim. Technol. Elementorg. Soedin. Polim.*, **1972**, 115.
702. V. I. Mar'in, A. D. Shebaldova, T. A. Bolshinskova, M. L. Khidekel' and I. V. Kalechits, *Kin. Katal.*, **14**, 613 (1973).
703. V. I. Mar'in and V. I. Labunskaya, *Issled. Obl. Sint. Katal. Org. Soedin.*, **1975**, 46.
704. V. I. Mar'in, A. D. Shebaldova and M. L. Khidekel', *Khim. Technol.*, **1976**, 43.
705. V. I. Mar'in, G. K. Kornienko, P. E. Matkovskii, A. D. Shebaldova and M. L. Khidekel', *Izvest. Akad. Nauk SSSR, Ser. Khim.*, **1979**, 699.
706. V. I. Mar'in, G. K. Kornienko, P. E. Makovskii, A. D. Shebaldova and M. L. Khidekel', *Izvest. Akad. Nauk SSSR, Ser. Khim.*, **1980**, 2269.
707. L. Markó, *J. Hung. Chem. Soc.*, **27**, 213 (1972).
708. P. R. Marshall and B. J. Ridgewell, *European Polymer J.*, **5**, 29 (1969).
709. K. Maruyama, K. Terada and Y. Yamamoto, *J. Org. Chem.*, **45**, 437 (1980).
710. T. Masuda, K. Hasegawa and T. Higashimura, *Macromolecules*, **7**, 728 (1974).
711. T. Masuda, S. Sasaki and T. Higashimura, *Macromolecules*, **8**, 717 (1975).
712. T. Masuda, K.-Q. Thieu, N. Sasaki and T. Higashimura, *Macromolecules*, **9**, 661 (1976).
713. T. Masuda, H. Kawai, T. Ohtori and T. Higashimura, *Polymer J.*, **10**, 269 (1978).
714. T. Masuda and T. Higashimura, *Macromolecules*, **12**, 9 (1979).
715. T. Masuda, H. Kawai, T. Ohtori and T. Hashiguchi, *Polymer J.*, **11**, 813 (1979).
716. T. Masuda, T. Ohtori and T. Higashimura, *Polymer Prepr.*, *Amer. Chem. Soc.*, *Div. Polymer Chem.*, **20**, 73 (1975).
717. T. Masuda, T. Mouri and T. Higashimura, *Bull. Chem. Soc. Japan*, **53**, 1152 (1980).
718. T. Masuda, Y. Kuwame and T. Higashimura, *Polymer J.*, **13**, 301 (1981).
719. T. Masuda, Y. Kuwame and T. Higashimura, *Polymer*, **23**, 744 (1982).
720. F. E. Massoth, *J. Catalysis*, **30**, 204 (1973).
721. F. E. Massoth, *J. Catalysis*, **36**, 164 (1975).
722. P. E. Matkovskii, G. N. Nesterenko, G. P. Startseva, L. N. Russiyan, V. N. Belova, A. H. Semenov, K. A. Brikenshtein and D. N. Sokolov, *Vysokomol. Soedin.*, *Ser. A*, **19**, 1836 (1977).
723. S. A. Matlin and R. G. Sammes, *J. Chem. Soc., Chem. Commun.*, **1973**, 174.
724. S. A. Matlin and R. G. Sammes, *J. Chem. Soc., Perkin Trans.*, *1*, **1978**, 624.
725. S. Matsumoto, K. Komatsu and K. Igarashi, *Amer. Chem. Soc.*, *Symp. Ser.*, **59**, 303 (1977).
726. B. Matyska, H. Balcar and M. Smutek, *React. Kinet. Catal. Lett.*, **20**, 193 (1982).
727. J. McGinnis, T. J. Katz and S. Hurwitz, *J. Amer. Chem. Soc.*, **98**, 605 (1976).
728. J. McGinnis, *Ph. D. Thesis, Diss. Abstr. Int. B*, **37**, 3965 (1977).
729. R. J. McKinney, T. A. Tulip, D. L. Thorn, T. S. Coolbough and F. N. Tebbe, *J. Amer. Chem. Soc.*, **103**, 5584 (1981).
730. J. Medema, C. Van Stam, V.H. J.DeBeer, A. J. A. Konings and D. C. Koningsberger, *J. Catalysis*, **53**, 386 (1978).
731. H. R. Menapace, N. A. Maly, J. L. Wang and L. G. Wideman, *Prepr.*, *Amer. Chem. Soc.*, *Div. Pet. Chem.*, **19**, 150 (1974).
732. H. R. Menapace, N. A. Maly, J. L. Wang and L. G. Wideman, *J. Org. Chem.*, **40**, 2983 (1975).
733. G. Metha, A. V. Reddy and A. Srikrishna, *Indian J. Chem.*, **20B**, 698 (1981).
734. G. Metha, A. Srikrishna, A. V. Reddy and M. C. A. Nair, *Tetrahedron*, **37**, 4543 (1981).
735. G. Metha and A. V. Reddy, *J. Chem. Soc., Chem. Commun.*, **1981**, 756.
736. G. Metha, *J. Chem. Ed.*, **59**, 313 (1982).

737. L. Michailov and H. J. Harwood, *Polymer Charact. Proc. Symp. 1970* **(Publ. 1971)**, 221.

738. F. W. Michelotti and W. P. Keaveney, *J. Polymer Sci., A*, **3**, 895 (1965).

738a. I. S. Millichamp and W. J. Feast, *5th Internat. Symp. Olefin Metathesis*, Graz, August 22—26, 1983.

739. Kh. M. Minachev, M. A. Ryashentzeva, G. V. Isagulyants and N. N. Rodjistvenskaya, *Izvest. Akad. Nauk SSSR, Ser. Khim.*, **1972**, 705.

740. Kh. M. Minachev, D. A. Kondrat'ev, B. K. Nefedov, A. A. Dergachev, T. N. Bondarenko, T. V. Alekseeva and T. B. Borovinskaya, *Izvest. Akad. Nauk SSSR, Ser. Khim.*, **1980**, 2509.

741. R. J. Minchak and H. Tucker, *Polymer Prepr., Amer. Chem. Soc., Div. Polymer Chem.*, **13**, 885 (1970).

742. R. J. Minchak and H. Tucker, *Polymer Prepr., Amer. Chem. Soc., Div. Polymer Chem.*, **22**, 247 (1981).

743. R. J. Minchak and H. Tucker, *Amer. Chem. Soc., Symp. Ser.*, **193**, 155 (1982).

744. M. T. Mocella, M. A. Busch and E. L. Muetterties, *J. Amer. Chem. Soc.*, **98**, 1283 (1976).

745. M. T. Mocella, R. Rovner and E. L. Muetterties, *J. Amer. Chem. Soc.*, **98**, 4689 (1976).

746. J. Mochida, Y. Ikeda, H. Fujitsu and T. Takeshita, *Chem. Letters*, **1975**, 1213.

747. A. J. Moffat and A. Clark, *J. Catalysis*, **17**, 264 (1970).

748. A. J. Moffat, M. M. Johnson and A. Clark, *J. Catalysis*, **18**, 345 (1970).

749. J. C. Mol, J. A. Moulijn and C. Boelhouwer, *J. Chem. Soc., Chem. Commun.*, **1968**, 633.

750. J. C. Mol, J. A. Moulijn and C. Boelhouwer, *J. Catalysis*, **11**, 87 (1968).

751. J. C. Mol, F. R. Visser and C. Boelhouwer, *J. Catalysis*, **11**, 114 (1970).

752. J. C. Mol, *Ph. D. Thesis*, University of Amsterdam, 1971.

753. J. C. Mol, *Chem. Weekbl. Mag.*, **1978**, 467.

754. J. C. Mol and E. F. G. Woerlee, *J. Chem. Soc., Chem. Commun.*, **1979**, 330.

755. J. C. Mol, *J. Mol. Catalysis*, **15**, 35 (1982).

755a. J. C. Mol and A. Hoogland, *5th Internat. Symp. Olefin Metathesis*, Graz, August 22—26, 1983.

756. A. Mortreux and M. Blanchard, *Bull. Soc. Chim. France*, **1972**, 1641.

757. A. Mortreux and M. Blanchard, *J. Chem. Soc., Chem. Commun.*, **1974**, 786.

758. A. Mortreux, E. Digny, D. Nyranth and M. Blanchard, *Bull. Soc. Chim. France*, **1974**, 241.

759. A. Mortreux, N. Dy and M. Blanchard, *J. Mol. Catalysis*, **1**, 101 (1975/1976).

760. A. Mortreux, J. C. Delgrange, M. Blanchard and B. Lubochinsky, *J. Mol. Catalysis*, **2**, 73 (1977).

761. A. Mortreux and M. Blanchard, *Rol. Koordinatsii v Katalize*, **1976**, 137.

762. A. Mortreux, F. Petit and M. Blanchard, *Tetrahedron Letters*, **1978**, 4967.

763. A. Mortreux, F. Petit and M. Blanchard, *J. Mol. Catalysis*, **8**, 97 (1980).

763a. A. Mortreux, M. Petit and F. Petit, *5th Internat. Symp. Olefin Metathesis*, Graz, August 22—26, 1983.

764. G. Motroni, G. Dall'Asta and I. W. Bassi, *European Polymer J.*, **9**, 257 (1973).

765. V. W. Motz and M. F. Farona, *Inorg. Chem.*, **16**, 2545 (1977).

766. V. W. Motz, *Ph. D. Thesis, Diss. Abstr. Int. B*, **38**, 2171 (1977).

767. J. A. Moulijn and C. Boelhouwer, *J. Chem. Soc., Chem. Commun.*, **1971**, 1170.

768. J. A. Moulijn, H. J. Reitsma and C. Boelhouwer, *J. Catalysis*, **25**, 434 (1972).

769. W. Mowat, J. Smith and D. A. Whan, *J. Chem. Soc., Chem. Commun.*, **1974**, 34.

770. E. L. Muetterties and M. A. Busch, *J. Chem. Soc., Chem. Commun.*, **1974**, 754.

771. E. L. Muetterties, *Inorg. Chem.*, **14**, 951 (1975).

772. E. L. Muetterties and E. Band, *J. Amer. Chem. Soc.*, **102**, 6572 (1980).

773. A. V. Mushegyan, V. K. Ksipteridis, R. K. Dzhylakyan, N. A. Gevorkyan and G. A. Chukhadzhyan, *Arm. Khim. Zh.*, **28**, 672 (1975).

774. A. V. Mushegyan, V. K. Ksipteridis and G. A. Chukhadzhyan, *Arm. Khim. Zh.*, **28**, 674 (1975).

775. A. A. Nadirov, M. N. Zhavoronkov and A. Z. Dorogochinskii, *Neftepererab. Neftekhim.*, **1976**, 25.

776. A. A. Nadirov, M. N. Zhavoronkov and A. Z. Dorogochinskii, *Zh. Fiz. Khim.*, **51**, 460 (1977).

777. M. Nagasawa, K. Kikukawa, M. Takagi and T. Matsuda, *Bull. Chem. Soc. Japan*, **51**, 1291 (1978).

778. R. Nakamura, H. Ida and E. Echigoya, *Chem. Lett.*, **1972**, 273.
779. R. Nakamura and E. Echigoya, *Nippon Kagaku Kaishi*, **1972**, 2276.
780. R. Nakamura, Y. Morita and E. Echigoya, *Nippon Kagaku Kaishi*, **1973**, 244.
781. R. Nakamura and E. Echigoya, *Bull. Japan Petrol Inst.*, **14**, 187 (1972).
782. R. Nakamura, Y. Morita and E. Echigoya, *Nippon Kagaku Kaishi*, **1974**, 420.
783. R. Nakamura, Y. Morita, H. Ida and E. Echigoya, *Nippon Kagaku Kaishi*, **1975**, 1138.
784. R. Nakamura, H. Ida and E. Echigoya, *Nippon Kagaku Kaishi*, **1975**, 1669.
785. R. Nakamura, S. Fukuhara, S. Matsumoto and K. Komatsu, *Chem. Lett.*, **1976**, 253.
786. R. Nakamura, S. Matsumoto and E. Echigoya, *Chem. Lett.*, **1976**, 1019.
787. R. Nakamura, H. Ida and E. Echigoya, *Nippon Kagaku Kaishi*, **1976**, 221.
788. R. Nakamura and E. Echigoya, *Chem. Lett.*, **1977**, 1227.
789. R. Nakamura and E. Echigoya, *Rec. Trav. Chim.*, **96**, M31 (1977).
790. R. Nakamura, K. Ichikawa and E. Echigoya, *Nippon Kagaku Kaishi*, **1978**, 36.
791. R. Nakamura, K. Ichikawa and E. Echigoya, *Nippon Kagaku Kaishi*, **1978**, 1602.
792. R. Nakamura, K. Ichikawa and E. Echigoya, *Chem. Lett.*, **1978**, 813.
793. R. Nakamura and E. Echigoya, *Chem. Lett.*, **1981**, 51.
794. R. Nakamura and E. Echigoya, *J. Mol. Catalysis*, **15**, 147 (1982).
795. R. Nakamura and A. Dedieu, *Nouveau J. Chim.*, **6**, 23 (1982).
796. N. S. Nametkin, V. M. Vdovin, E. D. Babich, V. N. Karl'skii and B. V. Kacharmin, *Dokl. Akad. Nauk SSSR*, **213**, 356 (1973).
797. N. S. Nametkin, V. M. Vdovin, E. D. Babich, B. V. Kacharmin, N. B. Bespalova and V. N. Karel'skii, *Dokl. Akad. Nauk SSSR*, **225**, 577 (1975).
798. G. Natta, G. Dall'Asta, G. Mazzanti, I. Pasquon, A. Valvassori and A. Zambelli, *Makromol. Chem.*, **54**, 95 (1962).
799. G. Natta, G. Dall'Asta, G. Mazzanti and G. Motroni, *Makromol. Chem.*, **69**, 163 (1963).
800. G. Natta, G. Dall'Asta and G. Mazzanti, *Angew. Chem.*, **76**, 765 (1964).
801. G. Natta, G. Dall'Asta and G. Motroni, *J. Polymer Sci.*, *B*, **2**, 349 (1964).
802. G. Natta, G. Dall'Asta and L. Porri, *Makromol. Chem.*, **81**, 253 (1965).
803. G. Natta, G. Dall'Asta, I. W. Bassi and G. Carella, *Makromol. Chem.*, **91**, 87 (1966).
804. G. Natta and I. W. Bassi, *European Polymer J.*, **3**, 33 (1967).
805. G. Natta and I. W. Bassi, *European Polymer J.*, **3**, 43 (1967).
806. G. Natta, I. W. Bassi, and G. Fagherazzi, *European Polymer J.*, **3**, 339 (1967).
807. G. Natta and I. W. Bassi, *J. Polymer Sci.*, *Part C*, **16**, 2551 (1967).
808. G. Natta and G. Dall'Asta in "Polymer Chemistry of Synthetic Elastomers", J. P. Kennedy and E. Thornquist (Eds.), Interscience Publ., New York, 1969.
809. N. V. Nekrasov, N. S. Ustinova, Yu. N. Usov and S. L. Kiperman, *Kinet. Katal.*, **23**, 649 (1982).
810. I. V. Nicolescu and O. Serban, *Rev. Chim.*, **23**, 589 (1972).
811. T. Nishiguchi, S. Goto, K. Sugisaki and K. Fukuzumi, *Yukagaku*, **29**, 15 (1980).
812. T. Nishiguchi, K. Fukuzumi and K. Sugisaki, *J. Catalysis*, **70**, 24 (1981).
813. S. M. Nosakova, *Nov. Aspekty Neftekhim. Sinteza*, **1978**, 43.
814. R. Nouguier, R. Mutin, J. P. Laval, G. Chapelet, J. M. Basset and A. Lattes, *Rec. Trav. Chim.*, **96**, M91 (1977).
815. E. A. Ofstead and N. Calderon, *Makromol. Chem.*, **154**, 21 (1972).
816. E. A. Ofstead, J. P. Lawrence, M. L. Senyek and N. Calderon, *J. Mol. Catalysis*, **8**, 227 (1980).
817. E. A. Ofstead, *Macromol. Synth.*, **6**, 69 (1977).
817a. E. A. Ofstead, M. L. Senyck and N. Calderon, *5th Internat. Symp. Olefin Metathesis*, Graz, August 22—26, 1983.
818. E. Ogata and Y. Kamiya, *Chem. Lett.*, **1973**, 603.
819. E. Ogata and Y. Kamiya, *Ind. Eng. Chem.*, *Prod. Res. Develop.*, **13**, 226 (1974).
820. E. Ogata, T. Sodesawa and Y. Kamiya, *Bull. Chem. Soc. Japan*, **49**, 1317 (1976).
821. J. Ogonowski, *Przem. Chem.*, **60**, 547 (1981).
822. R. F. Ohm, *Chem. Tech.*, **1980**, 183.
823. Y. Okamoto, T. Imanaka and S. Teranishi, *J. Catalysis*, **65**, 448 (1980).
824. T. Okuhara and K. Tanaka, *J. Catalysis*, **42**, 474 (1976).
825. A. A. Olsthoorn, *Ph. D. Thesis*, University of Amsterdam, 1974.
826. A. A. Olsthoorn and C. Boelhouwer, *J. Catalysis*, **44**, 197 (1976).

827. A. A. Olsthoorn and C. Boelhouwer, *J. Catalysis*, **44**, 207 (1976).
828. A. A. Olsthoorn and J. A. Moulijn, *J. Mol. Catalysis*, **8**, 147 (1980).
829. P. P. O'Neill and J. J. Rooney, *J. Chem. Soc., Chem. Commun.*, **1972**, 104.
830. P. P. O'Neill and J. J. Rooney, *J. Amer. Chem. Soc.*, **94**, 4383 (1972).
831. R. Opitz, L. Bencze, L. Markó and K.-H. Thiele, *J. Organomet. Chem.*, **71**, C3 (1974).
832. R. Opitz, K.-H. Thiele, L. Bencze and L. Markó, *J. Organomet. Chem.* **96**, C53 (1975).
833. I. A. Oreshkin, I.L. Red'kina, K. L. Makovetskii, E. I. Tinyakova and B. A. Dolgoplosk, *Izvest. Akad. Nauk SSSR, Ser. Khim.*, **1971**, 1123.
834. I. A. Oreshkin, B. A. Dolgoplosk, E. I. Tinyakova and K. L. Makovetskii *Dokl. Akad. Nauk SSSR*, **226**, 1351 (1976).
835. I. A. Oreshkin, L. I. Red'kina, I. L. Kershenbaum, G. M. Chernenko, K. L. Makovetskii, E. I. Tinyakova and B. A. Dolgoplosk, *European Polymer J.*, **13**, 447 (1977).
836. I. A. Oreshkin, K. L. Makovetskii, B. A. Dolgoplosk, E. I. Tinyakova, I. A. Ostrovskaya, I. L. Kershenbaum and G. M. Chernenko, *Vysokomol. Soedin.*, *B*, **19**, 55 (1977).
837. I. A. Oreshkin, L. I. Red'kina, I. L. Kershenbaum, G. M. Chernenko, K. L. Makovetskii, E. I. Tinyakova and B. A. Dolgoplosk, *Izvest. Akad. Nauk SSSR, Ser. Khim.*, **1977**, 2566.
838. I. A. Oreshkin, S. A. Smirnov, I. A. Kop'eva, B. A. Dolgoplosk and E. I. Tinyakova, *4-I Mezhdunar. Mikrosimpoz. Uspekhi v Obl. Mol. Polimeriz.*, *Ufa*, **1979**, 24.
839. T. Oshika, S. Murai and Y. Koga, *Kobunshi Kagaku*, **22**, 633 (1967).
840. T. Oshika, S. Murai and Y. Koga, *Chem. High Polymer Japan*, **22**, 211, (1968).
841. T. Oshika and H. Tabuchi, *Bull. Chem. Soc. Japan*, **41**, 211 (1968).
842. I. Ya. Ostrovskaya, L. I. Red'kina, K. L. Makovetskii and B. A. Dolgoplosk, *Izvest. Akad. Nauk SSSR, Ser. Khim.*, **1980**, 2650.
843. K. C. Ott, J. B. Lee and R. H. Grubbs, *J. Amer. Chem. Soc.*, **104**, 2942 (1982).
844. J. Otton, J. Colleuille and J. Varagnat, *J. Mol. Catalysis*, **8**, 313 (1980).
845. N. I. Pakuro, A. R. Gantmakher and B. A. Dolgoplosk, *Dokl. Akad. Nauk SSSR*, **223**, 868 (1975).
846. N. I. Pakuro, E. I. Agapova, A. L. Izyumnikov, A. R. Gantmakher and B. A. Dolgoplosk, *Dokl. Akad. Nauk SSSR*, **230**, 897 (1976).
847. N. I. Pakuro, A. R. Gantmakher and B. A. Dolgoplosk, *Dokl. Akad. Nauk SSSR*, **241**, 369 (1978).
848. N. I. Pakuro, A. R. Gantmakher and B. A. Dolgoplosk, *Vysokomol. Soedin., Ser. B*, **20**, 805 (1978).
849. N. I. Pakuro, A. R. Gantmakher and B. A. Dolgoplosk, *Internat. Symp. Ionic Polymerization, Ufa*, **1979**, 37.
850. N. I. Pakuro, A. R. Gantmakher and B. A. Dolgoplosk, *Izvest. Akad. Nauk SSSR, Ser. Khim.*, **1981**, 1052.
851. N. I. Pakuro, K. L. Makovetskii, A. R. Gantmakher and B. A. Dolgoplosk, *Izvest. Akad. Nauk SSSR, Ser. Khim.*, **1982**, 509.
852. G. Pampus, J. Witte and M. Hoffmann, *Rev. Gen. Caoutch. Plast.*, **47**, 1343 (1970).
853. G. Pampus, G. Lehnert and D. Maertens, *Polymer Prepr., Amer. Chem., Soc., Div. Polymer Chem.*, **13**, 880 (1972).
854. G. Pampus and G. Lehnert, *Makromol. Chem.*, **175**, 2605 (1974).
855. R. G. Pearson, *Top. Curr. Chem.*, **41**, 75 (1973).
856. F. Pennella, R. L. Banks and G. C. Bailey, *J. Chem. Soc., Chem. Commun.* **1968**, 1548.
857. F. Pennella and R. L. Banks, *J. Catalysis*, **31**, 304 (1973).
858. F. Pennella, R. B. Regier and R. L. Banks, *J. Catalysis*, **34**, 52 (1974).
859. F. Pennela, *J. Catalysis*, **69**, 206 (1981).
860. C. P. Pinazzi and D. Reyx, *Compt. Rend.*, **276**, 1077 (1973).
861. C. P. Pinazzi, J. Cattiaux, J. C. Soutif and J. C. Brosse, *J. Polymer Sci., Polymer Letters Ed.*, **11**, 1 (1973).
862. C. P. Pinazzi, J. Cattiaux and J. C. Brosse, *European Polymer J.*, **10**, 837 (1974).
863. C. P. Pinazzi, J. Cattiaux, A. Pleurdeau and J. C. Brosse, *Makromol. Chem.*, **175**, 1795 (1974).
864. C. P. Pinazzi, I. Guilmet and D. Reyx, *Tetrahedron Letters*, **1976**, 989.
865. C. P. Pinazzi, I. Campistron and D. Reyx, *Rec. Trav. Chim.*, **96**, M59 (1977).
866. C. P. Pinazzi, I. Campistron and D. Reyx, *Bull. Soc. Chim. France*, **1977**, 896.

867. C. P. Pinazzi, I. Campistron, M. C. Croissandeau and D. Reyx, *J. Mol. Catalysis*, **8**, 325 (1980).
868. I. I. Pis'man, V. V. Atlas and A. A. Bakhshi-Zadé, *Neftekhim.*, **12**, 216 (1972).
869. E. V. Pletneva, Yu. N. Usov and E. V. Skvortsova, *Neftekhim.*, **15**, 347 (1975).
870. E. V. Pletneva, Yu. N. Usov and E. V. Skvortsova, *Prevrasch. Uglevod. Kislotn.-Osnovn. Geterog. Katal.*, *Tezisy Dokl.*, **1977**, 173.
871. A. M. Popov, R. A. Fridman, E. S. Finkel'stein, N. S. Nametkin, V. M. Vdovin, N. A. Bashkirov, Yu. B. Kryukov and L. G. Liberov, *Izvest. Akad. Nauk SSSR, Ser. Khim.*, **1973**, 1423.
872. L. Porri, G. Natta and M. G. Galazzi, *Chimica e Industria*, **46**, 428 (1964).
873. L. Porri, R. Rossi, P. Diversi and A. Lucherini, *Polymer Prepr.*, *Amer. Chem. Soc.*, *Div. Polymer Chem.*, **13**, 897 (1972).
874. L. Porri, R. Rossi, P. Diversi and A. Lucherini, *Makromol. Chem.*, **175**, 3097 (1974).
875. L. Porri, P. Diversi, A. Lucherini and R. Rossi, *Makromol. Chem.*, **176**, 3121 (1975).
876. R. J. Puddephatt, M. A. Quyser and C. F. H. Tipper, *J. Chem. Soc., Chem. Commun.*, **1976**, 626.
876 a. F. Quignard, M. Leconte and J. M. Basset, *5th Internat. Symp. Olefin Metathesis*, Graz, August 22—26, 1983.
877. S. A. Radzinskii, A. P. Sheinker, T. Yu. Evdokymova and A. D. Abkin, *Vysokomol. Soedin.*, *B*, **23**, 736 (1981).
878. L. Ramain and Y. Trambouze, *Compt. Rend.*, **273C**, 1409 (1971).
879. A. K. Rappé and A. W. Goddard III, *J. Amer. Chem. Soc.*, **102**, 5114 (1980).
880. A. K. Rappé and W. A. Goddard III, *Nature (London)*, **285**, 311 (1980).
881. A. K. Rappé and W. A. Goddard III, *Pontential Energy Surf. Dyn. Calc. Chem. React. Mol. Energy Transf., Proc., Symp.*, **1980**, 661.
882. A. K. Rappé and W. A. Goddard III, *J. Amer. Chem. Soc.*, **104**, 448 (1982).
883. L. I. Red'kina, I. A. Oreshkin, K. L. Makovetskii, E. D. Babich and V. M. Vdovin, *Izvest. Akad. Nauk SSSR, Ser, Khim.*, **1980**, 1214.
884. L. I. Red'kina, K. L. Makovetskii and B. A. Dolgoplosk. *Vysokomol. Soedin.*, *B*, **23**, 227 (1981).
885. L. I. Red'kina, K. L. Makovetskii, I. L. Kershenbaum, U. M. Dzhemilev and G. A. Tolstikov, *Izvest. Akad. Nauk SSSR, Ser. Khim.*, **1981**, 113.
886. L. Reif and H. Höcker, *Makromol. Chem., Rapid Commun.*, **2**, 183 (1981).
887. L. Reif and H. Höcker, *Makromol. Chem., Rapid Commun.*, **2**, 745 (1981).
888. L. Reif, C. T. Thu and H. Höcker, *4th Internat. Symp. Olefin Metathesis*, Belfast, September 1—4, 1981.
889. W. Reimann, *Ph. D. Thesis*, Mainz, 1977.
890. D. Reyx, I. Campistron and P. Heiling, *Makromol. Chem.*, **183**, 173 (1982).
891. D. Reyx and M. C. Croissandeau, *Makromol. Chem.*, **183**, 1371 (1982).
892. H. Ridder, *Diplomarbeit*, TH Aachen, 1978.
893. H. Ridder, *Ph. D. Thesis*, TH Aachen, 1981.
894. K. Riebel, *Ph. D. Thesis*, Mainz, 1976.
895. R. E. Rinehard and H. P. Smith, *J. Polymer Sci.*, *B*, **3**, 1049 (1965).
896. S. M. Rocklage, J. D. Fellmann, G. A. Rupprecht, L. W. Messerle and R. R. Schrock, *J. Amer. Chem. Soc.*, **103**, 1440 (1981).
897. R. Rossi, P. Diversi, A. Lucherini and L. Porri, *Tetrahedron Letters*, **1974**, 879.
898. R. Rossi, *Chimica e Industria*, **57**, 242 (1975).
899. R. Rossi and R. Giorgi, *Tetrahedron Letters*, **1976**, 1721.
900. R. Rossi and R. Giorgi, *Chimica e Industria*, **58**, 517 (1976).
901. R. Rossi, *Chimica e Industria*, **58**, 517 (1976).
902. R. Rossi and R. Giorgi, *Chimica e Industria*, **58**, 518 (1976).
903. R. Rossi, *Gazz. Chim. Ital.*, **1976**, 106.
904. R. M. Rotchild, *Ph. D. Thesis, Diss. Abstr. Int. B.*, **36**, 2813 (1975).
905. H. Rudler, *J. Mol. Catalysis*, **8**, 53 (1980).
906. H. Rudler, F. Rose, M. Rudler and C. Alvarez, *J. Mol. Catalysis*, **15**, 81 (1982).
906a. K. J. Rutt and H. T. Dodd, *5th Internat. Symp. Olefin Metathesis*, Graz, August 22—26, 1983.

906b. H. Rudler, C. Alvarez and A. Parlier, *5th Internaz. Symp. Olefin Metathesis*, Graz, August 22—26, 1983.

907. A. V. Ryabukin, V. S. Fel'dblyum, M. E. Basner, I. V. Ishchuk, T. I. Baranova, O. P. Karpov, S. I. Kryukov and M. I. Farberov, *Neftekhim.*, **17**, 62 (1977).

908. A. V. Ryabukin, V. S. Fel'dblyum, S. I. Kryukov, T. I. Baranova, I. A. Gol'denberg and V. V. Makar'in, *Osnovn. Organ. Sintez. Neftekhim.*, **1976**, 121.

909. A. V. Ryabukin, V. S. Fel'dblyum, S. I. Kryukov and V. V. Makar'in, *Prom. Synt. Kauch.*, **1977**, 5.

910. T. Saegusa, T. Tsujino and J. Furukawa, *Makromol. Chem.*, **78**, 32 (1964).

911. H. Sakurai, Y. Kamiyama and Y. Nakadaira, *J. Organomet. Chem.*, **131**, 147 (1977).

912. R. G. Salomon, D. J. Coughlin and E. M. Easler, *J. Amer. Chem. Soc.*, **101**, 3961 (1979).

913. C. Sanchez, R. Kieffer and A. Kinnemann, *J. Prakt. Chem.*, **177**, 320 (1978).

914. C. Sanchez, R. Kieffer and A. Kinnemann, *J. Prakt. Chem.*, **127**, 329 (1978).

915. J. Sancho and R. R. Schrock, *J. Mol. Catalysis*, **15**, 75 (1982).

916. G. Sartori, F. Ciampelli and W. Cameli, *Chimica e Industria*, **45**, 1473 (1963).

917. G. Sartori, A. Valvassori and S. Faina, *Atti Acad. naz. Lincei, Rend., Classe Sci. Fis. Mat.*, **34**, 565 (1963).

918. H. Sato, Y. Tanaka and T. Taketomi, *Makromol. Chem.*, **198**, 1993 (1977).

919. Y. Sato, Y. Iwasawa and H. Kuroda, *Chem. Lett.*, **1972**, 1101.

920. V. Schneider and P. K. Frölich, *Ind. Eng. Chem.*, **23**, 1045 (1931).

921. J. Schopov and C. Jossifov, *Balkan Chemistry Days*, **Poster 3—92**, Varna, May 17—19, 1983.

922. R. R. Schrock, S. Rocklage, J. Wengrovius, G. Rupprecht and J. Fellmann, *J. Mol. Catalysis*, **8**, 73 (1980).

923. R. R. Schrock, S. Rocklage, J. Wengrovius, G. Rupprecht and J. Fellmann, *Prepr., Amer. Chem. Soc., Div. Pet. Chem.*, **1980**, 382.

924. R. R. Schrock, M. L. Listeman and L. G. Sturgeoff, *J. Amer. Chem. Soc.*, **104**, 4291 (1982).

924a. R. R. Schrock, *5th Internat. Symp. Olefin Metathesis*, Graz, August 22—26, 1983.

925. R. G. Schulz, *J. Polymer Sci., B*, **4**, 541 (1966).

926. K. W. Scott, N. Calderon, E. A. Ofstead, W. A. Judy and J. P. Ward, *Adv. Chem. Ser.*, **91**, 399 (1969).

927. K. W. Scott, N. Calderon, E. A. Ofstead, W. A. Judy and J. P. Ward, *Rubber Chem. Technol.*, **44**, 1341 (1971).

928. K. W. Scott, *Polymer Prepr., Amer. Chem. Soc., Div. Polymer Chem.*, **13**, 874 (1972).

928a. J. W. F. L. Seetz, B. J. J. van de Heisteeg, G. Schat, O. S. Akkerman and F. Bickelhaupt, *5th Internat. Symp. Olefin Metathesis*, Graz, August 22—26, 1983.

929. B. Sentürk and K. Hummel, *Chem. Acta Turcica*, **6**, 229 (1978).

930. B. Sentürk and K. Hummel, *Acta. Chem. Turcica*, **7**, 15 (1979).

931. K. Seyferth and R. Taube, *J. Organomet. Chem.*, **229**, 275 (1982).

932. A. M. Shapiro, Yu. V. Korshak, M. A. Tlenkopatchev and B. A. Dolgoplosk, *Dokl. Akad. Nauk SSSR*, **248**, 1173 (1979).

933. A. D. Shebaldova, V. I. Mar'in, I. V. Kalechits, T. P. Popova, V. A. Akhurin, G. G. Nasledskova and M. L. Khidekel', *V Sb., XI Mendeleevsk. S'ezd po Obshch. i Prikl. Khim., Ref. Dokl. i Soobshch.*, **1974**, 309.

934. A. D. Shebaldova, V. I. Mar'in, M. L. Khidekel', I. V. Kalechits and S. N. Kursanov, *Izvest. Akad. Nauk SSSR, Ser. Khim.*, **1976**, 2509.

935. A. D. Shebaldova, V. I. Mar'in, G. G. Nasledskova, M. L. Khidekel', T. A. Popova and V. A. Akhurin, *Neftekhim.*, **16**, 533 (1976).

936. A. D. Shebaldova, V. I. Mar'in and M. L. Khidekel', *Khim. Technol. Element. Soedin. Polym.*, **1976**, 47.

937. A. D. Shebaldova, M. L. Khidekel', P. E. Martkovskii, G. C. Kornienko and V. I. Mar'in, *13 Vses. Chugaev Sovesch. Khim. Kompleks. Soedin.*, **1978**, 453.

938. A. D. Shebaldova and V. I. Mar'in, *3rd Internat. Symp. Olefin Metathesis*, Lyons, September 10—12, 1979.

939. A. M. Sherman, *Ph. D. Thesis, Diss. Abstr. Int. B*, **36**, 5051 (1976).

940. F. K. Shmidt, E. A. Grechkina and V. G. Lipovich, *Kinet. Katal.*, **14**, 1080 (1973).

941. F. K. Shmidt, E. A. Grechkina, V. V. Saraev and V. G. Lipovich, *Neftekhim.*, **14**, 41 (1974).
942. F. K. Shmidt, E. A. Grechkina and V. P. Chumakova, *Fiz.-Khim. Issled. Vzaimodeistv. Solei Shcheloch. Met. Rasplav. Prod. Destr. Sapropel.* **1972**, 49.
943. F. K. Shmidt, E. A. Grechkina and Y. S. Levkovskii, *Fiz.-Khim. Issled. Vzaimodeistv. Solei Shcheloch. Met. Rasplav. Prod. Destr. Sapropel.* **1972**, 62.
944. F. K. Shmidt, Y. S. Levkovskii, B. V. Timashkova, V. B. Laurent'eva and E. A. Grechkina, *React. Kinet. Catal. Lett.*, **3**, 385 (1975).
945. F. K. Shmidt, E. A. Grechkina, Y. S. Levkovskii, O. I. Shmidt and V. B. Lavrent'eva, *Neftekhim.*, **18**, 23 (1978).
946. V. Siekermann, *Ph. D. Thesis*, TH Aachen, 1978.
946a. J. H. Sinfelt, *Ind. Eng. Chem.*, **62**, 66 (1970).
947. D. M. Singleton and D. J. Eatough, *J. Mol. Catalysis*, **8**, 175 (1980).
948. E. V. Skvortsova, E. V. Pletneva, L. A. Elovatskaya and N. S. Ustinova, *Isled. Obl. Sint. Katal. Org. Soedin.*, **1975**, 62.
949. A. S. Smirnov, I. A. Kop'eva, I. A. Oreshkin, E. I. Tinyakova and B. A. Dolgoplosk, *Dokl. Akad. Nauk SSSR*, **239**, 392 (1978).
950. S. A. Smirnov, I. A. Oreshkin and B. A. Dolgoplosk, *Dokl. Akad. Nauk SSSR*, **239**, 1375 (1978).
951. S. A. Smirnov, *Ph. D. Thesis*, **INKhS**, Akad. Nauk SSSR, Moscow, 1979.
952. S. A. Smirnov, I. A. Oreshkin, V. G. Zaikin and B. A. Dolgoplosk, *Dokl. Akad. Nauk SSSR*, **245**, 1400 (1979).
953. S. A. Smirnov, *Mech. Protses. Polym., Rabot. Molod. Uchen.*, M, **1980**, 3.
954. J. Smith, R. F. Howe and D. A. Whan, *J. Catalysis*, **34**, 191 (1974).
955. J. Smith, W. Mowat, D. A. Whan and E.A.V. Ebsworth, *J. Chem. Soc., Dalton. Trans.*, **1974**, 1742.
956. T. Sodesawa, E. Ogata and Y. Kamiya, *Nippon Kagaku Kaishi*, **1975**, 1046.
957. T. Sodesawa, E. Ogata and Y. Kamiya, *Bull. Japan Petrol Tnst.*, **18**, 162 (1976).
958. T. Sodesawa, E. Ogata and Y. Kamiya, *Bull. Chem. Soc. Japan*, **50**, 998 (1977).
959. T. Sodesawa, E. Ogata and Y. Kamiya, *Bull. Chem. Soc. Japan*, **52**, 1661 (1979).
960. J. P. Soufflet, D. Commereuc and Y. Chauvin, *Compt. Rend.*, **276**, 169 (1973).
961. A. N. Startsev, B. N. Kuznetsov and Yu. V. Ermakov, *React. Kinet. Catal. Lett.*, **3**, 321 (1975).
962. A. N. Startsev, E. A. Demin, B. N. Kuznetsov, E. M. Chalganov and Yu. I. Ermakov, *Izvest. Vyssh. Uchebn. Zaved. Khim. Khim. Tekhnol.*, **20**, 1167 (1977).
963. F. Stelzer, R. Thummer, D. Wewerka and K. Hummel, *Kolloid.-Z. u Z. Polymere*, **251**, 772 (1973).
964. F. Stelzer, R. Thummer and K. Hummel, *Colloid Polymer Sci.*, **255**, 664 (1977).
965. F. Stelzer, K. Hummel and R. Thummer, *Progr. Colloid. Polymer Sci.*, **66**, 411 (1979).
966. F. Stelzer, P. Heiling and K. Hummel, *3rd Internat. Symp. Olefin Metathesis*, Lyons, September 10—12, 1979.
967. F. Stelzer, C. Graimann and K. Hummel, *4th Internat. Symp. Olefin Metathesis*, Belfast, September 1—4, 1981.
968. F. Stelzer, C. Graimann and K. Hummel, *Colloid. Polymer Sci.*, **260**, 829 (1982).
969. H. W. Stern, *Erdöl Kohle*, **26**, 11 (1975).
970. A. V. Stolyarova, A. Ya. Vanyarkh and G. I. Kitaev, *Kauch. Rezina*, **1982**, 29.
971. W. H. J. Stork, J. G. F. Coolegem and G. T. Pott, *J. Catalysis*, **32**, 497 (1974).
972. W. H. J. Stork and G. T. Pott, *Rec. Trav. Chim.*, **96**, M105 (1977).
973. R. Streck, *Paper presented at the Ges. Deut. Chem. Meeting*, Köln, September 10, 1975.
974. R. Streeck, *4th Internat. Symp. Olefin Metathesis*, Belfast, September 1—4, 1981.
975. R. Streck, *J. Mol. Catalysis*, **15**, 3 (1982).
976. B. S. Strelchik, R. A. Fridman, E. S. Finkel'stein and V. M. Vdovin, *Izvest. Akad. Nauk SSSR, Ser. Khim.*, **1976**, 579.
977. V. V. Strelets, V. N. Tsarev and O. N. Efimov, *Kinet. Katal.*, **22**, 359 (1981).
978. A. I. Syatkovskii, T. T. Denisova, I. V. Ikonitskii and B. D. Babitskii, *J. Polymer Sci., Polymer Chem. Ed.*, **17**, 3939 (1979).
979. A. I. Syatkovskii, T. T. Denisova, N. A. Buzina and B. D. Babitskii, *Polymer*, **21**, 1112 (1980).

980. A. I. Syatkovskii, T. T. Denisova, E. L. Abramenko, A. S. Khatchaturov and B. D. Babitskii, *Polymer*, **22**, 1554 (1981).

981. A. I. Syatkovskii, T. T. Denisova, L. V. Gavrilova and B. D. Babitskii, *Kauch. Rezina*, **1931**, 6.

982. A. I. Syatkovskii, L. V. Gavrilova, T. T. Denisova and B. D. Babitskii, *3 Vses. Conf. Khim. Carben. Moscow*, **1982**, 80.

983. N. Taghizadeh, F. Quignard, M. Leconte, J. M. Basset, C. Larroche, J. P. Laval and A. Lattes, *J. Mol. Catalysis*, **15**, 219 (1982).

984. J. Takada, T. Otsu and M. Imoto, *Kogyo Kagaku Zaisshi*, **69**, 715 (1966).

985. T. Takagi, T. Hamaguchi, K. Fukuzumi and M. Aoyama, *J. Chem. Soc., Chem. Commun.*, **1972**, 838.

986. T. Takagi, K. Ichikawa, T. Hamaguchi, K. Fukuzumi and M. Aoyama, *Yukagaku*, **24**, 377 (1975).

987. T. Takagi, K. Ichikawa, K. Fukuzumi and T. Hamaguchi, *Yukagaku*, **24**, 518 (1975).

988. T. Takagi and K. Fukuzumi, *Trans. Met. Chem.*, **1**, 54 (1976).

989. T. Takahashi, *Kogyo Kagaku Zaisshi*, **73**, 728 (1970).

990. T. Takahashi, *Bull. Japan Petrol Inst.*, **14**, 40 (1972).

991. T. Takahashi, *Kogai Shigen Kenkyusho Iho*, **3**, 7 (1974).

992. T. Takahashi, *Nippon Kagaku Kaishi*, **1975**, 1058.

993. T. Takahashi, *Nippon Kagaku Kaishi*, **1975**, 2183.

994. T. Takahashi, *Kogai Shigen Kenkyusho Iho*, **4**, 63 (1974).

995. T. Takahashi, *Nippon Kagaku Kaishi*, **1976**, 1454.

996. T. Takahashi, *Nippon Kagaku Kaishi*, **1977**, 369.

997. T. Takahashi, *Nippon Kagaku Kaishi*, **1977**, 1524.

998. T. Takahashi, *Nippon Kagaku Kaishi*, **1978**, 418.

999. T. Takahashi, *Kogai Shigen Kenkyusho Iho*, **10**, 31 (1981).

1000. S. Tamagaki, J. R. Card and C. D. Neckers, *J. Amer. Chem. Soc.*, **100**, 6635 (1978).

1001. K. Tanaka, K. Tanka and K. Miyahara, *J. Chem. Soc., Chem. Commun.*, **1979**, 314.

1002. K. Tanaka and K. Miyahara, *Proc. 7th Internat. Congr. Catal.*, B, 1318 (1980).

1003. K. Tanaka, K. Tanaka and K. Miyahara, *Proc. 7th Internat. Congr. Catal.*, **B**, 1327 (1980).

1004. K. Tanaka, K. Miyahara and K. Tanaka, *Chem. Lett.*, **1980**, 623.

1005. K. Tanaka, K. Tanaka and K. Miyahara, *J. Chem. Soc., Chem. Commun.*, **1980**, 666.

1006. K. Tanaka, K. Miyahara and K. Tanaka, *Bull. Chem. Soc. Japan*, **54**, 3106 (1981).

1007. K. Tanaka, K. Miyahara and K. Tanaka, *Stud. Surf. Sci. Catal.*, **7**, 1318 (1981).

1008. K. Tanaka, K. Tanaka and K. Miyahara, *J. Catalysis*, **72**, 182 (1981).

1009. K. Tanaka, K. Miyahara and K. Tanaka, *J. Mol. Catalysis*, **15**, 133 (1982).

1009. a. K. Tanaka, T. Nakagawa, I. Yaegashi and K. Aomura, *5th Internat. Symp. Olefin Metathesis*, Graz, August 22—26, 1983.

1010. C. Tanielian, R. Kieffer and A. Harfouch, *Tetrahedron Letters*, 1977, **1977**, 4589.

1011. C. Tanielian, A. Kinnemann and T. Osparpucu, *Can. J. Chem.*, **57**, 2022 (1979).

1012. C. Tanielian, A. Kinnemann and T. Osparpucu, *Can. J. Chem.*, **58**, 2813 (1980).

1013. C. Tanielian, R. Kieffer and A. Harfouch, *J. Mol. Catalysis*, **10**, 269 (1981).

1014. E. K. Tarcsi, F. Ratkovics and L. Bencze, *J. Mol. Catalysis*, **1**, 435 (1976).

1015. R. Taube and K. Seyferth, *Z. Chem.*, **13**, 300 (1973).

1016. R. Taube and K. Seyferth, *Z. Chem.*, **14**, 284 (1974).

1017. R. Taube and K. Seyferth, *Z. Chem.*, **14**, 410 (1974).

1018. R. Taube and K. Seyferth, *Z. anorg. allgem. Chem.*, **437**, 213 (1977).

1019. R. Taube, K. Seyferth, L. Bencze and L. Markó, *J. Organomet. Chem.*, **111**, 215 (1976).

1019a. R. Taube and K. Seyferth, *5th Internat. Symp. Olefin Metathesis*, Graz, August 22—26, 1983.

1020. F. N. Tebbe, G. W. Parshall and D.W. Ovenall, *J. Amer. Cehm. Soc.*, **101**, 5074 (1979).

1021. F. N. Tebbe and R. L. Harlow, *J. Amer. Chem. Soc.*, **102**, 6149 (1980).

1022. H. H. Thoi, K. J. Ivin and J. J. Rooney, *4th Internat. Symp. Olefin Metathesis*, Belfast, September 1—4, 1981.

1023. H. H. Thoi, K. J. Ivin and J. J. Rooney, *J. Mol. Catalysis*, **15**, 245 (1982).

1024. H. H. Thoi, B. S. R. Reddy and J. J. Rooney, *J. Chem. Soc., Faraday Trans.*, **1**, 3307 (1982).
1025. H. H. Thoi, K. J. Ivin and J. J. Rooney, *Makromol. Chem.*, **183**, 1629 (1982).
1026. R. Thomas, J. A. Moulijn and F. P. J. M. Kerkhof, *Rec. Trav. Chim.*, **96**, M134 (1977).
1027. R. Thomas, M. C. Mittelmeijer-Hazeleger, F. P.J.M. Kerkhof, J. A. Moulijn, J. Medema and V. H. J. DeBeer, *Chem. Uses Molybdenum, Proc. Internat. Conf., 3rd*, **1979**, 85.
1028. R. Thomas, J. A. Moulijn, V. H. J. DeBeer and J. Medema, *J. Mol. Catalysis*, **8**, 161 (1980).
1029. R. Thomas, F. P. J. M. Kerkhof, J. A. Moulijn, J. Medema and V.H. J. DeBeer, *J. Catalysis*, **61**, 559 (1980).
1030. R. Thomas, *Ph. D. Thesis*, University of Amsterdam, 1981.
1031. R. Thomas and J. A. Moulijn, *J. Mol. Catalysis*, **15**, 157 (1982).
1032. E. Thorn-Csanyi and H. Perner, *Makromol. Chem.*, **180**, 919 (1979).
1033. E. Thorn-Csanyi, *3rd Internat. Symp. Olefin Metathesis*, Lyons, September 10—12, 1979.
1034. E. Thorn-Csanyi, H. Abendroth and H. Perner, *Makromol. Chem.*, **181**, 2081 (1980).
1035. E. Thorn-Csanyi, *Nach. Chem. Techn. Lab.*, **10**, 700 (1981).
1036. E. Thorn-Csanyi, *Angew. Makromol. Chem.*, **94**, 181 (1981).
1037. E. Thorn-Csanyi, *4th Internat. Symp. Olefin Metahesis*, Belfast, September 1—4, 1981.
1038. E. Thorn-Csanyi, C. Hannemann-Perner and H. Perner, *Makromol. Chem. Rapid Commun.*, **3**, 329 (1982).
1038a. E. Thorn-Csanyi and H. Timm, *5th Internat. Symp. Olefin Metathesis*, Graz, August 22—26, 1983.
1039. R. Thummer, F. Stelzer and K. Hummel, *Makromol. Chem.*, **176**, 1703 (1975).
1040. M. A. Tlenkopatchev, I. A. Kop'eva, N. A. Bychkova, Yu. V. Korshak, G. I. Timofeeva, E. I. Tinyakova and B. A. Dolgoplosk, *Dokl. Akad. Nauk SSSR*, **227**, 889 (1976).
1041. K. A. Torosyan, S. G. Babayan, A. V. Mushegyan and Zh. G. Martirosyan, *Prom.-st. Arm.*, **1977**, 68.
1042. C. Tosi, F. Ciampelli and G. Dall'Asta, *J. Polymer Sci., Polymer Phys. Ed.*, **11**, 529 (1973).
1043. W. I. Truett, D. R. Johnson, J. M. Robinson and B. A. Montague, *J. Amer. Chem. Soc.*, **82**, 2337 (1960).
1044. C. P. Tsonis, *Ph. D. Thesis, Diss. Abstr. Int. C*, **38**, 5370 (1978).
1045. C. P. Tsonis and M. F. Farona, *J. Polymer Sci., Polymer Chem. Ed.*, **17**, 185 (1979).
1046. C. P. Tsonis and M. F. Farona, *J. Polymer Sci., Polymer Chem. Ed.*, **17**, 1779 (1979).
1047. N. Tsuda and A. Fujimori, *J. Catalysis*, **69**, 410 (1981).
1047 a. W. Truda, T. Mori, W. Kosaka and Y. Sakai, *5th Internat. Symp. Olefin Metathesis*, Graz, August 22 — 26, 1983.
1048. J. Tsuji and S. Hashiguchi, *Tetrahedron Letters*, **1980**, 2955.
1049. J. Tsuji and S. Hashiguchi, *J. Organomet. Chem.*, **218**, 69 (1981).
1050. T. Tsujino, T. Saegusa, and S. Kobayashi and J. Furukawa, *J. Chem. Soc. Japan*, **67**, 1961 (1964).
1051. T. Tsujino, T. Saegusa and J. Furukawa, *Makromol. Chem.*, **85**, 71 (1965).
1052. A. Uchida, Y. Mukai, Y. Hamano and S. Matsuda, *Ind. Eng. Chem., Prod. Res. Develop.*, **10**, 369 (1971).
1053. A. Uchida, Y. Hamano, Y. Mukai and S. Matsuda, *Ind. Eng. Chem., Prod. Res. Develop.*, **10**, 372 (1971).
1054. A. Uchida, K. Kobayashi and S. Matsuda, *Ind. Eng. Chem., Prod. Res. Develop.*, **11**, 389 (1972).
1055. A. Uchida and S. Matsuda, *Kagaku No Ryoichi*, **26**, 243 (1972).
1056. A. Uchida, T. Ishikawa and M. Takagi, *Rec. Trav. Chim.*, **96**, M13 (1977).
1057. A. Uchida, M. Minenoya and T. Yamamoto, *J. Chem. Soc., Dalton Trans.*, **1981**, 1089.
1058. A. Uchida and K. Hata, *J. Chem. Soc., Dalton. Trans.*, **1981**, 1111.
1059. A. Uchida and K. Hato, *J. Mol. Catalysis*, **15**, 111 (1982).
1059a. A. Uchida and M. Kori, *5th Internat. Symp. Olefin Metathesis*, Graz, August 22—26, 1983.
1060. Y. Uchida, M. Hidai and T. Tatoumi,. *Bull. Chem. Soc. Japan*, **45**, 1158 (1972).
1061. T. Ueshima, S. Kobayashi and M. Matsuoka, *Japan Plastics*, **24**, 11 (1974).

1062. A. V. Ushakov, E. S. Finkel'stein, E. B. Portnykh and V. M. Vdovin, *Izvest. Akad. Nauk SSSR, Ser. Khim.*, **1981**, 2365.
1063. A. V. Ushakov, E. S. Finkel'stein, E. B. Portnykh and V. M. Vdovin, *Izvest. Akad. Nauk SSSR, Ser. Khim.*, **1981**, 2835.
1064. Yu. N. Usov, T. G. Vaistub, E. V. Pletneva and E. V. Skvortsova, *Neftekhim.*, **12**, 223 (1972).
1065. Yu. N. Usov, E. V. Skvortsova, L. A. Elovatskaya, E. V. Pletneva and T. G. Vaistub, *Neftepererab. Neftekhim.*, **1974**, 33.
1066. Yu. N. Usov, E. V. Skvortsova, E. V. Pletneva and T. G. Vaistub, *Izvest. Vyssh. Uchebn. Zaved. Khim. Khim. Technol.*, **17**, 898 (1974).
1067. Yu. N. Usov, E. V. Skvortsova, E. V. Pletneva and T. G. Vaistub, *Neftekhim.*, **14**, 196 (1974).
1068. Yu. N. Usov, E. V. Skvortsova, T. G. Vaistub and E. V. Pletneva, *Neftekhim.*, **14**, 368 (1974).
1069. Yu. N. Usov, E. V. Pletneva, E. V. Skvortsova and T. G. Vaistub, *5th Sb., Katal. Prevrasch. Uglevod.*, Irkutsk, **1974**, 66.
1070. Yu. N. Usov, L. A. Elovatskaya and E. V. Skvortsova, *5th Sb., Katal., Prevrasch. Uglevodorod. Irkutsk*, **1974**, 75.
1071. Yu. N. Usov, N. S. Ustinova and E. V. Skvortsova, *Neftekhim.*, **16**, 829 (1976).
1072. Yu. N. Usov, E. V. Skvortsova, N. S. Ustinova and T. G. Vaistub, *Katal. Prevrasch. Uglevod.*, **1976**, 133.
1073. Yu. N. Usov, E. V. Skvortsova and L. A. Elovatskaya, *Neftekhim.*, **18**, 193 (1978).
1074. N. S. Ustinova, B. S. Gutkov and T. B. Khrust, *3rd Vses. Conf. Mech. Katal. React.*, p-2, *Novosibirsk*, **1982**, 160.
1075. A. Vaghi, A. Castellan, J. C. J. Bart, N. Giordano and V. Ragaini, *J. Catalysis*, **42**, 381 (1976).
1076. P. B. Van Dam, M. C. Mittelmeijer and C. Boelhouwer, *J. Chem. Soc., Chem. Commun.*, **1972**, 1221.
1077. P. B. Van Dam and C. Boelhouwer, *React. Kin. Catal. Lett.*, **1**, 165 (1974).
1078. P. B. Van Dam and C. Boelhouwer, *React. Kin. Catal. Lett.*, **1**, 481 (1974).
1079. P. B. Van Dam, M. C. Mittelmeijer and C. Boelhouwer, *Fette, Seifen, Anstrichm.*, **76**, 264 (1974).
1080. P. B. Van Dam, M. C. Mittelmeijer and C. Boelhouwer, *J. Amer. Oil Chem. Soc.*, **51**, 389 (1974).
1081. F. H. M. Van Rijn and J. C. Mol, *Rec. Trav. Chim.*, **96**, M96 (1977).
1082. A. J. Van Roosmalen, P. Polder and J. C. Mol, *J. Mol. Catalysis*, **8**, 185 (1980).
1083. A. J. Van Roosmalen and J. C. Mol, *J. Chem. Soc., Chem. Commun.*, **1980**, 704.
1084. A. J. Van Roosmalen, D. Koster and J. C. Mol, *J. Phys. Chem.*, **84**, 3075 (1980).
1085. A. J. Van Roosmalen, *Ph. D. Thesis*, University of Amsterdam, 1980.
1086. J. M. Van Thiel and C. Boelhouwer, *Farbe Lack*, **80**, 1028 (1974).
1087. L. M. Vardanyan, Yu. V. Korshak, M. P. Teterina and B. A. Dolgoplosk, *Dokl. Akad. Nauk SSSR*, **207**, 345 (1972).
1088. L. M. Vardanyan, Yu. V. Korshak, S. A. Pavlova, G. I. Timofieva and B. A. Dolgoplosk, *Dokl. Akad. Nauk SSSR*, **208**, 1349 (1973).
1089. L. M. Vardanyan and S. S. Graukina, *Neftekhim. Sint. Vysokomol. Soedin.*, **1976**, 286.
1090. L. M. Vardanyan, Yu. V. Korshak, M. A. Markevich, N. A. Nechitalo, M. P. Teterina and B. A. Dolgoplosk, *Vysokomol. Soedin.*, **16**, 24 (1974).
1091. J. Vass, J. Pallos and L. Bencze, *Magy. Kem. Lapja*, **30**, 337 (1975).
1092. V. M. Vdovin, E. D. Babich, V. N. Karelskii, B. V. Kacharmin and N. B. Bespalova, *5th Sb., 11th Mendeleevsk. S'ezd Obsch. Prikl. Khim., Ref. Dokl. Soobsch.*, **1974**, 318.
1093. V. M. Vdovin, E. D. Babich, T. A. Butenko, N. B. Bespalova, N. A. Pritula, V. E. Fedorov and G. K. Fedorova, *Katal. Soderzh., Nenasen. Kompleksy, Simp. Taskent 1980*, Novosibirsk 1980, 18.
1094. E. Verkuijlen and C. Boelhouwer, *J. Chem. Soc. Chem. Commun.*, **1974**, 793.
1095. E. Verkuijlen and C. Boelhouwer, *Riv. Ital. Sost. Grasse*, **53**, 234 (1976).
1096. E. Verkuijlen and C. Boelhouwer, *Fette, Seifen Anstrichm.*, **78**, 444 (1976).
1097. E. Verkuijlen, F. Kapteijn, J. C. Mol and C. Boelhouwer, *J. Chem. Soc., Chem. Commun.*, **1977**, 198.

1098. E. Verkuijlen, R. J. Dirks and C. Boelhouwer, *Rec. Trav. Chim.*, **96**, M86 (1977).
1099. E. Verkuijlen and C. Boelhouwer, *Chem. Phys. Lipids*, **24**, 305 (1979).
1100. E. Verkuijlen, *J. Mol. Catalysis*, **8**, 107 (1980).
1101. E. Verkuijlen, *Ph. D. Thesis*, University of Amsterdam, 1980.
1102. J. Viale and J. M. Basset, *Rect. Kin. Catal. Lett.*, **2**, 397 (1975).
1103. D. Villemin and J. Levisalles, *3rd Internat. Symp. Olefin Metathesis*, Lyons, September 10—12, 1979.
1104. D. Villemin, *Tetrahedron Letters*, **1980**, 1715.
1105. D. Villemin and P. Cadiot, *Tetrahedron Letters*, **23**, 5139 (1982).
1106. J.-L. Wang and H. R. Menapace, *J. Org. Chem.*, **33**, 3794 (1968).
1107. J.-L. Wang, H. R. Menapace and M. Brown, *J. Catalysis*, **26**, 455 (1972).
1108. J.-L. Wang and H. R. Menapace, *J. Catalysis*, **28**, 300 (1973).
1109. L. Wang and W. K. Hall, *J. Catalysis*, **66**, 251 (1980).
1110. S. Warwel and W. Laarz, *Chem.-Ztg.*, **99**, 502 (1975).
1111. S. Warwel and A. Aivasidis, *Chem.-Ztg.*, **101**, 361 (1977).
1112. S. Warwel and W. Laarz, *Z. Naturforsch.*, **32B**, 1145 (1978).
1113. S. Warwel and P. Buschmeyer, *Angew. Chem.*, **90**, 131 (1978).
1114. S. Warwel and Buschmeyer and J. Heveling, *3rd Internat. Symp. Olefin Metathesis*, Lyons, September 10—12, 1979.
1115. S. Warwel, *Erdöl, Kohle, Erdgas, Petrochem.*, **1978/1979**, 437.
1116. S. Warwel and E. Janessen, *Chem.-Ztg.*, **106**, 266 (1982).
1117. S. Warwel, H. Ridder and W. Winkelmüller, *Angew. Chem.*, **94**, 718 (1982).
1117a. S. Warwel and W. Winkelmüller, *5th Internat. Symp. Olefin Metathesis*, Graz, August 22—26, 1983.
1118. E. Wasserman, D. A. Ben-Efraim and R. Wolovsky, *J. Amer. Chem. Soc.*, **90**, 3286 (1968).
1118a. K. Weiss and K. Hoffmann, *5th Internat. Symp. Olefin Metathesis*, Graz, August 22—26, 1983.
1119. P. A. Wender and J. C. Lechleiter, *J. Amer. Chem. Soc.*, **99**, 267 (1977).
1120. P. A. Wender and J. C. Hubbs, *J. Org. Chem.*, **45**, 365 (1980).
1121. P. A. Wender and L. J. Letendre, *J. Org. Chem.*, **45**, 367 (1980).
1122. P. A. Wender and S. L. Eck, *Tetrahedron Letters*, **1982**, 1871.
1123. C. Wengi, P. Deren, Z. Yuming, Z. Yulian, Y. Caiyum and S. Zhiquan, *Gaofenzi Tongxum*, **4**, 226 (1980).
1124. C. Wengi and Z. Yulian, *Xuaxue Xuebao*, **39**, 383 (1981).
1125. J. H. Wengrovius, R. R. Schrock, M. R. Churchill, J. R. Missert and W. J. Joungs, *J. Amer. Chem. Soc.*, **102**, 4515 (1980).
1126. J. H. Wengrovius, J. Sancho and R. R. Schrock, *J. Amer. Chem. Soc.*, **103**, 3932 (1981).
1127. J. H. Wengrovius and R. R. Schrock, *Organometallics*, **1**, 148 (1982).
1128. R. Westhoff and J. A. Moulijn, *J. Catalysis*, **46**, 414 (1977).
1129. D. Wewerka and K. Hummel, *Colloid. Polymer Sci.*, **254**, 116 (1976).
1130. L. G. Wideman, *J. Org. Chem.*, **33**, 4541 (1968).
1131. L. G. Wideman, H. R. Menapace and W. A. Maly, *J. Catalysis*, **43**, 371 (1976).
1132. G. B. Wills, J. Fathikalajahi, S. K. Gangwal and S. Tang, *Rec. Trav. Chim.*, **96**, M110 (1977).
1133. S. R. Wilson and D. E. Schalk, *J. Org. Chem.*, **41**, 3928 (1976).
1134. J. Witte and M. Hoffmann, *Makromol. Chem.*, **179**, 641 (1978).
1135. R. Wolovsky, *J. Amer. Chem. Soc.*, **92**, 2132 (1970).
1136. R. Wolovsky, N. Maoz and Z. Nir, *Synthesis*, **1970**, 656.
1137. R. Wolovsky and Z. Nir, *Synthesis*, **1972**, 134.
1138. R. Wolovsky and Z. Nir, *J. Chem. Soc., Chem. Commun.*, **1975**, 302.
1139. F. L. Woody, M. J. Lewis and G. B. Wills, *J. Catalysis*, **14**, 389 (1969).
1140. P. S. Woon and M. F. Farona, *J. Polymer. Sci., Polymer Chem. Ed.*, **12**, 1749 (1974).
1141. S. K. Wu, J. Lucki and J. F. Rabek, *Polym. Photochem.*, **2**, 73 (1982).
1141a. Xu Xiaoding, A. Andreini and J. C. Mol, *5th Internat. Symp. Olefin Metathesis*, Graz, August 22—26, 1983.
1142. A. I. Yablonskaya, D. A. Bol'shakov, L. A. Morozova, G. S. Paramanova, E. A. Shmulevich, V. F. Grigor'ev, R. A. Aronovich, V. S. Fel'dblyum and T. I. Baranova, *Sb. Nauchn. Trudy Monom. Sint. Kauch.*, **1976**, 41.

1143. A. I. Yablonskaya, D. A. Bol'shakov, R. A. Aronovich, V. S. Fel'dblyum, I. G. Fal'kov and V. F. Grigor'ev, *Kinet. Katal.*, **18**, 1615 (1977).
1144. A. I. Yablonskaya, A. D. Bol'shakov and L. A. Morozova, *Prom.-st. Sint. Kauch.*, **1980**, 6.
1145. T. Yamaguchi, Y. Tanaka and K. Tanabe, *J. Catalysis*, **65**, 442 (1980).
1146. D. L. Yang and Ch. Cheng Su, *Bull. Inst. Chem., Acad. Sing.*, **23**, 66 (1976).
1147. S. Yoshitomi, *Nenryo Kyokaishi*, **46**, 514 (1967).
1148. S. Yoshitomi, *Sekiyu Gakkai Shi*, **13**, 92 (1970).
1149. S. Yoshitomi, *Sekiyu to Sekiyu Kagaku*, **15**, 89 (1971).
1150. V. P. Yur'ev, G. Gailiunas, G. A. Tolstikov, V. I. Khvostenko and E. G. Galkin, *Zh. Obsch. Khim.*, **45**, 2741 (1975).
1151. A. H. Yusupbekov, Sh. T. Abdurashidova and S. S. Arislamov, *Sb. Nauchn. Trudy. Tashkent Polytechn. Inst.*, **326**, 17 (1981).
1152. L. M. Zemtsov, B. E. Davydov, G. P. Karpacheva, V. V. Khoroshilova and T. G. Samedova, *Kinet. Katal.*, **18**, 638 (1977).
1153. Y. Zhai, K. Tanaka and K. Aomura, *Bull. Fac. Eng. Hokkaido Univ.*, **107**, 83 (1982).
1154. M. N. Zhavoronkov and A. A. Nadirov, *Zh. Fiz. Khim.*, **52**, 196 (1978).
1155. M. N. Zhavoronkov, A. A. Nadirov and A. Z. Dorogochinskii, *Prevrasch. Uglevod. Kislotno.-Osnovn. Geterog. Katal., Vses. Konf.*, **1977**, 168.
1156. T. Zowade and H. Höcker, *Makromol. Chem.*, **165**, 31 (1973).
1157. E. A. Zuech, *J. Chem. Soc., Chem. Commun.*, **1968**, 1182.
1158. E. A. Zuech, W. B. Hughes, D. H. Kubicek and E. T. Kittleman, *J. Amer. Chem. Soc.*, **92**, 528 (1970).

C. Patents

1159. M. Abe, M. Nagata, S. Matsui, K. Yamamoto, I. Sakato, F. Imaizumi and K. Takahashi, **Japan. Kokai** 76 134,746, 22 Nov 1976.
1160. M. Abe, M. Nagata, T. Kotani, H. Ikeda and F. Imaizumi, **Japan. Kokai** 76 137, 752, 27 Nov 1976.
1161. S. Akutagawa, H. Kumobayashi and A. Komatsu, **Ger. Offen.** 2,409,194, 12 Sep 1974.
1162. S. Akutagawa, H. Kumobayashi and A. Komatsu, **U.S.** 3,865,888, 11 Feb 1975.
1163. M. A. Albright, **U.S.** 3,330,882, 24 Jan 1963.
1164. M. L. Alfonso, **U.S.** 3,280,207, 18 Oct 1966.
1165. M. L. Alfonso, **U.S.** 3,312,635, 04 Apr 1967.
1166. H. J. Alkema and R. Van Helden, **Brit.** 1,117,968, 26 Jun 1968.
1167. H. J. Alkema and R. Van Helden, **Brit.** 1,118,517, 3 Jul 1968.
1168. H. J. Alkema and R. Van Helden, **Brit.** 1,174,968, 22 Apr 1969.
1169. H. J. Alkema, D. Medema and F. Wattimena, **Ger. Offen.** 1,929,136, 11 Dec 1969.
1170. H. J. Alkema and R. Van Helden, **U.S.** 3,536,777, 27 Oct 1970.
1171. H. J. Alkema, D. Medema and F. Wattimena, **U.S.** 3,634,539, 11 Jan 1972.
1172. J. K. Allen, B. P. McGrath and B. M. Palmer, **S. African** 69 06,673, 19 Mar 1970.
1173. J. K. Allen, B. M. Palmer and B. P. McGrath, **Ger. Offen.** 2,230,490, 11 Jan 1973.
1174. J. K. Allen, B. M. Palmer and B. P. McGrath, **U.S.** 4,022,846, 10 May 1977.
1175. K. G. Allum and P. J. Robinson, **Ger. Offen.** 2,004,490, 22 Apr 1971.
1176. A. N. Anderson and N. C. Merckling, **U.S.** 2,721,189, 18 Oct 1955.
1177. F. Arai, M. Kira, S. Kokuryo and T. Ueshima, **Japan. Kokai** 76 101,060, 07 Sep 1976.
1178. F. Arai, M. Kira, T. Ueshima and T. Sato, **Japan. Kokai** 76 148,771, 21 Dec 1976.
1179. F. Arai and T. Ueshima, **Japan. Kokai** 77 65,566, 31 May 1977.
1180. R. P. Arganbright, **U.S.** 3,697,613, 10 Oct 1972.
1181. R. P. Arganbright, **U.S.** 3,702,827, 14 Nov 1972.
1182. R. P. Arganbright, **U.S.** 3,792,108, 12 Dec 1974.
1183. W. Ast and G. Rheinwald, **Ger. Offen.** 2,535,496, 17 Feb 1977.
1184. W. Ast, G. Rheinwald and R. Kerber, **Ger. Offen.** 2,535,562, 17 Feb 1977.
1185. V. V. Atlas, I. I. Pis'man and A. M. Bakhshi-Zadé, **U.S.S.R.** 253,797, 07 Oct 1969.
1186. C. J. Attridge, A. Morris and H. Thomas, **U.S.** 3,840,612, 08 Oct 1974.
1187. E. D. Babich, V. M. Vdovin, B. A. Dolgoplosk, B. V. Kacharmin, K. L. Makovetskii, N. S. Nametkin and I. A. Oreshkin, **U.S.S.R.** 464,182, 25 Dec 1976.

1188. B. D. Babitskii, T. T. Denisova, V. A. Kormer, I. M. Lapuk, M. I. Lobach, N. P. Siminova, K. S. Solov'ev, T. Ya. Chepurnaya and T. L. Yufa, U.S.S.R. 393,777, 20 Jan 1976.

1189. B. D. Babitskii, T. T. Denisova, V. A. Kormer, I. M. Lapuk, M. I. Lobach, N. N. Marasanova, N. P. Siminova, K. S. Solov'ev, T. Ya. Chepurnaya and T. L. Yufa, U.S.S.R. 445,304, 28 Feb 1978.

1190. B. D. Babitskii, T. T. Denisova, V. A. Kormer, I. M. Lapuk, M. I. Lobach, N. P. Siminova, K. S. Solov'ev, T. Ya. Chepurnaya and T. L. Yufa, Rom. 66,065, 31 Oct 1979.

1191. B. D. Babitskii, P. A. Vernov, T. T. Denisova, V. A. Kormer, B. S. Korotkevich, I. M. Lapuk, N. V. Lemaev, Yu. L. Lelyukhina and E. Ya. Mandel'shtam, U.S.S.R. 655,705, 05 Apr 1979.

1192. B. D. Babitskii, P. A. Vernon, T. T. Denisova, N. F. Kovalev, V. A. Kormer, B. S. Korotkovich, I. M. Lapuk, N. V. Lemaev and E. Ya. Mandel'shtam, U.S.S.R. 681,069, 25 Aug 1979.

1193. B. D. Babitskii, G. V. Broi-Karre, P. A. Vernon, V. M. Duletova, K. V. Kisin, V. A. Kormer, B. S. Korotkovich, Yu. A. Kuritsyn and I. M. Lapuk, U.S.S.R. 726,120, 05 Apr 1980.

1194. D. S. Banasiak, U.S. 4,247,417, 27 Jan 1981.

1195. D. S. Banasiak, U.S. 4,248,738, 03 Feb 1981.

1196. D. S. Banasiak, U.S. 4,262,156, 06 Feb 1981.

1197. D. S. Banasiak, U.S. 4,269,780, 26 May 1981.

1198. D. S. Banasiak, U.S. 4,291,187, 05 Sep 1981.

1199. R. L. Banks, Belg. 620,440, 21 Jan 1963.

1200. R. L. Banks, Belg. 633,418, 10 Dec 1963.

1201. R. L. Banks, Belg. 642,916, 27 Jul 1964.

1202. R. L. Banks, U.S. 3,261,879, 19 Jul 1966.

1203. R. L. Banks, U.S. 3,321,547, 23 May 1967.

1204. R. L. Banks, Ger. 1,280,869, 24 Oct 1968.

1205. R. L. Banks and J. R. Kenton, Belg. 713,190, 1968.

1206. R. L. Banks, U.S. 3,431,316, 04 Mar 1969.

1207. R. L. Banks and J. R. Kenton, Fr. 1,560,086, 14 Mar 1969.

1208. R. L. Banks, U.S. 3,442,969, 06 May 1969.

1209. R. L. Banks, U.S. 3,463,827, 26 Aug 1969.

1210. R. L. Banks, U.S. 3,538,181, 03 Oct 1970.

1211. R. L. Banks, U.S. 3,546,313, 08 Dec 1970.

1212. R. L. Banks, U.S. 3,631,118, 28 Dec 1971.

1213. R. L. Banks and R. B. Regier, Ger. Offen. 2,217,675, 09 Nov 1972.

1214. R. L. Banks, U.S. 3,658,929, 25 Apr 1972.

1215. R. L. Banks, U.S. 3,663,640, 16 May 1972.

1216. R. L. Banks, U.S. 3,696,163, 03 Oct 1972.

1217. R. L. Banks and F. Pennella, U.S. 3,729,525, 24 Apr 1973.

1218. R. L. Banks, U.S. 3,767,565, 23 Oct 1973.

1219. R. L. Banks, U.S. 3,785,956, 15 Jan 1974.

1220. R. L. Banks, U.S. 3,785,957, 15 Jan 1974.

1221. R. L. Banks, U.S. 3,883,606, 13 May 1975.

1222. R. L. Banks and J. R. Kenton, U.S. 3,996,166, 07 Dec 1976.

1223. R. L. Banks, U.S. 4,211,885, 08 Jul 1980.

1224. A. N. Bashkirov, Yu. B. Kryukov, R. A. Fridman, S. M. Nosakova and S. I. Beilin, U.S.S.R. 445,259, 21 Feb 1973.

1225. A. N. Bashkirov, R. A. Fridman, S. M. Nosakova and L. G. Liberov, U.S.S.R. 517,575, 15 Jun 1976.

1226. A. N. Bashkirov, R. A. Fridman, L. G. Liberov and S. M. Nosakova, U.S.S.R. 564,301, 05 Jul 1977.

1227. A. N. Bashkirov, L. G. Liberov, R. A. Fridman and S. M. Nosakova, U.S.S.R. 565,028, 15 Jun 1977.

1228. A. N. Bashkirov, E. I. Bogolepova and R. A. Fridman, U.S.S.R. 595,282, 28 Feb 1978.

1229. A. N. Bashkirov, R. A. Fridman, S. B. Mitina and L. G. Liberov, **U.S.S.R.** 607,831, 25 May 1978.
1230. A. N. Bashkirov, E. I. Bogolepova and R. A. Fridman, **U.S.S.R.** 639,883, 30 Dec 1978.
1231. A. N. Bashkirov, E. I. Bogolepova and R. A. Fridman, **U.S.S.R.** 644,766, 30 Jan 1979.
1232. J. M. Basset, J. Bousquet and R. Mutin, **Ger. Offen.** 2,511,359, 09 Oct 1975.
1233. J. M. Basset, J. Bousquet and R. Mutin, **U.S.** 4,080,398, 21 Mar 1978.
1234. J. M. Basset, G. Chapelet, A. Lattes, J. P. Laval, R. Mutin and R. Nouguier, **Ger. Offen.** 2,816,833, 12 Oct 1978.
1235. Bayer A. G., **Fr.** 2,002,570, 17 Oct 1968.
1236. Bayer A. G., **Fr.** 2,003,451, 07 Nov 1969.
1237. Bayer A. G., **Fr.** 2,003,536, 07 Nov 1969.
1238. Bayer A. G., **Fr.** 2,011,428, 27 Feb 1970.
1239. Bayer A. G., **Fr.** 2,014,139, 17 Apr 1970.
1240. Bayer A. G., **Fr.** 2,014,769, 17 Apr 1970.
1241. Bayer A. G., **Fr.** 2,065,300, 09 Oct 1970.
1242. Bayer A. G., **Fr.** 2,235,140, 27 Jun 1974.
1243. Bayer A. G., **Fr.** 2,245,691, 23 Aug 1974.
1244. Bayer A. G., **Fr.** 2,271,242, 16 May 1975.
1245. M. Beck and D. Theisen, **Fr.** 2,133,829, 17 Apr 1971.
1246. A. J. Bell, **U.S.** 3,772,255, 13 Nov 1973.
1247. A. J. Bell, **Ger. Offen.** 2,552,616, 02 Jun 1977.
1248. A. J. Bell, **Brit.** 1,483,163, 17 Aug 1977.
1249. A. J. Bell, **Fr.** 2,403,352, 23 Aug 1978.
1250. A. J. Bell, **Belg.** 869,987, 18 Dec 1978.
1251. L. Bencze and L. Markó, **Hung. Teljes** 8,274, 28 Jun 1974.
1252. L. Bencze and L. Markó, **Hung. Teljes** 9,421, 28 Feb 1975.
1253. A. J. Berger, **U.S.** 3,726,938, 10 Apr 1973.
1254. C. W. Blewett and W. R. Garret, Jr., **U.S.** 4,078,012, 07 Mar 1978.
1255. C. W. Blewett and W. R. Garret, Jr., **U.S.** 4,078,013, 07 Mar 1978.
1256. D. A. Bol'shakov, A. I. Yablonskaya, G. A. Stepanov, L. A. Morozova and N. F. Chernoknizhnaya, **U.S.S.R.** 480,442, 15 Aug 1975.
1257. D. A. Bol'shakov, A. I. Yablonskaya, G. A. Stepanov, L. A. Morozova, E. K. Lopatina, N. F. Chernoknizhnaya, E. A. Shmulevich and E. V. Orlov, **U.S.S.R.** 514,626, 25 May 1976.
1258. D. A. Bol'shakov, A. I. Yablonskaya, G. S. Paramonova, V. S. Fel'dblyum, A. M. Kutin, G. A. Stepanov and T. I. Baranova, **U.S.S.R.** 525,467, 25 Aug 1976.
1259. D. A. Bol'shakov, A. I. Yablonskaya, L. A. Morozova, E. V. Orlov, V. S. Fel'dblyum, A. M. Kutin, G. A. Stepanov and T. I. Baranova, **U.S.S.R.** 534,243, 05 Nov 1976.
1260. D. A. Bol'shakov, A. I. Yablonskaya, V. S. Fel'dblyum, T. I. Baranova, A. M. Kutin and G. A. Stepanov, **U.S.S.R.** 536,837, 30 Nov 1976.
1261. D. A. Bol'shakov, A. I. Yablonskaya, V. S. Fel'dblyum, A. M. Kutin, G. A. Stepanov, T. I. Baranova, E. A. Shmulevich and O. P. Karpov, **U.S.S.R.** 566,814, 30 Jul 1977.
1262. D. A. Bol'shakov, A. I. Yablonskaya, V. S. Fel'dblyum, T. I. Baranova and A. M. Kutin, **U.S.S.R.** 599,834, 30 Mar 1978.
1263. K. H. Bourne and C. J. L. Metcalfe, **U.S.** 1,145,503.
1264. C. P. C. Bradshaw and L. Turner, **Brit.** 1,121,806, 31 Jul 1968.
1265. C. P. C. Bradshaw, **Ger. Offen.** 1,900,490, 31 Jul 1969.
1266. C. P. C. Bradshaw, **Ger. Offen.** 1,909,951, 16 Oct 1969.
1267. C. P. C. Bradshaw, **Brit.** 1,180,459, 04 Feb 1970.
1268. C. P. C. Bradshaw, **Brit.** 1,252,799, 10 Nov 1971.
1269. C. P. C. Bradshaw and J. I. Blake, **Brit.** 1,266,340, 08 Mar 1972.
1270. C. P. C. Bradshaw, **Brit.** 1,282,592, 19 Jul 1972.
1271. C. P. C. Bradshaw, **U.S.** 3,723,563, 27 Mar 1973.
1272. H. H. Brintzinger, **Brit.** 2,000,153, 04 Jan 1979.
1273. British Petroleum Co., Ltd., **Brit.** 1,054,864, 08 Oct 1964.
1274. British Petroleum Co., Ltd., **Brit.** 1,064,829, 19 Nov 1964.
1275. British Petroleum Co., Ltd., **Brit.** 1,089,956, 23 Jul 1965.
1276. British Petroleum Co., Ltd., **Brit.** 1,093,784, 25 Aug 1965.

1277. British Petroleum Co., Ltd., **Brit.** 1,105,563, 23 Apr 1965.
1278. British Petroleum Co., Ltd., **Brit.** 1,105,564, 23 Apr 1965.
1279. British Petroleum Co., Ltd., **Neth. Appl.** 6,511,659, 09 Mar 1966.
1280. British Petroleum Co., Ltd., **Neth. Appl.** 6,514,985, 20 May 1966.
1281. British Petroleum Co., Ltd., **Neth. Appl.** 6,605,328, 24 Oct 1966.
1282. British Petroleum Co., Ltd., **Neth. Appl.** 6,605,329, 24 Oct 1966.
1283. British Petroleum Co., Ltd., **Neth. Appl.** 6,610,195, 24 Jan 1967.
1284. British Petroleum Co., Ltd., **Neth. Appl.** 6,610,196, 24 Jan 1967.
1285. British Petroleum Co., Ltd., **Neth. Appl.** 6,611,840, 27 Feb 1968.
1286. British Petroleum Co., Ltd., **Fr.** 1,485,859, 23 Jun 1967.
1287. British Petroleum Co., Ltd., **Fr.** 1,554,287, 17 Jan 1969.
1288. J. E. Burleigh and C. A. Uraneck, **U.S.** 3,640,986, 08 Feb 1972.
1289. R. L. Burnett, **U.S.** 3,856,876, 24 Nov 1974.
1290. N. Calderon, **Belg.** 679,624, 16 Apr 1965.
1291. N. Calderon and W. A. Judy, **Brit.** 1,124,456, 21 Aug 1968.
1292. N. Calderon and H. Y. Chen, **Brit.** 1,125,529, 28 Aug 1968.
1293. N. Calderon, **U.S.** 3,439,056, 15 Apr 1969.
1294. N. Calderon, **U.S.** 3,439,057, 15 Apr 1969.
1295. N. Calderon and W. A. Judy, **U.S.** 3,492,245, 14 Apr 1966.
1296. N. Calderon, **U.S.** 3,506,592, 14 Apr 1966.
1297. N. Calderon and Y. H. Chen, **U.S.** 3,535,401, 11 May 1966.
1298. N. Calderon, **U.S.** 3,597,406, 03 Aug 1971.
1299. N. Calderon, **U.S.** 3,932,373, 13 Jan 1976.
1300. N. Calderon and K. W. Scott, **Can.** 1,095,646, 10 Feb 1981.
1301. J. P. Candlin, A. H. Mawby and H. Thomas, **Ger. Offen.** 2,213,947, 25 Jan 1973.
1302. K. F. Castner, **U.S.** 4,038,471, 26 Jul 1977.
1303. K. F. Castner, **U.S.** 4,060,468, 29 Nov 1977.
1304. Charbonnage de France, **Fr.** 1,536,305, 10 Aug 1968.
1305. Charbonnage de France, **Fr.** 1,543,497, 25 Oct 1968.
1306. Charbonnage de France, **Fr.** 1,550,351, 20 Dec 1968.
1307. Charbonnage de France, **Fr.** 1,556,215, 07 Feb 1969.
1308. Charbonnage de France, **Fr. Addn.** 94,571, 12 Oct 1969.
1309. Charbonnage de France, **Fr. Addn.** 94,672, 03 Oct 1969.
1310. Charbonnage de France, **Fr.** 1,594,934, 17 Jul 1970.
1311. Charbonnage de France, **Ger. Offen.** 1,961,742, 18 Jun 1970.
1312. Charbonnage de France, **Fr.** 1,597,727, 07 Aug 1970.
1313. Charbonnage de France, **Fr.** 2,265,774, 26 Mar 1974.
1314. Y. Chauvin, J. P. Souflet, D. Commereuc and P.-N. Phung, **Ger. Offen.** 2,151,662, 04 May 1972.
1315. Chemische Werke Hüls, A. G., **Fr.** 2,296,651, 30 Dec 1975.
1316. Chemische Werke Hüls, A. G., **Fr.** 2,346, 378, 24 Mar 1977.
1317. M. Cherubin, E. Wilms and H. Arendsen, **Ger. Offen.** 2,360,981, 28 May 1975.
1318. D. L. Crain, **U.S.** 3,463,828, 22 Oct 1965.
1319. D. L. Crain, **Belg.** 688,784, 23 Jun 1967.
1320. D. L. Crain, **Belg.** 691,711, 23 Jun 1967.
1321. D. L. Crain, **Fr.** 1,515,037, 01 Mar 1968.
1322. D. L. Crain, **Fr.** 1,516,853, 15 Mar 1968.
1323. D. L. Crain, **U.S.** 3,575,947, 20 Apr 1971.
1324. D. L. Crain and R. E. Reusser, **U.S.** 3,658,927, 25 Apr 1972.
1325. D. L. Crain and P. R. Stapp, **U.S.** 3,658,931, 25 Apr 1972.
1326. D. L. Crain and R. E. Reusser, **U.S.** 3,660,516, 02 May 1972.
1327. G. Dall'Asta and G. Carella, **Ital.** 784,307, 02 Aug 1966.
1328. G. Dall'Asta and R. Manetti, **Ital.** 789,585, 21 Dec 1965.
1329. G. Dall'Asta and G. Carella, **Brit.** 1,062,367, 11 Feb 1965.
1330. G. Dall'Asta, **Belg.** 691,503, 20 Jun 1967.
1331. G. Dall'Asta, G. Motroni and G. Carella, **Fr.** 1,545,643, 15 Nov 1968.
1332. G. Dall'Asta and P. Meneghini, **Ital.** 859,384, 28 Mar 1969.
1333. G. Dall'Asta and R. Manetti, **Ital.** 932,461, 01 Jul 1971.

1334. G. Dall'Asta and G. Motroni, **Fr.** 2,149,810, 02 Aug 1971.

1335. G. Dall'Asta, V. Zamboni, P. Meneghini and U. Gennaro, **Ger.** 2,101,069, 22 Jul 1971.

1336. G. Dall'Asta, **Ital.** 913,463, 15 Mar 1972.

1337. G. Dall'Asta, **Ger.** 2,161,878, 06 Jul 1972.

1338. G. Dall'Asta and G. Motroni, **Ger.** 2,237,353, 15 Feb 1973.

1339. G. Dall'Asta, P. Meneghini and U. Gennaro, **Ger.** 2,252,082, 03 May 1973.

1340. J. W. Davison, **U.S.** 3,442,970, 06 May 1969.

1341. A. J. DeJong, H. J. Heijmen and R. Van Helden, **Ger. Offen.** 2,604,261, 19 Aug 1976.

1342. A. J. DeJong, R. Van Helden and R. S. Downing, **Ger. Offen.** 2,724,484, 22 Dec 1977.

1343. A. J. DeJong and H. J. Heijmen, **Ger. Offen.** 2,754,540, 15 Jul 1978.

1344. P. A. Delvin, E. F. Lutz and R. J. Patten, **U.S.** 3,627,739, 14 Dec 1971.

1345. K. F. Dickinson, **Eur. Pat. Appl. E. P.** 56,013, 14 Jul 1982.

1346. R. E. Dixon, **Belg.** 633,483, 11 Dec 1963.

1347. R. E. Dixon, **Can.** 755,135.

1348. R. E. Dixon, **U.S.** 3,485,890, 23 Dec 1969.

1349. R. E. Dixon and J. F. Hutto, **U.S.** 3,544,649, 01 Dec 1970.

1350. R. E. Dixon and P. D. Hann, **U.S.** 4,085,158, 18 Apr 1978.

1351. R. E. Dixon, **U.S.** 4,091,046, 23 Mar 1978.

1352. R. E. Dixon, **U.S.** 4,176,141, 27 Nov 1979.

1353. R. E. Dixon, **U.S.** 4,255,605, 10 Mar 1981.

1354. B. A. Dolgoplosk, I. A. Oreshkin, I. L. Kershenbaum, K. L. Makovetskii, E. I. Tinyakova and T. G. Golenko, **U.S.S.R.** 468,505, 25 Nov 1976.

1355. B. A. Dolgoplosk, I. A. Oreshkin, T. L. Kershenbaum, I. Ya. Ostrovskaya, K. L. Makovetskii, E. I. Tinyakova, G. M. Cherenko and V. V. Titov, **U.S.S.R.** 522,610, 25 Jan 1977.

1356. J. D. Donald, C. A. Uraneck and J. E. Burleigh, **Ger. Offen.** 2,137,524, 03 Feb 1972.

1357. G. Doyle, **Ger. Offen.** 2,047,270, 01 Apr 1971.

1358. G. Doyle, **U.S.** 3,686,136, 22 Aug 1972.

1359. G. Doyle, **U.S.** 3,849,513, 19 Nov 1974.

1360. E. D. M. Eades and P. R. Hughes, **U.S.** 3,652,704, 01 Aug 1969.

1361. E. D. M. Eades and P. R. Hughes, **Ger. Offen.** 1,942,635, 26 Feb 1970.

1362. Ya. T. Eidus, A. Lapidus, Kh. M. Minachev, Ya. I. Isakov and R. V. Aretisyan, **U.S.S.R.** 273,193, 15 Jun 1970.

1363. H. S. Eleuterio, **U.S.** 3,074,918, 20 Jun 1957.

1364. H. S. Eleuterio, **Ger. Offen.** 1,072,811, 07 Jan 1960.

1365. A. F. Ellis and E. T. Sabourin, **Ger. Offen.** 2,018,890, 07 Jan 1971.

1366. A. F. Ellis and E. T. Sabourin, **U.S.** 3,595,920, 27 Jul 1971.

1367. A. F. Ellis, **Ger.** 2,047,102, 18 Apr 1971.

1368. A. F. Ellis and E. T. Sabourin, **U.S.** 3,646,143, 29 Feb 1972.

1369. A. F. Ellis and E. T. Sabourin, **Ger. Offen.** 2,306,813, 30 Aug 1973.

1370. A. F. Ellis and E. T. Sabourin, **U.S.** 3,775,509, 27 Nov 1973.

1371. F. E. Farley, **U.S.** 3,647,906, 07 Mar 1972.

1372. V. Fattore, S. D. Milanese, M. Mazzei and B. Notari, **U.S.** 3,792,107, 21 Jul 1971.

1373. V. Fattore, M. Mazzei and B. Notari, **Fr.** 2,089,461, 11 Feb 1972.

1374. V. Fattore, M. Mazzei and B. Notari, **U.S.** 3,764,635, 09 Oct 1973.

1375. V. Fattore, M. Mazzei and B. Notari, **U.S.** 3,792,107, 12 Feb 1974.

1376. V. S. Fel'dblyum, T. I. Baranova and V. N. Petrushanskaya, **U.S.S.R.** 422,240, 05 Jul 1974.

1377. V. S. Fel'dblyum, V. N. Petrushanskaya and A. M. Kutin, **U.S.S.R.** 436,808, 25 Jul 1974.

1378. V.S. Fel'dblyum, V. N. Petrushanskaya, B. A. Dolgoplosk, E. I. Tinyakova, K. L. Makovetskii and I. A. Oreshkin, **U.S.S.R.** 520,377, 05 Jul 1976.

1379. P. Fitton and T. Whitesides, **U.S.** 3,855,338, 17 Dec 1974.

1380. G. Foster, **Ger. Offen.** 2,063,150, 08 Jul 1971.

1381. J. Furukawa, H. Kobayashi, K. Jinto and O. Hayashi, **Japan. Kokai** 77 151,400, 15 Dec 1977.

1382. P. A. Gautier, F.J.F. Van der Plas and P. A. Verbrugge, **Ger. Offen.** 2,208,058, 07 Sep 1972.

1383. P. A. Gautier, **U.S.** 3,776,974, 04 Dec 1973.

1384. T. W. Gibson, **Eur. Pat. Appl.** 21,496, 08 Jan 1981.
1385. R. E. Gilliland, D. K. McQueen and R. E. Dixon, **U.S.** 3,281,351, 25 Oct 1966.
1386. Goodyear Tire and Rubber Co., **U.S.** 3,535,401, 11 May 1966.
1387. Goodyear Tire and Rubber Co., **Ger. Offen.** 1,595,739, 14 Apr 1966.
1388. Goodyear Tire and Rubber Co., **Brit.** 1,124,456, 21 Aug 1968.
1389. Goodyear Rubber and Tire Co., **Ger. Offen.** 1,720,820, 27 Jul 1971.
1390. Goodyear Tire and Rubber Co., **Ger. Offen.** 2,049,026, 08 Apr 1971.
1391. Goodyear Tire and Rubber Co., **Fr.** 2,236,882, 27 Mar 1975.
1392. Goodyear Tire and Rubber Co., **Fr.** 2,241,571, 27 Mar 1975.
1393. Goodyear Tire and Rubber Co., **Fr.** 2,253,767, 27 Mar 1975.
1394. Goodyear Tire and Rubber Co., **Fr.** 2,265,775, 27 Mar 1975.
1395. Goodyear Tire and Rubber Co., **Fr.** 2,265,776, 27 Mar 1975.
1396. Goodyear Tire and Rubber Co., **Fr.** 2,271,237, 27 Mar 1975.
1397. Goodyear Tire and Rubber Co., **Fr.** 2,333,816, 01 Jul 1977.
1398. Goodyear Tire and Rubber Co., **Belg.** 836,404, 01 Apr 1978.
1399. A. Greco, F. Pirinoli and G. Dall'Asta, **Ger. Offen.** 2,406,554, 22 Aug 1974.
1400. A. Greco, F. Pirinoli and G. Dall'Asta, **U.S.** 3,956,178, 05 Nov 1976.
1401. M. Green and T. A. Kuc, **Ger. Offen.** 2,153,314, 27 Apr. 1972.
1402. P. Günther, W. Oberkirch and G. Pampus, **Ger. Offen.** 1,720,798, 08 Mar 1968.
1403. P. Günther and W. Oberkirch, **Ger. Offen.** 1,793,059, 29 Jul 1968.
1404. P. Günther, W. Oberkirch and G. Pampus, **Ger. Offen.** 1,812,338, 18 Jun 1970.
1405. P. Günther, W. Oberkirch and G. Pampus, **Ger. Offen.** 1,950,981, 22 Apr 1971.
1406. P. Günther, W. Oberkirch, F. Haas, G. Pampus and G. Marwede, **Ger. Offen.** 2,015,153, 21 Oct 1971.
1407. H. Haag, K.-H. Nordsiek, R. Streck and D. Zerpner, **Ger. Offen.** 2,500,025, 15 Jul 1976
1408. F. Haas, K. Nützel and H. J. John, **Ger.** 1,769,149, 11 Apr 1968.
1409. F. Haas, W. Graulich, K. Nützel and K. Dinges, **Ger. Offen.** 1,720,791, 13 Apr 1978.
1410. T. Hara, H. Horino and R. Kita, **Japan. Kokai** 74 01,699, 09 Jan 1974.
1411. T. Hara, H. Harino and R. Kita, **Japan. Kokai** 74 01,700, 09 Jan 1974.
1412. R. C. Harrison, **U.S.** 3,567,789, 10 Nov 1969.
1413. T. Hayakawa, S. Kobayashi and T. Seki, **Japan. Kokai** 72 42,701, 16 Dec 1972.
1414. L. F. Heckelsberg, **U.S.** 3,340,322, 05 Sep 1967.
1415. L. F. Heckelsberg, **U.S.** 3,365,513, 23 Jan 1968.
1416. L. F. Heckelsberg, **Belg.** 713,184, 1968.
1417. L. F. Heckelsberg, **Belg.** 713,185, 1968.
1418. L. F. Heckelsberg, **Belg.** 713,186, 1968.
1419. L. F. Heckelsberg, **Ger. Aus.** 1,768,124, 03 Apr 1968.
1420. L. F. Heckelsberg, **U.S.** 3,395,196, 30 Jul 1968.
1421. L. F. Heckelsberg and R. L. Banks, **S. African** 68 01,408, 31 Jul 1968.
1422. L. F. Heckelsberg, **S. African** 68 01,620, 15 Aug 1968.
1423. L. F. Heckelsberg, **S. African** 68 01,707, 15 Aug 1968.
1424. L. F. Heckelsberg, **S. African** 68 01,877, 20 Aug 1968.
1425. L. F. Heckelsberg, **S. African** 68 01,878, 20 Aug 1968.
1426. L. F. Heckelsberg, **U.S.** 3,418,390, 24 Dec 1968.
1427. L. F. Heckelsberg, **Fr.** 1,562,396, 04 Apr 1969.
1428. L. F. Heckelsberg, **Fr.** 1,562,397, 04 Apr 1969.
1429. L. F. Heckelsberg, **U.S.** 3,444,262, 13 May 1969.
1430. L. F. Heckelsberg, **U.S.** 3,445,541, 20 May 1969.
1431. L. F. Heckelsberg, **U.S.** 3,485,891, 23 Dec 1969.
1432. L. F. Heckelsberg, **U.S.** 3,546,311, 08 Dec 1970.
1433. L. F. Heckelsberg and R. L. Banks, **U.S.** 3,546,312, 08 Dec 1970.
1434. L. F. Heckelsberg, **U.S.** 3,676,520, 11 Jul 1972.
1435. L. F. Heckelsberg, **Brit.** 1,282,592, 19 Jul 1972.
1436. L. F. Heckelsberg, **U.S.** 3,707,581, 26 Dec 1972.
1437. L. F. Heckelsberg, **U.S.** 3,723,562, 27 Mar 1973.
1438. L. F. Heckelsberg and R. L. Banks, **Ger. Aus.** 1,788,124, 04 Jul 1974.
1439. L. F. Heckelsberg, **U.S.** 3,845,155, 29 Oct 1974.
1440. L. F. Heckelsberg and R. L. Banks, **U.S.** 3,870,763, 11 Mar 1975.

1441. L. F. Heckelsberg, **U.S.** 4,093,536, 06 Jun 1978.

1442. L. F. Heckelsberg, W. T. Nelson and S. Schiff, **U.S.** 4,319,064, 09 Mar 1982.

1443. G. J. Heiszwolf and R. Van Helden, **Ger. Offen.** 2,445,568, 27 Mar 1975.

1444. P. Hepworth, **Ger. Offen.** 2,231,995, 18 Jan 1973.

1445. P. Hepworth, **Ger. Offen.** 2,264,973, 18 Dec. 1975.

1446. J. L. Hérisson and Y. Chauvin, **Ger. Offen.** 2 008,246, 08 Oct 1970.

1447. V. A. Hodjemirov, V. A. Evdokimova and V. M. Cherednichenkov, **U.S.S.R.** 527,444, 03 Aug 1977.

1448. V. A. Hodjemirov, A. V. Nikolaev, V. A. Evdokimova and D. V. Drobot, **U.S.S.R-**565,709, 06 Sep 1977.

1449. V. A. Hodjemirov, **U.S.S.R.** 812,792,15 Mar 1981.

1450. Hoechst A. G., **Fr.** 2,006,777, 02 Jan 1970.

1451. Hoechst A. G., **Ger. Offen.** 1,768,267, 28 Jan 1970.

1452. E. H. Homeir, **U.S.** 3,761,537, 25 Sep 1973.

1453. H. Horino, T. Kita and J. Nakazato, **Japan. Kokai** 78 24,998, 24 Jul 1978.

1454. E. J. Howman and B. P. McGrath, **Brit.** 1,123,500, 14 Aug 1968.

1455. E. J. Howman, C. P. C. Bradshaw and L. Turner, **U.S.** 3.424,812, 28 Jan 1969.

1456. E. J. Howman, B. P. McGrath and K. V. Williams, **Brit.** 1,144,085, 05 Mar 1969.

1457. E. J. Howman and L. Turner, **U.S.** 3,448,163, 03 Jun 1969.

1458. R. J. Hoxmeier, **U.S.** 4,276,199, 30 Aug 1981.

1459. T. R. Hughes and R. P. Sieg, **U.S.** 3,178,676, 01 Jul 1970.

1460. T. R. Hughes, **U.S.** 3,718,576, 1970.

1461. T. R. Hughes, **U.S.** 3,773,845, 16 Jan 1970.

1462. T. R. Hughes, **Ger. Offen.** 2,049,164, 15 Apr 1971.

1463. T. R. Hughes, **Ger. Offen.** 2,101,740, 23 Sep 1971.

1464. T. R. Hughes, **U.S.** 3,775,505, 1973.

1465. T. R. Hughes, **U.S.** 3,784,662, 1973.

1466. T. R. Hughes, **U.S.** 3,793,251, 1973.

1467. T. R. Hughes, **U.S.** 3,808,285, 30 Apr 1974.

1468. T. R. Hughes, **U.S.** 3,914,330, 05 Mar 1973.

1469. W. B. Hughes and E. A. Zuech, **Fr.** 1,575,779, 25 Jul 1969.

1470. W. B. Hughes and E. A. Zuech, **U.S.** 3,562,178, 09 Feb 1971.

1471. W. B. Hughes and E. A. Zuech, **U.S.** 3,691,253, 17 Sep 1972.

1472. W. B Hughes and E. A. Zuech, **U.S.** 3,721,718, 20 Mar 1973.

1473. W. B. Hughes, **U.S.** 4,086,229, 25 Apr 1978.

1474. J. F. Hutto, R. E. Dixon, J. W. Davison, C. A. Ayres, C. A. Renberg and D. K. McQueen, **U.S.** 3,345,258, 04 Sep 1964.

1475. J. F. Hutto and E. R. Dixon, **U.S.** 3,565,969, 23 Feb 1971.

1476. I. C. I., Ltd., **Fr.** 2,120,509, 08 Dec 1972.

1477. I. C. I., Ltd., **Fr.** 2,196,329, 17 Aug 1973.

1478. H. Ikeda, S. Matsumoto and K. Makino, **Ger. Offen.** 2,447,687, 17 Apr 1975.

1479. H. Ikeda, S. Matsumoto, H. Enyo, N. Oshima and K. Komatsu, **Ger. Offen.** 2.526.020, 02 Dec 1976.

1480. H. Ikeda, S. Matsumoto and K. Makino, **Japan. Kokai** 77 47,900, 16 Apr 1977.

1481. H. Ikeda, S. Matsumoto, H. Enyo, N. Oshima and K. Komatsu, **U.S.** 4,039,491, 02 Aug 1977.

1482. H. Ikeda, S. Matsumoto and K. Makino, **U.S.** 4,068,063, 10 Jan 1978.

1483. H. Ikeda, S. Matsumoto, H. Enyo, N. Oshima and K. Komatsu, **Brit.** 1,508,871, 26 Apr 1978.

1484. H. Ikeda, M. Shuichi and K. Makino, **U.S.** 4,176,220, 27 Nov 1979.

1485. F. Imaizumi, K. Suzuki and M. Kawakami, **Japan. Kokai** 76 136,800. 26 Nov 1976.

1486. F. Imaizumi, R. Nakamura, K. Sekine and K. Harada, **Japan. Kokai** 76 95,500, 21 Aug 1976.

1487. F. Imaizumi, K. Enyo and T. Kotani, **Japan. Kokai** 77 63,299, 25 May 1977. 11 Apr 1777.

1488. F. Imaizumi, K. Enyo, Y. Nakata, K. Makino and T. Sato, **Japan. Kokai** 77 45,696, 11 Apr 1977.

1489. F. Imaizumi, K. Enyo, Y. Nakata, K. Makino and T. Sato, **Japan. Kokai** 77 45,697, 11 Apr 1977.
1490. F. Imaizumi, S. Matsumoto, T. Kotani, K. Enyo and K. Takahashi, **Japan. Kokai** 77 51,500, 25 Apr 1977.
1491. F. Imaizumi, K. Goto, K. Sekine and T. Sato, **Japan. Kokai** 77 52,999, 28 Apr 1977.
1492. F. Imaizumi, H. Nobuyo and M. Kawakami, **Japan. Kokai** 77 87,499, 21 Jun 1977.
1493. F. Imaizumi, H. Nobuyo and T. Kotani, **Japan. Kokai** 77 87,500, 21 Jul 1977.
1494. H. Inoue, **Japan. Kokai** 73 55,298, 03 Aug 1973.
1495. H. Inoue, H. Horino and R. Kita, **Japan. Kokai** 73 66,698, 12 Sep 1973.
1496. H. Inoue, H. Horino and R. Kita, **Japan. Kokai** 73 66,699, 12 Sep 1973.
1497. H. Inoue, H. Horino and R. Kita, **Japan. Kokai** 73 66,700, 12 Sep 1973.
1498. H. Inoue, H. Horino and R. Kita **Japan. Kokai** 74 25,099,066, Mar 1974.
1499. H. Inoue and H. Horino, **Japan. Kokai** 74 38,998, 11 Apr 1974.
1500. H. Inoue, M. Iwata and J. Nakazato, **Japan. Kokai** 74 41,496, 18 Apr 1974.
1501. H. Inoue, **Japan. Kokai**, 78 24,996, 24 Jul 1978.
1502. H. Inoue, H. Horino and T. Kita, **Japan. Kokai** 78 24,997, 24 Jul 1978.
1503. H. Inoue, M. Isoi, M. Yoda and M. Komatsu, **Eur. Pat. Appl.** 28,085, 08 May 1981.
1504. Inst. Fr. Petr. Carb. Lubrif., **Neth. Appl.** 6,613,407, 28 Mar 1967.
1505. V. Isaguliants, W. N. Rozhdestvenskaya and A. N. Kozlov, **U.S.S.R.** 266,766, 07 Aug 1977.
1506. M. Ishikami and Y. Inoue, **Japan. Kokai,** 77 93,408, 05 Aug 1977.
1507. T. Ishikawa, H. Uehara, N. Hirosuke, T. Narumiya, I. Ishituka and A. Onshi, **Japan. Kokai** 69 29,059, 27 Nov 1969.
1508. M. Iwata, **Japan. Kokai,** 73 55,300, 03 Aug 1973.
1509. M. Iwata, **Japan. Kokai,** 78 24,995, 24 Jul 1978.
1510. Japan Synthetic Rubber Co., Ltd., **Fr.** 2,252,314, 20 Jun 1975.
1511. Japan Synthetic Rubber Co., Ltd., **Japan. Kokai Tokkyo Koho** 80 11,691, 27 Mar 1980.
1512. J. R. Jones, **Ger. Offen.** 1,920,868, 30 Oct 1969.
1513. W. A. Judy, **Fr.** 2,016,360, 26 Aug 1968.
1514. W. A. Judy, **Fr.** 2,016,364, 26 Aug 1968.
1515. W. A. Judy, **Fr.** 2,029,747, 31 Jan 1969.
1516. W. A. Judy, **Ger. Offen** 2,000,245, 13 Aug. 1970.
1517. W. A. Judy, **U.S.** 3,746,696, 17 Jul 1973.
1518. T. Kamiya, K. Sataki, K. Sakamoto and T. Hamada, **Japan. Kokai** 81 08,042, 21 Feb 1981.
1519. Y. Kamiya and E. Ogata, **Japan. Kokai** 75 69,002, 09 Jun 1975.
1520. H. Kanazawa, H. Kuribayashi and T. Yamaguchi, **Japan. Kokai** 78 65,220, 09 Nov 1978.
1521. H. Kanazawa, H. Kuribayashi, T. Koyama and T. Yamaguchi, **Japan. Kokai** 78 65,221, 09 Nov 1978.
1522. H. Kanazawa, T. Koyama and T. Yamaguchi, **Japan. Kokai** 79 104,306, 06 Feb 1979.
1523. H. Kanazawa, T. Koyama and T. Yamaguchi, **Japan. Kokai** 79 104,307, 06 Feb 1979.
1524. F. Kanega, M. Shidara, F. Shyoji, H. Kogame and T. Fujii, **Japan. Kokai** 76 111,300, 01 Oct 1976.
1525. F. Kanega, M. Shidara, F. Shoji, H. Kogame and T. Fujii, **Japan. Kokai** 78 25,597, 27 Jul 1978.
1526. T. J. Katz and M. A. Shippey, **U.S. Pat. Appl.** 276,111, 21 May 1982.
1527. T. J. Katz, M. Nair and S. J. Lee, **U.S.** 4,334,048, 08 Jun 1982.
1528. J. R. Kenton, D. L. Crain and R. F. Kleinschmidt, **U.S.** 3,491,163, 20 Jan 1970.
1529. J. R. Kenton and J. E. Mahan, **U.S.** 3,658,930, 25 Apr 1972.
1530. M. L. Khidekel', A. D. Shebaldova, V. I. Mar'in, S. W. Kursanov and I. V. Kalechits, **U.S.S.R.** 491,397, 15 Nov 1975.
1531. L. Kim, T. E. Paxon and S. C. Tano, **U.S.** 4,266,085, 05 May 1981.
1532. J. Kita, **Japan. Kokai** 74 41,494, 18 Apr 1974.
1533. R. Kita, **Japan. Kokai** 73 17,600, 06 Mar 1973.
1534. R. Kita, **Japan. Kokai** 73 27,758, 25 Aug 1973.
1535. R. Kita, **Japan. Kokai** 74 25,048, 06 Mar 1974.
1536. R. Kita, **Japan. Kokai** 74 38,999, 14 Apr 1974.

1537. R. Kita and H. Inoue, **Japan. Kokai** 78 24,994, 24 Jul 1978.
1538. E. T. Kittleman and E. A. Zuech, **U.S.** 1,561,025, 21 Mar 1969.
1539. E. T. Kittleman and E. A. Zuech, **U.S.** 3,554,924. 12 Jan 1971.
1540. E. T. Kittleman and E. A. Zuech, **U.S.** 3,558,515, 26 Jan 1971.
1541. E. T. Kittleman and E. A. Zuech, **U.S.** 3,703,551, 02 Jan 1973.
1542. E. T. Kittleman and E. A. Zuech, **U.S.** 3,776,973, 04 Dec 1973.
1543. H. Knoche, **Ger. Offen.** 2,024,835, 10 Dec. 1970.
1544. H. Knoche, **U.S.** 3,855,340, 17 Dec 1974.
1545. M. Kobayashi, T. Kamishima and S. Kobayashi, **Japan. Kokai** 77 90,600, 29 Jul 1977.
1546. M. Kobayashi, T. Kamishima and S. Kobayashi, **Japan. Kokai** 77 95,800, 11 Aug 1977,
1547. M. Kobayashi and T. Ueshima, **Japan. Kokai** 77 98,799, 18 Aug 1977.
1548. M. Kobayashi, T. Kamishima and S. Kobayashi, **Japan. Kokai** 77 101,300, 25 Aug 1977.
1549. M. Kobayashi and T. Ueshima, **Japan. Kokai** 77 112,000, 20 Sep 1977.
1550. M. Kobayashi, T. Ueshima and S. Kobayashi, **Japan. Kokai** 77 115,898, 28 Sep 1977.
1551. M. Kobayashi, T. Ueshima and S. Kobayashi, **Japan. Kokai** 77 145,899, 28 Sep 1977.
1552. M. Kobayashi, T. Ueshima and S. Kobayashi, **Japan. Kokai** 77 121,099, 12 Oct 1977.
1553. M. Kobayashi, T. Ueshima and S. Kobayashi, **Japan. Kokai** 77 130,900, 02 Nov 1977.
1554. M. Kobayashi, T. Ueshima and S. Kobayashi, **Japan. Kokai** 78 02,598, 11 Jan 1978.
1555. M. Kobayashi, R. Haruo and T. Ueshima, **Japan. Kokai** 78 02,599, 11 Jan 1978.
1556. M. Kobayashi and S. Kobayashi, **Japan. Kokai** 78 09,900, 28 Jan 1978.
1557. M. Kobayashi and T. Ueshima, **Japan. Kokai** 78 75,300, 04 Jul 1978.
1558. M. Kobayashi, I. Terano, T. Kamishima and S. Kobayashi, **Japan. Kokai** 78 970,151, 24 Aug 1978.
1559. M. Kobayashi, T. Yokoyama and T. Ueshima, **Japan. Kokai Tokkyo Koho** 79 149,800, 24 Nov 1979.
1560. M. Kobayashi, K. Murayama and T. Ueshima, **Japan. Kokai Tokkyo Koho** 79 150,460, 26 Nov 1979.
1561. M. Kobayashi, T. Yokoyama and T. Ueshima, **Japan. Kokai Tokkyo Koho** 79 150,500, 26 Nov 1979.
1562. S. Kobayashi, Y. Kobayashi, T. Ueshima, Y. Tanaka and S. Kurosawa, **Brit.** 1,519,862, 02 Aug 1978.
1563. Y. Kobayashi, T. Ueshima, Y. Tanaka, S. Kurosawa and S. Kobayashi, **Ger. Offen.** 2,540,553, 17 Mar 1977.
1564. Y. Kobayashi, T. Ueshima and S. Kobayashi, **Japan. Kokai** 77 26,599, 20 Feb 1977.
1565. Y. Kobayashi, T. Ueshima and S. Kobayashi, **Japan. Kokai** 77 36,200, 19 Mar 1977.
1566. Y. Kobayashi, T. Ueshima and S. Kobayashi, **Japan. Kokai** 77 37,998, 24 Mar 1977.
1567. Y. Kobayashi, T. Ueshima and S. Kobayashi, **Japan. Kokai** 77 40,600, 29 Mar 1977.
1568. Y. Kobayashi, T. Ueshima and S. Kobayashi, **Japan. Kokai** 77 42,600, 02 Apr 1977.
1569. Y. Kobayashi, T. Ueshima and S. Kobayashi, **Japan. Kokai** 77 44,899, 08 Apr 1977.
1570. Y. Kobayashi, T. Ueshima and S. Kobayashi, **Japan. Kokai** 77 45,698, 11 Apr 1977.
1571. Y. Kobayashi, T. Ueshima and S. Kobayashi, **Japan. Kokai** 77 63,298, 25 Jul 1977.
1572. Y. Kobayashi, T. Kamishima and S. Kobayashi, **Japan. Kokai** 77 95,800, 11 Aug 1977.
1573. Y. Kobayashi, T. Kamishima and S. Kobayashi, **Japan. Kokai** 77 121,099, 12 Oct 1977.
1574. Y. Kobayashi, T. Kamishima and S. Kobayashi, **Japan. Kokai** 77 130,900, 02 Nov 1977.
1575. Y. Kobayashi, T. Kamishima and S. Kobayashi, **Japan. Kokai** 77 142,800, 28 Nov 1977.
1576. Y. Kobayashi, R. Jio and T. Ueshima, **Japan. Kokai** 78 17,699, 17 Feb 1978.
1577. Y. Kobayashi, T. Ueshima and S. Kobayashi, **U. S.** 4,080,491, 21 Mar 1978.
1578. Y. Kobayashi and T. Ueshima, **Japan. Kokai** 78 77,300, 08 Jul 1978.
1579. Y. Kobayashi, T. Terao, T. Ueshima and S. Kobayashi, **Japan. Kokai** 78 97,099, 24 Aug 1978.
1580. Y. Kobayashi, R. Jio and T. Ueshima, **Japan. Kokai Tokkyo Koho** 78 103,000, 07 Sep 1978.
1581. Y. Kobayashi, R. Jio and T. Ueshima, **Japan. Kokai Tokkyo Koho** 79 76,700, 19 Jun 1979.

1582. Y. Kobayashi, J. Nakamura, K. Murayama and T. Kashima, **Japan. Kokai** 79 117,557. 12 Sep 1979.

1583. T. P. Kobylinski and H. E. Swift, **Ger. Offen.** 2.060.160, 09 Jun 1971.

1584. T. P. Kobylinski, **U. S.** 3,644,559, 22 Feb 1972.

1585. T. P. Kobylinski and H. E. Swift, **U.S.** 3,725,496, 03 Apr 1973.

1586. S. Kokuryo, F. Arai and S. Kurosawa, **Japan. Kokai** 76 112,867, 05 Oct 1976.

1587. S. Kokuryo, J. Nakamura, H. Konuma and T. Ueshima, **Japan. Kokai** 76 131,568, 18 Nov 1976.

1588. S. Kokuryo, M. Kira and T. Ueshima, **Japan. Kokai** 76 145,562, 14 Dec 1976.

1589. S. Kokuryo, H. Konuma and J. Nakamura, **Japan. Kokai** 78 58,558, 26 Mar 1978.

1590. K. Komatsu, S. Matsumoto, S. Aotani and M. Takahashi, **Japan. Kokai** 76 11,900, 30 Jan 1976.

1591. V. A Kormer, B. D. Babitskii, T. L. Yufa and F. A. Poleyatova, **Fr.** 2,017,515, 22 May 1970.

1592. V. A. Kormer, B. D. Babitskii, M. Lobach, V. Markova and L. Churlyaeva **Ger. Offen.** 2,133,905, 13 Jul 1970.

1593. V. A. Kormer, T. L. Yufa, B. D. Banitskii, I. A. Poletaeva, N. P. Simanova, I. M. Lapuk, V. V. Markova, M. I. Lobach, N. F. Kovalev and G. V. Kholodniskaya, **Fr.** 2,176,056, 30 Nov 1973.

1594. V. A. Kormer, **Japan. Kokai,** 74 121.000, 19 Nov 1974.

1595. N. N. Korneev, I. V. Kalechits, Y. L. Lelynkhina, E. V. Chechegoeva, T. D. Ivanova, A. S. Chepkasova and N. M. Pugashova, **U.S.S.R.** 526,611, 30 Aug 1976.

1596. T. Kotani and Y. Kuwahara, **Japan. Kokai** 73 46,637, 02 Jun 1973.

1597. T. Kotani and Y. Kuwahara, **Japan. Kokai** 73 46,638, 03 Jun 1973.

1598. T. Kotani, S. Matsumoto, F. Imaizumi, K. Igarashi and K. Takahashi, **Japan. Kokai** 77 32,100, 10 Mar 1977.

1599. T. Kotani, S. Matsumoto, F. Imaizumi, K. Igarashi and K. Takahashi, **Japan. Kokai,** 77 39,800, 28 Mar 1977.

1600. T. Kotani, S. Matsumoto, F. Imaizumi, K. Igarashi and K. Takahashi, **Japan. Kokai,** 77 47,852, 16 Apr 1977.

1601. T. Kotani, S. Matsumoto, F. Imaizumi and K. Igarashi, **Japan. Kokai,** 77 126,500, 24 Oct 1977.

1602. T. Kotani, S. Matsumoto, K. Igarashi and K. Suzuki, **Ger. Offen.** 2,731,263, 26 Jan 1978.

1603. T. Kotani, S. Matsumoto, K. Igarashi and K. Suzuki, **Japan Kokai** 78 08,700, 26 Jan 1978.

1604. T. Kotani, S. Matsumoto, K. Igarashi and K. Suzuki, **Japan. Kokai** 78 34,890, 31 Mar 1978.

1605. T. Kotani, S. Matsumoto, F. Imaizumi, K. Igarashi and K. Takahashi **Japan. Kokai** 78 94,399, 18 Aug 1978.

1606. T. Kotani, S. Matsumoto and K. Igarashi, **Japan Kokai,** 78 94,400, 18 Aug 1978.

1607. T. Kotani, N. Katsuki, T. Ukaji and S. Matsumoto, **Japan. Kokai Tokkyo Koho** 78 102.363, 06 Oct 1978.

1608. T. Kotani, S. Matsumoto, K. Igarashi and K. Suzuki, **U.S.** 4,250,063 22 Jan 1979.

1609. T. Kotani, K. Takahashi and H. Tsuchikawa, **Japan. Kokai Tokkyo Koho** 79 99,198 04 Aug 1979.

1610. T. Kotani, K. Takahashi and Y. Okubo, **Japan. Kokai Tokkyo Koho** 79 126,299, 11 Oct 1979.

1611. W. R. Kroll and G. Doyle, **U. S.** 3,689,433, 05 Sep 1972.

1612. W. R. Kroll, G. Doyle and W. Ruhle, **U.S.** 3,691,095, 12 Sep 1972.

1613. D. H. Kubicek and E. A. Zuech, **U.S.** 3,558,520, 26 Jan 1971.

1614. D. H. Kubicek and E. T. Kittleman, **U.S.** 3,622.644, 23 Nov 1971.

1615. D. H. Kubicek and R. E. Reusser, **U.S.** 3,659,008, 25 Apr 1972.

1616. D. H Kubicek and E. A. Zuech, **U.S.** 3,670,043, 13 Jan 1972.

1617. D. H. Kubicek and E. A. Zuech, **U.S.** 3,703,561, 21 Nov 1972.

1618. D. H. Kubicek, **U.S.** 3,728,407, 17 Apr 1973.

1619. D. H. Kubicek, **U.S.** 3,888,940, 10 Jun 1975.

1620. D. H. Kubicek, **U.S.** 3,974,100, 10 Aug 1976.

1621. F. W. Küpper, R. Streck and K. Hummel, **Ger. Offen.** 2,051,798, 24 Apr 1972.

1622. F. W. Küpper, R. Streck, H. T. Heims and H. Weber, **Ger. Offen.** 2,051,799, 27 Apr 1972.
1623. F. W. Küpper and R. Streck, **Ger. Offen.** 2,326,196, 12 Dec 1974.
1624. F. W. Küpper and R. Streck, **Ger. Offen.** 2,363,705, 10 Jul 1975.
1625. F. W. Küpper and R. Streck, **U.S.** 3,974,231, 10 Aug 1976.
1626. F. W. Küpper and R. Streck, **Ger. Offen.** 2,531,959, 20 Jan 1977.
1627. F. W. Küpper and R. Streck, **Ger. Offen.** 2,533,247, 10 Feb 1977.
1628. F. W. Küpper and R. Streck, **U.S.** 4,016,220, 05 Apr. 1977.
1629. F. W. Küpper, **Ger. Offen.** 2,613,999, 06 Oct 1977.
1630. F. W. Küpper. **Ger. Offen.** 2,619,197, 17 Nov 1977.
1631. S. Kurosawa, T. Ueshima and S. Kobayashi, **Ger. Offen.** 2,452,461, 07 May 1975.
1632. S. Kurosawa, T. Ueshima, Y. Tanaka, S. Kobayashi and T. Saito, **Japan. Kokai** 75 61, 500, 22 May 1975.
1633. S. Kurosawa, T. Ueshima and S. Kobayashi, **Japan. Kokai** 75 103,600, 15 Aug 1975.
1634. S. Kurosawa, T. Ueshima and S. Kobayashi, **Japan. Kokai** 75 110,000, 29 Aug 1975.
1635. S. Kurosawa, T. Ueshima, Y. Tanaka and S. Kobayashi, **Ger. Offen.** 2,529,963, 15 Jan 1976.
1636. S. Kurosawa, M. Arioka, T. Ueshima and S. Kobayashi, **Ger. Offen.** 2,550,837, 20 May 1976.
1637. S. Kurosawa, T. Ueshima and S. Kobayashi, **Japan. Kokai** 77 45,699, 11 Apr 1977.
1638. S. Kurosawa, T. Ueshima and S. Kobayashi, **Japan. Kokai** 77 45,700, 11 Apr 1977.
1639. S. Kurosawa, T. Ueshima and S. Kobayashi, **Japan. Kokai** 77 47,899, 16 Apr 1977.
1640. S. Kurosawa, T. Ueshima, Y. Tanaka and S. Kobayashi, **U.S.** 4,022,954, 10 May 1977.
1641. S. Kurosawa, M. Kobayashi and T. Kamishima, **Japan. Kokai** 78 10,688, 31 Jan 1978.
1642. S. Kurosawa, T. Yokoyama and T. Ueshima, **Japan Kokai.** 78 12,000, 02 Feb 1978.
1643. S. Kurosawa, Y. Kobayashi and T. Ueshima, **Japan. Koaki** 78 12,999, 06 Feb 1978.
1644. S. Kurosawa, T. Yokoyama and T. Ueshima, **Japan. Koaki** 78 13,000, 06 Feb 1978.
1645. S. Kurosawa and T. Ueshima, **Japan. Kokai** 78 71,159, 24 Jun 1978.
1646. S. Kurosawa and T. Ueshima, **Japan. Kokai** 78 71,160, 24 Jun 1978.
1647. S. Kurosawa, T. Yokoyama and T. Ueshima, **Japan. Kokai** 78 97,100, 24 Aug 1978.
1648. S. Kurosawa, T. Yokoyama and T. Ueshima, **Japan. Kokai Tokkyo Koho** 78 101,052, 04 Sep 1978.
1649. S. Kurosawa, K. Iwasaki, R. Haruo and T. Ueshima, **Japan. Kokai Tokkyo Koho** 79 158,454, 14 Dec 1979.
1650. S. Kurosawa, K. Iwasaki, R. Haruo and T. Ueshima, **Japan. Kokai Tokkyo Koho** 79 158,455, 14 Dec 1979.
1651. Y. Kuwabara, N. Tagata, T. Kotani and Y. Naito, **Japan. Kokai** 73 56,778, 09 Aug 1973.
1652. Y. Kuwabara, N. Tagata, M. Iwama, Y. Naito and T. Kotani, **Japan. Kokai** 73 56,779, 09 Aug 1973.
1653. L. G. Larson and R. T. Wilson, **Belg.** 713,182.
1654. L. G. Larson, **U. S.** 3.546.314, 08 Dec 1970.
1655. M. J. Lawrenson, **Ger. Offen.** 2,340,081, 28 Feb 1974.
1656. M. J. Lawrenson, **U.S.** 3,974,233, 10 Aug 1976.
1657. G. Lehnert, D. Maertens and M. Zimmermann, **Ger. Offen.** 2,332,564, 27 Jun 1973.
1658. D. Maertens, J. Witte and M. Beck, **Belg.** 775,268, 14 Nov 1970.
1659. D. Maertens, G. Lehnert and G. Pampus, **Ger. Offen** 2,103,301, 17 Aug 1972.
1660. K. Makino, T. Teramoto, K. Komatsu and K. Enyo, **Japan. Kokai** 77 57,299, 11 May 1977.
1661. S. Malinowski, A. Gieysztov, A. Krzywiki, L. Chrapkowski and Z. Waligorski, **Pol.** 71,930, 15 Jan 1975.
1662. F. D. Mango, **U.S.** 3,424,811, 28 Jan 1969.
1663. F. D. Mango, **U.S.** 3,527,828, 08 Sep 1970.
1664. A. Marbach and C. Stein, **Eur. Pat Appl.** 46,095, 17 Feb 1982.
1665. P. R. Marshall and B. J. Ridgewell, **Fr.** 2,002,522, 17 Oct 1969.
1666. R. M. Marsheck, **Belg.** 678,016, 19 Sep 1966.
1667. R. M. Marsheck, **Brit.** 1,128,091, 25 Sep 1968.

1668. S. Matsumoto, R. Nakamura, S. Fukuhara, K. Suzuki and K. Komatsu, **Japan. Kokai** 75 22,100, 08 Mar 1975.

1669. S. Matsumoto, R. Nakamura, S. Fukuhara, K. Suzuki and K. Komatsu, **Ger. Offen.** 2,442,181, 28 May 1975.

1670. S. Matsumoto, R. Nakamura, S. Fukuhara and K. Komatsu, **Japan. Kokai** 78 14,120, 15 May 1978.

1671. B. P. McGrath and K. V. Williams, **Brit.** 1,159,055, 23 Jun 1969.

1672. B. P. McGrath and K. V. Williams, **Brit.** 1,159,056, 23 Jun 1969.

1673. B. P. McGrath, **Ger.** 1,940,433, 19 Feb 1970.

1674. B. P. McGrath, **Fr.** 1,602,364, 24 Dec 1970.

1675. B. P. McGrath, **Brit. Ammend.** 1,216,278, 19 Jul 1971.

1676. B. P. McGrath, **U.S.** 3,594,440, 20 Jul 1971.

1677. B. P. McGrath. **U.S.** 3,600,456, 17 Aug 1971.

1678. B. P. McGrath and C.P.C. Bradshaw, **U.S.** 3,637,891, 25 Jan 1972.

1679. B. P. McGrath and C. P. C. Bradshaw, **U.S.** 3,637,892, 25 Jan 1972.

1680. D. Medema, W. Brunmayer-Schilt and R. Van Helden, **Brit.** 1,193,943, 03 Jun 1970.

1681. D. Medema, W. Brunmayer-Schilt and R. Van Helden. **Ger. Offen.** 2,003,371, 24 Apr 1974.

1682. D. Medema and R. Van Helden, **U.S.** 3,647,908, 07 Mar 1972.

1683. D. Medema, H. J. Alkema and R. Van Helden, **U.S.** 3,649,709, 14 Mar 1972.

1684. H. R. Menapace, **U.S.** 3,636,126, 18 Jan 1972.

1685. K. Meyer, R. Streck and H. Weber, **Ger. Offen.** 2,242,794, 21 Mar 1974.

1686. Kh. M. Minachev, E. S. Mortikov, A. S. Leont'ev, T. S. Papko, A. A. Masloboev-Shvedov and N. F. Kononov, **U.S.S.R.** 521,007, 15 Jul 1976.

1687. R. J. Minchak, **Ger. Offen.** 2,324,932, 06 Dec 1973.

1688. R. J. Minchak, **Ger. Offen.** 2,409,694, 28 Feb 1974.

1689. R. J. Minchak, **Fr.** 2,249,905, 16 May 1974.

1690. R. J. Minchak, **Fr.** 2,306,224, 03 Apr 1976.

1691. R. J. Minchak, **Fr.** 2,307,830, 15 Apr 1976.

1692. R. J. Minchak, **Ger. Offen.** 2,613,562, 14 Oct 1976.

1693. R. J. Minchak, **Ger. Offen.** 2,616,079. 28 Oct 1976.

1694. R. J. Minchak and R. E. Beauregard, **U.S.** 4,025,708. 24 May 1977.

1695. R. J. Minchak, **U.S.** 4,069,376, 17 Jan 1978.

1696. R. J. Minchak, **U.S.** 4,110,528, 29 Aug 1978.

1697. R. J. Minchak, **U.S.** 4,138,448, 06 Feb 1979.

1698. R. J. Minchak, **U.S.** 4,262,103, 14 Apr 1981.

1699. J. Mochida, Y. Ikeda, H. Fujitsu and K. Takeshita, **Japan. Kokai** 77 31,004. 09 Mar 1977.

1700. Montecatini Edison S. p. A., **Brit.** 1,010,860, 11 Apr 1963.

1701. Montecatini Edison S. p. A., **Neth, Appl.** 6,617,575, 21 Dec 1965.

1702. Montecatini Edison S. p. A., **Ital.** 741,683, 16 Jan 1967.

1703. Montecatini Edison S. p. A., **Ger. Offen.** 1,299,868, 24 Jun 1969.

1704. Montecatini Edison S. p. A., **Ger. Offen.** 1,520,259, 06 Feb 1969.

1705. Montecatini Edison S. p.A., **Ger. Offen.**. 1,520,269, 30 Apr 1969.

1706. Montecatini Edison S. p. A., **Ger. Offen.** 1,520,271, 26 Feb 1970.

1707. Montecatini Edison S. p. A., **Ger. Offen.** 1,520,378, 22 Jan 1970.

1708. Montecatini Edison S. p. A., **Ger. Offen.** 1.570,940, 04 Sep 1969.

1709. Montecatini Edison S. p. A., **Ger. Aus.** 1,770,440, 26 Sep 1971.

1710. Montecatini Edison S. p. A., **Ger.** 1,520,303. 03 Aug 1972.

1711. D. P. Montgomery, **Ger. Offen.** 2,160,151, 20 Jul 1972.

1712. D. P. Montgomery, **U.S.** 3,707,579, 26 Dec 1972.

1713. D. P. Montgomery, R. N. Moore and R. W. Knox. **U.S.** 3,965,206, 22 Jun 1976.

1714. Y. Morita, H. Yoshiura and T. Isaka, **Japan. Kokai** 73 65,186, 08 Sep 1973.

1715. Y. Morita, H. Yoshiura and T. Isaka, **Japan. Kokai** 73 66,086, 11 Sep 1973.

1716. A. Morris, H. Thomas and C. J. Attridge, **Ger. Offen.** 2,213,948. 21 Dec 1972.

1717. A. Morris and H. Thomas, **U.S.** 3,816,340, 11 Jun 1974.

1718. Y. Murakoshi, J. Nakamura and S. Kokuryo, **Japan. Kokai Tokkyo Koho** 78 102,933, 07 Sep 1978.

1719. R. Mutin, J. M. Basset, J. L. Bilhou and J. Bousquet, **Ger. Offen.** 2,649,780, 12 May 1977.
1720. J. W. Myers and J. R. Harris, **U.S.** 4,102,939, 25 Jul 1978.
1721. A. A. Nadirov, M. N. Zhavoronkov and A. Z. Dorogochinskii, **U.S.S.R.** 649,693, 25 Feb 1979.
1722. T. Nagaoka, N. Kuwabara and T. Ueshima, **Japan. Kokai** 76 28,146, 09 Mar 1976.
1723. J. Nakamura, H. Konuma, S. Kokuryo, T. Ueshima and C. Tsuge, **Japan. Kokai** 75 142,661, 15 Nov 1975.
1724. J. Nakamura, S. Kurosawa and S. Kokuryo, **Japan. Kokai** 76 107,351, 19 Mar 1975.
1725. J. Nakamura, S. Kurosawa and S. Kokuryo, **Japan. Kokai** 76 109,950, 29 Sep 1976.
1726. J. Nakamura, S. Kurosawa and S. Kokuryo, **Japan. Kokai** 76 112,865, 05 Oct 1976.
1727. J. Nakamura, H. Konuma, S. Kokuryo and T. Ueshima, **Japan. Kokai** 76 112, 866, 29 Mar 1976.
1728. J. Nakamura and T. Ueshima, **Japan. Kokai** 77 13,548, 01 Feb 1977.
1729. J. Nakamura, Y. Murakoshi and T. Ueshima, **Japan. Kokai** 77 22,052, 19 Feb 1977.
1730. J. Nakamura, T. Ueshima and S. Kurosawa, **Japan. Kokai** 78 23,343, 03 Mar 1978.
1731. J. Nakamura, H. Konuma, S. Kokuryo, T. Ueshima and C. Tsujii, **U.S.** 4,105,608, 08 Aug 1978.
1732. J. Nakamura, S. Kokuryo and S. Kurosawa, **Japan. Kokai Tokkyo Koho** 78 102,959, 07 Sep 1978.
1733. R. Nakamura, S. Fukuhara, S. Matsumoto and K. Komatsu, **Japan. Kokai** 76 125, 317, 01 Nov 1976.
1734. R. Nakamura, K. Sekine, F. Imaizumi and M. Kawakami, **Japan. Kokai,** 76 131,812, 16 Nov 1976.
1735. S. Nakatomi and T. Yamamura, **Japan. Kokai,** 73 85,508, 13 Nov 1973.
1736. S. Nakatomi, T. Yamamura and Y. Akiba, **Ger. Offen.** 2,460,356, 17 Jul 1975.
1737. S. Nakatomi, T. Yamamura and Y. Akiba, **U.S.** 3 872,180, 18 Mar 1975.
1738. S. Nakatomi, Y. Kohtoku and T. Inoue, **Ger. Offen.** 2,428,820, 09 Jan 1975.
1739. S. Nakatomi, T. Yamamura and Y. Akiba, **U.S.** 3,943,186, 09 Mar 1976.
1740. J. Nakazato, R. Kita and H. Horino, **Japan. Kokai,** 73 102,899, 24 Dec 1973.
1741. G. Natta, G. Dall'Asta, G. Mazzanti, I. Pasquon, V. Valvassori and A. Zambelli, **Belg.** 619.877, 31 Oct 1962.
1742. G. Natta, G. Dall'Asta and G. Mazzanti, **Brit.** 1,010,860, 14 Apr 1963.
1743. G. Natta, G. Dall'Asta and G. Mazzanti, **Ital.** 733,857, 07 Jan 1964.
1744. G. Natta, G. Dall'Asta and G. Mazzanti, **Fr.** 1,394,380, 02 Apr 1965.
1745. G. Natta, G. Dall'Asta, G. Mazzanti, I. Pasquon, V. Valvassori and A. Zambelli, **Fr.** 1,392,142, 12 Mar 1965.
1746. G. Natta, G. Dall'Asta and G. Mazzanti, **Belg.** 667,392, 16 Nov 1965.
1747. G. Natta, G. Dall'Asta and G. Mazzanti, **Fr.** 1,425,601, 21 Sep 1966.
1748. G. Natta, G. Mazzanti, G. Dall'Asta and G. Motroni, **Ital.** 718,795, 02 Nov 1966.
1749. G. Natta, G. Dall'Asta and G. Mazzanti, **U.S.** 3,476,728, 04 Nov 1969.
1750. Nippon Petrochemicals Co., Ltd., **Fr. Demande** 2,314,164, 07 Jan 1977.
1751. Nippon Zeon Co., Ltd., **Japan. Kokai Tokkyo Koho** 81 157,448, 04 Dec 1981.
1752. K. H. Nordsiek and R. Streck, **Ger. Offen.** 2,131,354, 11 Jan 1973.
1753. E. N. Nowak and K. J. Frech, **U.S.** 3,952,070, 20 Apr 1976.
1754. E. N. Nowak and K. J. Frech, **U.S.** 4,046,832, 06 Sep 1977.
1755. K. Nützel, K. Dinges and F. Haas, **Ger. Offen.** 1,720,797, 07 Mar 1968.
1756. K. Nützel and F. Haas, **Fr.** 2,008,180, 09 May 1968.
1757. K. Nützel, K. Dinges and F. Haas, **Ger. Offen.** 1,919,046, 15 Apr 1969.
1758. K. Nützel, F. Haas and G. Marwede, **Ger. Offen.** 1,919,047, 15 Apr 1969.
1759. K. Nützel, F. Haas and G. Marwede, **Ger. Offen.** 1,919,048, 15 Apr 1969.
1760. W. Oberkirch, G. Pampus and P. Günther, **Ger.** 1,770,688, 21 Jun 1968.
1761. W. Oberkirch, P. Günther and J. Witte, **Belg.** 740,769, 25 Oct 1968.
1762. W. Oberkirch, P. Günther and G. Pampus, **Belg.** 742,391, 29 Nov 1968.
1763. W. Oberkirch, P. Günther and J. Witte, **Ger. Offen.** 1,805,158, 14 Jan 1971.
1764. W. Oberkirch, P. Günther and G. Pampus, **Ger. Offen.** 1,811,653, 17 Sep 1970.
1765. W. Oberkirch, P. Günther and G. Pampus, **Ger. Offen.** 1,957,025, 19 May 1971.
1766. W. Oberkirch, P. Günther and G. Pampus, **U.S.** 3,787,381, 22 Jan 1974.

1767. J. M. O'Connor, R. R. Smith and J. Lal, **U.S.** 3,859,263, 07 Jan 1975.
1768. E. A. Ofstead, **U.S.** 3,597,403, 27 Oct 1969.
1769. E. A. Ofstead, **U.S.** 3,666,742, 25 Sep 1968.
1770. E. A. Ofstead, **U.S.** 3,801,559, 25 Sep 1968.
1771. E. A. Ofstead, **Ger. Offen.** 2,509,577, 04 Dec 1975.
1772. E. A. Ofstead, **Ger. Offen.** 2,509,578, 02 Oct 1975.
1773. E. A. Ofstead, **U.S.** 3,935,179, 27 Jan 1976.
1774. E. A. Ofstead, **U.S.** 3,945,986, 23 Mar 1976.
1775. E. A. Ofstead, **U.S.** 3,997,471, 14 Dec 1976.
1776. E. A. Ofstead, **U.S.** 4,010,113, 01 Mar 1977.
1777. E. A. Ofstead, **U.S.** 4,020,254, 26 Apr 1977.
1778. E. A. Ofstead and K. W. Donbar, **Fr.** 2,402,670, 04 Jul 1978.
1779. E. A. Ofstead and M. L. Senyek, **Fr.** 2,402,671, 04 Sep 1978.
1780. E. A. Ofstead, **U.S.** 4,137,390, 30 Jan 1979.
1781. E. A. Ofstead and K. W. Donbar, **U.S.** 4,172,932, 30 Oct 1979.
1782. E. A. Ofstead and M. L. Senyek, **U.S.** 4,239,874, 16 Dec 1980.
1783. M. Ogura and K. Yamamoto, **Ger. Offen.** 2,310,018, 20 Sep 1973.
1784. J. L. O'Hara and C. P. C. Bradshaw, **S. African** 69 06,633, 31 Mar 1970.
1785. J. I. O'Hara and C. P. C. Bradshaw, **Brit.** 1,283,348, 26 Jul 1972.
1786. J. I. O'Hara and C. P. C. Bradshaw, **U.S.** 3,671,462, 20 Jun 1972.
1787. Y. Okubo, H. Tsubitani and T. Kotani, **Japan. Kokai** 79 126,299, 01 Oct 1979.
1788. J. R. Olechowski, **U.S.** 3,714,134, 30 Jan 1973.
1789. J. R. Olechowski, **U.S.** 3,720,654, 13 Mar 1973.
1790. J. R. Olechowski, **U.S.** 3,725,371, 03 Mar 1973.
1791. T. Oshika, S. Murai and Y. Koga, **Japan.** 68 5,894, 16 Jul 1964.
1792. T. Oshika, S. Murai and Y. Koga, **Japan.** 69 15,666, 02 Feb 1969.
1793. J. Otton, **Ger. Offen.** 2,819,598, 09 Nov 1978.
1794. B. M. Palmer and B. P. McGrath, **Ger. Offen.** 1,945,282, 16 Apr 1970.
1795. B. M. Palmer, **Brit.** 1,377,161, 11 Dec 1974.
1796. G. Pampus and J. Witte, **Ger. Offen.** 1,770,143, 06 Apr 1968.
1797. G. Pampus and J. Witte, **Fr.** 2,005,701, 12 Dec 1969.
1798. G. Pampus, J. Witte and M. Hoffmann, **Ger. Offen.** 1,954,092, 28 Dec 1969.
1799. G. Pampus, N. Schön, J. Witte and G. Marwede, **Ger. Offen.** 1,961,865, 10 Dec 1969.
1800. G. Pampus, J. Witte and G. Marwede, **Fr.** 2,040,131, 27 Mar 1969.
1801. G. Pampus, N. Schön, W. Oberkirch and P. Günther, **Ger. Offen.** 2,009,740, 30 Aug 1971.
1802. G. Pampus, J. Witte and M. Hoffmann, **Ger. Offen.** 2,016,471, 28 Oct 1971.
1803. G. Pampus, J. Witte and M. Hoffmann, **Ger. Offen.** 2,016,840, 21 Oct 1971.
1804. G. Pampus, J. Lehnert and J. Witte, **Ger. Offen.** 2,101,684, 20 Jun 1972.
1805. G. Pampus, G. Lehnert and J. Witte, **U.S.** 3,933,778, 20 Jan 1976.
1806. F. Pennella, **Belg.** 713,187, 1968.
1807. F. Pennella, **S. African** 68 01,533, 06 Aug 1968.
1808. F. Pennella, **U.S.** 3,544,647, 01 Dec 1970.
1809. F. Pennella, **U.S.** 3,549,722, 22 Dec 1970.
1810. F. Pennella, **U.S.** 3,655,304, 12 Aug 1968.
1811. F. Pennella, **U.S.** 4,287,378. 01 Sep 1981.
1812. E. F. Peters and B. L. Evering, **U.S.** 2,963,447, 06 Dec 1960.
1813. E. F. Peters and B. L. Evering, **Ger.** 1,214,001, 07 Apr 1966.
1814. J. E. Phillips, **U.S.** 3,236,912, 22 Feb 1966.
1815. Phillips Petroleum Co., **Neth. Appl.** 6,400,549, 27 Jul 1964.
1816. Phillips Petroleum Co., **Neth. Appl.** 6,410,487, 10 Mar 1965.
1817. P. H. Phung and G. Lefebvre, **Fr.** 1,594,582, 17 Jul 1970.
1818. G. C. Ray and D. L. Crain, **Belg.** 694,420, 22 Aug 1967.
1819. G. C. Ray and D. L. Crain, **Fr.** 1,511,381, 26 Jan 1968.
1820. G. C. Ray and D. L. Crain, **Ger. Aus.** 1,618,760, 19 Jun 1971.
1821. R. F. Raymond, **U.S.** 3,472,909, 14 Oct 1969.
1822. R. B. Regier, **U.S.** 3,652,703, 28 Mar 1972.
1823. R. B. Regier, **U.S.** 3,761,427, 25 Sep 1973.

1824. R. B. Regier, **U.S.** 3,792,106, 12 Feb 1974.
1825. R. B. Regier, **U.S.** 3,923,920, 02 Dec 1975.
1826. R. E. Reusser, **S. African.** 68 01,879, 20 Aug 1968.
1827. R. E. Reusser, **U.S.** 3,511,890, 12 May 1970.
1828. R. E. Reusser, **U.S.** 3,538,180, 03 Nov 1970.
1829. R. E. Reusser and D. L. Crain, **U.S.** 3,660,517, 02 May 1972.
1830. R. E. Reusser, **U.S.** 3,689,589, 05 Sep 1972.
1831. R. E. Reusser, **U.S.** 3,696,165, 03 Oct 1972.
1832. R. E. Reusser, **U.S.** 3,729,524, 24 Apr 1973.
1833. R. E. Reusser and S. D. Turk, **U.S.** 3,760,026, 18 Sep 1973.
1834. R. E. Reusser and S. D. Turk, **U.S.** 3,786,112, 15 Jan 1974.
1835. R. E. Reusser, **U.S.** 3,836,480, 12 Sep 1974.
1836. R. E. Reusser, **U.S.** 3,915,897, 28 Oct 1975.
1837. R. E. Reusser and W. B. Hughes, **U.S.** 4,180,524, 25 Dec 1979.
1838. F. Röhrscheid, **Ger. Offen.** 1,768,267, 28 Jan 1970.
1839. H. W. Ruhle, **Ger. Offen.** 2,062,448, 16 Sep 1971.
1840. H. W. Ruhle, **U.S.** 3,668,146, 06 Jun 1972.
1841. H. W. Ruhle, **U.S.** 3,832,417, 27 Aug 1974.
1842. A. V. Ryabukhin, S. I. Kryukov and V. S. Fel'dblyum, **U.S.S.R.** 694,485, 30 Oct 1979.
1843. M. R. Rycheck and F. Pennella, **U.S.** 4,188,501, 12 Feb 1980.
1844. R. J. Sampson and D. Jackson, **Ger. Offen.** 1,808,459, 12 Jun 1969.
1845. K. Satake, K. Sakamoto, T. Hamada and T. Kamei, **Japan. Kokai** 73 69,900, 24 Dec 1971.
1846. W. Schneider, **U.S.** 4,320,239, 16 Mar 1982.
1847. N. Schön, G. Pampus and J. Witte, **Ger. Offen.** 1,770,844, 10 Jul 1968.
1848. N. Schön, G. Pampus, J. Witte and D. Theisen, **Ger. Offen.** 2,006,776, 19 Aug 1971.
1849. K. W. Scott and N. Calderon, **Ger. Offen.** 2,058,198, 09 Jul 1971.
1850. K. W. Scott and N. Calderon, **U.S.** 4,010,224, 01 Mar 1977.
1851. M. Setsuraka, F. Yagaki, H. Takayama and F. Kinjo, **Japan. Kokai** 78 13,400, 10 May 1978.
1852. A. D. Shebaldova, G. G. Nasledkova, V. I. Mar'in, V. A. Akchurin, T. A. Popova and M. L. Khidekel', **U.S.S.R.** 486,774, 05 Oct 1975.
1853. A. D. Shebaldova, V. I. Mar'in, P. E. Makovetskii, M. L. Khidekel' and G. K. Kornienko, **U.S.S.R.** 733,714, 15 May 1980.
1854. Shell Internat. Res. Maatschapij, N. V., **Neth. Appl.** 6,804,643, 05 Apr 1967.
1855. Shell Internat. Res. Maatschapij, N. V., **Neth. Appl.** 6,814,835, 22 Apr 1969.
1856. Shell Internat. Res. Maatschapij, N. V., **Fr.** 2,132,099, 22 Dec 1972.
1857. Shell Internat. Res. Maatschapij, N. V., **Neth. Appl.** 7,014,132, 28 Mar 1972.
1858. Shell Internat. Res. Maatschapij, N. V., **Neth. Appl.** 7,111,882, 25 May 1972.
1859. Shell Internat. Res. Maatschapij, N. V., **Neth. Appl.** 7,413,164, 31 Jan 1975.
1860. F. T. Sherk, **U.S.** 3,296,330, 03 Jan 1967.
1861. M. Shioda and S. Matsumura, **Japan. Kokai** 75 146,650, 25 Nov 1975.
1862. M. Shitara, T. Shoji, H. Kougame and F. Kanega, **Japan. Kokai** 76 11,898, 30 Jan 1976.
1863. M. Shitara, T. Shoji, H. Kougame and F. Kanega, **Japan. Kokai** 77 33,000, 12 Mar 1977.
1864. F. K. Shmidt, E. A. Grechkina, V. G. Lipovich and I. V. Kalechits, **U.S.S.R.** 429,049, 25 May 1974.
1865. F. K. Shmidt, B. V. Timashkova, E. H. Kim. Yu. S. Levkovski and I. V. Kalechits, **U.S.S.R.** 725,697, 05 Apr 1980.
1866. J. N. Short, **U.S.** 3,995,095, 30 Nov 1976.
1867. Showa Denko Kobushiki Kaisha, **Fr.** 2,291,217, 12 Dec 1975.
1868. Showa Denko Kobushiki Kaisha, **Fr.** 2,323,709, 11 Sep 1975.
1869. R. P. Sieg. **U.S.** 3,676,522, 11 Jul 1972.
1870. R. P. Sieg, **U.S.** 3,752,862, 14 Aug 1973.
1871. R. P. Sieg, **U.S.** 3,761,535, 25 Sep 1973.
1872. D. M. Singleton, **Ger. Offen.** 1,937,495, 25 Jul 1968.
1873. D. M. Singleton, **U.S.** 3,530,196, 29 Jan 1970.
1874. D. M. Singleton, **Ger. Offen.** 2,035,963, 04 Feb 1971.

1875. D. M. Singleton, **U.S.** 3,637,893, 25 Jan 1972.

1876. D. M. Singleton, **U.S.** 3,829,523, 13 Aug 1974.

1877. L. H. Slaugh, R. Van Helden and J. J. H. M. Vincken, **Ger. Offen.** 2,425,685, 19 Dec 1974.

1878. P. R. Stapp and D. L. Crain, **Fr.** 1,557,672, 21 Feb 1969.

1879. P. R. Stapp and D. L. Crain **U.S.** 3,457,320, 1969.

1880. F. W. Steffgen, **U.S.** 3,634,538, 11 Jan 1972.

1881. C. Stein and P. LeDelliou, **Ger. Offen.** 2,607,921, 09 Oct 1976.

1882. R. Streck and H. Weber, **Ger. Offen.** 1,945,538, 03 Sep 1969.

1883. R. Streck and H. Weber, **Ger. Offen.** 2,027,905, 20 Jan 1972.

1884. R. Streck and H. Weber, **Ger. Offen.** 2,028,716, 16 Dec 1971.

1885. R. Streck and H. Weber, **Ger. Offen.** 2,028,935, 16 Dec 1971.

1886. R. Streck, H. K. Nordsiek, H. Weber and K. Meyer, **Ger. Offen.** 2,131,355, 11 Jan 1973.

1887. R. Streck and H. Weber, **Ger.** 2,157,405, 24 May 1973.

1888. R. Streck and H. Weber, **Ger.** 2,166,867, 22 Apr 1976.

1889. G. Suld, A. Schneider and H. K. Myer, Jr., **U.S.** 4,100,338, 11 Jul 1978.

1890. Sumimoto Chemical Co., Ltd., **Japan. Kokai Tokkyo Koho** 80 104,307, 09 Aug 1980.

1891. T. Suminoe, S. Aotani and M. Takahashi, **Japan. Kokai** 75 154,399, 12 Dec 1975.

1892. R. Takagi, H. Nakabayashi and S. Matsumura, **Japan. Kokai** 76 100,985, 06 Oct 1976.

1893. H. Takahashi and K. Abe, **Japan. Kokai.** 79 65,018, 31 Oct 1979.

1894. T. Takahashi, **Japan. Kokai** 75 160,303, 25 Dec 1975.

1895. T. Takahashi, **Japan. Kokai** 76 26,805, 05 Mar 1976.

1896. T. Takahashi, **Japan. Kokai** 76 32,502, 19 Mar 1976.

1897. T. Takahashi, **Japan. Kokai** 76 68,508, 14 Jun 1976.

1898. T. Takahashi, **U.S.** 4.024.201, 17 May 1977.

1899. Takasago Perfumery Co., Ltd., **Japan. Kokai** 81 77,243, 25 Jun 1981.

1900. Y. Tanaka, T. Ueshima and S. Kobayashi, **Japan. Kokai** 75 159,598, 24 Dec 1975.

1901. Y. Tanaka, T. Ueshima and S. Kurosawa, **Japan. Kokai** 75 160,400, 25 Dec 1975.

1902. N. Tani, H. Wakabayashi, M. Nishigachi, S. Huji, R. Tagachi and S. Matsumura, **Japan. Kokai** 78 24,400, 07 Mar 1978.

1903. L. P. Tenney and P. S. Lane, Jr., **U.S.** 4,136,247, 23 Jan 1979.

1904. L. P. Tenney, **U.S.** 4,136,248, 23 Jan 1979.

1905. L. P. Tenney and P. C. Lane, Jr., **U.S.** 4,136,249, 23 Jan 1979.

1906. L. P. Tenney and P. C. Lane, Jr., **U.S.** 4,178,424, 11 Dec 1979.

1907. K.-H. Thiele, W. Grahlert, K. Milowski, G. Wiegand and U. Langbein, **Ger. (East)** 93,025, 05 Oct 1972.

1908. Toa Nenryo K. K., **Japan. Kokai** 71 22,663, 22 Jan 1971.

1909. A. Toyoda, **Japan. Kokai** 73 14,799, 24 Feb 1973.

1910. A. Toyoda, **Japan. Kokai** 73 14,800, 24 Feb 1973.

1911. A. Toyoda, T. Matsui and S. Shiozaki, **Japan. Kokai** 73 49,900, 03 Jul 1973.

1912. T. Tsujino, A. Sano and H. Hayashi, **Japan. Kokai** 76 80,400, 13 Jul 1976.

1913. L. Turner, E. J. Howman and C. P. C. Bradshaw, **Brit.** 1,056,980, 01 Feb 1967.

1914. L. Turner and K. V. Williams, **Brit.** 1,096,200, 20 Dec 1967.

1915. L. Turner and K. V. Williams, **Brit.** 1,096,250, 20 Dec 1967.

1916. L. Turner and C. P. C. Bradshaw, **Brit.** 1,103,976, 21 Feb 1968.

1917. L. Turner and C. P. C. Bradshaw, **Brit.** 1,105,565, 06 Mar 1968.

1918. L. Turner, E. J. Howman and K. V. Williams, **Brit.** 1,106,015, 23 Jul 1965.

1919. L. Turner, E. J. Howman and K. V. Williams, **Brit.** 1,106,016, 23 Jul 1965.

1920. L. Turner, E. J. Howman and C. P. C. Bradshaw, **Brit. Ammend.** 1,054,864, 11 Dec 1970.

1921. L. Turner, C. P. C. Bradshaw and E. J. Howman, **U.S.** 3,526,676, 01 Sep 1970.

1922. L. Turner, E. J. Howman and C. P. C. Bradshaw, **U.S.** 3,641,181, 08 Feb 1972.

1923. L. Turner and K. V. Williams, **U.S.** 3,642,931, 15 Feb 1972.

1924. Y. Uchida and M. Hidai, **Japan. Kokai** 72 31,903, 14 Nov 1972.

1925. Y. Uchida, M. Hidai and T. Tatsumi, **Japan. Kokai** 72 31,904, 14 Nov 1972.

1926. Y. Uchida, M. Hidai and T. Tatsumi, **Japan. Kokai** 74 85,004, 15 Apr 1974.

1927. A. Ueda and K. Komuro, **Japan. Kokai Tokkyo Koho** 78 108,143, 20 Mar 1978.

1928. T. Ueshima, S. Kobayashi and M. Matsumura, **Ger. Offen.** 2,316,087, 04 Oct 1973.

1929. T. Ueshima and S. Kobayashi, **Japan. Kokai** 74 77,999, 26 Jul 1974.
1930. T. Ueshima, S. Kobayashi and Y. Tanaka, **Japan. Kokai** 74 130,500, 13 Dec 1974.
1931. T. Ueshima, Y. Tanaka, T. Yokoyama and S. Kobayashi, **Ger. Offen.** 2,414,840, 25 Jan 1975.
1932. T. Ueshima, T. Yokoyama and S. Kobayashi, **Ger. Offen.** 2,441,742, 13 Mar 1975.
1933. T. Ueshima, Y. Tanaka and S. Kobayashi,. **Ger. Offen.** 2,445,812, 03 Apr 1975.
1934. T. Ueshima, S. Kobayashi and S. Kurosawa, **Japan. Kokai** 75 61,452, 27 May 1975.
1935. T. Ueshima, Y. Tanaka, S. Kurosawa and S. Kobayashi, **Japan. Kokai** 75 71,800, 13 Jun 1975.
1936. T. Ueshima, S. Kurosawa, Y. Tanaka and S. Kobayashi, **Japan. Kokai** 75 112,500, 02 Sep 1975.
1937. T. Ueshima, Y. Tanaka and S. Kobayashi, **Japan. Kokai** 75 128,799, 11 Oct 1975.
1938. T. Ueshima, S. Kobayashi and M. Matsumura, **Japan. Kokai** 77 49,520, 17 Dec 1977.
1939. T. Ueshima and S. Kobayashi, **Japan. Kokai** 77 63,298, 25 May 1977.
1940. T. Ueshima, Y. Tanaka and S. Kobayashi, **U.S.** 4,028,482, 07 Jun 1977.
1941. Uniroyal, Inc., **U.S.** 3,135,016, 29 Jun 1966.
1942. Uniroyal, Inc., **Brit.** 1,131,160, 23 Oct 1968.
1943. C. A. Uraneck and W. J. Trepka, **Belg.** 705,673, 26 Apr 1968.
1944. C. A. Uraneck and W. J. Trepka, **Fr.** 1,542,040, 11 Oct 1968.
1945. C. A. Uraneck and J. E. Burleigh, **U.S.** 3,681,300, 01 Aug 1972.
1946. YuN. Usov. E. V. Skvortsova, T. G. Vaistub and N. S. Ustinova, **U.S.S.R.** 407,572, 10 Dec 1973.
1947. Yu. N. Usov, E. V. Skvortsova, E. V. Pletneva and N. N. Gutsienko, **U.S.S.R.** 701,695, 20 Dec 1979.
1948. Yu. N. Usov, E. V. Pletneva and E. V. Skvortsova, **U.S.S.R.** 755,294, 15 Aug 1980.
1949. R. Van Helden, W. F. Engel, J. H. Alkema and P. Krijger, **Brit.** 1,164,687, 18 Dec 1967.
1950. R. Van Heldan, C. F. Kohll and P. A. Verbrugge, **Ger. Offen.** 2,109,011, 16 Sep 1971.
1951. R. Van Helden, C. F. Kohll and P. A. Verbrugge, **U.S.** 3.728.414, 17 Apr 1973.
1952. P. A. Verbrugge, C. F. Kohll and R. Van Helden, **Ger. Offen.** 2,137,361, 03 Feb 1972.
1953. P. A. Verbrugge and G. I. Heiszwolf, **Fr.** 2,132,099, 22 Nov 1972.
1954. P. A. Verbrugge and G. I. Heiszwolf, **U.S.** 3,776,975, 04 Dec 1973.
1955. P. A. Verbrugge and G. I. Heiszwolf, **Neth. Appl.** 7,413,165, 15 Apr 1981.
1956. L. S. Veretenikova, R. N. Liubovskaya, A. D. Pomogailo, J. S. Kiashkina and M. L. Khidekel', **U.S.S.R.** 887,576, 02 Jan 1980.
1957. V. C. Vives, **U.S.** 3,786,111, 15 Jan 1974.
1958. V. C. Vives and R. E. Reusser, **U.S.** 3,792,102, 12 Feb 1974.
1959. V. C. Vives and R. E. Reusser, **U.S.** 3,878,262, 15 Apr 1975.
1960. H. Wakabayashi, T. Nakagawa. S. Fuji and S. Matsumura, **Ger. Offen.** 2,437,104, 20 Feb 1975.
1961. H. Wakabayashi, N. Tani and M. Nishigaki, **Japan. Kokai** 78 17,700, 17 Feb 1978.
1962. H. Wakabayashi, N. Tani and M. Nishigaki, **Japan. Kokai** 78 92,000 12 Aug 1978.
1963. H. Wakabayashi, N. Tani and M. Nishigaki, **Japan. Kokai Tokkyo Koho** 78 111,399, 28 Sep 1978.
1964. H. Wakabayashi, M. Nishigaki and N. Tani, **Japan. Kokai Tokkyo Koho** 78 115,763, 09 Oct 1978.
1965. J.-L. Wang, M. Brown and H. R. Menapace, **Ger. Offen.** 2,158,990, 06 Jul 1972.
1966. J.-L. Wang, **U.S.** 4,073,820, 14 Feb 1978.
1967. P. H. Washecheck, **U.S.** 3,696,134, 09 Jan 1970.
1968. Z. Weinstein, **Fr.** 2,298,560, 21 Jan 1975.
1969. K. V. Williams and L. Turner, **Brit.** 1,116,243, 06 Jun 1968.
1970. R. T. Wilson and L. G. Larson, **U.S.** 3,346,661, 10 Oct 1967.
1971. R. T. Wilson, **U.S.** 3,544,648, 01 Dec 1970.
1972. J. Witte, N. Schön and G. Pampus, **Ger. Offen.** 1,770,491, 21 Oct 1971.
1973. J. Witte, G. Pampus, N. Schön and G. Marwede, **Ger. Offen.** 1,957,026, 19 May 1971.
1974. J. Witte, G. Pampus and D. Maertens, **Ger. Offen.** 2,203,303, 02 Aug 1973.
1975. J. Witte and G. Lehnert, **Ger. Offen.** 2,357,193, 28 May 1975.
1976. R. Wolovsky, **Ger. Offen.** 2,108,965, 09 Sep 1971.

1977. R. Wolovsky, **Israeli** 33,970, 31 May 1973.
1978. K. Yagii, T. Yonezawa and T. Kotani, **Japan. Kokai Tokkyo Koho** 78 91,963, 12 Aug 1978.
1979. M. Yamaguchi and Y. Tsunoda, **Japan. Kokai** 73 22,683, 07 Jul 1973.
1980. M. Yamaguchi and Y. Tsunoda, **Japan. Kokai** 74 07,124, 19 Feb 1974.
1981. S. Yoshida and S. Kitagawa, **Japan. Kokai** 76 127,004, 05 Nov 1976.
1982. H. Yoshiura, T. Isaka and Y. Morita, **Japan. Kokai** 73 31,194, 24 Apr 1973.
1983. T. L. Yufa, I. A. Poleyatova, V. A. Kormer, B. D. Babitskii, M. I. Lobach V. V. Markova and L. A. Churlyaeva, **Belg.** 770,823,02 Feb 1972.
1984. T. L. Yufa, I. A. Poleyatova, V. A. Kormer, B. D. Babitskii, M. I. Lobach, V. V. Markova and L. A. Churlyaeva, **Ger. Offen.** 2,133,905, 20 May 1972.
1985. T. L. Yufa, I. A. Poleyatova, V. A. Kormer, B. D. Babitskii, M. I. Lobach, V. V. Markova and L. A. Churlyaeva, **Brit.** 1,338,155, 21 Nov 1973.
1986. T. L. Yufa, I. A. Poletaeva, V. A. Kormer, B. D. Babitskii, M. I. Lobach and V. V. Markova, **U.S.S.R.** 432,769, 15 Dec 1974.
1987. T. L. Yufa, I. A. Poletaeva, V. A. Kormer, B. D. Babitskii, M. I. Lobach, V. V. Markova and L. A. Churlyaeva, **Can.** 977,497, 04 Nov 1975.
1988. R. P. Zelinski and R. H. Gath, **U.S.** 4,012,566, 15 Mar 1977.
1989. D. Zerpner and R. Streck, **Ger. Offen.** 2,922,335, 11 Dec 1980.
1990. M. W. Zhavoronkov, A. Z. Dorogochinskii, A. A. Nadirov and A. L. Proskurnin, **U.S.S.R.** 694,484, 30 Oct 1979.
1991. Yu. M. Zhorov, E. V. Morozava and G. M. Panchenkov, **U.S.S.R.** 555,905, 30 Apr 1977.
1992. M. Zimmermann, G. Lehnert, D. Maertens and G. Pampus, **Ger. Offen.** 2,334,604, 30 Jan 1975.
1993. M. Zimmermann, G. Pampus and J. Witte, **Ger. Offen.** 2,334,606, 30 Jan 1975.
1994. E. A. Zuech, **Fr.** 1,561,026, 21 Mar 1969.
1995. E. A. Zuech, **Fr.** 1,575,778, 25 Jul 1969.
1996. E. A. Zuech, **Ger. Offen.** 2,017,841, 29 Oct 1971.
1997. E. A. Zuech, **U.S.** 3,558,518, 26 Jan 1971.
1998. E. A. Zuech, **U.S.** 3,590,095, 29 Jun 1971.
1999. E. A. Zuech, **U.S.** 3,778,385, 11 Dec 1973.
2000. E. A. Zuech, **U.S.** 3,865,892, 11 Feb 1975.
2001. E. A. Zuech, **U.S.** 3,981,940, 21 Sep 1976.
2002. E. A. Zuech, **U.S.** 4,010,217, 01 Mar 1977.

D. Supplementary Recent Publications

2003. A. J. Amass and M. Latfipour, *Polymer*, **24**, 1317 (1983).
2004. S. L. Buchwald and R. H. Grubbs, *J. Amer. Chem. Soc.*, **105**, 5490 (1983).
2005. R. Chemelli, F. Stelzer and K. Hummel, *J. Mass Spectr. Ion Phys.*, **48**, 55 (1983).
2006. M. Gilet, A. Mortreux, J. C. Folest and F. Petit, *J. Amer. Chem. Soc.*, **105**, 3876 (1983).
2007. R. H. Grubbs, *2nd IUPAC Symp. Organomet. Chem. Org. Synth.*, Dijon, August 28—September 1, 1983.
2008. G. J. Hamilton, K. J. Ivin, J. J. Rooney and L. C. Waring, *J. Chem. Soc., Chem. Commun.*, **1983**, 159.
2009. E. Janssen, Ph. D. Thesis, TH Aachen, 1982.
2010. A. A. Kadushin, R. K. Aliev, O.V. Krylov, A. A. Andreev, R. M. Edreva-Kardjeva and D. M. Shopov, *Kin. Catal.*, **23**, 228 (1982).
2011. F. Kapteijn, E. Homburg and J. C. Mol, *J. Chem. Thermodyn.*, **15**, 147 (1983).
2012. J. Kress and J. A. Osborn, *J. Amer. Chem. Soc.*, **105**, 6346 (1983).
2013. C. Larroche, J. P. Laval, A. Lattes, M. Leconte, F. Quignard and J. M. Basset, *J. Chem. Soc., Chem. Commun.*, **1983**, 220.
2014. J. B. Lee, K. C. Ott and R. H. Grubbs, *J. Amer. Chem. Soc.*, **104**, 7491 (1982).

2015. L. G. McCullough, M. L. Listemann, R. R. Schrock, M. R. Churchill and J.W. Ziller, *J. Amer. Chem. Soc.*, **105**, 6729 (1983).

2016. S. F. Pedersen and R. R. Schrock, *J. Amer. Chem. Soc.*, **104**, 7483 (1982).

2017. M. Petit, A. Mortreux and F. Petit, *J. Chem. Soc., Chem. Commun.*, **1982**, 1385.

2018. Yu. Ye. Shapiro, V. A. Efimov and B. S. Turov, *Vysokomol. Soedin., Ser. A.*, **25**, 506 (1983).

2019. Yu. Ye. Shapiro, N. P. Dozorova, B. J. Turov and V. A. Efimov, *Vysokomol. Soedin., Ser. A.*, **25**, 955 (1983).

2020. R. Spinicci, *IX Iberoam. Symp. Catal.*, Lisbon, July 16—21, 1984.

2021. K. Tanaka and K. Tanaka, *J. Chem. Soc., Chem. Commun.*, **1984**, 748.

2022. R. Taube and K. Seyferth, *J. Organomet. Chem.*, **249**, 365 (1983).

2023. B. Tinland, M. Leconte, J. M. Basset and F. Quignard, *J. Amer. Chem. Soc.*, **105**, 2924 (1983).

2024. M. A. Tlenkopachev, E. G. Avdeikina, I. V. Korshak, G. N. Bondarenko, B. A. Dolgo-plosk and D. F. Kutepov, *Dokl. Akad. Nauk S.S.S.R.*, **268**, 133 (1983).

2025. Yu. N. Usov, E. V. Pletneva and E. V. Skvortsova, *Kin. Katal.*, **24**, 1017 (1983).

2026. G. C. N. Van den Aardweg, R. H. A. Bosma and J. C. Mol, *J. Chem. Soc., Chem. Commun.*, **1983**, 262.

2027. A. J. Van Roosmalen and J. C. Mol, *J. Catalysis*, **78**, 17 (1982).

2028. S. Warwel, G. Hachen, H. Kirchmeyer, H. Ridder and W. Winkelmüller, *Tenside Deterg.*, **19**, 321 (1982).

2029. S. Warwel and V. Siekerman, *Makromol. Chem., Rapid Commun.*, **4**, 423 (1983).

2030. S. Warwel, H. Ridder and G. Hachen, *Chem.-Ztg.*, **107**, 115 (1983).

2031. K. Weiss and K. Hoffmann, *J. Organomet. Chem.*, **255**, C24 (1983).

2032. K. Weiss and P. Kindl, *Angew. Chem.*, **96**, 616 (1984); *Angew. Chem. Int. Ed.*, **23**, 629 (1984).

Subject Index